Toxicology, Survival and Health Hazards of Combustion Products

Issues in Toxicology

Series Editors:
Professor Diana Anderson, *University of Bradford, UK*
Dr Michael Waters, *Integrated Laboratory Systems Inc., N Carolina, USA*
Dr Timothy C Marrs, *Edentox Associates, Kent, UK*

Advisor to the Board:
Dr Alok Dhawan, *Ahmedabad University, India*

Titles in the Series:

1: Hair in Toxicology: An Important Bio-Monitor
2: Male-mediated Developmental Toxicity
3: Cytochrome P450: Role in the Metabolism and Toxicity of Drugs and other Xenobiotics
4: Bile Acids: Toxicology and Bioactivity
5: The Comet Assay in Toxicology
6: Silver in Healthcare
7: *In Silico* Toxicology: Principles and Applications
8: Environmental Cardiology
9: Biomarkers and Human Biomonitoring, Volume 1: Ongoing Programs and Exposures
10: Biomarkers and Human Biomonitoring, Volume 2: Selected Biomarkers of Current Interest
11: Hormone-Disruptive Chemical Contaminants in Food
12: Mammalian Toxicology of Insecticides
13: The Cellular Response to the Genotoxic Insult: The Question of Threshold for Genotoxic Carcinogens
14: Toxicological Effects of Veterinary Medicinal Products in Humans: Volume 1
15: Toxicological Effects of Veterinary Medicinal Products in Humans: Volume 2
16: Aging and Vulnerability to Environmental Chemicals: Age-related Disorders and their Origins in Environmental Exposures
17: Chemical Toxicity Prediction: Category Formation and Read-Across
18: The Carcinogenicity of Metals: Human risk through occupational and environmental exposure
19: Reducing, Refining and Replacing the Use of Animals in Toxicity Testing
20: Advances in Dermatological Sciences

21: Metabolic Profiling: Disease and Xenobiotics
22: Manganese in Health and Disease
23: Toxicology, Survival and Health Hazards of Combustion Products

How to obtain future titles on publication:
A standing order plan is available for this series. A standing order will bring delivery of each new volume immediately on publication.

For further information please contact:
Book Sales Department, Royal Society of Chemistry, Thomas Graham House, Science Park, Milton Road, Cambridge, CB4 0WF, UK
Telephone: +44 (0)1223 420066, Fax: +44 (0)1223 420247
Email: booksales@rsc.org
Visit our website at www.rsc.org/books

Toxicology, Survival and Health Hazards of Combustion Products

Edited by

David A. Purser
Hartford Environmental Research, Hatfield, UK
Email: david-purser@ntlworld.com

Robert L. Maynard
University of Birmingham, UK
Email: robertmaynard3@gmail.com

James C. Wakefield
Health Protection Agency, UK
Email: James.Wakefield@delphichse.com

ROYAL SOCIETY OF CHEMISTRY

THE QUEEN'S AWARDS FOR ENTERPRISE: INTERNATIONAL TRADE 2013

Issues in Toxicology No. 23

Print ISBN: 978-1-84973-569-8
PDF eISBN: 978-1-84973-748-7
ISSN: 1757-7179

A catalogue record for this book is available from the British Library

Published by The Royal Society of Chemistry,
Thomas Graham House, Science Park, Milton Road,
Cambridge CB4 0WF, UK

Registered Charity Number 207890

For further information see our web site at www.rsc.org

Printed in the United Kingdom by CPI Group (UK) Ltd, Croydon, CR0 4YY, UK

Preface

The toxic effects of combustion products are a major cause of human morbidity and mortality. We are all exposed, every day, to air pollutants produced by the combustion of organic materials, while significant numbers of people suffer severe injury and death from fires. The hazards range from continuous exposure to low concentrations of toxic vapours and particulates, both outdoors and indoors, throughout our lives, to single short-duration exposures to very high concentrations during fires.

In order to evaluate and find ways to mitigate these hazards it is not sufficient or practical to consider the toxicity of combustion products in isolation, but rather in the context of systems involving the source terms (fires or other combustion processes), the dispersal of effluent plumes and the dynamics of human exposure and toxicity.

In compiling this book we have been very fortunate to obtain chapters contributed by leading international experts in the relevant fields. In addition to benefitting from this eminent multiple authorship we have attempted to provide a comprehensive and coordinated guide, which we hope will be of value both to those wishing to form a general understanding of the subject, and to regulators, forensic investigators, clinicians and engineers involved in practical assessments of hazard and risks from combustion products. Chapter 1 provides an overview of the subjects covered, and throughout the book chapters on specific topics we have provided frequent cross-references to integrate between relevant sections of different chapters.

In the first section of the book we have addressed the formation and dispersal of combustion products. It is a common misconception that one can identify "toxicity" as a discrete property of specific substances such as wood or diesel fumes, but this cannot be farther from the truth. In reality the combustion products from individual fuels consist of a complex mixture of

Issues in Toxicology No. 23
Toxicology, Survival and Health Hazards of Combustion Products
Edited by David A. Purser, Robert L. Maynard and James C. Wakefield

Published by the Royal Society of Chemistry, www.rsc.org

many individual toxic substances in the form of vapours and particulates. The composition of the effluent plume depends partly on the elemental and molecular composition of the burning fuel, but very much upon the combustion conditions, so that the yields and concentrations of key toxic products, such as carbon monoxide, can vary by several orders of magnitude. Both the fire size and combustion conditions change considerably during fire development and the human exposure conditions further depend on the dynamics of air entrainment and plume dispersal. These issues are covered in the first section, including descriptions of fire physics and chemistry, models for the calculation of rates of formation and composition of fire plume from different fuels, and fire conditions and their dispersal both within buildings and in the outside environment.

The effects on exposed subjects in different locations relative to a fire also vary at different locations throughout an exposure. For fire victims inside buildings or vehicles the sequence of toxic hazards usually begins with the immediate pain and incapacitating effects of exposure to visually obscure smoke, containing a range of irritant acid gases and particulates, followed by asphyxiation from gases, including carbon monoxide and hydrogen cyanide, complicated by heat exposure and burns. Following exposure, another set of effects include chemical lung injury, burns, and neurological and cardiovascular effects. Evaluation of these hazards therefore requires assessment of the source term and exposure conditions, as well as the subsequent sequence of physiological and pathological effects. The toxicology of these effects and methods for their assessment are presented in the second section, with clinical aspects of toxicology and management in the third section.

Assessing the effects of exposure to widely dispersed fire effluent plumes or deposited pollutants in the outside environment presents another set of challenges. Unlike the life-threatening effects of exposures during fires inside buildings, the health effects of both single and repeated exposures to dilute smoke plumes or combustion products dispersed into the environment are more subtle, ranging from acute nuisance odour to long term health hazards such as cardio-respiratory diseases and carcinogenicity. Aspects of these are also described in terms of their formation and dispersal in the first section and their toxicity in the second section. The third, fourth and fifth sections discuss aspects of clinical management and assessment, examples of some specific large conflagration incidents and the public health aspects of fire incident management.

With diverse coverage, and edited and authored by recognised experts in the field, it is intended that this book will provide an essential text for those working in toxicology, combustion science, public health, safety engineering, forensic fire investigation and environmental research.

David Purser
Robert Maynard
James Wakefield

Acknowledgements

We would like to thank the authors for contributing their time and expertise. We would also like to thank the Royal Society of Chemistry commissioning editors and production team for their advice, patience and support during the preparation of the chapter manuscripts and their development to the final printed book.

Issues in Toxicology No. 23
Toxicology, Survival and Health Hazards of Combustion Products
Edited by David A. Purser, Robert L. Maynard and James C. Wakefield

Published by the Royal Society of Chemistry, www.rsc.org

Dedication

In Memoriam – David John Hall

It is with great sadness that we learnt of the death of Dr David Hall on 22nd February 2015. David has been a leading figure internationally in plume dispersal measurement and modelling, and in atmospheric sampling of dusts and aerosols. Formerly at Warren Spring Laboratory and the Building Research Establishment, since setting up "Envirobods" in 1999 David and his colleague Angela Spanton continued with plume dispersal research and studies into the dispersal of contaminants during major incidents. Among the many projects David was engaged in towards the end of this life was Chapter 6 "Dispersion of Fire Plumes in the Atmosphere", which provides a fitting memorial to his practical expertise and enthusiasm for this important subject.

Issues in Toxicology No. 23
Toxicology, Survival and Health Hazards of Combustion Products
Edited by David A. Purser, Robert L. Maynard and James C. Wakefield

Published by the Royal Society of Chemistry, www.rsc.org

Terms and Definitions

Burn: undergo combustion.

Boundary conditions: the value of a mathematical function at the boundary of a calculation domain either known or assumed to be known.

In context of fire calculations these might be surface temperatures, heat or ventilation fluxes or gas pressure. For chemically active surfaces it may include species production or annihilation.

Note: the concept may also be applied to fire tests in relation to physical features of the boundaries of a space containing a fire and influencing heat loss and ventilation. For example, the fire development and combustion conditions in a room enclosure fire depend on features of the boundary conditions, including the surface area and thermal properties of the walls and ceiling, the dimensions and locations of vents such as open doors or windows, wind velocity and direction, ambient temperature, humidity and pressure.

Combustion: exothermic reaction of a substance with an oxidizing agent.

Combustion product: solid, liquid or gaseous material resulting from combustion.

Equivalence ratio: fuel : air ratio divided by fuel : air ratio required for a stoichiometric mixture.

Exposure dose: measure of the maximum amount of a toxic gas or fire effluent that is available for inhalation, calculated by integration of the area under a concentration–time curve.

Note 1: for fire effluent, typical units are grams time minute per cubic metre (g min m^{-3}).

Issues in Toxicology No. 23
Toxicology, Survival and Health Hazards of Combustion Products
Edited by David A. Purser, Robert L. Maynard and James C. Wakefield

Published by the Royal Society of Chemistry, www.rsc.org

Note 2: for a toxic gas, typical units are microliters per litre ($\mu L\ L^{-1}$) at $T = 298$ K and $P = 1$ atm.

Fire: process of combustion characterized by emission of heat and fire effluent and usually accompanied by smoke, flame, glowing or a combination thereof.

Fire scenario: qualitative description of the course of a fire with respect to time, identifying key events that characterize the studied fire and differentiate it from other possible fires.

Note: it typically defines the ignition and fire growth processes, the fully developed fire stage, the fire decay stage and the environments and systems that impact on the course of the fire.

Fire effluent (also known as "smoke"): totality of gases and aerosols, including suspended particles created by combustion, or pyrolysis in a fire.

Fractional effective concentration (FEC): ratio of the concentration of an irritant to that concentration expected to produce a specified effect on an exposed subject of average susceptibility.

Note 1: as a concept, FEC may refer to any effect, including impairment of escape capability, incapacitation, lethality or other endpoints.

Note 2: when not used with reference to a specific irritant, the term FEC represents the summation of the FEC value for all measured irritants in a combustion atmosphere.

Fractional effective dose (FED): ratio of the exposure dose for a specific combustion product or mixture of products to that exposure dose predicted to produce a specified effect on an exposed subject of average susceptibility.

Note 1: as a concept, FED may refer to any effect, including incapacitation, lethality or other endpoints.

Note 2: when not used with reference of any specific substance, the term "FED" represents the summation of the FED values for all measured components of a combustion atmosphere.

Flame retardant: substance added, or treatment applied, to a material in order to suppress or delay the appearance of flame and/or reduce the flame-spread rate.

Heat release rate: rate of thermal energy production generated by combustion (essentially the burning rate).

Glowing combustion: combustion of a material in the solid phase without flame but with emission of light from the combustion zone.

Flaming combustion: combustion in the gaseous phase, usually with emission of light.

Flashover (stage of fire): transition to a state of total surface involvement in a fire of combustible materials within an enclosure.

Fuel-rich combustion: combustion in which the equivalence ratio is greater than unity.

Note: in ventilation-controlled fires the fuel/air mixture is fuel-rich and relatively high concentrations of pyrolysis products and incomplete combustion gases result.

Fuel-lean combustion: combustion in which the equivalence ratio is less than unity.

Note: in well-ventilated fires, the fuel/air mixture is fuel-lean and complete combustion tends to occur.

LC_{50} (lethal concentration 50%): the concentration statistically calculated to cause the deaths of one half of the animals exposed to a toxicant for a specified time. This may be expressed in terms of volume fraction/volume fraction (ppm, %) or mass/volume (g m^{-3}).

Note 1: care must be taken when comparing LC_{50} values of both the exposure duration and the post exposure period over which deaths were scored. In combustion toxicology, LC_{50} values have most commonly been measured using a 30 minute exposure period followed by a 14 day post exposure observation period.

Note 2: in combustion toxicology, an LC_{50} concentration may be related to a specific combustion product such as carbon monoxide, of may be related to the test material rather than its actual products, expressed in terms of mass charge or mass loss concentration. For example, the LC_{50} (mass loss) for wood decomposed under a specific combustion condition might be expressed as 100 g wood m^{-3}. This means that when 100 g of wood were decomposed in the furnace system and the products were dispersed into 1 m^3, exposure to the resultant combustion product atmosphere resulted in a 50% death rate.

Nominal atmosphere concentration mass charge (mass charge concentration): the mass of a test specimen exposed to heating in a furnace system divided by the volume of air into which the combustion products are dispersed (g m^{-3}).

Nominal atmosphere concentration mass loss (mass loss concentration): the mass loss (mass decomposed) of a test specimen exposed to heating in a furnace system divided by the volume of air into which the combustion products are dispersed (g m^{-3}).

Potency: the toxic potency is a measure of the amount of a toxic substance required to elicit a specific toxic effect—the smaller the amount required, the greater the toxic potency. Where the toxic effect is lethality, the potency can be expressed as the reciprocal of the LC_{50} concentration.

Pyrolysis: chemical decomposition of a substance by the action of heat (*cf.* thermal decomposition).

Note 1: pyrolysis is often used to refer to a stage or fire before flaming combustion has begun.

Note 2: in fire science, no assumption is made about the presence or absence of oxygen (but chemically it is possible to distinguished between endothermic pyrolysis under nitrogen or in air, the latter involving some degree of exothermic oxidation).

RD_{50} (respiratory depression 50%): statistically calculated concentration of a sensory irritant required to reduce the breathing rate of laboratory rodents (usually mice) by 50%.

Note: in combustion toxicology, an RD_{50} concentration may be related to a specific combustion product such as hydrogen chloride, or may be related to the test material rather than its actual products, expressed in terms of mass charge or mass loss concentration.

Smouldering combustion: combustion of a material without flame and without visible light.

Stoichiometric combustion: mixture of chemical reactants having proportions in accordance with the equation for a specified chemical reaction.

Note: in a fire context this means that the fuel : air equivalence ratio is at unity, so that there is exactly the right amount of oxygen available for complete combustion of the fuel.

Thermal decomposition: process whereby the action of heat or elevated temperature on an item cause changes to the chemical composition.

Yield: mass of a combustion product generated during combustion divided by the mass of the test specimen. Note: yield can be expressed as mass charge yield or mass loss yield (g product per g specimen).

Recovery fraction (also known as conversion efficiency): The yield of a specified fuel element in a combustion product as a fraction of the maximum yield obtainable from complete conversion of the fuel element to that in the product. This can be calculated from the measured yield of the product of interest relative to its notional yield from complete conversion. For example, if all the carbon in a fuel specimen is burned to carbon dioxide, the recovery fraction for carbon dioxide is 1. If only half the carbon is recovered as carbon dioxide, then the recovery fraction for carbon dioxide is 0.5.

Contents

Issues in Toxicology No. 23
Toxicology, Survival and Health Hazards of Combustion Products
Edited by David A. Purser, Robert L. Maynard and James C. Wakefield

Published by the Royal Society of Chemistry, www.rsc.org

CHAPTER 1

Overview of Combustion Toxicology

DAVID A. PURSER*[a] AND ROBERT L. MAYNARD[b]

[a] Hartford Environmental Research, Hatfield, UK; [b] University of Birmingham, UK, Email: robertmaynard3@gmail.com
*Email: david-purser@ntlworld.com

1.1 Introduction

There can be no doubt that the discovery of how substances can be caused to burn is the greatest of all human discoveries. Until the discovery of nuclear fission, combustion was the only known means of causing matter to release large amounts of stored energy. Combustion remains the overwhelmingly most important means of releasing such energy; indeed even nuclear power would be impossible without combustion: metals have to smelted and fabricated before nuclear reactors can be constructed. Just how important combustion is can be illustrated by considering a lump of coal, a piece of wood or a litre of oil: how could one release the energy within these materials without combustion? From very early times to the present, from the need for warmth in caves to the apparent need for high speed motor cars, combustion has played an essential role in man's development. The essence of the discovery: that combustion releases heat and that heat, a form of energy, can do work is known to everybody.

When organic substances burn, they release heat; they also release chemical products of combustion. The atoms forming the molecules of organic matter cannot be destroyed by combustion, but they can be caused to separate from their original combinations and to form other

Issues in Toxicology No. 23
Toxicology, Survival and Health Hazards of Combustion Products
Edited by David A. Purser, Robert L. Maynard and James C. Wakefield

Published by the Royal Society of Chemistry, www.rsc.org

combinations. Material comprising only molecules containing carbon, oxygen and hydrogen cannot, on burning, fail to produce molecules containing these elements. The principle products of combustion are carbon dioxide and water, but depending on the efficiency of the combustion process, other substances are produced: carbon monoxide, for example. Already we have identified a toxicologically active product of combustion. Carbon monoxide is a very poisonous gas and is responsible for a high percentage of deaths occurring in accidents involving fires. Combustion is often not very efficient and a range of gases and particulate materials are produced: smoke and ash. Some products remain at the source of the fire; others are carried into the atmosphere along with the heated air produced by the fire. Air pollution is produced. Even well regulated combustion of the type seen in modern internal combustion engines produces pollutants. Everybody has seen black smoke being emitted by old diesel powered vehicles; even the "cleanest" new motor car produces carbon dioxide.

The toxic effects of combustion products resulting from various sources are a major cause of morbidity and mortality. All people are exposed, every day, to air pollutants produced by combustion of organic material. Some are exposed to high concentrations: forest fires provide an example. Others are exposed to lower concentrations: levels of air pollution in the countryside of developed countries like the UK provide an example. Some people are exposed to potentially dangerous levels of pollution during their work: those working with diesel engines in confined spaces, those working as fire-fighters provide examples. And some are exposed as a result of accidents: those trapped in a burning building and those exposed to carbon monoxide being emitted by a faulty coke boiler provide obvious examples. Just how dangerous smoke from fires actually is may easily be forgotten. In fact it is very dangerous indeed: incapacitation by inhalation of smoke and consequent inability to escape is the major cause of deaths in fires. In the UK alone 200 deaths and 2500 injuries requiring hospital treatment are, each year, caused by accidental exposure to smoke from uncontrolled fires.[1]

Figures 1.1 and 1.2 show annual fire deaths and injuries per million population in the UK from smoke and burns, by far the majority of which occur in domestic dwellings. In the 1950s most deaths (7.2 per million) and injuries (35.7 per million) resulted from burns, with very few deaths (1.9 per million) and injuries (3.3 per million) attributable to toxic smoke exposure. Between the late 1950s and the early 1970s, although the incidence of injury and death from burns remained approximately constant, there was a more than fourfold increase in deaths and a fivefold increase in injuries from exposure to toxic smoke.

Although a number of factors may be involved, the main cause of this increase was considered to be changes in living styles in the average British home, and in particular the replacement of traditional materials used for the construction of upholstered furniture and bedding by man-made materials, especially polyurethane foam filling and synthetic covering materials. Not only did the incidence of flaming fires increase, but when fires occurred, fire

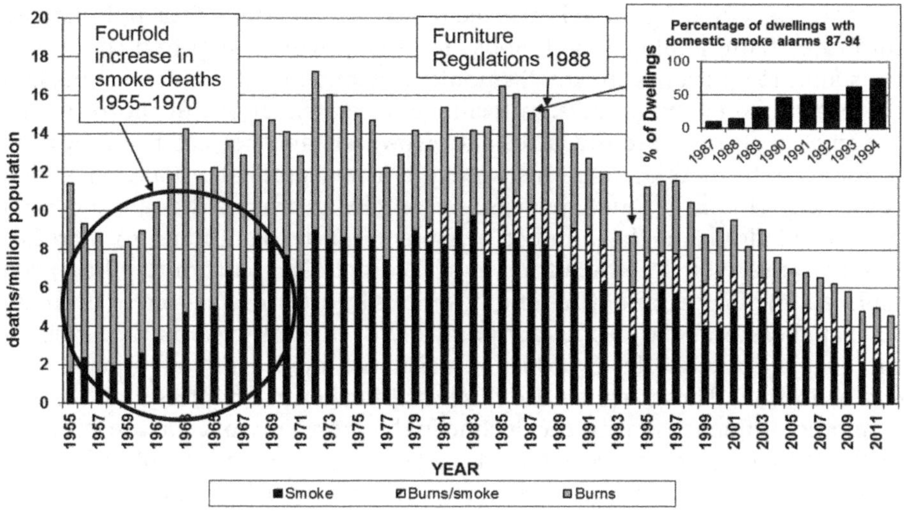

Figure 1.1 Annual fire deaths in the United Kingdom per million population from smoke exposure and burns between 1955 and 2012.[1]

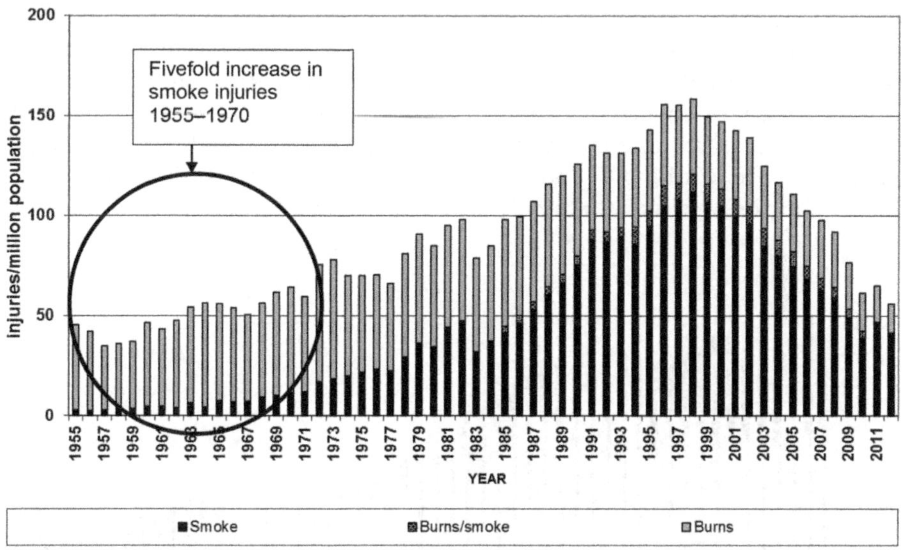

Figure 1.2 Annual fire injuries in the UK per million population from smoke exposure and burns between 1955 and 2012.[1]

growth was rapid and involved the production of large volumes of irritant smoke, containing high concentrations of particulates, carbon monoxide and hydrogen cyanide. The dense smoke was therefore much more likely to impede the escape of occupants, who were then rapidly overcome by asphyxiant gases and heat.

Concerns regarding these issues led to the introduction of the upholstered furniture flammability regulations in 1988, which required improved resistance to ignition from small smouldering and flaming sources. This had no immediate effect, since significant replacement of old furniture in the housing stock took around a decade. However, another safety innovation introduced from this time was smoke alarms, gradual uptake of which coincided with a gradual decrease in smoke deaths, although the total number of fires and the number of smoke injuries continued to increase. From around 2000, by which time much old upholstered furniture had been replaced, the incidence of fires and of serious injuries also started to decrease so that these and the death rate gradually decreased towards the levels of the 1950s. Toxic smoke exposure remains the man cause of injuries and deaths in fires. These aspects are discussed further in Chapters 2 and 15.

Another influence may be the decline in the prevalence of cigarette smoking in the UK over this period (Figure 1.3). This has important implications for both the incidence of acute fire deaths and injuries (since "smokers' materials" are reported as a major cause of fatal fires in the annual fire statistics[1]) and the long term morbidity and mortality from smoking-related diseases.[2,3] However, the decrease in incidence of smoking over the period up to 2000 in fact coincides with the increase in the numbers of fires in dwellings. It is possible that the further decline in smoking since 2000, coupled with the more recent unacceptability of smoking indoors, may be partly responsible for the decrease in the number of dwellings fires since this time.

This variety of causes of morbidity and mortality related to combustion products illustrates one aspect of their complexity: the wide range of toxic product concentrations of interest in evaluating effects, and the very wide range of exposure times that need to be considered. Figure 1.4 captures this,

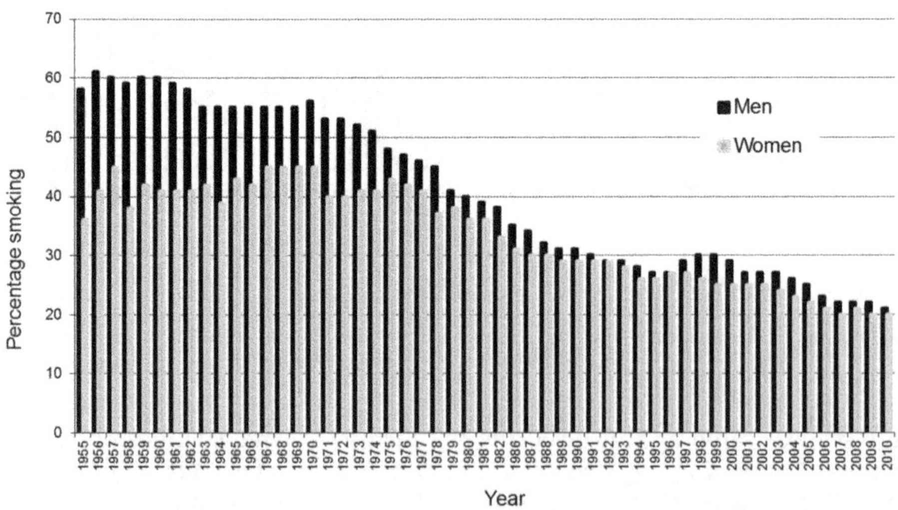

Figure 1.3 Prevalence of cigarette smoking in persons aged 16 and over, UK 1955–2010.[2,3]

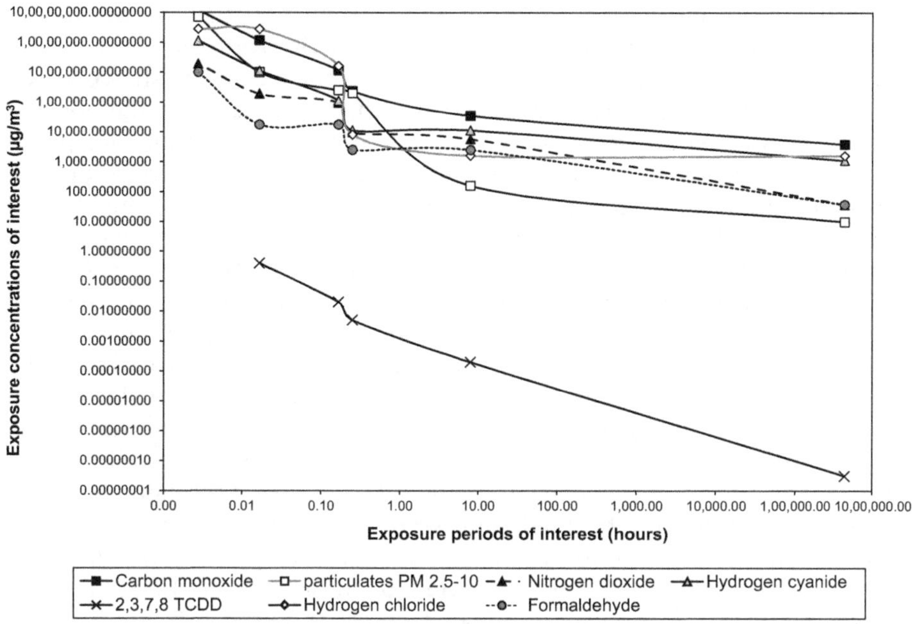

Figure 1.4 Ranges of exposure concentrations and exposure periods of toxicological significance for common components of combustion product atmospheres.

by showing the concentration ranges (expressed in µg m^{-3}) and exposure periods (expressed in hours) of interest for several toxic substances occurring in combustion products mixtures. The figure illustrates the enormous range of concentrations and times of interest involving over 16 orders of magnitude from pg m^{-3} concentrations of dioxins over a 50 year exposure period to over 100 g m^{-3} of carbon monoxide over periods as short as a few seconds.

The top left of Figure 1.4 shows concentrations of asphyxiant gases, such as carbon monoxide and hydrogen cyanide, acid gases and organic irritants, such as hydrogen chloride and formaldehyde, and smoke particulates, which can be present at concentrations up to percent levels by volume or g m^{-3} by mass during fires. In order to understand the effects on the survival of fire victims it is necessary to study the effects of smoke and irritants on escape capability and the incapacitating physiological effects of asphyxiant gases during critical periods of from a few seconds to a few minutes, in order to calculate time to loss of consciousness and death. For subjects rescued alive from fires it is also important to consider the more permanent effects of these gases on cerebral and cardiovascular function, and the effects of irritants and smoke particulates on lung function and pathology (complicated by burn injuries).

Moving on to exposure periods of approximately 1–100 hours, for situations such as acute exposures to diluted smoke plumes from wildfires, the concern relates mainly to effects from smoke particulates and irritants, including nuisance odour, mild eye and respiratory tract irritation and possibly more serious acute effects on vulnerable individuals.

Long term exposure to ambient air pollutants has been shown to be associated with a significant effect on health. In this context the concentrations of interest are very low, for example in the $\mu g\ m^{-3}$ range for pollutants, such as fine particulates (PM 2.5) or formaldehyde, and in the $pg\ m^{-3}$ range for dibenzodioxins. Studies reviewed by the Committee on the Medical effects of Air Pollutants (COMEAP) in 2009 and 2010[4,5] led to the conclusion that current levels of fine particles (PM 2.5) in the UK are responsible for 29 000 attributable deaths each year. A large proportion of material monitored as PM 2.5 comes, directly or indirectly, from combustion processes. No threshold of effect for such findings has been discovered. The effects leading to an increased risk of death involve those on the cardiovascular system and on the risk of lung cancer. It is interesting that effects on the respiratory system appear to be less important. Short term increases in ambient concentrations of air pollutants are also associated with increases in deaths and hospital admissions: in this case the respiratory system is affected, in addition to the cardiovascular system. The European Commission has funded a recent review of these effects.[6] In addition to these environmental exposures, tobacco smoking is the primary cause of preventable illness and premature death, accounting for approximately 100 000 death per year in the UK, representing 36% of all respiratory deaths, 28% of all cancer deaths and 14% or all circulatory deaths.[7]

With regard to some of these toxic substances, especially dioxins, the systemic health effects may result directly from inhalation exposure, but also from secondary ingestion following deposition of combustion products into water and taken up into foodstuffs, or even as a result of dermal exposure.[8] These issues are discussed in Chapters 3 and 12.

These few statistics illustrate the enormous ongoing adverse morbidity and mortality costs of exposure to combustion products.

1.2 Summary of the Main Aspects of the Subject Areas Addressed in the Book

In order to cover the different aspects of fires, combustion products and toxic health effects we have divided this book into five sections:

- The Science of Combustion.
- The Toxicity of Combustion Products.
- The Clinical Toxicology and Management of Combustion Product Exposures.
- Examples of Unusual Conflagrations.
- Public Health Aspects of Fire Management.

"The Science of Combustion" consists of five chapters addressing the chemical and physical aspects of combustion processes, fire development and fire types, the composition of combustion product "fire effluent" or "smoke" atmospheres, and the measurement and calculation of the yields and concentrations of toxic substances from a range of fuels in a variety of fire scenarios. An important aspect of fires is that the rates of production and composition of the combustion products vary considerably depending upon the combustion conditions. In Chapter 2, titled "Fire Types and Combustion Products", David Purser describes different fire types and fire stages in terms of fire scenario development, combustion conditions, the range of combustion products formed and the behaviour of different fuels depending on elemental and structural composition. In Chapter 3, titled "Estimating Yields and Quantities of Mass Releases of Toxic Products from Fires", he further develops this theme. The chapter describes methods for generating and measuring toxic products and yields under different fire conditions and presents yield data for a range of fuel types and fire scenarios. The chapter then explains how these can be used as input into engineering calculations for mass releases of combustion products both within and beyond enclosures such as buildings. Some of the technical issues with respect to valid test methods for burning materials under defined combustion conditions, and some challenges in sampling and measuring the composition of the combustion product atmospheres, are flagged up in this chapter. They are then described in more detail by T. Richard Hull and Anna Stec in Chapter 5, titled "Generation, Sampling and Quantification of Toxic Combustion Products". In Chapter 4, titled "Products of Combustion and Toxicity from Specific Types of Fires", James Wakefield describes a wider range of fire types and combustion products, including those involving industrial fires, from which specific industrial chemicals are released into the combustion product mixtures from the general building or vehicle materials. When large conflagrations such as these occur, large fire effluent plumes are released to the outdoor environment. In order to assess the hazards from such plumes, the mass releases from the source fire (calculated using methods such as those described in Chapter 3) are used as input to smoke plume dispersal calculation models, which are described by David Hall and Angela Spanton in Chapter 6.

Having generated our combustion products in the first section of the book, we next examine their toxicology in the following eight chapters. In Chapter 7 David Purser considers the application of animal models and human studies to prediction of combustion toxicity in humans. Combustion product atmospheres contain a complex mixture of gases and particulates interacting to produce a series of toxic effects on exposed subjects. Our understanding of these effects, which substances are the major actors and how they interact, is derived from a variety of sources. These include human fire incident investigations, animal exposures to combustion product atmospheres generated by burning a range of materials under different fire conditions, and exposures of rodents, primates and human volunteers to individual toxic substances (or specific mixtures of substances) known to occur in fire effluents. Prof. Purser explains how the main effects of acute exposure have been found to occur in a sequence consisting of immediate eye and upper respiratory tract irritancy followed by incapacitation due to the effects of asphyxiant gases during exposure, then followed by respiratory tract and lung inflammation some hours after exposure. These studies have shown that the major effects of exposure to fire effluents can be explained in terms or a small number of key irritant and asphyxiant substances. The quantification of the effects of these substances and their interactions is described and their incorporation into toxic potency calculation models for combustion product mixtures.

An important aspect of acute exposure to specific substances and evaluating their toxic effects is the concentration, exposure dose, toxicity relationships for different substances, and the extent to which they follow Haber's Law. Aspects of this are addressed in Chapter 7, and a detailed consideration of relationships to Haber's Law is provided by Robert Maynard and David Purser in Chapter 8, titled "Haber's Law and its Application to Combustion Products".

The remaining six chapters in this section consider the toxicology of specific substances, or groups of substances, of particular significance. Chapter 9, titled "Carbon Monoxide", by Robert Maynard, Isabella Myers and John Ross, and Chapter 10, titled "Hydrogen Cyanide – Physiological Effects of Acute Exposure during Fires", by David Purser contain detailed descriptions of the uptake dynamics, calculation of blood and tissue concentrations, and physiological and other toxic effects for these mayor asphyxiant gases. In Chapter 11 Ken Donaldson, Amanda Hunter, Craig Poland and Steve Smith describe the mechanism of action of combustion-derived nanoparticles. David Purser covers the health hazards from dioxins and carcinogens in Chapter 12, and James Wakefield those from irritant gases in Chapter 13.

Analysis of the life-threatening exposures of building occupants during fires is important for building design hazard assessments, understanding the condition of survivors at the fire scene and in the emergency room, and for forensic investigation of fatal fire incidents. For these applications it is necessary to be able to model and predict the sequence of effects on occupants during fire incidents and their severity. In Chapter 14, titled "Acute Effects of Combinations of Toxicologically Active Substances and Heat on Fire Victims in Buildings and During Exposures to Outdoor Smoke Plumes", David Purser describes the "fractional effective dose" modelling method with worked examples for actual fire scenarios.

The three chapters in the third section of the book cover clinical aspects of combustion toxicology and management of combustion product exposures. Timothy Marrs has contributed a chapter on the clinical management of cyanide poisoning, which complements Chapter 10 on hydrogen cyanide uptake and toxicology. David Baker has addressed the subject of irritant gases and acute lung injury, complementing descriptions of the toxicology of irritant combustion products in Chapter 7 and individual irritant gases in Chapter 13.

Although the severe acute health effects during and after exposure at fire scenes are well established, the significance of acute exposures to dilute outside smoke plumes from large fires is more controversial. Typically reported are nuisance odour, and minor eye and throat irritation, or breathing problems, but estimation of the predicted effects can be challenging. Wildland fires provide situations where populations, including the public and volunteer fire-fighters, may be exposed for periods from hours to several days to relatively dense smoke plumes from large areas of burning vegetation. Urban fire-fighters wear breathing apparatus during rescue and fire-fighting activities in buildings, but often have no respiratory protection in the immediate fire surroundings, when "damping down" during the last stages of a fire, or when investigating a contaminated post-fire scene. If there are significant acute or chronic health hazards from exposure to diluted smoke plumes they should show up in these groups. Jamie McAllister has reviewed this important subject in Chapter 17, titled "Health Effects in Groups Exposed to Wildland and Urban Fires".

The two chapters in the fourth book section consist of descriptions of some unusual major conflagrations. Thomas Waite, Catherine Keshishian and Virginia Murray in Chapter 18, titled "The Buncefield Fire", have contributed a chapter on this major incident, the largest fire in Europe since World War II, which had minimal health effects despite producing a large smoke plume over southern England for several weeks. Michaela Kendall, Mitchell Cohen and Lung-Chi Chen have described the consequences of the World Trade Center fires. The content of the effluent plume from this incident has led to considerable chronic health effects on exposed subjects. However, the main effects appear to be related not so much to the combustion products, which largely dispersed as a high altitude plume rather as at Buncefield, but owe more to the mineral dust and fibre plume released from the collapsing buildings. This illustrates an important point not covered elsewhere in the book, that mineral and related particulates can be carried up into energetic fire plumes during building fires, and then deposited in the surrounding areas depending on parameters such as the particle size, density and settlement velocity.

The final section consists of a chapter by Virginia Murray on the public health aspects of fire management in Chapter 20, titled "Providing Advice to Those Exposed to Combustion Products".

References

1. Fire Statistics, Great Britain. Published annually by communities and local government.
2. Trends in Smoking. Lung and asthma information agency, Fact sheet 98/2. St George's Hospital Medical School. London. 1998.
3. General Lifestyle Survey, Office for National Statistics, 2012.
4. Committee on the Medical Effects of Air Pollutants (COMEAP), *Long-term Exposure to Air Pollution: Effect on Mortality*, Department of Health, 2009.

5. Committee on the Medical Effects of Air Pollutants (COMEAP), *The Mortality Effects of Long-term Exposure to Particulate Air Pollution in the United Kingdom*, Department of Health, 2010.
6. World Health Organization, *Review of evidence on health aspects of air pollution – REVIHAAP. Technical Report*, World Health Organization Office for Europe, 2013.
7. Estimated premature deaths for each jurisdiction are as follows: England – 79 100; Scotland – 13 000; Wales – 5600; N. Ireland – 2300. Sources: Statistics on smoking: England, 2012; The NHS Information Centre for Health and Social Care, 2012; ScotPHO Smoking Ready Reckoner, 2011 edition; Welsh Government, Health Improvement; NIdirect – smoking.
8. D. A. Purser, Toxicity of fire retardants in relation to life safety and environmental hazards, in *Fire Retardant Materials*, ed. A. R. Horrocks and D. Price, Woodhead Publishing Ltd, Cambridge UK, 2001, pp. 69–127, Chapter 3.

The Science of Combustion

CHAPTER 2

Fire Types and Combustion Products

DAVID A. PURSER

Hartford Environmental Research, Hatfield, UK
Email: david-purser@ntlworld.com

2.1 Introduction

From the perspective of toxic fire hazard development and fire chemistry it is possible to classify fires into a range of basic types or stages depending on the nature of the burning fuel and the setting in which it is burning (the fire scenario).[1]

In any specific fire the toxic hazard at any location and time during a fire depends on:

- the mass burning rate of the fuel ($kg\ s^{-1}$);
- the yields of each of a set of toxic products ($kg\,kg^{-1}$ fuel burned);
- the air entrainment and volume into which the products are dispersed, which determines the concentrations of toxic products in the fire effluent.

Fire dynamics (mass burning rate and dispersal of the fire effluents) depend upon a range of variables related to the specific fire scenario, but the yields of toxic products depend upon three main aspects:

- The elemental composition of the fuel.
- The organic composition of the fuel.
- The combustion conditions.

Issues in Toxicology No. 23
Toxicology, Survival and Health Hazards of Combustion Products
Edited by David A. Purser, Robert L. Maynard and James C. Wakefield

Published by the Royal Society of Chemistry, www.rsc.org

With respect to the elemental composition, almost all fires involve organic materials consisting primarily of carbon and hydrogen, sometimes with oxygen, which are oxidised to a varying extent to produce water, carbon monoxide, carbon dioxide and a range of aliphatic and aromatic hydrocarbons and carbonaceous particulates. Many common fuels also contain "heteroelements", such as halogens, nitrogen, sulfur and phosphorus, which are often added as flame retardants. These have two effects on combustion product formation. Firstly, they introduce an additional range of products including acid gases and substituted organic compounds such as organic nitriles, organophosphorus products, and halogenated dibenzofurans and dioxins. Secondly, they influence the combustion conditions, thereby affecting the yields of other toxic products such as carbon monoxide, hydrogen cyanide, volatile organic compounds (VOCs), and smoke particulates.

The organic composition of the fuel also affects the mix of organic products evolved in fires. In general when organic materials are heated to a sufficiently high temperature to decompose thermally ($> \sim 250$–$300\,^{\circ}C$) they produce a range of common pyrolysis products including a series of saturated and unsaturated aliphatic hydrocarbons with a range of chain lengths and molecular masses from methane up to octane and beyond. The yields tend to be biased towards the lower molecular mass range and unsaturated compounds, with high yields of methane, ethene, and propene. Partial oxidation of these compounds forms aldehydes and ketones, of which the lower molecular mass aldehydes, especially formaldehyde and the unsaturated aldehydes acrolein and crotonaldehyde, are probably the most toxicologically significant due to their high irritant potency (see Chapter 7). In hot fire plumes some of these compounds undergo ring cyclisation, so that, for example, hexane forms cyclohexane, and then, through progressive further dehydrogenation, compounds, such as benzene, toluene, naphthalene, styrene, methyl styrene, and polyaromatic hydrocarbons, are formed.[1,2] Further dehydrogenation forms ever enlarging graphene-like plates culminating in carbon smoke particles. Partial oxidation forms compounds such as phenol. Analysis of thermal decomposition products from most organic materials usually shows strong peaks for benzene, naphthalene, styrene, methyl styrene, and phenol. These organic products, common to effluents from most materials, are then often enhanced by a range of organic decomposition products specific to the polymer being decomposed.[2] Thus effluent from polystyrene contains high yields of aromatic compounds, especially styrene monomer and soot particulates. Flexible polyurethane foams, common in upholstered furniture, decompose into polyol compounds (long chain alcohols), which tend to form a liquid pool fire, and toluene di-isocyanate, which polymerises with itself in the vapour phase to form irritant isocyanate "yellow smoke" particulates characteristic of furniture fires.[3,4] Pyrolysis of cellulosic materials releases proteoglycans.

The combustion conditions at different times and locations during a fire have a major effect on the yields of the combustion products. The most important determinants are whether the fire is flaming or non-flaming, and for flaming fires the fuel : air equivalence ratio [phi (ϕ)], expressed in terms of the actual fuel : air ratio divided by the stoichiometric fuel : air ratio.[5] In general, during well-ventilated flaming ($\phi < 1$), in which there is an excess of air to support combustion, non-flame retarded materials burn efficiently, with the main products being water, CO_2 and a small amount of powdery carbonaceous soot. Under non-flaming and fuel-rich flaming conditions [when the airborne mixture contains more fuel than the air required to burn it ($\phi > 1$)] combustion is inefficient, with high yields of CO, organic compounds and particulates. The particulates consist of carbon-rich particles, but also condensed organic liquid droplets, so that soot deposits tend to be sticky and oily as opposed to the dry, powdery soot from well-ventilated fires. Where halogens are present, the yields of dioxins and furans and their precursors (halogenated phenols and benzenes) are considerably increased by fuel-rich combustion conditions, and where nitrogen is present, high yields of hydrogen cyanide occur.[6,7]

Figures 2.1 and 2.2 show examples of the effects of fuel : air equivalence ratio on toxic product yields in flaming fires.[8,9] Figure 2.1 shows the yield of carbon monoxide from poly(methyl methacrylate) (PMMA) expressed as the ratio of CO yield at each equivalence ratio to that under well-ventilated combustion conditions. The CO yield increases by a factor of approximately 50 between well-ventilated flaming conditions ($\phi < 1$), and fuel-rich conditions ($\phi > 1$). The figure shows data from two different bench-scale combustion test methods designed to measure toxic product yields as a function of equivalence ratios, the ISO TS 19700 tube furnace[10] (see Chapter 3, Figure 3.2) and the ASTM E2058 flammability apparatus.[11] There is a good agreement between the results from the two methods and results from large-scale fire tests, indicating that the results are a fundamental property of the combustion conditions and therefore representative of effects occurring in full-scale fires. Figure 2.2 illustrates this by a direct comparison of the CO yields (CO g g^{-1} fuel burned) obtained using the ISO TS 19700 tube furnace, with those obtained in large-scale fires.[5] Such validation is important because a number of bench-scale test methods are in use for which the combustion conditions are poorly characterised, so that the toxic product yields show a poor correlation with those obtained from full-scale fires (see Chapter 5).

When materials are burned over a range of combustion conditions in terms of fuel : air equivalence ratios, the yields of different toxic products may remain approximately constant, or may increase or decrease depending on the product and how it is affected by the combustion conditions. Thus while the yield of CO, which is a product of incomplete combustion, increases under fuel-rich conditions, the yield of CO_2, which is the main end product of efficient combustion of fuel carbon, decreases approximately in proportion to the equivalence ratio. So, at an equivalence ratio of 2, when there is half the oxygen available for complete combustion of the fuel, the CO_2 yield is approximately halved. For halogens, which are released mainly in the form of hydrogen halides, the efficiency of conversion of fuel halide to hydrogen halide is generally high, so that for any

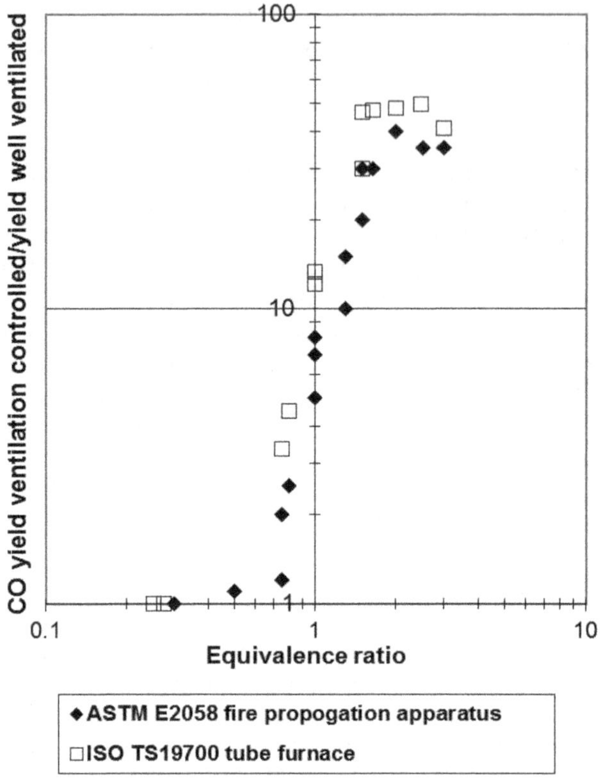

Figure 2.1 CO yield from flaming PMMA as a function of equivalence ratio in two bench-scale test methods: the ASTM E2058 fire propagation apparatus and the ISO TS 19700 tube furnace.[8]

individual material the yields are similar across a range of combustion conditions.[12]

Figure 2.3 illustrates the conversion of fuel carbon to the main product of combustion, carbon dioxide. The figure shows data for flaming combustion of a range of common polymeric materials as a function of the equivalence ratio in the ISO TS 19700 tube furnace.[13] The results are expressed in terms of the efficiency with which fuel carbon is converted to carbon dioxide. The results show that for most common non-flame retarded materials combusted under well-ventilated combustion conditions ($\phi<1$), 90–100% of fuel carbon (*i.e.*, recovery fraction 0.9–1.0) is released as carbon dioxide. Exceptions are the highly aromatic polymer polystyrene, and materials containing halogens as flame retardants (FRs) such as the FR-treated Bouclé fabric (consisting of a mixture of acrylonitrile, wool, polyester, and a flame retardant containing chlorine and bromine) and polyvinylchloride (PVC). These show inefficient combustion across the range of equivalence ratios. As combustion becomes fuel-rich at higher equivalence ratios, the

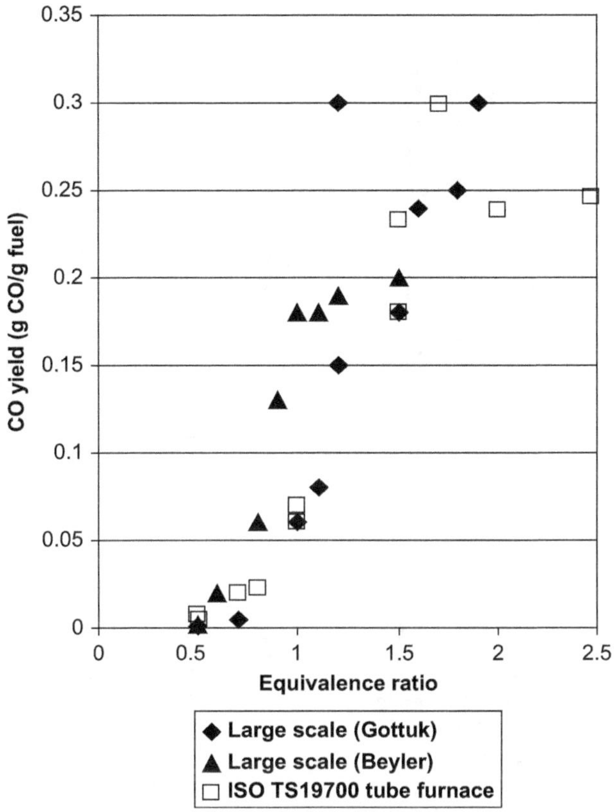

Figure 2.2 Validation of the ISO TS 19700 tube furnace data for CO yield from poly(methyl methacrylate) (PMMA) as a function of equivalence ratio by comparison with results from large-scale compartment fire tests.[8]

conversion to CO_2 decreases and the conversion efficiencies and yields of other products, such as CO, start to increase as shown in Figures 2.1 and 2.2.

Figure 2.4 illustrates the effects on a range of combustion products including CO, HCN, total mixed organics (expressed as CH_2), and smoke particulates from a nitrogen-containing polymer (polyamide 6). As the fuel : air ratio increases, the yields of oxidised products, such as nitrogen oxides, decrease (as does CO_2), but the yields of products of incomplete combustion increase including particulates, total organics (VOCs), and, in particular, hydrogen cyanide.[14] With regard to toxic nitrogen-containing substances, the yields of nitrogen oxides are low, even under well-ventilated combustion conditions, and the primary product is NO, which has a low toxic potency. This undergoes a slow, partial further oxidation to NO_2, which may make a minor contribution to lung irritation (see Chapter 7). However, under the fuel-rich combustion conditions occurring during most building fires, the release of fuel nitrogen as HCN increases considerably. In general,

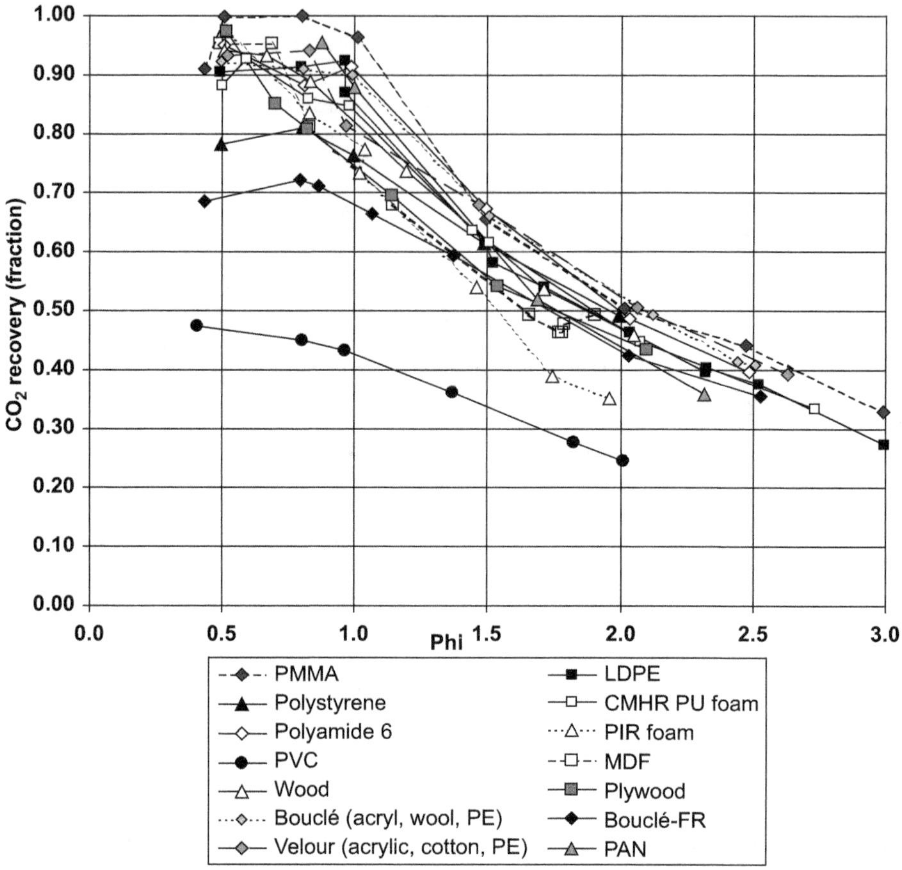

Figure 2.3 Carbon dioxide recovery fraction as a function of equivalence ratio for the flaming combustion of a set of common polymers in the ISO TS 19700 tube furnace.[13] LDPE, low-density polyethylene; CMHR PU, combustion modified high resilience polyurethane; PIR, polyisocyanurate; PAN, polyacrylonitrile.

for nitrogen-containing materials, the efficiency of conversion of fuel nitrogen to HCN is similar to the efficiency of conversion of fuel carbon to CO.[14] The yields of these substances then also depend on the carbon and nitrogen content of the fuel.

In addition to the major effect of non-flaming or flaming behaviour and the fuel : air ratio, oxygen concentration and temperature have also been found to affect toxic product yields for some polymers under fuel-rich combustion conditions.

The yields and rates of production of all of these products vary considerably in different fire scenarios, and during any particular incident, the nature of the fire may remain relatively constant, or more often develops through a series of stages, each involving a different set of combustion conditions. At any one time during a fire different areas of fuel may also burn under

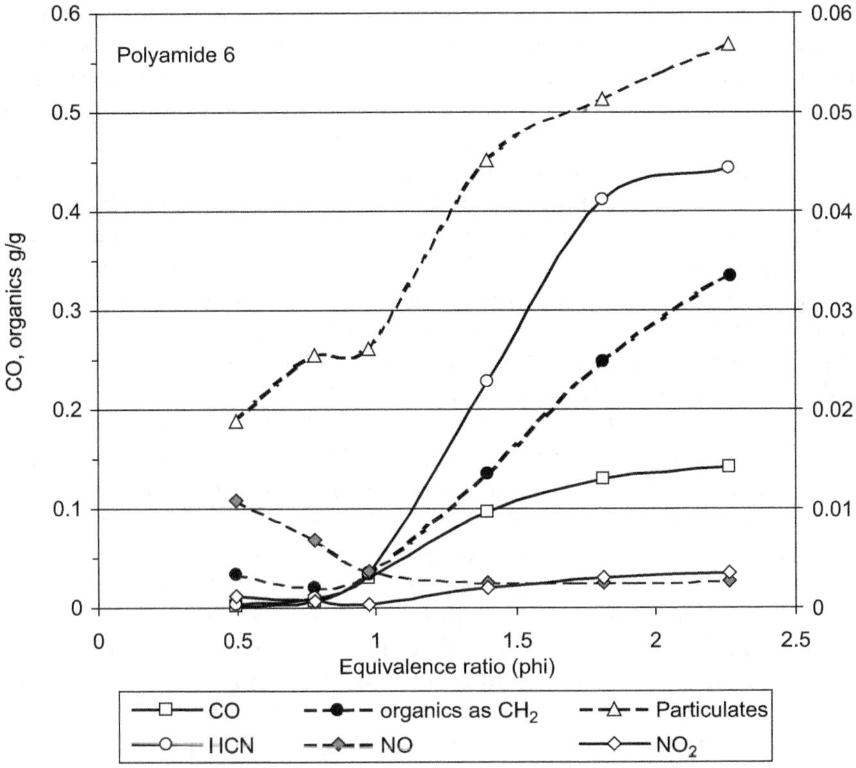

Figure 2.4 Yields of toxic products from flaming combustion of polyamide 6 in the ISO TS 19700 tube furnace.[13,15]

different conditions, evolving different combustion products into the effluent plume mixture. The main fire types may be classified as follows:

- Non-flaming and smouldering fires:
  - Pyrolysis in an inert atmosphere (such as under nitrogen).
  - Oxidative thermal decomposition (oxidative pyrolysis).
  - Smouldering.
- Early or well-ventilated flaming fires (fuel-controlled fires).
- Under-ventilated, fuel-rich (ventilation-controlled) fires:
  - Under-ventilated, pre-flashover fires.
  - Under-ventilated, post-flashover fires.

2.2 Non-flaming and Smouldering Fires

2.2.1 Non-flaming Thermal Decomposition

The initiating event in any fire is non-flaming thermal decomposition (pyrolysis) resulting from the application of heat, usually from an external source but sometimes through self-heating. The incident heat radiation from the source raises the surface temperature of the fuel. Solid fuels, such as natural and synthetic polymers, typically begin to decompose at temperatures exceeding approximately 200–300 °C by pyrolysis, an endothermic molecular scission releasing airborne volatile fuel vapours. The decomposition rate follows Arrhenius relationships, increasing with temperature, but varying considerably depending upon the fuel compositions. However, it is slow compared with that in flaming fires, so that non-flaming or smouldering conditions generally require several hours to generate toxicologically significant concentrations of combustion products in a building. For substances containing only carbon and hydrogen, such as polypropylene, the fuel vapours consist of a range of hydrocarbon fragments of varying molecular masses including those listed in the second column of Table 2.1.

When a fuel is heated in air some partial oxidisation occurs at the fuel surface, adding oxidised organic species (aldehydes, ketones, aliphatic alcohols and acids, phenol) and carbon oxides, to the products released. This process is exothermic, and may enhance the pyrolysis rate. For polypropylene, the oxidised species detected are shown in the third column of Table 2.1. When mice or primates were exposed to the products from polypropylene decomposed under nitrogen and therefore containing no oxidised organics, the products were found to be non-irritant and virtually non-toxic acutely, but when decomposed in air the atmospheres were found to be highly irritant, causing acute sensory and airway irritation during exposure and lung oedema and inflammation a few hours later (see Chapter 7). Carbon monoxide yields are also relatively high under these conditions. For polymers containing oxygen, including cellulosic materials (*e.g.*, wood, cotton), a wider range of products is released even under nitrogen pyrolysis, with accompanying irritancy, and even higher yields occur during decomposition in air. Cellulosic materials in particular produce relatively high yields of acrolein and formaldehyde: bonfire smoke provides a familiar example.[6,16] The thermal decomposition products include a mix of vapours, liquid droplets, and solid smoke particles. The particulates from most polymers are in the respirable size range, with a significant component in the micron or sub-micron range.

For materials containing nitrogen, the conversion of fuel nitrogen to nitrogen-containing products is generally low under non-flaming decomposition conditions, but depends upon the molecular structure of the polymer. Fuel nitrogen can be released as N_2, which is harmless, nitrogen oxides, ammonia, HCN, and organic nitriles. Polyacrylonitrile decomposes to release HCN even at relatively low temperatures.[6,14] Flexible polyurethane foams release nitrogen as organic isocyanates in the form of a highly irritant "yellow smoke",[3,17] which is dispersed, largely intact, away from the heated area, although some thermal decomposition can occur in heated regions above approximately 600 °C to produce small amounts of HCN.

Halogenated materials release chlorine and bromine mainly by dehydrohalogenation as HCl or HBr. In the fire effluent these are partitioned

Table 2.1 Pyrolytic and oxidative decomposition products of polypropylene at 500 °C showing percentage yields of major irritants.[6]

MS interpretation	Pyrolysis yield[b]/% ($\times10^{-1}$)	Oxidation yield[b]/% ($\times10^{-1}$)
Ethylene	10.4	8.1
Ethane	3.7	2.1
Propene	18.6	18.4
Cyclopropane	0.5	0.3
Formaldehyde[a]	—	33.2
Propyne	0.2	—
Acetaldehyde[a]	—	35.0
Butene	9.6	20.1
Cyclobutene	0.3	0.8
Methyl vinyl ether[a]	—	10.4
Acetone[a]	—	38.4
Butane	1.2	—
Methyl propane	0.4	—
Methyl butane	4.0	—
Butenone[a]	—	1.3
Methyl butene	29.7	12.9
Pentanol[a]	—	12.5
Cyclopentane	0.5	1.4
Pentadiene	1.3	—
Crotonaldehyde[a]	—	7.7
Ethylcyclopropane	0.1	—
Methyl vinyl ketone[a]	—	2.8
Methyl ethyl ketone[a]	—	4.7
Hexane	0.9	1.2
Cyclohexane	32.2	19.3
Hexadiene	3.7	2.2
Hexyne	—	1.3
Benzene	6.7	5.1
Methyl propyl ketone[a]	—	1.9
Pent-2-ene-4-one[a]	—	7.5
Phenol[a]	—	11.6
Toluene	2.4	16.1
Methyl cyclohexadiene	2.1	0.1
Xylene	6.1	0.2
Styrene	5.6	4.0

[a]Oxygen-containing products.
[b]Mass percentage conversion of polymer.

between free gases, acid mist associated with condensed water droplets and condensed onto smoke particles. While the acid gases act as eye and upper respiratory tract irritants, the droplets and particles are respirable, delivering acid deep into the lung. Bromine can also be released as bromine gas.

Table 2.2 shows the yields obtained of major combustion products from four materials decomposed under four different combustion conditions: non-flaming oxidative thermal decomposition (*i.e.*, heating in air), well-ventilated flaming ($\phi<1$), under-ventilated (fuel-rich) flaming ($\phi>1$) at 650 °C under conditions of enclosed building fires, and under-ventilated

Table 2.2 Toxic product yields (mass charge) and toxic potency of wood and 100% organic cable material (low density polyethylene) compared with a plasticised-PVC cable and a low smoke and fume (LSF) cable material containing aluminium hydroxide.

Material and decomposition condition	CO/ mg g^{-1}	CO_2/ mg g^{-1}	HCl/ mg g^{-1}	Organic carbon as CH_2/ mg g^{-1}	Particles/ mg g^{-1}	Smoke OD/g m^{-2} SEA m^2 kg^{-1}	LC_{50}/g m^{-3} mass charge	LC_{50}/g m^{-3} mass loss
Wood (*Pinus sylvestris*):								
Non-flaming 350 °C	48	87		14	152	82	34	23
Well ventilated ~650 °C	39	1494		34	2	3	77	77
Vitiated 650 °C	158	690		127	19	4	29	25
Post flashover ~825 °C	39	1165		69	7	3	52	48
LDPE:								
Non-flaming 350 °C	169	341		626	296	630	17	16
Well ventilated 650 °C	5	2584		133	49	167	70	70
Under ventilated 650 °C	149	1572		383	76	324	28	28
Post flashover 825 °C	112	1592		383	86	333	24	24
LSF:								
Non-flaming 350 °C	95	210		50	132	404	60	25
Well ventilated 650 °C	21	1651		0	23	112	168	100
Under ventilated 650 °C	87	578		390	28	129	73	35
Post flashover 825 °C	88	613		362	31	125	50	31
PVC (plastic + $CaCO_3$)								
Non-flaming 350 °C	9	108	345	18	255	672	28	12
Well ventilated 650 °C	69	1359	145	0	83	332	34	22
Under ventilated 650 °C	84	739	168	99	66	300	29	16
Post flashover 825 °C	170	834	405	0	56	341	18	9

(fuel-rich) flaming ($\phi > 1$) at 825 °C under conditions typical of large "post-flashover", fuel-rich building fires. The materials included Scots pine wood (*Pinus sylvestris*), low-density polyethylene (LDPE), low smoke and fume electrical cable insulation material (consisting of polyethylene 32%, vinyl acetate 8%, and hydrated alumina 60%), and a plasticised PVC cable

Table 2.3 Toxic product yields (mass charge) and toxic potency of cotton twill untreated and treated with borax/boric acid under non-flaming and flaming conditions at 400 °C and 700 °C at a mass charge concentration of 20 g m^{-3}.

Material and decomposition condition	CO/g g^{-1}	CO_2/g g^{-1}	Acrolein/ g g^{-1}, ×1000	Formaldehyde/ g g^{-1}, ×1000	Smoke OD/m	LC_{50} g m^{-3} mass charge	LC_{50} g m^{-3} mass loss
Cotton twill:							
Non-flaming 400 °C	0.21	0.75	2.10	24.66	0.16	23	21
Non-flaming 700 °C	0.33	0.70	5.83	17.74	0.06	14	14
Flaming 700 °C	0.04	1.61	0.00	14.40	0.00	57	55

insulation material containing 22% PVC, 28% $CaCO_3$, and 50% dibutyl phthalate. They were decomposed using the ISO TS 19700 tube furnace.[18]

For non-flaming decomposition in air, all four materials show relatively high yields of CO, organic products, and smoke particulates. PVC also shows high yields of HCl. The last two columns in Table 2.2 show the estimated rat LC_{50} concentrations for a 30 minute exposure period calculated using the method of Purser[19] (see Chapter 7). The LC_{50} concentrations are shown as the mass concentration of material m^{-3} expressed in terms of mass charge (the mass of material exposed to combustion conditions) and mass loss (the mass of material consumed during combustion), the difference being the mass of residue remaining after combustion. This is the accepted method for expressing the lethal toxicity of the mixed combustion product atmospheres from specific materials and relates to a standard 30 minute exposure period.

Table 2.3 shows data from a cellulosic (cotton twill) material decomposed in a tube furnace under non-flaming and flaming conditions.[16,20] Significant yields of acrolein and formaldehyde were measured, especially under non-flaming conditions.

Since non-flaming thermal decomposition depends upon an exterior heat source, it tends to be limited to a small fuel area and to be slow, so that although the yields of irritant products and CO can be relatively high, the mass production rate and resulting atmosphere concentrations are low. The hazard from such fires is generally low unless a transition to flaming occurs. Two exceptions involving more exotic toxic products are fluorocarbons and certain materials treated with flame-retardants containing phosphorus.

2.2.2 Fluorocarbons

For fluorocarbons the situation is much more complex[21] (see Chapter 7). Perfluorinated polymers like polytetrafluoroethylene (PTFE) are thermally stable, so that limited thermal decomposition occurs below around 450 °C, and flaming ignition requires temperatures of around 650 °C. When PTFE (for example, in the form of a cable insulation or a cookware coating) is heated, thermal decomposition starts to occur very slowly at temperatures above around 350 °C. The main product is the tetrafluoroethylene monomer, which is released into the vapour phase, where it re-polymerises to form high molecular mass particles (2000–4000 Daltons). The particles are initially in ultrafine ("nano") particle size ranges (0.01–0.15 μm). When inhaled they are deposited in the lung interstitium and are highly irritant to the lower respiratory tract, causing a respiratory pattern characteristic of stimulation of lung irritant receptors within a few minutes and fatal lung oedema within a few hours (see Chapter 7). However, when this aerosol ages over time scales of a minute or so, the ultrafine particles agglomerate to form much larger particles in the respiratory (1–2 μm) size range, which are harmless when inhaled. At temperatures above around 600 °C (non-flaming) and under flaming conditions, PTFE and other perfluoropolymers decompose to form mainly carbonyl fluoride, which hydrolyses to form HF. Another high toxic potency compound formed is perfluoroisobutylene (PFIB). Under these conditions the toxic effects include immediate painful sensory and lung irritancy, followed by lung inflammation and oedema. As discussed in Chapter 7, the lethal toxic potency of PTFE under conditions producing the ultrafine particulates is extremely high (rat 30 minute LC_{50} as low as to 0.17 g m^{-3} mass loss).

2.2.3 Fate of Phosphorus in Combustion Products

The fate of fuel phosphorus during thermal decomposition and flaming combustion is another complex issue. Phosphorus is used in elemental and organic form in various flame-retardant additives.[22] When heated these assist char formation, so a proportion may be retained in inorganic form in char, but a proportion is released into the gas phase, mostly as phosphorus pentoxide, which is rapidly hydrolysed to phosphoric acid.[20] There is a concern that some may be released as toxic organophosphorus products. One known example of such a product is trimethylolpropane phosphate (TMPP), a relatively thermostable caged biphosphosphorus ester, which is neurotoxic (see Chapter 7). This is formed by thermal decomposition of materials such as turbine engine lubricants containing a phosphate and trimethylol propane polyols.[20]

2.2.4 Brief Case Histories of Non-flaming Thermal Decomposition Incidents

Overheated Trouser Press

Hotels rooms often contain heated trouser presses. During a particular incident, the timer, thermal and electrical safety cut-outs failed, which resulted in the press heating clothing throughout the night while the room occupant slept. In the morning the occupant suffered a headache (presumably due to fumes and a small amount of CO) and a sore throat (from irritant thermal decomposition products), but was otherwise uninjured.

Overheated Heating Element in an Air-conditioning Humidifier

This incident occurred in a multi-storey office building in Birmingham. A water supply fault resulted in a water evaporation unit in an air-conditioning humidifier boiling dry in a roof-level plant room. This resulted in the thermal decomposition of the cover and insulation from a small length (approximately 0.5 m) of PVC insulated cable. The thermal decomposition products were released into the plant room, and then into the conditioned make-up air to a large sector of the building. Due to the enormous dilution factor the concentration of smoke products was extremely low, but when staff arrived for their morning shift, they became aware of an unpleasant odour, with a thin but visible smoke obscuration, especially on certain floors. After an hour or so the conditions were such that it became necessary to evacuate the building.

2.2.5 Smouldering

A special case of non-flaming decomposition that can be potentially more hazardous is smouldering decomposition. Smouldering decomposition occurs only with materials in which thermal decomposition occurs in a porous solid, into which air can penetrate beyond the surface. This therefore excludes flammable liquids or thermoplastic polymers, which melt on heating, forming a surface melt. Commonly, a carbonaceous char is formed in smouldering materials when heated. These include mainly cellulosic materials and polyurethane foams. Smouldering is an exothermic oxidation process, which can be initiated and driven by an externally applied heat source, or can be self-heating or self-sustaining. The most common hazard scenarios occur in domestic dwelling fires and in industrial or agricultural settings. In a domestic setting smouldering can occur in furniture and bedding or clothing, either as a result of overheating from radiant heat sources (for example, by hanging clothing to dry too close to a heater), or by contact with a discarded cigarette. In industrial and agricultural settings smouldering often results from interior self-heating of vulnerable bulk materials in heaps or silos. In stored natural materials, such as straw or wood chippings, heat can be generated in the interior by bacterial activity. Because the heat is slow to escape through the bulk material, the temperature can rise sufficiently to initiate thermal decomposition.

Once smouldering is initiated, a hot glowing char is often formed, in which the carbon oxidises exothermically in contact with air. The main products from the char are CO_2, some CO, and heat, but as the hot char decomposes, it heats the neighbouring material, initiating a non-flaming thermal decomposition. This releases high yields of smoke and organic decomposition products, leaving a further area of carbon char, which itself becomes glowing, thereby enlarging the decomposing volume.

The products from this process contain relatively high yields of CO and irritant organic species. A burning cigarette provides an example of these processes. The main products from the glowing tip are CO_2, CO, and some carbon smoke particulates. Immediately behind the hot tip, the tobacco and paper are pyrolised and partially oxidised releasing a range of organic products including highly irritant aldehydes. When air is inhaled through the cigarette, while material close to the char or with a low vapour pressure is combusted, volatile substances, such as nicotine, further from the hot zone are vaporised.

Hazards from Smouldering Fires

As with thermal overheat situations, smouldering begins and develops slowly, with low rates of thermal decomposition. Smouldering fire scenarios are therefore not particularly hazardous during the first hour or several hours after the first odour and smoke products become evident. The main hazard from such fires is that at some point a transition occurs to a rapidly growing flaming fire, which can become life threatening within minutes. Sometimes the rate of flaming fire growth following a period of overheating or smouldering can be more rapid than that occurring when the same fuel is ignited directly with a flaming ignition source.

Experiments involving smouldering in furniture and bedding in closed rooms show gradual contamination with smoke and a slowly increasing CO concentration over a period of several hours. In some cases CO concentrations approaching 1000 ppm or more (with quite dense smoke) can be achieved within two hours, which could be potentially hazardous to sleeping occupants if they were not aroused.[1] Fire incident statistics contain reports of people dying as a result of smouldering fires, but after the event it can be difficult to determine if there has been a period of smouldering followed by a brief (but lethal) period of flaming, which then self-extinguishes due to lack of ventilation. Based upon experimental data the author considers that deaths are rarely likely to result from simple smouldering fires.

Large-scale smouldering fires produce large amounts of dense, irritant smoke that is rich in toxic organic VOCs. Also, because the energy release is slow, the smoke from such fires has low buoyancy. The resultant smoke plumes tend to drift slowly away from an outdoor release site near ground level, with a low rate of mixing and air entrainment compared to a flaming fire source. The resulting plumes can present a considerable nuisance and a potential health hazard to the occupants of surrounding areas.

2.3 Well-ventilated Flaming Fires

Once flaming ignition of a fuel occurs, either directly or following a period of non-flaming decomposition, the subsequent fire growth is initially fuel dependent. Flaming involves positive feedback, in that some of the heat from the flame radiates back to the fuel surface, causing further pyrolysis and generating more vapour-phase fuel gases to support combustion. Flaming fires therefore exhibit a t^2 growth curve [the fire size in terms of heat release rate (Q, kW) is a function of the square of the burning time t in seconds]. The growth rate constant depends upon the physical and chemical properties of the fuel, but a simple classification used by fire engineers for different fire types is slow, medium, fast, and ultra-fast (Figure 2.5).[23]

Table 2.4 shows examples of generic design fire growth rates for different building fire scenarios. In general, fires in sparsely furnished enclosures, such as a picture gallery, would be predicted to be slow growing. Furnishings and other contents in typical building enclosures, including offices, hotel bedrooms or reception areas, or rooms in dwellings, tend to have medium fire growth rates, unless fast burning items, such as Christmas trees, are ignited. The contents of shops can vary, but fire growth can be fast in items and packaging stacked on shelving, especially if flammable liquids or aerosols become involved. With industrial storage or plant rooms, fires can grow at an ultra-fast rate, again depending upon the situation.[23]

Fires grow in this way as long as they can spread to include more unburned fuel and as long as they are freely ventilated. For a fire in the open,

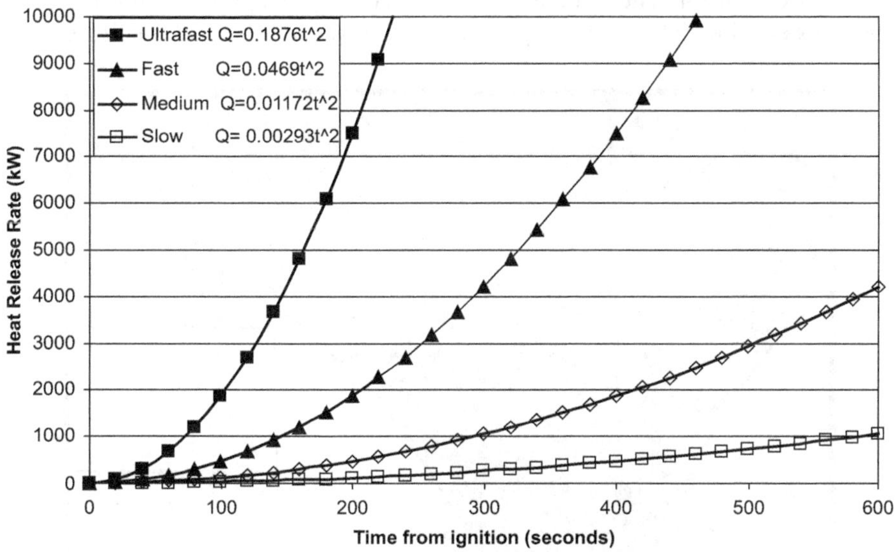

Figure 2.5 Fire growth curves used by engineers to estimate flaming fire growth for design purposes.

Table 2.4 Design fire growth rates.[23]

Picture gallery	Slow
Dwelling	Medium
Office	Medium
Hotel reception	Medium
Hotel bedroom	Medium
Shop	Fast
Industrial storage or plant room	Ultra-fast

the ultimate fire size is a function of the perimeter of the burning area, but fires in enclosures, such as a building, become limited by the size of the vents available for fresh air to enter and support combustion and for the fire effluent to escape.

For well-ventilated fires involving normally combustible materials (not flame-retarded), combustion is efficient, with almost all fuel carbon released as carbon dioxide and very low yields of most toxic products, as illustrated in Figure 2.3, Tables 2.2 and 2.3. Figure 2.6 illustrates conditions during typical well-ventilated flaming fires inside a large building. During the early stages, the fire plume containing the pyrolysed gaseous fuel products entrains a large mass of air, supporting combustion under lean fuel:air equivalence ratios in the $\varphi = 0.2–0.8$ range. The heated products are buoyant and rise within the building, entraining increasingly more air as the smoke plume rises, the total mass of air entrained being proportional to the plume height. The result is that as long as the fire remains small compared to the size of the fire enclosure, the toxic hazards to the building occupants are relatively minor. The smoke rises to the ceiling so a person at floor level is not exposed, the yields of toxic products are low, and dilution resulting from

Figure 2.6 Well-ventilated flaming fires in a building.

the mass of air entrained is high, so that the concentrations of toxic products in the smoke are very low. For the larger fire example shown, there is a possibility that building occupants on a high balcony level of the atrium space might be exposed to diluted smoke with a visibility of around 10 metres, but at such likely dilution levels the toxic hazards are minor.

In practice the combustion process in fires is always somewhat inefficient. When the temperature of some areas of the partially burned gaseous fuel in and around these turbulent diffusion flames decreases below a critical temperature, as a result of mixture with cold entrained air, the flame tip is extinguished: smoke particles and small amounts of partially decomposed organics escape without being fully combusted. This is called the "smoke point". Smoke formation depends partly on physical aspects of the combustion conditions, such as the disposition of the fuel, and partly on the fuel chemistry. Thus a pre-mixed flame (as from a Bunsen burner with the air vent open) for fuels such as methane or propane produces very little soot or smoke, while a diffusion flame spread over a wider area, with more turbulence and inefficient mixing with air, produces higher yields of smoke particles, especially for aromatic fuels such as benzene or polystyrene. When soot particles are present within a flame they are heated to temperatures high enough to emit radiation in the form of yellow light [as from a Bunsen burner with the air vent closed, or from a candle flame ($\sim$1000–1400 °C)]. Coal gas, once used for lighting in houses, contained ethylene, which burnt to produce incandescent carbon particles and thus generated light. A proportion of these particles may be preserved at the smoke point and released as smoke particles.

When the area of the flaming fuel becomes large, for example during major fires such as Buncefield (see Chapter 18), then this introduces further inefficiencies in combustion, because air may not penetrate and mix efficiently with the mass or airborne fuel. Another issue, especially with fires outdoors, such as wildfires or fires involving burning waste tips, is that some areas may consist of large flames with energetic and efficient flaming combustion of dry fuels, while other areas may involve smouldering or sporadic flaming, and materials such as green waste. The main fire area may then produce large flames with a relatively low smoke yield and a smoke plume rising high into the air with a very high dilution factor, which is then carried away from the local area. The subsidiary areas with small fires and non-flaming decomposition involve inefficient combustion, with high yields of toxic smoke vapours and particulates, producing large amounts of yellowish-grey smoke, containing higher concentrations of toxicants. Due to their low buoyancy, such smoke plumes remain close to ground level, with less air entrainment, and depending upon the wind conditions can be carried slowly away from the site to surrounding areas (Figure 2.7).

Despite these inefficiencies leading to release of smoke particulates and some toxic products in the smoke plumes, the general picture for well-ventilated flaming fires, both inside buildings and in the open, is of relatively efficient combustion, with low yields of toxic products and very high dilutions. Exposure to the resulting plumes is in most cases not acutely life threatening,

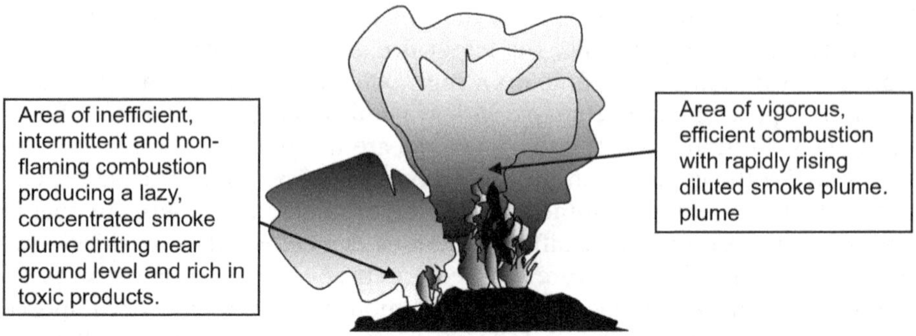

Figure 2.7 An outdoor fire with distributed fuels and areas of efficient and inefficient combustion.

but in some cases may present minor health and environmental hazards. The main exceptions are certain industrial fires releasing large quantities of specific toxic chemicals, which are carried in the smoke plume (see Chapter 4) and large wildfires. For wildfires, situations can occur in which local volunteers with limited respiratory protection attempt to suppress fires and work enveloped in smoke plumes from flaming and smouldering vegetation. In some situations such plumes can contain CO concentrations of up to approximately 2000 ppm in addition to irritant smoke particulates.

2.4 Ventilation-controlled (Fuel-rich) Flaming Fires

Although most flaming fires in enclosures (such as buildings, tunnels, and vehicles) begin as small, well-ventilated fires, they quickly grow to a size in which the combustion conditions and future fire development become influenced by the boundary conditions, especially the ventilation. The fire size then becomes limited by ventilation (ventilation-controlled), rather than by the available fuel, and the combustion process become fuel-rich and inefficient (vitiated). Most buildings and vehicles are enclosed for most of the time (with exterior doors and windows closed), so that the main ventilation available to support combustion is that inside the enclosure, at least during the early and middle periods of a fire. The two fire examples shown in Figure 2.6 are both fuel-controlled as shown and the larger fire in the large open mall is unlikely to become ventilation-controlled before the available fuel item is consumed (assuming the fire does not spread). The smaller fire is likely to become ventilation-controlled and fuel-rich as it grows due to the limited ventilation *via* the doorway and the low ceiling height. Fires such as this could occur in settings such as a shop unit in a shopping mall, a storage room in a factory, a vehicle, or a domestic lounge in a typical two-storey dwelling.

During the first few minutes the fire is well ventilated and the combustion products accumulate in the ceiling layer above the flames, but as the fire grows the smoke layer fills down from the ceiling level until the upper part of the flames penetrate the upper smoke layer. The mass of air entrained into the flames depends upon the height between the fire and the smoke layer, while the mass loss rate of the fuel increases as the fire becomes larger. The flames are also, in part, burning in the upper, oxygen-depleted, smoke layer. These processes result in inefficient, fuel-rich combustion with high yields of dense smoke and toxic chemical species. Since the smoke layer is now closer to the floor, any enclosure occupants are exposed to high concentrations of toxic fire effluent capable of causing incapacitation and death within a few minutes. Further development of the fire depends upon whether or not the door or windows of the room containing the fire are open. If the room is totally enclosed (for example, a storeroom or dwelling lounge) then the smoke layer descends to near floor level and the fire is likely to self-extinguish as the smoke layer oxygen concentration decreases to around 12–15%, leaving the room filled with a combination of dense smoke and asphyxiant gases (particularly CO and HCN) capable of causing death after a few minutes exposure. If, as in Figure 2.8, the fire enclosure door is open,

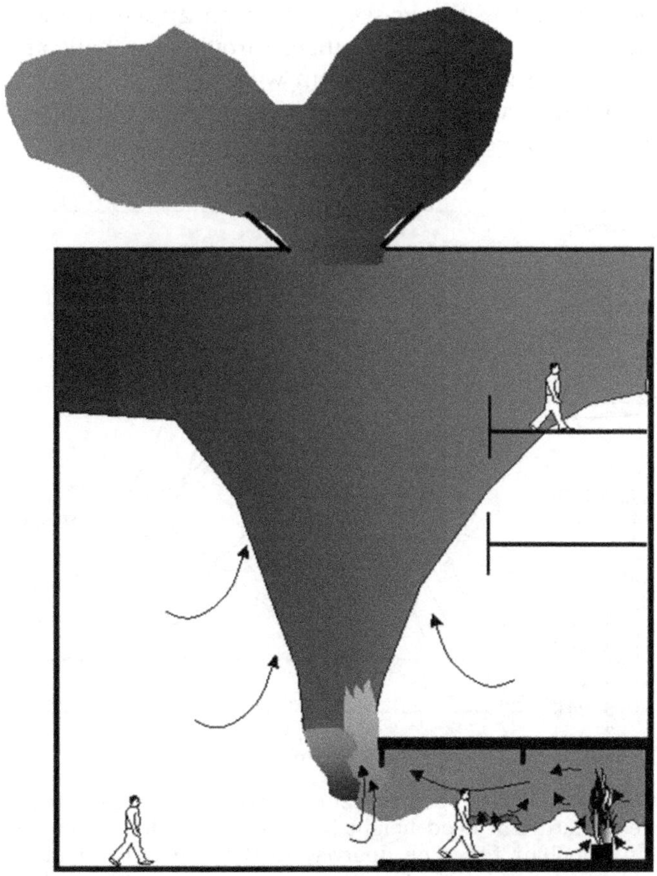

Figure 2.8 Under-ventilated (fuel-rich) fire in a shop unit.

then the vitiated smoke plume flows out through the doorway once the smoke layer descends to below the level of the doorway soffit. The buoyant smoke plume rises and entrains air as before, but the concentrations of smoke and toxic effluents are much higher than during the well-ventilated phase, and the mass production rate is higher because the fire is larger. Such fires then contaminate any continuous open areas with large volumes of toxic smoke. In this open room situation the ultimate size of the fire and depth of the smoke layer depend upon the size of the doorway opening. The conditions in the upper layer typically consist of temperatures up to around 600 °C, oxygen concentrations as low as 1% O_2, 18% CO_2, several percent CO, and >1000 ppm HCN (depending on the fuel composition).

This kind of scenario is the major cause of injury and deaths in fires. By far, the majority of annual smoke injuries and deaths in fires occur in domestic dwellings. Typically, such fires involve an item of upholstered furniture or bedding and the fire is limited to a small area around the item first ignited.

Figure 2.9 shows an example of such a fire in the open lounge of a typical UK domestic house. The fire was started in an armchair and the vitiated smoke plume flowed out of the lounge, through the hall, and filled the upstairs landing and an open bedroom with dense toxic smoke. The fire continued to burn, sustained by air entering through the lounge door from the hall, until the oxygen concentration in these areas became depleted.

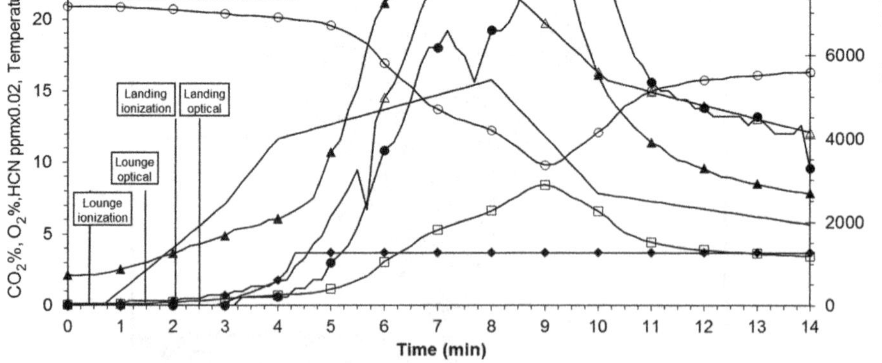

Figure 2.9 Example of time–concentration curves for smoke, toxic gases, and temperature at head height in the domestic lounge of a house during an armchair fire. The doorway to the hall is open but the house is otherwise enclosed. The time for activation of smoke detectors is shown.[1,7]

Once the oxygen concentration of the air entering the lounge at floor level decreased to 15% O_2 the fire self-extinguished.

2.5 Continuous Ventilation-controlled Pre- and Post-flashover Fires

The scenario in Figure 2.8 differs from that in the house fire (Figure 2.9), in that the fire plume flows out into a very large open space, enabling large volumes of air to enter the fire compartment and support combustion for an extended period. A similar situation occurs if a vent, such as a door or window, is opened on a lower floor. The fuel-rich toxic smoke plume from the fire flows to the outside, while fresh air enters at a lower level to support combustion. The conditions in this kind of scenario are illustrated in Figure 2.10.

In experimental studies of such fires, 100 kg fire loads were placed in the fire room leading to an open corridor (Figure 2.11).[7] Whereas the enclosed fires self-extinguished after consuming a few kilograms of fuel, these fires continued to burn until all the fuel was consumed. Under these conditions the smoke layer descends to provide continuous ventilation-controlled (under-ventilated) combustion. The toxic product yields and concentrations in the upper layer of the smoke plume flowing down the corridor are very high, while the oxygen is depleted almost completely to around 1% O_2, 18% CO_2, and 4% CO, as shown in Figure 2.10.

Under these conditions of continuous ventilation the temperature of the upper layer increases to around 600 °C or more. The size of the fire and the upper layer temperature is limited by the ventilation to the fire, which in these experiments was controlled by varying the fire room door opening

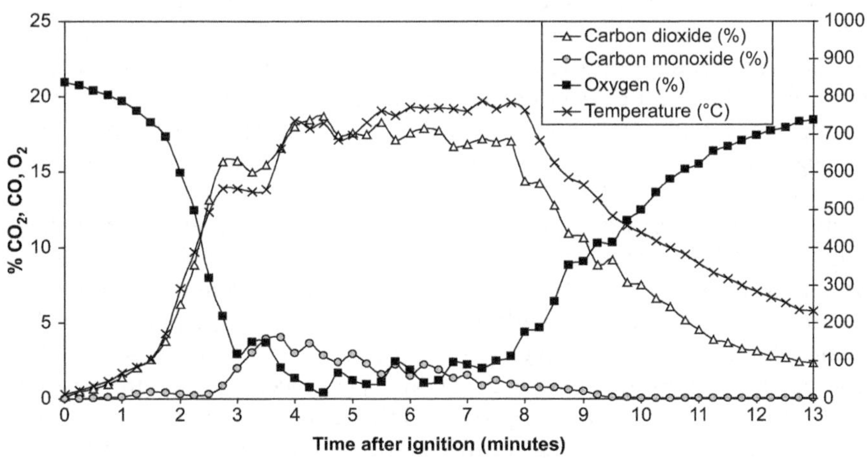

Figure 2.10 Composition and temperature of the upper layer of the fire room entrance for a wood fire in the full-scale open room-corridor rig shown in Figure 2.11.

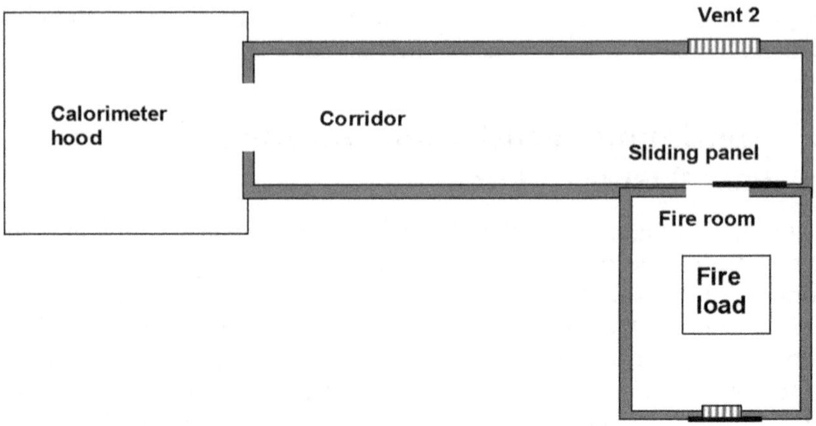

Figure 2.11 Full-scale room-corridor fire test rig.

using the sliding panel shown in Figure 2.11. If the ventilation is sufficient to raise the upper layer temperature by 550–600 °C, and the burning fuel package is large enough, or the hot smoke plume passes over fuel items remote from the original fire, then the downward heat radiation may be sufficient to pyrolyse and then ignite any remote fuel. All fuels in the area then become involved in the fire—a process known as flashover. The fire is still fuel-rich but larger and hotter than pre-flashover, thereby spreading and producing much larger volumes of toxic smoke. The higher temperatures also tend to increase the yields of toxic products. Once flashover occurs all combustible fuel surfaces within the enclosure are burning.

The scenario in Figure 2.8, involving a shop unit with an open doorway opening into a mall, would be likely to progress to flashover, especially if the front windows failed as the fire developed. Under these conditions the flammable toxic gases are preserved within the fuel-rich plume, but secondary flaming can occur at the interface with the air entering the fire and the plume leaving it. Also, as the hot effluent plume leaves the shop front and rises into the atrium, it mixes with fresh air, so that more vigorous secondary flaming occurs, partly consuming the fuel and toxic gases.

Such post-flashover effluent plumes can be observed pouring from large building conflagrations, when a number of exterior windows have failed, or there has been partial structural failure of parts of the building such as the roof. Under these conditions, with total involvement of large parts of the fuel contents of a building, very large fires of around 70–100 MW (2.5–4 kg s^{-1} for an average heat of combustion of 25 MJ kg^{-1}) can occur, with flames several metres high. Most of this burning fuel is initially consumed under relatively fuel-rich conditions within the building, but taking into account the secondary combustion outside the areas from which the fire is vented, the resulting smoke plume is likely to contain only moderately high yields and concentrations of toxic products, with a dark smoke plume.

The general picture is then that most flaming fires within buildings become fuel-rich and ventilation-controlled within a few minutes of ignition. If the building envelope remains intact and enclosed, such fires rapidly fill all contiguous open areas with smoke rich in toxic products within a few more minutes. Exposure to such smoke is likely to cause rapid incapacitation and be lethal within a few minutes. Such fires may then continue to burn slowly or self-extinguish due to lack of oxygen depending upon the "leakiness" of the building envelope. This situation is typical of fires in dwellings, but also occurs in larger buildings. Where there are large, high internal atrium spaces, or where vents to the exterior, such as several doors and windows, become open, then these fuel-rich fires are likely to progress to flashover, with total involvement of the fire enclosure contents. The very large volume of highly toxic effluent produced under these conditions can render large interior spaces lethal within seconds (as, for example, in the Dussledorf Airport[24] and Mont Blanc tunnel fires[25]), and release large volumes of toxic effluent into the outside environments.

2.6 Examples of Fire Incidents

For any fire incident, the hazards to exposed persons depend upon the fire development scenario. Important aspects are the rate of fire growth (fire growth curve), the ultimate spread, size and duration of the fire, the yields of toxic products, the heat release rate, and the dispersion of hot fire effluent through occupied areas. Fire incidents result in many injuries and deaths, so detailed investigations of specific incidents constitute an important source of information on fire hazards. The main limitation of such investigations is that the actual conditions at the time of exposure are not measured so must be inferred from a range of sources including investigation of the fire scene, estimation of the sequence of events and conditions from witness interviews, pathology studies of victims and, in major incidents, by reconstruction of the fire either by using physical experiments or computed fire simulations.[1,24–26]

While occasional, major incidents, usually in large public buildings or transport situations, have resulted in multiple injuries and deaths, leading to detailed investigations. The majority of fire injuries and deaths result from fires in domestic dwellings, which involve small numbers of people.

Figures 1.1 and 1.2 in Chapter 1 show the annual UK fire deaths and injuries. Both fire deaths and injuries increased considerably between the 1950s and 1980s, mainly due to a large increase in the number of people being overcome, injured or killed by exposure to toxic smoke. Since the peak in deaths and injuries, a number of changes have resulted in an improvement in the statistics. These include the introduction of upholstered furniture flammability regulations in 1988, the increasing use of automatic smoke detectors, and the decline in cigarette smoking in the home. Most fires in domestic dwellings, including fatal fires, consist of relatively small, under-ventilated fires, mostly limited to the item first ignited and occurring in enclosed spaces, but resulting in the spread of lethal smoke throughout the building.

Figure 2.12 shows the cumulative percentage of dwelling fires plotted against the percentage of the area of the room of fire origin damaged by fire.[27] This shows that for 80% of non-fatal fires, the area of fire damage was less than 20% of the area of the room in which the fire occurred, while even for fires involving fatalities, 60% had damage to less than 30% of the fire room area. Figure 2.13 shows the extent of fire spread beyond the room of fire origin, with no fire spread beyond the fire room for 93% of non-fatal fires and 67% of fatal fires. These findings confirm that the majority of fires remain small, although toxic smoke spreads throughout the buildings. Also, the extent of damage is assessed after the event so that even for fires that eventually spread, the fire size during the critical first few minutes, when most injuries and deaths occur, is considerably smaller than that observed later. Overall, approximately a third of fire deaths result from burns and two thirds from smoke exposures (see Chapter 1). Figures 2.14 and 2.15 show data for fatal cases in the London area (for the period 2002–2006, provided by London Fire and Rescue Service) with different ranges of percentage

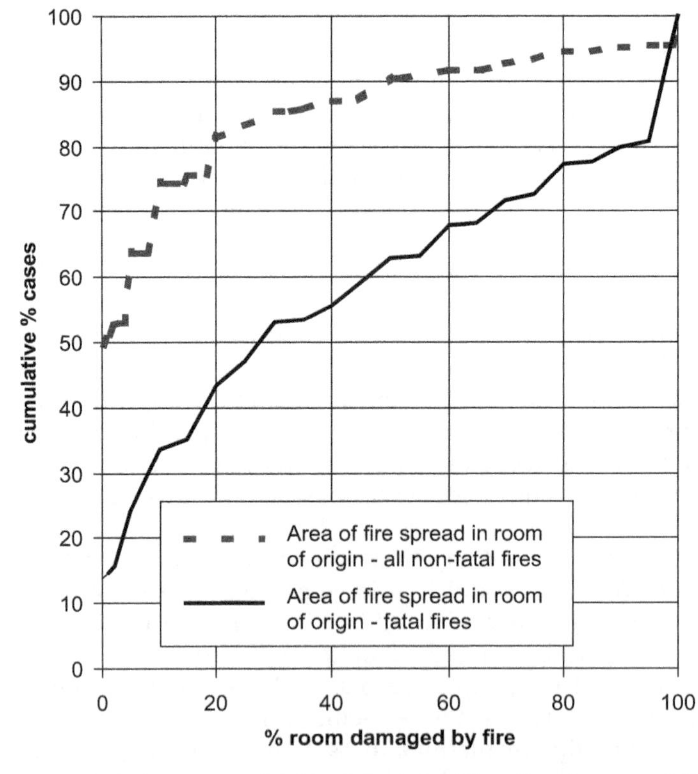

Figure 2.12 Cumulative % non-fatal and fatal dwelling fires by area of fire damage in fire room (1999).

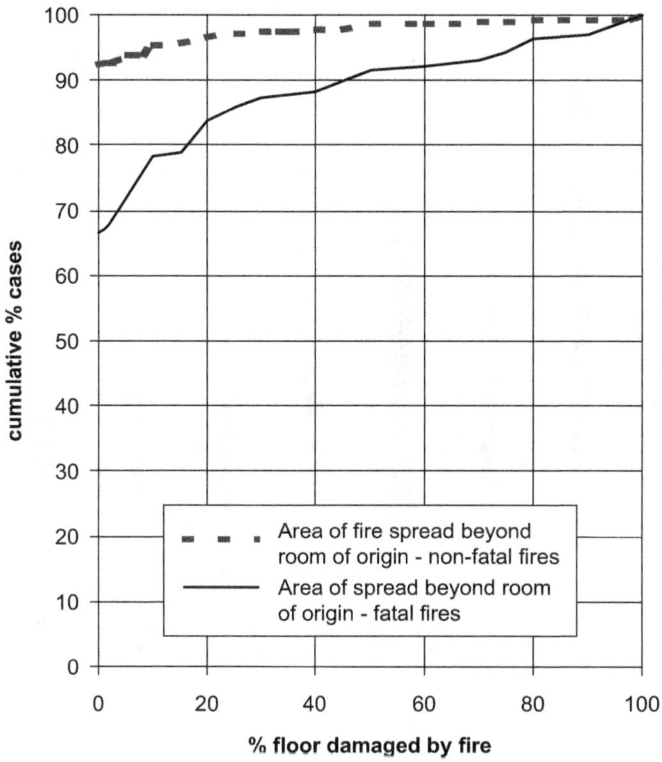

Figure 2.13 Cumulative % non-fatal and fatal dwelling fires by area of fire spread beyond room of origin (1999).

carboxyhaemoglobin (%COHb) and the incidence of burns and severe (potentially fatal) burns in each range.

Figure 2.14 is for victims dying in the room of fire origin. The %COHb achieved provides an indication of the approximate duration of exposure and the extent to which death is likely to have resulted from exposure to toxic smoke or from burns. A low %COHb ($<\sim$30% COHb), coupled with severe burns, indicates that the victims most likely died from burns before they had time to inhale a high dose of carbon monoxide. A high %COHb with a low incidence of severe burns indicates that the victims most likely died from exposure to toxic smoke and, in particular, carbon monoxide, with approximately 45–50% COHb and above considered to represent a fatal exposure (see Chapters 7 and 9). These results therefore confirm that for the majority of victims dying in the room of fire origin, burns are a major cause of death, but some subjects survive long enough to inhale a fatal dose of asphyxiant gases without suffering from burns. In contrast to this, Figure 2.15 shows that hardly any victims dying in rooms beyond the fire room have serious burns, and the majority have fatal %COHb concentrations, confirming that the spread of toxic smoke is the

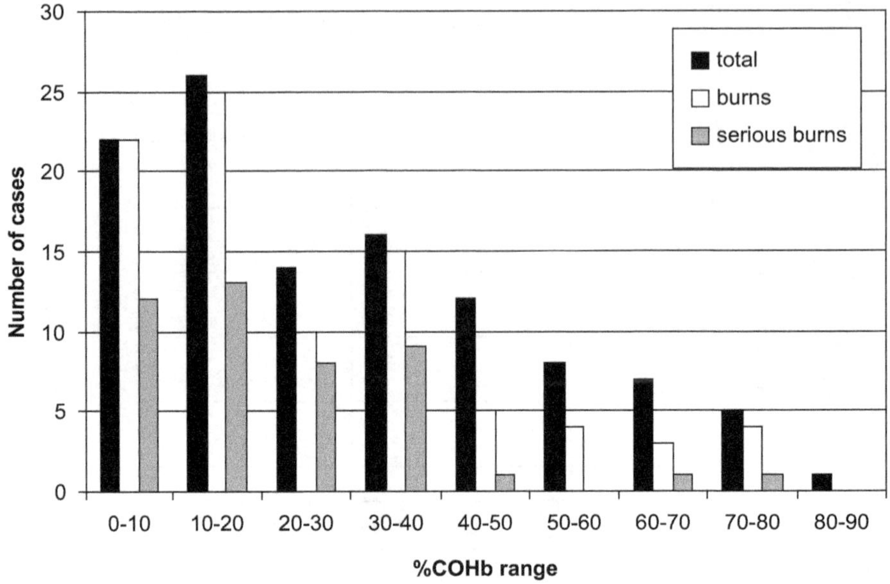

Figure 2.14 Distributions of %COHb, burns, and serious burns in 109 fire fatalities for deaths in the room of fire origin in London 2002–2006.

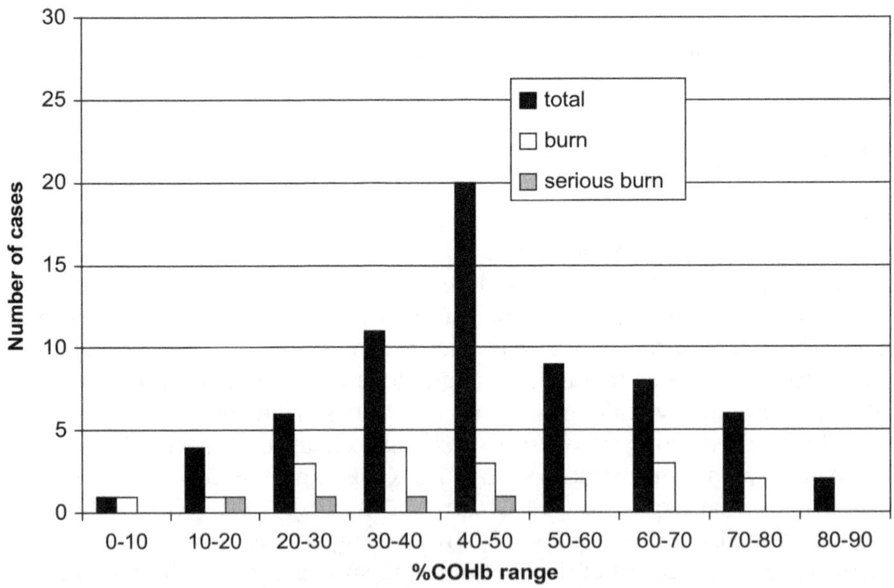

Figure 2.15 Distributions of %COHb, burns, and serious burns in 67 fire fatalities for deaths in rooms remote from that of the fire origin.
Data from London, 2002–2006, and a multi-fatality nursing home fire in Scotland.[28]

main cause of death. Figure 2.15 includes the London fire data and those from a single incident in Scotland (the Rosepark nursing home fire),[28] in which 14 persons died from toxic smoke exposure in locations remote from the fire enclosure. This incident is described in more detail in Chapter 14.

A number of major incidents have occurred in which fires have developed to a large size in one part of a structure, often progressing to flashover and producing large smoke plumes, containing high concentrations of asphyxiant gases, and spreading throughout a large single occupied enclosure, or penetrating multiple occupied enclosures. Examples of such incidents are two large hotel fires in the United States [MGM Grand (21st November 1980) and Dupont Plaza (31st December 1986)], the Summerland fire on the Isle of Man (2nd August 1983), Manchester Woolworths (8th May 1979) and Chesterfield department store fires (7th May 1993), King's Cross (London; 18th November 1987) and Daegu (South Korea; 18th February 2003) subway station/train fires, the Mont Blanc (24th March 1999) and Gotard (24th October 2001) road tunnel fires, the Kaprun (Austria) ski train fire (11th November 2000), the Manchester Airtours airplane fire (22nd August 1985), and the Düsseldorf Airport fire (11th April 1996). In all these cases, although some victims close to the fires may have suffered fatal burns before death, and other bodies may have become burned after death as the fires spread over periods of hours to days, the majority of victims were overcome and died as a result of exposure to toxic smoke, often some distance from the fire itself.

2.7 Practical Aspects

2.7.1 Toxicity and Toxic Hazard

Fires may be characterised in terms of the danger they pose to life by virtue of production of toxicologically active products: this is described as the toxic hazard of the fire. The toxic hazard of a fire is a system property of the full-scale fire scenario. It consists of a sequence of time-varying effects depending upon the changing combustion conditions and concentrations of different toxic substances, and the varying rates of development of a sequence of different toxic effects, caused by different types of toxic fire effluent components. For this reason it is not possible to make meaningful measurements of "toxicity" for any material in any small or even large-scale fire "toxicity" test. Rather, it is necessary to determine toxic hazards from measured or calculated exposure concentrations and exposure doses of different toxic species from full-scale fire scenarios over different time scales (see examples of such hazard analyses in Chapter 14).

Combustion product toxicity includes rapidly developing physiological effects and longer-term pathological effects. During fires it is the physiological effects that dominate, determining time to incapacitation, and hence survivability.[1] The danger to life from fires in buildings (or other enclosures) therefore depends upon the performance of a dynamic system involving interactions between the building, the fire, and the occupants. In a design context it is necessary to demonstrate that the available safe escape time (ASET) is greater than the required safe escape time (RSET) by an acceptable safety margin.[8] The ASET endpoint is the time when occupants are predicted to be incapacitated to the extent that they can no longer save themselves. An important consideration is that since all fires involve a set of toxic and physical hazards from smoke, toxic substances, and heat, the main hazard leading to injury or death may not be the hazard ultimately responsible for injury or death. Thus, the presence of irritant smoke during the early stages of a fire, although in itself not immediately life threatening, often results in occupants being delayed or impeded in escape attempts. This prolongs their exposure to asphyxiant gases or heat, which may then result in serious injury or death.

2.7.2 Acute Life-threatening Hazards in Building or Transport Fires

From examination of fire injury and death statistics and detailed investigations of individual fire incidents it is evident that that heat and burns are primary causes of injury and death mainly in situations in which victims are intimate with the fire, either by being in close proximity to a burning object or as a result of direct ignition of clothing. Victims of such incidents usually have severe burns but low blood carboxyhaemoglobin concentrations. When victims are exposed away from the immediate fire, they may still suffer burns as a result of heat radiation or contact with hot fire gases, but in most cases will have also have had time to achieve high blood carboxyhaemoglobin concentrations. However, the majority of fatal victims, and many injured fire victims, are overcome by exposure to toxic smoke away from the immediate fire area, either in the same room as the fire or often in a more remote part of a building and have no or limited burns. Typically, such victims are overcome after a period of smoke exposure and have high blood carboxyhaemoglobin concentrations. Sometimes, such victims dying from toxic smoke exposure can suffer post-mortem burns as a fire spreads, which can complicate the assessment of the contribution of burns, since it is not possible to distinguish directly between pre- and post-mortem burns.

Once a fire occurs in an occupied building, the developing hazards occur more or less in the following order:

1. Exposure to irritant and vision-obscuring smoke, immediately reducing escape capability. The severity of the effects on the eyes and upper respiratory tract depends on the exposure concentration.
2. Inhalation of asphyxiant gases (CO, HCN, CO_2, low oxygen) causing collapse and loss of consciousness within minutes when a sufficient dose has been inhaled.

3. Exposure to heat or flames resulting in body surface and/or respiratory tract burns (usually only if close to the fire scene, but can occur simultaneously with hazard 2).
4. Inhalation of irritants and particulates: minimal effects at the fire scene but can result in lung inflammation and oedema if a sufficient dose has been inhaled, or cardiovascular problems several hours to days after rescue.

Where subjects are exposed to smoke plumes in the open the main effects of concern are:

1. acute concentration-related nuisance or minor irritant effects from exposure to irritant smoke;
2. minor respiratory tract inflammation or other health effects resulting from exposure to inhaled irritants over periods of hours to days;
3. health hazards resulting from chronic exposures over long time scales.

Where smoke plumes release environmental toxicants into the outside environment the main effect of concern is:

4. secondary exposure (mainly *via* food or water) to environmental toxicants.

To determine the extent of exposure and time to any specific toxicological endpoints for any system it is necessary to determine:

- Time–concentration (or intensity) curves for the major toxic products, smoke, and heat from the fire at the breathing zone of those exposed, which in turn depend upon:
  - fire growth curve [mass loss rate of the burning fuel (kg s^{-1}) and its dispersal volume (kg m^{-3}) with time];
  - the yields of the major toxic products (kg kg^{-1}) and heat (kJ kg^{-1}).
- Concentration–time/physiological effects relationships of these products in terms of the physiological/toxic potency of the products and heat [the exposure concentration (kg m^{-3})], or exposure dose (kg m^{-3} min or ppm min) causing toxic effects (and the equivalent effects for heat and smoke obscuration) in terms of:
  - concentrations and doses (or heat intensity) likely to impair escape efficiency due to behavioural and/or physiological effects;
  - exposure concentrations or doses likely to cause incapacitation or prevent escape due to behavioural and/or physiological effects;
  - lethal exposure concentrations or doses;
  - exposure doses causing post-exposure health effects.

For acute exposures these endpoints can be calculated using physiological methods (fractional effective dose methods) and other toxicological assessment methods described in Chapter 14.

2.7.3 Determination of Time–Concentration Exposure Curves from Fire Incidents

In order to make an assessment of the toxic hazards from exposure to combustion products, whether from an acute life-threatening exposure or a chronic exposure over many years, it is important to determine the time–concentration curves for the toxic products at the breathing zone of an exposed subject. This may be achieved by direct measurement of the fire effluent either from a specific incident or by re-creation of a fire incident at full-scale as in Figures 2.9 and 2.10.

Another method is use of fire engineering and plume dispersal calculations as described in Chapter 3. Available methods range from simple calculations, as in BS 7974,[23] and the application of relatively simple zone computer models, such as the Consolidated Model of Fire and Smoke Transport (CFAST),[29] to the more sophisticated computational fluid dynamics models such as the Fire Dynamics Simulator (FDS).[30] With appropriate input data these can be used to simulate the developing fire conditions, enabling the composition of the fire effluent plume in a building at different times and locations to be calculated.

These methods can also be used to calculate the mass release rate and composition of effluent plumes dispersing outside buildings or from fires in the open. Plume dispersal models can then be used to calculate further dispersal, dilution, and deposition in the outside environments (see Chapter 6).

Another approach that can provide useful information about both fire conditions and their effects is to make measurements of effluent deposits, or the concentrations of key or marker toxicants in exposed subjects or environmental deposits. For fire victims, the blood carboxyhaemoglobin concentrations after an incident provides a measure of the integrated time–concentration exposure dose during the fire, not only to CO itself but also to other toxic fire products known to track with CO.[28] The extent of smoke deposition in the respiratory tract also provides some indication of the extent of exposure to fire effluents. The extent and severity of burns provides an indication of the heat exposure history.

All three approaches have advantages and limitations, and in practice they are most powerful when used in combination for the assessment of any specific fire incident.

2.7.4 Direct Measurement during Incidents or Full-scale Incident Reconstruction

Establishing the toxic hazard during an incident by direct measurement is challenging because the conditions change very rapidly during most fires and vary considerably at different locations relative to the source fire. Attempts have been made to measure toxic gases during fire incidents in buildings by equipping fire-fighters with sampling devices, and in some buildings, fire detection systems monitor parameters such as temperature, smoke density, and CO concentration.[25] To date, the results of such approaches have been limited. Fire-fighters are not usually present at a victim's location during the key exposure period, and when they do arrive they tend to be too occupied with essential activities to activate sampling equipment during the periods and in the locations needed. As sensor technology improves and more monitoring systems are integrated into buildings, more real-time data on fire effluent composition will be obtained during incidents.[31] Large conflagrations burning for several days present an opportunity to deploy sampling systems, but as with building fires, plume composition and dilution vary with changes in the fire and meteorological conditions. Post-incident forensic investigations involving measurements of the composition of soot deposits can provide useful information on the extent of environmental toxic releases of compounds such as dioxins.

For major fire incidents, the main method of investigation involves full-scale reconstruction of the incident or part of an incident location. In order to obtain realistic measurements of smoke and toxic gases it is essential to recreate not only the fire itself (involving the same fuels in the same disposition) but also the same fire boundary conditions as in the incident. This means that the physical dimensions and properties of the fire enclosure must be similar to those in the incident (similar size and thermal properties of the boundaries), and it is especially important that the ventilation conditions are the same as in the actual incident, an aspect often neglected. For example, as part of the investigation into a major incident in the United States (the Station Nightclub fire), a section of the building where the fire started and grew initially was recreated in full-scale, but in an open test rig that only comprised a small area of the actual fire enclosure and with somewhat different ventilation conditions.[32] Although tests such as this provide useful information, some doubt remains as to how well the actual conditions are simulated. In contrast, for the fatal fire in the Rosepark nursing home in Scotland, the Building Research Establishment (BRE) made as far as possible an exact reconstruction of the affected part of the building, including all relevant rooms and corridors with the same ventilation conditions as on the night of the fire.[33]

Although it is difficult to ensure an exact duplication of a previous fire, methods are now becoming available to validate the test conditions against those in the actual incident. One very valuable marker is CO. Calculation models exist for human subjects for both uptake of CO during exposure at a fire scene and for washout after rescue (allowing for different treatments with air, facemask oxygen, or oxygen by intubation). From the CO time–concentration curves in a reconstruction test it is possible to forward-calculate the predicted %COHb in exposed subjects up to the time of death or rescue. For the decedents it is possible to compare calculated %COHb concentrations with those measured post-mortem. For survivors it is possible to back-calculate the %COHb at the time of rescue from the concentrations measured on arrival at hospital using washout calculations.[28,33] The extent to which the forward- and back-calculated %COHb concentrations match then provides an estimate of the extent to which the conditions in the recreated fire matched those in the actual incidents (see Chapter 14).

Detailed full-scale reconstruction tests of this kind and other full-scale fire experiments, such as that shown in Chapter 14, Figure 14.11, are useful not only with respect to specific incidents, but because they can also provide information applicable to other fire scenarios, providing the generic scenario conditions in terms of fuel type, fire size, enclosure size, and ventilation conditions are similar. Thus, the exposure conditions in most domestic dwelling fires involving upholstered furniture or similar mixed fuels, in which the exterior doors and windows are closed, are likely to be generically similar to those in Figure 2.9, but significantly different from incidents in which exterior windows and doors are open throughout the fire (as in Figure 2.10). Table 2.5 shows approximate concentration ranges of major fire products in mixed fuel building fires in the effluent layer dispersed within building spaces contiguous with the fires, and in an effluent plume flowing outside through an open vent, such as a window, for which outside air entrainment into the rising plume results in a dilution factor of approximately 50–100. The main implication of these high dilution factors is that the concentrations of the most hazardous and life-threatening substances inside buildings—the asphyxiant gases: CO and HCN—are diluted to toxicologically insignificant concentrations in most situations, even in the zone immediately outside (within approximately 20 metres of the fire). However, due to their very large concentration ranges of toxic effects, particulates, irritant and odorous organics, and irritant acid gases are present at low but toxicologically significant concentrations. For mixed fuels the concentrations of irritants, and in particular acid gases, depend upon the fuel composition, but are likely to be sufficient to cause minor eye

Table 2.5 Approximate concentrations of toxic products in the smoke layers of fires inside the buildings and in outside smoke plumes in the vicinity of burning buildings.

Toxic product	Low range: near smouldering or small flaming fire	In outside plume, approximately 50–100 dilution	High range: in effluent plume from well-developed flaming fire	In outside plume, approximately 50–100 dilution
Carbon monoxide	100 ppm	1–2 ppm	40 000 ppm	400–800 ppm
Carbon dioxide	0.06%	0.04%	15%	1.5–3.0%
Oxygen depletion	Negligible	Negligible	19%	1.9–3.8%
Hydrogen cyanide	20 ppm	1–2 ppm	1500 ppm	8–15 ppm
NO_x	Negligible	Negligible	50 ppm	0.5–1 ppm
Total organics (VOCs)	~0.1 mg L^{-1}	~1–2 ppm	20 mg L^{-1}	2–4 mg L^{-1}
Acid gases	~10–100 ppm	0.1–1 ppm	10–1000 ppm	0.1–10 ppm
Particulates	~0.05 mg L^{-1}	1–2 μg L^{-1}	10 mg L^{-1}	1–2 mg L^{-1}

and respiratory tract irritation, while organic components add to the irritancy and nuisance odour at high dilution levels. As the smoke plume moves further away from the source, much greater dilution occurs (see Chapter 6).

2.7.5 Generation and Measurement of Combustion Product Effluents

Generation and measurement of the composition of combustion products is extremely challenging because the yields of individual toxic products are so dependent on the exact conditions under which they are generated and the effluent mixtures are so complex both physically and chemically. The main issues are as follows:

- The mix generated from any specific material varies considerably, both in terms of the identity and yields of different components, depending up the decomposition conditions. Detailed chemical measurements are often made using small bench-scale decomposition devices with little or no consideration of the extent to which the decomposition conditions in the apparatus represent those actually occurring in different types of fires.
- Combustion products contain mixtures of gases, vapours, liquid droplets, and solid particulates with a very wide size range. This presents challenging sampling conditions, in that components, such as acid gases and particulates, tend to be lost in sampling lines and on filters.
- Even after fire effluents have cooled sufficiently for the main reaction chemistry to become "frozen", certain changes in composition continue over different time scales. Vapours condense into liquid droplets, solid particles formed in the submicron range agglomerate to form complexes in the 1–10 micron range, with a proportion even in the millimetre size range. Gases and vapours become adsorbed onto the surfaces of particles, which affects their bioavailability and their deposition in different regions of the respiratory tract. Acid gases, organic vapours, and particulates become deposited onto building surfaces, reducing the inhalation hazard to occupants remote from a fire, but increasing the hazard during clean-up or post-fire occupation (sick building syndrome). Some components change chemically even after cooling. For example, NO, which has a low toxic potency, oxidises over time scales of minutes to NO_2, which is a potent lung irritant. Once dispersed into the outside atmosphere, further chemical and physical changes occur due to photochemistry. The composition of deposited dioxin congeners changes over time scales of years with significant implications for toxicity.[17]
- Sampling such complex atmospheres from large-scale fires, the outside environment, and even bench-scale apparatus also presents considerable difficulties. Reactive acid gases tend to be lost to the walls of the

sampling lines, either directly by reacting with metal tubing, by condensation with water, or by association with soot deposits. NO_2 diffuses through or reacts with sampling tubes other than Teflon.

Figure 2.16 illustrates the complexities involved in sampling two acid gases, HCN and HCl, from the upper layer during a furniture fire in an enclosed 30 m^3 room. The fire burned for 5 minutes before self-extinguishing due to lack of ventilation. The main method for sampling from the fire compartment was to continuously pump an effluent sample through a 3 metre PTFE-lined stainless steel sampling line heated to 120 °C. For acid gases, a sample was then taken *via* a T-piece through a heated glass fibre filter and Dreschel bottles containing sodium hydroxide solution. The sample was taken over a 5 minute period after the burning ceased. In order to plot the acid gas concentration curves during and after the flaming period, another sampling method involving 270 mL "grab" vessels was used. Ten evacuated glass grab vessels, each containing 20 mL sodium hydroxide solution, were connected to an automated valve system, enabling each to be opened in sequence to take a small sample over a few seconds. Each grab vessel was connected to the sampling line *via* a 150 mm long, 2 mm internal diameter stainless steel tube. In order to assess any potential losses

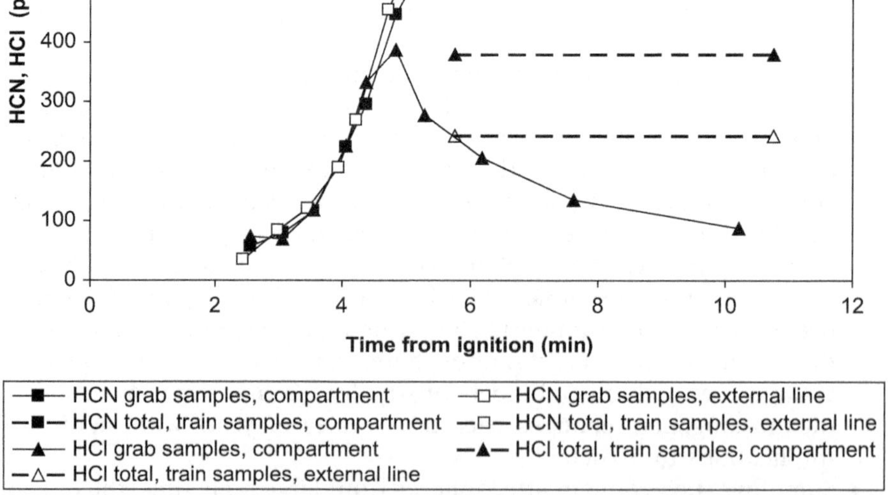

Figure 2.16 Sampling losses and partitioning of acid gases between gas and particulate phase.

along the sampling line, a duplicate set of sampling apparatus was placed inside the fire compartment directly in the hot upper smoke layer. This set consisted of an open face glass fibre filter holder and Dreschel bottle train, and a set of grab vessels. Figure 2.16 shows that for HCN the measured concentrations captured in the interior and exterior grab vessels were virtually identical, increasing to 550 ppm after 5 minutes burning and then remaining constant up to 10 minutes. The HCN results from the interior and exterior filter–Dreschel bottle sampling trains, representing the average HCN concentration over the period from 5–10 minutes, were also very similar to the grab vessel results and to each other (the interior sample being slightly higher than the others). From these results it was concluded that HCN remained in the gas phase, did not react with the sampling lines, and did not attach to particulates deposed in the lines. No HCN was recovered from the glass fibre filters.

In contrast to this, the grab vessel curve for HCl (sampled only inside the fire compartment) increased to a peak of 387 ppm at 5 minutes and then decreased over the next 5 minutes to 88 ppm. However, the interior filter–Dreschel bottle sample train gave an averaged value of 379 ppm over the 5–10 minute period, of which 90% was recovered from the filter. The exterior sample train gave an averaged concentration of 241 ppm HCl over the same period. From these results it was concluded that, over approximately 5–10 minutes following the fire, most HCl becomes associated with the particulate phase of the smoke, resulting in some losses in the sampling lines to the exterior of the test rig. For the grab vessels, once the HCl becomes associated with particulates, there is a tendency for losses to occur in the narrow stainless steel sampling line on each grab vessel, even in the hot upper layer inside the fire room, so that the measured concentrations in the grab vessels over the 5–10 minute period are only around 25% of those from the filter–Dreschel bottle system. However, during the early stages of the fire, when the fire is flaming and the HCl gas is freshly liberated from the fuel, most is in the vapour phase, so is efficiently trapped by the grab vessels. In other experiments, when freshly formed atmospheres were sampled from the tube furnace, a high proportion of HCl was trapped in the Dreschel bottle rather than the filter, while for more aged smoke, most was trapped on the filter.

Apart from demonstrating the considerable challenges in obtaining accurate measurements of actual fire atmosphere compositions, these findings may have implications for exposed human fire victims. During the first few minutes of exposure in a room fire, an important concern is the effect of the smoke density and irritancy of the fire effluent on the ability of an occupant to see, breathe, and escape from the fire. During this period the irritant effects of HCl vapour on the eyes and upper respiratory tract are important (see Chapter 7). If the victims remain in the fire room for a longer period, they inhale an increasing dose of irritant smoke into their lungs. The smoke particles are in a highly respirable size range and are deposed into the lower respiratory tract. If these particulates now carry an additional burden of HCl, in addition to organic irritants, they may further contribute to the development of lung oedema and inflammation.

- Once the difficulties of generation and sampling have been overcome, the chemical (and physical) analysis of the product mix is extremely challenging due to the enormous range of chemically different substances present, and to the low concentrations that are of toxicological significance for different substances. The potency of different irritant substances ranges over approximately six orders of magnitude, and the lethal toxic potencies over five orders of magnitude (see Chapter 7). This applies particularly to the organic components, which range from relatively low molecular mass gases, like methane, through to highly reactive species, such as acrolein and other aldehydes, to high molecular mass substances such as polycyclic aromatic hydrocarbons (PAHs) and dioxins. The substance mix covers a very wide range of concentrations, and while some substances (such as methane) can be present at percent levels but are not toxicologically significant, other substances, such as dioxins, can be of toxicological importance even at ng m^{-3} concentrations.
- The measured organic chemical composition of combustion atmospheres depends to some extent on the analytical methods used. While application of techniques, such as gas chromatography-mass spectroscopy (GC-MS), have revealed a wide range of products to be present, the results are often semi-quantitative, and very dependent upon the sampling techniques, type of column, and detection systems used. Generic analysis can therefore fail to detect compounds of toxicology interest, such as aldehydes, or due to the methods used, and may show severe limitations with respect to both low and high molecular mass components. When low concentrations of specific types of substances, such as organophosphorus compounds, halogenated organics (especially dioxins), or nitrogen-containing products, are of interest, the development of specially targeted and validated sampling and analysis methods is required.
- Historically, a number of research groups have attempted to measure the organic composition of combustion atmospheres for different reasons and using different approaches. During the 1970s and 1980s concerns that changes in the acute toxicity of fire atmospheres were responsible for the fourfold increase in fire deaths at this time led to groups, such as that of Dr David Woolley at the Fire Research Station, sampling and analysing the organic composition of combustion atmospheres from small-scale furnaces and full-scale fires.[2–4,6] Most of this work concentrated on the middle molecular mass range (ignoring the heavier fractions such as PAHs) and the main results (involving combined animal toxicity work) were that from the perspective of acute toxicity, the only aspect of interest, was the overall respiratory tract

irritancy of the products. With regard to irritant organic species, there was a definite correlation (as might be expected) between observed irritancy, the overall yields and concentrations of organic species in the total products and with the extent to which they were partially oxidised [containing aldehydes (especially unsaturated), organic acids, ketones, and phenol]. Due to the complexity of these mixtures, it proved impossible to calculate their overall irritancy from knowledge of the toxic potencies of the individual components. The observed effects (for example, on mice) were always greater than could be predicted from the composition (see Chapter 7). For these reasons the best approach with regard to estimation of acute irritant effects on exposed human subjects may be to use a generic method based upon the total organic content of an effluent atmosphere or the overall proportion of oxidised organics, rather than attempting to base predictions on a detailed chemical analysis.

- Other groups have been interested in chemical smoke analysis from the perspective of the possible effects of chronic exposure to low concentrations. A considerable effort was made by groups, such as that at the Laboratory of the Government Chemist, to measure the detailed composition of tobacco combustion products, combined with efforts by other groups to determine health-related toxicity such as airway disease and carcinogenicity. With regard to airway and lung diseases, the general picture appears to be somewhat similar to that for acute effects, in that the main criterion is the overall yields and concentrations of mixed organic compounds delivered into the respiratory tract and deposited in different parts of the lung, and on the overall content of mainly oxidised products such as aldehydes. For mutagenicity and carcinogenicity, a very wide range of compounds has been identified and measured both in fire smoke and tobacco smoke, and in this regard the health consequences of chronic exposure may be partly related to the mixture of carcinogens (including aldehydes, acrylonitrile, benzene, PAHs, nitrosamines), and possibly to some extent on specific substances such as benzo-*a*-pyrene. As with fire atmospheres, generation and sampling methods are issues. For example, side stream tobacco smoke (rising from a cigarette burning in an ashtray) is much more irritant than main stream cigarette smoke (that inhaled by a smoker taking a puff), and the chemical composition of the two components has been shown to be very different (see Chapter 12).
- A third category of investigations has been concerned with atmospheric pollution from sources such as unwanted fires, stack emissions, and vehicle fumes, either at the point of generation (stack measurements) or the point of exposure (street measurements). Most interest has been directed at the combustion particles (mainly in relation to size range rather than composition) and at individual acid gases including sulfur dioxide and nitrogen oxides.

- A fourth category relates to environmental contamination by halogenated organics. Again, this has been directed towards the source (*e.g.*, incinerator stacks) and the point of exposure (measurement of deposits in soils, water courses, vegetation, and human tissue).
- While progress is being made in all these areas, having attempted to characterise combustion atmosphere composition for many years, I find it a considerable frustration that while astronomers frequently describe the concentrations of CO and organic substances in nebulae many light years from Earth, I remain unconvinced that anyone has been able to make comprehensive and accurate measurements of the composition of laboratory combustion atmospheres. It is also possible to provide impressive lists of numbers for ranges of chemical substances using methods such as Fourier transform infrared (FTIR) analysis, but sometimes difficult to validate such results, as shown by publications of data in which the presence of inorganic gases are reported even when the necessary chemical elements are absent from the material burned.

References

1. D. A. Purser, Assessment of Hazards to Occupants from Smoke, Toxic Gases and Heat, *The SFPE Handbook of Fire Protection Engineering*, ed. P. J. DiNenno, National Fire Protection Association, Quincy, MA 02269, 4th edn, 2009, pp. 96–193.
2. W. D. Woolley and P. J. Fardell, Basic Aspects of Combustion Toxicology, *Fire Saf. J.*, 1982, **5**, 29.
3. W. D. Woolley, Nitrogen-Containing Products from the Thermal Decomposition of Flexible Polyurethane Foams, *Br. Polym. J.*, 1972, **4**, 27–43.
4. W. D. Woolley, S. A. Ames and P. J. Fardell, Chemical Aspects of Combustion Toxicology of Fires, *Fire Mater.*, 1979, **3**, 110–120.
5. W. M. Pitt, The Global Equivalence Ratio Concept and the Formation Mechanisms of Carbon Monoxide in Fires, *Prog. Energy Combust. Sci.*, 1995, **21**, 197–237.
6. D. A. Purser and W. D. Woolley, Biological Studies of Combustion Atmospheres, *J. Fire Sci.*, 1983, **1**, 118–145.
7. D. A. Purser, Toxic Product Yield and Hazard Assessment for Fully Enclosed Design Fires Involving Fire Retarded Materials, *Polym. Int.*, 2000, **47**, 1232–1255.
8. D. A. Purser, Influence of Fire Retardants on Toxic and Environmental Hazards from Fires, in *Fire retardancy of polymers. New strategies and mechanisms*, ed. T. R. Hull and B. K. Kandola, RSC Publishing, Cambridge, 2009, ch. 24, pp. 381–404.
9. D. A. Purser, ASET and RSET: Addressing Some Issues in Relation to Occupant Behaviour and Tenability, in *Proceedings of the 7th*

International Symposium on Fire Safety Science, International Association for Fire Safety Science, Boston, MA, 2003, pp. 91–102.
10. Controlled Equivalence Ratio Method for the Determination of Hazardous Components of Fire Effluents, *ISO/TS 19700*, International Organization for Standardization, Geneva, Switzerland, 2007.
11. A. Tewarson, Generation of Heat and Gaseous, Liquid and Solid Products in Fires, *The SFPE Handbook of Fire Protection Engineering*, ed. P. J. DiNenno, National Fire Protection Association, Quincy, MA 02269, 4th edn, 2009, pp. 109–194.
12. D. A. Purser, A. A. Stec and T. R. Hull, Effects of Material and Fire Conditions on Toxic Product Yields, in *Fire Toxicity*, ed. A. Stec and R. Hull, Woodhead, Cambridge UK, 2010, pp. 515–540.
13. D. A. Purser and J. A. Purser, The Potential for Including Fire Chemistry and Toxicity in Fire Safety Engineering, *BRE Report* No. 202804, 28th March 2003.
14. D. A. Purser and J. A. Purser, HCN Yields and Fate of Fuel Nitrogen for Materials Under Different Combustion Conditions in the ISO 19700 Tube Furnace and Large-scale Fires, 9th International Symposium on Fire Safety Science, University of Karlsruhe, Germany, 21st–26th September 2008.
15. D. A. Purser, A. A. Stec and T. R. Hull, Effects of the Material and Fire Conditions on Toxic Product Yields, in *Fire Toxicity*, ed. A. A. Stec and T. R. Hull, Woodhead, Cambridge, 2010, ch. 14, pp. 515–540.
16. T. Wright, Environmentally Friendlier Flame Retardant Systems, PhD Dissertation, Leeds University, Leeds UK, 1997.
17. D. A. Purser and P. Buckley, Lung irritance and inflammation during and after exposure to thermal decomposition products from polymeric materials, *Med., Sci. Law.*, 1983, **23**, 142–150.
18. D. A. Purser, P. J. Fardell, J. Rowley, S. Vollam and B. Bridgeman, An improved tube furnace method for the generation and measurement of toxic combustion products under a wide range of fire conditions, *Flame Retardants '94 Conference, London, UK, Proceedings*, Interscience Communications, London, January 1994, pp. 263–274.
19. D. A. Purser, Validation of additive models for lethal toxicity of fire effluent mixtures, *Polym. Degrad. Stab.*, 2012, **97**, 2552–2561.
20. D. A. Purser, Toxicity of fire retardants in relation to life safety and environmental hazards, in *Fire Retardant Materials*, ed. A. R. Horrocks and D. Price, Woodhead Publishing Ltd, Cambridge UK, 2001, ch. 3, pp. 69–127.
21. D. A. Purser, Recent developments in understanding the toxicity of PTFE thermal decomposition products, *Fire Mater.*, 1992, **16**, 67–75.
22. D. A. Purser, The combustion toxicology of anticholinesterases, in *Basic and Clinical Toxicology of Organosphosphates and Carbamates*, ed. Ballantyne and Marrs, Butterworth-Heinemann, Oxford, 1992, ch. 35, pp. 386–395.

23. Application of fire safety engineering principles to the design of buildings – Part 1: Initiation and development of fire within the enclosure of origin (Sub-system 1), PD7974-1:2003 British Standards Institution.
24. D. A. Purser, Design behavioural scenarios for escape behaviour modelling in tunnels and underground complexes. Advanced Research Workshop, Evacuation and Human Behaviour in Emergency Situations, University of Cantabria, Spain, 21st October 2011, Proceedings pp. 1–19.
25. D. A. Purser, Application of human behaviour and toxic hazard analysis to the validation of CFD modelling for the Mont Blanc Tunnel fire incident. Advanced Research Workshop: Fire Protection and Life Safety in Buildings and Transport Systems, University of Cantabria, Spain, 17 October 2009, Proccedings, pp. 23–57.
26. D. A. Purser, Structural fire engineering design: aspects of life safety, *BRE Digest*, 2004, 490.
27. Fire Statistics in the United Kingdom 2011, Communities and Local Government, London (published annually).
28. D. A. Purser, Fire safety and evacuation implications from behaviours and hazard development in two fatal care home incidents: Rosepark and Frampton House, Proceedings of the 5th International Symposium on Human Behaviour in Fire, Downing College, Cambridge, UK, 19–21 September 2012, Interscience, London, pp. 251–262.
29. W. W. Jones, R. D. Peacock, G. P. Forney and P. A. Ricke, CFAST – Consolidated Model of Fire Growth and Smoke Transport (Version 6), NIST Special Publication 1026, 2005.
30. K. McGrattan, S. Hostikka, J. Floyd, H. Baum and R. Rehm, Fire Dynamics Simulator (Version 5) Technical Reference Guide Volume 1: Mathematical Models, NIST Special Publications 1018-5, National Institute of Standards and Technology, Gaithersburgh, MA, July 2008, p. 92.
31. W. Jahn, G. Rein and J. L. Torero, A posteori modelling of the growth phase of Dalmarnock fire test one, *Build. Environ.*, 2011, **46**, 1065–1073.
32. W. Grosshandler, N. Bryner, D. Madrzykowski and K. Kuntz, Report of the Technical Investigation of the Station Nightclub Fire, *NIST NCSTAR 2*: Vol 1, National Institute of Standards and Technology, June 2005.
33. Experimental research for Scottish Building Standards Agency following the fire at the Rosepark Care Home, Glasgow, 31st January 2004, Building Research Establishment Ltd Fire Safety Framework, 6 October 2004, The Scottish Government.

CHAPTER 3

Estimating Yields and Quantities of Mass Releases of Toxic Products from Fires

DAVID A. PURSER

Hartford Environmental Research, Hatfield, UK
Email: david-purser@ntlworld.com

3.1 Modelling Conditions during Fire Incidents

In order to make hazard and risk assessments for potential fires, or for investigation of specific incidents, use is made of fire engineering and plume dispersal calculations. Available methods range from basic algebraic expressions, as in BS 7974,[1] to the application of zone computer models, such as the Consolidated Model of Fire and Smoke Transport (CFAST),[2,3] and the more sophisticated computational fluid dynamics (CFD) models such as the Fire Dynamics Simulator (FDS).[4,5] With appropriate input data these can be used to simulate the developing fire conditions, enabling the composition of the fire effluent plume in a building or outside environment to be calculated for different times and locations. Algebraic fire calculations make use of standard heat release rate input curves (such as those shown in Chapter 2, Figure 2.5) or curves derived from data obtained from experimental fires involving specific fuel packages. Fixed yields for combustion products are then assumed, and the air entrainment, dilution and spread of the effluent plume within and from building enclosures are calculated using appropriate formulae or equations.

Issues in Toxicology No. 23
Toxicology, Survival and Health Hazards of Combustion Products
Edited by David A. Purser, Robert L. Maynard and James C. Wakefield

Published by the Royal Society of Chemistry, www.rsc.org

For "zone" computer models the same basic algebraic expressions are used within the programme for the different fire parameters. A major simplification is made, that within each building enclosure the fire effluents are assumed to be dispersed evenly within an upper smoke layer, separated from a lower (essentially uncontaminated) air layer. Some models also include variables allowing for some exchange between the upper and lower layers. Computational fluid dynamics models are much more sophisticated in that the enclosures under consideration are divided into many cells and the movement of air, heat and combustion products and the combustion conditions from cell to cell are calculated for each successive short time period. The major disadvantage of CFD methods is that a single fire calculation, for even a simple enclosure, may require several days of computer time to complete. Also, the outcome can be critically dependent on the expertise with which the user sets up the initial cellular computation grid.

The advantages of using fire modelling methods to reconstruct or predict conditions during fire incidents are that they are considerably less expensive than carrying out full-scale experimental fire tests and that multiple simulations with different input data and boundary conditions can be used to examine a range of potential outcomes. Probabilistic assessments using event tree or Monte Carlo simulations can also be made.

The limitations of all modelling methods are that, although some aspects of fires, especially air entrainment and smoke movement within and from building enclosures, can be calculated with reasonable accuracy, the predicted conditions, especially in terms of the time–concentration curves for toxic products, are heavily dependent upon a range of input assumptions. Existing models are not fully capable of simulating the complexities of the actual fire source term (the processes whereby a fuel is pyrolised and then combusted), and while for a given fire they can predict the yields of "major species" such as CO_2 with reasonable accuracy, the prediction of "minor species", which includes essentially all the products of interest to toxicologists, is much less certain. Therefore, in order to carry out fire modelling simulations, the fire source input term (in terms of fire size and growth rate) is usually derived from standard fire curves or data from experimental fire tests involving similar fuels and fire scenarios. The yields of toxic products are also usually generic input terms derived from limited experimental test data and, as described in Chapters 2 and 5, in practice these can be extremely variable depending upon the exact combustion conditions.

3.2 Fire Modelling Basics

While it is considered beyond the scope of this chapter to describe different modelling methods in detail we shall consider some of the basic aspects relevant to the estimation of toxic hazards from fires that are essentially common to all models. The equations and data presented in the following sections represent relatively simple case examples. The reader is referred to the following source publications for more detailed information (*Application of Fire Safety Engineering Principles to the Design of Buildings*,[1] *SFPE Handbook of Fire Protection Engineering*[3,5] and *An Introduction to Fire Dynamics*[6]).

3.2.1 Source Fire and Fire Size

The starting point for engineering modelling of any fire is the fire growth curve, which is expressed in terms of the heat output (MW). This is usually an input term obtained from relevant test data or a set of standard heat release rate curves used mainly for building design purposes. Models are also under development for the prediction of fire growth from fundamental physical and chemical fuel characteristics. For non-flaming thermal decomposition it is possible to estimate pyrolysis rates from the applied heat radiation and the fuel characteristics such as heat of gasification. For smouldering fires a model has been developed by Quintiere describing pyrolysis rate by the expression:[1]

$$m = (0.1\ \text{g min}^{-1})t + (0.0185\ \text{g min}^{-3})t^2 \quad \text{for } 0 \leq t \geq 60\ \text{min} \tag{3.1}$$

$$m = (73\ \text{g min}^{-1}) \quad \text{for } 60 \leq t \geq 120\ \text{min} \tag{3.2}$$

where m is the smouldering pyrolysis rate (g min^{-1}; within limits as stated in BS7974, part 1, p. 27) and t is time (minutes).

For open burning (fuel-controlled flaming fires), the maximum fire size Q (MW) depends approximately on the area of fire and the heat release rate per m^2:

$$Q = Q''A_{\text{fire}} \tag{3.3}$$

where Q'' is the total heat release rate per unit area of the fire (kW m^{-2}) and A_{fire} is the fire area (m^2).

Table 3.1 shows values for Q'' for a variety of cases for fuel-bed-controlled fires.

For a growing fire spreading across an area of fuel, the fire size is approximately proportional to its area, and if the fire grows at a uniform rate,

Table 3.1 Suggested heat release rate per unit area for fuel-bed-controlled fires (from BS7974-1).

Occupancy or item	Heat release rate per unit area/kW m^{-2}
Shops[7]	550
Offices[7]	290
Hotel room[7]	250
Industrial[8]	90–620
Gasoline pool fire[9]	1833–2930
Mail bags, 1.5 m high[1] (App 2.4, Table A4)	400
Cardboard carton, 4.9 m high[1] (A4)	1700–4200
Polyethylene bottles in cardboard cartons[1] (A4)	6200–7600
Polyvinyl chloride (PVC) bottles in cartons[1] (A4)	3400–7000
Rigid polyurethane foam board stacked 4.6 m high[1] (A4)	1900–3200

the size is proportional to the square of time from ignition. This has proved to be a practical, simple engineering model for growing fires involving different common fuels used for building design ("design fires").[1]

$$Q = \alpha t^2 \tag{3.4}$$

where Q is the heat release rate (kW), α is the fire growth parameter (kW s^{-2}) and t is the time from ignition.

Design fires are classified into four main fire types and fire growth parameters in Chapter 2, Figure 2.5, with examples of fire scenarios in Chapter 2, Table 2.4.

It is also useful to consider the heat release rate (HRR) from individual objects or cases as listed in Table 3.2.

The peak HRR figures shown in Tables 3.1 and 3.2 are mainly for fuel-controlled burning, in which the ventilation is sufficient to support the maximum fire size that the fuel can generate. In building enclosures, the maximum fire size usually becomes limited after a few minutes of fire growth by the available ventilation. For fires in small, completely enclosed buildings, such as the domestic house fire shown in Chapter 2, Figure 2.9, the fire grows for a few minutes until it becomes ventilation limited, with a maximum size of around 500–1000 kW. This lasts for a minute or so before the fire dies down to a small size and is likely to self-extinguish, unless an exterior window or door is opened. When exterior vents are opened, especially one or more windows and a door, then a fire may grow to flashover, with total involvement of the fire enclosure contents. The size of the fire then depends on the size of the vents, and burns at a constant rate until the

Table 3.2 Examples of peak fire sizes for a range of individual items (fuel-bed-controlled burning) and building cases.

Item or case	Peak heat release rate/kW
Folded double sheet newspaper, 22 g[10]	4
Crumpled double sheet newspaper, 22 g[10]	17
Pillow (polyester fibre fill)[11]	33
Medium wastepaper basket with milk cartons[11]	100
TV set[11]	120–290
Wardrobe, plywood, 12.7 mm thick containing clothing and paper[11]	3100
Decorated Christmas tree, 2.86 m height[11]	3321
Items of upholstered furniture (CBUF)[12]	
Minor flaming of covers only	10–40
Gradually growing fire	300–500
Rapid involvement of covers and foam filling	700–2000
Small newsstand booth, 15–60 kg books[9]	3600–8000
Furnished living room[10]	4000–8000
World Trade Center 5 (fully involved, eight-storey office building)[13]	16 000
World Trade Center 1 or 2[10]	1 150 000

fuel is exhausted. Under these conditions the fire size can be calculated approximately as:[14]

$$Q = 0.09 A_w \sqrt{h \Delta H_c} \tag{3.5}$$

where A_w is the area of the opening (m^2), h is the height of the opening (m) and ΔH_c is the fuel heat of combustion (kJ kg^{-1}).

This may slightly overestimate the rate of burning since it assumes efficient (stoichiometric) combustion. During large building fires the heat release rate and hence the mass of effluents flowing out from the building increases considerably once one or more windows fail, with flashover and secondary combustion of the fire plume as it exits from the open vents. A further large increase in fire size occurs once structural failure occurs, involving roof and wall collapses that enable more air to reach the burning fuel.

3.2.2 Calculation of Burning Rates and Toxic Product Releases from the Heat Release Rate

The mass burning rate of a fire (m_f; the mass of fuel consumed per second, or mass loss rate of the fire)[14] is calculated from the HRR (Q in kW) as:

$$m_f = \frac{Q}{\chi H_c} \tag{3.6}$$

where m_f is the mass burning rate (kg s^{-1}), χ is the combustion efficiency (from 0 to 1) and H_c is the net chemical heat of combustion.

The net heat of combustion H_c is the net heat of complete combustion obtained by totally efficient combustion of the fuel. In practice, combustion is never 100% efficient in fires so the effective heat of combustion is given by the heat of combustion multiplied by a combustion efficiently factor between 0 and 1. During the early stages of well-ventilated fires, when the equivalence ratio (actual fuel : air ratio divided by the stoichiometric fuel : air ratio, phi or ϕ) is less than 1, the combustion efficiency is high at around 0.8–1 for non-flame-retarded fuels, but once a fire becomes ventilation-controlled the combustion efficiency depends upon the equivalence ratio, being around 0.5 for an equivalence ratio of 2. Figure 3.1 illustrates the relationship between equivalence ratio and combustion efficiency for some common fuels burned in the steady state tube furnace. As shown in Figure 3.1, another factor influencing combustion efficiency is the presence of flame retardants acting in the vapour phase—especially chlorine and bromine. PVC, which contains 57% chlorine, is an example of a material showing low combustion efficiency even under well-ventilated conditions. Table 3.3 shows the net heats of combustion and elemental compositions for a range of common natural and synthetic polymers tested in the ISO 19700 tube furnace. Table 3.4 shows the effective heats of combustion obtained under well-ventilated ($\phi < 1$) and ventilation-controlled ($\phi > 1$) combustion conditions.

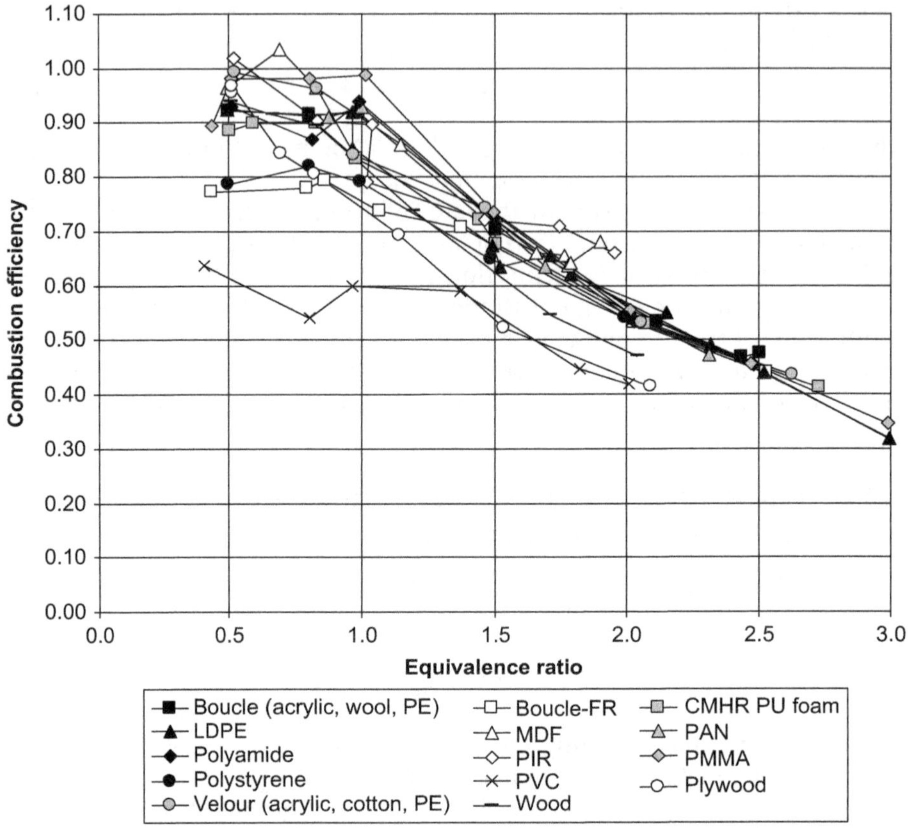

Figure 3.1 Combustion efficiency as a function of equivalence ratio for flaming combustion of a set of common polymers in the ISO TS 19700 tube furnace. PE, polyethylene; LDPE, low-density PE; FR, flame retarded; MDF, medium density fibreboard; PIR, polyisocyanurate; CMHR PU, combustion modified high resilience polyurethane; PAN, polyacrylonitrile; PMMA, poly(methyl methacrylate).

The mass release rate of each toxic product m_p (g s^{-1}) is then given as:

$$m_p = Y_p m_f \tag{3.7}$$

where Y_p is the mass loss yield of each product (g toxic product/g fuel mass consumed) and m_f is the mass loss rate of the fuel (g s^{-1}).

The yields of any specific toxic products from a particular material depend upon two main parameters: the mass fraction of the main chemical source elements in the fuel and the efficiency with which those fuel elements are converted to the toxic products containing those elements. The first parameter depends simply on the elemental composition of the combustible components of the fuel material. The second parameter depends primarily on the combustion conditions, but may also be affected by the molecular structure. For toxic products containing carbon, the yield therefore depends

Table 3.3 Composition of test materials.

Material	ΔH_c[a] kJ g^{-1}	Ψ_O[b] g g^{-1}	Elemental composition (%) C	H	O	N	Cl	Br	P	S
Boucle (acrylic/wool/polyester, 38/38/24; mixed fibre fabric)	26.5	2.02	63.1	6.4	16.7	12.89	<0.3	<0.5	—	0.94
Boucle (acrylic/wool/polyester, 38/38/24; FR back-coated)	25.0	1.91	59.0	6.3	16.1	10.83	0.95	6.09	—	0.76
Combustion modified high resilience (CMHR) polyurethane (PU) foam (FR)	24.5	1.87	56.45	7.67	24.1	8.22	2.53	—		—
Low density polyethylene (LDPE)	44.8	3.42	85.5	14.51						
MDF	16.9	1.35	47.90	6.13	41.66	3.69	0.62	<0.5	<0.01	—
MDF (FR board)		1.25	45.1	5.77	39.22	6.73	—	2.66	0.63	
Polyacrylonitrile (>85%), fabric	30.5	2.33	65.62	5.71	—	23.24				
Polyamide 6	30.5	2.33	63.68	9.79	14.14	12.4				
Polyisocyanurate (PIR), rigid foam	24.5	1.87	63.5	4.98	21.8	6.15	3.56	—	—	—
Poly(methyl methacrylate) (PMMA)	25.2	1.92	60.33	8.14	31.53					
Polystyrene	40.2	3.07	92.26	7.38						
PVC	16.8	1.28	38.44	4.84			56.73			
Plywood	17.8	1.36	46.32	5.80	47.56	0.32	—	—	—	—
Acrylic/cotton/polyester, 52/31/17, velour mixed fibre fabric	26.2	2.00	64.4	6.39	18.45	11.55	<0.3	<0.5	—	—
Wood (*Pinus sylvestris*)	18.1	1.38	49.2	6.44	44.22	0.14				

[a]ΔH_c = net heat of chemical combustion.
[b]Ψ_O = stoichiometric oxygen demand.

on the mass fraction of carbon in the burning fuel (or more specifically in the portion of the fuel consumed in the fire) and the efficiency of conversion of this fuel carbon to carbon in the toxic product of interest (also known as the recovery fraction).

3.3 Calculation and Measurements of Product Yields and Recovery Fractions

The yield of any specific product is given by:

$$Y_P = f_E R_E M_P / M_{A,E} n_E \tag{3.8}$$

where f_E is the mass fraction of the product element in fuel, R_E is the mass recovery fraction of fuel element E in the product, $M_{A,E}$ is the atomic mass, expressed in grams, of the element E, M_p is the molar mass of the product, n_E is the number of atoms of element E per molecule of the product, and M_E

Table 3.4 Toxic gas yields, effective heats of combustion and oxygen consumption under well-ventilated and under-ventilated combustion conditions for a range of common polymeric materials from the ISO TS 19700 tube furnace.[20]

Polymer	phi	Eff Ht/kJ g^{-1}	CO_2/ mg g^{-1}	CO/ mg g^{-1}	HC/ mg g^{-1}	O_2[d]/ mg g^{-1}	Soot/ mg g^{-1}	SEA/ m^2 kg	HCN/ mg g^{-1}	NO/ mg g^{-1}	NO_2/ mg g^{-1}	HCl/ mg g^{-1}	HBr/ mg g^{-1}	SO_2/ mg g^{-1}
Well-ventilated flaming: phi = 0.4–0.8														
LDPE	0.49	41.5	2836	15	85	3166	45	268						
Polystyrene	0.49	31.6	2644	61	82	2416	110	621						
Wood	0.51	16.9	1696	6	13	1293	5	12						
Plywood	0.52	17.3	1774	6	11	1324	3	1	0	2	1			
MDF	0.49	16.8	1680	7	24	1283	3	7	0	3	1			
PAN	0.88	30.4	2339	39	54	2320	25	104	8	2	1			
Polyamide 6	0.51	28.4	2216	3	34	2166	19	147	0	11	1			
PIR	0.52	24.6	2340	48	13	1874	33	75	3	2	1	69		
CMHR PU	0.59	25.3	2156	41	48	1928	28	154	4	3	1	9		
Boucle non-FR[a] (acrylic, wool, PE; 34/38/24)	0.50	24.4	2128	60	19	1861	26	103	1	7	1			12
Boucle FR	0.44	19.3	1486	130	81	1474	90	456	19	8	0	10	30	11
Velour[b] (acrylic, cotton, PE; 52/31/17)	0.52	26.3	2240	41	51	2005	19	84	2	4	0			
PVC[c]	0.40	10.7	667	177	70	815	32	163				447		

Fuel-rich (ventilation-controlled) flaming: phi = 1.5–2.0														
LDPE	1.71	29.4	1696	196	334	2242	85	668						
Polystyrene	1.99	21.8	1662	86	299	1664	179	820						
Wood	1.71	9.8	967	134	80	752	19	155						
Plywood	1.54	9.4	986	96	55	714	14	120	0	1	0			
MDF	1.66	8.9	870	113	62	681	19	150	3	1	1			
PAN	1.69	19.1	1271	130	235	1460	60	489	72	2	3			
Polyamide 6	2.03	16.3	1135	130	248	1246	51	413	41	3	3			
PIR	2.08	14.0	937	333	136	1068	72	495	20	1	2	57		
CMHR PU	2.07	14.9	1041	246	197	1134	59	403	14	1	2	5		
Boucle non-FR	2.12	14.2	1138	119	228	1080	104	594	35	1	2			4
Boucle FR	2.03	13.3	920	146	184	1016	100	611	25	2	1	3	28	8
Velour	2.06	14.0	1211	126	239	1071	84	526	34	2	1			
PVC	1.82	7.5	389	137	98	573	70	473				585		

[a]Boucle looped yarn mixed fabric (see Table 3.3). For other abbreviated fuel names see Table 3.3 and caption to Figure 3.1.
[b]Velour mixed fabric (see Table 3.3).
[c]Rigid, 100% PVC.
[d]Oxygen consumed (mg g^{-1}).

is the mass, expressed in grams, of element E per unit mass, expressed in grams, of material.

For example, the carbon content of medium density fibreboard (MDF) is shown in Table 3.3 as 38.44%. When this was burned at an equivalence ratio of 0.5 the recovery fraction of carbon in the form of CO was 0.225 (as shown in Figure 3.3). The yield of CO is therefore calculated as:

$$(0.3844 \times 0.225 \times 28.01)/(12.01 \times 1) = 0.202 \text{ g g}^{-1}$$

where:

0.3844 is the mass fraction of carbon in PVC;
0.225 is the recovery fraction of carbon in CO;
28.01 is the molar mass of CO;
12.01 is the atomic mass of carbon;
1 is the number of atoms of carbon in CO.

The recovery fraction (or conversion efficiency) is the yield of a specified fuel element in the product as a fraction of the maximum yield obtainable from complete conversion of the fuel element to that in the product. It can be calculated from the measured yield, Y_p, of the product of interest relative to its notional yield from complete conversion, Ψ_p. For a material containing element E, this corresponds to eqn (3.9):

$$R_E = Y_p/\Psi_p \tag{3.9}$$

where Y_p is the yield of product (g g^{-1} fuel mass loss) and Ψ_p is the stoichiometric yield of product (g g^{-1} fuel mass loss), *i.e.*, the notional maximum theoretical yield.

Smoke concentrations can be expressed in terms of the particulate concentrations (g m^{-3}) and particulate yields (g g^{-1}), or in terms of it light attenuation, which depends on the density of the smoke density.

An attenuation parameter used by engineers is the extinction coefficient α_k per meter path length derived from Bouguer's law:[15]

$$\alpha_k = \frac{1}{L} \times \ln(I_0/I) \tag{3.10}$$

where:

α_k is the light extinction coefficient, expressed as reciprocal metres (m^{-1});
I_0 is the intensity of incident light;
I is the intensity of transmitted light (at the detector);
L is the length, expressed in metres, of the light path through the smoke.

A yield-based term derived from this is the specific extinction area A_{SEA} (m^2 kg^{-1}). This represents the extinction coefficient obtained by decomposing one kg of fuel into a volume of 1 m^3. Another method often used to measure obscuration is the optical density per metre (D), which is similar to the extinction coefficient but expressed as the logarithm to the base 10 rather than the natural logarithm, so that:

$$D \times 2.303 = \alpha_k \tag{3.11}$$

The optical density per metre D is a useful parameter because visibility distance through smoke is approximately the reciprocal of the optical density, although this depends somewhat on the contrast and illumination of the target being viewed.

Fire dynamics models currently have a limited ability to calculate product yields or recovery fractions, especially for the "minor" products such as CO, HCN, irritants and particulates, which are the most important toxic species. Generic data may therefore be used as inputs to models or data measured in specific combustion tests, either large- or bench-scale, for the fuel materials of interest. Full-scale and large-scale fire tests are expensive to perform, so that although they may be used occasionally as reference scenarios or to replicate specific fire incidents, it is necessary to rely on bench-scale tests to generate the necessary data. A problem with this is that, as discussed in Chapter 5, most bench-scale tests have not been designed to replicate the range of specific combustion conditions existing in full-scale fires, and since different combustion conditions are known to have a considerable effect on toxic product yields, such tests are of questionable value.

Two test methods, the ISO TS 19700 steady state tube furnace[16] and the ASTM E2058 fire propagation apparatus,[17] have been developed to burn test specimens over a range of test conditions. The ISO TS 19700 test method has been designed to decompose test specimens under the full range of combustion conditions known to occur in fires, specified in terms of non-flaming/flaming decomposition, decomposition temperature and (for flaming combustion) the fuel : air equivalence ratio. Early versions of this apparatus have also been used to produce combustion product atmospheres for animal exposures for direct measurements of the toxic effects of irritant and asphyxiant combustion product mixtures, in combination with measurement of chemical composition[18] (see Chapter 7). The apparatus is shown in Figure 3.2. The principle of the method is the generation of an atmosphere of combustion products in the mixing chamber maintained at a constant "steady state" composition for an extended period to enable repeated and detailed chemical measurements to be made or to enable animal exposures. This is achieved by introducing the test specimen into the furnace in strip form at a constant rate under a constant stream of air, thereby maintaining constant decomposition conditions for the duration of a test run. The combustion products are expelled into a chamber where they are diluted with a secondary air supply to form the test atmosphere for chemical sampling or animal exposures. Different combustion conditions can be obtained in separate runs by varying the furnace temperature and fuel : air ratio. The concentration of combustion products passing through the mixing chamber then depends on the rate of decomposition of the specimen and the flow of secondary dilution air through the chamber.

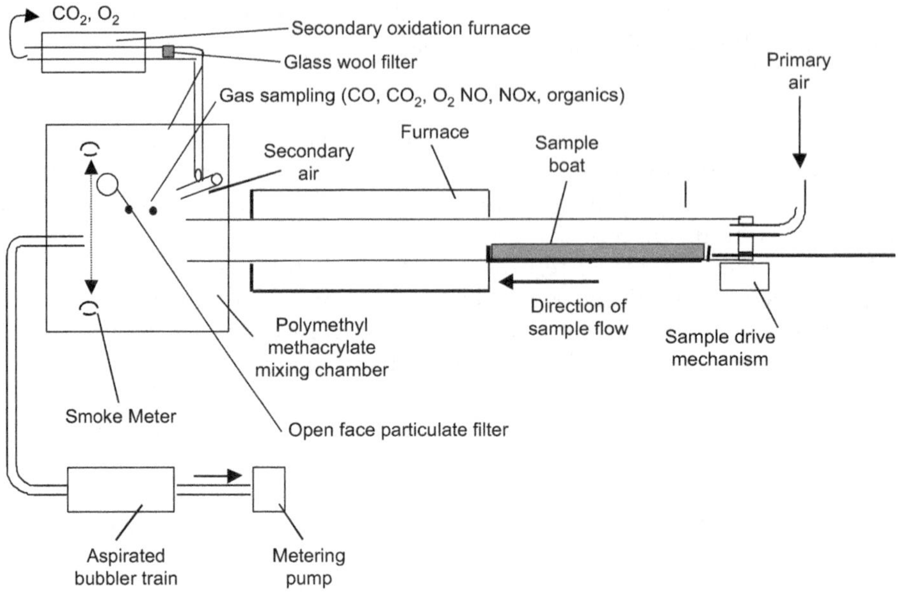

Figure 3.2 ISO TS 19700 steady state tube furnace apparatus.

The yield of each toxic product from any specific fuel can be measured in tests carried out under combustion conditions similar to those of the full-scale fire of interest. As stated, for flaming fires the yield from any particular material depends upon the equivalence ratio, and in particular whether combustion is well ventilated ($\phi < 1$) or ventilation controlled ($\phi > 1$).[19] When these yields are normalised in terms of the carbon or nitrogen content of the fuel they demonstrate the extent to which these fuel elements are converted to the different toxic products containing them. Figure 3.3 illustrates this in terms of the recovery fractions for CO as a function of equivalence ratio for flaming combustion of the set of common polymers listed in Table 3.3. For Figure 3.3, fuel carbon as CO is calculated as a proportion of the total fuel carbon released into the gas phase during combustion (*i.e.*, corrected for any fuel carbon remaining as char). The results show that for the majority of non-flame-retarded materials, the fuel carbon conversion to CO is very low under well-ventilated combustion conditions ($\sim$0.01) but then increases steeply for equivalence ratios above 1, often levelling off at equivalence ratios above 2.0. Expressing the data in terms of recovery fraction illustrates the extent to which the CO yields are influenced by the polymer composition. Under higher equivalence ratios, the highest CO conversions of up to 0.25–0.3 tend to be for substances with both carbon and oxygen in the molecule, such as cellulosic materials (medium density fibreboard and wood) or poly(methyl methacrylate). CO recoveries were somewhat lower for the aromatic polymer polystyrene and the aliphatic polymer polyethylene, neither of which contain oxygen. These experiments

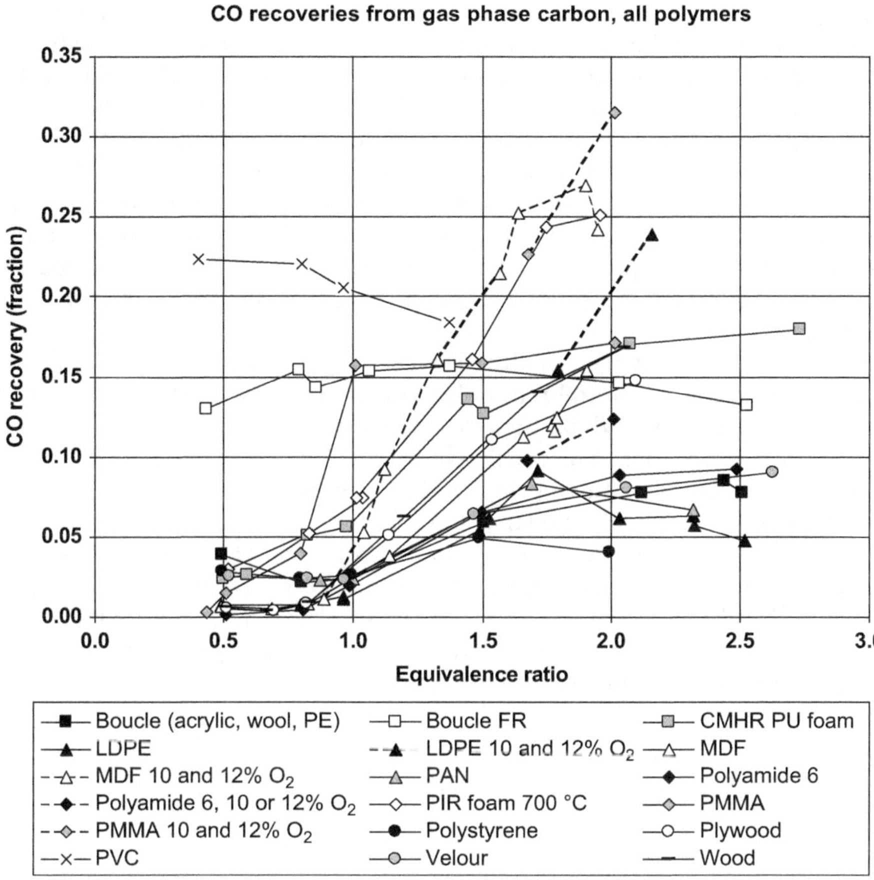

Figure 3.3 Recovery of carbon as CO as a proportion of fuel carbon released into the gas phase for flaming combustion in the ISO TS 19700 tube furnace at 650 °C (700 °C for PIR).[20]

were conducted at a furnace temperature of 650 °C (700 °C for polyisocyanurate). For these flaming combustion conditions, the temperature in the actual flames typically averages 1000–1400 °C but beyond the flame zone the temperature of the products passing through the furnace tube is close to the furnace temperature [representing the conditions beyond the flame zone in the hot upper layer of a full-scale building enclosure fire (see Chapter 2)]. At a higher temperature of 850 °C, characteristic of the upper layer in large post-flashover fires, the CO recoveries were increased for some materials. Another important determinant of CO yields and recoveries is the presence of flame-retardant additives, especially halogens and to some extent nitrogen and phosphorus.

When these materials burn, the halogens are released into the gas phase as halogen acid gases, which act as free-radical scavengers in the flame zone, thereby reducing combustion efficiency and promoting the formation of partly oxidised products including CO and HCN. In Figure 3.3, materials with a small but significant chlorine content, such as the CMHR PU foam and polyisocyanurate, show somewhat increased CO recoveries under well-ventilated combustion conditions compared with non-halogenated materials, while those with a high halogen content, such as PVC and Boucle-FR, show high CO recoveries across the range of equivalence ratios. Similar increases in recovery fractions and yields in relation to equivalence ratio, temperature and the presence of flame retardants have been measured for other toxic combustion products including HCN, organic products and particulates.

Figure 3.4 shows the relationship between the mass percentage of chlorine or bromine in a range of different polymers and the recovery fractions of CO, HCN, total organics and particulates when burned under well-ventilated combustion conditions. The results show an increasing trend for all products of inefficient combustion as the halogen content of the polymer increases, irrespective of the polymer type.

Figure 3.5 shows the recovery of fuel nitrogen as HCN for all nitrogen-containing polymers as a function of equivalence ratio when combusted at 650 °C.[20,21] At higher temperatures, the HCN recovery was slightly increased for the CMHR PU foam, but for polyamide it was increased considerably to 0.46, so that almost half the fuel nitrogen was released as HCN.

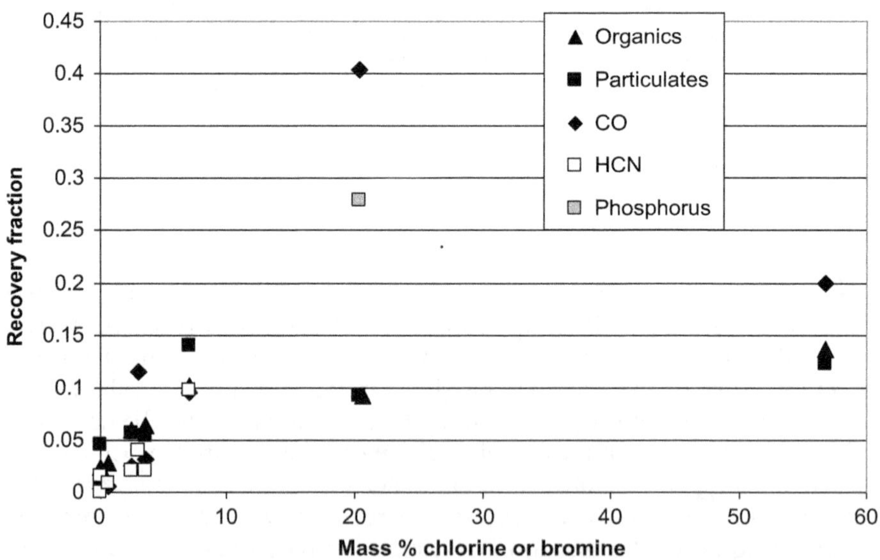

Figure 3.4 Recovery of fuel carbon, nitrogen and phosphorus to toxic products as a function of mass percentage chlorine or bromine in the original polymer under well-ventilated combustion conditions in the ISO TS 19700 tube furnace.[20]

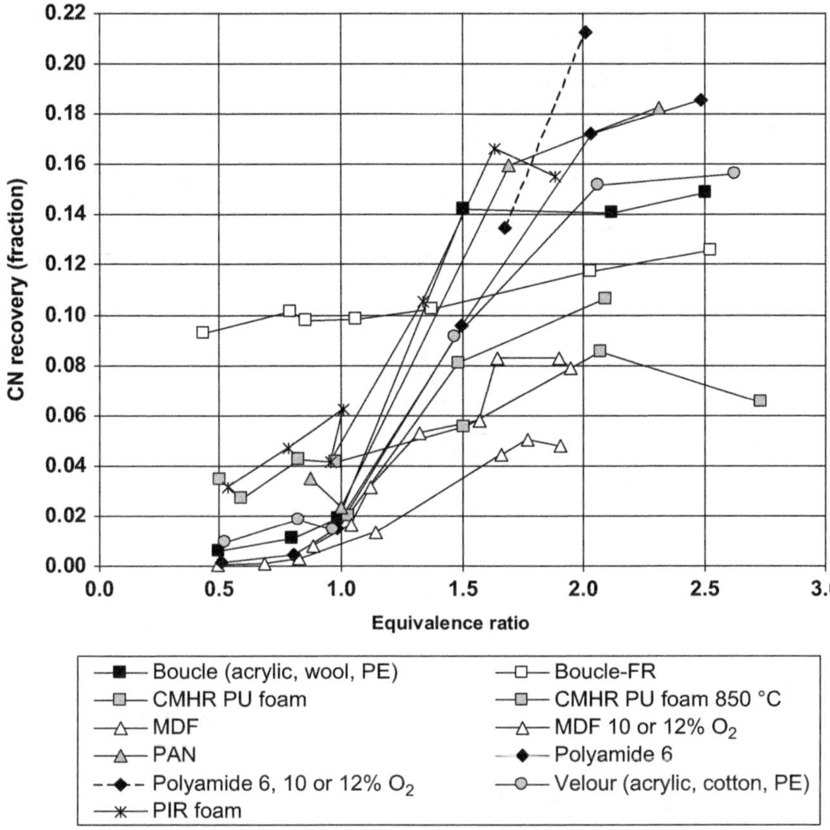

Figure 3.5 Recovery of fuel nitrogen as HCN for flaming combustion in the ISO TS 19700 tube furnace at 650 °C.

In general, recovery of fuel nitrogen as HCN under any specific combustion conditions was found to be similar to the recovery of fuel carbon as CO. The recoveries and yields of organic products also increased with equivalence ratio, as illustrated for polyamide in Chapter 2, Figure 2.4, and for poly(methyl methacrylate) in Figure 3.6. For most materials, there was also a small increase in particulate yields with equivalence ratio, but the main effect was on the particulate composition. Under well-ventilated combustion conditions the particulates tended to take the form of a dry and powdery soot, while for fuel-rich conditions the particulates contained a higher proportion of organic condensates, producing sticky soot deposits.

The main exceptions to the increase of product yields with equivalence ratio are those of carbon dioxide and oxides of nitrogen, which decrease as the equivalence ratio and/or fire retardants increase (see Chapter 2, Figures 2.3 and 2.4), and the yields of acid gases, which are generally similar across the range of combustion conditions (Figure 3.7), although the Boucle fabric (a fabric with a looped yarn) showed some decrease.

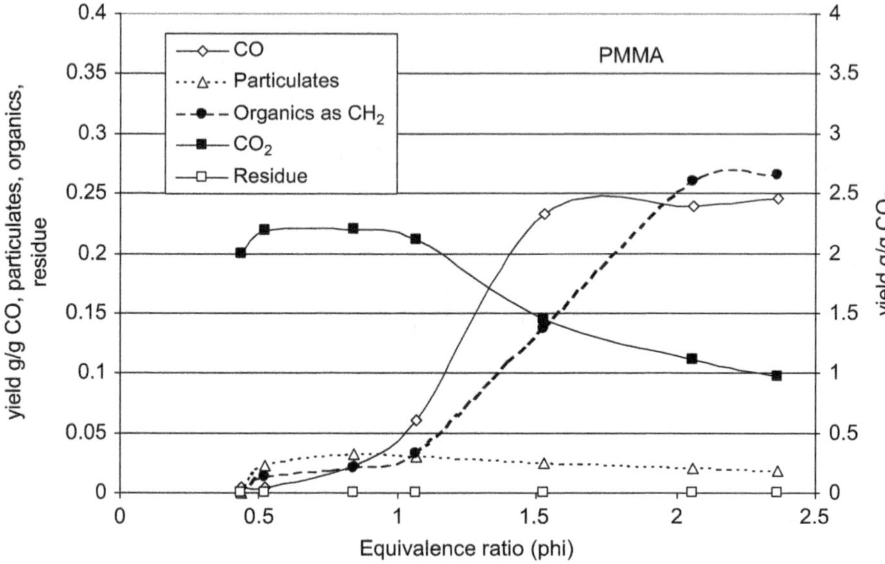

Figure 3.6 Comparison of the yields of major combustion products for PMMA under flaming combustion in the ISO TS 19700 tube furnace.

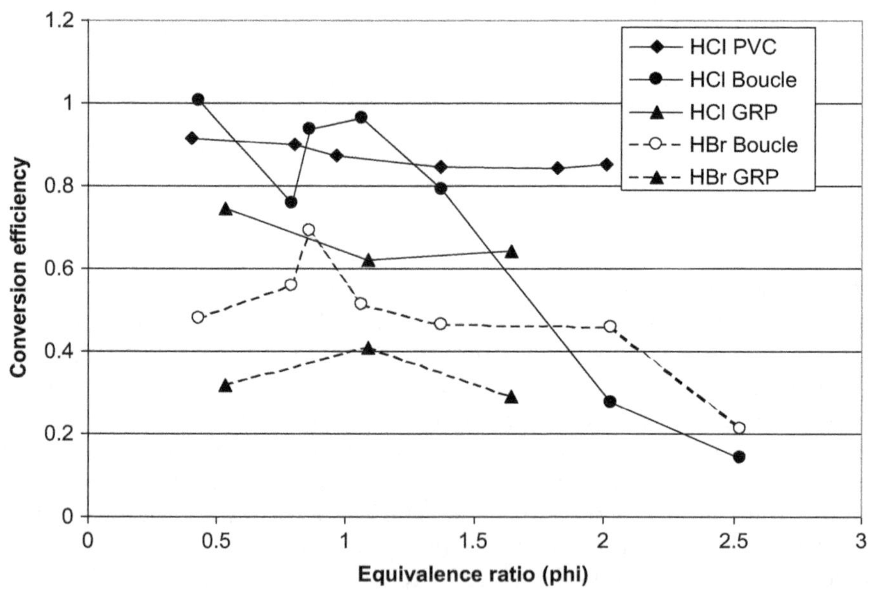

Figure 3.7 Conversion efficiency of fuel chlorine and bromine to gas phase halogens.[20] GRP, flame retarded Glass Reinforced Plastic.

Table 3.4 shows yield data for the main toxic products, total particulates and smoke obtained from all the materials listed in Table 3.3 using the ISO TS 19700 steady state tube furnace. In the table, yields are expressed as mg g^{-1}, so, for example, the mass production rate of CO from a ventilation-controlled fire involving low density polyethylene is given by:

$$m_{CO} = m_f \times 196/1000$$

where m_{CO} is the mass rate of carbon monoxide production (g s^{-1}) and m_f is the rate of burning of the fuel by mass (g s^{-1}).

Further yield data obtained using a different apparatus are shown in *The SFPE Handbook of Fire Protection Engineering*.[17]

The application of these methods enables the conditions to be calculated in different parts of a building during a fire, and in plumes from vents such as windows. For smoke plumes dispersing in the outside environment it is necessary to use the composition and production rate of smoke plumes from the source object (building) as inputs to plume dispersal models, taking into account the dynamics of the smoke plume development in relation to the meteorological conditions. This topic is dealt with in Chapter 6.

3.4 Using Individual Products as Markers for Estimating Concentrations of Other Components

Modelling the composition of smoke plumes both within buildings and outside can be very complex and challenging, requiring detailed knowledge of a wide range of parameters. A simpler method, which can be applied with reasonable effectiveness, especially for widely dispersed plumes, is to use one product that can be measured easily as a marker or tracer for other toxic products.

During fires, once the effluent has moved away from the fire source and cooled, the chemistry becomes effectively frozen, so that at least in the short term (and within certain limits) most products track together. Thus, in the house fire tests described, the ratios of the concentrations of smoke particulates, CO_2, oxygen depletion, CO and HCN measured in the fire room were similar to those remote from the fire in the upstairs bedroom. HCN is an example of a very toxic gas that is often not measured during fire tests or in fire victims due to the complexity involved (see Chapter 2); however, if the HCN and CO_2 concentrations are measured in one location, it is possible to make a good estimate of the HCN concentrations in other locations from the CO_2 concentration and the original CO_2 : HCN ratio.

Similar considerations apply to smoke density (or total particulate matter concentration). By measuring the ratio of smoke optical density (D) to the CO concentration in fires of any specific type, it is possible to make a reasonably good estimate of the CO concentration at a remote location, if the smoke density is estimated or measured. For example, in building fires involving fuel-rich combustion, when the smoke density (D) reaches approximately 0.5 m^{-1} (equivalent to approximate 2 m visibility) not only is this predicted to present direct difficulties for escaping occupants, but the CO concentration can be predicted to be sufficiently high to cause incapacitation within a few minutes' exposure. Further from the fire, if D decreases to 0.1 m^{-1} (representing approximately 10 m visibility), then it can be stated with confidence that the associated CO concentration would be insufficient to cause problems to an exposed subject with up to a 1 hour exposure period. Such an approach can be especially valuable for outside fire plumes during incidents.

In order to apply both this approach and the more complex modelling approach it is essential to obtain some information on the relative concentrations and yields of key toxic products in the source fire. This can be achieved by carrying out combustion tests on the materials involved under appropriate fire conditions. In the ISO TS 19700 tube furnace, measurements are made of a range of relevant parameters including the particulate yield, smoke optical yield, CO_2, CO, HCN total organics and acid gases. By comparing the relative yields of any of these products to the concentrations of marker species, such as smoke density, CO or CO_2, measured at any remote location it is then possible to estimate the likely extent of exposure to subjects at that location.

For materials that have not been tested in this way it is even possible to make reasonable estimates of the likely yields of key toxic products under specified combustion conditions, providing the elemental composition of the fuel is known. In particular, as stated, the efficiency of conversion of fuel nitrogen to hydrogen cyanide has been shown to be similar to the efficiency of conversion of fuel carbon to carbon monoxide, so that if the carbon and nitrogen contents of the fuel are known and the CO concentration has been measured, then it is possible to estimate the concentration of HCN by calculation.[21,24]

Table 3.5 shows some data for the yields of some potentially toxic products from burning African Savannah biomass in wildfires. The data were compiled by Christian *et al.* from laboratory experiments and reported field data.[22] The authors found that yields of different products varied to some extent depending upon the fuel type and the combustion efficiency, not so much as a function of the equivalence ratio, since all fires were well ventilated, but depending on the extent of flaming and smouldering combustion in the fuel bed. Thus, the yields of CO and organic products were lower for grass, which flamed efficiently, than for peat or secondary leaf litter, which tended to involve a mix of smouldering and flaming. In other experiments, higher yields of a succession of organic products were obtained when fuels were ramp heated to a point where smouldering occurred but not flaming.[23]

3.5 Validation of Fire Test Data for Application to Fire Hazard Assessments

As described, the yields of different toxic substances in fires are very dependent on the combustion conditions as well as the fuel composition. Where some form of large- or bench-scale fire test is used to generate fire effluents it is important to ensure, as far as possible, that the combustion conditions in the test are representative of those anticipated in the actual

Table 3.5 Molar emission ratios relative to CO_2 (mmol mol^{-1}) and yields (g kg^{-1}) for combustion products in smoke from fires in African Savannah fuels.[22]

Compound	Emission ratio/mmol mol^{-1} relative to CO_2	Yield/g kg^{-1}
Carbon dioxide	1000	1689
Carbon monoxide	66.4	71.4
PM 2.5	—	20.0 (2.9–61.6)
Methane	3.52	2.17
Nitrogen oxides, NO_x as NO	3.04	3.50
Ethylene	1.14	1.23
Acetic acid	1.06	2.44
Formaldehyde	0.97	1.12
Methanol	0.96	1.18
Sulfur dioxide	0.85	2.09
Hydrogen cyanide	0.57	0.59
Ammonia	0.46	0.30
Acetaldehyde	0.45	0.76
Formic acid	0.39	0.69
Acetylene	0.29	0.29
Phenol	0.23	0.83
Acetol	0.22	0.62
Glycolaldehyde	0.21	0.4
Propylene	0.20	0.32
Ethane	0.19	0.22
Methyl vinyl ether	0.11	0.24
Furan	0.085	0.21
Acetone	0.085	0.19
Benzene	0.069	0.21
1,3-Butadiene	0.035	0.073

incident scenario of interest. It is also useful to attempt to obtain some indication of the likely variability of the yields occurring in both the tests and the full-scale fire scenario. In this chapter and Chapter 2, data have been presented for full-scale fire tests representing specific building fire scenarios,[24–26] from large-scale fire tests, such as the ISO 19705 room scale test method,[27–29] to a bench-scale test method (the ISO TS 19700 test method).[16,20,21,30] Of these different methods, each full-scale building fire scenario test can be considered as representative of the fire conditions, toxic product concentrations and yields occurring in one example case of that full-scale scenario. The results therefore provide direct data on the conditions to which an occupant could have been exposed if they had been in the building at the time of the specific fire test. Repeat experiments for these scenarios using identical or similar fuels and test conditions also provide some indication of the variations that may occur in fire development and conditions, which can be significant, even when the fuels and test protocols are identical. Large-scale fire tests represent one level of abstraction from the full-scale scenario reconstruction, in that they replicate some but not all of the features of the full-scale scenario. For example, the full-scale house fire tests[24] and the Rosepark reconstruction experiments[25,26] recreated the full-scale building enclosures and ventilation conditions for actual fire scenarios or incidents. The ISO 9705 room tests, while they enable items such as an armchair or area of wall lining to be tested at full-scale, do not replicate exactly the dimensions and, in particular, the ventilation conditions in the full-scale building scenario, which can result in differences in the fire conditions and development. For a bench-scale test, such as the ISO TS 19700 test, it is not possible to replicate the complexities of the full-scale or large-scale fuel disposition and fire scenario, but it may be possible to replicate the main parameters of the decomposition conditions affecting fuels or fuel effluents at different locations and stages during the full-scale fire, enabling prediction of yields under specified conditions. On the other hand, it is possible to conduct some bench-scale and even large-scale tests in such a manner as to provide highly misleading data.

In order to ensure that the data used for any fire hazard assessment are useful and relevant to the fire scenario of interest it is important to consider carefully the major characteristics of the full-scale fire scenario of interest and ensure that any test data or calculation methods are relevant. For example, the ISO TS 19700 tube furnace method has been developed specifically to reproduce the combustion conditions known to occur in different types and stages of full-scale fires in terms of flaming or non-flaming behaviour, temperature and fuel : air equivalence ratio. The results for a range of fuels have been validated against sets of large- and full-scale fire test data for the same or very similar fuels to ensure that the test results are not artefacts of the bench-scale method, but represent real phenomena occurring in full-scale fires and provide comparable yield data for toxic products.

Figure 3.8 shows plots of CO yield data obtained by the author from MDF for flaming combustion over a range of equivalence ratios in the ISO TS 19700 test (SSTF) and two different large-scale compartment fire test methods (ISO 19705 room and a half-scale room-corridor test).[20] Of these the ISO19705 room is a large-scale standard method, but consists of a simple room with an open doorway. The room-corridor rig is considered to be more representative of an area of an actual building. The results show good agreement between the three methods. Figure 3.9 shows a comparison between the tube furnace test and large-scale compartment fire tests for six different fuels over a range of equivalence ratios.

For each set of large-scale fire test data (such as those in Figure 3.8) a trend line curve was fitted. The difference of each test point from the trend line curve was then measured and standard deviations calculated (left-hand range for each fuel) in Figure 3.9. This then provides an indication of the expected variability of the large-scale test results. On the right for each fuel is shown the average deviation and standard deviation also from the bench-scale test data. This provides an indication of the average difference and variability of the bench-scale test results as a predictor of the large-scale fire CO yields. The results show that the bench-scale test data provide good predictions of the averages and ranges of the large-scale results, thereby validating the use of the bench-scale test data for this purpose.

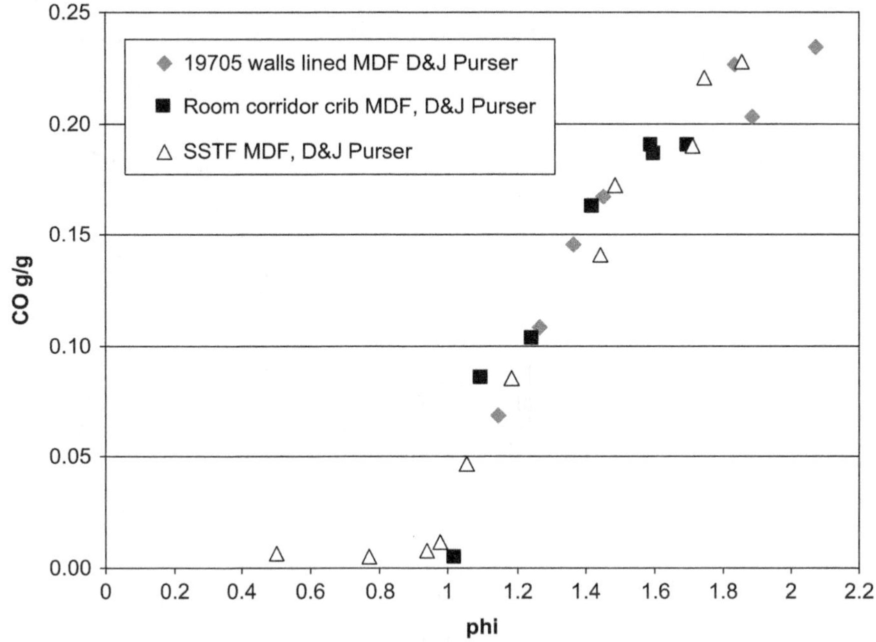

Figure 3.8 Comparisons of CO yields obtained from a cellulosic fuel (MDF) in the ISO TS 19700 tube furnace (SSTF) and two large-scale test methods, the ISO19705 room test and a room-corridor test.[20]

Another issue is that the measured yields of the different individual substances depend on the sampling and analytical methods used. The effects of sampling losses on results are illustrated in Chapter 2, Figure 2.16 for hydrogen chloride. Acid gases, such as HCl and NO_2, can be subject to considerable losses on the walls of test chambers and in sampling systems, or may be adsorbed onto particles. Hydrogen cyanide on the other hand has been found not to be subject to such losses (see Chapter 2). Analytical issues can be a particular problem for organic effluent components. The range of products detected can be very dependent on the methods used and their sensitivity. For example, Table 3.5 shows results from a detailed analysis for biomass fires, with low yields reported for a number of substances. However, there is no mention of two important irritants know to be a feature of burning cellulosic biomass: acrolein and crotonaldehyde (see Chapter 2, Table 2.3). These issues are discussed in more detail in Chapter 5.

3.6 Estimation of Combustion Products Capable of Causing Acute and Chronic Health Hazards or Environmental Contamination

Outdoor environmental and health hazards from fires depend to some extent on the same basic variables as those inside buildings. For acute

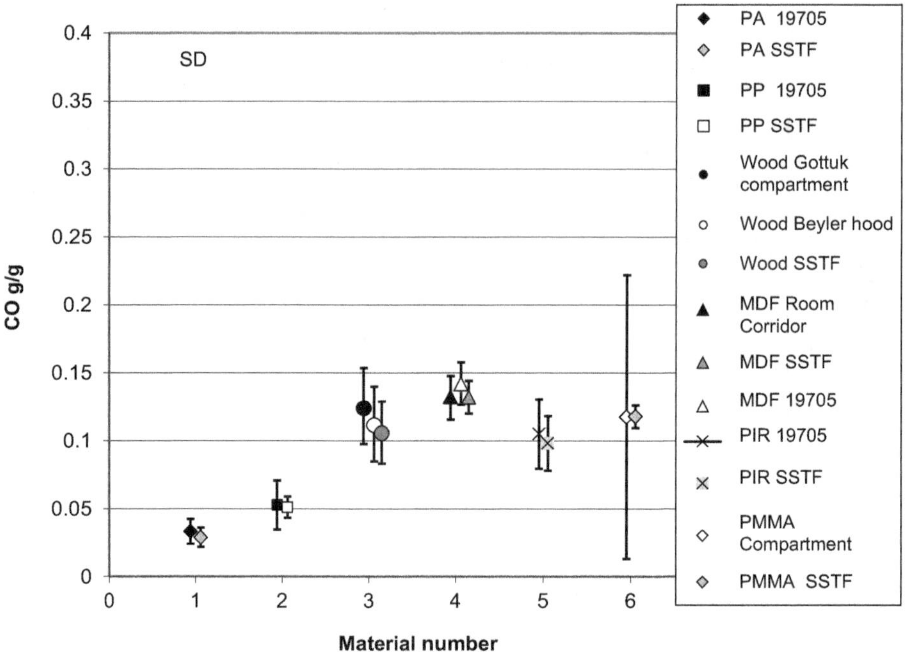

Figure 3.9 Means and standard deviations (SD) for CO yields measured in the ISO TS 19700 test and large-scale compartment fire tests for six different fuels. For each fuel the left-hand range is for the large-scale test data and the right-hand range is for the bench-scale test data.

exposures, the time–concentration curves and exposure doses of mixed toxic products are the main considerations, while for chronic exposure hazards the cumulative exposure dose over periods of years may be the main consideration. Environmental hazards depend upon a range of parameters including the deposition, uptake, degradation, bioaccumulation and persistence or substances released in combustion products.

Major differences from large fires inside buildings are the much larger dilution factors and exposure time-scales involved. For acute exposures, the main concerns are particulates, volatile organics and irritants including acid gases. These are capable of causing nuisance odour, irritancy, sore throats or respiratory symptoms in vulnerable populations. For chronic exposures, these substances and their derivatives affected by photochemistry are implicated in the long-term development of respiratory and cardiovascular diseases. The carcinogens in combustion products are also of concern. For human exposures to environmental toxins, bioaccumulation and systemic toxicity of a range of substances, particularly potential carcinogens and endocrine disruptors, also need to be considered.

The concentrations of these agents in outside fire plumes can be determined by measurements and calculations in the same way as for large building fires, but the concentrations and exposure doses important as potential health hazards are much lower. Chapter 2, Tables 2.1–2.3 and Table 3.5 show some data for products from materials commonly involved in building fires and wildfires. These give indications of the likely yields and relative yields of particulates, irritant acid gases and a range of organic products. In general, the presence of acid gases depends upon the elemental composition of the fuel. The presence of a range of organic species, particularly partly oxidised organic species, has been found to be indicative of irritancy. Thus, fire effluents from smouldering or inefficient flaming combustion, with relatively high yields of aldehydes and organic acids, especially low molecular mass (formaldehyde) and unsaturated aldehydes (acrolein and crotonaldehyde) and isocyanates, have been found to be irritant to mice and primates (see Chapter 7). However, the measured irritancy of effluent mixtures to mice has been found to be much greater than can be accounted for in terms of measured constituents. Fire effluent mixtures, especially from inefficient combustion, also contain a range of known carcinogens (see Chapter 2, Table 2.1 and Table 2.5, and Chapter 12). These include formaldehyde, benzene, polycyclic aromatic hydrocarbons, 1,3-butadiene, 1,2-dichloroethane, halogenated polychlorinated biphenyls (PCBs) and dioxins, acrylonitrile, nitrosamines, carcinogenic metals and fibres.

3.7 Environmental Contamination by Dioxins and Furans from Halogenated Materials

Combustion of any material containing chlorine or bromine produces a range of products that are not immediately hazardous in a fire, but which may contaminate a building interior or the wider environment with toxic substances hazardous to health over prolonged periods such as PCBs, and halogenated dioxins and furans.

With respect to unwanted fires there are two particular issues:

1. The extent to which the production, use and disposal of organohalogen compounds leads to general levels of environmental contamination by toxic flame-retardant additives, dioxins and dibenzofurans, which constitute a risk to public health.
2. The extent to which exposure of people to combustion products from halogenated materials during and after fires in buildings constitutes a risk to their long-term health.

These issues are reviewed in Chapter 12 and in *Fire Retardant Materials*.[31] The general extent of environmental contamination by dioxins and furans in North America and Europe has decreased considerably since a peak over the period from approximately 1955 to 1980 (see Chapter 12, Figure 12.1). Up to this time three main sources were dioxins as contaminants in PCBs used in transformers, penta-chlorinated phenols (used as wood preservatives) and municipal waste incineration (municipal solid waste, MSW). Other sources were industrial and hospital incinerators, crematoria, steel processing, coal fires, leaded petrol combustion in motor vehicles, certain herbicides and stubble burning. Since this time, controls on these and other minor sources have resulted in a rapid decline to low levels. Despite the large and increasing tonnage of PVC and other halogenated materials produced annually, this has not led to signs of increasing environmental contamination. The yields of halogenated dioxins and furans are highly dependent upon the combustion conditions, with high yields associated with inefficient combustion conditions occurring during some fire incidents, which can result in local contamination, and burning of halogenated waste other than in appropriately designed incinerators. It is considered that acute exposure to populations in areas surrounding fire incidents is unlikely to present a significant health hazard (except where fires in specific industrial plants might result in large releases (as at Seveso)), but can result in significant local contamination in soot deposits at fire sites. Potential occupational hazards to fire investigators and fire fighters during post-fire investigations and clean-up can be minimised by simple hygiene precautions (avoiding dermal contact or oral intake and the inhalation of contaminated soot deposits).

References

1. Application of fire safety engineering principles to the design of buildings – Part 1: Initiation and development of fire within the enclosure of origin (Sub-system 1). PD7974-1:2003 British Standards Institution.
2. W. W. Jones, R. D. Peacock, G. P. Forney and P. A. Ricke, *CFAST – Consolidated Model of Fire Growth and Smoke Transport (Version 6)*, NIST Special Publication 1026, 2005.
3. W. D. Walton, D. J. Carpenter and C. B. Wood, Zone Computer Fire Models for Enclosures, in *SFPE Handbook of Fire Protection Engineering*, ed. P. J. DiNenno, National Fire Protection Association (NFPA), Quincy, MA, 4th edn, 2008, ch. 3–7. pp. 222–228.
4. K. McGrattan, S. Hostikka, J. Floyd, H. Baum and R. Rehm, *Fire Dynamics Simulator (Version 5) Technical Reference Guide Volume 1: Mathematical Models*, NIST Special Publications 1018-5, National Institute of Standards and Technology, Gaithersburgh, MA, July 2008, p. 92.
5. K. McGrattan and S. Miles, Modeling Enclosure Fires using Computational Fluid Dynamics (CFD), in *SFPE Handbook of Fire Protection Engineering*, ed. P. J. DiNenno, National Fire Protection Association (NFPA), Quincy, MA, 4th edn, 2008, ch. 3–8, pp. 229–246.
6. D. Drysdale, *An Introduction to Fire Dynamics*, 3rd edn, 2011, Wiley.
7. D. D. Evans, Ceiling Jet Flows SFPE Handbook of Fire Protection Engineering, National Fire Protection Association (NFPA), Quincy, Massachusetts, 1st edn.

8. C. R. Theobold, Studies of fires in industrial buildings, *Fire Prev. Sci. Technol.*, 1977, **17**.
9. W. K. Chow, G. Zou, H. Dong and Y. Gao, Necessity of carrying out full-scale burning tests for post-flashover retail shop fires, *Int. J. Eng. Perform.-Based Fire Codes*, 2003, **5**, 20–27.
10. J. Milke, Kodur and C. Marrion, World Trade Center Building Performance Study: Data Collection, Preliminary Observations and Recommendations. Overview of fire protection in buildings. Appendix A. Overview of Fire Protection in Buildings, *FEMA 403*, September 2002.
11. V. Babrauskas, Heat Release Rates, *SFPE Handbook of Fire Protection Engineering*, National Fire Protection Association (NFPA) Quincy, Massachusetts, 4th edn, 2008.
12. *Fire Safety of Upholstered Furniture – The Final Report on CBUF Research Programme*, ed. B. Sundstrom, Interscience Communications, London, 1996.
13. K. J. LaMalva, Failure analysis of the World Trade Center 5 Building, MSc Thesis, Worcester Polytechnic Institute, May 1 2007, THES JRB-599.
14. R. Chitty and J. Fraser-Mitchell, *Fire Safety Engineering: A Reference Guide*, Building Research Establishment, 2003.
15. G. W. Mulholland, Smoke Production and Properties, in *The SFPE Handbook of Fire Protection Engineering*, ed. P. J. DiNenno, National Fire Protection Association, Quincy, MA 02269, 4th edn, 2009, pp. 291–302.
16. *Controlled Equivalence Ratio Method for the Determination of Hazardous Components of Fire Effluents*, *ISO/TS 19700*, International Organization for Standardization, Geneva, Switzerland, 2007.
17. A. Tewarson, Generation of Heat and Gaseous, Liquid and Solid Products in Fires, in *The SFPE Handbook of Fire Protection Engineering*, ed. P. J. DiNenno, National Fire Protection Association, Quincy, MA 02269, 4th edn, 2009, pp. 109–194.
18. D. A. Purser and W. D. Woolley, Biological Studies of Combustion Atmospheres, *J. Fire Sci.*, 1983, **1**, 118–145.
19. D. A. Purser, Influence of fire retardants on toxic and environmental hazards from fires, in *Fire Retardancy of Polymers. New Strategies and Mechanisms*. ed. T. R. Hull and B. K. Kandola, RSC Publishing, Cambridge, 2009, ch. 24, pp. 381–404.
20. D. A. Purser and J. A. Purser The potential for including fire chemistry and toxicity in fire safety engineering, BRE report No. 202804. 28th March 2003.
21. D. A. Purser and J. A. Purser, HCN yields and fate of fuel nitrogen for materials under different combustion conditions in the ISO 19700 tube furnace, *Fire Safety Science – Proceedings of the ninth international symposium*, International Association for Fire Safety Science, 2008, pp. 1117–1128.
22. T. J. Christian, B. Kleiss, R. J. Yokelson, R. Holzinger, P. J. Crutzen, W. M. Hao, B. H. Sahar and D. E. Ward, Comprehensive laboratory measurements of biomass-burning emissions: 1. Emissions from

Indonesian, African and other fuels, *J. Geophys. Res.*, 2003, **108**, 4719–4731, DOI: 10.1029/2003JD003704.
23. R. J. Yokelson, R. Susott, D. E. Ward, J. Reardon and D. W. T. Griffith, Emissions from smoldering combustion of biomass measured by open-path Fourier transform infrared spectroscopy, *J. Geophys. Res.*, 1997, **102**, 18867–18877.
24. D. A. Purser, Toxic Product Yield and Hazard Assessment for Fully Enclosed Design Fires Involving Fire Retarded Materials, *Polym. Int.*, 2000, **47**, 1232–1255.
25. Experimental research for Scottish Building Standards Agency following the fire at the Rosepark Care Home, Glasgow, 31st January 2004. Building Research Establishment Ltd. Fire Safety Framework. 6 October 2004. The Scottish Government.
26. D. A. Purser, Fire safety and evacuation implications from behaviours and hazard development in two fatal care home incidents: Rosepark and Frampton House. *Proceedings of the 5th International Symposium on Human Behaviour in Fire, Downing College, Cambridge, UK*, Interscience, 19–21 September 2012, London, pp. 251–262.
27. ISO9505:1993. Fire tests – Full-scale room test for surface products, International Organization for Standardization, Geneva, Switzerland, 1993.
28. B. Andersson, F. Markert and G. Holmstedt, Combustion products generated by hetero-organic fuels on four different test scales, *Fire Saf. J.*, 2005, **40**, 439–465.
29. P. Blomqvist and A. L. Lonnermark, Characterization of the combustion products in large-scale fire tests: comparison of three experimental configurations, *Fire Mater.*, 2001, **25**, 71–81.
30. D. A. Purser, P. J. Fardell, J. Rowley, S. Vollam and B. Bridgeman, An improved tube furnace method for the generation and measurement of toxic combustion products under a wide range of fire conditions, *Flame Retardants '94 Conference, London, UK, Proceedings*, Interscience Communications, London, January 1994, pp. 263–274.
31. D. A. Purser, Toxicity of fire retardants in relation to life safety and environmental hazards, in *Fire Retardant Materials*, ed. A. R. Horrocks and D. Price, Woodhead Publishing Ltd, Cambridge, UK, 2001, ch. 3, pp. 69–127.

CHAPTER 4

Products of Combustion and Toxicity from Specific Types of Fires[†]

JAMES C. WAKEFIELD

Health Protection Agency, Near Didcot, Chilton OX11 0RQ, UK
Email: James.Wakefield@delphichse.com

4.1 Introduction

A comprehensive review of the toxicology of combustion products was conducted to support the Health Protection Agency (HPA) strategic goals through Programme 2: 'to protect against the adverse health effects of acute and chronic exposure to chemicals, poisons and other environmental hazards'.

The Chemical Hazards and Poisons Division (CHaPD) frequently has to advise on the health effects arising from incidents due to fires. The purpose of this chapter, which is adapted from a previous review for the HPA, is to consider the combustion products from different types of fire and summarise their toxicity. Following smoke inhalation, toxicity may result either from thermal injury, or from the toxic effects of substances present. This review only considers the latter.

[†]The views expressed here are those of the author and should not be taken as those of the UK Department of Health or those of the Health Protection Agency (England and Wales).

Issues in Toxicology No. 23
Toxicology, Survival and Health Hazards of Combustion Products
Edited by David A. Purser, Robert L. Maynard and James C. Wakefield

Published by the Royal Society of Chemistry, www.rsc.org

The pyrolysis and combustion of materials can result in the generation of many toxic smoke products, as described in Chapters 2 and 3. These cause irritation and asphyxiation, which can result in incapacitation or death at the fire scene. Systemic toxic effects may occur following acute exposures, which can lead to injury or death. Some of the common toxic chemicals that may be present in a fire effluent include asphyxiant gases, such as carbon monoxide (CO) and hydrogen cyanide (HCN), irritant gases, such as hydrogen chloride (HCl), oxides of nitrogen (NO_x), acrolein and phosgene, and complex molecules, such as polycyclic aromatic hydrocarbons (PAHs). As described in Chapters 2 and 3, the amounts of toxic products evolved during combustion vary with the type of combustion, the availability of oxygen, the temperature and the materials involved. Therefore, the conditions of combustion will affect the severity of the adverse health effects in those exposed to the products of combustion.

This is a complex area and there is the potential for the generation of a huge range of pyrolysis products depending on the nature of the fire and the conditions of burning. Although each fire will have individual characteristics and need to be considered on a case by case basis, there are commonalities, particularly with regard to the most important components relating to toxicity. This review aims to identify generalisations that may be made regarding the toxicity of common products present in fire smoke, with respect to the combustion conditions (temperature, oxygen availability, *etc.*) and materials involved, focusing largely on the adverse health effects to humans following acute exposure to these chemicals in smoke.

Consideration is given first to the general types of fire with respect to ventilation and temperature, which are key factors governing the type of pyrolysis products formed. This is followed by a section giving consideration of the important toxic substances that may be produced, divided into general classes, such as asphyxiant gases, irritant gases (inorganic and organic) and complex organic compounds, and also includes consideration of specially sensitive groups in the population. This is followed by a section on the characteristics of fires from different combustible materials with regard to the type of toxic substances produced. Consideration is then given to how smoke and the pyrolysis products produced may be dispersed. Finally, there is a conclusions section.

4.2 Types of Fires

4.2.1 Effects of Thermal Breakdown Conditions

The thermal breakdown of materials can occur under a number of different conditions, which can affect how complete the degradation of the material may be and has a considerable effect on the composition of the products. Thermal degradation can be classified into the main fire types described in detail in Chapters 2 and 3 and summarised below. All fuel materials may be decomposed under these different fire conditions, but while some combustion products are common to almost all fires, others are specific to particular fuel types, depending on their elemental and molecular compositions.

4.2.2 Pyrolysis

The initial stage of any combustion process is pyrolysis, which can be described as the thermal breakdown and chemical conversion of materials. When any fuel is exposed to a heat source of sufficient intensity, the surface temperature increases. Liquid fuels with a high vapour pressure may evaporate initially without decomposing. Some solid fuels (including thermoplastic polymers) may melt, while others (including cellulosic materials and thermoset polymers) remain solid. When the temperature of the fuel reaches a sufficiently high temperature, thermal decomposition begins to occur, involving molecular bond scission and the release of a range of volatile breakdown products.[1] This process is endothermic, so that the rate of decomposition and the yields of decomposition products increase with temperature, typically becoming significant at temperatures above approximately 300–400 °C depending upon the fuel composition. In air, the pyrolysis process is facilitated by partial oxidation of the pyrolysis products, which is exothermic, and this results in a further range of partially oxidised combustion products. The range of individual products and their proportions may also change as the temperature increases.

Oxidative pyrolysis under externally applied heat produces high yields of a wide range of organic composition products, many of which are irritant and/or systemically toxic. However, decomposition is limited to the area heated and the rate of decomposition is very slow compared with flaming fires.

4.2.3 Smouldering

Smouldering is a slow form of combustion occurring mainly in char-forming cellulosic materials such as wood and cotton. In smouldering, the exothermic oxidative pyrolysis of the solid fuel becomes sufficient for self-sustained combustion to occur. Smouldering, once initiated, spreads through the fuel, progressively heating adjacent areas and releasing pyrolysis products.[1,2]

4.2.4 Flaming Combustion, Well-ventilated

When the concentration of gaseous fuel products, the fuel:air ratio and temperature are sufficient, flaming combustion occurs. For liquid and solid fuels, the heat from the flame must also be sufficient to maintain the supply of fresh combustible gaseous products by pyrolysis at the fuel surface. During the early stages, most flaming fires are well-ventilated, in that the mass of oxygen in the air entrained into the fire plume is more than enough for complete combustion of the fuel gases (fuel : air equivalence ratio less than 1). Under these conditions, combustion of most fuels consisting of carbon, hydrogen and oxygen (those not containing halogens or other flame-retardant additives) is relatively efficient, so that the main products are carbon dioxide, water and heat, but low yields of smoke particulates and products of incomplete combustion are also formed, and some pyrolysis products may escape the flame zone and be entrained into the fire effluent plume.[1,3,4] Where fuels contain other elements, such as sulfur, halogens or phosphorus, these are released mainly as acid gases. Fuel nitrogen is released mainly as N_2, but a small proportion is released as NO, which gradually oxidises further to NO_2. In large fires, very small amounts of nitrogen oxides may also be formed from atmospheric nitrogen. Halogens, phosphorus and nitrogen are commonly used in various combinations as flame-retardant components or additives in common polymers. Those acting as gas phase flame retardants, especially halogens, reduce the efficiency of combustion, resulting in the release of high yields of products of incomplete combustion, including CO, HCN, partially oxidised organics and smoke particulates, even under well-ventilated combustion conditions.[1]

4.2.5 Flaming Combustion, Ventilation-controlled

Fires in enclosures, such as buildings, often grow until their size becomes limited by the mass of oxygen supplied by the available ventilation compared to the amount of fuel available for combustion. Under these conditions the fire plume becomes fuel-rich (equivalence ratio > 1), so that combustion becomes inefficient and high yields of products of incomplete combustion are formed in the smoke effluent plume. Most fires in confined spaces (such as in buildings) become ventilation-controlled fairly early on in the stages of combustion. Ventilation-controlled fires may be either small enclosed fires, or large post-flashover fires. Flashover is a term often used to describe the stage in a fire when all of the combustible gases and materials in the vicinity ignite simultaneously at high temperature. The fuel-rich, low oxygen concentration conditions present in these types of fires lead to the production of large amounts of fire effluents containing high yields of carbon monoxide, carbon dioxide, hydrogen cyanide, other organic and inorganic gases, and smoke.[1–3]

4.2.6 Composition of Smokes

The composition of fire effluent (smoke) in a combustion environment depends on a large number of variables and can be extremely complex due to the dynamic nature of a fire. Some of the principal factors that can affect smoke composition include the nature of the fuels involved (chemical composition, structure and formulation), the stage of combustion (smouldering, flaming or post-flashover), the temperature of combustion and the availability of oxygen and ventilation in the fire environment (see Chapters 2 and 3).[1,5,6]

4.3 Common Toxic Combustion Products

4.3.1 Different Categories of Toxicants

The gaseous products formed during the combustion of most organic materials can be classified into two main categories on the basis of their toxicity. The first category relates to fire smoke components that have asphyxiant properties and also oxygen depletion due to the fire itself. The second category relates to the smoke components that cause irritation, either as sensory irritants affecting the eyes and upper respiratory tract (nose, mouth and throat), or as pulmonary irritants (affecting the lungs), although in many cases sensory and pulmonary irritation may be present simultaneously.[1,3] A third category may occasionally be used to describe products that may give rise to toxicity unlike that covered by the two main categories. There are, however, few documented cases of specific toxic combustion products that fall into this third category.[7] Also considered in this third category are toxic chemicals that may be present at the site of the fire and do not undergo decomposition in the combustion atmosphere. This may be an important consideration in scenarios such as a fire at a specialised chemicals factory.

4.3.2 Asphyxiant Gases

Asphyxiant gases produced during combustion can give rise to narcosis due to depression of the central nervous system. Exposure to these combustion products at sufficient concentration or duration of exposure can lead to unconsciousness and eventually death due to tissue hypoxia.[1,7,8] The principal asphyxiants produced during the combustion of organic materials are carbon monoxide, hydrogen cyanide and carbon dioxide together with low-oxygen conditions.[1,3,4,9] The effect of these can interact producing additive effects resulting in higher toxicity. Table 4.1 gives occupational exposure limits and acute emergency guideline levels for the substances considered.

Table 4.1 Worker exposure limits (WEL) and acute emergency guideline levels (AEGL) for asphyxiant gas combustion products.

	WEL[a]/ppm[32]		AEGL – 2/AEGL – 3/ppm[b,33]				
	LTEL[c]	STEL[d]	10 min	30 min	60 min	4 h	8 h
CO[e]	30	200	420/1700	150/600	83/330	33/150	27/130
HCN	–	10	17/27	10/21	7.1/15	3.5/8.6	2.5/6.6
CO_2	5000	15 000	Data not available				

[a]WEL = workplace exposure limits.
[b]AEGL – 2 = acute exposure guideline level – 2 (the level of the chemical in air at or above which there may be irreversible or other serious long-lasting effects or impaired ability to escape), AEGL – 3 = acute exposure guideline level – 3 (the level of the chemical in air at or above which the general population could experience life-threatening health effects or death).
[c]LTEL = long term exposure limit (8 h time weighted average).
[d]STEL = short term exposure limit (15 min reference period).
[e]interim AEGL.

Chemical asphyxiants prevent the normal uptake of oxygen by tissues by interfering with specific aspects of oxygen delivery and metabolic processes. Thus, carbon monoxide and hydrogen cyanide are chemical asphyxiants. Simple asphyxiants are physiologically inert gases that, if inhaled, displace oxygen from the alveoli and lead to hypoxia. Nitrogen and methane are considered to be simple asphyxiants. Breathing a reduced concentration of oxygen also has this effect, but is not considered a simple asphyxiant. The reduced concentration of oxygen is due to the presence of other gas, assuming the barometric pressure is unchanged. The effects of carbon dioxide are more complex, but when present in very high concentrations it can be considered as an asphyxiant.

The physiological effects of asphyxiant gases individually and their interactions when present in combination, and the derivation of calculation models to predict time to incapacitation in humans are described in Chapter 7. Chapters 9 and 10 present detailed reviews of the toxicology and physiology of carbon monoxide and hydrogen cyanide respectively. Chapter 14 describes the application of calculation models to the evaluation of acute effects of combinations of toxicologically active substances and heat on fire victims in buildings and during exposures to outdoor smoke plumes.

The following sections provide brief summaries of the effects of the major chemical asphyxiants.

4.3.2.1 *Carbon Monoxide (CO)*

Carbon monoxide (CO) is the most common asphyxiant product in most fire environments and is formed during both smouldering and flaming combustion of all organic materials.[1,7,9] The production of CO in a fire is dependent upon the availability of oxygen in the combustion environment, with an increase in CO formation with decreasing availability of O_2 (see Chapters 2 and 3). The production of CO is therefore greater in cases of ventilation-controlled combustion than with well-ventilated combustion. The CO yields and rates of production increase considerably as the fuel : air equivalence ratio increases over the range from 1 to approximately 2 in ventilation-controlled fires. Where there is sufficient ventilation for a fire to pass to flashover, the production of CO increases significantly due to the combination of the combustion becoming ventilation-controlled and a the rapid increase in the mass burning rate (see Chapter 2).

The toxicity of CO (reviewed in Chapter 9) is due mainly to a combination of a reduction in the oxygen-carrying capacity of the blood and impairment of the release of the oxygen that is carried into the tissues, resulting in a tissue hypoxia.[1,3,7,10] Hypoxia following exposure to CO results from the conversion of haemoglobin to carboxyhaemoglobin (COHb) due to competition between O_2 and CO for the haem-binding sites.[7] The main factor for the toxicity of CO is widely regarded to be that the affinity of haemoglobin for CO is around 200–250 times greater than the affinity for O_2.[1,3,6,11] As such, once COHb is formed it will not be readily displaced by O_2. Therefore, only partial conversion of haemoglobin to COHb significantly reduces the oxygen-carrying capacity of the blood.[1,3,7] The formation of COHb also leads to a left shift in the dissociation curve of oxyhaemoglobin: the oxygen that is bound will have a greater affinity for the haemoglobin and will be released less readily at the tissues.[7,11] This alteration in oxygen dissociation from haemoglobin resulting from COHb formation further increases the likelihood of hypoxia. The concentration of COHb in the blood will be elevated in most cases in individuals exposed to a combustion atmosphere, with the concentration of COHb being dependent upon the duration of exposure, the activity level (respiratory minute volume) of the subject and the concentration of CO in the fire environment. The presence of COHb in low concentrations in the blood of individuals exposed to a fire atmosphere may be difficult to interpret, since environmental factors (*e.g.*, an urban environment or tobacco smoking) may be associated with raised levels of COHb.[3] Most fatalities following inhalation of CO have reported the concentration of COHb in the blood to be greater than 50%.[1,3,12,13] During fatal exposures, the COHb levels continue to increase in a comatose subject until breathing and circulation cease. The COHb levels measured after death range from around 50% to in excess of 90% COHb in different individuals with a mode around 75% COHb (see Chapter 7, Figure 7.1). However, while the proportion of exposed subjects surviving acute CO exposure is high following rescue and treatment for %COHb levels up to 45% COHb, it is low once levels exceed 50% COHb.[1,14]

A COHb concentration in the range of 30–45% is likely to be associated with cause of death being due to a combination of both CO poisoning and other factors such as the presence of additional toxic combustion products. For a fatality in which there is a COHb concentration of less than 30%, the main cause of death is likely to be due to effects other than CO poisoning.[15]

Acute health effects resulting from CO-induced hypoxia at concentrations below that causing lethality can include neurological effects such as headache, dizziness, confusion, disorientation, loss of coordination, memory loss, fainting, cerebral oedema and coma.[3,14,16] These neurological effects of CO exposure in a fire environment may hinder the ability to perform tasks, recognise danger and escape from a hazardous fire situation.[3,14] Neurological symptoms following severe acute toxicity may appear 2–40 days post exposure, including lethargy, irritability and lack of concentration, and possible severe effects, including dementia and psychosis, which may not all be related to CO-induced hypoxia.[13] Inhalation of CO is also likely to give rise to metabolic acidosis.[3,16] The heart is particularly sensitive to the effects of CO, and acute exposure may give rise to cardiovascular effects including reduced myocardial function, hypotension, vasodilation, cyanosis, cardiac arrhythmias, shock, circulatory failure and cardiac arrest.[17,18] Pregnant women, the foetus *in utero* and new-born infants are at an increased risk from CO exposure. CO readily crosses the placenta and binds to foetal haemoglobin with a higher affinity than for maternal haemoglobin. CO is also cleared from foetal blood much slower than from maternal blood, resulting in a 10–15% increase in COHb formation in the foetus relative to the mother.[13]

4.3.2.2 Hydrogen Cyanide (HCN)

Any organic material containing carbon and nitrogen will produce hydrogen cyanide (HCN) during combustion under most conditions. Polymers, such as nylons, polyurethanes and polyacrylonitrile, are known to be prominent sources of HCN on thermal decomposition or combustion (see Chapters 2 and 3).[1,3,19,20] The yield of HCN from the combustion of materials containing nitrogen is dependent upon the temperature and availability of oxygen in the fire environment. In an oxygen-limited environment, the generation of HCN is initiated at high temperatures, with the yield increasing with further increases in temperature. In air, the evolution of HCN begins at much lower temperatures than in an anaerobic environment, with the yield increasing with equivalence ratio and temperature (see Chapter 3). The generation of HCN under specific combustion conditions has been shown to be approximately proportional to the nitrogen content of the polymer involved in the combustion, but the conversion efficiency of polymer nitrogen to HCN, and hence the overall yield, varies somewhat between different polymers (see Chapter 3).[1,21,22]

Following exposure and systemic uptake of HCN, it undergoes dissociation in the blood to form the cyanide ion. The cyanide ion is readily distributed within the body and is responsible for the toxicity of HCN by reducing the cellular utilisation of oxygen (cellular respiration). The effects of exposure to HCN are described in Chapters 7 and 10, and clinical aspects of treatment in Chapter 15. The cyanide ion binds to cytochrome oxidase, which is a principal enzyme involved in the utilisation of oxygen in most cells throughout the body, and inhibits it by forming a cytochrome oxidase–cyanide complex. The inhibition of cytochrome oxidase results in a rapid onset of cytotoxic hypoxia and loss of cellular function. The cardiac and cerebral tissues are particularly susceptible to the effects of cyanide on cellular respiration. The most common cause of death from HCN intoxication is due to depression of the respiratory system resulting from the cytotoxic hypoxic effect of the cyanide ion on the central nervous system, but effects on the cardiovascular system may also be a cause of death.[7] Some features of acute exposure to hydrogen cyanide at below fatal concentrations may include headache, nausea, dizziness, confusion, muscle weakness, loss of coordination, hyperventilation, cardiac arrhythmia, bradycardia, rapid loss of consciousness and coma.[1,3,7,14,23] These acute effects of HCN inhalation may impede the escape from a fire environment (see Chapter 14).[1,14,19]

The concentration of HCN fatal to humans following inhalation is dependent upon the duration of exposure. It has been widely reported that a concentration of 130 ppm for 30 minutes is likely to be fatal, a concentration of 180 ppm HCN is likely to be fatal after just 10 minutes and a HCN concentration of 270 ppm is considered to be immediately fatal.[23,24] A blood cyanide concentration of greater than 1 μg mL^{-1} in blood samples taken post mortem from fire fatalities is considered to suggest significant toxicity of HCN. Blood cyanide levels of 3 μg mL^{-1} or greater are considered to be lethal levels of cyanide.[1,7] However, measurements of blood cyanide levels can be problematic and analysis should be cautious with respect to factors including the time between sample removal and analysis, and the storage method.[25,26]

The majority of fire fatalities cannot be attributed to inhalation of HCN alone, since in a combustion atmosphere where HCN is being generated many additional toxic products are also likely to be evolved. CO is produced in all fires involving organic materials and there is likely to be an interaction between CO and HCN, causing hypoxia by two separate mechanisms. In a combustion environment where both CO and HCN are present at sub-lethal doses, the combination of the two vapours has an additive effect, which can prove to be fatal.[7,27,28] HCN also causes rapid incapacitation, so that victims may be first overcome by HCN then remain in the fire to inhale a fatal dose of CO (see Chapter 14). Hyperventilation resulting from acute inhalation of HCN could also give rise to increased toxicity of CO, reducing the time to death by increasing the amount of CO respired.[3] Therefore, in many cases of fire fatalities resulting from smoke inhalation it is difficult to isolate a cause of death to one chemical in particular from COHb and blood cyanide measurements, as it is likely to be due to sequential and a combined effect from both chemicals.

4.3.2.3 Carbon Dioxide (CO_2)

Carbon dioxide (CO_2) is generated in all fires involving organic materials, although its rate of production is largely dependent upon the availability of oxygen in the fire environment. As the level of oxygen present in a fire environment diminishes, there is a shift from production of CO_2 to CO. In this respect, the highest rate of generation of CO_2 occurs when combustion is complete due to sufficient ventilation.[6]

An increase in the amount of CO_2 inhaled at concentrations such as those likely to be generated during the early stages of a fire is not considered to cause significant toxicity on its own. However, the inhalation of CO_2 in a fire atmosphere will give rise to physiological effects that enhance the toxicity of other combustion products. An increase in CO_2 concentration will stimulate the rate and depth of respiration, increasing the respiratory minute volume.[1,7,14] This leads to an increase in the amount of any toxicants present that are inspired over a given period. A concentration of 2% CO_2 has been shown to increase the respiratory minute volume by 50%; whilst 10% CO_2 may increase the volume of air respired in one minute by as much as tenfold.[1,7] Human volunteers exposed to 7.5% CO_2 for 15 minutes have also reported difficulty breathing, headache, sweating, increased heart rate, restlessness, disorientation and visual distortion.[29] Prolonged exposures over a few minutes to similar concentrations of CO_2 (7 to 10%) can cause the onset of severe distress, nausea and may lead to a loss of consciousness.[1,4]

The presence of CO_2 has been shown to potentiate the toxicity of CO by increasing the formation of COHb due to an increase in the volume of CO inspired.[1,3,30] Additionally, exposure to elevated levels of CO_2 will give rise to respiratory acidosis, which in conjunction with metabolic acidosis caused by CO will result in severe acidosis, with a prolonged recovery period following cessation of exposure.[30]

The general situation with respect to exposure during fire incidents is that during the early stages the main effects of inhaled CO_2 are to increase ventilation and the uptake of CO and HCN, resulting in loss of consciousness due to hypoxia before the inhaled CO_2 concentration exceeds 7%, which is itself sufficient to induce collapse within a few minutes.

4.3.2.4 Low Oxygen Concentration

In a combustion atmosphere, as oxygen is consumed by the fire, the level of oxygen is depleted, particularly if the fire is in a closed environment. The depletion of oxygen below normal levels (21%) in a fire environment can give rise to adverse health effects. Therefore, a reduction in oxygen levels should also be considered to be a toxic product of combustion.[1,7,14] A reduction in the concentration of oxygen of only 4% (17% O_2) can lead to an impairment of motor co-ordination. Further reduction of the level of oxygen in the combustion environment to around 14 to 10% can lead to fatigue and an increased likelihood of making faulty judgements. These effects of low oxygen concentration could seriously hinder escape from a hazardous fire situation. If the level of oxygen falls below approximately 10%, it is likely that an exposed individual will become unconscious and, to prevent fatality, will require immediate removal to fresh air or treatment with oxygen.[14,17,19,31]

The effects of a reduction of the oxygen concentration in a combustion environment are due to hypoxic hypoxia caused by a decrease in the partial pressure of oxygen (PO_2) in the arterial blood.[1] Depletion of oxygen in a fire environment could potentially enhance the toxicity of the asphyxiant gases produced during combustion. As CO and HCN also give rise to hypoxia, a reduction in the level of atmospheric oxygen will be likely to reduce the concentration at which significant toxicity may be observed compared to a normal oxygen environment due to a further reduction in the availability of oxygen for cellular respiration. Low oxygen concentration in addition to raised carbon dioxide levels has a marked effect on breathing. Physical exertion, such as may be required to escape a hazardous fire environment, will increase the individual's demand for oxygen, which may also accelerate the onset of hypoxia.[1,7]

4.3.3 Irritant Gases

The combustion of most commonly used materials ranging from natural sources, such as wood, to synthetic plastics and polymers will result in the generation of irritant gases. Therefore, irritant gases are present in most fire atmospheres, irrespective of whether the combustion is smouldering or flaming (see Chapters 2 and 3). The irritant gases evolved and the rate of generation may, however, depend upon the fuel composition and the mode of combustion, relative to non-flaming or flaming, decomposition, temperature (especially for flaming fires) and the ventilation conditions. Irritant gases produced during combustion can be divided into two main classes relating to their chemical composition as either inorganic acid gases or organic irritants. For acid gases the conversion efficiencies of fuel elements, such as halogens, to acid gases in the combustion products is generally high and similar for all combustion conditions. The yields of nitrogen oxides from fuel nitrogen are higher under well-ventilated flaming combustion conditions, which also involve some oxidation of atmospheric nitrogen. The yields of organic irritants are very dependent on the combustion conditions. Under non-flaming decomposition conditions in air, yields of organic irritants (mostly partly oxidised substances such as aldehydes and ketones) are high. Under flaming conditions, the yields depend on the combustion efficiency. Under well-ventilated flaming conditions, for fuels low in gas phase flame inhibitors (especially halogens), combustion efficiency is high and yields of irritants are generally low. Where combustion efficiency is low, due to fuel-rich ventilation conditions or the presence of flame inhibitors, yields of organic irritants are high (see Chapters 2, 3 and 7).[1]

The injury following exposure to an irritant gas depends upon the chemical involved, its concentration, the exposure duration and its solubility. However, the initial effect of exposure to these irritant gases is likely to be sensory irritation.[1,7,11,34] Irritation of the eyes will cause pain and stinging of the eyes, initiation of a blinking reflex and lacrimation.[1,7,34] The severity of sensory irritation is dependent upon the concentration of the irritant present, and is independent of the exposure duration. An individual exposed to irritant gases in a combustion atmosphere with the effect of stinging or burning of the eyes and throat may shut their eyes and hold their breath to alleviate the irritation, hindering their ability to escape from the hazard.[1,7] An additional characteristic sign of exposure to irritant gases is a burning sensation of the mucous membranes of the upper respiratory tract including the nose, mouth and throat.[1,7,11] Pulmonary irritation will commonly occur following sensory irritation due to inhalation of the irritant gas into the lungs. This irritation of the lungs gives rise to bronchoconstriction, coughing and breathing difficulties. Unlike sensory irritation, the severity of pulmonary irritation is dependent upon both the concentration and the duration of exposure to the irritant gas. Exposure to high concentrations of irritant gases can cause inflammation of the lung tissues, pulmonary oedema and could potentially be fatal in a period of between 6 and 48 hours after removal from the exposure.[1,7]

Some irritant products commonly occurring in combustion products and their physiological and toxic effects are reviewed in Chapter 13. The results of human and animal exposure studies on individual irritant substances and irritant combustion product mixtures are reviewed in Chapter 7. Aspects related to the yields, rates of production and chemistry of combustion product irritants are reviewed in Chapters 2 and 3. The derivation and application of calculation models for predicting the effects of irritant exposures are presented in Chapter 14.

4.3.4 Complex Molecules

The combustion of many organic materials, including fossil fuels, natural polymers such as wood and synthetic polymers such as plastics, will not only give rise to the asphyxiant and irritant gas compounds mentioned previously. In many cases, the combustion of organic materials, particularly if it is incomplete, may also give rise to more complex molecules in the smoke plume and may typically include longer carbon chains and multiple carbon rings. The acute toxicity of these compounds is generally low and may not pose a direct health hazard during exposure. However, some of these compounds, in particular those from the polycyclic aromatic hydrocarbon groups, are recognised mutagens and carcinogens. The risks from a single (acute) exposure are very small (and unquantifiable). Others give rise to concern because of possible effects on the reproductive system (*e.g.*, dioxins).[35–37]

4.3.4.1 Polycyclic Aromatic Hydrocarbons (PAHs)

Polycyclic aromatic hydrocarbons (PAHs) are a large group (over 100) of organic compounds containing a minimum of two fused benzene rings. Some of the most commonly known PAHs include benzo[*a*]pyrene, naphthalene and anthracene.[36] PAHs may be evolved in the fire effluent from all combustion processes, with the largest quantities likely to be generated during the slow, incomplete combustion of organic materials.[35,36] PAHs are present as complex mixtures rather than as single individual compounds.

There is very little information regarding the adverse health effects following an acute exposure to PAHs, either from experimental data on individual compounds or from mixtures. Much of the data regarding the acute toxicity of PAHs relates to reports of accidental exposure of naphthalene.[36] Naphthalene may give rise to headaches, nausea, vomiting, confusion, profuse sweating, nose, throat and eye irritation, and corneal damage following acute exposure to vapours. The onset of acute haemolytic anaemia, particularly in individuals with glucose-6-phosphate deficiency, is a characteristic feature of acute exposure to naphthalene, which may be delayed post exposure.[35,38]

The health effects following exposure to mixtures of PAHs focuses on the fact that many of the compounds are known to have the potential to be carcinogenic, based on animal experiments or data from occupational exposures to mixtures of PAHs (*e.g.*, coke oven workers).[35] In most cases the carcinogenic compounds also have mutagenic potential and they are considered 'genotoxic carcinogens'. An exception is naphthalene, which is an animal carcinogen, but has no significant mutagenic potential. Some commonly known PAHs, which are classified by the International Agency for

Research on Cancer (IARC) as being probably carcinogenic to humans (group 2a), include benzo[*a*]pyrene, dibenz[*a,h*]anthracene and benz[*a*]-anthracene.[39] PAHs that are classified as possibly carcinogenic to humans (group 2b) include naphthalene, indeno[1,2,3-*cd*]pyrene and benzo[*b*]-fluoranthene, whilst compounds including chrysene, fluorine and anthracene are not classifiable as to their carcinogenicity in humans (group 3).[39] The experimental data relating to carcinogenicity of PAHs is mainly from chronic (long term) exposure studies in animals. Any risks from a single acute exposure are likely to be very small. Aspects of the carcinogenicity of combustion products are considered further in Chapters 11, 12, 17 and 19.

4.3.4.2 Dioxins and Dibenzofurans

Polychlorinated dibenzo-*p*-dioxins (PCDDs) and polychlorinated dibenzofurans (PCDFs) are large groups (75 dioxin isomers, 135 furan isomers) of polychlorinated tricyclic aromatic compounds that have very similar structures. PCDDs and PCDFs are formed mainly during the incomplete combustion of any materials containing carbon, oxygen and chlorine, and are therefore commonly found as emissions in most fire effluents.[40–42] The pyrolysis of organochlorine polymers, such as polyvinylchloride, has been shown to give rise to the formation of both PCDDs and PCDFs at temperatures ranging from 500 to 700 °C.[43–45] Once formed, PCDDs and PCDFs are relatively resistant to thermal decomposition and are only destroyed after heating to 800 °C for a prolonged period.[41] PCDDs and PCDFs present in combustion effluents are most likely to be present as complex mixtures. The amounts of dioxins and dibenzofurans produced during combustion will depend largely on the material involved and the combustion conditions. Data on the total emissions of PCDDs and PCDFs from municipal solid waste incinerators suggest that the amounts are very small, between the range of a few to several thousand ng m^{-3}.[41] However, it has been demonstrated that the formation of PCDDs and PCDFs from the combustion of chlorine-containing plastics increased dramatically in the presence of copper (Cu) in the form of electrical wire.[46] The combustion of plastics containing polychlorinated biphenyls (PCBs) was known to be a source of PCDFs and PCDDs, although the use of PCBs has been restricted since the late 1970s.[40] However, fires involving PCB-containing materials, such as electrical transformers and capacitors manufactured prior to the beginning of the 1980s, could potentially form PCDFs and PCDDs in the combustion effluent. The most commonly known of the PCDDs is 2,3,7,8-tetrachlorodibenzo-*p*-dioxin (2,3,7,8-TCDD) as it is one of the most toxic and most extensively studied of the dioxins.[47] Many of the other isomers are very much less potent than TCDD and have low toxicity, *e.g.*, the fully chlorinated derivative. The toxicity of mixtures of TCDD is quoted in terms of TCDD equivalents, with TCDD being the standard. The most toxic PCDFs are those which have chlorine atoms substituted at the 2, 3, 7 and 8 carbon positions, particularly 2,3,4,7,8-pentachlorodibenzofuran (2,3,4,7,8-pentaCDF) and 2,3,7,8-tetrachlorodibenzofuran (2,3,7,8-TCDF).[48]

The most commonly documented adverse health effect following exposure to dioxins is the gradual onset of chloracne, which may appear within days or even months post exposure.[40,41,47] Chloracne is typically characterised by comedones, epidermal cysts and inflamed papules with hyperpigmentation.[41] The distribution of chloracne differs from that typically observed with adolescent acne and is more commonly associated with areas of contact. Some additional adverse health effects that may occur following acute exposure to dioxins include transient hepatotoxicity, hypertension, hyperlipidaemia and possible peripheral and central neurotoxicity.[41] Studies in firefighters involved in PCB transformer fires (considered to be the worst case scenario, with respect to dioxin formation) did not detect any observable health effects after 1 year post exposure, although PCDD levels were slightly elevated.[40] Data from animal studies indicate that dioxins have adverse effects on the reproductive system, including teratogenicity, and that the most sensitive and consistent effect is upon the developing reproductive system in the male offspring. Exposure to chlorinated dibenzofurans appears to give rise to similar adverse effects as that observed following exposure to 2,3,7,8-TCDD, however, the toxicity of these dibenzofurans and other chlorinated dioxins is of much lower potency. There have been no reports of human fatalities due solely to acute exposure to dioxins.[47] However, 2,3,7,8-TCDD has been evaluated as carcinogenic to humans (group 1) by the IARC.[49] Polychlorinated dibenzo-*p*-dioxins other than 2,3,7,8-TCDD and polychlorinated dibenzofurans are not classifiable as to their carcinogenicity to humans (group 3).[49] However, TCDD and other dioxins do not have significant mutagenic properties and prolonged exposure is likely to be necessary for the promoter effects to induce carcinogenicity. Risks from single exposure, if any, would be expected to be very low.

Further consideration of dioxins is presented in Chapter 12.

4.3.4.3 Isocyanates

The thermal decomposition of polyurethane foams and plastics is known to evolve isocyanates in the combustion effluent due to dissociation of the urethane monomer.[50,51] Polyurethane foams are widely used and may therefore be present as materials in many combustion situations. Flexible polyurethane foams are used in many common domestic materials, including furniture, and bedding and carpet underlay, while semi-flexible foams are employed widely in motor vehicle interiors and rigid foams are commonly used as insulation for central heating tanks and pipes and in appliances such as refrigerators.[52] The amount of isocyanates formed in combustion effluents is expected to be greater from polyurethane foams that contain unreacted isocyanate. The combustion of polyurethane will not only yield isocyanates, but is also likely to result in the formation of greater yields

Table 4.2 Acute emergency guideline levels (AEGL) for methyl isocyanate.

	AEGL – 2/AEGL – 3/ppm[a,34]				
	10 min	30 min	60 min	4 h	8 h
Methyl isocyanate	0.40/1.2	0.13/0.4	0.067/0.2	0.017/0.05	0.008/0.025

[a]AEGL – 2 = acute exposure guideline level – 2 (the level of the chemical in air at or above which there may be irreversible or other serious long-lasting effects or impaired ability to escape), AEGL – 3 = acute exposure guideline level – 3 (the level of the chemical in air at or above which the general population could experience life-threatening health effects or death).

of aromatic compounds such as benzene and toluene.[50,52] The isocyanates that may be present as products of combustion will vary depending upon the composition of the material involved and the combustion conditions, but may include relatively simple compounds, such as methyl isocyanate, or more complex compounds such as toluene diisocyanate.

Methyl isocyanates are severe irritants to eyes and mucous membranes. Studies in volunteers indicated eye irritation and lachrymation in all individuals exposed to 1 ppm for 5 minutes (see Table 4.2 for AEGLs). Nose and throat irritation was reported in some at this exposure level. A large amount of data is available on the effect of methyl isocyanates resulting from the Bhopal accident, which resulted in over 250 deaths. The most frequently reported symptoms were burning/watering of the eyes, coughing, respiratory distress from pulmonary congestion, nausea, muscle weakness and central nervous system (CNS) involvement secondary to hypoxia. However, insufficient details are available regarding the levels of exposure producing these effects.[53] There is also evidence that methyl isocyanate produces allergic sensitisation reactions. Animal studies indicate that exposure of pregnant animals to 2 ppm produces foetotoxic effects (reduced body weight), whilst having no effects on the maternal animals, indicating that an acute exposure may harm the unborn child.[53]

Toluene diisocyanate is severely irritating to the eyes and throat with exposure to 0.5 ppm for 30 minutes producing lachrymation. Exposure to 1.3 ppm was intolerable for 10 minutes; several hours later cold-like symptoms with a cough persisted. Studies in pregnant animals indicated no adverse effects on development at exposure levels that were not toxic to maternal animals.[53,54] Data from experimental exposure have clearly established that toluene diisocyanate is a potent allergic sensitiser following inhalation exposure, producing an asthma-like reaction. Persons previously sensitised would be expected to have such reactions following exposure to very low levels of toluene diisocyanate.[55]

Inhalation of isocyanates can lead to the delayed onset of potentially fatal pulmonary oedema, which may develop 12 to 48 hours post exposure.[53,54]

4.3.4.4 Fluoropolymers and Perfluoroisobutylene (PFIB)

The toxicity of combustion products from perfluorinated polymers, such as polytetrafluoroethylene (PTFE), covers a wide range of potencies and the mix of toxic products are even more dependent on the exact combustion conditions than for other materials. The combustion chemistry and yields of toxic substances are described in Chapter 2 and the toxic effects are described in Chapter 7. The basic findings are that under non-flaming decomposition at temperatures above approximately 450 °C, fluorocarbons decompose into tetrafluorethylene monomer, which re-polymerises in the gas phase to form an ultrafine particulate in the nanoparticle size range. This deposits in the lung interstitium and within a few minutes produces rapidly fatal lung oedema and inflammation in rats. The toxic potency of this particulate is extremely high, giving PTFE a mass loss lethal concentration (LC_{50}) of 0.017 g m^{-3} following 30 minutes' exposure. This represents the toxicity when this mass of PTFE is decomposed in a furnace, of which only a small (unknown) proportion is in the form of the actual particulates. Traces of other toxic substances are also present but, under these conditions, are at concentrations well below those capable of making a significant contribution to toxicity.[55] A further complication is that once released into the air the ultrafine particles rapidly agglomerate to form larger particles in the respirable size range, which have a much lower toxic potency, so that the toxicity of fluoropolymer fumes has been found to decrease considerably within a few minutes of formation.[55] The main effect reported in humans has been polymer fume fever, although one workplace exposure death has been reported.[56]

At temperatures above 650 °C and under flaming combustion conditions the ultrafine particulate is not formed and the toxic potency of the products decreases by a factor of up to approximately 100. Under these conditions the toxicity of the combustion products can be understood in terms of a mix of products consisting mainly of carbonyl fluoride, which hydrolyses to hydrogen fluoride, and low yields of perfluoroisobutylene.[55]

Perfluoroisobutylene (PFIB) is one of the toxic products formed during the thermal decomposition of fluorine-containing polymers, such as PTFE, at temperatures greater than 475 °C.[57–59] PFIB may therefore be present as a product in the effluent following thermal degradation of polymers, such as PTFE, at relatively low combustion temperatures.

Inhalation of PFIB is extremely irritating to the respiratory tract. The adverse health effects of exposure to PFIB are similar to those of phosgene, however, the toxic potency of PFIB is approximately ten times greater than phosgene.[60,61] In addition to the sensory and respiratory irritant effects, exposure to vapours of PFIB may give rise to headache, cough, chest pain and dyspnoea. Inhalation of sufficient amounts of PFIB can also lead to the onset of potentially fatal pulmonary oedema, which may be delayed up to 8 hours post exposure.[60,62]

4.3.4.5 Particulate Matter (PM)

Particles of organic and inorganic matter are likely to be released during all types of fires involving organic materials, particularly under conditions of incomplete combustion. Particulate matter (PM) is usually comprised of a complex mixture of different particle types and particles sizes. Many of the types of PM derived from combustion are unlikely to cause adverse health effects.[63] However, some PM, particularly the smaller more respirable particles, can cause significant adverse effects.[64] Smoke particles and their toxic effects are described in Chapter 11. Those particles that are below 10 μm in diameter (commonly denoted as PM_{10}) are able to pass through the upper respiratory tract and can be deposited within the airways. Particles with diameters less than 2.5 μm ($PM_{2.5}$) may be respired deeper within the lungs and can be deposited within the alveoli.[65] Ultrafine particles (including those from fluorocarbons described in the previous section) are defined as particles that have a diameter of less than 100 nm in at least one dimension. Ultrafine particles have the potential to migrate from the site of deposition, through membranes, into the blood and are then able to infiltrate other target organs.[63,64]

Knowledge of the health effects of PM_{10} comes mainly from epidemiology studies relating to the effect of air pollution. A large number of such studies have shown that PM_{10} levels are associated with effects on health even at very low levels. These include increased daily deaths, increased admissions to hospital in patients suffering from heart and lung disorders, and a worsening of conditions in those with asthma. It is believed that these effects on the respiratory and cardiovascular system are caused predominately by the very small particles usually referred to as ultrafine (or nanoparticles), *i.e.*, those smaller than 100 nm in diameter.

Upon deposition in the lungs, some PM and especially the ultrafine particles produce free radicals that may provoke oxidative stress. This oxidative stress can contribute to inflammation in the lung tissue and may exacerbate any pre-existing lung condition, such as asthma,[64,66] or in the case of fluoropolymer particulate, a fatal lung inflammation.[55] Ultrafine particles that are able to migrate from the lungs into the blood have also been implicated in causing fume fevers and adverse effects on the cardiovascular system.[67] There are two main theories as to how effects on the cardiovascular system may be induced. One is that the particulates set up an inflammatory response in the interstitium of the lung and this provokes an increase in the likelihood of blood to clot or atherosclerotic plaques to rupture, which could result in myocardial infarction.[64,67] This hypothesis is true for sustained exposure, such as ambient air pollution, but may play a less significant role following exposure to ultrafine particles in smoke. An alternative hypothesis points to a reflex effect on the heart with receptors in the airways and lungs.[67]

In contrast to many of the other combustion products highlighted in this review, the degree of adverse effects observed following exposure to PM is not directly related to the external exposure concentration, but may be more closely related to the size and surface area of the particles deposited at the target organ.[68] PAHs are some of the most common toxic compounds present in combustion-derived PM, which may also include nitrogen-substituted polycyclic aromatic hydrocarbons (nitro-PAHs) and nitro-PAH lactones, many of which are highly mutagenic and carcinogenic[36,37] (see Chapter 12). Particulate matter may also include contamination with metals such as lead, arsenic, cadmium and nickel.[69,70]

4.3.5 Health Issues for Vulnerable Groups

Individuals who are most at risk from exposure to hazardous combustion products are those that have pre-existing respiratory diseases such as asthma or chronic obstructive pulmonary disease. The presence of an existing respiratory condition increases the susceptibility of the individual to the adverse effects of exposure to asphyxiant gases such as CO.[16] Acute exposure to smoke containing mixtures of asphyxiant and irritant gases is therefore likely to exacerbate these conditions.

Pregnant women are particularly at risk following exposure to smoke, as unborn infants are particularly susceptible to carboxyhaemoglobin due to CO. Following exposure to CO, the foetal circulation would be expected to have a greater concentration of carboxyhaemoglobin than the maternal circulation due to differences in the uptake and elimination of CO, involving differences in the affinity of foetal haemoglobin for CO. This increased level of COHb in the foetus compared to the mother may cause potentially serious hypoxia to the foetus at COHb levels that are less harmful to the mother. Newborn infants and children may also be at increased risk of adverse effects, such as hypoxia, respiratory irritation and pulmonary oedema, following exposure to hazardous combustion products. Infants are more susceptible to toxicity since the hazardous products present in the fire effluent are likely to result in toxicity to infants and children at lower concentrations than those required to cause similar effects in adults. Other combustion products, such as methyl isocyanates, give rise to concern because of adverse effects following *in utero* exposure.

Elderly individuals exposed to hazardous combustion products would also be expected to be at greater risk of potentially life threatening health effects due to conditions associated with age such as reduced in lung function. Any reduction in the utilisation of oxygen therefore increases the individual from being at risk of hypoxia resulting from exposure to asphyxiant gases, such as CO and HCN, and low oxygen concentration.

Infants, children and elderly individuals may also be less able to escape from an environment containing hazardous combustion products, thereby increasing the duration of exposure and thus increasing the potential for adverse effects.

Individuals who smoke are also likely to be at greater risk to toxicity following inhalation of combustion products as the baseline level of COHb is likely to be greater than would be observed in non-smokers, although this is partly offset by an increased haemoglobin concentration. Therefore, the inhalation of combustion products at concentrations that do not cause significant toxicity to non-smokers may cause considerable toxicity to individuals that smoke regularly.[3]

4.4 Hazardous Combustion Products Formed by Fuel Type

4.4.1 General and Case Specific Considerations

As stated, the products formed during combustion processes can vary greatly depending upon the material involved. The compounds generated during each individual fire scenario should be viewed on a case by case basis due to the large number of variables affecting the products formed. However, general predictions of the products most likely to present a hazard to health can be made if the materials involved are known (see Table 4.3).

4.4.2 Fires Involving Polymeric Materials

Polymeric materials, such as plastics, resins, fibres and foams, are likely to produce significant quantities of carbon monoxide upon combustion. The presence of a high level of carbon monoxide is, however, likely to be of most concern within the immediate vicinity of the fire, particularly if the fire is enclosed within a building and is ventilation-controlled.

The combustion of polymers that contain large quantities of nitrogen, such as nylons, polyurethanes and polyacrylonitriles, are likely to yield significant amounts of hydrogen cyanide (HCN), nitrogen oxides (NO_x) and ammonia. However, HCN, NO_x and ammonia are not exclusively formed from the combustion of nitrogen-containing plastics, but may also be present to a lesser extent during the combustion of non-nitrogen containing plastics due to the incorporation of atmospheric nitrogen. In this case, the incorporation of atmospheric nitrogen to form HCN and NO_x will be largely dependent upon its availability in the combustion environment and the temperature.

Many plastics contain a considerable proportion of halogens, which can be released as irritant gases or as more complex molecules. Hydrogen chloride (HCl) is released in large quantities during the combustion of plastic, such as polyvinylchloride, with the majority of the chlorine present in the plastic being released as HCl. In addition to HCl, some chlorinated plastics may also release phosgene and small amounts of PCDDs and PCDFs. HBr is likely to be generated during the combustion of many plastics used for electrical applications due to the addition of bromine as a flame retardant. Fluorine-containing polymers, such as PTFE, are known to liberate HF upon combustion, and if the combustion is incomplete can also generate PFIB and toxic ultrafine particulates, which are extremely potent irritants. Many more recent flame retardant plastics have been produced that do not contain halogens, but instead contain phosphorous. The combustion of these phosphorous-containing plastics yields oxides of phosphorous such as P_2O_5 and phosphoric acid. The combustion of polyurethane foams commonly used in furnishings has been shown to generate isocyanates and their derivatives.

Table 4.3 Hazardous combustion products generated with respect to the material involved.

Material involved	Fire zone[a]	CO	HCN	HCl/ HBr/HF	NO_x	SO_2	P_2O_5	Organic irritants, *e.g.*, acrolein or formaldehyde	Inorganic irritants, *e.g.*, phosgene or ammonia	PAHs	Complex/exotic, *e.g.*, PCDDs/ PCDFs/isocyanates/ PFIB	PM
Polymeric materials	1	+++	+++	+++	++	+	+	++	+	++	++	+++
	2	±	±	+	+	±	−	++	+	±	++	++
Wood	1	++	−	−	+	−	−	++	−	+	+	+++
	2	−	−	−	±	−	−	+	−	±	+	±
Rubber/tyres	1	+++	+	+	+	+++	+	++	+	++	++	+++
	2	±	±	+	±	+++	±	++	±	±	++	++
Oil/petrol	1	++	−	−	−	±	−	++	−	++	++	+++
	2	−	−	−	−	±	−	++	−	±	+	++

[a]Zone 1 relates to the immediate vicinity/compartment of the fire. Zone 2 relates to the location immediately outside the source of the fire.

Key: +++, likely to be present in very high quantities; ++, likely to be present in high quantities; +, likely to be present; ±, may be present at low level; −, unlikely to be present.

As the composition of polymers, such as plastics, resins, fibres and foams, can vary greatly, this will influence the products formed and it may not be possible to identify all of the types of plastic that may be involved during an individual combustion situation.

In general, for fires involving plastics the greatest hazard to public health outside the immediate vicinity of the fire will be due to inorganic and organic irritant gases, particulate matter and other more complex organic compounds. The asphyxiant gases generated during plastic combustion are most likely to be of greatest concern to health within the same compartment/building as the fire.

4.4.3 Fires Involving Wood

The combustion of cellulosic materials, such as wood, either as vegetation, such as in forest fires, or that used in construction and furnishings will be likely to lead to the formation of organic irritants such as acrolein and formaldehyde. However, depending upon the conditions, relatively small amounts of PAHs, particulate matter and more complex exotic molecules may be formed, particularly during incomplete combustion, all of which may present a concern to public health.

Asphyxiant gases, such as carbon monoxide, will be generated during the combustion of wood, although its presence is only likely to be of concern in the immediate vicinity of the source of the fire. The formation of CO from the combustion of wood in the open environment, such as that seen in the case of forest fires, is less likely to be a concern than in building fires due to the greater availability of oxygen and the greater potential for dispersion of the fire effluent.

In addition to the hazardous products from the combustion of wood, burning of wood incorporating preservatives, such as the heavy-metal-containing preservative chromated copper arsenate (CCA), may lead to the liberation of heavy metals as oxides such as chromium trioxide and arsenic trioxide.[71]

The main hazard to public health following exposure to products from wood fires is due to the generation of organic irritant gases. These irritant compounds are more likely to be present at toxicologically significant concentrations in the effluent plume away from the source than gases, such as CO, which are only expected to be present in significant quantities in the immediate vicinity/compartment of the fire.

Due to the presence of halogens in wood, combustion also produces halogenated dioxins and dibenzofurans.

4.4.4 Fires Involving Rubber and Tyres

Fires that involve large quantities of rubber, such as tyre fires, may give rise to the generation of significant yields of sulfur dioxide (SO_2) due to the high sulfur content resulting from the vulcanisation process.

The combustion of rubber is also likely to give rise to the formation of CO, organic irritants, inorganic irritants, PAHs, some complex organic molecules and particulate matter. Some rubber compounds contain organophoshate-based flame retardants, which upon combustion may additionally yield phosphorous pentoxide (P_2O_5).

The compounds most likely to pose a hazard to the health of individuals outside the immediate vicinity of the fire might be expected to be irritants, such as SO_2, and organic irritants such as acrolein.

4.4.5 Fires Involving Oil and Petrol

Fires involving oil or petrol in an external environment might be expected to undergo extensive combustion due to the high temperature and availability of oxygen, with carbonaceous particles being a prominent product. In this fire situation the generation of particulate matter may therefore be significant. Such a fire would also be expected to result in the generation of some PAHs, other complex compounds and organic irritants, which would be present in the greatest quantities at the combustion source. If the temperatures are high enough, these more complex organic chemicals may be completely broken down. The health hazard that these compounds pose will be reduced as the distance from the source increases, but could be of concern for individuals directly exposed to the plume.

The combustion of petrol or oil will most likely lead to the generation of irritants and particulate matter in the effluent plume, which may be expected to be the greatest hazard to the health of individuals outside the immediate vicinity of the source.

Smoke plumes from oil and petrol fires also contain carcinogenic compounds, such as benzene, resulting partly from release as a constituent of the fuel and partly formed from ring cyclisation in the flames (see Chapters 2, 3 and 12).

4.4.6 Fires Involving Hazardous Chemicals (Chemical, Pesticide Manufacturer and Storage)

This scenario involves a fire at a facility concerned with the manufacture or storage of hazardous chemicals such as an industrial chemical or pesticide manufacturer. In such a situation the hazardous materials present may be evolved unchanged during the fire. Specific examples of hazardous chemicals that may either be present unchanged or as hazardous decomposition products in a smoke plume include organophosphorous and organochlorine pesticides. The compounds and products arising during combustion at a location where hazardous chemicals, such as pesticides, may be present will be largely dependent upon the specific chemicals present at the source of the fire. Pesticides containing chlorine, for example, are unlikely to undergo complete decomposition under most combustion conditions, giving rise to the presence of hazardous organochlorine compounds in the smoke plume.

In general, if the quantity of these specialised chemicals in a fire situation was small and that any emissions were well dispersed, there is unlikely to be any significant additional risk to that arising from the smoke from any large building fire.

4.4.7 Fires Involving Asbestos

Large scale fires in which the fabric of the building may contain asbestos, *e.g.*, from asbestos cement roofing, give rise to significant concern by the public regarding exposure to asbestos. However, there is considerable data to show that, providing appropriate clean-up procedures are followed, there is no significant public health risk resulting from the asbestos.[72]

4.5 Smoke Behaviour

The generation of hazardous compounds are likely to pose less of a health hazard for individuals as the distance from the source of the fire increases due to dispersion and conversion of the products as the smoke plume travels further from the source. The area affected by the plume of smoke varies, sometimes rapidly, depending upon the meteorological conditions including wind speed, wind direction and atmospheric stability. This plume can be modelled by the Met Office using modelling software such as CHEMET (see Chapter 6).

Some hazardous combustion products are produced in large quantities during limited ventilation such as CO and NO, but are readily converted into CO_2 and NO_2, respectively, as the availability of oxygen in the effluent plume increases. CO is therefore likely to only present a significant hazard to health for individuals who are in the same compartment (such as the room or building) or immediate vicinity of the fire source (zone 1), but NO_2, which has a much higher toxic potency than NO, is likely to present a health hazard beyond zone 1. For individuals in zone 1, the fire environment may become lethal within a matter of minutes due to heat and the rapid generation of toxic compounds at life threatening concentrations. In zone 1, the most important hazardous combustion products are the asphyxiant gases, CO, HCN and CO_2, and the low availability of oxygen, as these may be fatal in just a few minutes. The hazard from these asphyxiant gases can however enhanced by the presence of irritant gases and a high optical density of the smoke since this may impede escape from this zone, thereby prolonging the exposure duration. The asphyxiant gases produced from a well-developed fire contained within a building may, however, be fatal to individuals present in a location distant to the source of the fire. As the concentration of gases produced from the fire increases, the smoke will disperse throughout the building. As the concentration of the compounds within the smoke increases steadily, a lethal concentration may be reached without an individual becoming aware (which can often occur if the individual is asleep). For this reason individuals may be found in locations within the same compartment, but distant from the source of the fire, that have been overcome by hypoxia, with no sign of fire damage in that location. Due to the immediate hazards of a combustion environment, such as heat, flames and lack of structural integrity, in addition to the presence of asphyxiant gases at lethal concentrations, zone 1 of a fire is most likely to be of major concern to the emergency services.

The exposure of individuals to the fire effluent in the zone outside the immediate fire zone (zone 2) would be expected to be of most concern to public health. This zone is likely to cover many more people than zone 1 and it may not be immediately obvious that these individuals are at risk. The combustion products present in this area are less likely to pose an immediate danger to life, but should be considered as hazardous. Asphyxiant gases, such as CO, are likely to be present in the effluent at much lower concentrations than would be seen in zone 1 and are therefore likely to be less of a hazard to health, unless individuals are directly in contact with the effluent. The major immediate hazard to public health in zone 2 is, therefore, expected to be exposure to irritants and particulates generated in the effluent. Low concentrations of the irritant gases may cause significant irritation of the eyes and respiratory tract, which may affect a large number of people who are not directly exposed to the effluent plume. The adverse effects resulting from exposure to these irritants are likely to be completely resolved following removal from the exposure, with no long term sequelae. However, the generation of more complex products, such as PAHs, dioxins, dibenzofurans and particulate matter, are of concern, but are likely to present a significantly greater hazard from long term or repeated exposure than following a large single acute exposure. Any risk of carcinogenicity from PAHs or dioxins from a single exposure is likely to be very low.

4.6 Main Conclusions

The prediction of toxic combustion products is a complex area and there is the potential for the generation of a huge range of pyrolysis products depending on the nature of the fire and the conditions of burning. Although each fire will have individual characteristics and will ultimately need to be considered on a case by case basis, there are commonalities, particularly with regard to the most important components relating to toxicity.

- Asphyxiant gases (carbon monoxide, hydrogen cyanide and carbon dioxide) and low oxygen concentration are most likely to be of concern to individuals in zone 1 (within the fire compartment/building), but are less likely to pose a major hazard to public health or individuals in zone 2 (outside the immediate compartment of the fire) due to dispersion and dilution.
- The most common potential hazards to individuals in zone 2 from all considered sources are:
  - organic irritants, including acrolein and formaldehyde;
  - acid gases, including hydrogen halides, NO_x, sulfur and phosphorous oxides;

- ○ complex molecules, including dioxins, dibenzofurans and isocyanates;
- ○ particulate matter.
- Scenarios in which the combustion is considered to be incomplete, due to low temperature, lack of ventilation and absence of flaming, would be expected to form the greatest quantities of hazardous combustion products.

These generalisations may assist in rapidly identifying which hazardous combustion products are likely to be of most concern to public health during a fire depending upon the materials involved.

References

1. D. A. Purser, Assessment of Hazards to Occupants from Smoke, Toxic Gases and Heat, *The SFPE Handbook of Fire Protection Engineering*, ed. P. J. DiNenno, National Fire Protection Association, Quincy, MA 02269, 4th edn, 2009, pp. 96–193.
2. T. J. Ohlemiller, Smoldering combustion, *SFPE Handbook of Fire Protection Engineering*, ed. P. J. DiNenno, D. Drysdale, C. L. Beyler and W. D. Walton, 3rd edn, 2002.
3. J. C. Norris and B. Ballantyne, Toxicology and implications of the products of combustion, in *General and Applied Toxicology*, ed. B. Ballantyne, T. Marrs and T. Syversen, Macmillan Reference Ltd, London, 2nd edn, 1999, pp. 1915–1933.
4. Department of Health (DoH), *Health advisory group on chemical contamination incidents; Smoke toxins*, 1996.
5. B. C. Levin and E. D. Kuligowski, Toxicology of fire and smoke, in *Inhalation Toxicology*, ed. H. Salem and S. A. Katz, Taylor & Francis, Boca Raton, 2nd edn, 2006, pp. 205–228.
6. A. Tewarson, Ventilation effects on combustion products, *Toxicology*, 1996, **115**(1–3), 145–156.
7. G. E. Hartzell, Overview of combustion toxicology, *Toxicology*, 1996, **115**(1–3), 7.
8. D. A. Purser, The evolution of toxic effluents in fires and the assessment of toxic hazard, *Toxicol. Lett.*, 1992, **64/65**, 247–255.
9. D. A. Purser and K. R. Berrill, Effects of carbon monoxide on behavior in monkeys in relation to human fire hazard, *Arch. Environ. Health*, 1983, **38**(5), 308–315.
10. R. L. Maynard and R. Waller, Carbon Monoxide, in *Air Pollution and Health*, ed. S. T. Holgate, J. M. Samet, H. S. Koren and R. L. Maynard, Academic Press, 1999, pp. 749–796.
11. T. Prien and D. L. Traber, Toxic smoke compounds and inhalation injury–a review, *Burns*, 1988, **14**(6), 451–460.
12. H. L. Kaplan and G. E. Hartzell, Modeling of toxicological effects of fire gases: I. Incapacitating effects of narcotic fire gases, *J. Fire Sci.*, 1984, **2**, 286–305.

13. Health Protection Agency (HPA), *Carbon Monoxide, Toxicological Overview*, in *HPA Compendium of Chemical Hazards*, 2007.
14. D. A. Purser, Asphyxiant components of fire effluents, in *Fire Toxicity*, ed. A. Stec and R. Hull, Woodhead, Cambridge UK, 2010, ch. 4, pp. 118–198.
15. H. Maeda, K. Fukita, S. Oritani, K. Nagai and B. L. Zhu, Evaluation of post-mortem oxymetry in fire victims, *Forensic Sci. Int.*, 1996, **81**(2–3), 201–209.
16. International Programme on Chemical Safety (IPCS), Carbon monoxide, *Environmental Health Criteria 213*, WHO, Geneva, 1999.
17. Institute for Environment and Health (IEH), *IEH assessment on Indoor Air Quality in the Home (2): Carbon Monoxide*, Medical Research Council, 1998.
18. R. D. Stewart, The effect of carbon monoxide on humans, *Annu. Rev. Pharmacol.*, 1975, **15**, 409–423.
19. D. A. Purser and W. D. Wooley, Biological studies of combustion atmospheres, *J. Fire Sci.*, 1983, **1**, 118–144.
20. E. Bertol, F. Mari, G. Orzalesi and I. Volpato, Combustion products from various kinds of fibres: Toxicological hazards from smoke exposure, *Forensic Sci. Int.*, 1983, **22**, 111–116.
21. D. A. Purser, Toxic product yield and hazard assessment for fully enclosed design fires involving fire retarded materials, *Poly. Int.*, 2000, **49**, 1232–1255.
22. D. A. Purser and J. A. Purser, (2008) HCN yields and fate of fuel nitrogen for materials under different combustion conditions in the ISO 19700 tube furnace. Fire Safety Science – Proceedings of the ninth international symposium. International Association for Fire Safety Science. Fire Safety Science 9, pp. 1117–1128. DOI: 10.3801/IAFSS.FSS.9-1117.
23. World Health Organisation (WHO), Hydrogen Cyanide and Cyanides: Human Health Aspects, *Concise International Chemical Assessment Document*, 2004, vol. 61.
24. G. Kimmerle, Aspects and methodology for the evaluation of toxicological parameters during fire exposure, *J. Fire Flammability/Combust. Toxicol. Suppl.*, 1974, **1**, 42.
25. B. Ballantyne, Changes in blood cyanide as a function of storage time and temperature, *J. Forensic Sci. Soc.*, 1976, **16**(4), 305–310.
26. J. L. McAllister, R. J. Roby, B. Levine and D. Purser, Stability of Cyanide in Cadavers and in Postmortem Stored Tissue Specimens: A Review, *J. Anal. Toxicol.*, 2008, **32**, 1–9.
27. S. J. Moore, I. K. Ho and A. S. Hume, Severe hypoxia produced by concomitant intoxication with sublethal doses of carbon monoxide and cyanide, *Toxicol. Appl. Pharmacol.*, 1991, **109**(3), 412–420.
28. P. Lundquist, L. Rammer and B. Sorbo, The role of hydrogen cyanide and carbon monoxide in fire casualties: a prospective study, *Forensic Sci. Int.*, 1989, **43**(1), 9–14.
29. D. E. Busby, Carbon dioxide toxicity, *Space Clin. Med.*, 1968, **1**, 381–419.

30. B. C. Levin, M. Paabo, J. L. Gurman, S. E. Harris and E. Braun, Toxicological interactions between carbon monoxide and carbon dioxide, *Toxicology*, 1987, **47**(1–2), 135–164.
31. I. N. Einhorn, Physiological and toxicological aspects of smoke produced during the combustion of polymeric materials, *Environ. Health Perspect.*, 1975, **11**, 163–189.
32. Health and Safety Executive (HSE), EH40/2005, Workplace exposure limits. 2005, The Stationary Office: London.
33. U.S. Environmental Protection Agency (USEPA). Acute exposure guideline levels. Available at http://www.epa.gov/oppt/aegl/pubs/chemlist.htm 2007 (cited 2007, Dec).
34. D. A. Purser, Hazards from smoke and irritants, in *Fire Toxicity*, ed. A. Stec and R. Hull, Woodhead, Cambridge UK, 2010, ch. 3, pp. 51–117.
35. International Programme on Chemical Safety (IPCS), Selected Non-heterocyclic Polycyclic Aromatic Hydrocarbons, *Environmental Health Criteria 202*, WHO, Geneva, 1998.
36. J. Lewtas, Air pollution combustion emissions: Characterization of causative agents and mechanisms associated with cancer, reproductive, and cardiovascular effects, *Mutat. Res.*, 2007, **636**(1–3), 95–133.
37. International Agency for Research on Cancer (IARC), Polynuclear Aromatic Compounds, Part 1, Chemical, Environmental and Experimental Data. IARC Monographs on the Evaluation of Carcinogenic Risk to Humans, 1983, 32.
38. Agency for Toxic Substances and Disease Registry (ATSDR), *Toxicological profile for naphthalene, 1-methylnaphthalene and 2-methylnaphthalene*, US department of Health and Human Services, Atlanta, US, 2005.
39. International Agency for Research on Cancer (IARC), Overall Evaluations of Carcinogenicity: An Updating of IARC Monographs Volumes 1 to 42, IARC Monographs on the Evaluation of Carcinogenic Risk to Humans, 1987, Supplement 7.
40. International Programme on Chemical Safety (IPCS), Polychlorinated dibenzo-p-dioxins and dibenzofurans, *Environmental Health Criteria 88*, 1989, WHO, Geneva.
41. D. Pelclova, P. Urban, J. Preiss, E. Lukas, Z. Fenclova, T. Navratil, Z. Dubska and Z. Senholdova, Adverse health effects in humans exposed to 2,3,7,8-tetrachlorodibenzo-*p*-dioxin (TCDD), *Rev. Environ. Health*, 2006, **21**(2), 119–138.
42. B. R. Stanmore, The formation of dioxins in combustion systems, *Combust. Flame*, 2004, **136**, 398–427.
43. S. V. Levchik and E. D. Weil, Overview of the recent literature on flame retardancy and smoke suppression in PVC, *Polym. Adv. Technol.*, 2005, **16**(10), 707.
44. W. F. Carroll Jr, Is P. V. C. in house fires the great unknown source of dioxin?, *Fire and Mater.*, 1996, **20**(4), 161.
45. National Institute for Occupational Safety and Health (NIOSH), DHHS (NIOSH) Publication No. 86-111. Current Intelligence Bulletin 45.

Polychlorinated Biphenyls (PCB's): Potential Health Hazards from Electrical Equipment Fires or Failures, 1986, United States Department of Health and Human Services.
46. T. Nakao, O. Aozasa, S. Ohta and H. Miyata, Formation of toxic chemicals including dioxin-related compounds by combustion from a small home waste incinerator, *Chemosphere*, 2006, **62**(3), 459–468.
47. Agency for Toxic Substances and Disease Registry (ATSDR), *Toxicological profile for chlorinated dibenzo-p-dioxins*, US department of Health and Human Services, Atlanta, US, 1998.
48. Agency for Toxic Substances and Disease Registry (ATSDR), *Toxicological profile for chlorodibenzofurans*, US department of Health and Human Services, Atlanta, US, 1994.
49. International Agency for Research on Cancer (IARC), Polychlorinated dibenzo-para-dioxins and polychlorinated dibenzofurans. IARC Monographs on the Evaluation of Carcinogenic Risk to Humans, 1997, 69.
50. M. Paabo and B. C. Levin, A review of the literature on the gaseous products and toxicity generated from the pyrolysis and combustion of rigid polyurethane foams, *Fire Mater.*, 1987, **11**(1), 1.
51. H. Tinnerberg, M. Spanne, M. Dalene and G. Skarping, Determination of complex mixtures of airborne isocyanates and amines. Part 3. Methylenediphenyl diisocyanate, methylenediphenylamino isocyanate and methylenediphenyldiamine and structural analogues after thermal degradation of polyurethane, *Analyst*, 1997, **122**(3), 275–278.
52. M. M. Esperanza, A. N. Garcia, R. Font and J. A. Conesa, Pyrolysis of varnish wastes based on a polyurethane, *J. Anal. Appl. Pyrol.*, 1999, **52**, 151.
53. B. Cohrssen, Cyanides and nitriles. Chapter sixty-one, in *Patty's Toxicology*, ed E. Bingham, B. Cohrssen and C. H. Powell, John Wiley & Sons, Inc, New York, 5th edn, 2001, vol. 3.
54. International Programme on Chemical Safety (IPCS), Toluene 2,4-diisocyanate (TDI). *Poisons Information Monograph: PIM 534*, 1997, WHO, Geneva.
55. D. A. Purser, Recent developments in understanding the toxicity of PTFE thermal decomposition products, *Fire Mater.*, 1992, **16**, 67–75.
56. C. H. Lee, Y. L. Guo, P. J. Tsai, H. Y. Chang, C. R. Chen, C. W. Chen and T.-R. Hsiue, Fatal acute pulmonary oedema after inhalation of fumes from polytetrafluoroethylene (PTFE), *Eur. Respir. J.*, 1997, **10**, 1408–1411.
57. L. W. Smith, R. J. Gardner and G. L. Kennedy Jr, Short-term inhalation toxicity of perfluoroisobutylene, *Drug and Chem. Toxicol.*, 1982, **5**(3), 295.
58. R. S. Waritz, An industrial approach to evaluation of pyrolysis and combustion hazards, *Environ. Health Perspect.*, 1975, **11**, 197–202.
59. H. Wang, R. Ding, J. Ruan, B. Yuan, X. Sun, X. Zhang, S. Yu and W. Qu, Perfluoroisobutylene-induced acute lung injury and mortality are heralded by neutrophil sequestration and accumulation, *J. Occup. Health*, 2001, **43**(6), 331.

60. J. Patocka and J. Bajgar, Toxicology of perfluoroisobutylene, *ASA Newsl.*, 1998, 22.
61. B. Jugg, J. Jenner and P. Rice, The effect of perfluoroisobutene and phosgene on rat lavage fluid surfactant phospholipids, *Human Exp. Toxicol.*, 1999, **18**(11), 659.
62. M. P. Maidment and D. G. Upshall, Retention of inhaled perfluoroisobutene in the rat, *J. Appl. Toxicol.*, 1992, **12**(6), 393.
63. K. Donaldson, L. Tran, L. A. Jimenez, R. Duffin, D. E. Newby, N. Mills, W. MacNee and V. Stone, Combustion-derived nanoparticles: A review of their toxicology following inhalation exposure, *Part. Fibre Toxicol.*, 2005, **2**, 10.
64. R. Duffin, N. L. Mills and K. Donaldson, Nanoparticles – A thoracic toxicology perspective, *Yonsei Med. J.*, 2007, **48**(4), 561.
65. U.S. Environmental Protection Agency (U.S. EPA), Air quality criteria for particulate matter. 2004, National Center for Environmental Assessment-RTP Office, Office of Research and Development: Research Triangle Park, NC.
66. N. Stenfors, C. Nordenhall, S. S. Salvi, I. Mudway, M. Soderberg, A. Blomberg, R. Helleday, J. O. Levin, S. T. Holgate, F. J. Kelly, A. J. Frew and T. Sandstrom, Different airway inflammatory responses in asthmatic and healthy humans exposed to diesel, *Eur. Respir. J.*, 2004, **23**(1), 82.
67. Committee on the Medical Effects of Air Pollutants (COMEAP), Cardiovascular Disease and Air Pollution; A report by the Committee on the Medical Effects of Air Pollutants, 2006, Department of Health.
68. U.S. Environmental Protection Agency (U.S. EPA), Air quality criteria for particulate matter. 2004, National Center for Environmental Assessment-RTP Office, Office of Research and Development: Research Triangle Park, NC.
69. European Commission (EC), DG XI, Working Group on Lead, Air Quality Daughter Directives – Position Paper on Lead, 1997.
70. European Commission (EC), DG Environment, Working Group on Arsenic, Cadmium and Nickel Compounds, Ambient Air Pollution by As, Cd and Ni compounds – Position Paper, 2001.
71. K. Lundholm, D. Bostrom, A. Nordin and A. Shchukarev, Fate of Cu, Cr, and As During Combustion of Impregnated Wood with and without Peat Additive, *Environ. Sci. Technol.*, 2007, **41**(18), 6534–6540.
72. Health Protection Agency (HPA), The Public Health Significance of Asbestos Exposures from Large Scale Fires, HPA-CHaPD-003, 2007, Health Protection Agency.

CHAPTER 5

Generation, Sampling and Quantification of Toxic Combustion Products

T. RICHARD HULL* AND ANNA A. STEC

Centre for Fire and Hazard Science, University of Central Lancashire, Preston, Lancashire PR1 2HE, UK
*Email: trhull@uclan.ac.uk

5.1 Introduction

Analytical chemistry has made tremendous advances over the last 20 years: instrumental sensitivities for most analytes have increased by several orders of magnitude. Although it is not usually possible to detect a single molecule, the number of molecules required for positive identification has fallen from parts per million to parts per billion or trillion. These improvements have occurred over a period when advances in combustion toxicity have been rather slow.

There are several reasons for this.

1. The practical difficulties associated with quantifying the toxic components present in a fire effluent.
2. A generally poor understanding of fire toxicity—without properly quantifying the fire conditions, the yields of even major toxic products appear random and irreproducible.

Issues in Toxicology No. 23
Toxicology, Survival and Health Hazards of Combustion Products
Edited by David A. Purser, Robert L. Maynard and James C. Wakefield

Published by the Royal Society of Chemistry, www.rsc.org

3. A regulatory focus on control of flammability and rate of heat release to control fire hazards rather than focus on the toxicity of the combustion products.
4. The range of components present in a fire effluent, and a lack of knowledge relating to the toxicity of all the individual components, makes quantification of fire toxicity more expensive, while the cost of assessment of the toxicity of every detectable component would be almost limitless.

This chapter is limited in scope to address the first point and aspects of the second point—the practical difficulties of replicating the combustion conditions and quantifying the toxic components present in fire effluents. These can be subdivided into three main areas.

1. **Generation of the effluent.** Earlier chapters have shown that fire effluent toxicity is not a material property, but a combination of the material's chemistry and the combustion conditions under which it is burnt. In general, larger fires show a wider range of fire conditions than small laboratory-scale test methods. For example, the oxygen concentration in the effluent from a fire in a domestic dwelling can fall as low as a few percent, while in some test methods, claimed to be suitable for assessing fire toxicity, the flame is extinguished when the oxygen concentration falls below 12%.
2. **Sampling and pre-concentration of the combustion effluent.** Combustion effluents are usually mixtures of gaseous and aerosol particles, containing significant quantities of condensable water. Many of the gaseous species are water soluble and can dissolve in water droplets when condensation has occurred, or at cold spots in the sample line; they can be absorbed on to the walls of the test apparatus, the sampling lines and onto aerosol particles; if a filter is used, then gaseous analytes will adsorb and desorb onto the particulate matter on the filter; if pumps or other apparatus need to be present before the analyser, they may also absorb certain analytes. When analysing for minor components, such as polycyclic aromatic hydrocarbons (PAHs), aldehydes *etc.*, methods such as sorption tubes or solid phase micro-extraction (SPME) may be used to pre-concentrate the analyte. In these cases the results are highly dependent on the affinity of each analyte for the sorbent. Also, the enclosed metal chambers used for some laboratory-scale methods have much greater surface : volume ratios compared with full-scale building enclosures. This can result in a rapid decrease in the concentrations of key products due to deposition and/or reaction with the chamber walls.[1]
3. **Analysis of the effluent.** A large number of methods have been described for analysis of most components of a fire effluent.[2–4] Some of these methods rely on grab sampling, providing a snapshot at a particular point in time, while others use continuous analysis.

Unfortunately there is no single method, or even combination of methods, able to fully quantify all the toxicants present in a combustion effluent. Specific methods for individual gases [particularly oxygen (O_2), carbon dioxide (CO_2), carbon monoxide (CO) and hydrogen cyanide (HCN)] are very robust, and these gases have relatively few sampling problems. Methods such as Fourier transform infrared (FTIR) and gas chromatography-mass spectrometry (GC-MS) are capable of identifying larger numbers of species in a single experiment, but both suffer from calibration problems and overlap, where several toxic species obscure the presence of others.

The aim of this chapter is to present the state-of-the-art in generation, sampling and analysis of toxic components of fire effluents. The challenges will be described to alert the unwary reader to potential pitfalls, rather than to present an exhaustive compendium of methods of generation, sampling and analysis that may be applied to combustion effluents.

5.2 Generation of Combustion Effluents

Burning behaviour and toxic product yields depend most strongly on a few factors: physical aspects of the fire scenario, such as shape and orientation of the fuel; material composition;[5] temperature; and oxygen concentration.[6] The generalised development of a fire has been used to classify fire growth into stages, from smouldering combustion and early well-ventilated flaming, through to fully-developed under-ventilated flaming.[7] Although smouldering (oxidative pyrolysis) can generate toxicologically significant quantities of effluent (for example, smouldering cotton, or polyurethane foam in a small enclosed space), the rate of reaction, and hence the amount of toxic species generated, is much lower than in flaming fires and typically only affects people in close proximity to the smouldering object. Well-ventilated flaming fires are generally small, so extinguishment or escape is still feasible and any fire effluent will accumulate below the ceiling, but above head height. However, as fires grow, and the upper layer descends, they become ventilation-controlled and rapidly change from well-ventilated to under-ventilated in enclosures such as buildings. These fires are larger and therefore produce greater volumes of effluent, affecting occupants over a much larger part of any building (see Chapter 2). In the laboratory assessment of fire hazard, for flammability, 21% oxygen is normally considered the worst case scenario, whereas for fire toxicity, under-ventilated flaming and oxygen concentrations below 10% are the worst case scenario. This arises because the yields of the major toxic products [particularly the asphyxiants, carbon monoxide (CO) and hydrogen cyanide (HCN)] will be much greater in under-ventilated conditions (see Chapter 3).

Combustion systems for energy release normally premix the fuel and oxidant, for example, by allowing air access to the base of the flame, giving a clean burn and high combustion efficiency. In contrast, almost all unwanted fires are diffusion flames, with inefficient mixing of fuel and oxygen. The interior of large diffusion flames are always under-ventilated, because oxygen cannot penetrate the flame. For a particular material, at a certain flame height, these products of incomplete combustion escape through the tip of the flame and smoke appears at a characteristic "smoke point height", releasing unburnt hydrocarbons and carbon monoxide.[8] When flames penetrate the descending, oxygen-depleted, upper layer, combustion becomes increasingly inefficient. Thus, for any larger fire there will always be a significant yield of CO, hydrocarbons and smoke. Indeed data from large-scale fires[9,10] in enclosures, such as a room, show much higher levels of the asphyxiants, CO and HCN, under conditions of developed flaming. It is therefore essential to the assessment of toxic hazard from fire that each fire stage can be adequately replicated and preferably the individual fire stages treated separately.

5.2.1 Assessment of Combustion Toxicity

The yields of combustion products from laboratory-scale methods depend most strongly on the fuel : air ratio and whether the decomposition is flaming or non-flaming. High yields of CO and HCN from under-ventilated fires are responsible for most deaths, but this under-ventilated burning is the most difficult to create on a laboratory scale. A review of the extensive research predicting the carbon monoxide evolution from flames of simple hydrocarbons has shown the importance of the equivalence ratio ϕ in defining the fire condition, see eqn (5.1).[11] The dependence of fire toxicity and toxic product yields on the ventilation conditions in fires emphasises the need for reliable replication of fire conditions.

$$\phi = \frac{\text{actual fuel to air ratio}}{\text{stoichiometric fuel to air ratio}} \tag{5.1}$$

An equivalence ratio of 0.5 represents a well-ventilated scenario, typical of an early growing fire, while a ratio of 2 corresponds to the under-ventilated stage responsible for high yields of toxic effluents. When $\phi = 1$ the theoretical amount of air is available for complete combustion to carbon dioxide (CO_2) and water *etc.* The equivalence ratio approach has been refined and subdivided: "global" equivalence ratio has been distinguished from "plume" equivalence ratio to improve the prediction of yields of carbon monoxide and other products of incomplete combustion, where the effluent mixes with fresh air and forms a reactive hot layer. According to Pitts, this condition arises when the hot layer temperature exceeds 625 °C.[11] In many fire tests the equivalence ratio can be determined directly: for fuels of known composition, from the burning rate or mass loss rate and the air flow rate, provided all the air passes through the fire plume (or the reactive hot layer). Where the material is of unknown composition, the equivalence ratio can be deduced from the oxygen consumption during well-ventilated burning. For a laboratory-scale method to differentiate between the stages of a fire, an indicator of fire condition, such as the equivalence ratio or the combustion efficiency, must be quantified so that the individual fire stages can be identified and the product yields correlated to each stage.

Most fire deaths and injuries actually occur in residential fires, although assessment of fire toxicity is currently focused on areas where escape is restricted, such as aeroplanes, railway carriages and passenger ships, which include requirements to quantify the fire toxicity of internal components. In most countries, there are no regulations covering the fire toxicity of building components, or for most road vehicles including goods vehicles in tunnels. In China and Japan, there are specific restrictions on the use of materials with high fire toxicity in high risk applications such as tall buildings. An increasing number of jurisdictions permit the alternative performance-based design approaches to fire safety. Reliable rate of heat release, fire effluent toxicity and smoke generation data are all essential components of such an assessment. The general approach, described in ISO 13571,[12] is to ensure that the available safe escape time (ASET), before escape routes become obscured by smoke or filled with toxic gases, exceeds the required safe escape time (RSET). Various apparatus and protocols for quantifying fire effluent toxicity in different jurisdictions and industries have been critically reviewed.[13] The quantification of fire toxicity is being led by the ISO TC92 Subcommittee 3 – *Fire Threat to People and the Environment*, where the steady state tube furnace (SSTF; currently ISO/DIS 19700) is being developed specifically as the means of quantifying toxic product yields from burning materials on a laboratory scale. Meanwhile, the transport industries have tried to adopt the smoke density chamber (SDC) ISO 5659-2[14] for quantification of toxic product yields[15,16] using conventional or Fourier transform infrared (FTIR) gas analysis, despite significant problems of reproducibility. It has been suggested that the reproducibility problems arose from the single point measurement (the tip of the probe may be in the centre of the plume, below it or, if mixing is more efficient, the upper layer may be recirculated through the flame), or because the timing of the effluent sampling may cause instabilities (for example, an initial proposal to sample after 8 minutes was replaced by a proposal to sample when the smoke density reached its maximum). The reproducibility of a revised protocol is being investigated based on continuous sampling of the fire effluent. In addition, the European rail standard[16] also specifies use of the French Railway test (NF X 70-100; NFX)[17] for smaller components. Although at an earlier stage of development, the controlled atmosphere cone calorimeter (CACC) is being considered within ISO as an apparatus for quantifying heat release rates from under-ventilated flaming and has been proposed as a potential candidate for quantifying toxic products yields from different fire conditions. The fire propagation apparatus (FPA; ISO 12136) was developed by the US insurance company Factory Mutual (now FM Global), but has been standardised for heat release measurement[18] and for engineering assessments of CO and CO_2 yields.

5.2.2 Laboratory-scale Methods

Laboratory-scale methods used for generation of toxic fire effluents should be capable of reproducing individual fire stages or combustion conditions for input into models of combustion toxicity. Full-scale fires simultaneously involve different fire stages in different places, which are changing with time. However, bench-scale methods that allow the combustion conditions to change during the test are much more difficult to relate to full-scale fires, because the duration of each condition is unknown and the behaviour of fires changes on scale-up. Most laboratory-scale methods have non-constant combustion conditions, such as those in closed chambers exposed to a constant source of heat, including the smoke density chamber (ISO 5659-2),[14] and static tube furnace tests, such as the NF X 70-100.[17] Those with constant combustion conditions are more suited to producing data suitable for comparison and modelling: the steady state tube furnace (ISO/DIS 19700)[19] has been specifically designed to achieve this. Intermediate between these two approaches are those that can produce quasi-steady combustion conditions, such as the cone calorimeter (ISO 5660-1)[20] with a non-standardised controlled atmosphere attachment (CACC) and the fire propagation apparatus (ISO 12136).[18] The difficulty of replicating the conditions of fully developed, under-ventilated flaming on a laboratory scale is caused by several practical problems. ϕ depends on the mass loss rate of the specimen and the available air; for most methods, one or both are unknown; ϕ will be increased by an unknown factor if products are recirculated into the flame zone. Apparatuses where ϕ changes rapidly allow little time for sampling and measurement of mass loss and effluent composition, with resultant errors and uncertainties. Progressive changes in the composition of a static specimen (for example, due to char formation) provide additional complexity. The reactions of under-ventilated flaming occur in both the flame zone and in the hot upper layer. Only the SSTF has a heated reaction zone that replicates the hot layer. The applied heat flux must be large enough for burning to continue at oxygen concentrations as low as 5%. In some laboratory-scale apparatuses the heat flux is constant and often insufficient to sustain flaming at such low oxygen concentrations; further, an unknown quantity of fresh air bypasses the fire plume, so the ventilation condition remains undefined. Some fire models, such as the cone calorimeter, fire propagation apparatus and smoke density chamber, use the temperature of the radiant heater to preselect the radiant heat flux, and then check this using a radiant heat flux meter. Others, such as the NF X 70-100, and the ISO 19700 SSTF use the furnace temperature setting to ensure a consistent radiant heat flux. The radiant heat flux in the ISO 19700 apparatus has been measured[21] and is 40 kW m^{-2} in the centre of the furnace at 650 °C and 78 kW m^{-2} at 825 °C. The following sections describe each method briefly, and then present experimental data comparing these different methods, relating them to large-scale test data.

5.2.2.1 The Smoke Density Chamber (SDC)

The most widely used fire-test apparatus, stipulated in smoke regulations in most countries of the developed world, is the smoke density chamber (ISO 5659-2),[14] shown in Figure 5.1. Its widespread availability has encouraged its adaptation to toxic gas generation and assessment, for which it is much less well-suited. The standard specifies four test conditions, which it is assumed are representative of different fire scenarios, by varying the heat flux in the presence or absence of a pilot ignition source. Of the conditions specified here, 25 kW m^{-2} without piloted ignition (presumably corresponding to oxidative pyrolysis), 25 kW m^{-2} with piloted ignition (corresponding to early, well-ventilated flaming) and 50 kW m^{-2} without piloted ignition (corresponding to developed, under-ventilated flaming) are considered. The fourth condition (50 kW m^{-2} with piloted ignition) was introduced in the 2012 version after the experimental work was completed. The sample is a 75×75 mm^2 solid sheet and the standard for smoke measurement states that the results are only valid at the thickness tested (typically 1–4 mm). For a fixed chamber volume (0.51 m^3), assuming complete combustion, the sample thickness and combustible fuel content (stoichiometric oxygen demand) will dictate the ventilation condition, thus a thin sample, or one with a high content of inert filler, will burn under well-ventilated conditions with minimum toxic products, while a thicker sample with a high combustible fuel content might be expected to produce a high yield of CO and other products of incomplete combustion. The protocol has been modified as a toxicity test by the mass transport industries, in the aircraft,[22] maritime[15] and railway tests.[16] Some of these methods attempt to address the transition through the fire stages by monitoring the formation of toxic gases as a

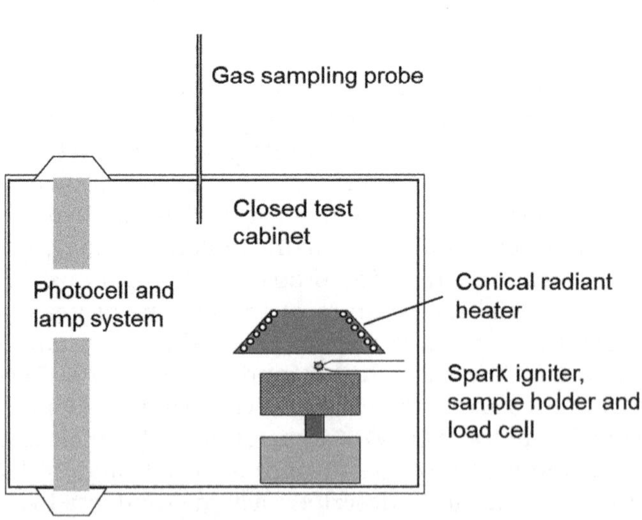

Figure 5.1 The smoke density chamber (ISO 5659-2) showing a sampling probe for fire smoke toxicity.

function of time as the oxygen concentration falls and the fire condition changes from well-ventilated to under-ventilated. However, unlike a real fire, the heat flux remains constant, and so when the oxygen concentration falls, the flame may be extinguished.

A particular issue with the calculation of toxic product yields is that although it is possible to measure the concentrations of gases at the sampling point near the chamber ceiling, since the chamber contents are not mixed, the concentrations at other locations in the chamber (especially at lower levels) are unknown. Since the heated buoyant combustion products collect and fill down from the top of the chamber it is likely that combustion product concentrations will be lower at lower heights in the chamber.

5.2.2.2 The Controlled Atmosphere Cone Calorimeter (CACC)

The cone calorimeter (ISO 5660)[20] is probably the most widely used apparatus for measurement of flammability properties such as ignitability and heat release rate.[23] It can be used for testing samples 100 mm×100 mm and up to 25 mm thick, in both the horizontal and vertical orientation. The open cone calorimeter replicates the early well-ventilated stage of flaming where a fire would be too small to produce enough toxicants to cause harm, except in very small enclosures. However, several non-standard modifications of the apparatus have been reported, enclosing the fire model combustion zone in a controlled-ventilation chamber in an attempt to replicate oxygen-depleted conditions. In these modifications, the controlled atmosphere cone calorimeter (CACC),[24] shown in Figure 5.2, has its fire model enclosed in a heat resistant glass chamber (400 mm high with a 300 mm×300 mm base) so that the air flow may be controlled. In the tests discussed here, the effluent continued to burn as it emerged from the chamber (secondary flaming in Figure 5.2), ultimately giving well-ventilated flaming. In others, under reduced oxygen concentrations, the fuel lifts from the surface, but ignition does not occur.[25] Hietaniemi[26] used the controlled atmosphere cone calorimeter, but argues correctly, we believe, that an instantaneous "effective" global equivalence ratio ϕ_{eff} should be used rather than an averaged local equivalence ratio, based on the oxygen supply to the chamber, because, in some experiments, substantial secondary flaming occurred outside the test chamber, such that the amount of oxygen available to combustion exceeded the amount that was fed to the enclosed chamber. The calculation of ϕ_{eff} takes into account both the primary oxygen supply inside the chamber and the secondary oxygen supply above the chamber.

5.2.2.3 The Fire Propagation Apparatus (FPA)

This fire propagation apparatus[18] (Figure 5.3) is similar in principle to the cone calorimeter but the fire zone is contained within a 172 mm diameter circular vertical silica tube, allowing better mixing of the fuel oxidant and

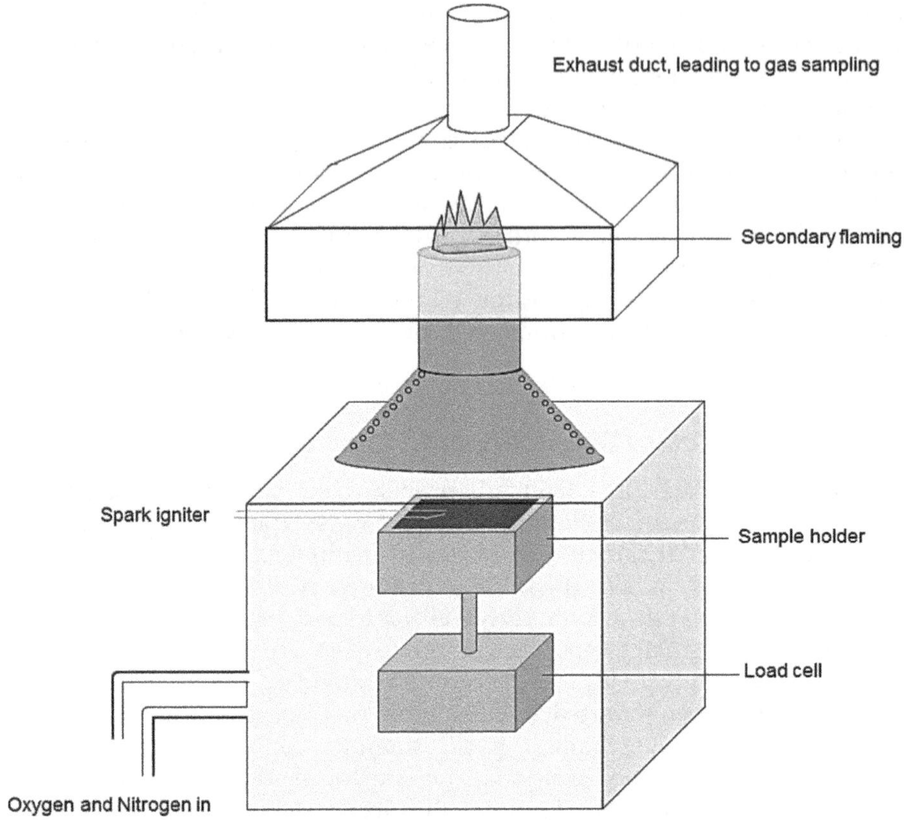

Figure 5.2 The controlled atmosphere cone calorimeter.

avoiding contact with the heaters, which are outside the tube. Horizontal test specimens, typically 100 mm^2 and up to 25 mm thick, or vertical specimens 100 mm wide, 305 mm high and up to 25 mm thick, are exposed to thermal radiation and a pilot flame. The effluent flows through an instrumented duct and the rate of heat release is determined from the CO_2 and CO generation and oxygen depletion rates. The effluent yields can be directly related to individual fire stages through the equivalence ratio. It has been used to quantify the generation of toxic products under different ventilation conditions and data has been published describing CO and hydrocarbon yields from various polymers.[27]

5.2.2.4 *The Non-dynamic Tube Furnace (NFX)*

In the standard tube furnace thermal degradation method[17] a small static specimen is thermally decomposed in a furnace with a metered air flow, which drives the effluent to the sampling system or gas measurement devices. Mass loss is obtained by weighing the sample before and after the test.

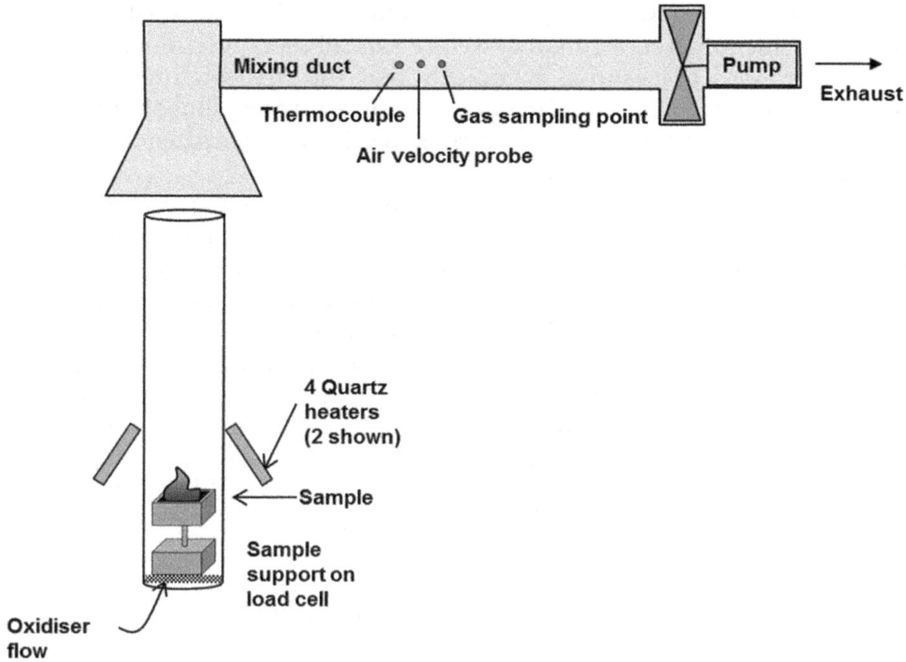

Figure 5.3 The fire propagation apparatus (ISO 12136).

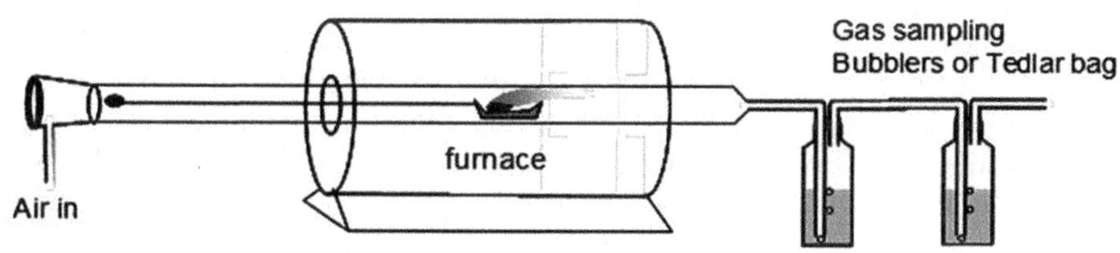

Figure 5.4 The non-dynamic tube furnace test, NF X 70-100.

Determination of the equivalence ratio is not possible unless the rate of pyrolysis is known, for example, from on-line analysis. Since different materials pyrolyse at different rates and have different stoichiometric oxygen requirements, using the same air flow will result in burning at equivalence ratios that vary with furnace temperature and material.

The NF X 70-100 (NFX) method shown in Figure 5.4 was developed to quantify the toxicity of materials and products used in railway vehicles. The static tube furnace uses 1 g (or 0.1 g for low density materials) of sample in a crucible, inserted into the middle of a silica furnace tube of diameter 44 mm and thermally decomposed in air flowing at 2 L min^{-1}. Furnace temperatures of 400, 600 and 800 °C are used. Although used as a standard test, without correlation of the conditions to particular fire stages, it would appear that a temperature of 400 °C corresponds to pyrolysis without ignition; a temperature of 600 °C auto-ignites the sample with a fairly slow rate of burning, which may correspond to well-ventilated flaming; a temperature of 800 °C pyrolyses the sample so quickly that the fire condition becomes under-ventilated. In the current work, the three furnace temperatures have been considered to represent oxidative pyrolysis, well-ventilated and under-ventilated conditions, respectively.

The method is easy to use and uses simple equipment with specified operating conditions of temperature and air flow. The greatest sensitivity in the test, leading to problems of repeatability and reproducibility, occur when the sample is close to ignition and, in some cases, flaming combustion occurs, while in other cases it does not. This makes the results very susceptible to differences in furnace design.[28] Alongside the SDC, this is part of EN TS 45545-2 and the method is increasingly used for fire toxicity testing of materials used for railway vehicles. The lack of requirement for flaming to be observed leaves the assignation of fire stages to be assumed. A practical problem is that several replicate test-runs are often needed to obtain sufficient volume of sample for complete gas analysis.

5.2.2.5 The Steady State Tube Furnace (SSTF)

The steady state tube furnace,[19] shown in Figure 5.5, feeds the sample (typically around 25 g pellets or granules) into its hot zone at a fixed rate, under a controlled air supply, inside a horizontal silica tube of diameter 48 mm, allowing adequate mixing of fuel and oxidant. It forces combustion by driving the sample into a furnace of increasing heat flux at a fixed rate, so that, by running several tests with the same material with different ventilation conditions, each fire stage can be replicated by steady state burning. The products generated in the flame zone then pass through the heated furnace tube, maintaining a high temperature, as in the upper layer of a compartment fire. The toxic product yields may be quantified from the gas concentrations and mass feed rate during the steady state burn period after the initial peak. It has been designed to generate data for input to fire hazard

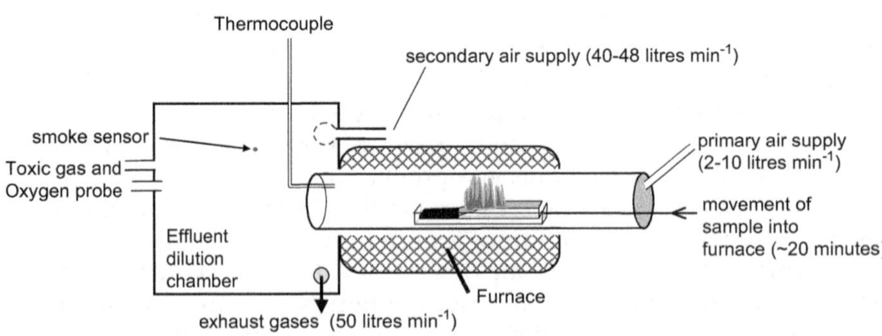

Figure 5.5 The steady state tube furnace apparatus, ISO 19700.

assessments, using the methodology in ISO 13344[29] and ISO 13571,[12] particularly in relation to the ISO fire stages.[7] The sample is spread evenly in a long silica boat over a length of 800 mm and fed into a tube furnace at a typical rate of 1 g min^{-1} with flowing air. Secondary air is added in a mixing chamber to give a total gas flow of 50 L min^{-1}.

5.2.3 Reference Data from Large-scale Fire Tests

There is only a limited amount of suitable large-scale data available where the fire condition is both known (in terms of equivalence ratio) and flaming was under-ventilated. One of the best data sets comes from the TOXFIRE project,[26,30–32] where a phi meter[33] was used to establish the fire condition inside a lightweight concrete room with dimensions in accordance with the ISO 9705 room test.[34] The ISO 9705 room had one opening, 0.8 m wide, 2.0 m high, centrally located at one end. The ventilation conditions inside the room were controlled by sealing the lower part of the opening with slabs of non-combustible fibreboard, giving a reduced opening of height 0.89, 0.68, 0.56 and 0.45 m to create under-ventilated conditions. The fuel was put into square pans of 1.2 m^2 for polypropylene (PP) and 1.4 m^2 for polyamide 6.6 (PA 6.6). Four tests were carried out, burning 60 kg of PP and between 55 and 75 kg of PA 6.6, using a load cell to measure the mass loss. A phi meter determines the equivalence ratio of the fire by adding oxygen to the fire effluent, passing the mixture over a catalyst at 900 °C then measuring the resulting oxygen concentration. Corresponding data have been reported for the different fire conditions using the bench-scale methods described above.

5.3 Comparison of Yields from Laboratory-scale Tests and Large-scale Reference Data

5.3.1 Comparisons Based on Equivalence Ratio

Steady state and quasi steady state methods allow the equivalence ratio to be determined. If large-scale fire toxicity test data has also been recorded as a function of equivalence ratio, direct comparison is possible. For both PP and PA 6.6 burning under the conditions described, negligible residue remains, and therefore the mass charge and mass loss yields will be indistinguishable. Figure 5.6 shows a comparison for the CO yield from PP as a function of equivalence ratio, using controlled atmosphere cone calorimeter data published as part of the TOXFIRE project,[26] steady state tube furnace data[35] and data from the fire propagation apparatus (actually reported for low density polyethylene, LDPE, which is isomeric with PP),[27] compared to the large-scale data.[30] It is apparent from all tests that the inherent variation in burning behaviour gives much larger scatter in under-ventilated conditions than in well-ventilated conditions. In general, this shows a significant increase in the CO yield as the fire develops from well-ventilated to

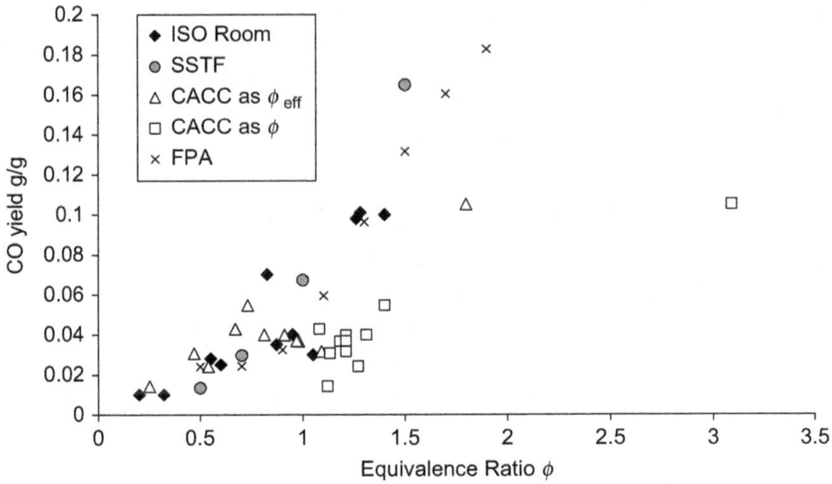

Figure 5.6 Comparison of CO yield for polypropylene for the large-scale (ISO Room), steady state tube furnace, fire propagation apparatus and controlled atmosphere cone calorimeter, as a function of equivalence ratio ϕ.

under-ventilated. Yields from the controlled atmosphere cone calorimeter are shown as average values as a function of equivalence ratio (ϕ)[26] and corrected values of the equivalence ratio (ϕ_{eff}), where the oxygen depletion that resulted from flaming occurring outside the enclosure, above the cone heater and detailed in reference,[26] was taken into account in the equivalence ratio calculation. With a single exception, the controlled atmosphere cone calorimeter failed to replicate the higher CO yields associated with large-scale behaviour.

Figure 5.6 shows acceptable agreement between the ISO room data and the steady state tube furnace, the controlled atmosphere cone calorimeter and fire propagation apparatus data, which actually generated greater extremities of well- and under-ventilated burning than occurred in the ISO room. However, in the results from the controlled atmosphere cone calorimeter, there is only a single data point (at $\phi = 3.2$ or $\phi_{eff} = 1.8$) that shows the higher yields of carbon monoxide associated with under-ventilated combustion, suggesting that it is not easy to replicate this condition in the CACC. Data for the controlled atmosphere cone calorimeter generally shows lower CO yields in under-ventilated combustion, and after the correction for combustion outside the chamber, the apparatus is unable to replicate the more toxic conditions of under-ventilated burning responsible for most fire deaths, and therefore has limited value as a tool for assessment of fire toxicity.

Figure 5.7 compares the published CO yields from the large-scale ISO room tests for PA 6.6 under a range of ventilation conditions with those from the controlled atmosphere cone calorimeter,[26] steady state tube furnace[35] and the fire propagation apparatus.[27] The SSTF data include results from

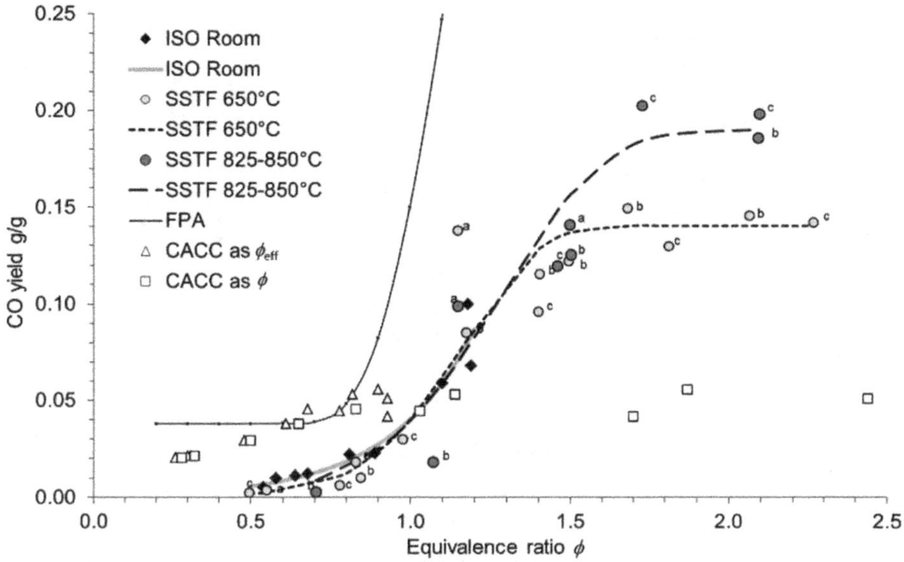

Figure 5.7 Comparison of CO yield for aliphatic polyamides from the steady state tube furnace, controlled atmosphere cone calorimeter and fire propagation apparatus with ISO Room, as a function of equivalence ratio ϕ. Letters indicate data source and material for SSTF: (a) polyamide 6.6,[36] (b) polyamide 6.6,[6] (c) polyamide 6.[37]

four data sets: (a) Blomqvist,[36] (b) Stec[6] and (c) Purser.[37] Of the three SSTF data sets, (a) was obtained using the same batch of polyamide 6.6 as was used for the ISO 9705 room in the TOXFIRE project; (b) was also polyamide 6.6, but obtained from a different source; (c) was for polyamide 6 (which had identical chemical composition and a slightly different structural formula to PA 6.6). Again, this shows a significant increase in the CO yield as the fire moves from well-ventilated to under-ventilated, accompanied by an increase in scatter, using all methods. CO yields in the SSTF were found to be temperature sensitive, with somewhat higher yields at 825–850 °C than at 650 °C. During the ISO room tests at the higher equivalence ratios, the upper layer temperatures were 750–1000 °C, corresponding more with the higher temperature SSTF results. The SSTF results are very similar from all three aliphatic polyamide samples. It is notable that the controlled atmosphere cone calorimeter results, based on the averaged local equivalence ratio ϕ reported by Hietaniemi,[26] show higher CO yields in well-ventilated conditions and, crucially, lower CO yields in under-ventilated conditions. Again, the data have been corrected by calculation of the effective equivalence ratio ϕ_{eff}, which aligns the yields to the well-ventilated stages of the large-scale data.[26] For polyamides, the FPA does show a large increase in CO yields from low levels, as ϕ approaches and exceeds unity, however, the corresponding yields are shifted to lower ϕ values compared with the large-scale and ISO 19700 test data.

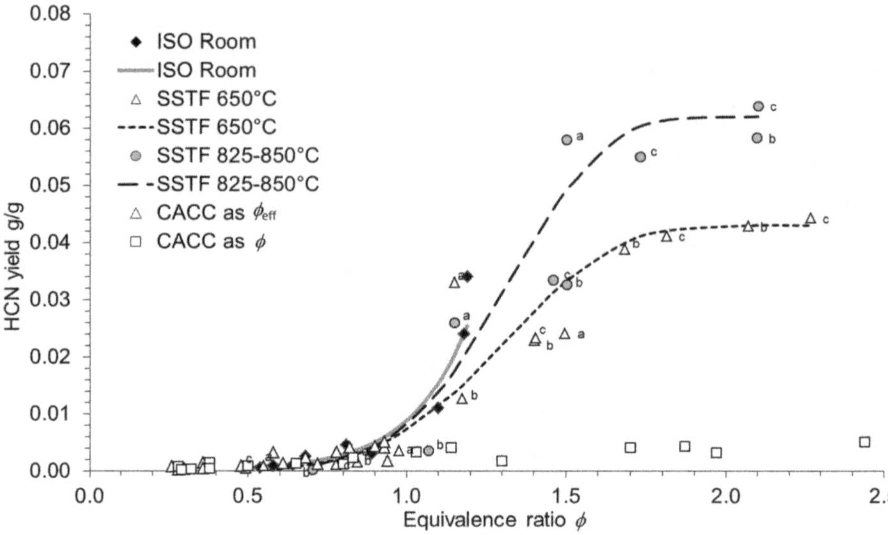

Figure 5.8 Comparison of HCN yield for aliphatic polyamides from the steady state tube furnace and controlled atmosphere cone calorimeter with ISO Room, as a function of equivalence ratio ϕ. Letters indicate data source and material for SSTF: (a) polyamide 6.6,[36] (b) polyamide 6.6,[6] (c) polyamide 6.[37]

Figure 5.8 shows the same ISO room experiments compared for the HCN yield to the steady state tube furnace and the controlled atmosphere cone calorimeter. Again, this shows good agreement between SSTF and large-scale data, alongside the sensitivity of HCN yields from polyamides to the furnace temperature in under-ventilated conditions. This illustrates the failure of the controlled atmosphere cone calorimeter to replicate the conditions of under-ventilated combustion even more dramatically. While the controlled atmosphere cone calorimeter data fit well with the large-scale data up to $\phi = 1$, they do not replicate the higher CO and HCN yields for under-ventilated conditions, where $\phi > 1$. Unfortunately, Tewarson has not reported a similar correlation for HCN yield from PA 6.6 from the fire propagation apparatus. It is apparent that the controlled atmosphere cone calorimeter also fails to replicate the higher yields of the other major asphyxiant, hydrogen cyanide, in under-ventilated fires. Indeed, the ~50 fold increase in HCN yield as the fire condition changes from well-ventilated to under-ventilated is not seen using the CACC. Again, correcting the data to ϕ_{eff} merely shifts it to equivalence ratios below 1.0.

Thus, the steady state tube furnace and fire propagation apparatus can adequately replicate large-scale burning, but the controlled atmosphere calorimeter cannot. Two recent reports of the use of the CACC for assessment of fire toxicity describe experimental programs investigating a range of fuels. In one, six materials, including polyvinyl chloride (PVC), polyisocyanurate foam and a glass-reinforced polyester containing brominated flame retardant, are described,[38] but the only results reported are the HCl yields. In the other paper[39] some carbon monoxide yields are reported for the burning of an undefined "sandwich composite material", which shows higher CO yields (up to 0.07 g g^{-1}) in well-ventilated conditions (the oxygen concentration was 21% and 14%) and much lower (0.03 g g^{-1}) for oxygen concentrations of 0, 5 and 9%. Based on these results, the authors concluded that "no gas phase oxidation can occur below 10% O_2: the formation of CO tends to decrease." This is in direct contradiction with the observations from large-scale fires where CO yields above 0.2 g g^{-1} are common in oxygen concentrations below 10%, presumably because burning in the CACC cannot be sustained at the applied heat flux. Indeed, upper layer oxygen concentrations as low as 1.2% O_2 have been measured.[40] In full-scale compartment fires, oxygen concentrations exceeding 10% would only occur at lower heights where the flames are formed; above the smoke layer they will be forcing combustion to occur at high equivalence ratios (as replicated in the SSTF). It appears that there is a significant body of work necessary before the CACC can replicate the higher yields of toxic products associated with under-ventilated flaming in full-scale fires.

5.3.2 Comparisons Based on Generic Combustion Conditions

For non-constant combustion conditions, although the equivalence ratio cannot be obtained, yields of major toxic products are available and loosely assigned to fire condition. Steady state tube furnace (SSTF), fire propagation apparatus (FPA) and large-scale ISO room fire test data have been compared to yields from the static tube furnace (NFX) and smoke density chamber (SDC) tests for LDPE/PP and PA 6.6.[41]

Figure 5.9 compares the CO yield for LDPE/PP and PA 6.6 in the ISO room, SSTF and FPA in well-ventilated and under-ventilated conditions to those in the NFX and SDC tests. It shows very high CO yields for the well-ventilated conditions for LDPE and PA 6.6 for the NFX test, and better agreement of CO yields for the SDC. For the under-ventilated condition, for LDPE both the NFX and SDC tests underestimate the CO yield, while for PA 6.6 the NFX shows good agreement for CO, while the SDC shows the poorest agreement.

Figure 5.10 compares the HCN yield for PA 6.6 in the ISO room and SSTF in well-ventilated and under-ventilated conditions to those in the NFX and SDC tests. This also shows high HCN yields for well-ventilated conditions in the NFX test, but lower HCN yields for the SDC. In contrast, in under-ventilated conditions, the NFX test shows good agreement with the SSTF, while the SDC fails to replicate the high HCN yields associated with under-ventilated flaming.

Overall, the NFX test overestimates the yields from well-ventilated fires, but underestimates the yields from under-ventilated fires; the SDC replicates well-ventilated flaming, but not under-ventilated fire conditions.

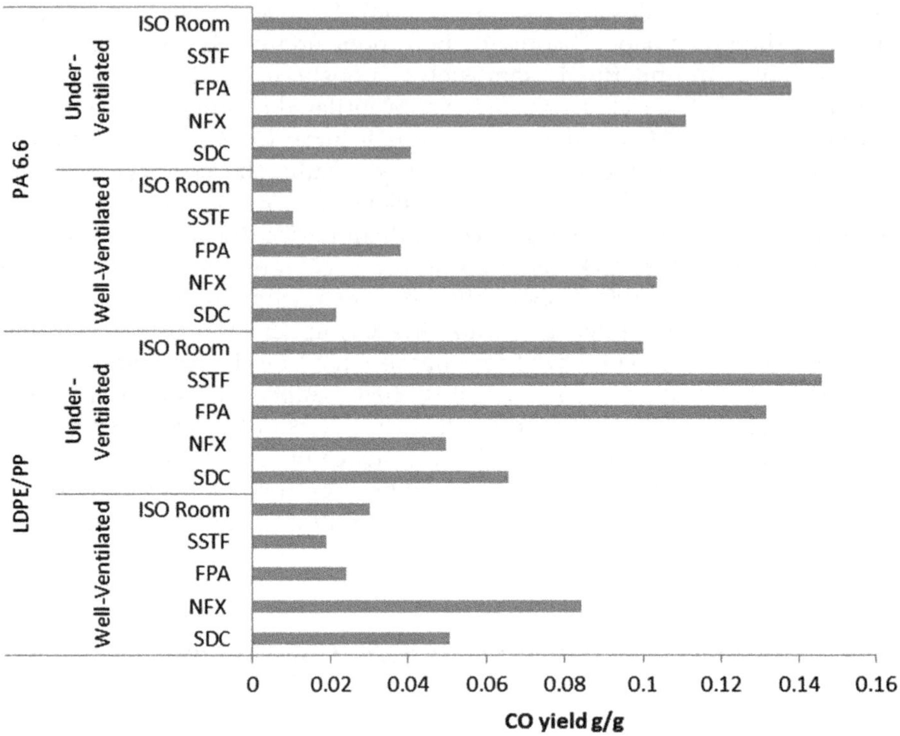

Figure 5.9 CO yield in different laboratory-scale tests and fire conditions.

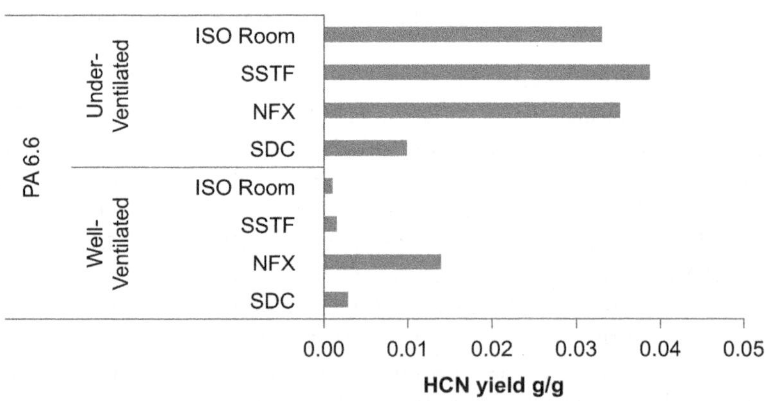

Figure 5.10 HCN yield in different laboratory-scale tests and fire conditions.

5.3.3 Attempted Replication of Under-ventilated Flaming in the Smoke Chamber

The results described in the preceding section led to a further investigation into the capability of the smoke chamber to replicate under-ventilated fires

Table 5.1 Predicted and actual effect of sample thickness on CO yields in SDC.

	Predicted for 100% mass loss			Actual			
Thickness/mm	Fire condition	ϕ Predicted	Corresponding CO yield/g g^{-1}	Mass loss/%	ϕ (Estimated from mass loss)	CO yield/g g^{-1}	O_2 at peak smoke/%
5	Well-ventilated	0.61	0.02	95.7	0.54	0.07	12.0
10	Under-ventilated	1.22	0.08	57.6	0.67	0.04	12.8
15	Under-ventilated	1.83	0.17	30.4	0.54	0.04	14.4

by burning different thicknesses of LDPE.[41] The equivalence ratio was calculated assuming complete combustion and Tewarson's equation[27] was used to predict the CO yield, and these predictions were compared to the actual results, shown in Table 5.1. When the tests were conducted under the most severe regime in the smoke chamber protocol, at 50 kW m^{-2}, as the atmosphere became oxygen depleted, instead of under-ventilated burning, the samples extinguished. Although the thicker samples were 10 and 15 mm, only the upper 5–7 mm actually burnt, giving an equivalence ratio and CO yields corresponding to well-ventilated flaming. Thus, like the controlled atmosphere cone calorimeter, it was not possible to replicate under-ventilated burning in the smoke chamber.

5.3.4 Problems with the Smoke Chamber for Assessment of Fire Toxicity

In the smoke chamber most of the toxic gas production should occur towards the end of the test, when the oxygen is most depleted, provided flaming combustion is maintained. However, it is this stage where it is most difficult to maintain flaming combustion. While the smoke chamber is well-suited to the measurement of visual obscuration during the early stages of flaming, where the smoke can be quantified non-destructively, and is independent of the smoke layer height and distribution, using the smoke chamber to quantify toxic gas generation presents a number of problems.

The ventilation condition, or equivalence ratio, corresponding to the time when the products are generated is unknown, because an unknown fraction of the available oxygen will be accessible to the fire plume. Indeed, the circulation within the chamber will depend primarily on the heat release rate. When sampling a stratified fire effluent from a fixed point, such as a tube 20 cm below the chamber roof, this could be too low, so missing the main effluent plume altogether, or sample from its centre, without dilution. The use of a fan, in some variants of the smoke density chamber standard, disrupts the burning process considerably and results in recirculation of all the fire effluent through the flame zone. The act of sampling the fire effluent, and clean-up prior to gas analysis, will change its composition. If the fire gases are sampled in real time, it may be possible to isolate individual fire stages, but then the effluent may change during analysis, or may not be returned to the chamber after analysis. If the analyte is returned to the chamber, it will usually have been filtered, removing both the particulates and the gases absorbed onto them, to protect the analysers, while acid gases may have deposited onto the sampling line. Similarly, the high surface area of the (steel) chamber wall will promote deposition and disproportionately larger losses of particulates and certain analytes, such as hydrogen chloride (HCl) or hydrogen bromide (HBr). However, of greatest significance in quantification of the toxicity of fire effluents is that the most important under-ventilated conditions cannot be replicated. The apparent inability of the smoke density chamber to replicate under-ventilated flaming renders it unsuitable for the assessment of fire effluent toxicity. Typically, in well-ventilated flaming there would be 0.02 g carbon monoxide per 1 g burning polymer and 0.002 g hydrogen cyanide per 1 g nitrogen-containing polymer, while in under-ventilated flaming there would be typical yields of 0.2 g g^{-1} CO and 0.06 g g^{-1} HCN, *i.e.*, increases of factors of 10 and 30, respectively.

It is most unfortunate that the widespread availability of the smoke density chamber in test laboratories has led to a practice where it is also considered as a prime candidate for assessment of fire toxicity and has already been adopted by the mass transport industries to quantify fire hazards on aircraft, trains and ships. In particular, the failure of the protocol to replicate under-ventilated flaming prevent it from being able to discriminate against those materials that produce very toxic effluents in a developed fire from those producing less toxic effluents. Indeed, the current procedure adopted by the International Maritime Organisation[15] references ISO 19702, the standard relating to the use of FTIR for quantifying fire gas toxicity, and ISO 5659-2,[14] with no mention of how they should be combined.

5.4 Sampling and Pre-concentration

5.4.1 Challenges

Combustion effluents contain a mixture of permanent, unreactive gases [nitrogen (N_2), O_2, CO, CO_2 *etc.*], reactive gases [hydrogen chloride (HCl), hydrogen bromide (HBr), nitric oxide (NO), nitrogen dioxide (NO_2), sulfur dioxide (SO_2) *etc.*], organic vapours, water vapour, and liquid and solid particulate matter (tar droplets and soot particles). Fortunately, around 80% of the volume is occupied by N_2, so changes in the volume of effluent (for example, if the CO_2:CO ratio changes) have only a minor effect on the measured concentrations. Unfortunately, many of the reactive gases, water, organic vapours and particulate matter interfere with the operation of, or even damage, the analytical instrumentation and need to be removed without changing the concentration of the target analyte.

5.4.2 Materials

The ideal materials for minimising interaction with the analyte (particularly acid gases) are glasses and polymers with very low surface energy. The polymer with the lowest surface energy is polytetrafluoroethylene (PTFE), comprising a $-(CF_2-CF_2)_n-$ chain. However, this molecule is very rigid and has such high crystallinity it starts to decompose before melting, and so cannot be extruded (although precipitated PTFE can be sintered into solid objects, which are porous to most gases). Modification of its crystallinity with randomly located C_3F_7-O- side branches produces perfluoroalkoxy (PFA) polymers with lower melting points, so, unlike PTFE these are melt processable, but still have very low affinity for analyte gases. Other alternatives include fluorinated ethylene propylene (FEP) copolymers $-(CF_2-CF_2)_n-(CF_2-CF(CF_3))_m-$ and polyvinylidene fluoride (PVDF) $-(CH_2-CF_2)_n-$. Lower cost silicone rubbers have relatively low affinity for most combustion gases and are also used. Glass also has a low affinity for most combustion effluents, with the exception of hydrogen fluoride. Materials with polar functionality, such as polyamides and PVC, are unsuitable, owing to their ability to selectively adsorb analytes.

For sampling directly from a flame, stainless steel is used most widely. The passivating chromium oxide layer on the surface reduces the surface energy, which, together with the heat, minimises adsorption of analyte onto the tube walls. The materials used in the construction of pumps located before the analyser in the sampling train are also of concern. (It is often a requirement to run a gas analyser at atmospheric pressure, so the pump is located between the sample probe and analyser.) These can include metals, rubbers, polyamides *etc.*, unless the pump has been specifically designed to be in inert and coated with low energy surfaces. If grab sampling is employed, using bags or evacuated vessels, these must be inert to the analyte in question. Bags made from polyvinyl fluoride (PVF; Tedlar) $-(CH_2-CHF)_n-$ are relatively inert to most fire effluents and have low permeability, and are widely used. Glass is the most suitable material for evacuated vessels, although other products have come on to the market recently, such as stainless steel canisters, which are chemically treated to render their surfaces inert.

5.4.3 Heated Lines

Since most combustion effluents contain enough water that cooling to room temperature causes droplets to condense, and most acid gases, such as HCl, HBr and NO_2, are highly soluble in water, it is usually necessary to heat the sampling lines above 100 °C to prevent loss of these analytes. In addition, if the line is to be heated it is essential to ensure that there are no cold spots, where moisture would otherwise accumulate. Commercial heated lines, containing a stainless steel tube lined with PTFE, provide a suitable means of transferring the combustion effluent to analytical instruments, such as an FTIR. However, particular care must be taken to ensure that all ancillary connections, filters, pressure gauges *etc.* are heated to at least the temperature of the sampling line. Temperatures of 150 to 180 °C are generally recommended. Significant losses of acid gases have been found to occur even in heated lines (see Chapter 3).

5.4.4 Filtration

Almost all combustion effluents contain particulate matter that will gradually deposit along the sampling lines and, if it is not removed within the analytical instrumentation, it causes temporary or permanent damage. It is generally necessary to filter all effluents prior to the analysis, but this causes three problems.

1. Many toxic species are absorbed onto the surface of the particles. This can either occur during the free movement of the fire effluent, or once the particulate matter is trapped on the filter. As the filter loading increases, so all gases must pass over the particulate matter to get to the analyser, maximising their chance of adsorption. As aerosol particle toxicants can often travel deeper into the lung than gas molecules, they have significantly fewer collisions with the mucus membrane lining the upper respiratory tract.
2. As the adsorption sites become occupied and equilibrium is established between the particulate and gas phase, so those toxicants may also be released, delaying the response from the flame zone to the analyser. This is particularly problematic for large or non-steady state flaming.
3. If there is further build-up of particulate matter on the filter (often the case when burning sooty materials), this will lead to partial or complete blockage of the filter. If there are any potential sources of leakage in the sampling line downstream of the filter, fresh air may leak in and dilute the combustion gases. If the pump is situated after the analyser, when the filter blocks the pressure may fall well below atmospheric, giving erroneous results.

Typically, in the analysis of fire effluents, particulate filters with a 1 μm pore size are used to protect the analysers. The formation of soot particles is an agglomeration of molecular species, such as polycyclic aromatic hydrocarbons, initially as spherical particles, which then agglomerate like a string of beads, until large enough to be trapped by the filter. When a 3 m heated line was used with a large 1 μm heated cartridge filter at the inlet end, it was also necessary (as evidenced by the particulate deposit) to have a second 1 μm filter at the outlet end of the heated line, as soot particle formation and agglomeration continued to occur after the first filter during transport along the 3 m heated line.

5.4.5 Pre-concentration

Since the undiluted combustion effluent will be around 80% N_2, up to 15% CO_2, 5 to 10% water vapour and up to 10% O_2, and these species are of little toxicological interest, it is often better to remove some of them prior to analysis. For water soluble components, such as acid gases, bubblers are frequently used to extract these components from the effluent. An advantage of bubblers without glass frits is that particulates can also be trapped in aqueous solutions and water soluble components leach out from the particles. It is generally prudent to put two or more bubblers in series to quantify the degree of carryover. It is also necessary to include a filter near the end of the line, since a significant proportion of fine particles is carried through the bubblers and could deposit in the pump or flow meter. It is, of course, essential to accurately determine the volume flowing through the bubblers, since the quantity of an analyte will be directly proportional to the volume sampled.

Volatiles trapped using evacuated glassware can be isolated from permanent gases by condensation using liquid nitrogen. A convenient design of bulb and extractable liquid trap is shown in Figure 5.11. For minor organic toxicants, such as aldehydes, ketones and aromatics, as well as more exotic toxicants, such as halogenated dibenzodioxins and dibenzofurans, polycyclic aromatic hydrocarbons, organophosphates *etc.*, various methods are available for separation from permanent gases. In general, these can be very useful for identification (qualitative analysis) but less valuable for quantification (quantitative analysis) of their concentrations because of the differential selectivity of individual analytes to the sorbent matrix. The methods can involve selective trapping of the analyte mixture on a sorbent for subsequent analysis. The sorbent, such as a non-volatile liquid coated onto an

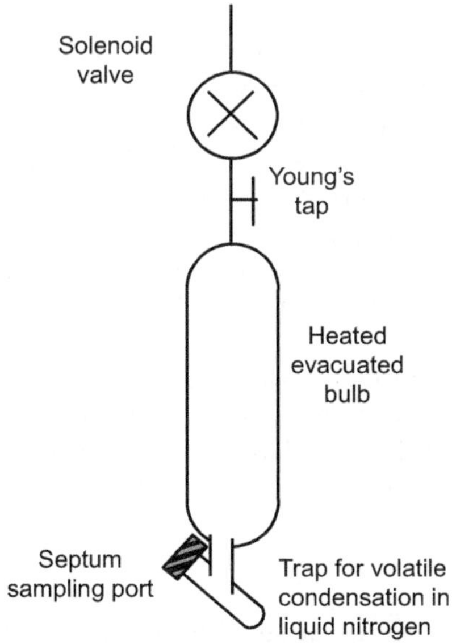

Figure 5.11 Glass vacuum sampling vessel, showing the trap at the base for cooling condensable vapours using liquid nitrogen and the septum sampling port for direct injection to GC-MS.

inert support, is selected for trapping a particular group of compounds and will show variable selectivity for the analytes in question. This effectively means that the trapping of certain compounds will be very effective, while others will be able to pass through the sorbent, so that they will only be detected if present in high concentrations. For this reason these sorption pre-concentration methods are not really suited to quantitative analysis without careful calibration using internal standards.

The choice of sorbent for such pre-concentration has been reviewed elsewhere.[42] These fall into two broad categories: those used with solvent desorption, such as silica gel, activated charcoal, Carboxens® carbon molecular sieve (CMS) materials, porous polymers and carbon molecular sieves; and those used with thermal desorption techniques, such as Tenax TA, Chromosorb 106, graphitised carbons and carbon molecular sieves. Thermal desorption sorbents may be purchased pre-loaded into "sorption tubes" or may be prepared in the laboratory. Commercial thermal desorption instruments then act as the interface for loading into a gas chromatograph (GC) or more often a gas chromatograph-mass spectrometer. The obvious disadvantage of single sorbents may be offset to some extent by using multi-bed tubes, containing sections of a few different sorbents.[43]

5.4.6 Solid-phase Micro-extraction

Solid-phase micro-extraction (SPME) combines sampling and pre-concentration of analytes for direct transfer to GC-MS. Volatiles pass over a short length of fine fibre coated with sorbent liquid protruding from a special syringe needle. The fibre is retracted and the end of the syringe may be capped. The syringe is then "injected" through the inlet septum of the GC into the heated zone of the injector, where the volatiles are desorbed directly into the carrier gas stream. Improved separations may be achieved by cryogenic focusing inside or outside the GC instrument.

SPME affords good sensitivity from an easy to use, lightweight and compact device, generally giving a linear signal/concentration response. However, it is less suitable for lower volatility analytes, is very susceptible to sampling and injector temperatures, and the sample is relatively unstable (>1 h) for storage.

The effects of environmental variables, such as the presence of water vapour, on sampling and recovery are uncertain. However, SPME appears most useful in the analysis of canister samples or the head-space analysis of solvent-extracted sorbent samples.[44]

5.4.7 Particulates

Particular sampling problems arise when quantifying the particulates present in fire effluents. Unlike most fire gases, whose concentration is frozen at a certain temperature, particulates continue to agglomerate, from PAHs able to pass through a filter, to soot particles that are trapped. Even at ambient temperatures, the number of particles is decreasing while their size is increasing. Thus, the distance from the fire source will affect the particulate concentrations and particle size distributions. Further, the gas sampling speed will also affect the particle size distribution, such that isokinetic sampling should be used to ensure that all particle sizes are sampled equally.[45]

5.5 Quantification

Permanent gaseous combustion products can generally be quantified continuously in real time using a combination of commercially available non-dispersive infrared (NDIR) analysers (CO_2, CO and hydrocarbons), electrochemical cells (CO and O_2) and paramagnetic analysers (O_2), giving direct readouts of concentration based on a two point ("zero and span") calibration. Their outputs are generally easy to log using suitable analogue to digital converters interfaced to a personal computer. Other species are generally analysed off-line using analyte-specific methods, and frequently collected in bubblers for subsequent analysis. For example, hydrogen cyanide can be quantified by derivatisation, forming a purple compound by reaction with chloramine-T, which can be detected spectroscopically, avoiding interferences. Water soluble acid gases may be determined titrimetrically, or using high performance ion chromatography (HPIC). Specific aldehydes and other oxygenated products may be quantified in bubbler solutions using derivitisation methods coupled with high performance liquid chromatography (HPLC).

In contrast to the continuous analysis of permanent gases, where the concentration is measured directly, the total volume of effluent flowing through the bubblers must be accurately known and used in subsequent calculations to determine the gas phase concentrations in the fire effluent.

5.5.1 Fourier Transform Infrared Spectroscopy

FTIR has been hailed as the all-embracing solution to fire effluent analysis. The Fourier transform technique allows for short scanning times, improving the temporal resolution. The acquisition of complete spectra every few seconds provides the opportunity to record all the compounds present in a fire effluent as a function of time. The use of a heated sampling line and heated gas cell is essential to keep water in the vapour phase and avoid losses of water soluble gases. The gas phase infrared spectra of most small molecules are rather elegant, almost symmetrical patterns. Figure 5.12 shows both the fundamental vibrational frequency and the rotational fine structure of HCl. Note the way the lines at the high energy end (greater than 3000 cm^{-1}) are closer together, while the lines at the low energy end are furthest apart, resulting from centrifugal distortion of the HCl molecule at high rotational energies. At a sufficient resolution, this gives every molecule a unique spectrum from which, in isolation, most molecules could be easily

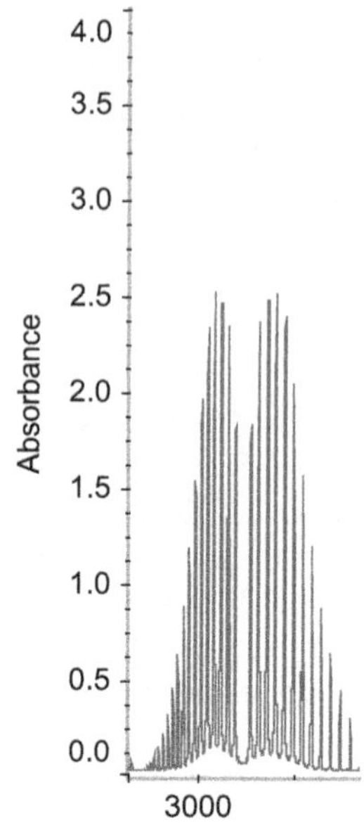

Figure 5.12 FTIR spectra of HCl showing the vibrational band composed of the rotational fine structure (satellite peaks of around 25% of the full height result from absorption by $H^{37}Cl$).

identified. This makes it ideal for industrial process gas analysis, where five to ten target analytes are programmed into the system and there is a degree of confidence that other gases will not be present.

Unfortunately, combustion effluents contain a large number of different molecules, present at very different concentrations, and so are usually dwarfed by water and carbon dioxide. Even at 200 to 300 °C, water does not exist as a single molecular species H_2O, but as its dimer $(H_2O)_2$, trimer $(H_3O)_3$, tetramer $(H_4O)_4$ and so forth. Each will have a unique spectrum. The result is an intense noisy region in the spectrum, which overlaps many toxicologically significant analytes. Worse, combustion effluents usually contain mixtures of known and unknown compounds. The latter are also likely to interfere with the analysis of the former.

In addition, despite the faster scanning times of modern FTIR spectrophotometers, they are rarely adequate for capturing crucial toxicological information. Sampling and analysis of gases by FTIR requires a good understanding of the problems of gas sampling and analysis, and the nature of the FTIR technique. A schematic of a simple sampling system is shown in Figure 5.13.

In order to benefit from the detailed fine structure of the spectra of individual toxicants, it is essential to record spectra at 0.5 cm^{-1} resolution or better. Most of this fine structure is lost at 4 cm^{-1}. Further, it is necessary to use long (folded) path length gas cells (typically 4 m path length, of volume

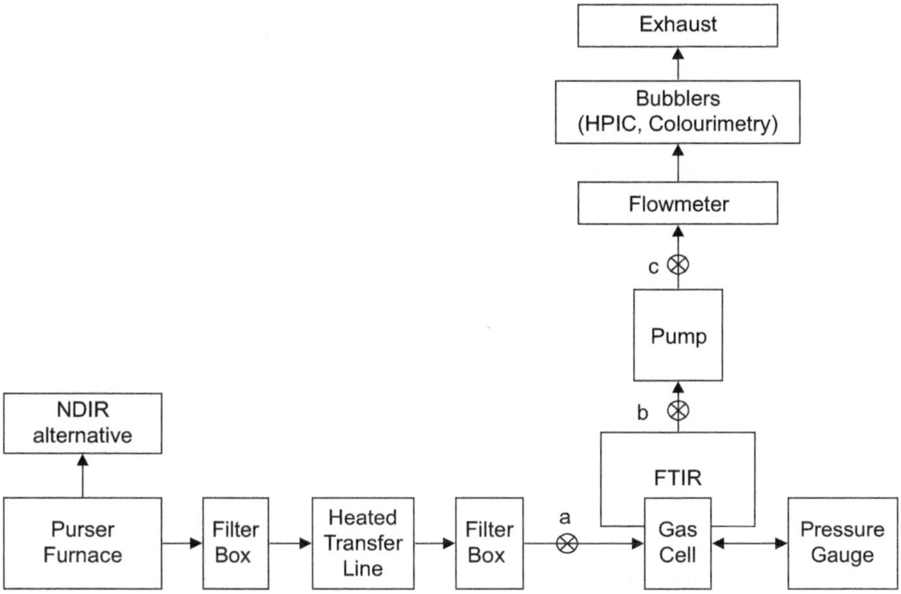

Figure 5.13 Schematic of FTIR gas sampling system and complementary analysis.

around 1 L) and high sensitivity mercury cadmium telluride (MCT) detectors (requiring cooling with liquid nitrogen).

Quantification is aggravated by the non-linear response of the infrared absorption to analyte concentrations for most toxicants. These problems have been compounded by the lack of investment in data analysis software for gas phase spectra by the major FTIR instrument manufacturers. One estimate suggests that it takes a fire laboratory 6 months to calibrate an FTIR instrument for the major gases quantified in fire effluents. Some commercial enterprises offer off-site calibration services for FTIR gas cells, for which there is no guarantee of validity within the laboratory FTIR instrument, particularly as the signal deteriorates through use. Others offer a service where the entire instrument (gas cell and FTIR spectrophotometer) are calibrated off-site before delivery.

An example spectrum is presented in Figure 5.14, below a diagram showing overlapping regions. There are many other molecules with absorptions of infrared within the same region. Looking at Figure 5.14 it is evident that HCl and HBr spectral regions overlap, as well as water overlapping parts of the acrolein, NO_2 and HCN spectra.

Figure 5.15 shows a simple example of the spectra of a combustion effluent obtained by burning Nylon 6.6 in the steady state tube furnace at 650 and 750 °C under well-ventilated conditions. Peaks for water, CO_2, CO, methane (CH_4) and HCN are all present in sufficient intensity to be clearly visible. Other toxicants are hidden behind the stronger peaks of the major components.

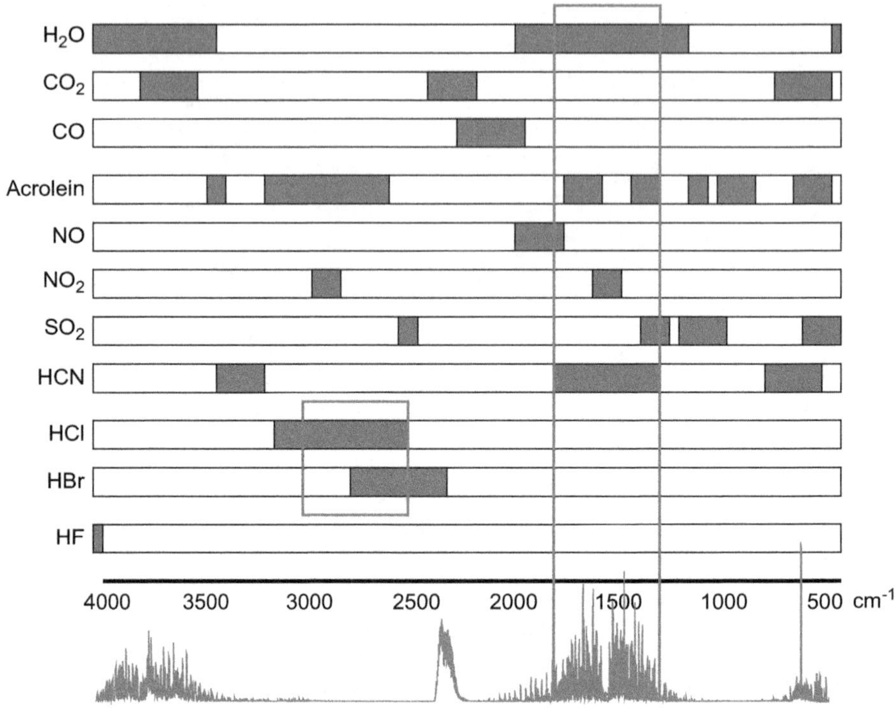

Figure 5.14 Top: absorption regions from different species present in fire gases; bottom: the gas phase spectrum of CO_2 and water, showing regions of overlap.

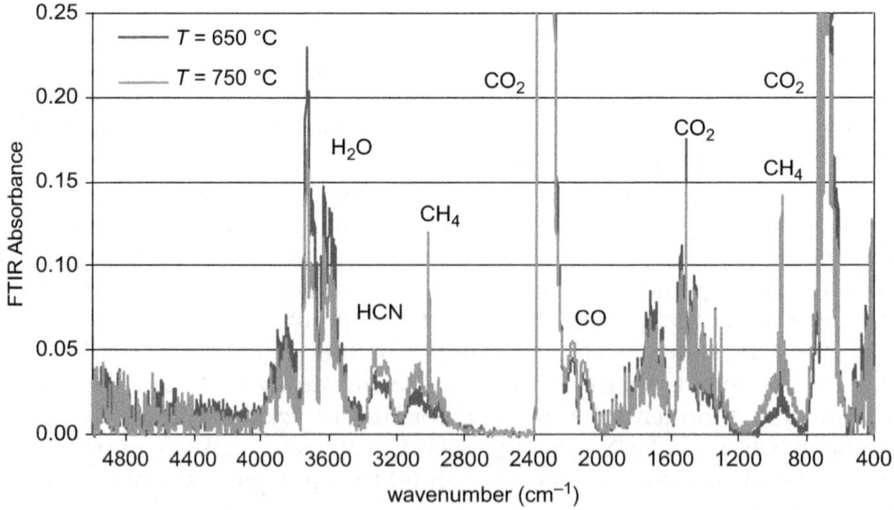

Figure 5.15 Nylon 6.6 products burned in the SSTF at 650 °C and 750 °C.

5.6 Conclusions

Fire effluent toxicity is the largest killer and greatest cause of injury in fires. For the purposes of fire hazard assessment, fire toxicity must be determined as a function of both the material and of fire conditions, particularly temperature and oxygen availability in the fire zone. The problem of assessing the fire toxicity can thus be divided into three components, or stages: generation, sampling and analysis.

A significant part of this chapter reviews the bench-scale methods available for generating fire effluent, so as to replicate real fire behaviour. There is a limited amount of large-scale data that reports toxic product yields as a function of equivalence ratio (or other quantifiable measure of ventilation condition). The steady state tube furnace and the fire propagation apparatus show acceptable agreement with large-scale data over the range of fire conditions. The controlled atmosphere cone calorimeter and the smoke density chamber fail to replicate the high yields of asphyxiant gases, CO and HCN, which occur in under-ventilated fires (the fire stage most responsible for injury and death). This shows that they are generally unsuitable for assessment of toxic hazards of fire effluents. The non-dynamic tube furnace replicates the toxic product yields intermediate between well-ventilated and under-ventilated flaming, seemingly independent of the pyrolysis temperature. Only the SSTF and FPA are capable of generating data under controlled conditions, allowing the relationship between equivalence ratio and the toxic product yields to be investigated in order to provide suitable inputs to engineering based assessment of toxic fire hazard.

Since fire effluents will contain around 80% nitrogen and a cocktail of other components, several techniques have been proposed for sampling and analysis of fire effluents. Unfortunately, used in isolation, none of these provide a satisfactory solution, although combined use can eliminate many of the drawbacks of the individual methods. The aim of this section of the chapter was to ensure that the reader is aware of the potential pitfalls before embarking on an extended programme of work.

The analysis of fire effluents also presents a number of challenges and no single technique is suitable for robust analysis of all the toxic components in a fire effluent (although some are certainly being marketed as though this were not the case)! Although FTIR has great potential, the spectrum must be recorded using a long (4 m) path length and a heated gas cell with resolution of at least 0.5 cm^{-1}. Then, only with the help of sophisticated, user-friendly software (or time-consuming visual inspection) can the spectra be deconvoluted in order to quantify the individual toxic products present.

References

1. K. T. Paul, Feasibility Study to Demonstrate the Potential of Smoke Hoods in Simulated Aircraft Fire Atmospheres: Development of the fire model, *Fire Mater.*, 1989, **14**, 43–58, DOI: 10.1002/fam.810140203.

2. ISO 19701:2013 Methods for sampling and analysis of fire effluents.
3. ISO 19702:2006 Toxicity testing of fire effluents – Guidance for analysis of gases and vapours in fire effluents using FTIR gas analysis.
4. ISO 29904:2013 Fire chemistry – Generation and measurement of aerosols.
5. T. R. Hull, R. E. Quinn, I. G. Areri and D. A. Purser, Combustion toxicity of fire retarded EVA, *Polym. Degrad. Stab.*, 2002, **77**, 235–242, DOI: 10.1016/S0141-3910(02)00039-3.
6. A. A. Stec, T. R. Hull, K. Lebek, J. A. Purser and D. A. Purser, The effect of temperature and ventilation condition on the toxic product yields from burning polymers, *Fire Mater.*, 2008, **32**, 49–60, DOI: 10.1002/fam.955.
7. ISO 19706:2007, Guidelines for assessing the fire threat to people.
8. G. H. Markstein, Radiant emission and smoke points for laminar diffusion flames of fuel mixtures, *Symp. (Int.) Combust.*, 1988, **21**, 1107–1114.
9. Y. Alarie, Toxicity of fire smoke, *Crit. Rev. Toxicol.*, 2002, **32**, 259–289, DOI: 10.1080/20024091064246.
10. D. A. Purser, Toxic product yields and hazard assessment for fully enclosed design fires, *Polym. Int.*, 2000, **49**, 1232–1255, DOI: 10.1002/1097-0126(200010)49:10<1232::AID-PI543>3.0.CO;2-T.
11. W. M. Pitts, The global equivalence ratio concept and the formation mechanisms of carbon monoxide in enclosure fires, *Prog. Energy Combust. Sci.*, 1995, **21**, 197–237, DOI: 10.1016/0360-1285(95)00004-2.
12. ISO 13571:2012 Life threat from fires – Guidance on the estimation of time available for escape using fire data.
13. T. R. Hull and K. T. Paul, Bench-scale assessment of combustion toxicity-A critical analysis of current protocols, *Fire Saf. J.*, 2007, **42**, 340–365, DOI: 10.1016/j.firesaf.2006.12.006.
14. ISO 5659-2:2012 Plastics – Smoke generation – Part 2: Determination of optical density by a single-chamber test.
15. 2010 Fire Test Procedure Code, Maritime Safety Committee, (MSC 87/26/Add.3) Annex 34, Part 2 Smoke and Toxicity Test, International Maritime Organisation, London.
16. EN 45545-2:2013 Railway applications. Fire protection on railway vehicles. Requirements for fire behaviour of materials and components.
17. NF X 70-100, Analysis of pyrolysis and combustion gases. Tube furnace method. Part 1, Methods of analysis of gas generated by thermal degradation. Part 2, Method of thermal degradation using tube furnace.
18. ISO 12136:2011, Reaction to fire tests – Measurement of material properties using a fire propagation apparatus.
19. ISO/DIS 19700:2013 Controlled equivalence ratio method for the determination of hazardous components of fire effluents – the steady state tube furnace.
20. ISO 5660-1:2002 Fire tests – Reaction to fire – Part 1: Rate of heat release from building products (cone calorimeter method).
21. A. A. Stec, T. R. Hull and K. Lebek, Characterisation of the steady state tube furnace (ISO TS 19700) for fire toxicity assessment, *Polym. Degrad. Stab.*, 2008, **93**, 2058–2065, DOI: 10.1016/j.polymdegradstab.2008.02.020.

22. EN 2826:2011 Aerospace series – Burning behaviour of non metallic materials under the influence of radiating heat and flames – Determination of gas components in the smoke; ABD 0031 Fire-Smoke-Toxicity (FST) Test Specification (Airbus Industries); Boeing BSS 7239, Test method for toxic gas generation by materials on combustion.
23. B. Schartel and T. R. Hull, Development of fire-retarded materials – interpretation of cone calorimeter data, *Fire Mater.*, 2007, **31**, 327–354.
24. V. Babrauskas, W. H. Twilley, M. Janssens and S. Yusa, Cone calorimeter for controlled-atmosphere studies, *Fire Mater.*, 1992, **16**, 37–43.
25. M. Christy, R. Petrella, and J. Penkala, Controlled-atmosphere cone calorimeter, *Fire and Polymers II: Materials and Tests for Hazard Prevention*, 1995, vol. 599, pp. 498-517.
26. J. Hietaniemi, R. Kallonen and E. Mikkola, Burning characteristics of selected substances: Production of heat, smoke and chemical species, *Fire Mater.*, 1999, **23**, 171–185, DOI: 10.1002/(SICI)1099-1018(199907/08)23:4<171::AID-FAM680>3.0.CO;2-C.
27. A. Tewarson, Generation of heat and chemical compounds in fires, *SFPE Handbook of Fire Protection Engineering*, Quincy, MA, 3rd edn, 2002, pp. 3–82.
28. Y. LeTallec, D. A. Smith, J. Hunter, and F. R. Groenfeld, Evaluation of toxicants for materials used in military vehicles and equipment, 10th International Fire Science and Engineering Conference (Interflam), 2004, pp. 631–637.
29. ISO 13344:1996, Estimation of lethal toxic potency of fire effluents.
30. P. Blomqvist and A. Lonnermark, Characterization of the combustion products in large-scale fire tests: Comparison of three experimental configurations, *Fire Mater.*, 2001, **25**, 71–81, DOI: 10.1002/fam.761.
31. B. Andersson, F. Markert and G. Holmstedt, Combustion products generated by hetero-organic fuels on four different fire test scales, *Fire Saf. J.*, 2005, **40**, 439–465, DOI: 10.1016/j.firesaf.2005.03.002.
32. J. Heitaniemi, R. Kallonen and E. Mikkola, Fires at Chemical Warehouses – A cone calorimeter study on the burning characteristics and fire effluent composition of selected chemical compounds, VTT Research Notes 1810, VTT, Espoo, Finland, 1997.
33. V. Babrauskas, W. J. Parker, G. Mulholland and W. H. Twilley, The phi meter: A simple, fuel-independent instrument for monitoring combustion equivalence ratio, *Rev. Sci. Instrum.*, 1994, **65**, 2367–2375, DOI: 10.1063/1.1144690.
34. ISO 9705:1993, Fire tests – Full-scale room test for surface products.
35. T. R. Hull, A. A. Stec, K. Lebek and D. Price, Factors affecting the combustion toxicity of polymeric materials, *Polym. Degrad. Stab.*, 2007, **92**, 2239–2246, DOI: 10.1016/j.polymdegradstab.2007.03.032.
36. P. Blomqvist, T. Hertzberg and H. Tuovinen, A small-scale controlled equivalence ratio tube furnace method – Experience of the method and link to large scale fires, *Proceedings of the 11th International Interflam Conference*, Interscience Communications Ltd, London, 2007, pp. 391–402.

37. D. A. Purser and J. A. Purser, HCN yields and fate of fuel nitrogen for materials under different combustion conditions in the ISO 19700 tube furnace and large-scale fires, *Fire Saf. Sci.*, 2009, 1117–1128, DOI: 10.3801/IAFSS.FSS.9-1117.
38. C. Gomez, M. Janssens, and A. Zalkin, Measuring yields of toxic gases from materials during different stages of fire development, *Proceedings of the 12th International Conference on Fire and Materials*, Interscience Communications, 2011.
39. D. M. Marquis, E. Guillaume, A. Camillo, M. Pavageau, T. Rogaume, Usage of controlled-atmosphere cone calorimeter to provide input data for toxicity modeling, *Proceedings of the 12th International Conference on Fire and Materials*, Interscience Communications, 2011.
40. D. A. Purser, J. A. Rowley, P. J. Fardell and M. Bensilum. *Fully Enclosed Design Fires For Hazard Assessment In Relation To Yields Of Carbon Monoxide And Hydrogen Cyanide Interflam '99 Proceedings*, 1999, vol. 2, pp. 1163–1169.
41. K. Kaczorek, MPhil Thesis, University of Central Lancashire, 2009.
42. M. R. Ras, F. Borrull and R. M. Marce, Sampling and preconcentration techniques for determination of volatile organic compounds in air samples, *TrAC, Trends Anal. Chem.*, 2009, **28**, 347–361.
43. M. Harper, Sorbent trapping of volatile organic compounds from air, *J. Chromatogr. A*, 2000, **885**, 129–151, DOI: 10.1016/S0021-9673(00)00363-0.
44. K. Elke, E. Jermann, J. Begerow and L. Dunemann, Determination of benzene, toluene, ethylbenzene and xylenes in indoor air at environmental levels using diffusive samplers in combination with headspace solid-phase microextraction and high-resolution gas chromatography-flame ionization detection, *J. Chromatogr. A*, 1998, **826**, 191–200.
45. ISO 29904:2013 Fire chemistry – Generation and measurement of aerosols.

CHAPTER 6

Dispersion of Fire Plumes in the Atmosphere

DAVID. J. HALL AND ANGELA M. SPANTON*

Envirobods Ltd, Stevenage, Hertfordshire, UK
*Email: amspanton@gmail.com

6.1 Introduction

One of the main routes of exposure of the work force and general populace to material from fires is from the dispersing plume of fire products and other related material released by the fire. Its behaviour and the distributed concentration of materials from the fire as it disperses are thus critical to exposure estimates. The dispersion, resultant distribution and exposure of the workplace, populace and the natural environment to air pollutants in general are long-studied and highly researched subjects. That of exposure to fire products is rather less so; fire plumes have specific, rather specialised properties that distinguish them from the more general run of dispersing contaminants and which directly affect their dispersion.

In view of limited space, this chapter concentrates on these specific properties of fire plume dispersion and how they affect local exposure. There already exists a large literature on more general aspects of dispersion to which the reader is referred for essential basic information. A brief description of basic dispersion and a bibliography of useful introductory sources are given in the Appendix.

Issues in Toxicology No. 23
Toxicology, Survival and Health Hazards of Combustion Products
Edited by David A. Purser, Robert L. Maynard and James C. Wakefield

Published by the Royal Society of Chemistry, www.rsc.org

The main distinctions of fire plumes and their dispersion are:

- Fire sources are normally at or near the ground and commonly associated with buildings, arising from flush exits on the building surface or from within partially destructed structures.
- There are varying amounts of associated heat in the fire plume, which will probably pass through a cycle from initiation to growth and then exhaustion or suppression. At the maximum in the combustion cycle the heat release may be very large, often limited only by the availability of oxygen. Generated heat in the range 100–1000 MW is perfectly possible in large fire plumes (see Chapters 2, 3 and 18). The heat generates a buoyant rise in the plume, which can be very large. Because of their usually low source height, buoyant rise is critical in fire plumes as it directly affects the plume concentration locally at the ground and, consequently, the exposure of those in the vicinity of the source and the surrounding area.
- The initial surface area of a fire plume source is usually relatively large and its size and distribution directly affect the plume behaviour.
- The high temperatures usually associated with fires result in large associated ambient air density differences, which can generate high local airspeeds around or within the fire. There may also be mechanical damage to structures, internal collapse and internally generated explosions. This combination of high levels of internally generated mechanical energy and high local air velocities can generate significant levels of initial momentum in the plume discharge.
- The high local air velocities, collapsing structures and local small explosions can also eject quite large pieces of material, 'large particles', into the fire plume, which may not be a direct part of the combustion process. They can have a size far in excess of conventional combustion products and can be dispersed over significant distances in the fire plume.
- Fire plumes are often not coherent; parts of them may detach from the whole and disperse separately.
- It is usually difficult to estimate with any accuracy the heat release associated with a fire plume, which is critical to calculating its buoyant rise. Similarly, it is usually difficult to estimate the 'source term', the rate of emission of any toxic or debilitating contaminants. Both of these parameters are critical to estimating the level of exposure around the source. They are dealt with in other chapters of this book.

Most of these features of fire plumes require some specific treatment as they are not usually dealt with by conventional dispersion models. They are discussed here in some detail.

In general, factors such as the size and area distribution of fire sources and the manner of their release, directly from the ground or associated with buildings, control the initial conditions of the fire plume. They are important as they control the initial spreading of the plume, and its initial buoyant rise, in the area near the source where plume concentrations are high and exposure is at its greatest. At long distances these source term effects tend to disappear as the plume eventually forgets the detailed conditions of its release. Plume buoyancy from the fire heat release is, however, normally conserved indefinitely. Its effects thus persist for much longer distances, though eventually even these also disappear.

6.2 Fire Plume Buoyancy Scaling and Basic Properties

It is helpful to understand the relationship between wind speed, scale and heat release in fire plumes and the self-similarity in their behaviour that results. The main governing parameter of fire plume behaviour in the atmosphere is the unitless 'dimensionless buoyancy flux', which is defined as

$$\frac{F}{U^3 L} \tag{6.1}$$

where L is a suitable length scale (for example, the height of an associated building or the area of the fire source) and U is a reference wind speed (typically taken at building height or some other fixed height). F is the 'flux of buoyancy' at the source, defined as

$$g\frac{\Delta\rho}{\rho_a}\frac{V}{\pi} \tag{6.2}$$

where g is the gravitational acceleration. ρ is the density of the fire gas $\Delta\rho/\rho_a = (\rho_a - \rho)/\rho_a$, which is the density difference of the fire gases relative to ambient (ρ_a) and V is the volume rate of release of the fire gases. π appears in eqn (6.2) as a scaling factor from the original definition of eqn (6.2) by Briggs,[1] based on his dimensional analysis.

The 'flux of buoyancy' F has units of $m^4\ s^{-3}$ and is essentially the rate of discharge of the buoyant force in the plume, the larger it is the greater the buoyancy in the plume. Conveniently for the present discussion, it is possible to define F in terms of an equivalent heat release Q (in MW), where for ambient air and conventional combustion products

$$F = 8.9Q \tag{6.3}$$

where F is in metric units of $m^4\ s^{-3}$ as above. The dimensionless buoyancy flux, eqn (6.1), is ubiquitous (sometimes in disguised form) in fluid mechanical flows involving buoyancy and is a base parameter in Briggs' plume rise equations (which are in almost universal use).

Basically, any similar arrangement of a buoyant discharge will have a common appearance and dispersion pattern for a similar value of the dimensionless buoyancy flux. A plot of the relationship between heat release, wind speed and scale is shown in Figure 6.1, taken from Hall and Spanton,[2] for a length scale (in this case an approximate associated building height) of

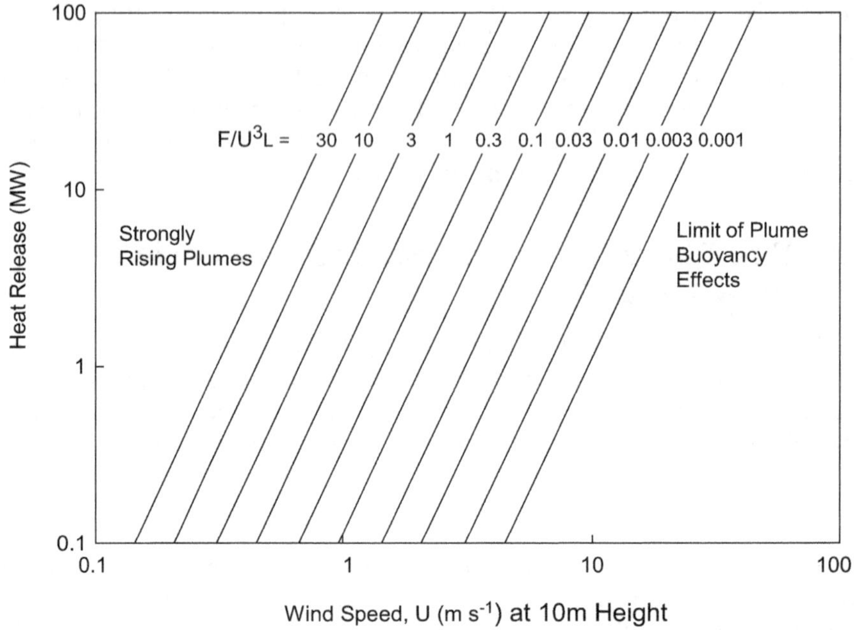

Figure 6.1 Relationship between scale, heat release and wind speed for buoyant plume scaling. Based on a characteristic scale (the building height) of 10 m. (For comparison, the 50th percentile wind speed at a height of 10 m in the central UK is about 3 m s^{-1}, the 90th percentile is about 10 m s^{-1} and the 99th percentile is about 20 m s^{-1}.)
Taken from Hall and Spanton, ADMLC/2003/1.[2]

10 m. Each line for a constant value of the dimensionless buoyancy flux on the plot represents a range of wind speeds and source buoyancies for which a plume of similar behaviour, relative dispersion and appearance results. Since the definition of the dimensionless buoyancy flux has U^3 in its denominator, fire plume behaviour, particularly its buoyant rise, is very sensitive to wind speed. The buoyant rise falls approximately in proportion to the increase in wind speed.

In terms of the effects of plume behaviour, the results of small scale wind tunnel experiments have shown that the lowest value of the dimensionless buoyancy flux in Figure 6.1, 0.001, corresponds to ground-based plumes showing no significant buoyancy effects; they are not distinguishable from neutrally buoyant plumes. The highest value of the dimensionless buoyancy flux, 30, corresponds to plumes with high buoyancy, which normally (though not always) rise very strongly from the source. Note that this range of conditions covers a dimensionless buoyancy flux range of 30 000 : 1. Note also in Figure 6.1 that the most buoyant release conditions require a combination of low wind speed and large heat release.

Some sense of the appearance and behaviour of fire plumes with different buoyancy levels can be seen in Figures 6.2, 6.3 and 6.4. Figure 6.2, from

Figure 6.2 Kerosene pool fire from the field experiments of Yates.[3] Pool about 18 m×12 m, heat release about 72 MW. Wind speed 5 m s^{-1} at 10 m height. Dimensionless buoyancy flux (based on a 10 m reference height) approx. 0.5.
(Taken from A.N. Yates, Dispersion of Smoke Plumes from Pool Fires. Ph.D. Thesis, University of Greenwich, 1996.)

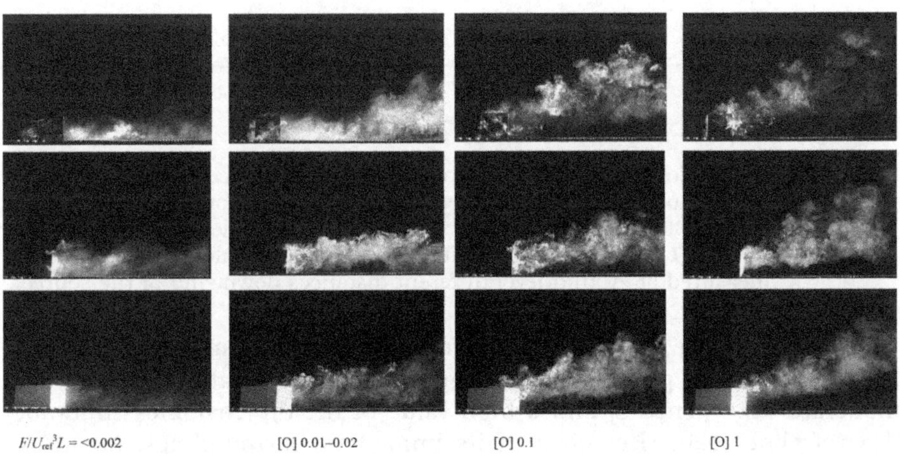

Figure 6.3 Flow visualisation of buoyant plumes discharged from a building surface. From small scale wind tunnel experiments of Hall and Waters.[4] Upper row: discharge from upwind face of building. Centre row: discharge from downwind face of building. Bottom row: discharge from downwind face of building skewed at 45° to the flow direction. Values of the dimensionless buoyancy flux (based on a 60 m building height) appended.
Crown copyright held by The National Archives, Department of Trade and Industry, 1970–1974. Reference: FV 12/87.

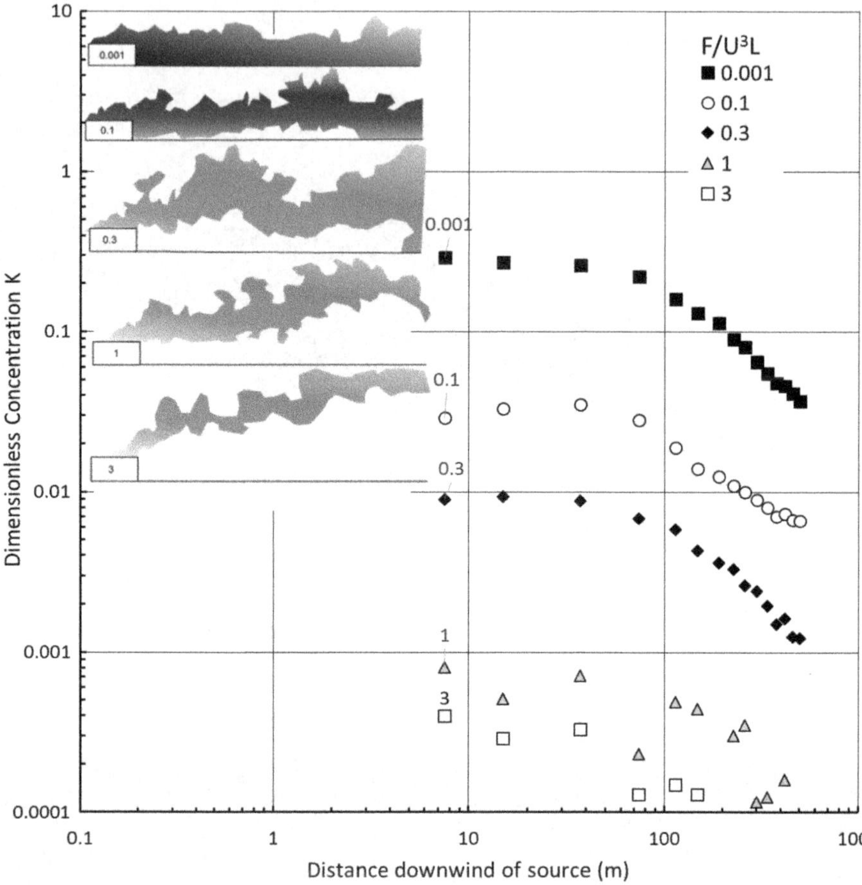

Figure 6.4 Illustrations of buoyant plumes rising from the centre of small scale wind tunnel experiments representing a warehouse building shell, 30 m × 30 m × 10 m height, at five levels of dimensionless buoyancy flux F/U^3L (from Figure 6.1) with plots of dimensionless concentration K, measured at ground level different distances downwind of the source.

Yates,[3] shows the smoke plume from an experimental ground-based kerosene fire. The plume, with a large heat release (of about 72 MW) in a moderate wind (5 m s^{-1}) and a high value of the dimensionless buoyancy flux (of about 0.5), rises clear of its immediate surroundings. Figure 6.3 shows visualised plumes released from different faces of a large (60 m × 120 m × 60 m high) building in reduced scale wind tunnel experiments by Hall and Waters.[4] Local downwash and enhanced dispersion due to the presence of the building reduced the plume rise immediately downwind for buoyancies at the lower values of the dimensionless buoyancy flux. However, at the highest buoyancy fluxes (above about 0.1), the buoyancy forces in the plume tended to overcome the building downwash and the plume rose clear of the building wake. The buoyant rise was then comparable to that of the pool fire in open terrain in Figure 6.2. Buoyant plume interactions with the wind flows around buildings can be complex. An interesting feature in the right hand images in Figure 6.3 is the effect of the strong trailing vortex shed from the skewed upwind edge of the building, a common feature of building flows. The vortex is similar to the trailing vortices shed from aircraft wing tips. At the highest plume buoyancy the trailing vortex has entrained the plume, but the plume's own high buoyancy has lifted the vortex with it. The pronounced swirling of the plume could be clearly seen in the experiments. Figure 6.4, illustrates (inset) the distribution pattern of plumes of varying buoyancy rising from the core of a small warehouse building (30 m $\times$ 30 m $\times$ 12 m height) whose roof has collapsed, leaving the building shell containing the fire. These were also based upon data from small-scale wind tunnel experiments.[5] The plume in the upper image is neutrally buoyant, with the buoyancy increasing in the lower images. It can be seen that, in this case, the approximate boundary condition for the plume starting to leave the ground, minimising exposure levels to the plume at the ground around the building, is for a dimensionless buoyancy flux of about 0.3, with the images for the two higher levels of dimensionless buoyancy flux, 1 and 3, showing it lifting clear of the ground.

Figures 6.2, 6.3 and 6.4 also show clearly the essentially unsteady nature of the dispersion. In a steady wind the concentration pattern in the plume has stable time-averaged properties, thus for example, the distribution of contaminant concentration across the plume both laterally and vertically is usually close to the Gaussian form. However, there are large short term variations in concentration at any point in the plume. This is a normal property of dispersion in the atmosphere, but is not normally calculated in most conventional numerical dispersion models.

6.3 Fire Plume Dispersion Scaling for Contaminant Concentration

Fire plumes disperse in the atmosphere in essentially the same way as all other dispersing material, and scale in the same way. In neutrally buoyant releases, for a specific release arrangement the relative dispersion pattern in the atmosphere is fixed if the meteorological conditions (wind speed and direction, surface roughness and atmospheric stability) are constant. Contaminant concentrations within the plume are then related to the contaminant emission rate, wind speed and scale by the dimensionless concentration K, defined as

$$K = \frac{CUL^2}{Q} \tag{6.4}$$

where C is the concentration in some part of the plume, U is the wind speed at some reference height as before, Q is the rate of emission of the pollutant and L is a suitable length scale for the process, as before.

The distribution of values of K through the plume remains the same, irrespective of the contaminant release rate, wind speed and scale of the release.

If this is rearranged to give

$$C = K\frac{Q}{UL^2} \tag{6.5}$$

it can be seen immediately that the concentration in a plume is directly proportional to the emission rate of the contaminant Q, and inversely proportional to the wind speed U and the square of the scale L. The distribution of values of K within the plume can be found by measurement (in full scale or in small scale wind tunnel experiments for example) or by calculation using dispersion models. Dispersion models usually apply eqn (6.5) directly to the calculation for fixed values of the scale, wind speed and contaminant emission rate to provide concentration patterns for specific conditions. However, in doing so they mask the essential relationships between the plume release scale and meteorological conditions described here.

In buoyant fire plumes this relationship still holds for specific values of the dimensionless buoyancy parameter, which fixes the shape and rise of the plume. These basic parameters hold good for full scale buoyant plume discharges in similar conditions, and in both numerical and small scale models of fire plumes. It is in this way that data from small scale wind tunnel experiments, from which most of the examples given here are derived, is converted to full scale conditions. They thus form a useful basic way of defining the behaviour and contaminant dispersion patterns in fire plumes.

6.4 Effects of Source Buoyancy, Shape and Size on Fire Plume Dispersion

Most fires occupy some significant surface area either on the ground (as in Figure 6.2), or are associated with a burning building (as in Figures 6.3 and 6.4). The area may be quite large, as in a burning field or forest fire for example (though these quite often burn in strips), or on a whole industrial site. Alternatively, they may be associated with liquid pools of moderate area from hydrocarbon liquid spillages or a burning mixture from a fractured pipe or valve, which will have a limited, relatively small, source area. Fires arising in buildings may go through a complex cycle of development, during which their effective source size may vary considerably. Dispersion and heat release from fires arising in buildings are discussed in Section 6.5.

The effect of plume buoyancy on fire plume dispersion is illustrated here with experimental data from small scale wind tunnel experiments. Figure 6.4 shows values of dimensionless concentration K measured at the ground in the small scale wind tunnel experiments at different values of dimensionless buoyancy flux (F/U^3L) from 0.001 to 3, for which the plume dispersal patterns are illustrated. The substantial decrease in concentration at the ground with increasing buoyancy (by a factor of about 1000) shows the importance of plume height in the near field.

The size and shape of fire sources significantly affect their dispersion and buoyant rise, so need some further consideration. Increasing the source area of a discharging plume generally reduces concentrations at short distances, since the released material is initially spread more widely across the wind. At longer distances, beyond about twenty source widths across the wind, these source effects start to disappear and the source conditions become less important to the dispersion. Many conventional numerical models will account to some degree for the effect of increased source size on dispersion in this way. However, there is also a marked effect of source size on buoyant plume rise, which is also critically important to contaminant exposure in the near field. This is accounted for less commonly.

The effect of source size and shape on buoyant plume rise and dispersion from different rectangular sources was measured by Hall and Walker.[6] These experiments included square sources of different sizes and rectangular sources of different lengths, spread either along or across the wind. An array of plume tracer gas concentration sampling points was placed along the ground and laterally and vertically at two downwind distances. Plots were then made of the height of the centre of the plume at the nearest distance of vertical measurements for varying width or length, whichever is the greater, for three source buoyancies. The results showed a limited difference between plumes from square sources or spread across the wind over the same width; both showed a marked and broadly similar reduction in plume height with increasing lateral spread. Contrastingly, the data for sources spread along the wind show a different behaviour, with a small but slight increase in the plume height for increasing source length. The reasons for the latter behaviour, for sources spread along the wind, has yet to be fully explained. The same effect was also observed in some cases in Hall *et al.*'s[5] experiments. In the Hall and Waters data,[4] one explanation is simply that some of the buoyant discharge was further from the measurement point, allowing for further plume rise over this extra distance; however it does not seem to be a sufficient explanation for the quite marked effects shown in some cases. One possibility is that the longitudinal vorticity that develops in nearly all buoyant plumes is enhanced in some way in these cases, thus enhancing the plume rise.

The effects on plume rise of distributing sources laterally can be very marked in some cases. Figure 6.5 shows illustrations of plume dispersion at a dimensionless buoyancy flux of 0.3 for a discharge from the roof of a model of a warehouse in which the discharge was through a varying number of roof lights assumed to have burned through in the course of a fire. Discharging the plume though one roof opening produced a plume rising clear of the building. Redistributing the plume through 15 separate openings spread across the wind largely destroyed the plume rise.[5] This behaviour is to some extent predictable by treating the individual discharges as separate plumes

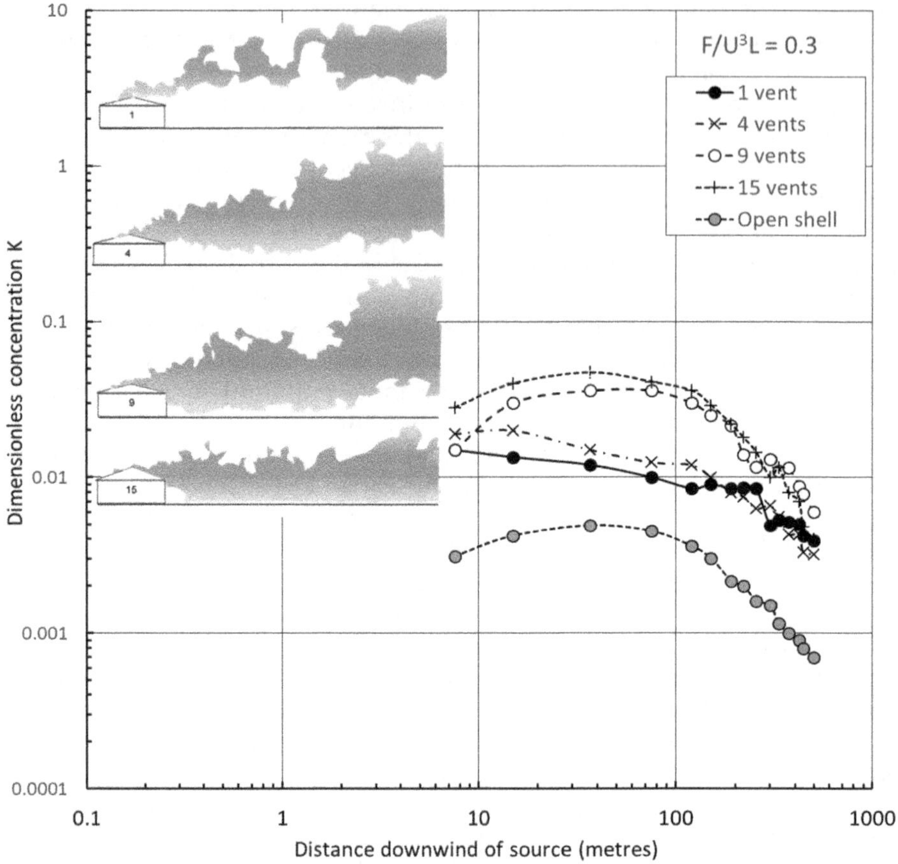

Figure 6.5 Illustrations of buoyant plumes distributed across varying numbers of roof openings from small scale wind tunnel experiments representing a 100 m wide warehouse building with up to 15 roof openings, set normally across the wind, with plots of dimensionless concentration K measured at ground level different distances downwind of the source. Dimensionless buoyancy flux, $F/U_{ref}^3L = 0.3$.

(for which the individual dimensionless buoyancy flux was about 0.02. However, this is not done in most numerical fire modelling, where the plume is usually treated as a single entity.

Figure 6.5 also shows related values of dimensionless concentration *K* at the ground for the illustrated plumes. As the number of vents and thereby the distribution of the fire plume across the building increases so the rise of the plume decreases and the concentrations at ground level increase.

This effect of number of vents on plume dispersal occurs consistently for small buildings but can be more varied with larger buildings such as those illustrated in Figure 6.5. For larger buildings the dispersal pattern is affected by the additional lateral dispersion in the building wake as the discharge is distributed laterally. This progressive lateral spread occurs with large but not small buildings, where complete lateral mixing always occurs. With the large building the additional reduction in concentration due to increasing the number of discharge openings may counterbalance the increases in concentration due to reduced plume rise. For a fully open building shell the plume dispersal dynamics are somewhat different (not similar to a large number of discrete vents). Figure 6.5 includes a case for a fully open building shell, where the ground level concentrations are lower even than those for the single vent case.

These interactions can lead to different effects on plume dispersal at different discharge buoyancies. Figure 6.6 shows the effect of distributing the fire plume across the building for three discharge buoyancies, values of F/U^3L of 0.001 (neutrally buoyant), 0.1 and 0.3 (the latter value matching the buoyant plume visualisation images of Figure 6.5). For the neutrally buoyant release, the ground level concentrations closer to the

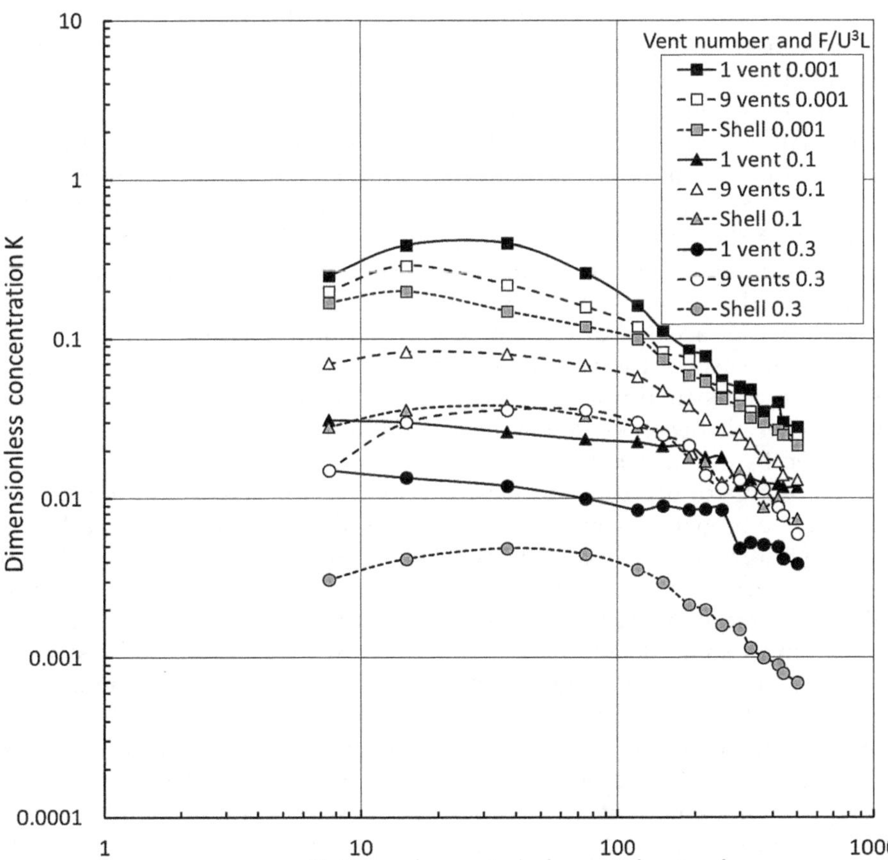

Figure 6.6 Effect of three different dimensionless buoyancy fluxes, $F/U_{ref}{}^3L = 0.001$, 0.1 and 0.3 on plume dispersal for the 100m wide warehouse building illustrated in Figure 6.5 with up to 15 Roof Openings and open shell.

source are in fact higher for the single vent and decrease with the number of vents, although the effect is relatively small and reduces to a small difference in concentration at the longest range of the measurements. At 0.1 and 0.3 buoyancy flux values there is a significant increase in ground level concentration with increasing vent number, with quite a large effect for the highest plume buoyancy, increasing concentrations at the ground by about a factor of five when redistributing the discharge from a single opening to all fifteen openings across the wind. This difference remained over the whole range of downwind distances of the measurements. These effects of the relationships between building size, number of open vents and buoyancy are discussed in reference 5.

6.5 Fire Plume Heat Release from Buildings

The heat content of fire plumes depends on both the availability of combustible material and of combustion air to allow the combustion to occur. It is common for fires to start in closed buildings, where the fire development may be controlled by the building ventilation, which limits the amount of oxygen available for combustion. In the initial stages of a fire, the available combustion air may be limited to that obtained by ventilation through small gaps in the structure or limited ventilation openings (see Chapters 2 and 3). The full progression of fires normally damages the structure of the containing building, initially by burning out roof lights (common in warehouses) or doors and windows. This process allows an increase in the ventilation rate and of the fire combustion rate and heat release, further accelerating the process of building damage and the subsequent increase in the ventilation rate. Eventually, there may be major damage to the structure by the roof or wall(s) collapsing, when access to combustion air becomes largely unlimited and the rate of combustion is controlled by other factors.

It is possible to use conventional building natural ventilation practice to estimate the probable heat release from a closed or partially closed building. Building natural ventilation is driven by a mixture of buoyancy forces, due to internal/external temperature differences, and wind pressures on the building external surfaces. ASHRAE[7] covers the fundamentals of building ventilation in some detail. Typically, buoyancy-driven ventilation occurs at low wind speeds, above which wind-driven ventilation becomes dominant. The nominal wind speed at which this change occurs (gradually) varies somewhat with the building shape, height and the degree to which it is sheltered from the wind by surrounding structures, but is typically around 2–4 m s^{-1}. The same ventilation processes occur with fires in buildings, but the very much larger internal/external temperature differences that occur inside buildings with fires (hundreds of degrees rather that tens of degrees in conventional ventilation) result in buoyancy-driven ventilation remaining dominant up to quite high wind speeds.

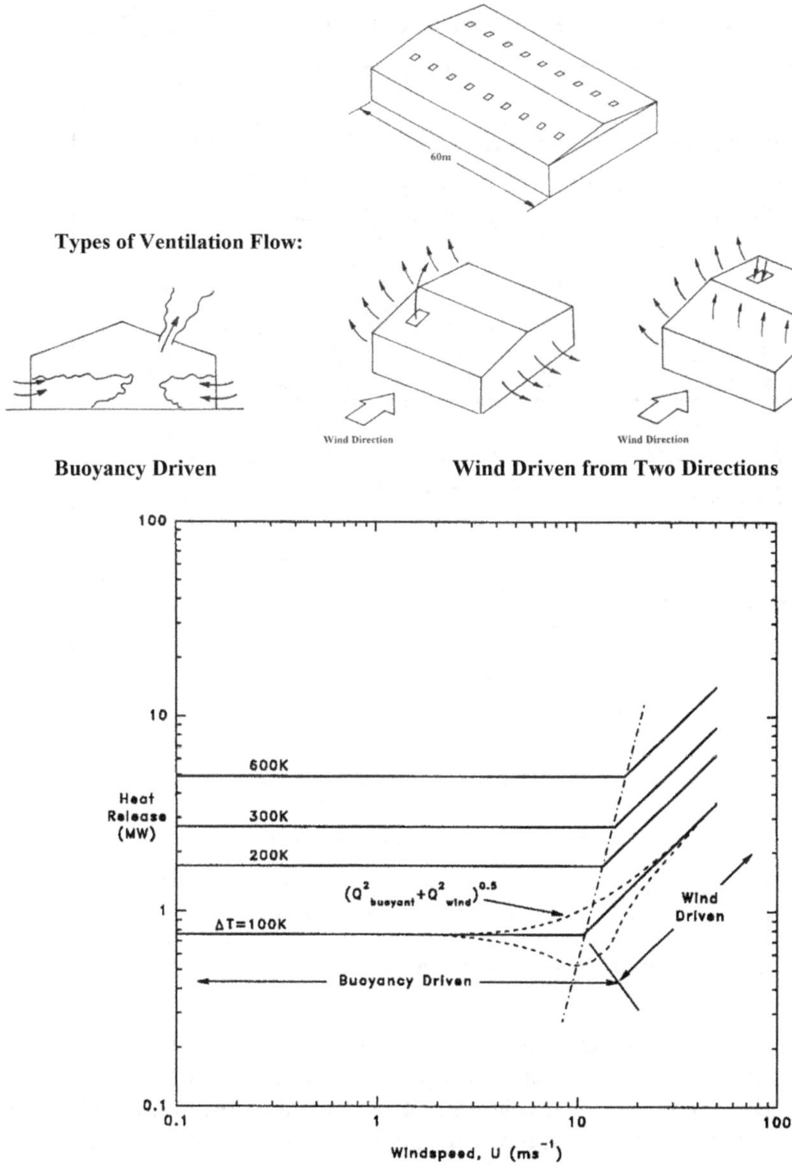

Figure 6.7 Upper: types of ventilation flow from a 60 m × 30 m × 12 m high warehouse with a single roof light burned through. Lower: calculated heat release at varying internal temperatures.

Results of an example calculation of the heat release from a warehouse with plastic roof lights burning through are shown in Figures 6.7 and 6.8. Details of the calculations are given in Hall *et al.*[5] Figure 6.7 shows the calculated heat release for different internal temperatures in a specific

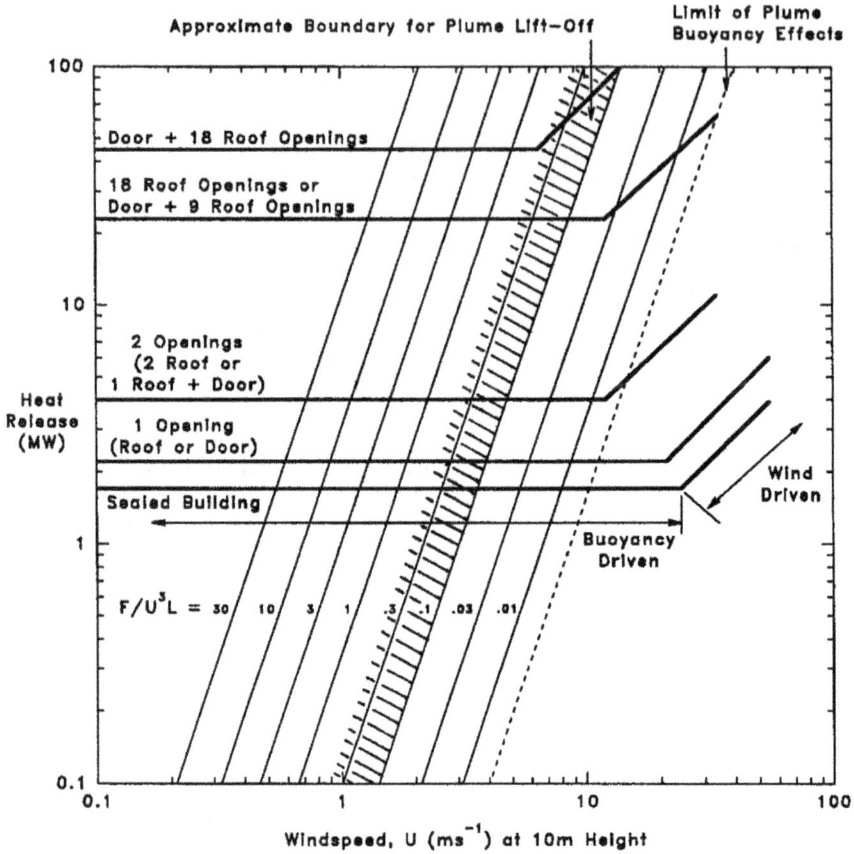

Figure 6.8 Estimated upper bounds of heat release for various plume discharge conditions in varying wind speeds. For a warehouse 60 m × 30 m × 12 m height with 16 plastic roof lights, burning through in progression, and an assumed fire gas temperature of 300 K above ambient.

condition of a warehouse with one roof light burned through. Typical buoyancy- and wind-driven ventilation patterns are shown in the top of the figure and the calculated heat release in the bottom of the figure. The heat release for a fixed internal temperature is constant at low wind speeds as the wind plays little part in the ventilation. At higher wind speeds the ventilation rate becomes approximately proportional to the wind speed once wind-driven ventilation dominates. Ventilation in the intermediate state, where both buoyancy- and wind-driven ventilation act, can be variable depending on whether the two ventilation flows reinforce or oppose one another. The broken lines on the calculated ventilation for a 100 K internal/external temperature difference show the sort of variation in the ventilation rate that can occur here. The calculated heat releases in Figure 6.7 are quite modest, typically 1–3 MW until the temperature differences exceed 300 K. Also notable in Figure 6.7 is the relatively high wind speed, around 10 m s^{-1} (around the 90th percentile UK wind speed at 10 m height), beyond which wind-driven ventilation becomes dominant.

Figure 6.8 shows the result of further calculations for the same scenario, but for a fixed internal/external temperature difference of 300 K with a building passing through steadily increasing stages of destruction. It can be seen that a fair degree of destruction of the building is necessary before a high heat release occurs. With the 'sealed' building (ventilated through gaps, cracks and small ventilators) or with only a few openings, the calculated heat release is limited to about 2 MW.

6.6 Buoyant Fire Plume 'Lift-off'

There has been some interest in the literature in the conditions for which 'lift-off' occurs. This is the boundary condition of buoyancy for plumes released on the ground or on buildings lifting clear of the building surface. This is important in limiting the exposure to the plume near its source. The concept was originally proposed by Briggs in an unpublished note in about 1974 and was investigated experimentally by a number of other workers. Briggs' original definition was in the form of a dimensionless 'lift-off' parameter L_p, at the point of the plume lifting from the ground, defined as

$$L_p = g\frac{\Delta\rho}{\rho}\frac{H}{u_*^2} \tag{6.6}$$

where H is a vertical dimension associated with the plume and u_* is the friction velocity of the atmospheric boundary layer.

Briggs suggested a critical value for L_p for 'lift-off' of around 20–30. It will be noted that eqn (6.1) and (6.6) are not the same: the dimensionless buoyancy flux, eqn (6.1), is a term related to the plume source conditions, describing the whole plume dispersion pattern, and Briggs' eqn (6.6), a form of Richardson's number, is related to specific conditions at some point along the plume. Nonetheless, they must have related values corresponding to the 'lift-off' condition. Hall *et al.*[8] related Briggs' L_p to a value of the dimensionless buoyancy flux for conditions associated with a building of height H and found an equivalent value of the dimensionless buoyancy flux of about 0.1–0.3 to Briggs' value of 20–30 for L_p. This is shown as a shaded region in Figure 6.8. Reference to the visualised plumes in Figures 6.3 and 6.4 shows that this estimate is close to a 'lift-off' condition, where the lower edges of the plumes can be seen to be just clearing the ground. It does not, however, correspond to a sharp boundary in exposure levels at the ground. Figure 6.9 shows measured plume concentrations at the ground at different fixed distances downwind of plumes emitted from buildings; the upper plot is for the 60 m×120 m×60 m high building, visualised in Figure 6.3, and the lower plots are for the warehouse

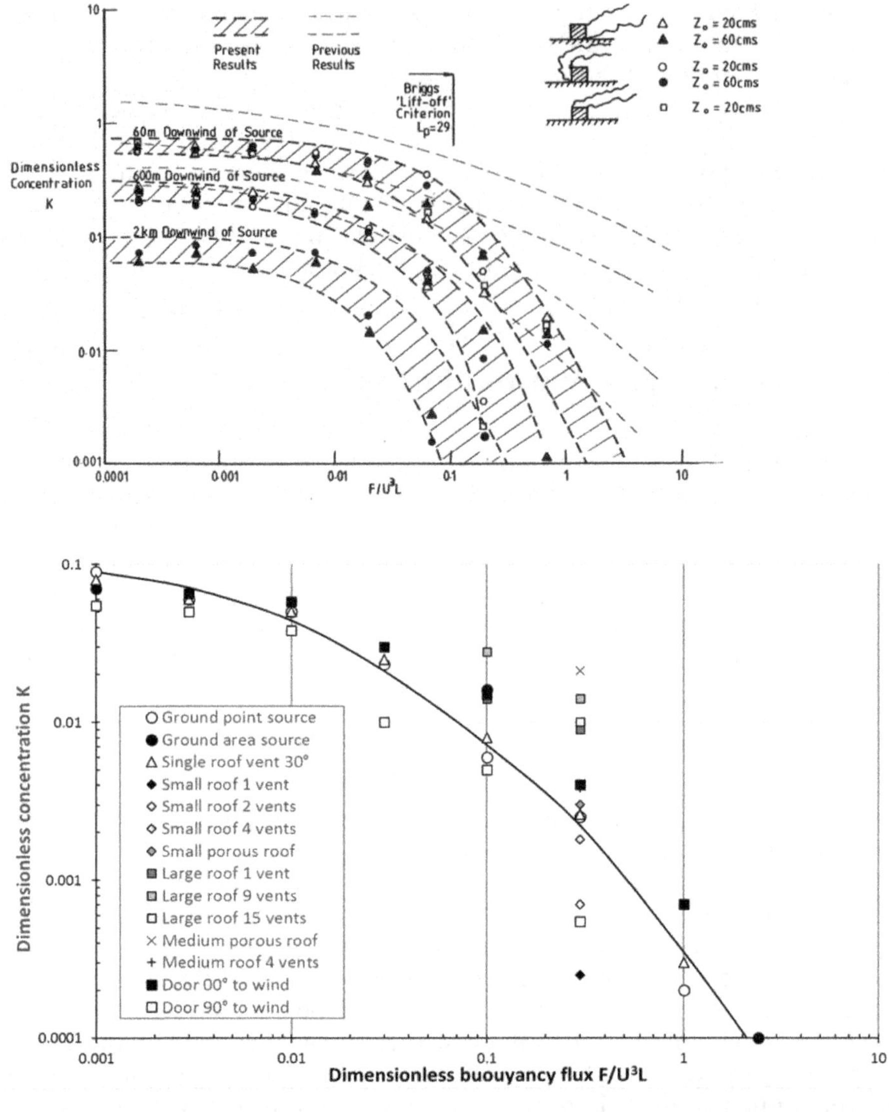

Figure 6.9 Variation in maximum plume concentrations on the ground at fixed downwind distances with the dimensionless buoyancy flux from releases at the ground and on buildings matching the visualised plumes of Figures 6.3, 6.4 and 6.5.
From Hall and Waters.[9] Copyright held by The National Archives, Department of Trade and Industry, 1970–1974, reference FV 12/87 n[2].

model plumes visualised in Figure 6.4 and Figure 6.5.[5] It can be seen in Figure 6.9 that there is no step change in concentration at the ground as the 'lift-off' criterion is passed. However, it does correspond to a marked change in the effect of the dimensionless buoyancy flux on concentration at the ground. Below the 'lift-off' criterion the effect of plume buoyancy on concentrations at the ground is limited, showing small effects of plume buoyancy. Above it there is a significantly more rapid reduction in concentration at the ground with increasing plume buoyancy. The same criterion seems to apply approximately to area sources at the ground. The lower part of Figure 6.9 shows that for all sources ground concentrations are similar and high for buoyancy fluxes up to around 0.01–0.03. Between 0.03 and 0.3 the ground concentrations decrease considerably, but are sensitive to various source characteristics such as source area, the number of roof vents, size and orientation of the building relative to wind direction and the presence and orientation of open doors. At higher buoyancy fluxes of around 1 the ground concentrations tend to be lower and somewhat less variable. Plume heights from Hall and Walker's data[6] for all their sources was analysed on the basis of a dimensionless buoyancy parameter by using the buoyancy flux per unit width in eqn (6.1). This produced the results shown in the upper part of Figure 6.10, which shows a plot of plume centre height at a fixed distance against dimensionless buoyancy flux per unit width of the emission. On this basis of analysis there is a common form of the data, with a clear break in plume centre height from the ground at values of the dimensionless buoyancy flux above about 0.1, but some effect around 0.01–0.03. In practice the plume centre must be at a significant height before the lower part of the plume starts to clear the ground. This did seem to correspond again to a dimensionless buoyancy flux per unit width around 0.1–0.3. The trend lines for different geometries show that plume rise increases most for long buildings (orientated in the same direction as the wind), somewhat less for square buildings and still less for wide buildings.

There has been a further analysis of the 'lift-off' criterion in relation to releases associated with buildings by Hanna *et al.*,[9] who applied it to buoyant releases on buildings. Their final correlation was of the form

$$\frac{cU_{\mathrm{ref}}R^2}{q} = \frac{\exp\left(-6\left(\frac{F}{U_{\mathrm{ref}}{}^3 W}\right)^{0.4}\right)}{\left(0.037 + 0.03\left(\frac{x}{H}\right)^2 + \left(\frac{F}{U_{\mathrm{ref}}{}^3 W}\right)^2\left(\frac{x}{H}\right)^4 + \left(\pi\frac{\sigma_y\sigma_z}{R^2}\right)\right)^{1/3}} \tag{6.7}$$

where c is the concentration at the ground, W is the building width, H is the building height and R is a length scale, taken for buildings as $H^{2/3}W^{1/3}$. σ_y and σ_z are the standard deviations of the plume Gaussian concentration distribution [see Appendix, eqn (A6.1)]. The wind speed reference U_{ref} is at the nominal building height. It is also assumed that $U_{\mathrm{ref}}/u_* = 10$, so that the surface is moderately rough.

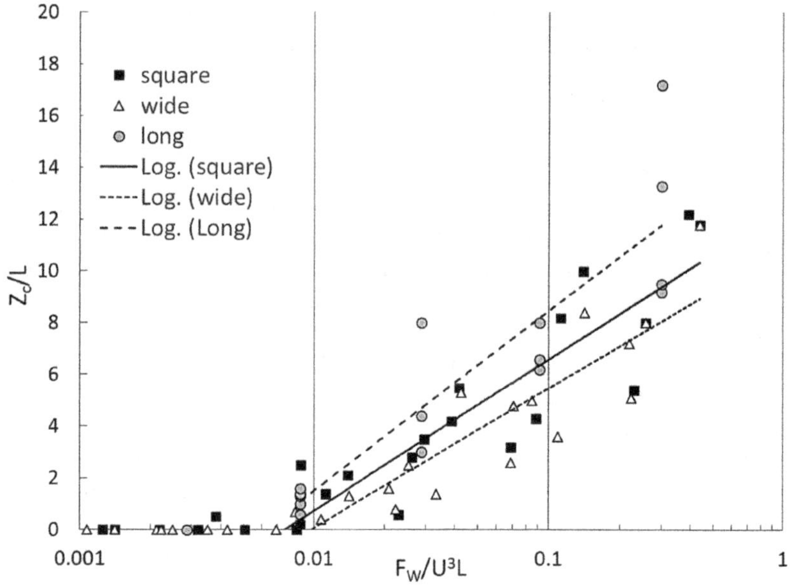

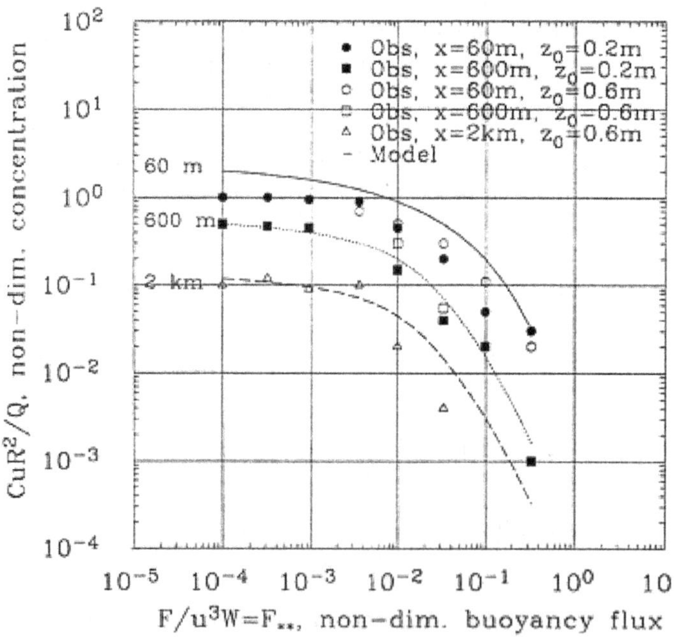

Figure 6.10 Examples of standardised methods of estimating fire plume rise from area sources at the ground and on buildings. Upper: plume height from ground-based sources of varying shape, reduced to a Common dimensionless buoyancy flux based on buoyancy flux per unit width. Lower: application of Hanna *et al.*[14] (eqn (6.7)) to Hall and Water's Data.[4]

For releases at the surface in open terrain, eqn (6.7) could be reduced to

$$\frac{cU_{\text{ref}}}{q} = \frac{\exp\left(-6\left(\frac{F}{U_{\text{ref}}{}^3 W}\right)^{0.4}\right)}{\left(\left(\frac{F^{1/3}x^{2/3}}{U_{\text{ref}}}\right)^6 + \left(\pi\sigma_y\sigma_z\right)^3\right)^{1/3}} \tag{6.8}$$

In both eqn (6.7) and (6.8), there is a dependence on the width of the source, which seems to be necessary as noted in the discussion earlier.

A comparison by Hanna *et al.*[9] of eqn (6.7) with Hall and Waters data[4] (shown in Figure 6.9 in its original form) is shown in the lower plot of Figure 6.10, where the comparison with the experimental data is quite reasonable.

6.7 Fire Plume Dispersion in Urban Areas

There is very little information in the literature on the effects of urban areas on the dispersion of buoyant fire plumes either from theory or from experimental data. Despite this lack of any positive direction on the behaviour of fires in urban areas, it is possible to speculate on some of the effects based on what is known of the main characteristics of urban flow and dispersion. These depend on the urban array area density (the plan area occupational density). For low urban area densities (below about 15–20%), the individual building wakes can still be identified, so that the plume entrainment is probably similar to that for individual buildings, but within a modified wind field. The main effects are firstly reduced wind speeds and increased sheltering within the urban building array and, secondly, high levels of vertical mixing and enhanced turbulence up to about twice the mean urban array height. At higher area densities, above about 20%, the individual building wakes start to merge, wind speeds near the ground reduce considerably and there is a more ready transfer of material laterally from one building wake to another. There has been a recent review of urban dispersion and wind fields by Robins and Macdonald,[10] and earlier by Hall *et al.*[11]

This combination of wind sheltering and rapid vertical mixing in urban areas provides opposing forces, respectively encouraging and inhibiting buoyant plume rise from the ground or from building surfaces. Wind sheltering reduces local wind speeds and encourages plume rise, while the rapid vertical mixing within the urban canopy (up to 2–3 building heights) encourages their more rapid vertical dispersion back down to the ground. The same dichotomy was discussed by Hall *et al.*[8] and by Hall and Waters[4] in relation to plume rise associated with isolated building wakes.

Overall, with plumes interacting with single building wakes, the effect of the building in the available experiments has been to reduce plume rise and increase concentrations at the ground compared with a ground-based source of the same size in open terrain. This implies that the rapid lateral and vertical mixing in the building wake reduced the relative buoyant force in the enlarged plume in excess of any enhancement in plume rise due to the reduced local wind speed. Further, the region of the building from which the plume was discharged (as distinct from its width across the wind) had little effect on concentrations at the ground.

There seems no particular reason why this behaviour of buoyant plume interaction with single buildings should be different in principle within an urban array, though the typical wind flow patterns within the array suggests that some of these characteristics will be enhanced. Firstly, the array will probably encourage initial lateral spreading of the plume beyond any associated single building, thus tending to reduce the plume rise. Secondly, the higher levels of turbulence within the building canopy, compared with those due to a single building, will enhance the initial plume dispersion, further inhibiting plume rise and encouraging generation of a low buoyancy partitioned fraction of the plume that remains at the surface. It also seems probable that lift-off would be delayed by this behaviour and a residual element of the plume would remain at the surface as surface roughness increased; Section 6.10 discusses this latter behaviour in more detail.

6.8 Particle Dispersion in Fire Plumes

Most fires are associated with the production of a large number of particles, which are discharged in the fire plume and the question of how they disperse in comparison with the hot gaseous plume, the main subject of consideration so far, must be considered. The problem is understood, but not particularly well investigated or described.

There are, broadly, two types and size ranges of particles discharged in fire plumes. The first are the direct products of combustion or partial combustion, for example, smoke, fumes (condensed particles from vaporised solid materials) and their agglomerated products. These are generally of quite small size, with aerodynamic diameters from sub-micron up to perhaps 10–20 μm. The second are particles generated by mechanical processes occurring during fires, the collapse of structures or small explosions that eject particles and small items of debris into the fire plume; these particles may be of the raw materials of the fire, from the building's structure or partially combusted materials. The smaller end of the size range of these mechanically produced particles can be of the micrometre size, similar to those of the fire combustion products from fine powders stored in the fire area or dust from collapsing masonry for example, up to quite large pieces of material. The core velocity of a large fire plume can be relatively high due to the high buoyancy forces usually involved; for example, the plume gas velocities through the warehouse building roof openings discussed in Section 6.4 were estimated to be around 10 m s^{-1} for a fire plume temperature of 300 K above ambient. Thus, any particle with a falling speed below this can be carried out of a building in the fire plume. This can include quite large pieces of material of low density; practical examples of large particulate materials from major fires that have deposited on to the local populace include pieces of charred wood (some still alight), asbestos, burned roofing felt, construction materials and charred plastic (from storage bags) up to a centimetre across, contaminated with toxic agrochemicals.

It is easiest to examine the dispersion of particles in fire plumes in terms of their inertial properties. To first order, these can be described in terms of the falling speed of the particle v_f. This is normally described in terms of the particle aerodynamic diameter *d*, which is the diameter of the sphere of unit density that has the same falling speed as the particle in question, whatever its shape or effective density. A plot of particle falling speeds against aerodynamic diameter is given in the upper part of Figure 6.11. The upper value of the aerodynamic diameter on the plot is about 4 mm (the size of a large rain drop), which has a falling speed of about 10 m s^{-1}.

There are two scaling parameters that approximately describe particle behaviour in a discharging plume. The first is the slope (the falling angle) α of the particle path in a uniform wind of speed *U*, where,

$$\alpha = \frac{v_f}{U}. \tag{6.9}$$

The second is the particle inertia, most conveniently described here in terms of the particle stop distance D_s, where

$$D_s = \frac{v_f U}{g} \tag{6.10}$$

The stop distance indicates the length scale of particle inertial effects in relation to the scale of the flow and dispersion pattern of the fire plume. Thus, the ratio

$$\frac{D_s}{H} = \frac{U^2 \alpha}{gH} \tag{6.11}$$

defines the relative significance of particle inertia to particle dispersion in the plume. *H* here is some characteristic length scale of the flow, such as the height of an associated building in the near field of a plume or the depth of the plume at some point. The ratio D_s/H is effectively a Stokes number for the flow and it can be seen from eqn (6.11) that its value increases with both the wind speed and the angle of fall of the particles. In general, if v_f/U and D_s/H are low then the particle dispersion may be little or no different from that of a gas. If they are high, the particle dispersion may be significantly altered, the rates of particle dispersion may be diminished and losses to the ground increased. Some examples are shown in Table 6.1 of the particle properties in a range of wind speeds that give rise to specific angles of fall.

Some sense of the effects of the basic particle parameters on plume concentrations at the ground can be seen in the results of some small scale wind tunnel experiments by Hall *et al.*,[12] shown in the lower part of Figure 6.11,

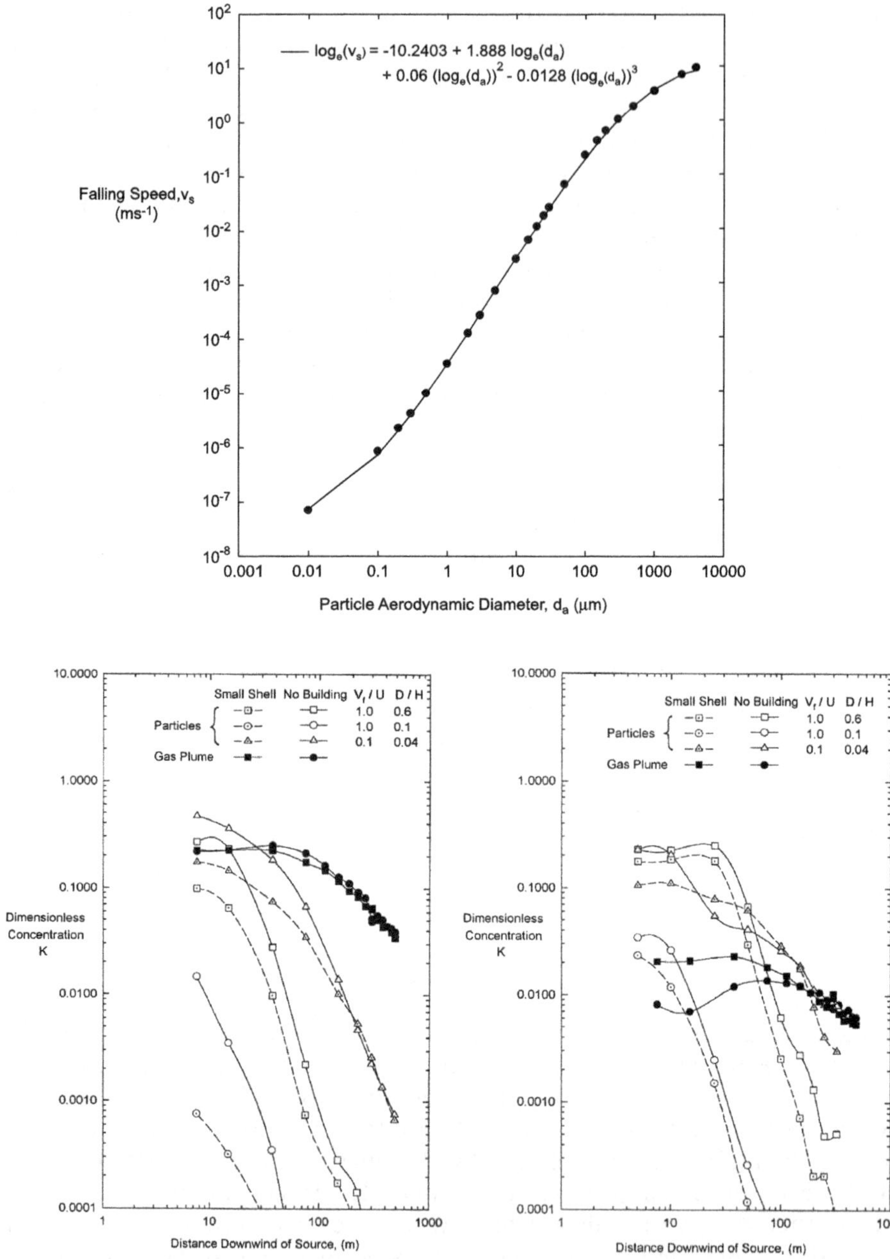

Neutrally buoyant plume. $F/U_{ref}^3L = 0$

Buoyant plume. $F/U_{ref}^3L = 0.1$

Figure 6.11 Upper: Particle falling speeds. Lower: Concentrations at the ground of gases and particles from a warehouse fire plume.
From Hall *et al.*[12] Reprinted with permission from Elsevier. Copyright 1998.

Table 6.1 Some particle aerodynamic properties for different wind speeds and particle sizes.

Wind speed/m s^{-1}	Particle parameters	Particle trajectory slope (v_f/U) 0.01	0.1	1
1	v_f (m s^{-1})	0.01	0.1	1
	D_s (m)	0.001	0.01	0.1
	d_a (μm)	18	58	280
2	v_f (m s^{-1})	0.02	0.2	2
	D_s (m)	0.004	0.05	0.4
	d_a (μm)	25	88	500
5	v_f (m s^{-1})	0.05	0.5	5
	D_s (m)	0.025	0.25	2.5
	d_a (μm)	40	160	1250
10	v_f (m s^{-1})	0.1	1	10
	D_s (m)	0.1	1	10
	d_a (μm)	58	277	4000
20	v_f (m s^{-1})	0.2	2	20
	D_s (m)	0.4	4	–
	d_a (μm)	88	500	–

which followed the small scale warehouse fire plume dispersion experiments on discharged gases (described earlier) with some directly comparable experiments using particles of varying inertia additionally discharged into the buoyant plume. The plots show measurements of concentration at the ground of particles discharged in plumes of neutral and moderate buoyancy, compared with those of gases in the same plume. The experiments show markedly different dispersion characteristics between gases and particles with relatively large inertia in the same plume.

A more general expression of the effects of particle inertia on dispersion can be found in the diagram in Figure 6.12, which shows a plot of particle fall angle against wind speed. The solid lines on the plot are for fixed values of D_s/H and the broken lines are of particle falling speeds that correspond to specific sets of values of fall angle and wind speed on the plot. These lines are drawn as straight on the plot as a convenient first order approximation. In practice this assumes that the whole flow is in the Stokes flow regime, which is not formally true at the large fall angles and wind speeds shown on parts of the plot. The 'upper wind speed limit' of 10 m s^{-1} marked on the plot is nominal; it is the approximate 90th percentile UK wind speed at 10 m height.

Figure 6.12 is divided into a number of particle behaviour regimes. The shaded area to the bottom left of the plot is the regime of low falling speeds and low particle inertia. Particles in this regime disperse in a manner not distinguishable from that of gases; they include most combustion products and inhalable particles. The shaded area to the top right of the plot encompasses a region where particles are unlikely to be ejected in a fire, mainly due to falling speeds in excess of 10 m s^{-1}. The intermediate unshaded area is of relatively large particles with high inertia and/or falling speeds and

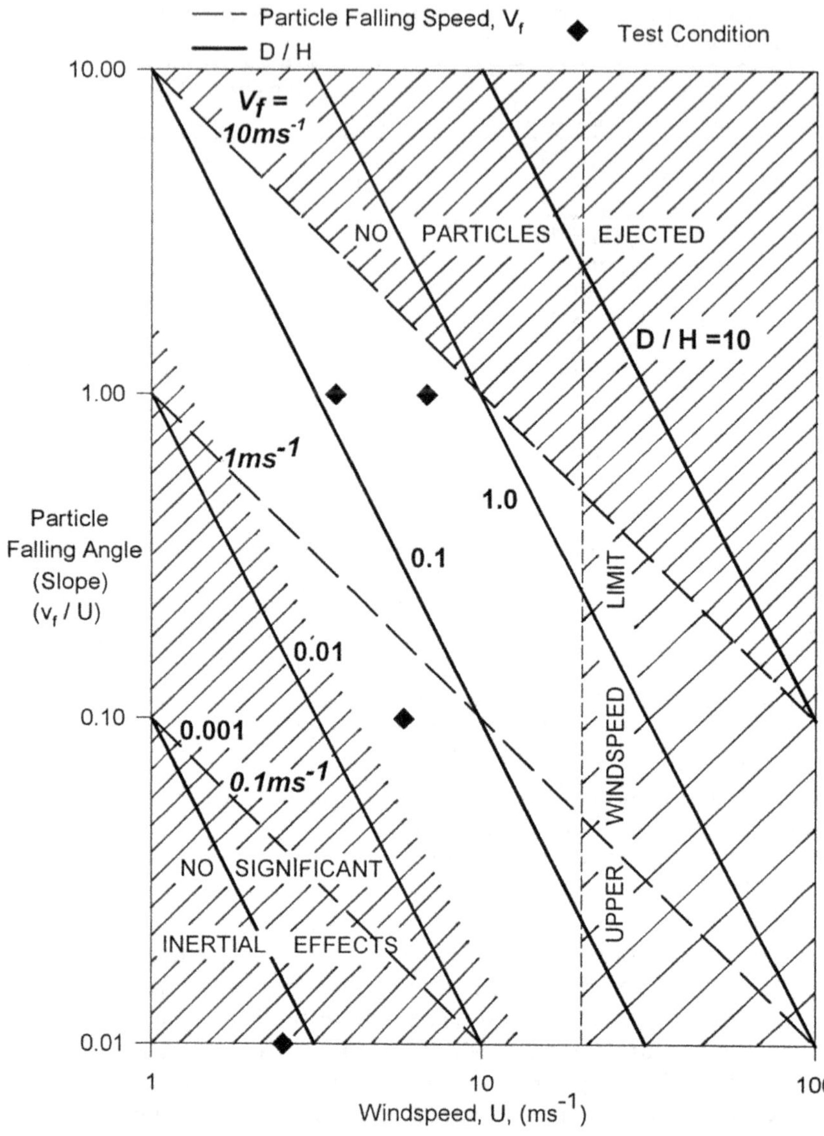

Figure 6.12 Diagram of the aerodynamic characteristics of particles discharged in fire plumes. The equivalent conditions for the experiments in Figure 6.14 are shown as black diamonds.
From Hall *et al.*[12] Reprinted with permission from Elsevier. Copyright 1998.

angles, which can be ejected with the fire plume. Particles of this sort have high deposition velocities so are rapidly depleted in the early stages of the plume dispersion. They also do not perfectly follow the main air flow and plume motion and have modified (largely reduced) rates of dispersion as their inertia may be high relative to the atmospheric turbulence that drives the plume dispersion. Their behaviour is difficult to predict.

6.9 Deposition to the Ground from Fire Plumes

Toxic contaminants from fire plumes deposited to the ground and other surfaces may leave a long lasting residue, which needs consideration, both due to uptake through the food chain and from direct human contact. Both gases and particles can be deposited by washout in rainfall (wet deposition) or directly on the ground (dry deposition) and the deposition rates can be estimated using conventional deposition modelling methods. Their application in the present context is described briefly below. More detailed descriptions of deposition modelling can be found in the first instance in Jones[13] and in the review by Hall *et al.*[14]

Wet deposition in rainfall is different for gases and particles. In wet deposition of contaminant gases there is normally a Henry's law equilibrium between the gaseous and dissolved contaminant fractions in the rainfall. Since this equilibrium usually develops fairly quickly, the deposition to the ground can be calculated to first order from the local liquid/vapour equilibrium at the ground. Thus, if there is no contaminant concentration at the ground (as from an airborne plume) there is no gaseous wet deposition. There are more complex conditions that may alter the liquid/vapour equilibrium beyond discussion here; the reader is referred to the reviews above. Particle washout is usually assumed to be permanent, due to collision between particle and raindrop, and is based on a rate of removal depending on rainfall rate and particle size. Thus, particle washout is cumulative during the raindrop's path through the contaminant plume, irrespective of its height above the ground.

The rate of contaminant dry deposition D to the ground of both gases and particles is given by

$$D = C\nu_{\mathrm{d}} \tag{6.12}$$

where D is the deposition rate, C is the contaminant concentration (particle or gas) close to the ground and ν_{d} is the deposition velocity.

Note that this is a rate of removal of the contaminant to the ground. The concentration C can be found from dispersion modelling calculations, small scale wind tunnel modelling or field measurement. The deposition velocity ν_{d} is more difficult to determine; it is different for gases and particles and varies with the meteorological conditions and the nature and state of the surface. The current state of the art is to calculate deposition velocities using 'resistance' models, 'resistance' R being the inverse of deposition velocity ν_{d}. This is a convenient way of describing the succession of processes controlling deposition; for gases these are the transfer of contaminant close to the surface by a turbulent diffusion (aerodynamic resistance), its molecular diffusion through the laminar sublayer (sublayer resistance) and its final loss to the surface (surface resistance). The deposition velocity is then obtained from

$$v_d = \frac{1}{R(\text{total})} = \frac{1}{R_a + R_b + R_s} \tag{6.13}$$

where R_a, R_b and R_s are the aerodynamic, sublayer and surface resistances, respectively. For gases the sum of the aerodynamic and sublayer resistances is approximately

$$R_a + R_b \approx \frac{13}{u_*} \tag{6.14}$$

The surface resistance depends on the nature of the surface; it may vary between zero, for highly soluble or surface reactive gases, to infinity (that is, no deposition), for insoluble and unreactive gases. For gases with zero surface resistance it can be seen from eqn (6.14) that the deposition velocity is then proportional to the wind speed and increases with the surface roughness, which also changes u_*.

For all but the very smallest particles, the sublayer resistance is very high (controlled effectively by the Brownian motion rather than by molecular diffusion as with gases), but the deposition is augmented by the particle gravitational settling velocity v_f. Eqn (6.13) is then modified for particles to the form

$$v_d = \frac{1}{R_a + R_b + R_a R_b v_f} + v_f \tag{6.15}$$

The surface resistance for particle deposition is usually taken as zero, once a particle contacts the surface it is permanently removed, so does not appear in eqn (6.15). The value of v_f may be large compared with the aerodynamic components, as with the very large particles discussed earlier in this section, in which case gravitational settling dominates the deposition. Alternatively, it may be relatively small, as with the micron and sub-micron elementary fire products, in which case deposition is dominated by the molecular and turbulent transfer mechanisms that control gaseous deposition. Particle deposition velocity is thus a result of a number of combining processes of (wind-driven) molecular and turbulent transfer, (wind-driven) inertial impaction at the surface and gravitational settling. This leads to a complex pattern of deposition velocities depending on particle size, wind speed and surface roughness. Figure 6.13[15] shows an example of the well-known plot of particle deposition velocities in relation to the controlling parameters. The contributions of wind speed and surface roughness (*via* u_*) to the deposition velocity can be seen. It can also be seen that there is a minimum in deposition velocity for particle sizes between about 0.1 μm and 1 μm. This is also a characteristic of all filtration systems

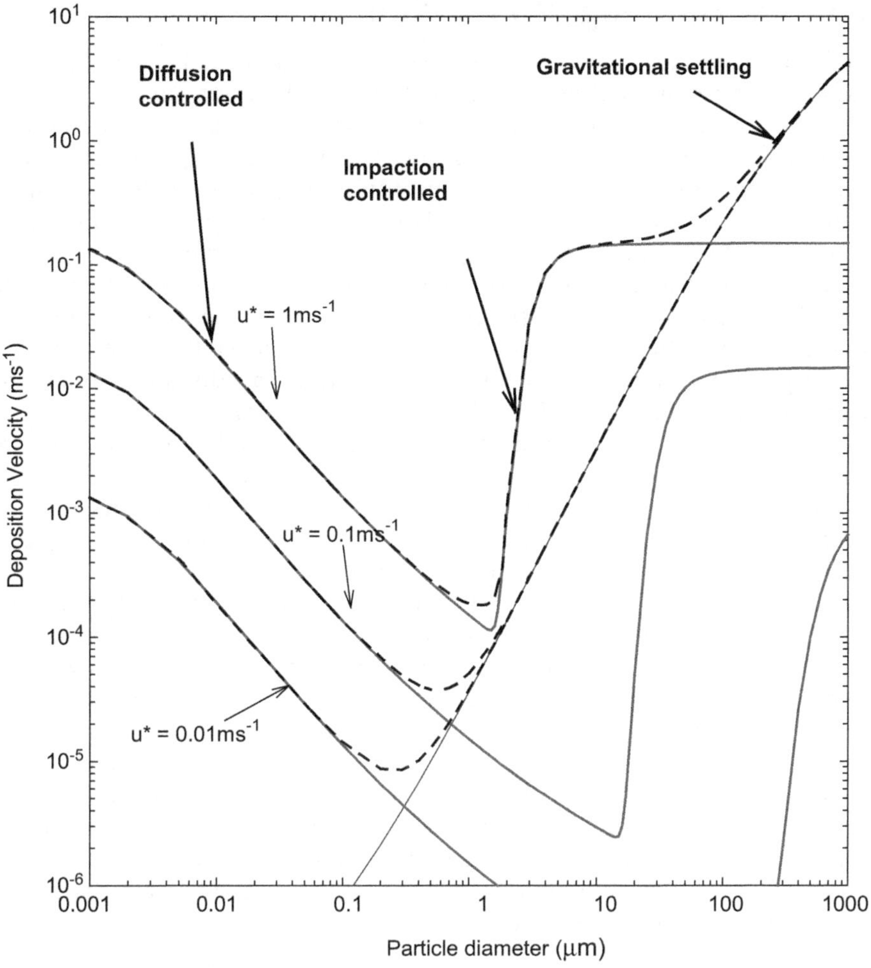

Figure 6.13 The variation of particle deposition velocity with aerodynamic diameter and the different processes governing the deposition. The broken line shows the overall deposition velocity. From Hall and Spanton.[15]

that rely on material being carried by an airflow through a physical filter, and of the deposition of particulate material in the lung. For larger particles the contributions of inertial impaction and gravitational settling become large, and for smaller particles the contribution of molecular diffusion becomes large. Comparison of Figure 6.13 with the diagram of Figure 6.12 shows that the diffusion-controlled particle deposition is related to conditions at the bottom left of Figure 6.12 and the inertial and gravitational settling-controlled deposition is related to conditions on the upper right part of Figure 6.12.

6.10 Partitioning of Fire Plumes

The list of distinctive properties of fire plumes in Section 6.1 noted that fire plumes are often not coherent and that parts of them may detach from the whole and disperse separately. This phenomenon has been observed in small scale experiments and in the field, where strongly rising buoyant plumes from the ground or from buildings appeared to leave a small residue on the ground. This residue dispersed separately as a ground-based plume with little buoyancy. Though having only a small fraction of the main plume's content, it may produce relatively high concentrations at the ground in the near field compared with those from a strongly rising, very buoyant body of the main plume (see Chapters 2 and 3). Its contribution to local exposure and deposition may thus be disproportionate, though this behaviour is usually ignored in dispersion calculations.

The matter is not well understood and is discussed more fully in Hall *et al.*[5] Briefly, there are a number of mechanisms that generate this behaviour. Plumes generated in association with buildings may have an inhomogeneous structure, with sub-plumes of material emitted from separate openings or fissures in the structure that do not connect with the main plume. Further, the complex wind flow patterns around buildings may detach parts of the main plume. Even with ground-based fire plumes in open terrain, the buoyancy-driven air flow over the body of the fire has very low velocities at the outer edges of the air inflow over the fire and parts of the plume may be detached in this region by turbulent eddies in the wind flow over the fire. There is also a mechanism in strongly rising buoyant plumes, which generates vertically oriented vortices connecting the base of the plume to the ground; these vortices carry elevated plume material back down to the ground. In combination, these phenomena appear to be able to generate persistent levels of plume material at the ground, even when the main body of a fire plume is above the ground.

6.11 Conclusions

The generation and dispersal of fire plumes into the atmosphere and the extent of direct exposure of populations, and of ground deposition of contaminants, are highly dependent on a set or interactions. These include the structural characteristics and scale of the source building and surroundings, the locations and sizes of vents, the size of the fire and the meteorological conditions, particularly in terms of wind velocity and direction.

In general, factors such as the size and area distribution of fire sources and the manner of their release, directly from the ground or associated with buildings, control the initial conditions of the fire plume. These parameters and the wind speed are important as they control the initial spreading of the plume, and its initial buoyant rise, in the area near the source where plume concentrations are high and exposure at its greatest.

At long distances these source term effects tend to disappear as the plume eventually forgets the detailed conditions of its release. Plume buoyancy from the fire heat release is, however, normally conserved indefinitely. Its effects thus persist for much longer distances, though eventually even these also disappear.

The results of wind tunnel and field experiments described, the data obtained and calculation expressions derived from them, provide a basis for calculation of fire plume generation and dispersal and of ground level exposure concentrations at different locations and distances from the source. They also give an indication of the likely variations in ground level exposures resulting from interactions between the different variables.

Appendix

A Brief Introduction to Dispersion Modelling

Dispersion modelling attempts to predict the spread of emissions of pollutants from various sources and hence to calculate the concentration of the pollutants at receptor locations. One of the most widely used methods is the Gaussian dispersion model, which is usually a good representation of observed behaviour behaviour (as shown in Hall and Walker[6]). In this model it is assumed that the vertical and lateral concentration profiles are Gaussian in distribution. Diagrams of Gaussian plume dispersion are shown in Figure 6.14 for two types of discharge of interest here: a buoyant plume rising from a ground-based source and a non-buoyant plume dispersing at the ground from a ground-based source. The mean plume cross section is elliptical and the lateral and vertical distribution of concentration is Gaussian, centred about the plume centreline. In the lower diagram, for the non-buoyant ground-based plume, the plume 'centreline' is at the ground and the vertical concentration distribution is then a half-Gaussian with its maximum concentration at the ground, as the plume folds upon itself; the part of the plume that would be below the ground is 'reflected' above the ground. Concentrations within the ground-based plume are then double those of its airborne equivalent due to the 'reflection' of the 'below-ground' part of the plume adding to the above-ground part.

In this type of model the main concern is to determine the variation of the standard deviations of the Gaussian concentration distributions vertically and laterally with distance from the source. Expressions for the values of the standard deviations of these distributions, σ_y and σ_z, have been developed from experimental data. Earlier dispersion models (*e.g.*, R91[16] and ISC[17]) used empirical equations to calculate the vertical and lateral spreads, as a function of downwind distance, for different atmospheric stability categories defined by time of year, time of day and wind speed. Newer, '2nd generation' models, for example, ADMS[18] and

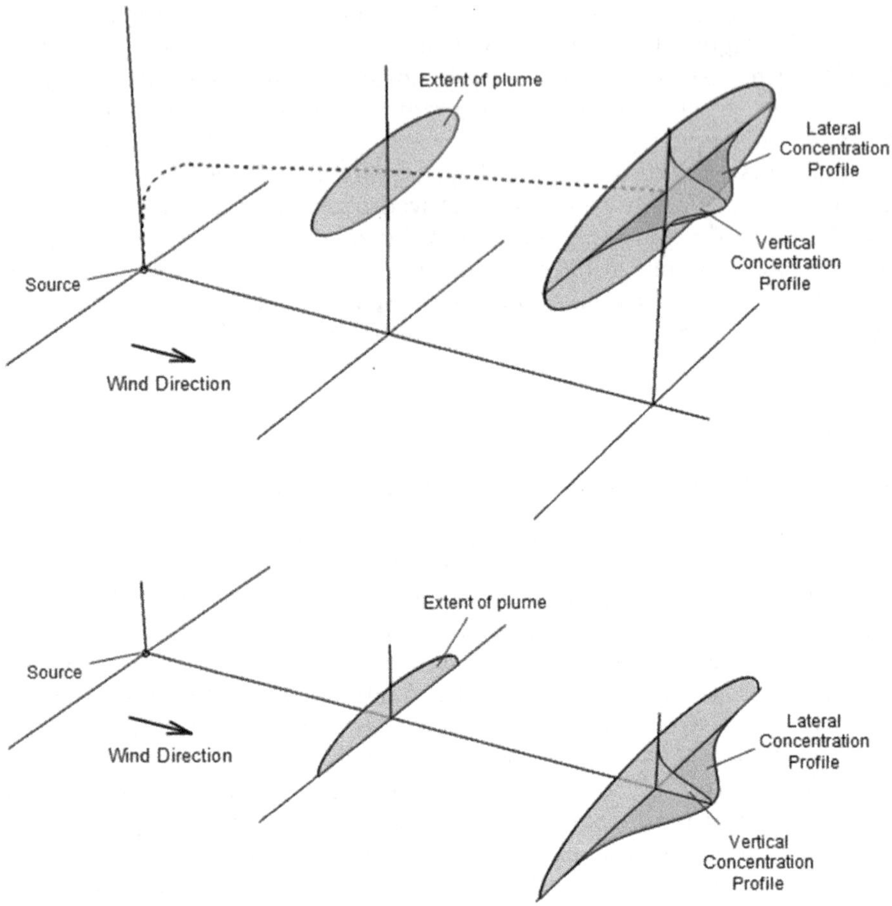

Figure 6.14 Diagrams of Gaussian plume dispersion. Upper: buoyant plume rising from a ground-based source. Lower: non-buoyant plume dispersing at the ground from a ground-based source.

AERMOD,[19] use the boundary layer height and the Monin–Obukhov length scale to determine stability and profiles of atmospheric turbulence from which the dispersion parameters, σ_y and σ_z, are calculated. This avoids the discrete set of stability conditions of the earlier models and more closely resembles atmospheric behaviour. Dispersion downwind is assumed to be proportional to the wind speed U, with the concentration being inversely proportional to the wind speed as in eqn (6.4) and (6.5). The concentration C at a point with coordinates (x, y, z) relative to the position of the source is then given by:

$$C = \frac{Q}{2\pi U \sigma_y \sigma_z} \exp\left(\frac{-y^2}{2\sigma_y^2}\right)\left[\exp\left(\frac{-(z-H)^2}{2\sigma_z^2}\right) + \exp\left(\frac{-(z+H)^2}{2\sigma_z^2}\right)\right] \qquad \text{(A6.1)}$$

where H is the height of the plume centreline above the ground. The left hand term on the right hand side scales the maximum concentration in the centre of the plume and the exponential terms scale the Gaussian distribution of concentration away from the plume centre. The two exponential terms for the vertical distribution are to allow for reflections from the ground as described above; further refinements can also be added to this basic equation to account for reflections from the top of the boundary layer.

The rise of buoyant plumes is normally calculated separately using simple plume rise models, for example, those due to Briggs and Moore, though as noted in the main text, these do not model the reduced plume rise from area sources or those associated with buildings common in fire plumes.

More detailed descriptions of dispersion modelling and the related meteorology can be found in the following resources.

Dispersion modelling:

M. R. Beychok, *Fundamentals of stack gas dispersion*, Published by Milton R. Beychok, 2233 Martin St., Unit 205, Irvine CA 92612, USA, 1994, ISBN 0-9644588-0-2.

R. H. Clarke, *A model for short and medium range dispersion of radionuclides released to the atmosphere*, National Radiological Protection Board, UK, Report NRPB-R91, 1979.

S. R. Hanna, G. A. Briggs and R. P. Hosker, *Handbook on atmospheric diffusion*, Published by Technical Information Center, U.S. Dept. of Energy (Oak Ridge, TN), 1982, ISBN 0870791273.

J. A. Jones, *A procedure to include deposition in the model for short and medium range atmospheric dispersion of radionuclides*, National Radiological Protection Board, UK, Report NRPB-R122, 1981.

J. A. Jones, *Models to allow for the effects of coastal sites, plume rise and buildings on dispersion of radionuclides and guidance on the value of deposition velocity and washout coefficients*, National Radiological Protection Board, UK, Report NRPB-R157, 1983.

D. B. Turner, *Workbook of atmospheric dispersion estimates*, Lewis Publishers (CRC Press), 1994, ISBN 1-56670-023-X.

Meteorology:

P. S. Arya, *Introduction to micrometeorology*, Academic Press, 2001, ISBN 0-12-059354-8.

J. C. Kaimal and J. J. Finnigan, *Atmospheric boundary layer flows – Their structure and measurement*, Oxford University Press, 1994, ISBN0-19-506239.

R. B. Stull, *Meteorology for scientists and engineers*, Brooks/Cole, 2000, ISBN 0-534-37214-7.

As noted in the main text, there has also been a more wide-ranging review of fire plume dispersion by Hall and Spanton.[2]

References

1. G. A. Briggs, Plume Rise, US Atomic Energy Commission, AEC Critical Review Series TID-25075, Library of Congress Catalogue Card No: 72-603261, November 1969.
2. D. J. Hall and A. M. Spanton, A Review of Models for Dispersion Following Fires, Envirobods Ltd Report for the UK Atmospheric Dispersion Modelling Liaison Committee, ADMLC/2003/1, Available at http://www.admlc.org.uk/ar03-04.htm (accessed May 2015), April, 2005.
3. A. N. Yates, Dispersion of Smoke Plumes from Pool Fires, PhD Thesis, University of Greenwich, 1996.
4. D. J. Hall and R. A. Waters, Further Experiments on a Buoyant Emission from a Building, Warren Spring Laboratory, Report No. LR567(PA), July 1986, Held by: The National Archives – Department of Trade and Industry, 1970–1974 Reference FV 12/87.
5. D. J. Hall, V. Kukadia, S. Walker and G. W. Marsland, Plume Dispersion from Chemical Warehouse Fires, BRE Report No. CR 56/95, September 1995, Available as a PDF from the authors.
6. D. J. Hall and S. Walker, Plume Rise from Buoyant Area Sources at the Ground, Building Research Establishment, Report No 80921, February 2000.
7. ASHRAE ASHRAE Handbook – Fundamentals, ASHRAE, 2009.
8. D. J. Hall, C. F. Barrett and A. C. Simmonds, Wind Tunnel Model Experiments on a Buoyant Emission from a Building, Warren Spring Laboratory, Report No. LR355(AP), 1980.
9. S. R. Hanna, G. A. Briggs and J. C. Chang, Lift-Off of Ground-Based Buoyant Plumes, *J. Hazard. Mater.*, April 1998, **59**, 123–130.
10. A. G. Robins and R. W. Macdonald, Review of Flow and Dispersion in the Vicinity of Groups of Buildings, Atmospheric Dispersion Liaison Committee – Annual Report 1998/99, J. A. Jones, National Radiological Protection Board, NRPB R322: 53–157, 2001.
11. D. J. Hall, A. M. Spanton, R. Macdonald and S. Walker, A Review of Requirements for a Simple Urban Dispersion Model, Building Research Establishment, Report No. CR 77/96, 1996.
12. D. J. Hall, V. Kukadia, S. Walker and G. W. Marsland, Deposition of Large Particles from Warehouse Fire Plumes – A Small Scale Wind Tunnel Model Study, Building Research Establishment, Report No. CR 42/96, February 1996, Also in *J Haz Mat.*, **59** (1) 13–29, 1998.
13. J. A. Jones, Models to Allow for the Effects of Coastal Sites, Plume Rise and Buildings on Dispersion of Radionuclides and Guidance on the Value of the Deposition Velocity and Washout Coefficients, National Radiological Protection Board, Report No. NRPB-R157, December 1983.
14. D. J. Hall, A. M. Spanton and C. B. Powlesland, Review of Modelling Methods of Near-Field Acid Deposition, Environment Agency, R&D Technical Report P4-083/3/TR, May 2005, Available from the Environment Agency web site or the authors.

15. D. J. Hall and A. M. Spanton, A Wet and Dry Deposition Model for the UDM, Envirobods Ltd, Technical Report No. 04/04, February 2004.
16. R. H. Clarke, A model for short and medium range dispersion of radionuclides released to the atmosphere, National Radiological Protection Board, UK, Report NRPB-R91, 1979.
17. USEPA User's guide for the Industrial Source Complex (ISC3) Dispersion Models: Volume I, User Instructions, U.S. Environmental Protection Agency, Research Triangle Park, NC, EPA-454/B-95-003a, 1995, http://www.epa.gov/scram001/userg/regmod/isc3v1.pdf (accessed April 2015).
18. D. J. Carruthers, R. J. Holroyd, J. C. R. Hunt, W. S. Weng, A. G. Robins, D. D. Apsley, D. J. Thomson and F. B. Smith, UKADMS: A New Approach to Modelling Dispersion in the Earth's Atmospheric Boundary Layer, *J Wind Eng. Ind. Aerodynamics.*, 1994, **52**, 139–153.
19. A. J. Cimorelli, S. G. Perry, A. Venkatram, J. C. Weil, R. J. Paine, R. B. Wilson, R. S. Lee, W. D. Peters, R. W. Brode and J. O. Paumier, AERMOD – Description of Model Formulation, USEPA document, EPA-454/R-03-004, September 2004, Available at http://www.epa.gov/scram001/7thconf/aermod/aermod_mfd.pdf (accessed May 2015).

The Toxicology of Combustion Products

CHAPTER 7

Application of Animal Models and Human Studies to Prediction of Combustion Toxicity in Humans

DAVID A. PURSER

Hartford Environmental Research, Hatfield, UK
Email: david-purser@ntlworld.com

7.1 Introduction

7.1.1 General

This chapter considers how a combination of human and animal exposures has been used to:

- determine the main physiological and pathological effects of exposure to toxic fire effluents, how they impair escape and cause incapacitation during exposure, acute injuries and deaths during and after exposure and chronic health effects;
- identify the main toxic species responsible for these effects;
- determine the relationships between exposure concentration, dose and severity for each toxic substance individually, and in the combinations occurring in fire effluents.

Issues in Toxicology No. 23
Toxicology, Survival and Health Hazards of Combustion Products
Edited by David A. Purser, Robert L. Maynard and James C. Wakefield

Published by the Royal Society of Chemistry, www.rsc.org

This has involved a combination of:

- fire incident investigations;
- studies of effects of individual toxic gases on humans in industrial accidents;
- experimental studies on humans, non-human primates and rodents, involving individual toxic gases, specific toxic gas mixtures and fire effluents containing mixtures of toxic gases and particulates from a range of burning materials.

The toxic effects of exposure to combustion products range from acute lethality (resulting from acute exposure of a few minutes duration to very high concentrations of toxic products during fire incidents) to chronic health effects (resulting from many years' exposure to low concentrations of combustion products in the outdoor and indoor environments and from tobacco smoke).

This chapter considers the development of toxic hazards in fires and how the main toxic hazards (consisting of irritancy and asphyxiation) have been established from studies of fire incidents and fire victims, and from experimental studies of mechanisms of incapacitation using non-human primates and rodents. Consideration is then given to the advantages and limitations of different approaches using humans and each different animal species for measurement of sensory irritant, pulmonary irritant and asphyxiant potencies in terms of predicting effects in humans.

Studies on non-human primates and rodents have been used to establish that the physiological and pathological effects of fire effluent mixtures from the vast majority of materials can be explained in terms of a small number of key asphyxiant gases and irritants. Animal exposure studies have been carried out on individual fire gases, on mixtures containing different combinations of individual gases, and on the mixed effluents obtained during combustion of different materials. The results of these studies have then been used for the development and validation of mathematical models for the prediction of the toxic effects of the mixtures of major toxic gases and particulates in humans. These take into account the effects of each individual component and the interactions between them and predict the overall physiological and pathological effects during and after an exposure.

7.1.2 Thermal Decomposition, Combustion and Fire Effluents

As described in Chapter 2, fire effluents consist of a complex mixture of gases, vapours, liquid and solid particulates containing hundreds of substances in varying ratios, the composition being highly dependent upon the combustion conditions and also the elemental and organic composition of materials burned. When materials are heated in air they initially undergo non-flaming thermal decomposition (pyrolysis), producing a wide range of organic fragments of differing molecular masses and inorganic gases such as CO and CO_2. Depending upon the oxygen content of the original material and the extent of oxidation at the heated fuel surface, these may include a variety of partially oxidised organic species. When the temperature and concentration of gaseous pyrolysis products and the air supply are sufficient, a transition to flaming combustion occurs. Under these conditions the bulk of the organic pyrolysis products are fully oxidised to carbon dioxide and water, but depending upon the ventilation conditions (fuel : air ratio), a proportion are released as partially oxidised species including CO, HCN, volatile and condensed organics, carbonaceous soot particles and acid gases. Fire effluents (smoke) therefore consist of the mixture of solid, liquid and gaseous products from any fire, which may result from non-flaming decomposition or flaming combustion or a mixture of both. The composition of an effluent atmosphere depends upon the yields of all toxic substances produced by each fuel component and the extent of dilution at the exposure location. The toxicity of a fire effluent atmosphere may be considered in terms of the measured effects during exposure to concentrations of individual toxic components and their interactions. On a simpler basis it may also be expressed in terms of the effects of the mixed effluents evolved from a particular material decomposed under specified decomposition conditions. An example of the latter could be the lethal toxicity of wood decomposed under well-ventilated flaming combustion conditions expressed as LCt_{50} g min m^{-3}. In this context the exposure concentration is expressed in terms of the mass of wood decomposed into a given volume (g wood mass loss m^{-3}). This is multiplied by the exposure period to provide an expression of "exposure dose". This then represents the concentration of mixed effluents available for inhalation by an exposed subject over a period of time. In this simple form it does not consider aspects such as uptake, metabolism and excretion of specific toxic substances, which are taken into account for more sophisticated analyses.

7.1.3 Exposure Concentration, Time and Dose Interactions for Combustion Product Mixtures and Fractional Effective Dose Methodology

The mathematical models applied to combustion product mixtures are known as fractional effective dose (FED) models.[1,2] The fractional effective dose of a toxic combustion product (or set of products) is calculated as the dose to which a subject is exposed over a period of time expressed as a fraction of the dose require to cause a given toxicity endpoint such as incapacitation or death. The simplest forms of FED models have been applied to bench-scale combustion toxicity tests used to measure lethal exposures (LCt_{50}) in rats, typically for a constant 30 minute exposure period plus a 14 day post-exposure observation period. Thus, for carbon monoxide the rat lethal concentration (LC_{50}) is 5400 ppm for a 30 minute exposure giving an LCt_{50} of 162 000 ppm min, while for HCN the rat LC_{50} is 165 ppm for a 30 minute exposure giving an LCt_{50} of 4950 ppm min.[3,4] For any specific exposure concentration (C), the FED is then given by the actual $C \times t$ ("*Ct* product", where t is time) exposure dose as a fraction of the lethal *Ct* product exposure dose. In general terms:

$$\text{FED} = \frac{\text{Exposure dose over time t } (i.e., \ Ct)}{\text{Effective Ct dose to cause incapacitation or death}} \tag{7.1}$$

This normalised method for expressing exposures in terms of their toxicological effects is particularly useful for examining the contributions of mixture components to overall toxicity. For example, for an exposure to a mixture containing an FED of 0.5 for each of two toxic gases (such as CO and HCN), it is possible to determine the extent to which the observed lethality can be understood in terms of each agent acting independently or whether there are additive, antagonistic or synergistic interactions between the different components.

The main application of FED methodology to rat lethality studies in 30 minute bench-scale tests has therefore been to examine the extent to which different components of fire effluent mixtures interact and for the development of validated calculation methods to predict toxic potency for a measured set of combustion products.[5–8] This can then be used to determine the extent to which toxic potency of the combustion products from any particular material can be understood in terms of the measured components, or whether some toxic component is present with unpredicted specific toxicity or high toxic potency not normally associated with combustion products (sometimes known as a "supertoxicant"). To the extent that the toxicity of combustion product mixtures can be understood in terms of a small number of key toxic substances, it is then possible to calculate the predicted toxic potency for the combustion product mixture from any burning material. For example, it is possible to calculate the mass of wood that would need to be decomposed in a 1 m^3 volume chamber under specified combustion conditions for a lethal outcome after a 30 minute exposure (which is 25 g m^{-3} for under-ventilated flaming fire conditions). This could then be compared with the mass required for another polymer, such as polyamide, under the same combustion conditions (1.4 g m^{-3}).[7]

The other main advantage of using the FED method is for application to full-scale fires, in which the exposure concentrations of the different components change rapidly with time. For this application, the concentration of each toxic gas is averaged over successive short time periods during a fire (for example, each minute) and the exposure doses for each toxic gas during each time period are calculated and combined (for example, using an additive model). The exposure doses for successive periods are then summed until the total FED = 1, at which time the chosen endpoint is predicted to occur.[2] In the international standard on this method (ISO13571),[2] the principle of the model in its simplest form is expressed in the form:

$$\mathrm{FED} = \sum_{i=1}^{n} \sum_{t_1}^{t_2} \frac{C_\mathrm{i}}{(Ct)_\mathrm{i}} \Delta t, \tag{7.2}$$

where C_i is the average concentration of toxic component "i" over the chosen time period Δt and $(Ct)_\mathrm{i}$ is the concentration×time exposure dose required for the chosen toxicity endpoint.

For toxic substances obeying Haber's rule over timescales of interest during fires (so that for any specific toxicity endpoint $Ct = k$), then the Ct denominator is a constant (k). For substances deviating significantly from Haber's rule, different expressions of the general form $C^n t = k$ can be used. The extent to which physiological and pathological effects of different toxic combustion products deviate from Haber's rule is discussed later in this chapter and in more detail in Chapter 8.

The FED method is applied to time-varying concentrations in fires mainly for the purpose of estimating time to incapacitation for human fire victims, usually over periods of a few minutes (up to approximately an hour). For this application, while some products (such as CO) can be considered approximately to follow Haber's rule over these timescales, so that constant Ct product denominators can be used, for the important asphyxiant gases, HCN and low oxygen hypoxia, and for the effects of CO_2, the concentration–time effect relationships are more complex so that a series of physiological expressions has been derived.[1,9] These are presented in detail in Chapter 14 dealing with the calculation of acute effects in human fire victims.

The FED concept is also applied to estimation of the "fractional exposure concentrations" (FEC) of irritants, for which the chosen endpoint is sensory irritancy sufficient to impair escape capability. For sensory irritancy it is considered that effects occur almost instantaneously after exposure and depend upon the exposure concentration rather than the exposure dose.[10–12] For rodent studies (mainly using mice) it is possible to measure the RD_{50} concentration (the concentration resulting in a 50% decrease in breathing rate) and by expressing the exposure concentration of any irritant as a fraction of the RD_{50} concentration, it is possible to examine the interactions between different components in a mixture of irritants (so, for example, it is possible to examine the extent to which FECs of 0.5 for a mixture of HCl and HBr would produce an additive FEC = 1).

The development and validation of FED "N-gas" models enabling calculation of lethal toxic potency for fire effluents from materials to rats (LC_{50} concentrations) *via* chemical analysis of effluent atmospheres is also considered in this chapter. Two materials forming exceptions are also discussed for which unpredicted toxic effects have been measured in animal studies that were not predicted from chemical analysis.

Consideration is then given to the contributions to overall toxic potency from different individual chemical species in effluent mixtures from different materials, decomposed under non-flaming and flaming combustion conditions [well-ventilated, under-ventilated (or "vitiated") and post-flashover under-ventilated]. The range of lethal toxic potencies to rats for individual toxic species and fire effluent mixtures under different decomposition conditions is described. The final section examines the similarities and differences between humans and different animal models for toxic effects and how these should be taken into account when extrapolating between species.

7.2 Development of Toxic Hazards in Fires and Health Hazards from Chronic Exposures

Fires tend to develop rapidly, giving rise to conditions likely to be lethal to any building occupants within a few minutes, mainly as result of exposure to toxic fire effluent but sometimes to heat (see Chapter 2). Before lethal conditions develop, exposure to fire effluent may first impair the ability of occupants to escape and then cause incapacitation, preventing escape. Once a person is incapacitated, they are likely to die some minutes later from the lethal effects of heat, toxic gases or structural failure of building elements. Although the lethal effects and lethal exposure doses of fire effluents are of interest, the most important determinants of survival in fires are the effects that might slow or otherwise impair escape and the time to incapacitation (the point at which occupants are no longer able to save themselves).[1,2,13] For persons sheltering in a temporary refuge within a building, the most important consideration is the time by which exposure to heat or toxic gases causes incapacitation.

Fire hazard and fire safety therefore depend upon the outcome of two parallel time lines: the available safe escape time (ASET), which is the time from ignition of the fire to the development of incapacitating conditions, and required safe escape time (RSET), which is the time required for occupants to reach a place of safety.[13–15]

Exposure to toxic fire effluent influences both these parameters. Studies of human and animal survivors and victims of fire incidents have revealed the following major hazards occurring in fires more or less in the order shown.[1,9]

- Behavioural effects of seeing fire or smoke (reluctance to enter smoke-filled escape routes or move past flames).
- Physiological, behavioural and pathological effects of direct exposure to optically obscure, irritant smoke containing high concentrations of toxic combustion products:
  - Difficulty of finding escape routes and slow movement speed due to effects of smoke preventing clear vision.
  - Further impairment of vision and eye pain due to immediate, concentration-related effects of sensory irritant smoke products on eyes—blepharospasm (reflex closure of eyes due to pain produced by stimulation of corneal receptors).

- Impairment and eventual prevention of evacuation due to immediate, concentration-related painful sensory irritant effects of smoke on upper respiratory tract (mouth, nose, throat) and airways (bronchoconstriction, chest pain).
- Uptake over a period of several minutes of asphyxiant gases leading to incapacitation and collapse once a critical exposure dose has been inhaled.
- Impairment and eventual prevention of evacuation due to pain from heat exposure, with incapacitation as soon as burns reach a critical level.
- Death during exposure at the fire scene (or after rescue, usually within a few hours) resulting from exposure to asphyxiant gases, exacerbated by respiratory tract irritancy and effects of heat or burns to respiratory tract.
- Death during exposure at the fire scene (or within a few days of rescue) due primarily to heat exposure and/or body surface burns.
- Death within a few hours of rescue from lung oedema and inflammation due to inhalation of irritant fire effluent gases and particulates (sometimes complicated by thermal airway damage).
- Death within a few days of rescue due to bronchopneumonia.
- Increased risk of heart attack or stroke (usually between a few hours to several days after rescue) resulting from effects of fine particles and asphyxiant gases on blood circulation (cardiac infarct and/or cerebrovascular incidents) or from stress.
- Long-term health problems including neuropathology, sensitisation and reactive airways disease syndrome (RADS).

• Acute or chronic exposure to low concentrations of combustion products from a diluted smoke plume, in the open, from indoor or outdoor environmental air pollution from combustion products, or deliberate self-exposure from tobacco smoke may result in the following:
    - Nuisance odour and mild eye and respiratory tract irritancy during exposure.
    - Bronchoconstriction or acute rhinitis if already sensitised to any component of the smoke.
    - Mild signs and symptoms of respiratory tract irritancy during the next day or so (sore eyes and throat, acute bronchitis).
    - Following continued subacute–chronic exposure: ciliastasis, squamous metaplasia of larynx and trachea, inflammatory lung cell infiltration, development of chronic obstructive pulmonary disease.
    - Possible continued risk of cardiovascular disease due to systemic uptake of ultrafine and nano-soot particles.
    - Increased risk of lung and other systemic cancers from exposure to carcinogenic smoke components: metals, radioactive isotopes, fibres,

volatile organics (such as formaldehyde, acrylonitrile, benzene, isocyanates and others) and less volatile organics [polycyclic aromatic hydrocarbons, polychlorinated biphenyls (PCBs), halogenated dioxins and dibenzofurans, nitrosamines].

This chapter considers mainly the acute incapacitating physiological effects of exposure to combustion products during fires and the effects on the lungs occurring during the hours and first few days following rescue from a single incident. Consideration is also given to the minor acute health effects of low-concentration exposures. The long-term health effects from chronic low-concentration exposures are considered in Chapters 11 (particulates) and 12 (dioxins and other carcinogens).

7.3 Identification of Toxic Species in Fire Effluents and Assessment of Contributions to Toxic Hazards

The identity and physical form of substances occurring in fire effluents has been measured during fire incidents, full-scale fire experiments and using a variety of bench-scale combustion test methods. Information on the toxic effects of exposure has been obtained from investigations of fire incidents and from experimental studies. The latter have involved human and animal exposures to individual fire gases and animal exposures to combustion product atmospheres generated in bench-scale combustion toxicity tests. An important set of key studies on the incapacitating effects of fire gases and combustion product mixtures was carried out using non-human primates. Larger data sets involving studies of a wider range of materials are available from experimental studies using rodents, although these are somewhat limited in terms of the physiological measurements made. The following sections describe the results obtained from these different approaches and the extent to which, when combined, they provide an understanding of toxic hazards from combustion products.

7.4 Fire Incident Investigations, Associated Fire Tests and Pathology Studies

7.4.1 Smoke Irritants

Everyone has experienced the painful effects of exposure to irritant smoke from fires involving garden waste, cooking or cigarette smoke, so it is not surprising that fire victims report that the irritancy of smoke, in addition to simple visual obscuration, affected their ability to escape. In studies of occupant behaviour in smoke during fire incidents, Wood in the UK[16] and Bryan in the US[17,18] found that around 30% of people turned back rather than continued through smoke-logged areas and the average smoke density at which people turned back was at a "visibility" distance of 3 m (optical density m^{-1} 0.33, extinction coefficient 0.76). In the UK study, 91% of those turning back having entered smoke did so when visibility was less than approximately 4 m, representing 62% of those exposed to smoke at this visibility or lower, while only 11% of those exposed to smoke with a visibility >4 m turned back (5% for visibility >9 m). In an experiential setting Jin[19,20] found that walking speed through irritant wood smoke was much more reduced than that through non-irritant smoke of a given optical density. The ability to see in irritant smoke decreased sharply at a smoke density exceeding a certain level. "The subjects could only keep their eyes open for a short time and tears ran so heavily that they could not see the words on the signs". Both smoke density and irritancy were found to affect walking speeds, the speed decreasing very rapidly even at low smoke density levels. The subjects could not keep their eyes open and they walked with a zigzag pattern or along the wall.[20]

Smoke containing acid gases, such as smoke from burning polyvinyl chloride (PVC), which contains high concentrations of HCl, is reported as being particularly painful. Very low smoke levels in minor incidents involving PVC cable insulation have resulted in buildings being evacuated due to eye and respiratory tract irritation. The penetration of irritant smoke vapours and particulates causes a chemical pneumonitis in some fire victims, resulting in a transient hypoxic crisis. Low arterial blood oxygen concentrations develop several hours after rescue due to pulmonary oedema and inflammation. This can be fatal at this stage, or resolve after around 24 hours. Recovery may then be rapid or there may be a progression to a fatal bronchopneumonia. Fire survivors with significant smoke exposure reported a slight hoarseness and productive cough for a few days after an incident. At three month follow up they are generally free of respiratory symptoms.[21] However, studies of fire incidents do not really enable the identification of important irritant substances or provide a method for predicting the severity of effects.

7.4.2 Asphyxiant Gases

Although a number of irritant and asphyxiant gases are involved in incapacitation preventing escape, the majority of people injured or dying in fires have high blood carboxyhaemoglobin concentrations (%COHb; see Chapter 2). Since the %COHb distribution is somewhat similar in both fire fatalities and victims of CO gas poisoning it is considered that CO is the main ultimate cause of most "smoke" deaths in fires. This is shown by comparing the %COHb distribution for non-fire CO poisoning deaths in Figure 7.1 with the %COHb distribution for smoke deaths in Figure 7.2. For both data sets the mode post-mortem %COHb is around 75% COHb, but the somewhat higher proportions of fatalities at lower %COHb levels in non-burned fire victims may indicate some contribution from other asphyxiants.[22]

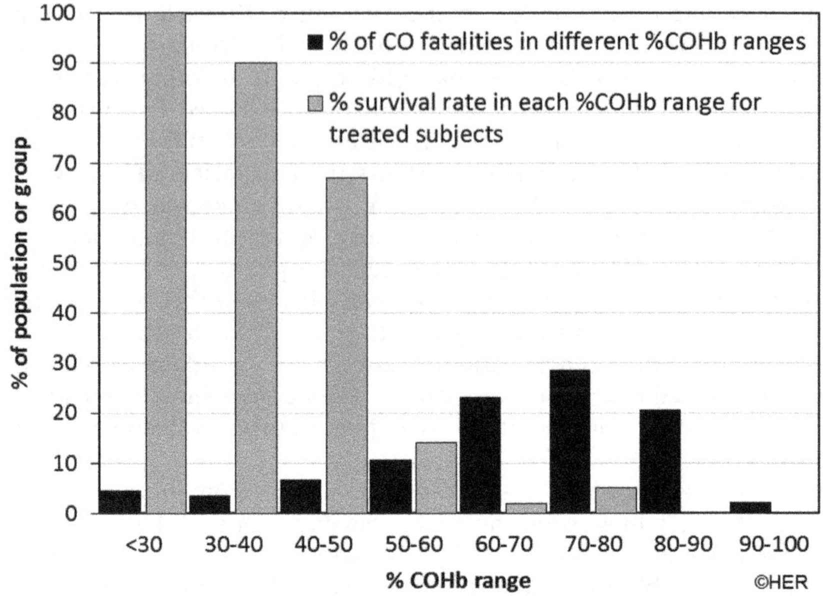

Figure 7.1 Distribution of %COHb in fatal non-fire CO poisoning cases (after Nelson)[22] compared with proportions of survivors and fatalities in different %COHb ranges from a sample of 260 CO poisoning cases (after Pach).[23]

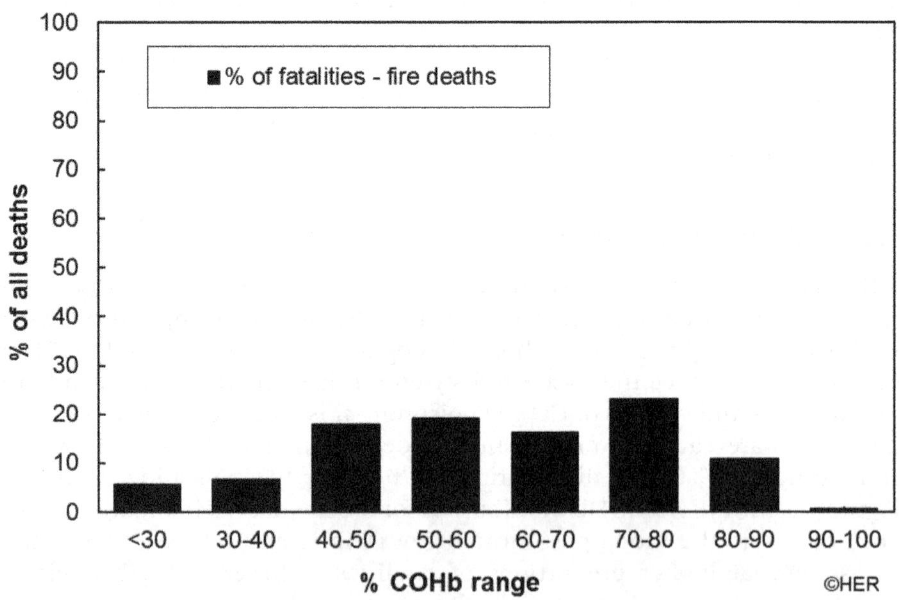

Figure 7.2 Distribution of fatal %COHb in non-burned fire cases (after Nelson).[22]

An important consideration is that there is a considerable difference between the post-mortem %COHb measured in fatalities and the maximum %COHb survivable in victims rescued alive. Once a subject is incapacitated by CO during an incident, uptake continues until respiration and circulation cease, which can result in final blood concentrations exceeding 90% COHb in some cases, as shown in Figures 7.1 and 7.2, but this does not mean that they could have survived if rescued before this time. A key consideration is the average maximum level survivable after rescue and treatment. This is illustrated in Figure 7.1 from data by Pach,[23] which shows the proportion surviving following treatment of CO exposure victims having achieved different %COHb when rescued. This shows an approximate 100% survival rate below 30% COHb, 68% survival for subjects with 40–50% COHb but only about 5% survival above 60% COHb. From the data in Figures 7.1 and 7.2 for CO alone it is proposed that the average lethal level is around 50% COHb, with a threshold just below 40% COHb. For fire victims dying during incidents from smoke exposure (without burns) there are few cases with %COHb levels below approximately 30–40% COHb (see Chapter 2). Although %COHb levels at death might be expected to vary somewhat with age and health status, the differences are small, so that the distribution of %COHb levels in CO and fire fatalities is only slightly left shifted for the over 60 age group. Also, when %COHb levels in fatalities were compared with their pre-existing health status, no relationship was found with pre-existing respiratory or general health, but some tendency for death occurring at slightly lower %COHb levels was found for subjects with pre-existing cardiovascular disease. Carbon monoxide toxicity is described in detail in Chapter 9.

Since concentrations of smoke particulates and other toxic gases track together with CO in fire effluents, the presence of soot in the lungs and carboxyhaemoglobin in the blood of fire victims are markers for the individual's exposure to a range or other smoke toxicants. Elevated blood cyanide concentrations have been found in a proportion of fire victims, but always accompanied by elevated %COHb levels.[24] Cyanide is less stable in blood than CO,[25] and due to the dynamics of uptake and dispersal of cyanide in the blood and body fluids, it is difficult to determine its importance in producing incapacitation at the fire scenes from post-mortem studies[26–28] (see Chapter 10). In experimental house fires involving domestic furniture, HCN concentrations of 1000 ppm were present with CO concentrations of around 10 000 ppm, so that HCN is likely to be a major factor in causing rapid incapacitation (loss of consciousness), thereby limiting time available for escape in many fire incidents.[29,30] When blood samples were taken from fire survivors at the time of rescue from fire scenes in Paris, high blood cyanide concentrations were measured.[31] Similar findings were reported for the Manchester Airtours fire victims.[32]

Exposures to high concentrations of carbon dioxide and/or to low concentrations of oxygen are difficult to identify from studies of survivors and fatalities since no evidence of such exposures is likely to be available from analysis of blood samples. Carbon dioxide is always present at high concentrations during fire experiments, and in fire victims its main effect is to cause hyperventilation, thereby increasing the rate of uptake of other fire gases. With regard to low oxygen hypoxia the effects are more difficult to predict. During enclosure fire experiments (such as that shown in Chapter 2)[29,30] it is generally found that oxygen depletion at head height is relatively minor by the time victims are predicted to be severely affected by CO and HCN.[9] In some fires, the oxygen concentration in the upper layer of fire effluent plumes can be as low as 1%, as well as being at temperatures of several hundred degrees centigrade.[29] If a fire victim were to inhale an atmosphere such as this, they would be expected to collapse unconscious within seconds. There are some reports of observations that could be consistent with this effect: victims in fire incidents have been observed to collapse following inhalation of a single breath of smoke. This situation is likely to occur when a person opens the door of a closed room in which a fire has burned for some time and has significantly depleted the local oxygen concentration.

Other potential asphyxiant gases present in fire effluents are oxides of nitrogen. During well-ventilated fires, NO is formed at concentrations considered harmless. Over periods of several minutes in the ageing smoke, a small proportion of this is oxidised to highly irritant NO_2 but concentrations measured have not been high enough to make a significant contribution to asphyxia or lung irritancy, except in specific incidents such as fires involving nitrate film[33] (see Chapter 2).

The general indications from fire incident studies and full-scale fire experiments are that a short list of asphyxiant gases is important in most fires, including CO, HCN, CO_2 and low O_2. Of these, CO is present in all fires and almost always found at high concentrations in the blood of fire survivors and fatalities with serious smoke exposure in fires. CO is also likely to be the major cause of deaths during fires, but it is likely that both HCN and CO are important as causes of incapacitation. Their uptake is enhanced by the presence of CO_2. Irritants and simple hypoxia may be important in some fires as additional causes of overall hypoxia, and all may be additive with CO as ultimate causes of deaths from asphyxia during fires. Based upon the %COHb levels in fire fatalities it is possible to estimate the fatal exposure doses of CO for human fire victims. By comparing clinical status with measured %COHb concentrations in fire survivors it is possible to make some estimates of CO exposure doses and %COHb concentrations causing incapacitation during fires, however, wide variations are reported in practice.[34]

7.5 Studies Using Animals in Conjunction with Bench-scale and Large-scale Fire Toxicity Tests

7.5.1 Considerations Related to Fire Models and Animal Models Used

Most studies of the toxicity of fire effluent atmospheres using animals involve exposure to combustion atmospheres generated using bench-scale methods, although a few studies involving exposure of rodents to effluent samples taken from large-scale fires have been reported.

For bench-scale experiments involving animals there are two major considerations with respect to their relevance to toxic hazards experienced by humans in full-scale fires:

- The extent to which the yields of toxic products and exposure concentrations in the bench-scale tests represent those occurring in different types and stages of full-scale fire scenarios involving human exposures.
- The extent to which the animal models provide qualitative and/or quantitative data relevant to the physiological and pathological effects on human fire victims during and after exposure in fire incidents.

Most animal exposure studies were performed during the 1970s and 1980s and few attempts were made to relate the decomposition conditions and product mixes generated in small-scale combustion generation equipment with those occurring during full-scale fires. Most exposures were to products generated by non-flaming oxidative decomposition of material specimens heated in air and only a few under the flaming decomposition conditions occurring in the vast majority of accidental fires involving human fire victims (see Chapter 2). In particular, very few experiments reproduced the under-ventilated ("vitiated") flaming conditions occurring in most full-scale fires in buildings or vehicles. This limits the work as a predictive tool for full-scale fire hazards. However, where animal exposures were combined with measurements of the concentrations of the major fire gases and particulates, they do give some indication of the extent to which the toxicity of complex fire effluent mixtures can be expressed in terms of that small number of major fire gases, and the extent to which other atmosphere components may also be important. In practice, the fire effluent mixtures produced by non-flaming decomposition are by far the most complex in terms of the numbers of chemical products at significant yields and concentrations. So if the toxic potency of these rich, complex mixtures can be quantified in terms of a few major toxic components, then toxic potency calculations for the simpler mixtures from flaming combustion should be much easier to achieve. Also, by screening a relatively large number of natural and synthetic polymeric materials, these bench-scale toxicity studies using rodents provide an indication of the extent to which the overall toxic potencies of fire effluent atmospheres from different materials are similar, and of instances where more extreme or unusual toxic potency has been observed from burning specific materials.

The other major limitation of many of these experiments was the toxicity model used. As described, the major toxic hazards considered important during fires are the rapidly developing physiological effects of sensory irritation and incapacitation due to effects of asphyxiants. Most animal tests involving rats and mice used simple measurements of lethality (LC_{50} concentrations), usually involving 30 minute exposure periods, sometimes followed by a 14 day observation period. Such methods produced a crude average of the overall lethal toxic potency from asphyxiants during exposure, and lung inflammation following exposure, but gave no indication of time or exposure dose to incapacitation. Other important limitations are the physiological differences between rodents and humans. For this reason, the relationships between exposure concentration and time to death, or number of deaths in any specific test, are only very approximately relevant to predicted effects in humans. Exposures of rodents (usually mice) are useful to measure the sensory irritant potencies of both individual irritants known to occur in combustion products and to combustion product mixtures produced by thermal decomposition of different polymers. Continuous measurements of respiratory pattern and especially respiratory rate are used on the basis that the severity of sensory irritancy to the eyes and upper respiratory tract is considered to be proportional to the decrease in respiratory rate. The RD_{50} concentration (respiratory depression 50%, the concentration resulting in a 50% decrease in breathing rate) then provides an indication of irritant potency, while observations of time course of response development indicate the rapidity or development of the response and the extent to which it is concentration- or dose-related. The few experiments involving exposure of primates to fire effluent atmospheres with detailed physiological and behavioural measurements of effects leading to incapacitation are more directly applicable to humans.[1,35]

7.6 Findings from Primate Exposure Studies

The UK fire statistics showed a four-fold increase in deaths between the 1950s and 1970s resulting from exposure to toxic smoke in fires (see Chapter 2). Following this finding a series of experiments was commissioned involving sub-lethal exposures of macaque monkeys to individual asphyxiant fire gases (CO, CO_2, HCN and low oxygen hypoxia) and a set of mixed fire effluent atmospheres generated from a range of polymeric materials.[35–39] These experiments were used to investigate the physiological mechanisms by which fire effluents impede escape and cause incapacitation in human fire victims. Physiological parameters were measured before, during and after 30 minute exposures under dynamic steady-state exposure conditions (*i.e.*, the individual gas or effluent mixture flowed continuously through the chamber at a constant rate, in order to maintain a constant concentration of fresh gas or fumes). Combustion product atmospheres were generated using an early version of the ISO19700 tube furnace,[40] whereby strips of material were introduced at a constant rate into a tube furnace under a stream of air. The effluents were passed into a mixing chamber where they were diluted with air. The animals were exposed to this atmosphere *via* a facemask or in a whole-body chamber. Physiological parameters indicating respiratory, cardiovascular and neurophysiological status were measured continuously. These included respiratory flow, tidal volume and frequency, heart rate and electrocardiogram, electrocorticogram, nerve conduction velocity and auditory evoked potentials. Clinical signs and body reflexes were also monitored, and blood samples were taken at intervals for carboxyhaemoglobin and blood cyanide measurements. Observations were continued over a 14 day recovery period, followed by lung pathology investigations in some cases.

For fire effluent exposures, wood (*Pinus sylvestris*), polyacrylonitrile and flexible polyurethane foam were pyrolysed under nitrogen at 300 °C, 600 °C and 900 °C. A rigid polyurethane foam, nylon 6, polystyrene and polypropylene were then decomposed under non-flaming oxidative decomposition conditions in air at 440–500 °C. Polypropylene was also decomposed under flaming and pyrolytic decomposition conditions and polystyrene under pyrolysis. The fire effluent atmospheres were monitored continuously for CO, HCN, CO_2, O_2, smoke optical density and total particulates. The fire effluents contained significant concentrations of asphyxiant gases (CO and HCN) and smoke particulates. Gas chromatography-mass spectrometry (GC-MS) investigations[41] showed a rich mix of aliphatic and aromatic hydrocarbons generated under pyrolysis, with a significant additional content of partially oxidised species (including acrolein, formaldehyde, crotonaldehyde, acetaldehyde, phenol and aliphatic acids) formed under non-flaming oxidative decomposition conditions. Flaming combustion consumed most of the organic content. Flexible polyurethane foam decomposed into isocyanate-derived "yellow smoke" at the lower temperatures, but at 900 °C formed a relatively smoke-free effluent.

In addition to these experiments on resting primates, two sets of experiments used animals conditioned to carry out a behavioural task in a chamber. The animals were conditioned to press a lever in response to an audible signal, after which they had a few seconds to reach the other end of the chamber where a food reward was presented for a brief period. The task was repeated at 3–5 minute intervals throughout a 30 minute exposure and post-exposure period.[36] The animal's performance was measured throughout exposure while they maintained an active state. Exposures were to different concentrations of carbon monoxide and to fire effluent atmospheres generated from polypropylene under pyrolytic and non-flaming oxidative thermal decomposition conditions.[1]

In other experiments a conditioned escape paradigm was used to examine effects of 5 minute exposures of baboons to different concentrations of acrolein or HCl gas. After 5 minutes' exposure the animals were conditioned to press a lever to release the door catch and escape from the exposure chamber.[42]

The results from the individual gas experiments were:

- Carbon monoxide had no physiological or behavioural effects until the blood COHb concentration reached approximately 30% (for active animals) or 40% (for animals at rest). Free moving animals were unaffected until they suddenly became lethargic and, during a few seconds, passed through a brief period of apparent intoxication to collapse and loss of consciousness. Resting animals were unaffected

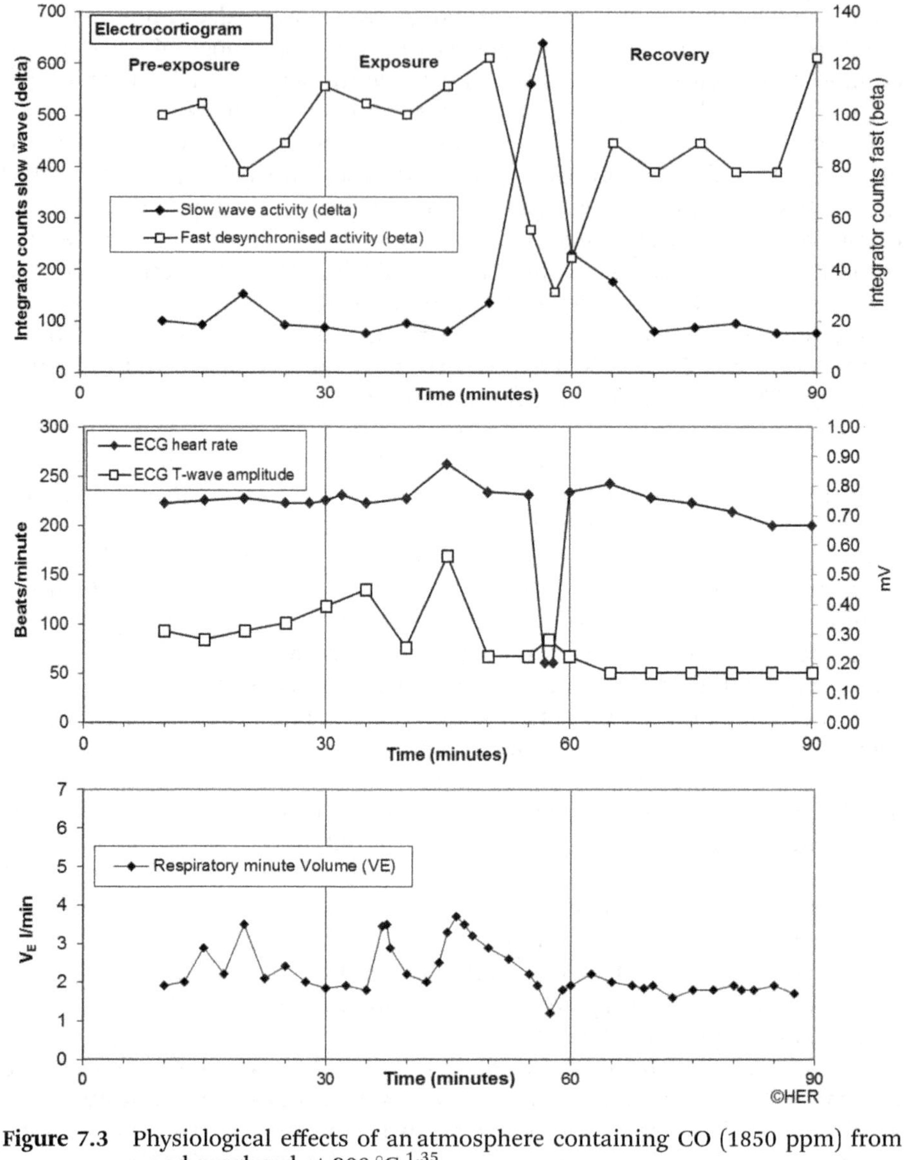

Figure 7.3 Physiological effects of an atmosphere containing CO (1850 ppm) from wood pyrolysed at 900 °C.[1,35]

until their consciousness suddenly became impaired (Figure 7.3). As the animals then became unconscious there was a decrease in heart rate with occasional arrhythmias, a large increase in slow-wave brain activity, a decrease in respiration and failure of withdrawal reflexes. This occurred at an exposure dose of 27 000 ppm min for free moving animals and 36 000 ppm min for resting animals. For lightly active, but larger baboons (~20 kg compared with 3–4 kg macaques) incapacitation occurred at an exposure dose of 34 350 ppm min.[42]

- Hydrogen cyanide also had no immediate effects, but after a few minutes (depending upon the concentration) the animals started to hyperventilate. A positive feed-back resulted such that the increased ventilation increased the rate of uptake of HCN, which further stimulated respiration until the animals lost consciousness and then showed similar signs to those exposed to CO (see Chapter 10 and Figure 7.4). There was then a large decrease in respiration, so the rate of HCN

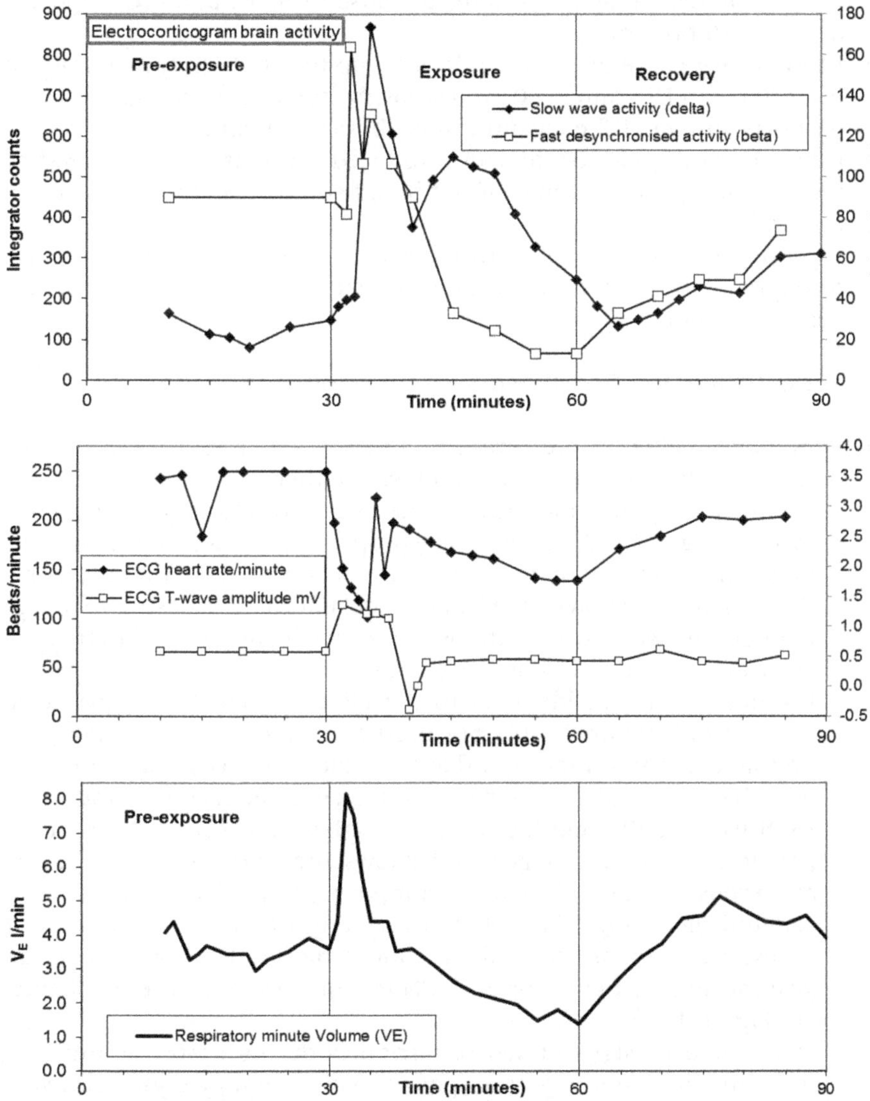

Figure 7.4 Physiological effects of an atmosphere containing HCN (163 ppm) from polyacrylonitrile pyrolysed at 600 °C.[1,38]

uptake decreased and the blood cyanide concentration stopped increasing. Animals sometimes recovered slightly at this point, and then gradually passed deeper into unconsciousness. A rapid recovery occurred a few minutes after the end of the exposure, but there was little decrease in whole blood cyanide concentrations over periods of up to 1 hour.[27,38] Incapacitation occurred within a few minutes during exposure to high HCN concentrations (150–200 ppm) (*Ct* exposure dose = 270 ppm min), while lower concentrations of 80–100 ppm required approximately 30 minutes to cause incapacitation (*Ct* exposure dose = 2610 ppm min).[1]

- Baboons were able to escape efficiently up to very high concentrations of 15 000 ppm HCl and 2780 ppm acrolein, but severe irritancy occurred post-exposure at HCl concentrations above 1000 ppm.
- Exposure to 5% carbon dioxide produced an approximate three-fold increase in $\dot{V}_E$ (the volume of air breathed each minute) and minor cerebral effects.
- Exposure to 10% oxygen caused cerebral depression and lethargy throughout most of a 30 minute exposure with semi-consciousness being apparent.

The results from the fire effluent exposures were:

- Physiological effects were similar to those of the main toxic gases present in the effluent atmosphere and were either:
    - Dominated by and identical to the effects of CO alone (*e.g.*, wood pyrolysed at 900 °C, flaming polypropylene). Atmospheres were almost smoke-free with most fuel carbon released as CO and some CO_2. Figure 7.3 shows the physiological effects of exposure to a high temperature wood smoke atmosphere, which are identical to those of exposure to an equivalent CO gas mixture.
    - Dominated by and identical to the effects of HCN (*e.g.*, pyrolysed polyacrylonitrile or flexible polyurethane foam at 900 °C, rigid polyurethane foam under non-flaming oxidative decomposition). Atmospheres were also almost smoke-free, with high concentrations of HCN with small amounts of other nitriles, and some CO from the polyurethane foams. Figure 7.4 shows the physiological effects of an exposure to thermal decomposition products from polyacrylonitrile pyrolysed at 600 °C, which are virtually identical to those of exposure to an equivalent concentration of HCN gas in air (see the similar figure for an HCN gas mixture exposure in Chapter 10, Figure 10.1).
    - Dominated by signs of sensory irritancy during exposure and lung inflammation after exposure (polyurethane foam pyrolysed at 300 °C and 600 °C, wood pyrolysed at 300 °C).
    - The effects of CO and HCN mixtures were consistent with the toxicity of the two gases being somewhat additive.

- Where CO and irritants were present, there was some hyperventilation resulting from stimulation of lung irritancy receptors but, although $\dot{V}_E$ was increased, the uptake of CO was not increased. Ventilation–perfusion ratio changes may have limited CO uptake.
- The particulate isocyanate-derived "yellow smoke" produced when flexible polyurethane foam was pyrolised was highly irritant, producing respiratory changes during exposure and lung inflammation and oedema some hours later.[11,35]
- Pyrolysed polypropylene gave a hydrocarbon mist containing a wide range of aliphatic and aromatic pyrolysis products, but the atmosphere was innocuous, producing no signs of irritancy or other toxic effects in the animals, and no disruption of the behavioural task performance. Polypropylene decomposed under non-flaming oxidative conditions gave a highly irritant smoke, producing a distinctive respiratory pattern with hyperventilation, and signs of lung oedema and inflammation some hours after exposure. Behavioural task performance was inhibited at 1.85 mg L^{-1} mass loss concentration (*i.e.*, exposure to the effluent mixture produced by decomposition of 1.85 mg polypropylene into each litre of chamber air).

These primate studies confirmed the effects reported in fire fatalities and survivors. These were that the only important acute effects of exposure to fire effluents from common polymeric materials were asphyxia from the mixed asphyxiant gases present (CO, HCN, CO_2 and low oxygen hypoxia) and irritant effects, consisting mainly of sensory irritancy at the time of exposure and lung inflammation and oedema after exposure. The effects of CO and HCN were considered somewhat additive as later confirmed in experiments with rats.

The irritant effects were predictable in general terms from the chemical composition of the effluents, in that atmospheres rich in partially oxidised organic species, such as unsaturated aldehydes or isocyanate-derived compounds, were highly irritant, while atmospheres low in organic content, or lacking partially oxidised organic species, were of low irritancy.

Figure 7.5 shows the respiratory pattern in a macaque exposed to the irritant thermal decomposition products from wood (*Pinus sylvestris*)

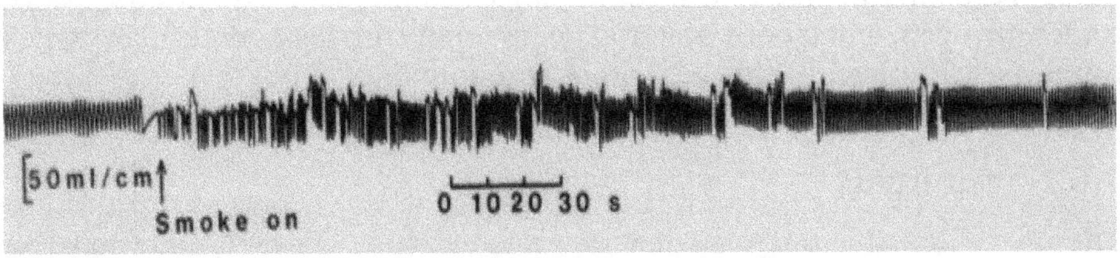

Figure 7.5 Effects on respiration pattern of a macaque monkey following exposure to smoke from wood pyrolysed at 300 °C (trace shows tidal volume).

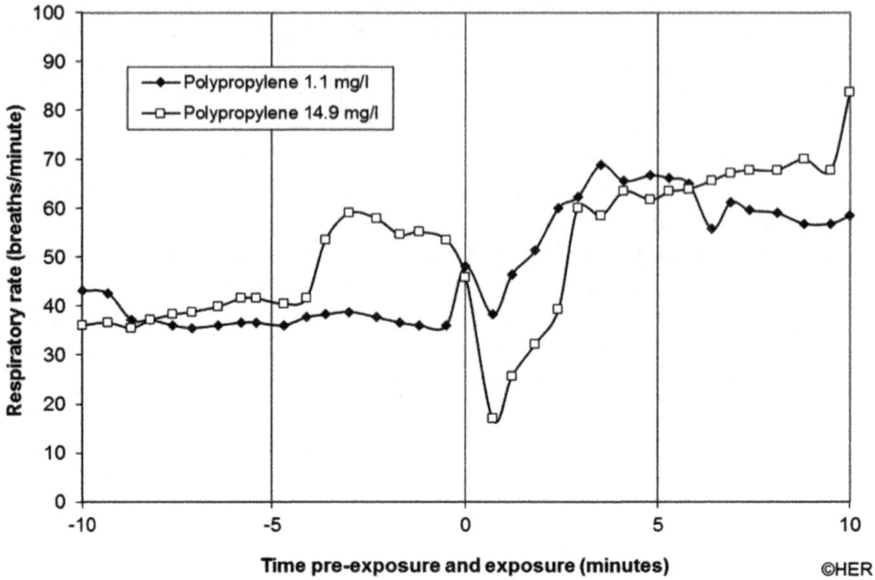

Figure 7.6 Respiratory rate in a macaque monkey before and during exposures to irritant thermal decomposition products from non-flaming polypropylene.[1,11]

pyrolysed at 300 °C in an early version of the ISO19700 tube furnace. There is initial evidence of a trigeminal-mediated respiratory depression reflex with pauses at end inspiration, giving way to a continued hyperventilation due to stimulation of lung vagal receptors.

Polypropylene, when decomposed under non-flaming oxidative decomposition conditions in air, provides another good example of these effects, involving an initial respiratory rate suppression followed by hyperventilation (Figure 7.6).

When the furnace temperature was increased to 700 °C, so that flaming combustion occurred, a "cleaner" atmosphere was formed with a lower smoke and organic content and most of the fuel carbon released was as oxides of carbon. The resulting atmosphere was of low irritancy but caused asphyxia due to the carbon monoxide content. Irritant atmospheres were produced by both wood and flexible polyurethane foam when decomposed under nitrogen, since both materials contained sufficient molecular oxygen to release significant yields of irritant compounds.[11,35]

7.6.1 Summary

These results illustrate a general relationship between thermal decomposition conditions and fire effluent composition (see Chapter 2). For any material, the most irritant, organic-rich atmospheres are obtained from non-flaming oxidative decomposition. Well-ventilated flaming results in efficient combustion, so that organic species are oxidised to form carbon oxides and irritants are absent, or present at very low yields (unless gas phase fire retardants additives are present). For under-ventilated (vitiated) combustion conditions the organic content and irritancy increases somewhat, but less so than under non-flaming conditions, since both the number of species present and their yields are lower.[41,43] If any unexpected acute toxic effects are likely to occur from organic chemical species in fire effluent mixtures, they are almost certain to be found in the organic-rich vapour and particulate effluents from non-flaming oxidative decomposition. For flaming combustion conditions, the main toxic effects can be predicted to be those of the major asphyxiant gases, plus a degree of irritancy, especially from products of vitiated combustion, or from acid gases. None of the materials used in the primate studies were fire-retarded, but increased irritancy is to be predicted from fire-retarded materials, mainly from the acid gases released (halides, sulfur oxides and phosphoric acid) under all decomposition conditions, and also because fire retardants acting in the gas phase (especially halogen fire retardants) reduce the efficiency of combustion under flaming conditions, thereby also increasing the content of potential organic irritants. The relationships between combustion conditions and toxic product yields in fires are considered in detail in Chapters 2 and 3.

The sub-lethal exposures of primates to individual fire gases and fire effluent mixtures have provided useful information on the mechanisms of incapacitation of human fire victims and quantitative data on the uptake and effects of toxic fire gases. These have been used to develop predictive methods for calculating time to incapacitation in human fire victims (see Chapter 14). However, the materials studied were limited to a small number typically used in domestic furnishings and the range of combustion conditions was limited mainly to non-flaming decomposition. Although human fire victims are exposed to these combustion conditions in some fire incidents, the main cause of injury and death is from flaming fires and flaming fires are also the source of most environmental air pollution. Since this work was done, extensive studies have been made of the combustion products from full-scale fire incidents and from a wide range of materials in small-scale combustion tests designed to produce different fire conditions. As stated, the main differences in effluent composition between the non-flaming conditions used for most of the primate experiments and flaming fires are that the organic content, including organic irritants, is considerably lower in flaming fires, while the yields and concentrations of asphyxiant gases, including CO, CO_2 and HCN, can be higher. In particular, the yields and concentrations of CO_2 are likely to be an order of magnitude higher. When attempting to extrapolate from the results of these early primate and rodent combustion toxicity studies to predict effects on human fire victims it is therefore important to take into account the effects of increasing the CO_2 concentration from around 0.2–0.4% by volume in these studies to 5–15%, which is typical in large-scale fires.[29] From human and primate data on the effects of hypercapnia it is considered that the main effect of inhaled CO_2

during the early stages of fires is to cause hyperventilation, increasing the rate of uptake of other toxic products in the smoke, while possibly providing some degree of amelioration of CO and low oxygen hypoxia resulting from shifts of the haemoglobin oxygen dissociation curve.[1,28] After a period of exposure, the acidosis, circulatory changes and direct toxic effects of hypercapnia can cause distress and incapacitation at concentrations exceeding 5% CO_2. In most fire incidents, whilst inhaled CO_2 is likely to increase the rate of uptake of other gases, incapacitation from the effects of HCN and CO are usually predicted before the direct toxic effects of CO_2 inhalation become a dominant factor. Hypoxia from lowered oxygen concentrations can also be a significant factor in some fire exposures.

Another characteristic of flaming fires is the nature of the particulates, which tend to contain more carbon soot particles and lower concentrations of condensed organic vapours than for non-flaming fires. The yields of polycyclic aromatic hydrocarbons (PAHs) and dioxins may also be higher in flaming fires. In order to study the toxic effects of fire effluents from a wide variety of materials it has been necessary to use rodent models, as described in the next section.

7.7 Estimates of Sensory Irritant Potency of the Products of Combustion from Natural and Synthetic Polymers under Different Fire Conditions using Rodents, and the Contribution to Overall Irritant Potency from Major Irritant Gases

7.7.1 Use of Rodents

Rodents have been used to measure both sensory and lung irritancy from a wide range of individual substances and mixed combustion products from many different materials. Although these studies provide a convenient and effective model for studying irritancy, it is important to consider some key inter-species differences when extrapolating from effects in rodents to predict effects in humans.

7.7.2 Studies using the Rodent Respiratory Rate Depression (RD_{50}) Test

The rodent respiratory rate depression test, and in particular measurement of the mouse RD_{50} concentration, is a useful quantitative method for comparing the relative sensory irritant potencies of individual irritant gases and fire effluent mixtures.[10,44–47] Although the basic trigeminal reflexes involved are very similar in humans and mice, the test method does not rely on the mouse and human responses being identical, but rather that the intensity of the mouse response can be used as a predictive scalar for the severity of effects in humans. Most individual irritant chemicals are highly irritant to humans at the mouse RD_{50} concentration. So this method (in conjunction with other available data) can be used to estimate concentrations of individual irritants or irritant fire effluent mixtures likely to seriously impair escape of human fire victims.

Figure 7.7 shows the relationship between the mouse RD_{50} concentration and reported effects in humans for a variety of common irritant chemicals. The black squares compare the concentration of each gas reported to be painfully irritant to humans with the mouse RD_{50} concentration. The open triangles compare workplace maximum tolerated exposure concentrations for the same set of irritant chemicals (the 1980 workplace threshold limit values [TLV-TWA], for which human tolerance levels were set on the basis of irritancy) against $0.03\times$the mouse RD_{50} concentrations. Both show a reasonably good agreement between the human and mouse data, validating

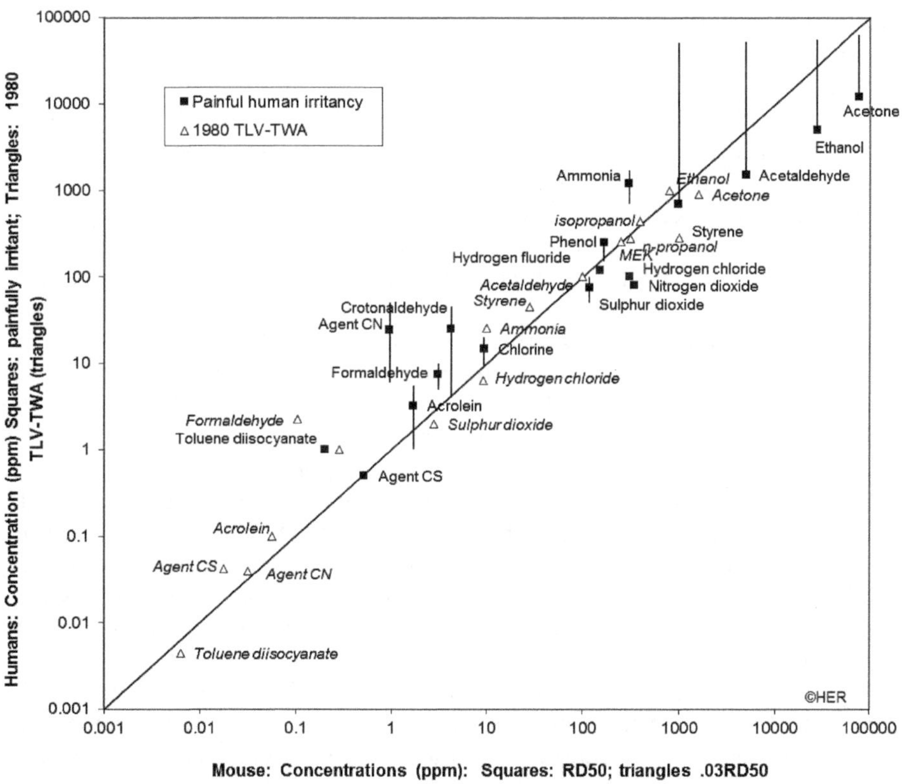

Figure 7.7 Relationships between two irritancy endpoints in humans and mice. Black squares compare mouse RD_{50} concentrations with concentrations reported to be painfully irritant to humans. Open triangles compare $0.03\times$mouse RD_{50} concentration with workplace tolerance levels for irritancy (1980 TWA-TLV).
Data from Purser[1] and Alarie.[44]

the method by demonstrating that the mouse RD_{50} model is quite a good predictor of irritancy in humans, and that the mouse RD_{50} concentration, in general, represents a concentration likely to be painfully irritant to humans. For most irritants, an upper limit for tolerance in the workplace is approximately 1/30 of a concentration causing severely painful irritancy.[44]

As an example, the mouse RD_{50} for the highly irritant unsaturated aldehyde acrolein is 1.7 ppm, while the concentrations reported as painfully irritant to humans are 1–5.5 ppm, which are between 0.6 and 2.3 times the mouse RD_{50}. Multiplying the mouse RD_{50} by 0.03 to predict a tolerable human workplace hygiene level for irritancy gives a figure of 0.051 ppm. This compares with the 1980 TLV-TWA of 0.1 ppm.

Figure 7.7 also illustrates the enormous irritant potency range of over approximately six orders of magnitude for different chemicals, from the most irritant substances, such as toluene diisocyanate, to low potency chemicals such as acetone. These low potency irritants (such as ethanol and acetone) require much higher concentrations to elicit a similar painfully irritant effect. For the four least potent substances, the reported painfully irritant concentrations for humans were reported as "greater than", as illustrated by the range bars on Figure 7.7.

Although the RD_{50} concentration is an indication of likely severe irritancy in humans, experimental and incident data are needed to estimate the severity of effects on human escape capability. Table 7.1 shows potency data for irritants occurring in combustion products (with the exemption of the riot control agents, CS and CN, included for comparison). The table compares the mouse RD_{50} with concentrations reported as severely irritant in humans. Also shown are the LC_{50} concentrations normalised to a 30 minute exposure period (assuming Haber's rule). The right hand column shows the LC_{50}/RD_{50} ratio. Since, for most of these substances, lethality is due to lung inflammation, this ratio provides a useful indication of the difference between exposure concentrations causing sensory irritancy and those causing lung tissue damage. For agent CS, the high ratio of 529 indicates that a concentration causing severe pain presents a low health hazard, while the very low ratio for NO_2 indicates that it is very dangerous at noticeably irritant concentrations. NO_2 has a low aqueous solubility, resulting in low absorption and stimulation of the nasal receptors, with a high level of penetration into the deep lung, resulting in oedema and inflammation.

Although Figure 7.7 and Table 7.1 provide some indication of the severity of irritant effects of individual chemicals on humans, an important aspect in relation to fires is the evaluation of irritancy for combustion product effluent mixtures from different materials. Figure 7.8 shows RD_{50} test run data for a combustion product effluent mixture, in terms of the average respiratory rate for groups of four mice exposed to thermal decomposition products from a PVC cable jacket material (PVC-J) at a mass loss concentration of 0.72 g m^{-3} (approximately twice the RD_{50} concentration of 0.34 g m^{-3}). The material was decomposed in an early version of the ISO/TS 19700 tube furnace.[40,46]

Table 7.1 Sensory and pulmonary irritancy of combustion products.[d]

Irritant	RD_{50} mouse[a]/ppm	Severe sensory irritancy in humans/ppm	30 minute LC_{50} mammal[c]/ppm	LC_{50}/RD_{50}
	0.1–1.0			
Toluene diisocyanate	0.20	1.0[48]	100[48]	500
o-Chlorobenzylidene malonitrile (CS)[b]	0.52	0.5[12]	150–400[49,50]	529
α-Chloroacetophenone (CN)[b]	0.96	6–50[12]	300–400[51,52]	365
	1.0–10			
Acrolein	1.7	1–5.5[45,48]	140–170[42,51,52]	91
Formaldehyde	3.1	5–10[45,48]	700–800[52,53]	242
Chlorine[b]	9.3	9–20[45]	100[53]	11
	10–100			
Crotonaldehyde	4.2[54]	4–45[48]	600–1500[48,51,56]	156
Acrylonitrile		>20[48]	4000–4600[52]	
Penteneone			1000[52]	
Phenol	166[55]	>50[48]	400–700[52]	3
	100–1000			
SO_2	117	50–100[45,48]	300–500[52,53]	3
NH_3	303	700–1700[58]	1400–8000[52]	16
HF	151[57]	120[48]	900–3600[52]	13
HCl	309	100[45,48]	1600–6000[51,52]	12
HBr		100[48]	1600–6000[51]	
NO_2	349	80[45,48]	60–250[52,53]	0.4
Styrene	980	>700[48]	10 000–80 000[52]	46
Acetaldehyde	**1000–10 000**			
	4946	>1500[48]	20 000–128 000[52]	15
	10 000–100 000			
Ethanol	27 314	>5000[48]	400 000[52]	15
Acetone	77 516	>12 000[48]	128 000–250 000[52]	3

[a] RD_{50} concentrations from Alarie,[44] except where otherwise indicated.
[b] Substances not detected in combustion atmospheres.
[c] LC_{50} concentrations have been normalised to a 30 minute exposure time according to Haber's rule.
[d] The potential for causing sensory irritation spans six orders of magnitude, while that for causing death spans approximately three orders of magnitude. For substances in the table down to NO_2, death is likely to be due to lung irritation, while for the remainder from styrene to acetone, death is likely to be due to asphyxiation.

It is noteworthy that the effects on breathing rate are almost instantaneous at this highly irritant concentration, the rate decreasing to around 50% of the pre-exposure rate within 4 seconds and settling at 30% of the pre-exposure level within 1.5 minutes. Figure 7.9 shows a plot of the RD_{50} determination from the relationship between percentage respiratory rate depression and log exposure concentration for thermal decomposition products from this material.

These experiments were used to measure the irritant potency of effluent from different materials, in order to rank them in comparison to each other and in relation to the irritancy of common irritant chemical species. Another aim has been to establish the extent to which the overall irritancy of fire

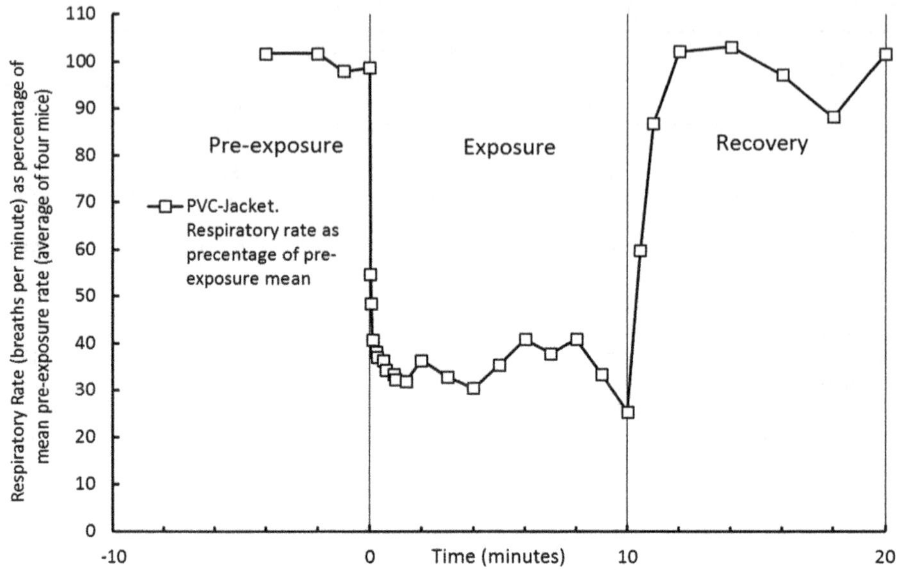

Figure 7.8 RD_{50} test: respiratory rate depression in mice exposed to thermal decomposition products from PVC cable jacket.

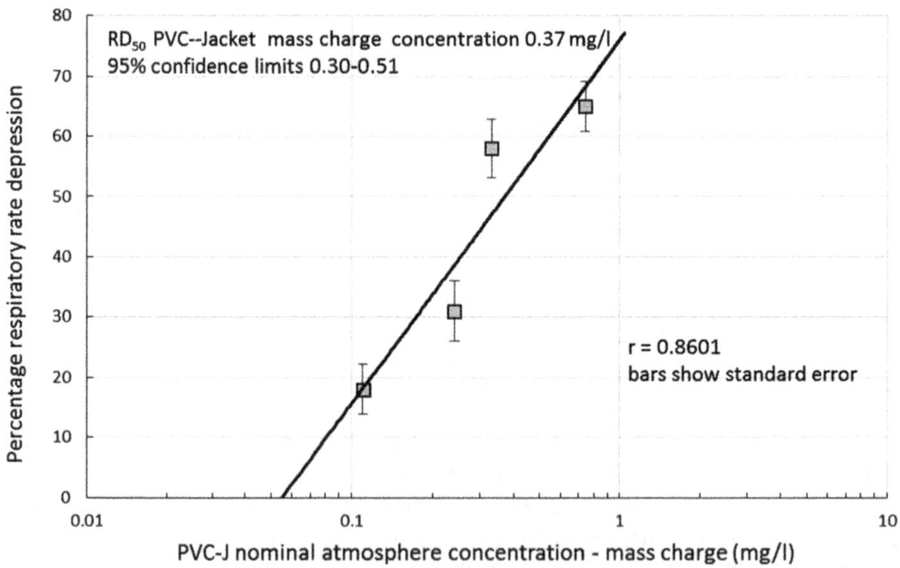

Figure 7.9 Relationship between breathing rate and log exposure concentration in mice for the same material as in Figure 7.8 (PVC-J).

effluent mixtures can be explained in terms of the individual chemical components present and to develop a predictive calculation model for sensory irritant potency. A default model assumption is that each irritant in a mixture contributes in an additive manner to the overall irritancy depending on the concentrations and potencies of each irritant component. In order to sum the effects of the different components, the fractional effective concentration approach is used whereby the fractional irritant concentration (FIC) of each component is expressed in terms of a fraction for which the numerator is the concentration of the irritant present in the atmosphere and the denominator RD_{50} concentration for each irritant present.

The overall FIC for exposure to mice an irritant mixture is then as follows:

$$\begin{aligned} FIC = {} & FIC_{HCl} + FIC_{HBr} + FIC_{HF} + FIC_{SO_2} + FIC_{NO_2} + FIC_{CH_2CHO} \\ & + FIC_{CH_2O} + \Sigma FIC_x \end{aligned} \tag{7.3}$$

Where ΣFIC_x is the sum of the FICs for any other irritants present.

If the irritant components were additive, an overall FIC of unity would represent a mixture concentration resulting in a 50% decrease in respiratory rate of exposed mice. This method is commonly used to assess the potential effects of mixtures of toxic gases for industrial hygiene purposes and has the merit that it is reasonably conservative. There are few data on the effects of mixtures of specific irritants, but it seems reasonable to assume direct additivity for different halogen acid gases. Although different chemical irritants substances stimulated different irritant receptors, there is evidence that effects are additive in terms of nerve stimulation, and even that heat pain and irritancy are additive (for example, when eating spicy food and taking a hot drink).

The method is also used to assess effects on human fire victims. Sensory irritant effects occur on a continuum, increasing in severity in proportion to the log of the exposure concentration, but it is considered feasible to recognise thresholds levels for effects on humans related to fires:

- A concentration capable of seriously impairing escape capability and movement speed.
- A concentration capable of causing incapacitation, such that the subject effectively cannot move.

For these two endpoints the same FIC equation is used with different denominators for each term depending upon the desired endpoint (see Chapter 14).

7.7.3 Findings from Combustion Irritancy Tests

Alarie and associates have carried out mouse RD_{50} determinations for a wide range of individual chemicals[44,45,53,59] and also for combustion product mixtures generated using the University of Pittsburgh (UPITT) furnace test method.[60–62] Unfortunately, the usefulness of UPITT data are limited by the nature of the decomposition method. The method employed ramp heating of a sample of material in a large crucible under a stream of air. The effluent was carried through a chamber in which groups of four mice were exposed nose-only and their respiratory pattern measured. The specimen passed through a number of stages involving a period of increasing non-flaming oxidative decomposition as it heated up, followed in some cases by flaming combustion if ignition occurred. Since the mass loss rate of the fuel was not measured and there was limited analysis of the atmosphere it is difficult to relate the effects observed to the yields and concentrations of thermal decomposition products evolved. It is also difficult to relate the furnace conditions to those occurring during actual full-scale fires[1] (see Chapter 2).

An early version of the ISO 19700 tube furnace apparatus was used by Purser for measuring the RD_{50} of a range of materials. For this method, as described in Chapter 2, the decomposition products are evolved continuously from a strip of material introduced into the furnace tube at a constant rate under a constant stream or air, giving a constant atmosphere composition throughout a test. The method enables a wide range of combustion conditions to be set up in different experiments, specified in terms of continuous flaming or non-flaming behaviour and fuel : air equivalence ratio. The concentrations of effluents in the exposure chamber can be characterised in terms of a wide range of measured atmosphere components and the mass loss or mass charge concentration of the material under investigation. Figures 7.8 and 7.9 illustrate the results from one such set of experiments on a PVC cable jacket and insulation materials.[46]

Most of these experiments using mice were for non-flaming oxidative decomposition of the test materials, but a few were under flaming conditions. Figure 7.10 ranks the materials in terms of RD_{50} concentration,[9] so that the shorter the bar the more irritant the atmosphere. Results are expressed as concentrations in g m^{-3} mass loss (standard error bars are shown). As irritancy scalars, the lower three bars show individual irritants common in fire effluents and considered as likely major individual fire effluent irritant species. The low molecular weight unsaturated aldehydes (acrolein and crotonaldehyde) are among the most potent irritant species known to occur in fire effluents (RD_{50} concentration 0.004 g m^{-3}), while hydrogen chloride has an RD_{50} concentration of 309 ppm (0.468 g m^{-3}). These polymeric materials were highly irritant when decomposed under non-flaming oxidative thermal decomposition conditions (RD_{50} 0.05–1.38 g m^{-3}), (apart from a phenolic resin/fibreglass composite material with a low irritancy [RD50 ~9.1]). The most irritant materials were the fumes given off by the binder oil in the fibreglass insulation blanket materials used as insulation in aircraft fuselages (RD_{50} 0.05 g m^{-3}), low density polyethylene (RD_{50} 0.05 g m^{-3}) and silicone rubber (RD_{50} 0.06 g m^{-3}). Plasticised PVC laminate, Redux adhesive and polypropylene (not shown in Figure 7.10) were also highly irritant (RD_{50} 0.1 g m^{-3}). However, when several of these materials were decomposed under flaming conditions, the irritant potency was considerably reduced. For these single experimental runs under flaming conditions the results are only approximate, but were generally at least an order of magnitude less irritant than when the same materials were decomposed under non-flaming oxidative conditions.

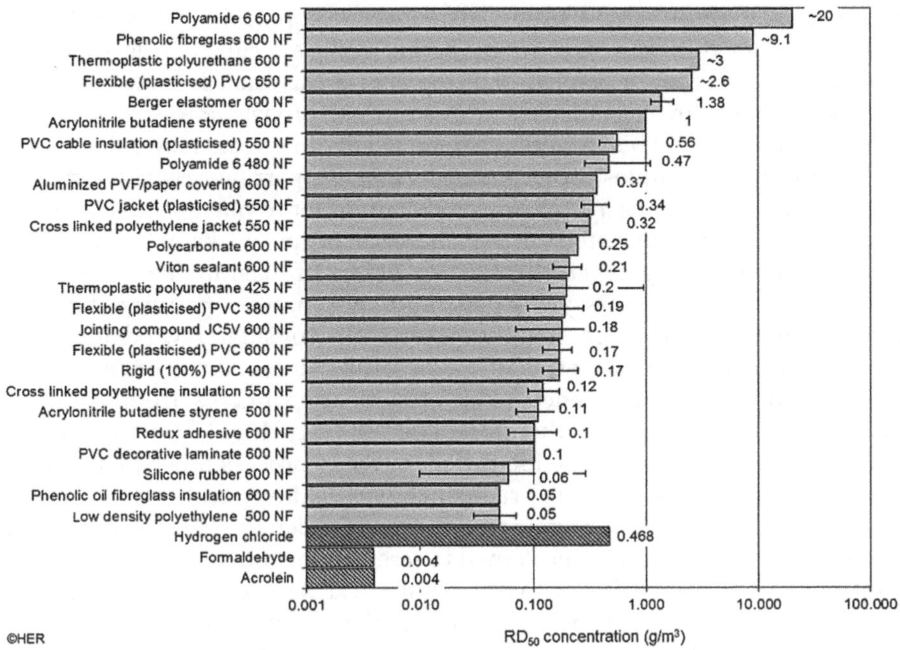

Figure 7.10 Mouse RD_{50} concentrations measured under non-flaming oxidative (NF) and flaming (F) combustion conditions for a range of materials using and early version of the ISO 19700 tube furnace method.[9]

Almost all materials therefore produce highly irritant atmospheres under non-flaming oxidative thermal decomposition conditions, when relatively high yields of organic substances, including irritant species, are known to occur. The irritancy is reduced under flaming conditions although materials such as PVC remain somewhat irritant due to the release of HCl.

Some attempt has been made to determine the extent to which the observed irritancy can be explained in terms of the effluent composition using gas chromatography-mass spectrometry and ion chromatography to characterise the atmospheres. For the most irritant materials, such as polyethylene, approximately 30% of the mass loss is in the form of organic products, giving a mass loss concentration of total organics of around 0.015 g m^{-3}. The most irritant substances detected in these atmospheres are acrolein and formaldehyde, but in order to account for the observed irritant potency the total mass loss concentration of these substances would need to be 0.004 g m^{-3}, representing approximately 26% of the total organic mass loss. In practice the bulk of the organic mass loss is known to be in the form of low molecular weight aliphatic hydrocarbons (methane, ethane, ethene, propane *etc.*), as well as small amounts of benzene, phenol and styrene, and traces of other compounds including aldehydes.

The fraction of the mass loss as formaldehyde and acrolein in combustion products is very small. For example, the highest yields of these compounds are obtained from non-flaming oxidative decomposition of cellulosic materials and, during tests on cotton the highest formaldehyde yield was 0.25 mg g^{-1} and that of acrolein 0.05 mg g^{-1}. If the main irritant substance produced from fibreglass insulation or from polypropylene was produced at similar low yields and had a similar low molecular mass, then it would need to be several hundred times more potent as a sensory irritant than substances such as formaldehyde, acrolein or agent CS in order to obtain the RD_{50} mass loss concentration measured in the mice. There are no known substances with sensory irritant potencies within two orders of magnitude of this.

From the organic mass yield data and the analyses performed it was evident that the observed irritant potency is considerably greater than could be accounted for in terms of the organic irritants detected (assuming a simple additive FEC model). The effluent atmospheres from common polymeric materials must therefore contain very highly potent irritant compounds that remain to be identified, or some process other than simple chemical irritancy, such as free radical activity, is involved.[1,47]

It was also possible to determine the extent to which the measured irritancy of effluents from PVC can be explained in terms of the measured HCl and what contribution organic irritants make. Figure 7.11 shows the contribution of hydrogen chloride to the sensory irritancy of thermal decomposition products from a plasticised PVC under non-flaming oxidative thermal decomposition conditions at 380 °C.[63] At the RD_{50} concentration for this material the HCl concentration was 49 ppm, while that for HCl gas alone is 296 ppm. The mixed thermal decomposition products from the PVC were therefore six times more irritant than could be accounted for in terms of the HCl content (the difference presumably being due to organic irritants in the atmosphere in addition to the HCl). Under flaming conditions at 650 °C an approximate 50% decrease in

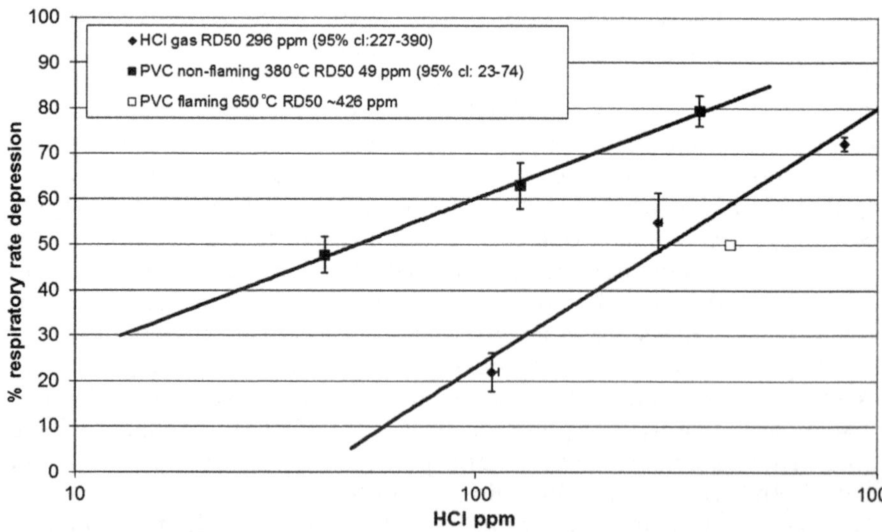

Figure 7.11 Concentration–effect relationships for atmospheres containing HCl in different forms.

respiratory rate was obtained for an HCl concentration of 426 ppm, which is reasonably close to the HCl RD_{50} concentration of 296 ppm.

Under flaming conditions most of the observed irritancy of PVC can therefore be accounted for in terms of HCl.

The overall irritant effect of a combination of chemical species in a fire effluent mixture can therefore be measured using mice without the need to identify and measure them all chemically. Chemical analysis of combustion products has proven a poor basis for predicting irritancy, since in every case under non-flaming conditions the observed irritant potency to the mice was considerably greater than could be accounted for in terms of the concentrations of known irritant acid gases and organic irritants. Irritancy was much reduced under flaming conditions and may be more easily related to a small number of irritant chemical substances.

Although there is a need for more research on the irritant potencies of mixtures of individual irritant species and to determine the contribution of individual organic irritants to the overall irritant potencies of thermal decomposition products, it is considered that the simple additive FIC model is a reasonable approach for estimating the overall irritant potencies of mixtures in humans. From the observations made in mice, primates and humans, and the chemical analysis of fire effluent mixtures, it is likely that in full-scale flaming fires the most important irritants are likely to be acid gases, with some contribution from the organic irritants, particularly under vitiated combustion conditions. Although different sensory irritants stimulate nociceptors by different biochemical mechanisms, it is likely that the effects of different chemical species will be broadly additive up to a particular level, above which the pain response is likely to reach saturation.[63]

7.7.4 Development Time and Concentration–Dose Relationships of Sensory Irritant Responses

As a fire hazard sensory irritancy is most important in terms of the extent to which it impairs or prevents escape in fire victims, or causes distress during exposure at lower concentration, outside, exposures. The hazard assessment models described in the previous section are based upon the concept that sensory irritant effects on the eyes and upper respiratory tract occur almost instantaneously on exposure and that the severity is proportional to the log of the exposure concentration. This is considered to the most meaningful approach because:

- At concentrations causing severe irritant signs and symptoms, the effects in humans are reported as occurring immediately or within a few seconds (for example, the effects of side-steam cigarette or bonfire smoke in the eyes or nose).
- Figures 7.5 and 7.6 demonstrate this instantaneous effect on respiratory pattern in primates exposed to irritant wood and polypropylene smoke.
- Figure 7.8 shows the rapid decrease in respiratory rate in mice exposed to PVC smoke.

An alternative would be some form of $C^n t$ dose model, but while such a dose-based model is useful for the prediction of effects on lung pathology and lethality, it does not fit well with the transient respiratory rate depression in primates (including humans) and the maintained respiratory rate depression plateau in mice (see Figure 7.8), as opposed to an effect increasing with accumulated dose.[47] More importantly, it does not fit with reported human symptoms (for example, during experiments with agent CS), with the most severe symptoms occurring immediately on exposure, then maintained more or less constant or somewhat decreasing with exposure duration.[12,64]

Despite these findings it has been found that at least for some irritant substances the development of the respiratory rate depression reflex in mice is not instantaneous, but takes some time to develop during a continuous exposure, thereby demonstrating at least some degree of dose relationship.

This effect is illustrated in Figure 7.12, which was obtained by fitting development rate curves to experimental data for respiratory rate depression in mice. These were exposed to hydrogen chloride gas using a "square wave" exposure system with instantaneous exposure to different maintained concentrations. At progressively higher exposure concentrations, not only is the extent of respiratory rate depression greater, but the peak response is also reached more quickly. In particular, at severely irritant concentrations close to the RD_{50} concentration or greater, the time to a 50% rate decrease

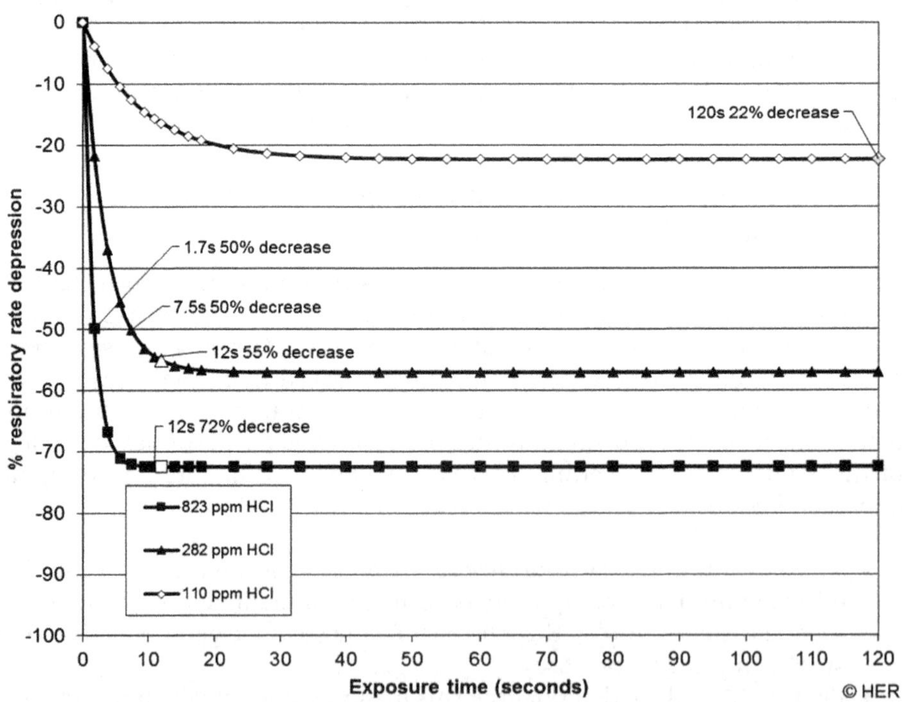

Figure 7.12 Time–response curves for percentage respiratory rate depression in mice exposed to hydrogen chloride gas.[10]

is reduced to a few seconds.[47] Attempts to model a mechanism for this response pattern gave a good representation by assuming a developing equilibrium between an accumulating inhaled dose at the site of action (nasal trigeminal receptors) balanced by some form of absorption or detoxification at a rate proportional to the concentration at that site.[47] One important factor reducing the rate of accumulation is the respiratory rate depression reflex itself, which can result in a decrease in inhaled minute volume of up total of 90%. The time constants for the development of the response plateau vary with different inhaled irritants, as shown in Table 7.2, depending on aspects such as their absorption in the nasal mucosa and the extent to which they are particle-bound.

In addition to the time required for the receptor response, the mouse respiratory rate decrease response is a neuro-muscular reflex, which takes a significant time to develop depending upon the neural pathway involved, and may not necessarily reflect the timing and severity of pain sensation. It is noteworthy that the time to achieve a 50% rate decrease is somewhat longer for the thermal decomposition atmospheres than for HCl gas, even though the PVC decomposition products contain some HCl. One reason for the more prolonged response time may be that a proportion of the irritant organic vapours and acid gases in thermal decomposition atmospheres are adsorbed onto respirable smoke particles and into condensed liquid droplets. This results in greater pulmonary deposition, reducing the proportion deposited in the nasal passages. Also, for those particles deposited in the nose, it may slow the release of active substances for absorption by the nasal epithelium and receptors.

In general the sensory irritant response to inhaled thermal decomposition products is very rapid at painfully irritant concentrations; this is consistent with human experience. On the basis of these overall findings from experiments in humans and mice it is considered that the use of an instantaneous maximal, maintained concentration-related response in humans is the most realistic, especially at irritant concentrations causing significant behavioural impairment.

Table 7.2 Mouse RD_{50} concentrations for two individual irritant substances and two thermal decomposition product atmospheres from materials decomposed under non-flaming oxidative (NF ox.) conditions in an early version of the ISO 19700 tube furnace.

		Time to 30% rate decrease at RD_{50} concentration/ seconds	~Time to 50% rate decrease at RD_{50} concentration/ seconds
α-Chloroacetophenone	8.59 μg L^{-1}	40	172
HCl gas	295 ppm	3.2	15.5
PVC NF ox., 380 °C[a]	0.26 mg L^{-1}	13	56
LDPE NF ox., 500 °C[a]	0.06 mg L^{-1}	14.9	69

[a]PVC, flexible polyvinyl chloride; LDPE, low density polyethylene.

7.8 Estimates of Lethal Toxic Potency for Natural and Synthetic Polymers under Different Fire Conditions using Rodents and Contribution to Overall Toxic Potency from Major Toxic Gases

7.8.1 Small-scale Combustion Toxicity Tests using Rodents

Small-scale toxicity tests have been carried out for a range of materials using rats to measure LC_{50} concentrations from combustion products, under both non-flaming oxidative decomposition conditions and well-ventilated flaming combustion conditions. These results can therefore be used to compare the lethal toxicity of different materials decomposed under both sets of conditions. The primate experiments already described have established that, at least for a limited range of materials, the main incapacitating toxic effects can be understood in terms of asphyxiant gases and irritants. Where rodent experiments include measurement of the major toxic gases present it should also be possible to estimate the contributions to overall lethal toxic potency using LC_{50} data in rats for each gas. In order to do this it has been necessary to obtain data on the lethal exposure doses of each individual toxic gas and on how mixing the gases affects toxicity. These data have been used to develop calculation models to allow for the interactions between the different toxic gas mixtures present. The basic models used for this are called N-gas FED models and are described in the next section. Because the standard exposure period for combustion toxicity experiments is 30 minutes, the N-gas models use 30 minute LC_{50} concentrations for the gases considered. This includes deaths both during exposure and during a 14 day post exposure period, which is important to allow the effects of gases causing lung inflammation to be captured.

7.8.2 N-gas Models for Predicting Lethality (LC_{50} Concentrations) in Rats Following a 30 minute Exposure

There are two main FED N-gas models using similar concepts, differing somewhat in which factors are considered and in the ways the various interactions are handled. The models estimate FED for lethality for rats in terms of the concentration of each toxic gas present, expressed as a fraction of the LC_{50} concentration. The fractions of lethal doses of almost all gases are assumed to be additive. So, for example, if half the LC_{50} concentration of CO is present with half the LC_{50} concentration of HCN, then exposure of rats to this mixture for 30 minutes will, on average, result in the deaths of half the animals. In terms of overall lethality, the effects of asphyxiant gases (CO, HCN and low oxygen hypoxia) are considered to be additive with the effects of irritant gases (acid gases, such as HCl and NO_2, and organic irritants such as formaldehyde). It is recognised that CO_2 modifies the toxicity of other gases and this is treated differently in different models. It is also recognised that NO_x has some protective effect against HCN toxicity due to methaemoglobin formation (see Chapter 10).

7.8.2.1 Levin N-gas Model for Predicting FED for Lethality in Rats

The N-gas model developed by Levin *et al.*[8,65] is based upon a number of LC_{50} experiments on individual asphyxiant and irritant gases, and experiments on various concentration mixtures of two or three gases. The main precepts of the model are:

1. Fractions of lethal doses of almost all gases are directly additive.
2. Based upon combination experiments of CO and CO_2, it is considered that CO_2 enhances the lethal toxicity of CO, with a maximum effect at a concentration of 5% CO_2. Above 5% CO_2 the enhancement of CO toxicity decreases. This effect is handled by the use of constants m and b, which represent the slope and intercepts of the combination gas toxicity curve.[65]
3. Corrective terms are added for the protective effect of NO_2 on HCN toxicity (but NO is not considered).[8] Versions of the model are also developed for predicting deaths during exposure as well as for combined deaths during and after exposure.

The current version of the N-gas model[3] for total deaths (during and after exposure) is as follows:

$$\mathrm{FED} = \frac{m[\mathrm{CO}]}{[\mathrm{CO_2}] - b} + \frac{21 - [\mathrm{O_2}]}{21 - \mathrm{LC_{50}O_2}} + \left(\frac{[\mathrm{HCN}]}{\mathrm{LC_{50}HCN}} \times \frac{0.4[\mathrm{NO_2}]}{\mathrm{LC_{50}NO_2}}\right) + \frac{0.4[\mathrm{NO_2}]}{\mathrm{LC_{50}NO_2}} + \frac{[\mathrm{HCl}]}{\mathrm{LC_{50}HCl}} + \frac{[\mathrm{HBr}]}{\mathrm{LC_{50}HBr}} \tag{7.4}$$

The constants used for the different terms are presented in Table 7.3.

Table 7.3 Constants and LC_{50} concentrations for gases used in the Levin N-gas model.

Gas	LC_{50} concentrations for 30 minute exposures plus 14 day post-exposure period
For $[CO_2] \leq 5\%$	$m = -18$ and $b = 122\,000$
For $[CO_2] > 5\%$	$m = 23$ and $b = -38\,600$
Hypoxia (oxygen depletion)	$21 - 5.4 = 15.6\%$ depletion
HCN	150 ppm
HCl	3700 ppm
HBr	3000 ppm
NO_2	200 ppm

7.8.2.2 Purser Model for Predicting FED for Lethality in Rats

The Purser rat lethality model[1,5–7] is based upon the same general additive concept as the N-gas model and makes use of the rat LC_{50} data for individual gases and gas interactions obtained by mainly by Levin *et al.*[3] and Hartzell *et al.*[66]
The key precepts of the model are:

1. Fractions of lethal doses of all gases except CO_2 are directly additive.
2. The main effect of CO_2 is considered to be a multiplicative effect on the rate of uptake of other gases depending upon the extent of CO_2-driven hyperventilation. In addition it is considered that once animals are incapacitated, CO_2-induced respiratory acidosis enhances the metabolic acidosis already present, providing an additive toxicity factor.
3. Low oxygen hypoxia is usually a minor factor in small-scale rodent toxicity experiments and can be ignored unless oxygen concentrations are allowed to decrease below 12%. At low levels, a non-linear additive term can be used. An exponential function has been developed to allow for the effects of oxygen at low concentrations.
4. A correction for the protective effect of NO and NO_2 on HCN toxicity due to methaemoglobin formation can be made if necessary and for the additive effect of other nitriles present.
5. It is considered important to make allowance for the effect of all inorganic acid gases present and for organic irritants.

The estimated FED for lethality in rats for a 30 minute exposure to any defined set of toxic gas concentrations is calculated as follows:

$$\mathrm{FED} = \left(\frac{[\mathrm{CO}]}{\mathrm{LC_{50}CO}} + \frac{[\mathrm{CN}] - [\mathrm{NO}x]}{\mathrm{LC_{50}CN}} + \sum_{i=1}^{n}\frac{[\mathrm{I_A}]_i}{\mathrm{LC_{50}I_{A\mathit{i}}}} + \sum_{i=1}^{n}\frac{[\mathrm{I_O}]_i}{\mathrm{LC_{50}I_{O\mathit{i}}}}\right) \times V\mathrm{CO_2} + A + \frac{1}{H} \tag{7.5}$$

where:

I_{Ai} represents each irritant acid gas present (summed for *n* acid gases);
I_{Oi} represents each irritant organic substance present (summed for 0 to *n* organics);
VCO_2 is a multiplication factor for CO_2 driven hyperventilation $= 1 + (\exp(0.14\times[CO_2]) - 1)/2$;
A is an acidosis factor $= ([CO_2]\times 0.05) - 0.02$;
H is a hypoxia function $= \exp(8.13 - 0.54\times[21 - O_2])$; this can be omitted if $[O_2]$ is not less than 12%;
[CN] represents the concentration of cyanide;
[NO*x*] represents the summed concentration of NO and NO_2.

Where data on concentrations of organic irritants are lacking it is recommended that a contribution to the overall FED should be derived from an estimate of the total yield of organic products. The FED component for organic irritants (FED_{org}) is then estimated as:

$$FED_{org} = (\text{mass loss concentration of organic material} \times \text{fraction of mass loss carbon present as organic carbon})/35 \quad (7.6)$$

FED_{org} is then substituted for the organic irritant term in eqn (7.5). This term is empirically derived from the contribution to overall lethal toxicity attributable to the organic fraction during rat exposure to non-flaming wood and non-flaming flexible polyurethane foam.[1]

If these data are unavailable, a FED_{org} of 0.35 is used for non-flaming decomposition, 0.105 for vitiated or inefficient combustion and 0.035 for well-ventilated combustion at the LC_{50} mass loss concentration.

If the concentrations of the irritants present and their lethal exposure doses are known, then the equation can be solved fully. Where unknown irritants are present the equation enables the maximum LC_{50} to be predicted based upon the asphyxiant gases and a generic expression for overall irritants.

The LC_{50} concentration values used for the Purser LC_{50} FED model are shown in Table 7.4.

The VCO_2 exponential function is derived from the measured increase in the rate of uptake of CO and other inhaled toxic gases in humans and rats to increased CO_2. This is an exponential function of the inhaled CO_2 concentration (over the range 0–10% CO_2), as shown from experiments in humans.[1] In rats, inhalation of 5.25% CO_2 produced a 1.54 times increase in the rate of CO uptake, so an exponential function has been derived intersecting this point. In situations where a 30 minute exposure is lethal, animals become comatose after approximately 15 minutes so that hyperventilation ceases. For this reason, the increased rate of uptake applied for half the exposure time. In addition to the hyperventilation effects, CO_2 is known to cause

Table 7.4 LC_{50} concentrations for gases used in the Purser LC_{50} FED model.

Gas	LC_{50} concentrations for 30 minute exposures plus 14 day post-exposure period/ppm
CO	5705
HCN	165
HCl	3800
HF	2900
HBr	3000
SO_2	400
NO_2	170
Hypoxia (using function)	$21 - 5.4 = 15.6\%$ depletion

acidosis and this respiratory acidosis, combined with hypoxia-induced metabolic acidosis, results in an additive toxic component, which has been estimated according to the function shown. The data on which the derivation of these functions was based were obtained from ref. 8 and 65.

The hypoxia function is also based upon experimental data from humans and rats. From human data it is known that there is an exponential function between percentage oxygen depletion and time to and severity of effects, with minimal effects down to 12% O_2.[1] In rats, the measured LC_{50} concentration for a 30 minute exposure is 5.4% oxygen (15.6% oxygen depletion).[3,8] For the hypoxia term an exponential function has therefore been derived providing a minimal FED at concentrations above 12% O_2, an FED of 1 at 5.4% O_2 and increasing FED values at lower concentrations.

The derivation of these terms for the lethal FED contributions of mixed organics, carbon dioxide and low oxygen concentrations provides a toxicologically and physiologically based approach to the calculation of lethal toxic potency for complex fire effluent mixtures. The extent to which the equation including these terms provides an improved model for the calculation of lethal toxic potency from polymer effluent mixtures has been tested by application to the experimentally derived rat lethal toxicity database for combustion product mixtures, as described in the next section.

7.8.3 Application of LC_{50} N-gas FED Models to Rat Lethality Data from Small-scale Combustion Toxicity Tests

Several different combustion toxicity test methods have been used with rats, mostly in the 1970s and 1980s to estimate the lethal toxic potency of the effluents from a range of different natural and synthetic polymeric materials. Few of these studies involved LC_{50} concentration measurements in combination with comprehensive measurement of the chemical composition of the atmospheres or the mass loss concentrations of the materials tested. Also, few have been conducted under combustion conditions that can be readily correlated with those in full-scale fires, and for most methods the composition of the test atmosphere and the decomposition conditions changed continuously during a test run. Issues relating to the test decomposition conditions are discussed in detail in Chapter 2, but three methods have provided some data that can be used to determine LC_{50} concentrations under reasonably constant and well-defined combustion conditions, coupled with measurements of the concentrations of a least a few of the major toxic gases. Where these data are used to indicate the possible LC_{50} concentrations in full-scale fires, it is essential to examine carefully the relationship between the decomposition conditions in the test and those in the large-scale fire scenarios of interest. However, in order to examine the validity of the rat lethality LC_{50} N-gas models, the exact decomposition conditions are less important. The main requirement is that a mixed fire effluent of relatively constant composition is produced containing a reasonably typical mix of fire effluent gases and particulates, including those used in the models.

The largest database of rodent (mouse) lethality data for different materials was obtained using the University of Pittsburgh toxicity test method.[60,61] Unfortunately, the method involved ramp heating of a static sample under a stream of air and is of limited value, because the conditions, identities and yields of products change continuously throughout the test exposure as the specimen is heated and the decomposition mode changes from non-flaming to flaming. The effluent composition and sample mass measurements were also limited and the animals were not retained for a 14 day post-exposure observation period. The results obtained by Levin *et al.*[67] using the then National Bureau of Standards (NBS) cup furnace method are more useful. This method involved decomposing polymer samples of different mass in a crucible furnace under non-flaming oxidative thermal decomposition conditions at a temperature 25 °C below the auto-ignition temperature of the specimen and also under flaming conditions at 25 °C above the auto-ignition temperature. The effluent passed directly into an enclosed 200 L volume chamber to which groups of six rats were exposed nose-only. Exposure began as the sample was dropped into the crucible: the concentration of the combustion products increased for a time as the specimen decomposed and then remained relatively stable. However, some changes in atmosphere composition occurred during this post-combustion period due to deposition of smoke particles, acid gases and condensates on the chamber walls. Further decomposition of some atmosphere components (significant for polyurethanes and fluoropolymers) occurred as the chamber atmosphere was re-circulated through the heated furnace. For flaming exposures there was also a period during which products of non-flaming decomposition were generated before ignition occurred, but this was usually quite short. Exposure concentrations of the major asphyxiant gases (CO, CO_2, O_2 and HCN) were measured continuously and the exposure concentrations were averaged over the 30 minute exposure period. Animal deaths were scored during the exposure period plus a 30 minute post-exposure period (scored as within exposure deaths) and also for the exposure period plus a 14 day post-exposure period. In general this means that within-exposure period deaths are likely to be due primarily to the effects of asphyxiant gases, while exposure plus 14 days includes both asphyxiant deaths occurring during and immediately after exposure and deaths resulting primarily from lung oedema and inflammation caused by irritants, which typically occur over a period of from a few hours after exposure to 14 days. The test material exposure concentrations are reported on a mass charge basis. This means that the exposure concentration is calculated as the mass of material dropped into the crucible divided by the chamber volume. Where allowance is made for the mass of any residue left after the test, it is possible to calculate the mass loss concentration, which is the mass actually decomposed divided by the chamber volume. In practice, for most materials there is little or no residue, so that the two measures are similar.

In later experiments the crucible furnace was replaced by a radiant furnace.[68] For this method the specimen, in the form of a flat strip, is irradiated from above by two angled radiant heaters. The effluent rises through a slit between the heaters into the enclosed 200 L chamber, while replacement air circulates back into the air space between the sample and the heater *via* two slits at each end of the heated area. The main decomposition mode is flaming combustion and, as with the crucible furnace method, the initial decomposition is under non-flaming conditions until the specimen ignites, followed by a period of flaming until the specimen is extinguished, and a subsequent period during which off-gassing may occur from any residue. The basic exposure procedure is the same as for the cup furnace method, involving a total 30 minute exposure period over which the gas concentrations are averaged. The mass loss rate of the specimen is also measured and the exposure concentrations are reported on both a mass charge and mass loss basis.

Another method for which some useful rat lethality data have been generated is the DIN 53436 tube furnace method.[69,70] This method represents an early version of the concept used for the ISO 19700 test method, in which the sample in the form of a strip is decomposed continuously in a stream of air, producing a dynamic steady-state decomposition condition with fresh combustion products flowing through the exposure chamber to maintain a constant atmosphere composition. The rats are exposed nose-only over a 30 minute period (plus a 14 day post-exposure observation period). This method has mostly been used for non-flaming decomposition, although some flaming decomposition experiments have been performed.[71] The test atmosphere composition measurements made were somewhat limited.

For the NBS cup furnace method and the National Institute of Standards and Technology (NIST) radiant method, the flaming combustion conditions vary depending upon the size and behaviour of the specimen. In relation to the description of fire types in Chapter 2, the combustion conditions in these tests can be classified approximately based upon the $CO_2:CO$ ratios obtained. In practice (for non-fire retarded materials), for different materials and specimen masses, they vary over a range from well-ventilated, through an equivalence ratio of around unity, to slightly vitiated combustion conditions. For the DIN 53436 experiments the conditions were generally rather vitiated flaming at high furnace temperatures more or less simulating post-flashover vitiated combustion conditions. However, as stated, the important consideration is that these combustion product atmospheres did contain mixtures of toxic fire gases and particulates generically typical of those occurring in fire combustion products, thereby providing an opportunity to determine the extent to which the individual atmosphere components acted in isolation or interacted to produce the overall toxic potency of the overall combustion product mixture.

In the following section the data from these tests have been used to compare measured rat LC_{50} concentrations (exposure plus 14 days) with LC_{50} concentrations calculated from the measured atmosphere compositions using the Levin N-gas LC_{50} calculation method and the Purser LC_{50} calculation method.[7] The materials tested and the toxic product concentrations at the measured rat LC_{50} concentrations for each material are shown in Table 7.5.

Table 7.5 LC_{50} concentrations (exposure plus 14 days, mass charge) for materials tested in the NBS cup furnace, NIST radiant furnace and DIN 53436 tube furnace apparatus.

Test material and abbreviation[a]	Mass conc./ g m^{-3}	CO_2/ %	CO/ ppm	O_2[b]/ % dep	HCN/ ppm	HCl/ ppm	NOx	Org[c]/ % term
NBS cup furnace non-flaming								
Acrylonitrile butadiene styrene (ABS)	30.9	0.53	680	0.63	164	—	—	0.35
Douglas fir (DFIR)	22.8	0.69	2700	1.00	—	—	—	0.347
Flexible polyurethane foam (FPU)	35.0	0.40	1126	1.00	11	—	—	0.869
Modacrylic (MOD)	5.3	0.53	435	1.75	249	—	—	0.35
Polyphenylsulfone (PPS)	9.5	0.51	4465	0.91	—	—	—	0.35
Polyvinylchloride (PVC)	30.3	0.46	640	0.98	—	2308	—	0.18
PVC/zinc ferrocyanide (PVC zn ferrocyn)	11.3	0.53	1141	0.98	—	—	—	0.108
Red oak (REDO)	30.3	0.73	2515	0.98	—	—	—	0.35
Wool	25.1	0.70	979	1.19	301	—	—	0.35
NBS cup furnace flaming								
ABS	19.3	1.06	1448	1.13	124	—	—	0.035
DFIR	39.8	3.71	3400	4.70	—	—	—	0.000
FPU	20.3	2.10	1040	2.20	86	—	—	0.035
MOD	4.4	0.40	339	1.75	180	—	—	0.035
Polystyrene (PS)	38.9	1.95	1323	1.81	—	—	—	0.79
REDO	56.8	4.43	2102	4.22	—	—	—	0.134
Rigid polyurethane foam (RPU)	13.3	1.20	1689	1.03	126	—	—	0.035
Wool	28.2	2.50	705	1.19	116	—	—	0.035
NIST radiant furnace flaming								
Douglas fire rad (DFIRrad)	56.0	3.71	3233	4.11	—	—	—	0.007
Rigid polyurethane foam (RPUrad)	22.4	2.10	1800	2.55	133	—	—	0.000
Flexible polyurethane foam (FPUrad)	52.0	5.31	1392	6.74	53	—	—	0.035
PVC (PVCrad)	26.8	1.36	2000	1.90	—	2400	—	0.063
Vinyl fabric (VFrad)	32.0	1.81	3020	2.42	8	647	12.8	0.035
Melamine (Melrad)	12.5	0.99	401	1.24	111	—	47.0	0.035
DIN tube furnace, 850 °C vitiated fl.								
Scot's pine (*Pinus sylvestris*; SP)	29.2	1.70	5515	2.00	—	—	—	0.090

[a]For each material name an abbreviation has been added for use in later figures and tables, *e.g.*, ABS represents acrylonitrile butadiene styrene.
[b]Percentage oxygen depletion.
[c]Calculated as mass loss concentration of organic polymer material × fraction of mass loss carbon as organic carbon/35. If data unavailable: non-flaming 0.35, flaming 0.035 vitiated flaming 0.105.

The measured and calculated LC_{50} concentrations are listed in Table 7.6. Figure 7.13 shows the measured rat LC_{50} concentrations plotted against those calculated using the Levin (N-gas) method. Linear regression lines have been fitted for the non-flaming and for the flaming data. The extent to

Table 7.6 LC_{50} concentrations (g m^{-3}) (exposure plus 14 days, mass charge) for materials tested in the NBS cup furnace, NIST radiant and DIN 53436 apparatus.

	Non-flaming LC_{50} g m^{-3}			Flaming LC_{50} g m^{-3}		
	Actual	Purser	Levin	Actual	Purser	Levin
NBS cup furnace						
ABS	30.9	20.8	25	19.3	17	17.4
DFIR	22.8	25.8	50.2	39.8	40.9	39.2
FPU	35	30.3	108.6	20.3	21.3	22.5
MOD	5.3	2.8	2.9	4.4	2.9	3.5
PPS	9.5	8.3	12.6	—	—	—
PS	—	—	—	38.9	32.1	91
Polytetrafluoroethylene (PTFE)	—	—	—	—	—	—
PVC	20	21.4	25.4	17.3	17.3	17.4
PVC Zn ferrocyn	11.3	9.2	9.3	—	—	—
REDO	30.3	35	54.6	56.8	60.4	73.2
RPU	—	—	—	13.3	11	11.4
Wool	25.1	10.6	11.3	28.2	25.2	28.8
NIST radiant furnace						
DFIRrad				56	58.9	58.3
RPUrad				22	16.8	18.7
FPUrad				52	46.3	46.4
PVCrad				26	22.8	24.4
VFrad				32	32.4	35
Melrad				12.5	14.4	14.2
DIN tube furnace, 850 °C						
SP				29.2	23.9	27.4

which the regression lines overlay the 45° actual LC_{50} line gives an indication of the predictive power of the N-gas calculation method. Figure 7.14 shows the results for the Purser LC_{50} calculation method plotted against the same rat LC_{50} data. Where a calculated line lies to the left of the 45° line, this indicates that the calculation method underestimates the toxic potency of the mixed combustion products, while a line laying to the right indicates that the calculation method overestimates the toxic potency.

The results show quite a good relationship between the actual rat LC_{50} concentrations and those calculated using the Levin N-gas LC_{50} calculation method ($R^2 = 0.6913$) for materials decomposed under flaming conditions, which is improved if one outlier (flaming polystyrene) is removed ($R^2 = 0.9392$). For non-flaming decomposition data, the fit is less convincing ($R^2 = 0.5455$), underestimating the observed toxic potency. The results for the Purser rat LC_{50} model show a good fit for the flaming data ($R^2 = 0.9707$) and a reasonably good fit for the non-flaming data ($R^2 = 0.7004$).

The models are intended to be general and therefore applicable to all forms of combustion atmosphere (both flaming and non-flaming). Pooling the non-flaming and flaming data gives the following linear fit equations for the two different calculation models:

$$\text{Levin N-gas: } y = 1.235x - 1.052 \qquad R^2 = 0.531$$
$$\text{Purser } LC_{50}\text{: } y = 1.026x - 2.498 \qquad R^2 = 0.919$$

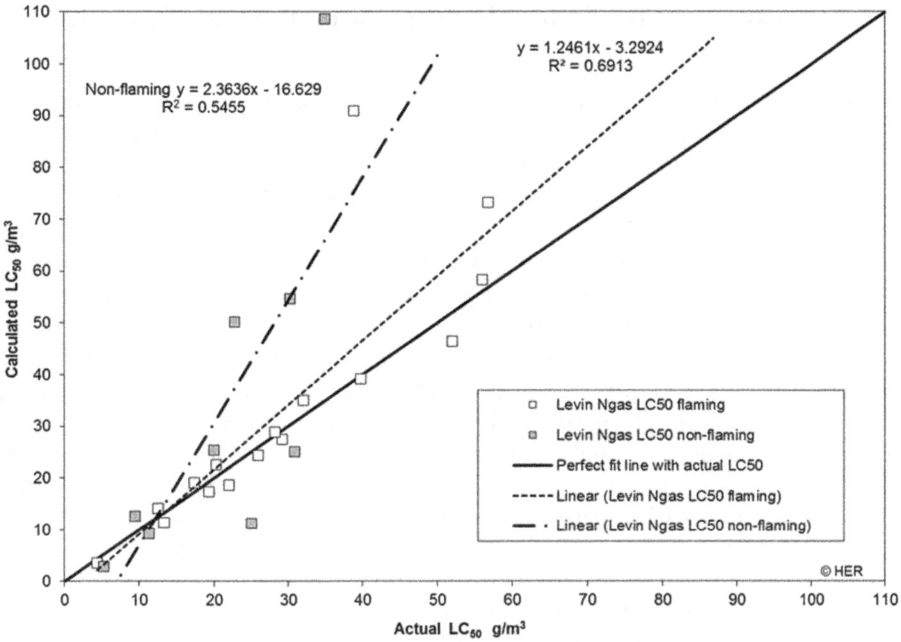

Figure 7.13 LC_{50} concentrations calculated using the Levin N-gas model compared with measured rat LC_{50} concentrations for different materials decomposed under non-flaming and flaming combustion conditions.[7]

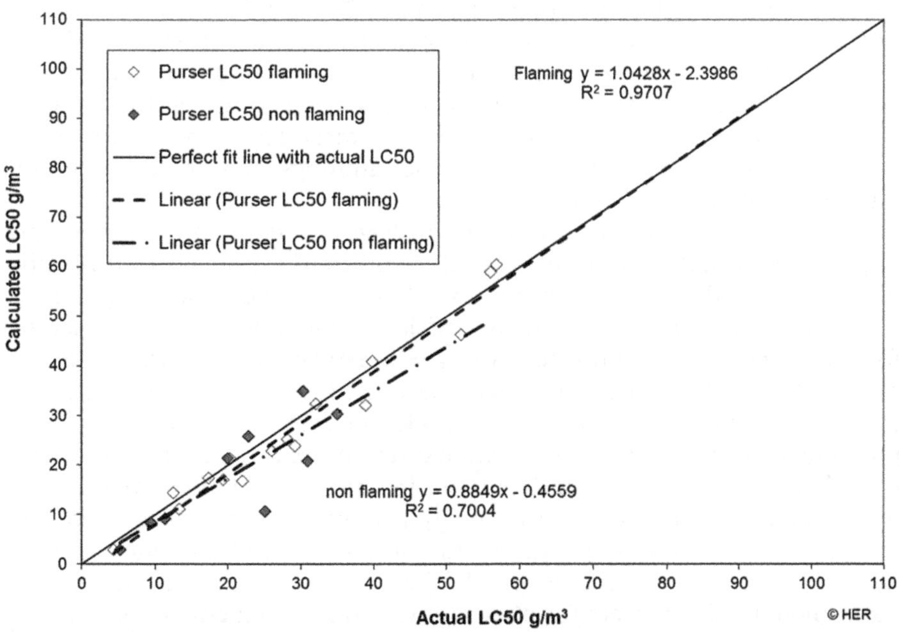

Figure 7.14 LC_{50} concentrations calculated using the Purser rat LC_{50} model compared with measured rat LC_{50} concentrations for different materials decomposed under non-flaming and flaming combustion conditions.[7]

7.8.4 Comparison of Models and Confirmation of Additive Toxicity

Both toxicity models therefore provide reasonably good predictions of actual rat LC_{50} concentrations using measured concentrations of a small number of key toxic gases for a range of common polymeric materials with a variety of chemical compositions. This is especially true for flaming decomposition, for which the composition of the atmospheres is much simpler than under non-flaming conditions and consists mainly of the common asphyxiant gases (for polymers containing C, H, O and N), with the addition of some irritant acid gases (for polymers containing N, F, Cl, Br, P or S), but low concentrations of organic species.[1] Since both models are based primarily on an additive FED concept, it also demonstrates that the toxic effects of these gases can legitimately be considered as primarily additive and, in particular, that the lethal effects of the asphyxiant gases present during exposure can be considered approximately additive with the lethal effects of irritants on the lung to some extent during exposure and also during the post exposure period up to 14 days (mainly during the first 48 hours after exposure). This is consistent with Pauluhn's findings of a reasonable agreement between N-gas calculations and experimental data obtained using the DIN43536 apparatus.[72]

The importance of this is that it demonstrates that for a range of common of materials (with the exception of per-fluorinated polymers and particular phosphorus-containing polymers) the toxic effects of fire effluent mixtures can be predicted from this small set of gases using relatively simple additive models and that (especially for flaming combustion conditions) there is no evidence for any major unknown acute toxic effects other than the asphyxiant and irritant effects predicted from these few mixed gases. These results therefore confirm the more detailed findings from the primate work for a much wider range of materials tests.

Comparing the two models, the Levin N-gas model is more empirically based, using a set of 30 minute rat LC_{50} studies of the effects of individual fire gases and different gas mixtures. These have been used to provide a set of additive FED fractions, with functions generated to derive constants for the effects of CO_2 and interactions between NO_2 and HCN. The Purser model also assumes basic additivity between the FEDs of the main toxic gases, but differs from the Levin model in attempting to use a more physiologically based approach, with a different treatment for the effects of CO_2, low oxygen hypoxia and organic irritants. The Purser model gives more weight to the effect of CO_2 on $\dot{V}_E$ (the volume of air breathed each minute) so that increasing hyperventilation at higher CO_2 concentrations increases the rate of uptake of all the other toxic gases, whereas the Levin model only corrects for effects of CO_2 on CO toxicity. Less weight is given to the hypoxic effects of oxygen depletion in the Purser model, with little effect until oxygen concentrations approach 5–6%, as observed experimentally. The greater weight given to hypoxia in the Levin model compensates to some extent for the lower weight given to CO_2. Another important difference is that the Levin model ignores the effects of organic irritants. From effects in humans and primates, and the results in the rats and mice (particularly the post-exposure deaths),[1] it is clear that the mixed organic irritants produced make an important contribution to overall lethal toxicity. This is particularly the case under non-flaming oxidative thermal decomposition conditions, but also to some extent under vitiated flaming combustion conditions, especially where the organic atmosphere component is high in relation to that of the asphyxiant gases. It is difficult to obtain fully quantitative measurements of the concentrations of all the potentially irritant organic compounds in fire effluent mixtures and, based upon the mouse irritancy studies, it is considered that a number of important irritant species remain unidentified.[1,2] Such GC-MS studies as have been performed have shown that the mix of compounds formed under specific decomposition conditions are often relatively similar for different materials, containing a range of substances common to most materials, with the addition of a number of substances more specifically related to the molecular structure of the material.[1] On this basis, the approach taken in the Purser model is to use a generic term for the total organic content of the atmosphere, calculated from the organic fuel mass loss multiplied by the fraction of fuel carbon present in the form of organic carbon in the fire effluent.

This term accounts for most of the differences between the LC_{50} concentrations predicted by the Levin and Purser models, giving a better fit to the data for the Purser model especially for non-flaming decomposition conditions, when approximately a third of the observed toxicity can be attributed to the effects of organic irritants on the lung. For these experiments carried out on rats, the effect of organic irritants is much lower under flaming conditions, with the exception of polystyrene (which produces a dense smoke under flaming conditions that is rich in particulates and organic irritants including styrene monomer) and PVC (producing HCl). However, most of these experiments were conducted under relatively efficient combustion conditions and it is considered that at higher equivalence ratios the contribution from organic irritants is likely to be somewhat increased.

Figure 7.15 shows the percentage decrease in LC_{50} (*i.e.*, increase in toxicity) between exposure plus 30 minutes and exposure plus 14 days. This represents the increase in toxicity resulting from post-exposure toxic effects due mainly to lung inflammation caused by inhalation of irritants. For most materials, toxicity increased significantly during the post-exposure period under non-flaming decomposition conditions and also for a number of materials under flaming decomposition conditions.

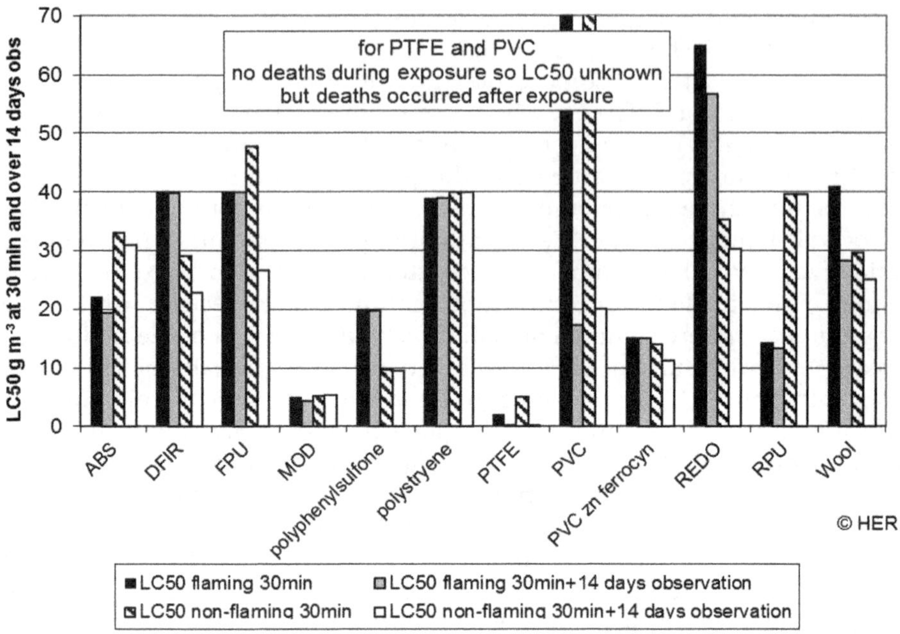

Figure 7.15 Percentage decrease in LC_{50} concentration (*i.e.*, toxicity increase) over 14 day post-exposure period.

7.8.5 Contribution of Different Toxic Gases to Overall Lethal Toxic Potency from Different Materials

Figures 7.16–7.19 show the extent to which different toxic species contribute to the overall toxic potency of different materials for the main different fire types described in ISO 19706[73] calculated using the Purser LC_{50} FED model. The chemical atmosphere composition data are from the NBS cup furnace, NIST radiant and mainly from the ISO19700 tube furnace.[74,75]

Figure 7.16 shows non-flaming oxidative thermal decomposition data (mostly with the NBS cup furnace) plotted as toxic potency ($1/LC_{50}$), so that the higher the bar the greater the toxicity. The shading shows the contribution to the overall toxic potency from each atmosphere component. Under these conditions toxic potencies are relatively high, ranging from 0.029 to 0.37 (LC_{50} 34.8–3.8 g m^{-3}). The main toxic components depend somewhat on the composition of the polymer, but in all cases pyrolysis and partial oxidation produce significant contributions from CO and a range of organic irritants. For nitrogen-containing polymers some HCN is formed and makes a major contribution to toxic potency, although the yield is generally quite low other than from polyacrylonitrile. Halogens are released as acid gases and make a significant contribution if they are present as a high mass percentage of the original polymer, as for PVC formulations (see Chapter 3).

For well-ventilated flaming (Figure 7.17) the scale is expanded five-fold because toxic product yields and toxic potencies are generally low [ranging

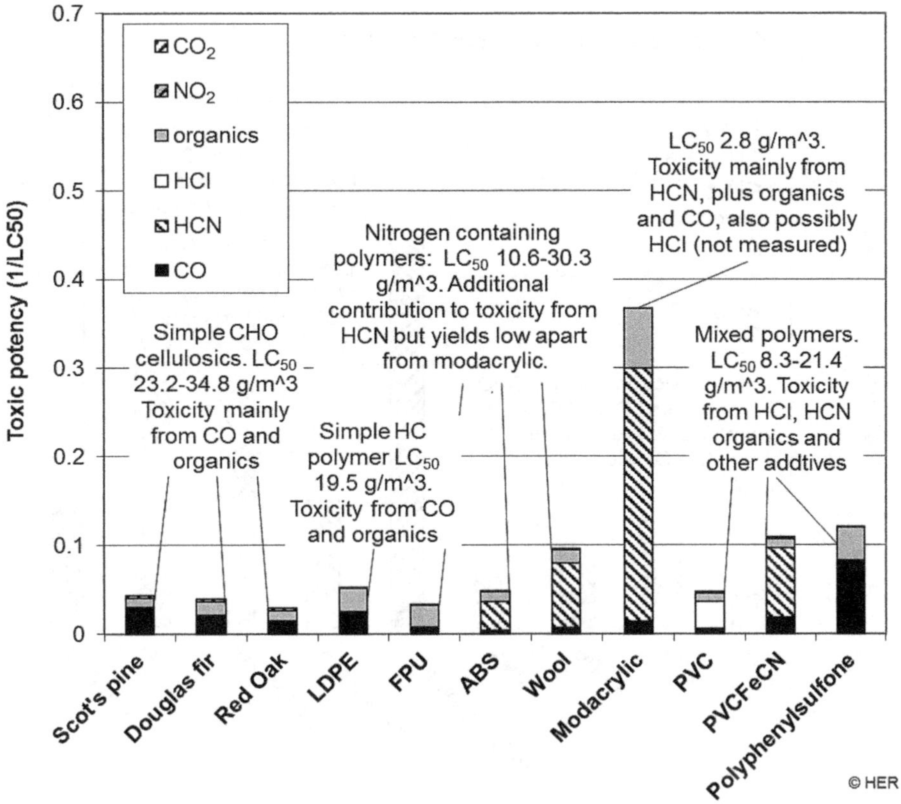

Figure 7.16 Toxic potencies ($1/LC_{50}$) for non-flaming oxidative decomposition conditions. LDPE, low density polyethylene.

from around 0.008 to 0.13 (LC_{50} 118–7.9 g m^{-3})]. A variety of different components contribute to the overall potencies, but for well-ventilated combustion CO_2 is always important as a cause of hyperventilation. In this sense it magnifies the toxicity of the other components by increasing their rate of uptake. It also makes a direct toxic effect at high concentrations due to acidosis. For nitrogen-containing materials, HCN is important, but NO_2 also makes a contribution, since the yields of nitrogen oxides are highest under well-ventilated combustion conditions. CO is also important for some materials, and for materials containing chlorine, both CO and HCl are significant components of the overall toxicity (due to inefficient combustion resulting from the gas phase effect of chlorine on oxidative free radicals in the flame). Hypoxia and organic irritants also make a contribution at high mass loss concentrations.

As fires in buildings grow and become under-ventilated (Figure 7.18), the yields of toxic gases and overall toxic potencies are considerably higher than for well-ventilated combustion conditions (toxic potency 0.03–0.48; LC_{50} 31.8–2.1). Carbon monoxide is a significant toxic component in all cases, with a contribution from CO_2 and organic irritants. For nitrogen-containing

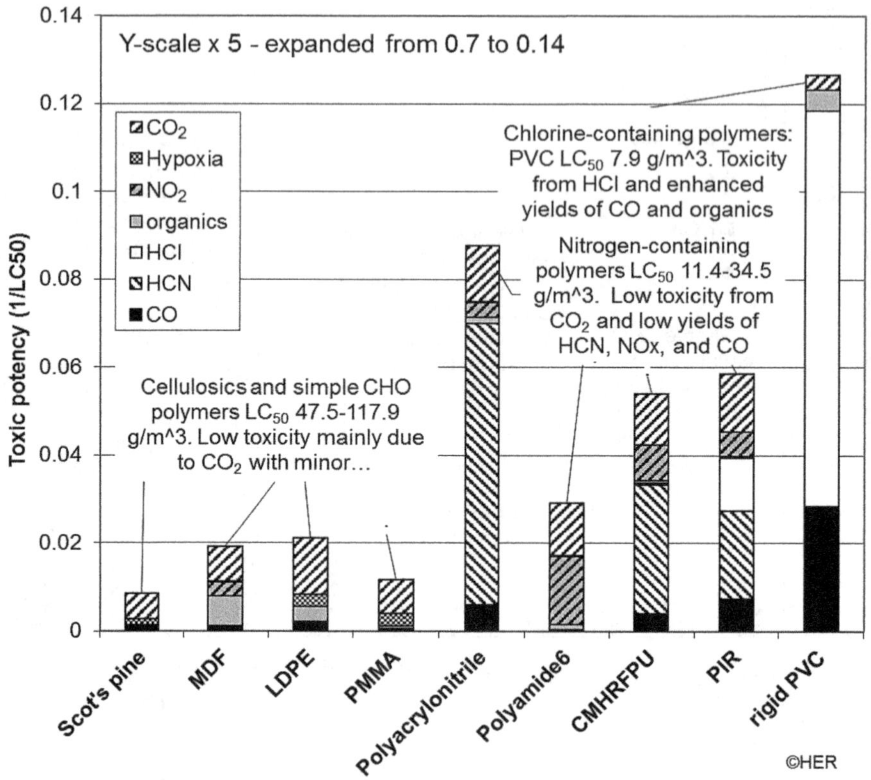

Figure 7.17 Toxic potencies ($1/LC_{50}$) for well-ventilated flaming conditions. MDF, medium density fibreboard; PMMA, polymethylmethacrylate; CMHRFPU, combustion modified high resilience flexible polyurethane foam, PIR rigid polyisocyanurate foam.

materials the toxicity is dominated by HCN, which is produced at high yields, with a small contribution from NO_2 and for PVC, HCl and CO are the main toxic components with a small contribution from organic irritants. Polystyrene toxicity is heavily influenced by irritant organic particulates.

At higher temperatures representing post-flashover under-ventilated combustion conditions (Figure 7.19), toxic potencies and patterns are similar to those under pre-flashover, under-ventilated combustion conditions, but yields of CO and HCN can be somewhat higher, so that toxic potencies range from 0.04 to 0.67 (LC_{50} 23.9–1.5 g m^{-3}). High yields of HCN can be formed from polyacrylonitrile and polyamides (see Chapter 3).

In this set of materials only one (the fluoropolymer, polytetrafluoroethylene) showed toxic effects under certain specific non-flaming conditions not accounted for in terms of simple asphyxiant gases and organic irritants.

Excluding fluoropolymers, the lethal toxic potency (mass loss LC_{50} concentrations of fire effluent mixtures from different materials) ranges from polyamide-6 decomposed under post-flashover, vitiated flaming conditions

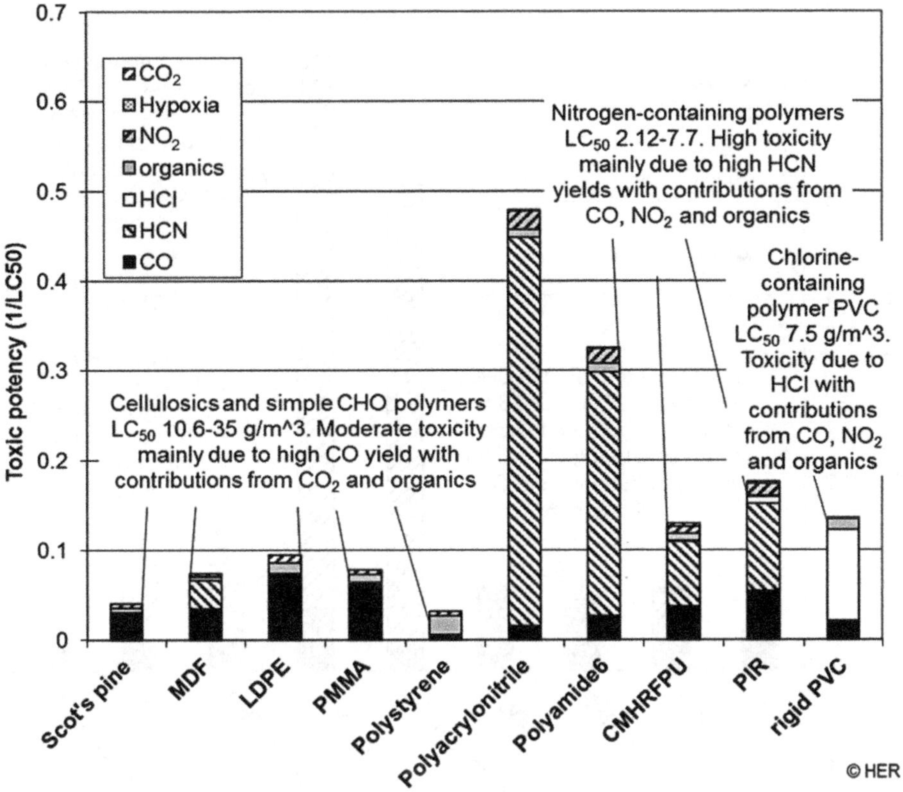

Figure 7.18 Toxic potencies ($1/LC_{50}$) for under-ventilated flaming conditions.

(equivalence ratio >1.5) (LC_{50} 1.51 g m^{-3}) to Scot's pine and similar cellulosic materials decomposed under well-ventilated flaming conditions at low equivalence ratios (0.5–0.8), giving an LC_{50} concentration of 119 g m^{-3}. In practice the lowest possible mass loss toxic potency (highest possible LC_{50} concentration) for any flammable material is limited to somewhere in the 120–150 g m^{-3} range, because at these loadings CO_2 and oxygen depletion are in the lethal range, giving a range of approximately two orders of magnitude (factor of 79) between the most toxic and least toxic effluent atmospheres for common materials. In general the toxic potencies are lower under well-ventilated combustion conditions, and higher under non-flaming and under-ventilated conditions, as illustrated in Figures 7.17 and 7.18. Table 7.7 shows approximate LC_{50} concentrations and approximate lethal exposure doses for common classes of materials under the four main fire conditions.

Flaming combustion conditions in compartment fires begin by being well-ventilated and then gradually become more vitiated as the fire grows and the ventilation becomes limiting. Individual materials in fires are therefore first decomposed in well-ventilated conditions at low equivalence ratios, and the equivalence ratio increases as combustion becomes under-ventilated. The effect of this on lethal toxic potency is illustrated in Figure 7.20,[76] which

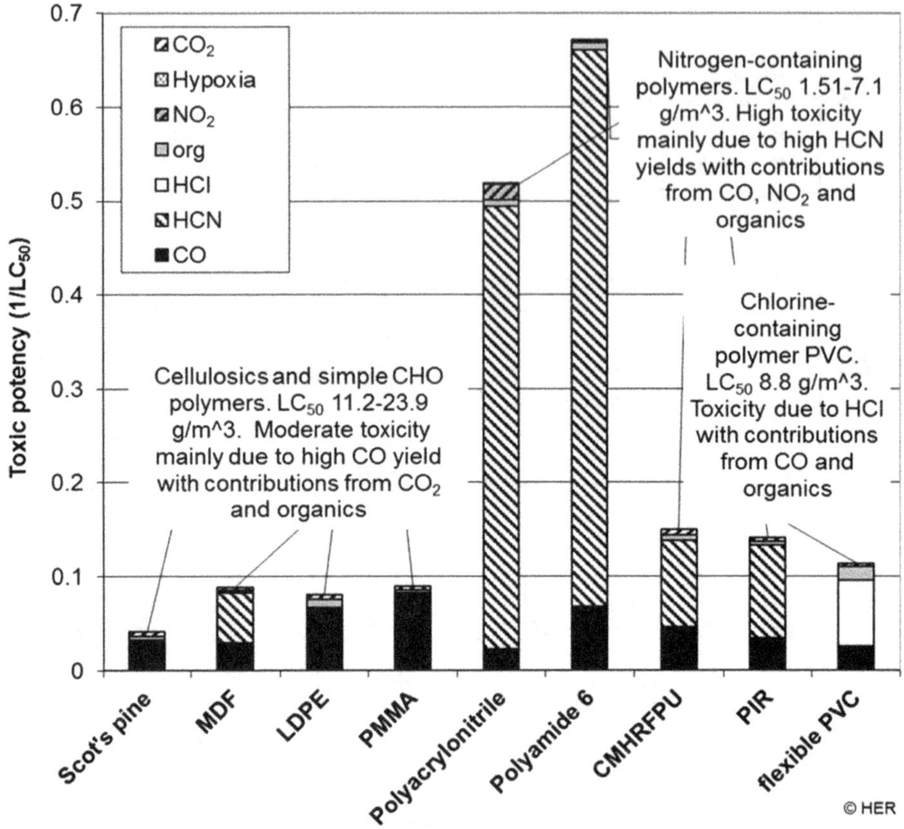

Figure 7.19 Toxic potencies ($1/LC_{50}$) for post-flashover under-ventilated flaming conditions. MDF, medium density fibreboard.

shows plots of calculated LC_{50} against equivalence ratios for a set of materials combusted in the ISO TS19700 tube furnace. Most materials show a considerable increase in toxic product yields and overall toxic potency (lower LC_{50} concentration) as the equivalence ratio increases. Materials containing halogens as fire retardants show a more constant toxicity over the fuel : air equivalence ratio range. When some materials were decomposed at higher temperatures, or under lowered oxygen concentrations (10 or 12% oxygen), the toxic potency was significantly increased.

7.8.5.1 Fluoropolymers

Thermal decomposition and combustion products from fluoropolymers show a great range of toxic potencies, depending upon the precise thermal decomposition or combustion conditions. A detailed account is given in Purser.[77] At decomposition temperatures above 650–700 °C polytetrafluoroethylene (PTFE) burns with a flame. Exposure results in post-exposure airway

Table 7.7 Approximate LC_{50} concentrations (g m^{-3}) and lethal exposure doses (LED) (g m^{-3} min) for common material classes under different fire conditions.[a]

Material	Non-flaming		Well-ventilated flaming		Under-ventilated flaming		Post-flashover, under-ventilated flaming	
	LC_{50}	LED	LC_{50}	LED	LC_{50}	LED	LC_{50}	LED
Fluoropolymers	0.017	0.51	~8	240	~8	240	5.4	162
Polyamide,	17	510	17	510	3.4	54	1.8	54
Modacrylic, polyacrylonitrile	5	150	12	240	4.4	132	1.0	60
Wool	25	750	28	840	4	120	4	120
Rigid polyurethane foam	40	1200	15	450	7	210	7	210
Flexible polyurethane foam	27	810	13	390	7	210	7	210
PVC	7	240	7	240	7	240	8	270
CHO polymers	17	510	70	2100	20	600	12	360
Cellulosics	24	720	119	3570	25	750	15	450

[a]Approximate rat LC_{50} concentrations and LEDs for 30 minutes exposure plus 14 days observation.

and lung inflammation, which can be understood in terms of known toxic products, including carbonyl fluoride, which hydrolyses to form HF, and perfluoroisobutylene. This results in rat LC_{50} or approximate lethal concentration (ALC) values (30 minutes exposure + 14 days observation) in the 5–50 g m^{-3} mass loss range. Figure 7.21 shows the effects on respiratory pattern in a rat during exposure to products from PTFE decomposed under non-flaming decomposition conditions at 600 °C in a tube furnace. Under these conditions the main toxic products are similar to those under flaming conditions, resulting in a powerful respiratory rate depression indicative of upper airway sensory irritation.

Below 400 °C, fluoropolymers are relatively stable, without significant decomposition or toxicity, but between 400 and 650 °C the toxicity is highly dependent upon the polymer type and exact decomposition conditions. If decomposition is non-flaming and the products are rapidly dispersed, perfluorinated polymers produce tetrafluorethylene vapour, which polymerises to form a high molecular mass polytetrafluorethylene condensate in the form of a highly toxic ultrafine particulate. This has a very high toxic potency if inhaled. Freshly formed particulates are approximately 0.01 μm (10 nm) in diameter, and thus in the nanoparticle size range. Figure 7.22 shows the respiratory pattern during exposure to a very low mass loss concentration of PTFE containing large numbers of these particles but very low concentrations of other toxic species. The hyperventilatory pattern with occasional pauses is characteristic of irritant effects on lung vagal nociceptors and develops after a few minutes during an exposure.

The ultrafine particles are formed in large numbers in fresh fume, but agglomerate rapidly, forming particles larger by a factor of ten or more within a minute or so. When rats were exposed to this aged fume atmosphere

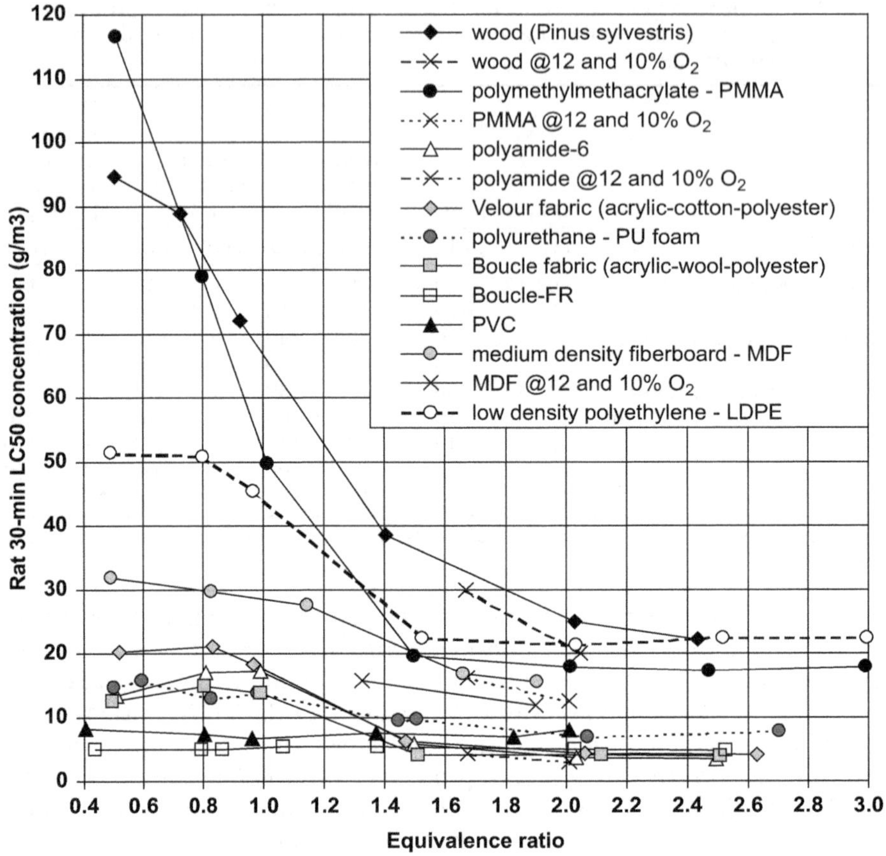

Figure 7.20 LC_{50} concentrations (g m^{-3}) as a function of equivalence ratio under flaming combustion conditions.[76]

it was found to be harmless. The conditions of generation and exposure are therefore critical to the atmosphere composition and toxicity. When rats are placed in small chambers close to continuously generated ultrafine fume, they are seriously affected within a few minutes, but when placed in larger chambers, those close to the fume source are badly affected, while others placed approximately 1 m away are unaffected. Although both groups of rats are exposed to the same total mass concentration of fumes, those further away from the source inhale the agglomerated particulate, which is non-toxic.

Extreme toxic potency has also been observed under conditions where a fluoropolymer, such as PTFE or fluorinated ethylene propylene (FEP), is heated for several hours at temperatures around 420 °C under a stream of air, with rats exposed to the fresh particulate stream. Other conditions resulting in extreme toxicity involve heating at temperatures between 450 °C and 650 °C in an enclosed space (such as the NBS cup furnace apparatus),

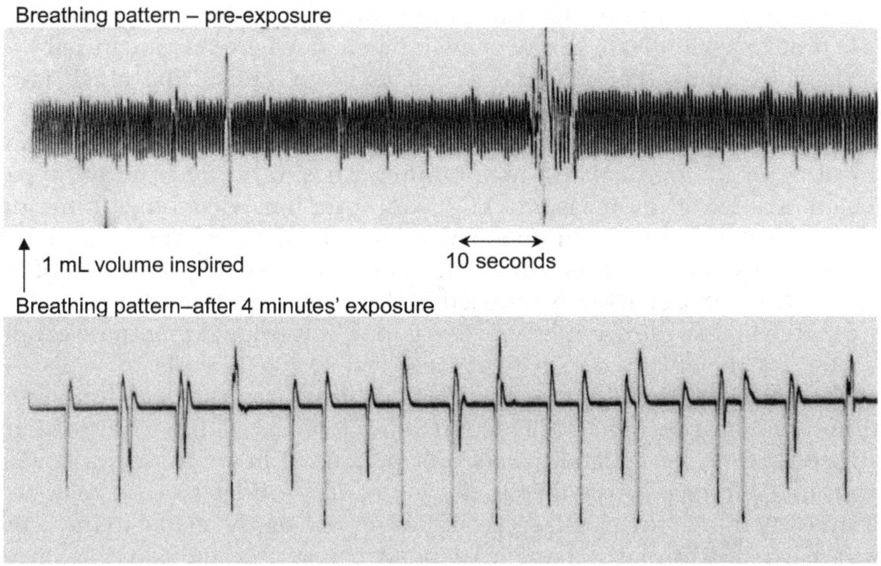

Figure 7.21 The effects on the breathing pattern of a rat exposed to a respiratory tract irritant (PTFE) decomposed at 600 °C in the DIN 53436 apparatus (4.85 mg L^{-1}).

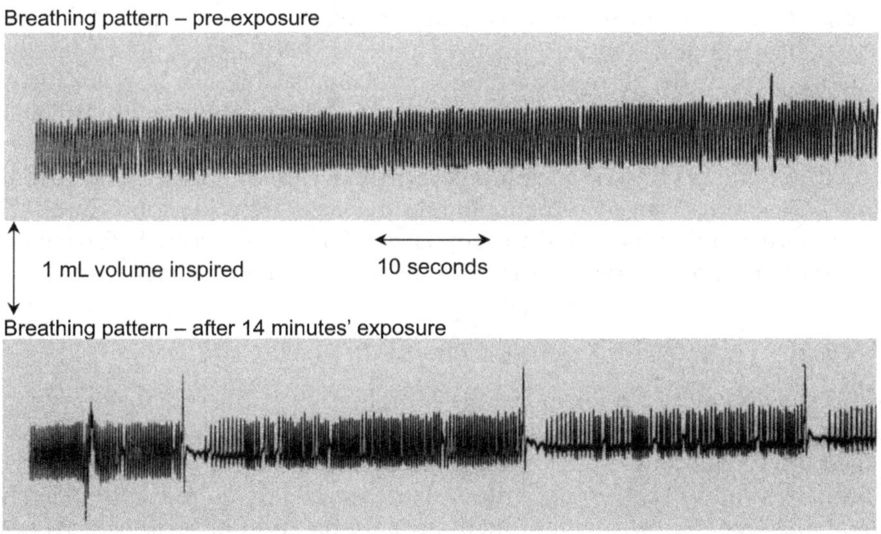

Figure 7.22 The effect on the breathing pattern of a rat exposed to a lung irritant (PTFE) decomposed at 575 °C in the NBS cup furnace (0.054 mg L^{-1}).

such that particulate fume is formed, released into the atmosphere and then recirculated back through a heated zone. In this way the ultrafine toxic fume is formed and continuously regenerated in the hot zone, so that the rats are exposed continuously to freshly generated toxic fumes.

Under temperatures or decomposition conditions where the particulates have time to agglomerate before being inhaled, or where they are mixed with particulates from other materials, such as wood smoke, the much lower toxicity is more related to those of the normal toxic products, with LC_{50} in the 0.5–5 g m^{-3} range, but the greater the opportunity for recirculation and reheating of the particulates, maintaining exposure to fresh ultrafine particulate, the lower the measured LC_{50} concentrations, down to a minimum of around 0.017 g m^{-3} fluoropolymer mass loss. Since the LC_{50} of perfluoroisobutylene vapour is 0.047 g m^{-3} and only a small proportion of the mass of decomposed PTFE is released in this form, and since the LC_{50} of HF and carbonyl fluoride are 1.88 and 1.96 g m^{-3}, it is evident that these cannot account for the observed toxicity at these very low LC_{50} levels.

When inhaled, the ultrafine particles penetrate into the deep lung interstitium and an inflammatory reaction starts to occur within a few minutes (Figure 7.22), which is fatal to rats over periods of hours to days following exposure. The mechanism of toxicity is not fully understood. It may be a simple function of the particle size and/or may be partly related to the active surface chemistry of the particles (stable peroxy free radicals have been detected on the particle surface), or partly as a result of the particle size favouring delivery to and uptake by the lung interstitial cells. Similar effects have since been reported for many types of ultrafine particles, particularly those containing transition metals, which are known to become involved in oxidation reactions involving free radical formation (see Chapter 11). However, with a PTFE mass loss LC_{50} concentration of 0.017 g m^{-3}, of which only a proportion is in the form of ultrafine particulates, this has a much higher toxic potency than other ultrafine particulates. The extreme toxicity particulate is not formed from fluoropolymers containing hydrogen in the molecular structure, such as ethylene tetrafluoroethylene (ETFE) copolymer. The extreme toxicity resulting from these nanoparticles released from fluoropolymers and other sources raises potential toxicity issues for other materials formulated with nanoparticle additives if these are released in fire effluents.[77]

7.8.5.2 Trimethylolpropane Phosphate (TMPP)

Materials containing phosphorus in combination with trimethylolpropane polyol[78,79] can give rise to another unusual toxic effect. The toxic agent is a caged biphosphorous ester, which is released into the vapour phase and is thermally stable. Exposed rats suffer seizures similar to elliptic episodes. The agent has been shown to inhibit brain gamma-aminobutyric acid (GABA) receptors, which in turn limit general levels of inhibitory activity in the brain.[80] TMPP formation during combustion was originally identified from a flexible polyurethane foam formulated using a trimethylolpropane polyol and containing a phosphate fire retardant additive. This was never marketed, but the same problem occurs with turbine lubricants used in certain marine applications and aircraft engines, which are also formulated with trimethylolpropane polyols and tricresyl phosphate additives. Fortunately, the toxic ester (TMPP) is not formed from other polyols such as pentaerythritol. Although the pattern of toxicity differs from the irritancy and asphyxia observed from other fire effluents, the toxic product is formed at relatively low yields, so that the lethal toxic potency is not significantly different from those of other materials. Combustion products from other phosphate-containing materials have not produced acute neurotoxic effects. The main combustion product from phosphates is phosphoric acid, which has some irritant potency, although somewhat high lethal toxicity was observed in DIN furnace experiments involving a phosphate fire retardant used for treating cotton.[81] Since some phosphate esters are very toxic, further research on combustion products of phosphate-containing polymers would be useful.

7.9 Dose–Effect Relationships and Extrapolation to Humans

7.9.1 General

In the previous sections it has been proposed that the acute toxic effects of exposure to combustion products can be categorised mainly in terms of acute sensory irritancy experienced within seconds of exposure, asphyxiant effects occurring over timescale of minutes and lung inflammation occurring over a timescale of approximately 2–36 hours. It has also been established using animal models that the acute asphyxia can be understood in terms of the combined (mainly additive) effects of a small number of gases (CO, HCN, CO_2 and low oxygen hypoxia), while lethality can be modelled in terms of the additive effects of a these asphyxiant gases, a small number of acid gases and the overall organic content. Lethal toxic potency for combustion product mixtures is calculated by summing the fractions of a 30 minute LC_{50} concentration present for each component. In practice this represents the sums of the normalised *Ct* exposure doses for each gas measured during 30 minute exposures of rats. The extent to which a similar result would be obtained for longer or shorter exposure periods depends upon the extent to which, in rats and humans, the lethal exposure doses and additive effects of the components follow Haber's rule (see Chapter 8).

Sensory irritant potency for combustion product mixtures has been shown to be related to the content of inorganic acid gases and to the presence of partially oxidised organic substances, although it has not been possible to predict the intensity of sensory irritant effects in mice from an analysis of all the identifiable inorganic and organic components. Sensory irritant effects are generally considered to be concentration-related and occurring almost instantaneously on exposure in humans. The respiratory depression reflex response can also be shown to be rapidly developing in mice, but is not instantaneous, so that questions have been raised regarding the extent to which sensory irritancy is in fact a concentration- or dose-related phenomenon.

The calculation of time and dose to incapacitation from asphyxiant gases in humans is very important in terms of assessing escape and survival during fire incidents, in which differences of a few seconds often determine survival or death, so it is important to consider the physiological effects and interactions of these gases, and similarities and differences between different animal models and humans.

In combustion toxicology, a common approach is to express toxicity in terms of the *Ct* exposure dose (ppm min or g m^{-3} min) for incapacitation or lethality rather than in terms of the inhaled dose or concentration of the toxicant at the target organ. This causes potential problems when extrapolating from animals, especially rodents to humans, particularly in relation to effects of relative body size.

In terms of the toxicity there are two main possibilities:

- The concentration of the toxic agent at the target organ increases directly in relation to the inhaled dose, so that toxicity follows Haber's rule.
- The concentration of the toxic agent at the target organ depends on the dynamics of uptake *versus* excretion or detoxification, so that the target organ concentration tends to increase linearly at first at a rate depending upon the inhaled concentration, ventilation ($\dot{V}_E$, L min^{-1}) and body size, but then tends to a maximum as equilibrium between uptake and removal is reached.

The main problems in extrapolating from animal models to humans are as follows:

- Size differences.
- Body structure differences.
- Biochemical and physiological differences.

7.9.2 Effects of Allometric Relationships on Rates of Uptake of Asphyxiant Gases and Time to Incapacitation

Allometric relationships (relationships between the size of each part of an animal, or each animal relative to others) need to be considered when comparing effects in animals and humans. The effect of body surface area : volume ratio on metabolic rate and respiration is reflected in the respiratory minute volume per unit bodyweight, which is a major determinant of the rate of uptake of toxic gases, as shown in Figure 7.23, which is an expression of Guyton's formula,[82] whereby

$$\text{Volume of air breathed per minute } (\dot{V}_E,\ \text{mL}) = 2.18 \times \text{bodyweight in grams}^{0.75} \quad (7.6)$$

As Figure 7.23 shows, the volume of air inhaled per minute per kg bodyweight in a mouse is greater than that in a human by the factors shown in Table 7.8.

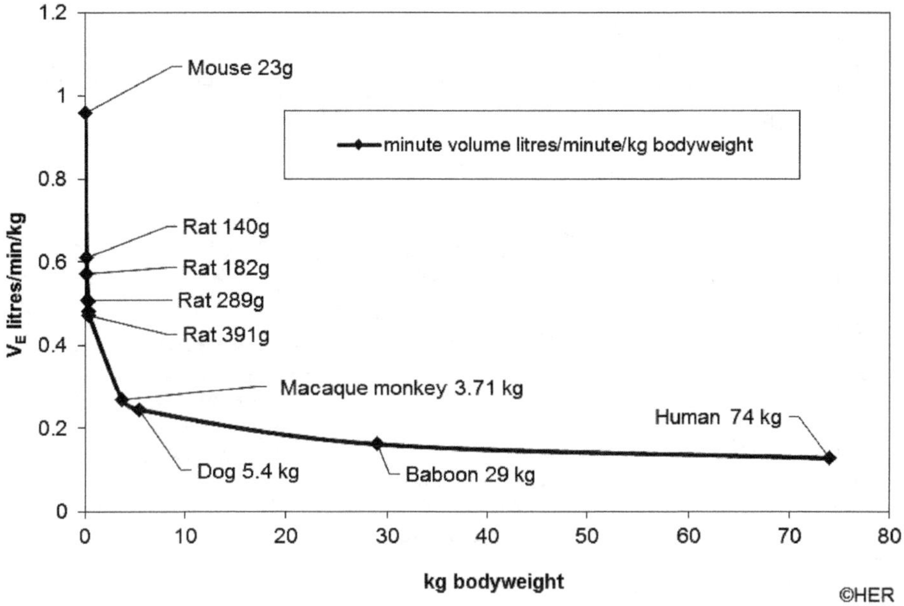

Figure 7.23 Relationship basal respiratory minute volume (V_E) in L min^{-1} and bodyweight for different mammal species.

Table 7.8 V_E ratios for different species.

Species	V_E factor
Human	1.0
Baboon	1.3
Dog	1.9
Monkey	2.1
Rat	4.0
Mouse	7.5

This is reflected in large differences in the rate of uptake of CO in rats and humans exposed to the same CO concentrations, as described in Purser[28] and illustrated Figure 7.24.

Figure 7.24 shows plots of blood carboxyhaemoglobin concentration (%COHb) against time for continuous exposures to two CO concentrations, 1200 ppm and 3000 ppm, in rats and humans. Predicted times to incapacitation are also shown. For rats, %COHb and time to incapacitation were calculated from the fitted curve developed by Kaplan and Hartzell.[83] For humans, CO uptake as %COHb was calculated using the Coburn Forster Kane equation.[84] A single experimentally determined point is also shown for macaque monkeys exposed to 1200 ppm CO. For humans and monkeys at rest, incapacitation occurs as confusion and loss of consciousness at approximately 40% COHb,[35,37,85,86] while for rats incapacitation occurred at

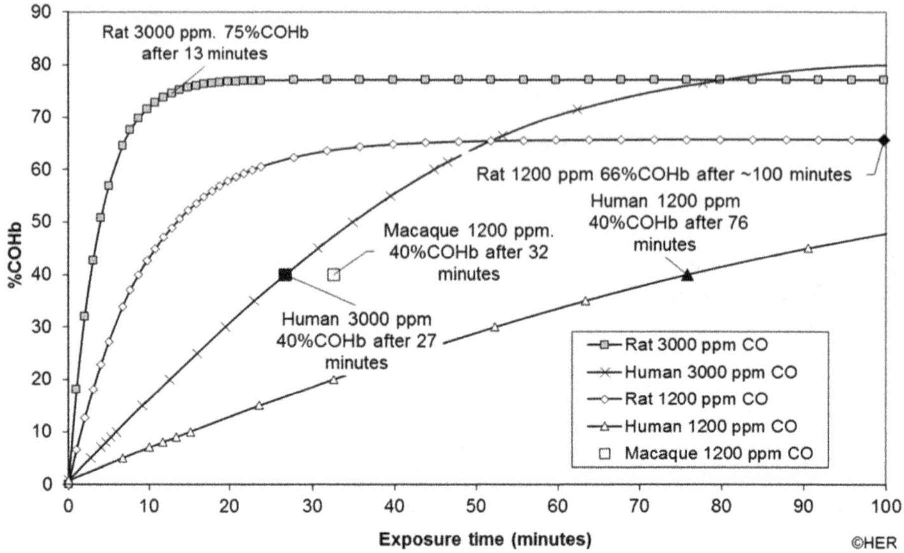

Figure 7.24 Rate of uptake of CO as %COHb, time to and %COHb at incapacitation for rats, humans and macaque monkeys at 1200 and 3000 ppm CO exposure concentrations.

65–76% COHb. However, at 1200 ppm CO the rat uptake curve levels off at around 65% COHb as the blood carboxyhaemoglobin concentration approaches equilibrium with the inhaled CO concentration. In practice the true equilibrium concentration is slightly higher at around 72% COHb, so the uptake model shown is not quite a perfect fit, as discussed in Hartzell *et al.*,[87] and the actual %COHb levels at incapacitation in rats can be somewhat higher. Uptake curves for continuous exposure to 3000 ppm CO are also shown, reaching 40% COHb after 2.7 minutes in rats with incapacitation at 76% COHb after 13 minutes. This is achieved much more rapidly than at 1200 ppm (after ~100 minutes) since it is below the point at which equilibrium is reached and the uptake curve levels off. Incapacitation is predicted after 27 minutes for humans at around 40% COHb.

The rate of uptake of CO is therefore initially much more rapid in the rat than in humans and monkeys, as predicted from the differences in $\dot{V}_E$ per unit body mass. At 1200 ppm CO the time to reach 40% COHb for rats is 8.76 minutes, compared to 32 minutes for monkeys and 76 minutes for humans. The blood carboxyhaemoglobin concentrations therefore reach equilibrium values much more rapidly in rats than in humans. However, rats can tolerate much higher %COHb concentrations before incapacitation occurs (where incapacitation in humans and monkeys is represented by loss of consciousness, and in rats by failure of leg flexion shock avoidance response). At rest, the rate of uptake in rats to 40% COHb is approximately ten times more rapid than in humans, which is even greater than a factor of four difference between the basal $\dot{V}_E$ levels. Of course, the rate of uptake in humans and primates also varies with the level of activity and hence $\dot{V}_E$, as discussed in Purser,[28] so that, in practice, time to incapacitation in a resting monkey is similar to that in a lightly active adult human, while a primate provides a more direct uptake model for an infant human of similar body size.

Time to incapacitation is shorter for rats exercising in an activity wheel than for resting animals when it is measured using the leg flexion avoidance response.[83] There is some evidence that the physiological effects of CO are not simply a function of the %COHb level achieved, but also of the time for which it is maintained, as in humans.[88,89] It is important to consider these differences in uptake rates and sensitivities to hypoxia in different species when comparing incapacitating and lethal doses.

Comparing effects between species in terms of exposure dose ignores aspects of uptake and distribution of toxic substances in the body. In practice significant deviations from Haber's rule occur for several important fire gases (especially CO over concentrations and timescales approaching equilibrium, and also for HCN, low oxygen hypoxia and CO_2 at all concentrations). Table 7.9 shows approximate *Ct* exposure doses for incapacitation and death from CO in different species. For humans at rest the CO exposure dose for incapacitation is around twice that in a macaque as predicted from the difference in uptake rates. But it is also approximately twice that in a rat, which is a somewhat coincidental result of the combination of more rapid uptake rates and a higher incapacitating dose threshold in the rat. Under conditions of light activity the exposure doses for incapacitation are similar in all four species; however, the human data are for fast walking, while the primate data are for less vigorous intermittent movement in a shuttle box chamber. For humans the database of %COHb concentrations in fatal victims of CO poisoning shows a mode of 75% COHb, but survival is rare above 50% COHb, see Figure 7.1.[22] This puts the fatal exposure dose for humans at approximately 60 000–190 000 ppm min when a subject is active and then collapses, and up to 110 000–240 000 if at rest throughout exposure.

For larger animals, such as macaques and humans, uptake of CO to a blood COHb concentrations predicted to cause incapacitation (30–40%

Table 7.9 *Ct* product exposure doses for incapacitation and death by CO for different species at rest and during light activity.

	Incapacitation		Fatal	
	CO at rest/ ppm min	CO light activity/ ppm min	CO at rest/ ppm min	CO light activity/ ppm min
Human, 70 kg	80 000–100 000	30 000–35 000	~110 000–240 000	~60 000–190 000
Baboon, ~20 kg		34 000		
Macaque, 3–4 kg	38 000–40 000	27 000		
Rat, ~300 g	30 000–40 000	22 000–36 000	162 000	

COHb) is approximately linear at inhaled CO concentrations significantly greater than the equilibrium concentration of 400 ppm, so that the *Ct* exposure dose for incapacitation at a given level of activity is constant. This is illustrated for macaque monkeys in Figure 7.25. The animals were placed in a chamber and trained to perform a behavioural task keeping them physically active while being exposed for up to 30 minutes to different constant CO concentrations. The task consisted of obtaining a food reward presented at one end of the chamber for a few seconds following a lever press at the other end of the chamber. The figure shows the time to incapacitation (loss of consciousness at ~30% COHb) for different CO concentrations between 900 and 8000 ppm. Between these levels loss of consciousness occurred at a constant exposure dose of approximately 27 000 ppm min, irrespective of the inhaled concentration.[36]

For hydrogen cyanide (also shown in Figure 7.25), the dynamics of uptake into the blood and dispersal throughout the body are more complex than for CO (see Chapter 10). At high concentrations, incapacitation occurs at much lower *Ct* exposure doses than for lower concentrations. The *Ct* exposure dose for incapacitation in a resting macaque is approximately 2610 ppm min for a 30 minute exposure period,[1,90,91] which compares with approximately 3000 ppm min for a rat.[83] Again, it is likely that the similarity is something of a coincidence. For low oxygen hypoxia the 30 minute LC_{50} concentration in

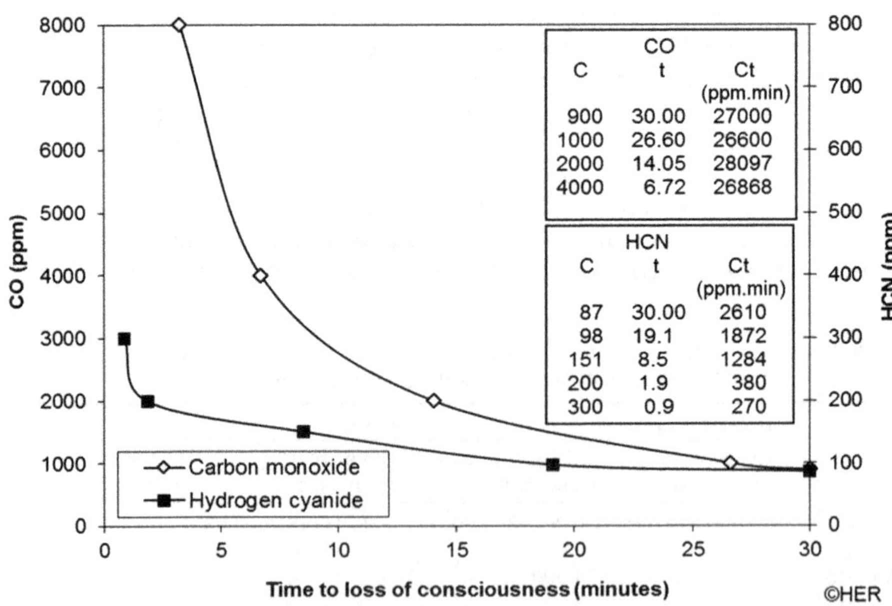

Figure 7.25 Relationships between time to incapacitation and concentration for CO and HCN exposures in primates. Time and concentration are equivalent for CO, but for HCN a small increase in concentration causes a large decrease in time to incapacitation[1] (see Chapter 10.)

rats is 5.4% (representing a depletion of 15.6%),[92,93] while around 8% O_2 (representing a depletion of 13%) is fatal in humans within around 12 minutes.[94] The effects of hypoxia expressed in these terms are non-linear, so that exposure doses for incapacitation are much smaller at high levels of depletion than at lower levels. Rats are certainly more tolerant of hypoxia than humans.

Human data on exposure doses for lethal lung damage from irritants are limited, but assumed to be approximately similar to those in rodents. When extrapolating to humans for safety purposes it is common to apply a factor of ten to allow for interspecies differences. Since the 30 minute lethal exposure dose for hydrogen chloride in rats is 114 000 ppm min, it could be considered potentially dangerous to expose humans to more than 11 400 ppm min or approximately 380 ppm for 30 minutes. This aspect is considered in more detail in Purser.[47]

7.9.3 AEGL Values for Irritant Gases Released in Combustion Products

The Acute Exposures Guidelines (AEGL)[95] provide guidance for the assessment of likely acute effects of human exposures to a range of individual substances likely to be released during industrial accidents, including a range of irritant gases occurring in combustion product mixtures (see Chapter 13). The AEGL publications are useful because the documentation includes detailed reviews of available toxicity data from both human and animal studies, with carefully justified opinion on extrapolation between and within species, and application to the selection of guideline ceiling levels for exposure periods from 10 minutes to 8 hours. The AEGL guideline levels are for three endpoints in exposed humans are shown in Table 7.10.

All three levels have application to different exposure situations involving combustion products, but in relation to incapacitating effects of sensory irritants on escape capability the AEGL-2 values are most relevant. The AEGL 2 values for some irritant fire gases are shown in Table 7.11.

Table 7.10 Acute Exposure Guideline Levels (AEGLs) of hazardous substances.

AEGL-1 is the airborne concentration of a substance above which it is predicted that the general population, including susceptible individuals, could experience notable discomfort, irritation, or certain asymptomatic nonsensory effects. However, the effects are not disabling and are transient and reversible upon cessation of exposure.

AEGL-2 is the airborne concentration of a substance above which it is predicted that the general population, including susceptible individuals, could experience irreversible or other serious, long-lasting adverse health effects or an impaired ability to escape.

AEGL-3 is the airborne concentration of a substance above which it is predicted that the general population, including susceptible individuals, could experience life-threatening adverse health effects or death.

Table 7.11 AEGL-2 values for different exposure durations of irritant gases occurring in combustion products (ppm).

Exposure period/hours	0.167	0.5	1	4	8
Substance					
HCl	100	43	22	11	11
HBr	100	43	22	11	11
HF	95	34	24	12	12
NO_2	20	15	12	8.2	6.7
SO_2	0.75	0.75	0.75	0.75	0.75
Formaldehyde	14	14	14	14	14
Acrolein	0.44	0.18	0.1	0.1	0.1

AEGL-1 and AEGL-2 make use of low-level human exposure physiological signs and animal toxicity data, especially histopathological changes to the respiratory tract. For sensory irritancy, use is made of human experience and mouse RD_{50} data on the basis that such effects are not dose-related, but concentration-related and maintained. The concentration predicted to cause significant sensory irritancy therefore sets a ceiling on the exposure levels over the entire range of exposure periods from 10 minutes to 8 hours. Where significant pathological effects are reported at lower exposure concentrations for different exposure periods these are applied to reduce the ceiling exposure concentrations (and thereby the exposure doses) for longer exposure periods.

The AEGL-3 values provide most hard data for consideration of exposure duration–dose response relationships. Different forms of $C^n t$ relationship are used as the main basis for setting the ceiling concentrations for the different exposure times, depending upon the effects of individual toxicants. The most reliable animal data are usually obtained using 4 hour exposure periods, although this is rather long for most fire incidents, for which the 10 and 30 minute exposure periods are most relevant.

Figure 7.26 shows a plot of the AEGL-3 *Ct* products for the irritant gases, normalised to the 4 hour exposure values. This therefore illustrates the extent to which, taking all available data into account, the AEGL committee estimated that the effects of exposure to these gases deviated from Haber's rule over the period up to 8 hours. The steeper the slope, the greater is the estimated deviation from Haber's rule. For most gases it was considered that short, high concentration exposures might be more hazardous than equivalent longer exposures.

7.9.4 Effects of Differences in Body Structure

Allometric differences in body structure and proportions, physiological and biochemical differences all affect the extrapolations from effects in rodents to predict effects in humans. There are considerable differences between rodents and primates, including humans, in the structure and function of the upper respiratory tract. Rodents have well-developed nasal passages with

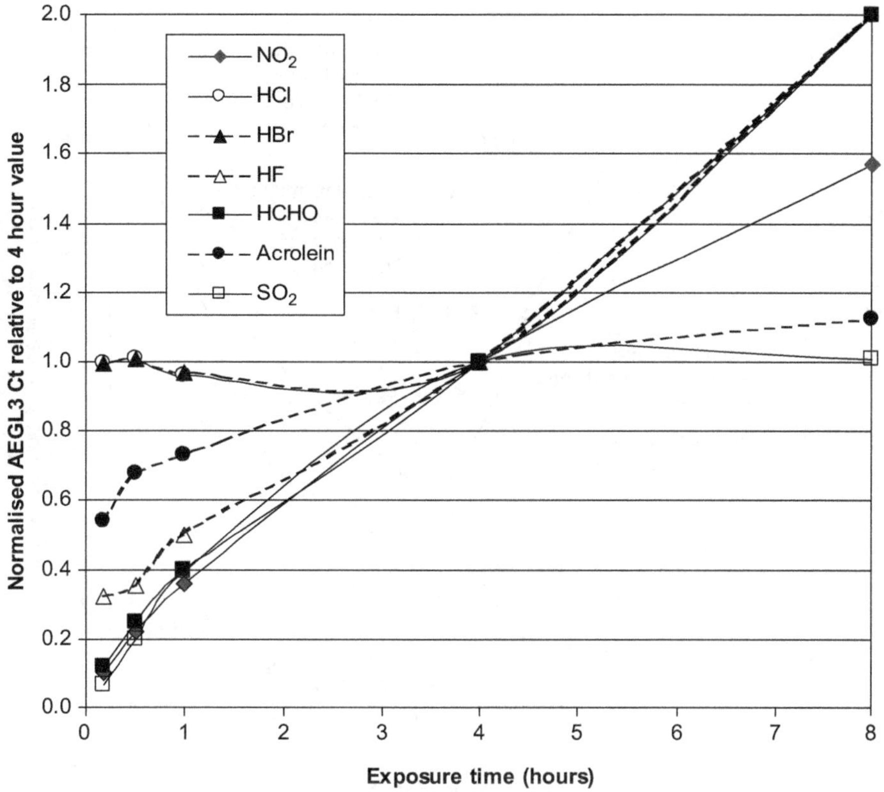

Figure 7.26 AEGL-3 *Ct* product exposure doses for irritant gases normalised to the 4 hour values.

a larger relative surface area compared to primates, and are obligate nose-breathers, unlike primates, which can use both nasal and oral breathing especially under stress or exercise. Water-soluble acid gases are therefore absorbed more readily in the upper respiratory tract of rodents than primates, resulting in differences in the physiological response to inhaled acid gases shown in Figures 7.5 and 7.21. The rodent sensory irritant response of maintained respiratory rate depression contrasts with the primate (including human) response of transient rate depression followed by vagally-induced hyperventilation.

The effects of differences in nasal structure and respiratory pattern between different species are illustrated in Figure 7.27. At rest, a mouse inhales 5.5 times the volume of air each minute per unit bodyweight as an adult human, so it might be predicted that they would be more sensitive to a specific inhaled concentration of a sensory or lung irritant. However, due to their very large nasal cavities compared with primates and especially humans, the nasal surface area per gram bodyweight is 33 times greater in the mouse than in humans. The airflow coming in to contact with each

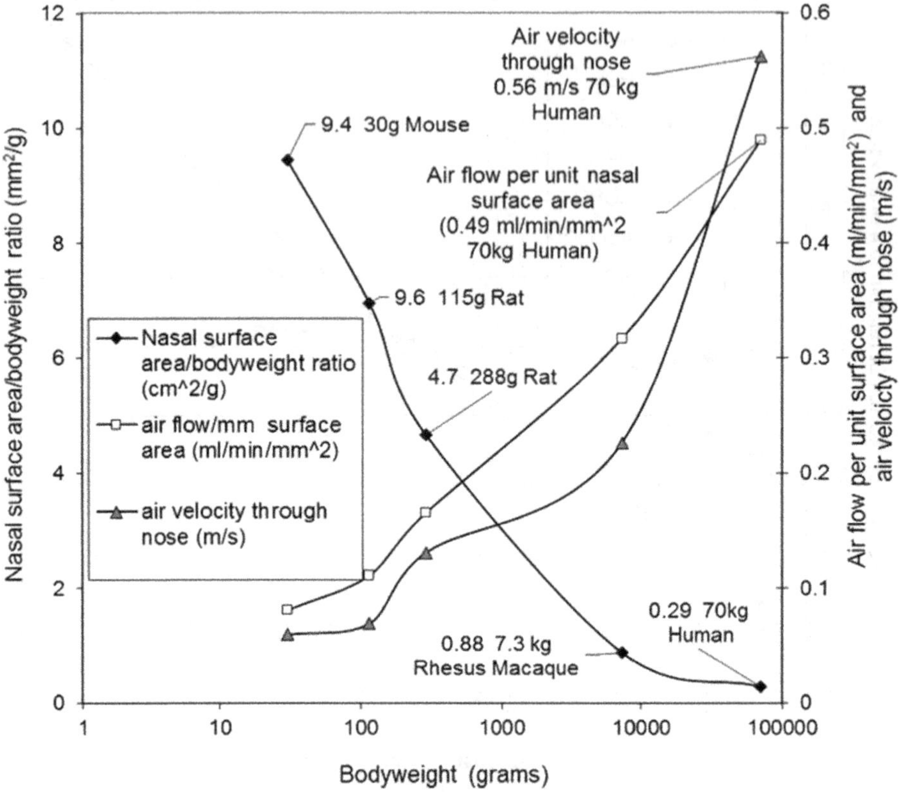

Figure 7.27 Comparison of nasal surface area : bodyweight ratio and air velocity through nose and air flow per unit nasal surface area in different species at resting respiration.

square millimetre of nasal surface area is six times greater in humans than mice (0.49 mL min^{-1} mm^{-2} in humans compared with 0.08 mL min^{-1} mm^{-2} in mice). The air velocity through the nose is nine times greater in humans than in mice (0.56 m s^{-1} for humans compared with 0.06 m s^{-1} in mice). These aspects may go some way to explaining the more efficient nasal scrubbing of water-soluble gases in rodents (larger relative surface and slower air velocity), while at the same time producing a more marked sensory irritant response than in primates. It may also explain why the time to reach an equilibrium sensory response is greater in rodents. In primates the larger airflow coming into contact with each unit surface area promotes rapid equilibrium. The slower airflow and larger relative surface area in rodents requires a longer period for equilibrium to develop. Similar issues arise in the large airways, which have a larger relative surface area in rodents relative to body mass. Figure 7.28 shows that each millimetre of trachea length enables more absorption in mice than humans. These effects are further exaggerated by the strong decrease in respiratory rate and minute volume in

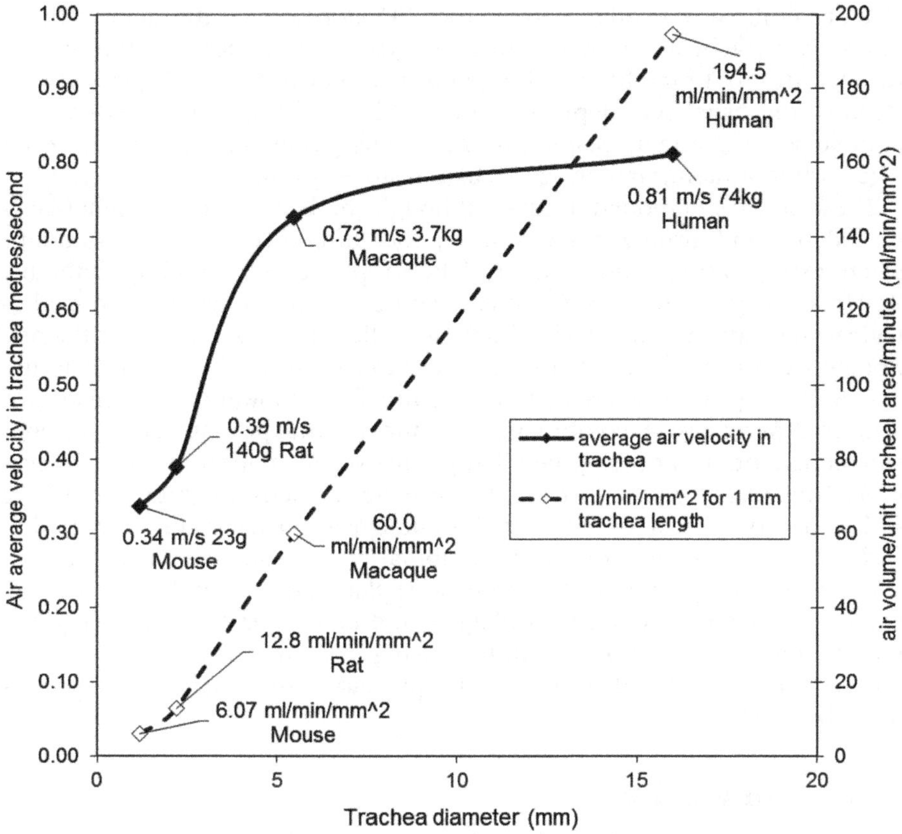

Figure 7.28 Comparison of air velocity in trachea and air flow per unit tracheal area in different species at resting respiration.

rodents when inhaling sensory irritants compared with the hyperventilation response in primates. The result is that rodents have a considerable degree or protection from penetration of acid gases into the lung compared with primates. Alarie showed that when the nose is bypassed by introducing irritant gases directly into the trachea of rodents, the LC_{50} concentrations (resulting from lung inflammation) are considerably lower.[53]

Rats and mice are able to reduce their respiration by up to around 90% when inhaling an irritant atmosphere. They tolerate the resultant hypoxia well, partly by reducing their peripheral circulation (similar to the diving response in marine mammals). This provides a protective effect, reducing the uptake of toxic gases into the lungs. When rats were exposed to 3000 ppm CO with 1000 ppm HCl the time to incapacitation due to CO intoxication was increased from 13 minutes with CO alone up to 22.2 minutes in the presence of HCl. The lethal exposure time was increased from approximately 65 minutes to 109.5 minutes.[87] This does not occur in primates, so that for macaque monkeys exposed to CO in the presence of irritants, the rate of uptake was similar to that from CO alone.[11,35] Introducing irritant gases *via* a tracheal cannula in mice to bypass the upper respiratory tract resulted in rapid breathing with a pattern similar to that in Figure 7.22 instead of the usual rate depression (as in Figure 7.21), and increased lung pathology.[53] Enhancing respiration by forcing guinea pigs to exercise on a wheel while inhaling irritant gases also resulted in increased toxicity.[96]

These studies demonstrate that, although the basic toxicity mechanisms in rodents and primates are similar, the differences can be important when extrapolating from rodent data to primates, including humans. Rodents are good models for investigating the physiological and pathological mechanisms of toxicity from fire effluent gases and fire effluents. Rodent exposure doses can also give an indication of likely incapacitating and lethal exposure doses in humans, provided allowances are made for species differences. A combination of human and primate data are considered essential to develop detailed predictive calculation methods effects of asphyxiant fire gases. In order to estimate the sensory irritant and lung inflammatory effects of irritants (physiological, behavioural and pathological) it is necessary to use a combination of data from all available sources, including human data from workplace exposes and industrial accidents, mouse RD_{50} data, physiological and behavioural studies using primates, and lung pathology studies using rodents. The application of methods for assessment of time to incapacitation in human fire victims is described in Chapter 14.

7.10 Conclusions

Studies of the physiological and toxicological effect of exposure to combustion product atmospheres have involved fire incident investigations and experimental exposures to fire gases and combustion product mixtures in humans and laboratory animals. These have demonstrated that the main effects and their severity (consisting of sensory irritancy, asphyxiation and post-exposure lung inflammation) can be understood in terms of a small number of irritant and asphyxiant gases and particulates. For sensory irritancy, overall severity of exposure to combustion product mixtures is reasonably well represented by directly additive models for the contribution from different components. This also basically applies to the interactions between asphyxiant gases, with allowances for deviations from Haber's rule, especially for HCN and low oxygen hypoxia, and for the effects of the hyperventilatory stimulus by inhaled CO_2 on the rates of uptake of other asphyxiant gases. Rodent studies of combustion product toxicity have confirmed that, for a wide range of polymers, overall lethal potency (including both within exposure and post-exposure deaths up to 14 days) is consistent with fractional additive effects for all asphyxiant and irritant atmosphere components. Only two cases of unpredicted toxic effects have been found, consisting of the extreme potency for lung inflammation of ultrafine fluoropolymer particulates, and the neurotoxic effects of exposure to trimethylolpropane phosphate evolved from fuels containing trimethylolpropane polyol additives and phosphates.

Using these physiological and toxicological data from the range of sources described, it has been possible to develop calculation models for the assessment of incapacitating and lethal effects on humans from the concentrations of a small number of key irritant and asphyxiant gases and particulates in fire effluents, as presented in Chapter 14.

References

1. D. A. Purser, Assessment of Hazards to Occupants from smoke, toxic gases and heat, in *SFPE Handbook of Fire Protection Engineering*. ed. P. J. DiNenno, National Fire Protection Association, Quincy MA 02269, USA, 4th edn, 2008, section 2, ch. 6, pp. 96–193.
2. ISO 13571 Life-threatening components of fire – Guidelines for the estimation of time to compromised tenability in fires, International Organization for Standardization, Geneva, Switzerland, 2nd edn, 2012.
3. B. C. Levin, M. Paabo, J. L. Gurman and S. C. Harris, Effects of Exposure to Single or Multiple Combinations of the Predominant Toxic Gases and Low-Oxygen Atmospheres Produced in Fires, *Fundam. Appl. Toxicol.*, 1987, **9**, 236–250.
4. G. E. Hartzell, D. N. Priest and W. G. Switzer, Modelling of Toxicological Effects of Intoxication of Rats by Carbon Monoxide and Hydrogen Cyanide, *J. Fire Sci.*, 1985, **3**, 115–128.
5. D. A. Purser, The evolution of toxic effluents in fires and the assessment of toxic hazard, *Toxicol. Lett.*, 1992, **64/65**, 247–255, DOI: 10.1016/0378-4274(92)90196-Q.
6. Determination of the Lethal Toxic Potency of Fire Effluents, *ISO 13344*, 1996.
7. D. A. Purser, Validation of additive models for lethal toxicity of fire effluent mixtures, *Polym. Degrad. Stab.*, 2012, **97**, 2552–2561.
8. B. C. Levin, New research avenues in toxicology: 7-gas N-gas model, toxicant suppressants and genetic toxicology, *Toxicology*, 1996, **115**, 89–106, DOI: 10.1016/S0300-483X(96)03497-X.
9. D. A. Purser, Behavioural Impairment in Smoke Environments, *Toxicology*, 1996, **115**, 25–40.
10. D. A. Purser, The application of exposure concentration and dose to evaluation of the effects of irritants as components of fire hazard. Interflam 2007. 3–5th September 2007 Royal Holloway College, Egham, UK. Proceeding, Interscience Communications, Greenwich, UK, pp. 1033–1041.
11. D. A. Purser and P. Buckley, Lung Irritance and Inflammation During and After Exposure to Thermal Decomposition Products from Polymeric Materials, *Med. Sci. Law*, 1983, **23**, 142–150.
12. F. W. Beswick, P. Holland and K. H. Kemp, Acute Effects of Exposure to Orthochlorobenzylidene Malonitrile (CS) and the Development of Tolerance, *Br. J. Ind. Med.*, 1972, **29**, 298–306.

13. D. A. Purser, 2002, ASET and RSET: addressing some issues in relation to occupant behaviour and tenability. 7th International Symposium on Fire Safety Science, Worcester Polytechnic Institute – Worcester Massachusetts, USA 16–21 June 2002. Fire Safety Science – Proceedings of the seventh international symposium. International Association for Fire Safety Science, 2003, pp. 91–102.
14. The application of fire safety engineering principles to fire safety design of buildings – Part 6: Human Factor: Life safety strategies – Occupant evacuation behaviour and conditions PD 7974-6:2004, British Standards Institution.
15. Fire safety engineering, Technical information on methods for evaluating behaviour and movement of people, ISO/TR 16738, International Organisation for Standardization, Geneva, 2009.
16. P. G. Wood, The Behavior of People in Fires, *Fire Research Note 953*, Fire Research Station, Watford, UK, 1972.
17. J. L. Bryan, Smoke As a Determinant of Human Behavior in Fire Situations (Project People), *NBS-GCR-77-94*, U.S. Department of Commerce, National Bureau of Standards, Washington, DC, 1977.
18. J. L. Bryan, in *SFPE Handbook of Fire Protection Engineering*, NFPA, Quincy, MA, 3rd edn, 2002, pp. 315–341.
19. T. Jin, Studies of Emotional Instability in Smoke from Fires, *J. Fire Flammability*, 1981, **12**, 130–142.
20. T. Jin, Visibility and Human Behavior in Fire Smoke in *SFPE Handbook of Fire Protection Engineering*, ed. P. J. DiNenno *et al.*, National Fire Protection Association, Quincy, MA, 4th edn, 2008, pp. 54–66.
21. D. A. Purser and M. Kuipers, 2004 Interactions between buildings, fire and occupant behaviour using a relational database created from incident investigations and interviews. 3rd International Symposium on Human Behaviour in Fire. Europa Hotel, Belfast, 1st–3rd September 2004, Proceedings, pp. 443–456, Interscience Communications, London, UK.
22. G. L. Nelson, Carbon Monoxide and Fire Toxicity: A Review and Analysis of Recent Work, *Fire Technol.*, 1998, **34**, 38–58.
23. J. Pach *et al.*, Analysis of Predictive Factors in Acute Carbon Monoxide Poisonings, *Folia Med. Cracov.*, 1978, **20**, 159–168.
24. R. A. Anderson, I. Thompson and W. A. Harland, The Importance of Cyanide and Organic Nitriles in Fire Fatalities, *Fire Mater.*, 1979, **3**, 91–99.
25. J. L. McAllister, R. J. Roby, B. Levine and D. Purser, Stability of cyanide in cadavers and in postmortem stored tissue specimens: A review, *J. Anal. Toxicol.*, 2008, **32**, 1–9.
26. D. A. Purser, P. Grimshaw and K. R. Berrill, Intoxication by Cyanide in Fires: A Study in Monkeys Using Polyacrylonitrile, *Arch. Environ. Health*, 1984, **39**, 394–400.
27. D. A. Purser, *Determination of Blood Cyanide and Its Role in Producing Incapacitation in Fire Victims*, Royal Society of Chemistry Meeting, Huntingdon, 1984.

28. D. A. Purser, Asphyxiant components of fire effluents, in *Fire Toxicity*, ed. A. Stec and R. Hull, Woodhead, Cambridge UK, 2010, ch. 4, pp. 118–198.
29. D. A. Purser, Toxic product yield and hazard assessment for fully enclosed design fires involving fire retarded materials. Fire Retardant Polymers, 7th European Conference. University of Greenwich, UK 8th-10th September 1999. Published in Polymer International 49: 1232–1255, 2000.
30. D. A. Purser, J. A. Rowley, P. J. Fardell and M. Bensilum, Fully Enclosed Design Fires for Hazard Assessment in Relation to Yields of Carbon Monoxide and Hydrogen Cyanide, *Interflam '99, Eighth International Fire Science and Engineering Conference*, Edinburgh, Proceedings pp. 1163–1169, Interscience Communications, London (June–July 1999).
31. F. J. Baud, P. Barriot, V. Toffis *et al.*, Elevated Blood Cyanide Concentrations in Victims of Smoke Inhalation, *N. Engl. J. Med.*, 1991, **325**, 1761–1766.
32. D. F. King, *Aircraft Accident Report 8/88*, UK Department of Transport, Air Accidents Investigation Branch, HMSO, London, UK, 1988.
33. D. A. Purser and J. A. Purser, HCN yields and fate of fuel nitrogen for materials under different combustion conditions in the ISO 19700 tube furnace and large-scale fires, *9th International Symposium on Fire Safety Science*, University of Karlsruhe, Germany, 21st-26th September 2008.
34. D. A. Purser, Fire safety and evacuation implications from behaviours and hazard development in two fatal care home incidents: Rosepark and Frampton House, *Proceedings of the 5th International Symposium on Human Behaviour in Fire, Downing College, Cambridge, UK*, Interscience, London, 19–21 September 2012, pp. 251–262.
35. D. A. Purser and W. D. Woolley, Biological Studies of Combustion Atmospheres, *J. Fire Sci.*, 1983, **1**, 118–145.
36. D. A. Purser and K. R. Berrill, Effects of Carbon Monoxide on Behaviour in Monkeys in Relation to Human Fire Hazard, *Arch. Environ. Health*, 1983, **38**, 308–315.
37. D. A. Purser, A bioassay model for testing the incapacitating effects of exposure to combustion product atmospheres using cynomolgus monkeys, *J. Fire Sci.*, 1984, **2**, 20–36.
38. D. A. Purser, P. Grimshaw and K. R. Berrill, Intoxication by Cyanide in Fires: A Study in Monkeys Using Polyacrylonitrile, *Arch. Environ. Health*, 1984, **39**, 394–400.
39. D. A. Purser and P. Grimshaw, The incapacitative effects of exposure to the thermal decomposition products of polyurethane foams, *Fire Mater.*, 1984, **8**, 10–16.
40. Controlled Equivalence Ratio Method for the Determination of Hazardous Components of Fire Effluents," *ISO/TS 19700*, International Organization for Standardization, Geneva, 2007.
41. W. D. Woolley and P. J. Fardell, Basic Aspects of Combustion Toxicology, *Fire Saf. J.*, 1982, **5**, 29.

42. H. L. Kaplan, A. F. Grand, W. G. Switzer, D. S. Mitchell, W. R. Rogers and G. E. Hartzell, Effects of Combustion Gases On Escape Performance of the Baboon and the Rat, *J. Fire Sci.*, 1985, **3**(4), 228–244, DOI: 10.1177/073490418500300401.
43. W. D. Woolley, S. A. Ames and P. J. Fardell, Chemical Aspects of Combustion Toxicology of Fires, *Fire Mater.*, 1979, **3**, 110–120.
44. Y. Alarie, Bioassay for Evaluating the Potency of Airborne Sensory Irritants and Predicting Acceptable Levels of Exposure in Man, *Food Cosmet. Toxicol.*, 1981, **19**, 623–626.
45. L. Kane, C. S. Barrow and Y. Alarie, A Short-Term Test to Predict Acceptable Levels of Exposure to Airborne Sensory Irritants, *Am. Ind. Hyg. Assoc. J.*, 1979, **40**, 207–209.
46. M. M. Hirschler and D. A. Purser, Irritancy of the smoke (non-flaming mode) from materials used for coating wire and cable products, both in the presence and absence of halogens in their chemical composition, *Fire Mater.*, 1993, **17**, 7–20.
47. D. A. Purser, Hazards from smoke and irritants, in *Fire Toxicity*, ed. A. Stec and R. Hull, Woodhead, Cambridge UK, 2010, ch. 3, pp. 51–117.
48. Documentation of the Threshold Limit Values for Substances in Workroom Air, *American Conference of Governmental Industrial Hygienists*, Cincinnati, 1980.
49. C. L. Punte, J. T. Weimer, T. A. Ballard and J. L. Wilding, Toxicologic studies on *o*-Chlorobenzylidene Malononitrile, *Toxicol. Appl. Pharmacol.*, 1962, **4**, 656–662.
50. B. Ballantyne and S. Calloway, Inhalation Toxicology and Pathology of Animals Exposed to *o*-Chlorobenzylidene Malonitrile, *Med. Sci. Law*, 1972, **12**, 43–65.
51. L. Kane, C. S. Barrow and Y. Alarie, A Short-Term Test to Predict Acceptable Levels of Exposure to Airborne Sensory Irritants, *Am. Ind. Hyg. Assoc. J.*, 1979, **40**, 207–209.
52. *Registry of Toxic Effects of Chemical Substances*, National Institute for Occupational Safety and Health, Washington, DC, 1982.
53. Y. Alarie, *Proceedings of the Inhalation Toxicology Symposium, Upjohn Company*, Ann Arbor Science (The Butterworth Group), Ann Arbor, MI, 1980.
54. W. H. Steinhagen and C. S. Barrow, Sensory irritation structure-activity study of inhaled aldehydes in B6C3F1 and Swiss Webster mice, *Toxicol. Appl. Pharmacol.*, 1984, **72**, 495–503.
55. J. C. DeCeaurriz, J. C. Micillini, P. Bonnet and J. P. Guenier, Sensory irritation caused by various industrial airborne chemicals, *Toxicol. Lett.*, 1981, **9**, 137–143.
56. H. Salem and H. Cullumbine, Inhalation Toxicities of Some Aldehydes, *Toxicol. Appl. Pharmacol.*, 1960, **2**, 183–187.
57. Acute Exposure Guideline Levels for Selected Airborne Chemicals Volume 4 (2004) Hydrogen Fluoride p 123. Subcommittee on Acute Exposure Guideline Levels, Committee on Toxicology Board on

Environmental Studies and Toxicology. National Academic Press, Washington.
58. T. J. Cole, J. E. Cotes, G. R. Johnson, H. deV. Martin, J. W. Reed and M. J. Saunders, Ventilation, Cardiac Frequency and Pattern of Breathing During Exercise in Men Exposed to *o*-Chlorobenzylidene Malonitrile (CS) and Ammonia Gas in Low Concentrations, *Q. J. Exp. Physiol.*, 1977, **62**, 341–351.
59. Y. Alarie, Sensory Irritation by Airborne Chemicals, *CRC Crit. Rev. Toxicol.*, 1973, **2**, 299.
60. R. C. Anderson, P. A. Croce, F. G. Feeley and J. D. Sakura, Study to Assess the Feasibility of Incorporating Combustion Toxicity Requirements into Building Materials and Furnishing Codes of New York State, *Reference 88712*, Arthur D. Little, Cambridge, MA, 1983.
61. C. S. Barrow, H. Lucia, M. F. Stock and Y. C. Alarie, Development of Methodologies to Assess the Relative Hazards from Thermal Decomposition Products of Polymeric Materials, *Am. Ind. Hyg. Assoc. J.*, 1979, **40**, 408–423.
62. Y. Alarie and R. C. Anderson, Toxicologic and Acute Lethal Hazard Evaluation of Thermal Decomposition Products of Synthetic and Natural Polymers, *Toxicol. Appl. Pharmacol.*, 1979, **51**, 341–362.
63. D. A. Purser, The application of exposure concentration and dose to evaluation of the effects of irritants as components of fire hazard, Interflam 2007, Royal Holloway College, University of London, UK 3-5th September 2007, Proceedings, pp. 1033–1046.
64. C. L. Punte, E. J. Owens and P. J. Gutentag, Exposures to ortho-Chlorobenzylidene Malonitrile, *Arch. Environ. Health*, 1963, **6**, 366–374.
65. B. Levin, J. Gurman, M. Paabo, L. Baier and T. Holt, Toxicological effects of different time exposures to fire gases: carbon monoxide or hydrogen cyanide or to carbon monoxide combined with hydrogen cyanide or carbon dioxide, Proceedings of the 9th Joint Panel Meeting of the UNJR Panel on Fire Research and Safety, US National Bureau of Standards, Gaithersbuerg, MD, Report NBSIR 88-3753, p. 368, 1988.
66. G. E. Hartzell, A. F. Grand and W. G. Switzer, Modeling of Toxicological Effects of Fire Gases: VI. Further Studies on the Toxicity of Smoke Containing Hydrogen Chloride, *J. Fire Sci.*, 1987, **5**, 368–391, DOI: 10.1177/073490418700500602.
67. B. C. Levin, A. J. Fowell, M. M. Birky, M. Paabo, S. Stolte and D. Malek, Further Development of a Test Method for the Acute Inhalation Toxicity of Combustion Products, *NBSIR 82-2532*, National Bureau of Standards, Washington, DC, 1982.
68. V. Babrauskas, B. C. Levin, R. G. Gann *et al.*, *Toxic Potency Measurement for Fire Hazard Analysis*, NIST Special Publication 827, 1991.
69. DIN, Producing Thermal Decomposition Products from Materials in an Air Stream and Their Toxicological Testing, *DIN 53 436*, Deutsches Institut für Normung, Berlin, Germany.

70. H. Klimisch, H. W. Hollander and J. Thyssen, Comparative Measurements of the Toxicity to Laboratory Animals of Products of Thermal Decomposition Generated by the Method of DIN 53 436, *J. Combust. Toxicol.*, 1980, **7**, 209–230.
71. J. M. Jounay, C. Boudene and R. Truhout, The physiogram as a method for the evaluation of combustion products in controlled ventilation experiments, Polymer Conference Salt Lake City, 1976.
72. J. Pauluhn, A Retrospective Analysis of Predicted and Observed Smoke Lethal Toxic Potency Values, *J. Fire Sci.*, 1993, **11**, 109–130.
73. Guidelines for Methodology for Assessing the Fire Threat to People, ISO 19706, ISO, Geneva, Switzerland, 2007.
74. D. A. Purser and J. A. Purser, The potential for including fire chemistry and toxicity in fire safety engineering, BRE Project Report No. 202804, 26th March 2004, Building Research Establishment, Garston Watford UK.
75. D. A. Purser and J. A. Purser, HCN yields and fate of fuel nitrogen for materials under different combustion conditions in the ISO 19700 tube furnace and large-scale fires, *Fire Saf. Sci.*, 2009, **9**, 1117–1128, DOI: 10.3801/IAFSS.FSS.9-1117.
76. D. A. Purser, Influence of fire retardants on toxic and environmental hazards from fires, in *Fire retardancy of polymers. New strategies and mechanisms*, ed. T. R. Hull and B. K. Kandola, RSC Publishing, Cambridge, 2009, ch. 24, pp. 381–404.
77. D. A. Purser, Recent Developments in Understanding the Toxicity of PTFE Thermal Decomposition Products, *Fire Mater.*, 1992, **16**, 67–75.
78. J. H. Petajan, K. L. Voorhees, S. C. Packham, R. C. Baldwin, I. N. Einhorn, M. L. Grunnet, B. G. Dinger and M. M. Birky, Extreme Toxicity from Combustion Products of a Fire-Retarded Polyurethane Foam, *Science*, 1975, **187**, 742–744.
79. D. A. Purser, Toxicity of fire retardants in relation to life safety and environmental hazards, in *Fire Retardant Materials*, ed. A. R. Horrocks and D. Price, Woodhead Publishing Ltd, Cambridge UK, 2001, ch. 3, pp. 69–127.
80. W. Y. Kao, Q. Y. Liu, W. Ma, G. D. Richie, J. Lin, A. F. Nordholm, J. Rossi, J. L. Barker, D. A. Stenger and J. J. Pancrazio, Inhibition of spontaneous GABAergic transmission by trimethylolpropane phosphate, *Neurotoxicology*, 1999, **5**, 843–849.
81. R. Kallonen, A. von Wright, L. Tikkanen *et al.*, The toxicity of fire effluents from textiles and upholstery materials, *J. Fire Sci.*, 1985, **3**, 145–160.
82. A. C. Guyton, Measurement of the respiratory volumes of laboratory animals, *Am. J. Physiol.*, 1947, **150**, 70–77.
83. H. L. Kaplan and G. E. Hartzell, Modeling of Toxicological Effects of Fire Gases: I. Incapacitating Effects of Narcotic Fire Gases, *J. Fire Sci.*, 1984, **2**(4), 286–305, DOI: 10.1177/073490418400200404.

84. R. F. Coburn, R. E. Forster and P. B. Kane, Consideration of Physiological Variables That Determine the Blood Carboxyhaemoglobin Concentrations in Man, *J. Clin. Invest.*, 1965, **44**, 1899–1910.
85. R. D. Stewart, The effects of carbon monoxide on man, *J. Combust. Toxicol.*, 1974, **1**, 167–176.
86. P. R. Sayers and S. J. Davenport, Review of Carbon Monoxide Poisoning, *Public Health Bulletin 195*, U.S. Government Printing Office, Washington D.C., 1930.
87. G. E. Hartzell, H. W. Stacy, W. G. Switzer, D. N. Priest and S. C. Packham, Modeling of Toxicological Effects of Fire Gases: IV. Intoxication of Rats by Carbon Monoxide in the Presence of an Irritant, *J. Fire Sci.*, 1985, **3**(4), 263–279, DOI: 10.1177/073490418500300403.
88. D. A. Purser, Interactions Among Carbon Monoxide, Hydrogen Cyanide, Low Oxygen Hypoxia, Carbon Dioxide and Inhaled Irritant Gases, in *Carbon Monoxide Toxicity*, ed. D. G. Penney, CRC Press, Boca Raton, 2000, ch. 7, pp. 157–192.
89. R. A. M. Meyers and S. R. Thom, Carbon monoxide and cyanide poisoning, in *Hyperbaric Medical Practice*, ed. E. P. Kindwall, Best Publishing Co, 1st edn, 1995, ch. 18, pp. 344–372.
90. D. A. Purser, P. Grimshaw and K. R. Berrill, Intoxication by Cyanide in Fires: A Study in Monkeys Using Polyacrylonitrile, *Arch. Environ. Health*, 1984, **39**, 394–400.
91. D. A. Purser, *Determination of Blood Cyanide and Its Role in Producing Incapacitation in Fire Victims*, Royal Society of Chemistry Meeting, Huntingdon, 1984.
92. B. C. Levin, M. Paabo, J. L. Gurman and S. C. Harris, Effects of Exposure to Single or Multiple Combinations of the Predominant Toxic Gases and Low-Oxygen Atmospheres Produced in Fires, *Fundam. Appl. Toxicol.*, 1987, **9**, 236–250.
93. B. C. Levin, New research avenues in toxicology: 7-gas N-gas model, toxicant suppressants and genetic toxicology, *Toxicology*, 1996, **115**, 89–106.
94. U. C. Luft, A*via*tion Physiology—The Effects of Altitude, in *Handbook of Physiology*, American Physiology Society, Washington, DC, 1965, 1099–1145.
95. National Advisory Committee for Acute Exposure Guideline Levels for Hazardous Substances (NAC/AEGL). Acute Exposure Guideline Levels for Selected Airborne Chemicals Subcommittee on Acute Exposure Guideline Levels, Committee on Toxicology Board on Environmental Studies and Toxicology. National Academic Press, Washington.
96. D. E. Malek and Y. Alarie, Ergometer within a whole-body plethysmograph to evaluate performance of guinea pigs under toxic atmospheres, *Toxicol. Appl. Pharmacol.*, 1989, **101**, DOI: 10.1016/0041-008X(89)90282-2.

CHAPTER 8

Haber's Law and its Application to Combustion Products[†]

ROBERT L. MAYNARD[a] AND DAVID A. PURSER[b]

[a] University of Birmingham, UK; [b] Hartford Environmental Research, Hatfield, UK
*Email: robertmaynard3@gmail.com; david-purser@ntlworld.com

8.1 Introduction

The concentration–dose–effect relationships of the individual toxic substances and mixtures in combustion products depend on a set of parameters, including the dynamics of uptake, target organ concentration and activity, duration of exposure, metabolism and excretion. For some key toxic combustion gases, such as carbon monoxide, hydrogen cyanide and hydrogen chloride, measurement of these parameters has enabled the development of accurate calculation models to predict their effects on exposed subjects, as described in the other chapters in this book. However, for the majority of individual substances known to occur in combustion products, knowledge relating to these key parameters is limited, especially over the full ranges of exposure concentrations and exposure durations of interest. Also,

[†]Much of the material included in this chapter has already been published in the Health Protection Agency's *Chemical Hazards and Poisons Report*, April 14th 2009, and is reproduced here with permission of the editor of that publication, Professor Virginia Murray.

Issues in Toxicology No. 23
Toxicology, Survival and Health Hazards of Combustion Products
Edited by David A. Purser, Robert L. Maynard and James C. Wakefield

Published by the Royal Society of Chemistry, www.rsc.org

some health and safety applications require simple assessment methods and guideline exposure criteria that can readily be derived and applied to emergency response guidelines and workplace exposure limits.

One of the simplest and most practical approaches is to base predictions on Haber's Law and its various derivatives, which have stood the test of time, being in use for almost a century. The simple basis of this approach is to treat the subject as a "black box", considering only the exposure concentration for the airborne toxicant, the exposure time and a specified toxic endpoint such as incapacitation or lethality.

This chapter describes the history of Haber's Law, the derivation of some of its variants and their application to some combustion gases. Consideration is given to some limitations, especially with respect to the prediction of the effects of acute exposures to high concentrations of irritant and asphyxiant gases in fires.

8.2 History of Haber's Law

Fritz Haber, one of Germany's greatest scientists of the late 19th and early 20th centuries, was born in Breslau in 1868. His family were prosperous chemical and dye merchants. An early interest in chemical experiments at home led to Haber attending courses at Berlin and Heidelberg universities. These courses were followed by a short time in the family business, but this seems to have been unhappy and Haber went back to university. He flourished: organic chemistry led to thermodynamics and electro-chemistry, and by 1905, with his book "*The Thermodynamics of Technical Gas Reactions*" in print, he was an acknowledged leader of German chemistry. Arguments with Nernst (physiologists will recall the Nernst equation!) followed, but with the Haber Process for synthesising ammonia for the production of fertilisers and explosives, Haber swept to a Nobel Prize in 1918 (awarded in 1919). An early discovery had related to the strength of C–C as compared with C–H linkages in aliphatic and aromatic compounds and this led to Haber's Rule:

> *"As regards thermal stability the C–C link is stronger than the C–H link in aromatic compounds and weaker in aliphatic compounds."*

Haber's Rule, Haber's Process; next came Haber's Law.

In 1914, Haber placed the resources of the Kaiser Wilhelm Institute, which he directed, at the service of the German War Ministry. Explosives were needed and the Haber Process provided the essential ammonia. Then, in late 1914, Haber turned to chemical warfare. Nernst was already active in the field and working on irritants, but Haber's drive and enthusiasm again swept him to the front: he introduced chlorine as a chemical warfare agent for a large-scale and very effective attack on April 22nd 1915, and in late 1916 he was appointed Chief of the Chemical Warfare Service with responsibility for all aspects of the subject from research to training. A range of agents were developed, including sulfur mustard in 1917.

Although Haber felt that use of chemical warfare agents was justified, his wife Clara could not accept it and was so distressed that she committed suicide by shooting herself on 2nd May 1915 following an argument on the subject. The post-war period was also unhappy for Haber: he failed to extract enough gold from seawater to pay off the German war debt, he fell foul of the Nazi regime (he was of Jewish origin) and despite continuing to produce first class work, was effectively exiled in 1933. He died in 1934.

8.3 Haber's Law

Haber's toxicological work on chemical warfare agents led him to suggest that for a given animal species and a given chemical, the likelihood of mortality could be expressed as

$$Ct = w, \tag{8.1}$$

where C is the exposure concentration (mg m^{-3}), t is the duration of exposure (minutes) and w is the mortality product (Tödlichsckeitproduckt) or lethal index (LI).

This equation is often referred to as Haber's Rule, though Haber's Law seems more appropriate given that Haber's Rule, see above, was applied to organic bond strengths.

Thus, if the exposure to x mg m^{-3} for 10 minutes caused death, so would exposure to $0.5x$ mg m^{-3} for 20 minutes.

This equation seemed to apply well to compounds said to produce irreversible effects by local reactions, *e.g.*, chlorine and phosgene, or compounds for which the half-life for detoxification or excretion is long with respect to the exposure duration (as described by Klimisch *et al.*).[1] Some compounds producing systemic poisoning fitted the equation less well, where the rate of detoxification and excretion prevented accumulation until a certain threshold concentration was inhaled, when the rate of uptake exceeded the maximum rate of removal. Accumulation then proceeds at a rate dependent upon the difference between the inhaled concentration and the threshold concentration (the rate of uptake and the constant rate of removal). An "elimination factor" e was included to allow for this, as described by Flury and Zernik (1931):[2]

$$(C - \mathrm{e})t = w \tag{8.2}$$

Unfortunately, this simple principle does not always hold true. In particular, some volatile substances (such as CO) are both taken up and excreted *via* the lungs and others, such as hydrogen cyanide, have more complex uptake and distribution dynamics (see Chapters 7, 9 and 10). Haber himself stated that HCN did not follow his law in a series of lectures presented in 1920–1923.[3] For CO the rate of uptake depends on the difference between the concentration inhaled and that in the body, giving an exponential uptake so that:

$$C(1 - \mathrm{e}^{-tk}) = w \tag{8.3}$$

This is the basis for the Coburn–Forster–Kane (CFK) equation[4,5] describing the uptake of CO in humans (see Chapters 7 and 9). This relationship approaches the linear Haber's Law when the concentration C in the atmosphere is high with respect to the concentration in the body required to cause incapacitation or death, so that for short exposures to high CO concentrations, uptake is approximately linear. This effect is illustrated by the results from CO exposure experiments in primates described in Chapter 7 (Figure 7.25).

Somewhat similar relationships occur for short exposures to substances where the inhaled substance accumulates initially at a particular site at a rate depending upon the inhaled concentration, but is rapidly detoxified or disperses from the initial site at a rate depending upon the accumulated concentration. This appears to be the case for the acute sensory irritancy to the cornea and upper respiratory tract epithelium resulting from inhalation of acid gases such as HCl. Exposure causes a reflex decrease in respiratory rate (especially in mice), the extent of which is proportional to the logarithm of the inhaled concentration. As described in Chapter 7, this response develops very rapidly (as does the upper respiratory tract pain response in humans) and can be predicted in terms of a calculation model with these uptake and dispersal features.[6] A similar uptake and response pattern occurs with short exposures of a few minutes to high concentrations of hydrogen cyanide above approximately 100 ppm. As described in Chapter 7 for primate exposures, time to incapacitation is highly dependent on the exposure concentration of HCN, with a *Ct* product exposure dose for incapacitation (loss of consciousness) of 270 ppm min at an inhaled concentration of 300 ppm compared with a *Ct* product of 2610 ppm min at an inhaled concentration of 87 ppm (Chapter 7, Figure 7.25). A detailed analysis of uptake and circulation dynamics, and physiological effects in primates (presented in Chapter 10) shows that, over these short time scales, time to incapacitation is related to the plasma cyanide concentration, the time–concentration profile of which varies considerably under different exposure conditions.

The effect of acid gases on respiration is a major parameter influencing the toxicity profile of such compounds. In mice, exposure to strongly irritant concentrations of acid gases readily absorbed in the upper respiratory tract results in a decrease in respiratory rate and minute volume of up to 90% of pre-exposure levels (see Chapter 7). This therefore decreases the rate of uptake of the toxicant. On the other hand HCN acts initially as a respiratory stimulant, so that exposure of primates to HCN atmospheres produced a factor of two to three increase in minute volume over a period of a few minutes, resulting in an increase in HCN uptake and a transient peak in plasma HCN.

In contrast to these more complex concentration–dose–effect relationships over short exposure times, deviating considerably from Haber's Law, the inflammatory effects of irritants acid gases, such as HCl on the lungs and the systemic effects of lower concentration exposures to hydrogen cyanide,

Table 8.1 AEGL-3 (life threatening) Acute Exposure Guideline Levels for hydrogen chloride.

Exposure period	10 min	30 min	60 min	4 h	8 h
AEGL-3/ppm	620	210	100	26	26
Ct/ppm hour	104	105	100	104	208

Table 8.2 Hydrogen cyanide and tabun (the nerve agent GA).

Hydrogen cyanide/ppm					
	10 min	30 min	60 min	4 h	8 h
AEGL-1	2.5	2.5	2.0	1.3	1.0
AEGL-2	17	10	7.1	3.5	1.0
AEGL-3	27	21	15	8.6	6.6
Agent GA (tabun)/ppm (mg m^{-3} data shown in brackets)					
	10 min	30 min	60 min	4 h	8 h
AEGL-1	0.0010 [0.0069]	0.0060 [0.0040]	0.00042 [0.0040]	0.00021 [0.0014]	0.00015 [0.0010]
AEGL-2	0.0013 [0.087]	0.0075 [0.050]	0.0053 [0.0035]	0.0026 [0.0017]	0.0020 [0.0013]
AEGL-3	0.11 [0.76]	0.057 [0.038]	0.039 [0.26]	0.021 [0.14]	0.015 [0.10]

are much more in agreement with Haber's Law. For HCl the exposure dose for lung inflammation follows the classic Haber equation [eqn (8.1)]. This is reflected in the US Acute Exposure Guideline Level (AEGL-3)[7] values for life threatening adverse health effects for HCl shown in Table 8.1, in that the *Ct* exposure dose is constant at ~103 ppm hour for exposure periods between 10 minutes and 4 hours.

Chapter 7, Figure 7.26 shows the extent to which *Ct* product AEGL-3 levels vary with exposure time for seven different irritant vapours, with significant deviations from Haber's Law in several cases, in that short, high concentration exposures were generally considered to be more hazardous than longer exposures to the same *Ct* product exposure dose.

For HCN over longer time scales, metabolism in the liver becomes a significant factor, so that eqn (8.2) provides a better model. The AEGL values for HCN, shown in Table 8.2, indicate that higher exposure doses can therefore be tolerated over longer time scales (for example, AEGL-3 of 10.5 ppm hour over 30 minutes but 52.8 ppm hour over 8 hours).

8.4 Further Developments of the Haber's Law Model

The equation $Ct = w$ describes a rectangular hyperbola. This relationship can be plotted as a straight line very simply by writing:

$$C = w\frac{1}{t} \tag{8.4}$$

If C is plotted against 1/t, a straight line of gradient w will be produced. We may recall that the equation for a straight line is:

$$y = mx + b. \tag{8.5}$$

In our equation, $C = y$, $w = m$, $1/t = x$ and b is the intercept on the y axis—in our case b is zero as the line passes through the origin.

American work led to dispute regarding Haber's figures for the lethal index of many chemical warfare agents. In dogs, for example, the LI for phosgene was ten times greater than that reported by Haber in cats (4500 as compared with 450 mg m^{-3} min). Flury (1921)[8] noted that compounds that were rapidly detoxified in the body did not follow Haber's Law, and in 1934 Flury and Wirth[9] reported that for some compounds C seemed more important that t. In 1940, Bliss[10] revived an equation,

$$(C - C_0)^n\ t = k, \tag{8.6}$$

proposed by Ostwald and Dernoschek in 1910,[11] before Haber's work was undertaken. In this equation C_0 was the threshold concentration and n and k were constants. Toxicologists have focused on C in Haber's Law and the general form:

$$C^n t = k \tag{8.7}$$

has come to be widely used. n can vary from 1 to about 7, being high for irritant compounds. This equation lends itself to logarithmic transformation:

$$C^n t = k$$

$$n \log C + \log t = \log k \tag{8.8}$$

$$\log t = -n\ \log C + \log k. \tag{8.9}$$

This too represents a straight line, the gradient being $-n$ and the intercept on the y axis being $\log k$.

Application to the measured lethal toxicity of hydrogen chloride to rats over periods of up to 1 hour illustrate how applying a $C^n t$ model may provide an improved fit to the data. Although the lethal exposure dose for hydrogen chloride in rats is regarded as approximately following Haber's Law this is not strictly the case for exposure times common in relation to fires, as illustrated in Figure 8.1 from Purser,[6] which plots data from Hartzell *et al.*[12] on the LCt_{50} for HCl in rats for different exposure times of from 5 to 60 minutes (plus 14 days post-exposure observation period). The LCt_{50} for 60 minute exposures is approximately twice that for a 5 minute exposure.

Figure 8.2 shows a plot of $\log C$ *versus* $\log t$ for the HCl rat lethality data giving an intercept for k of 6.55 and a slope for the n value of 1.4.

Applying the derived figure of n in Figure 8.3 gives a somewhat better fit for $C^{1.4}t = k$ than for $Ct = k$.

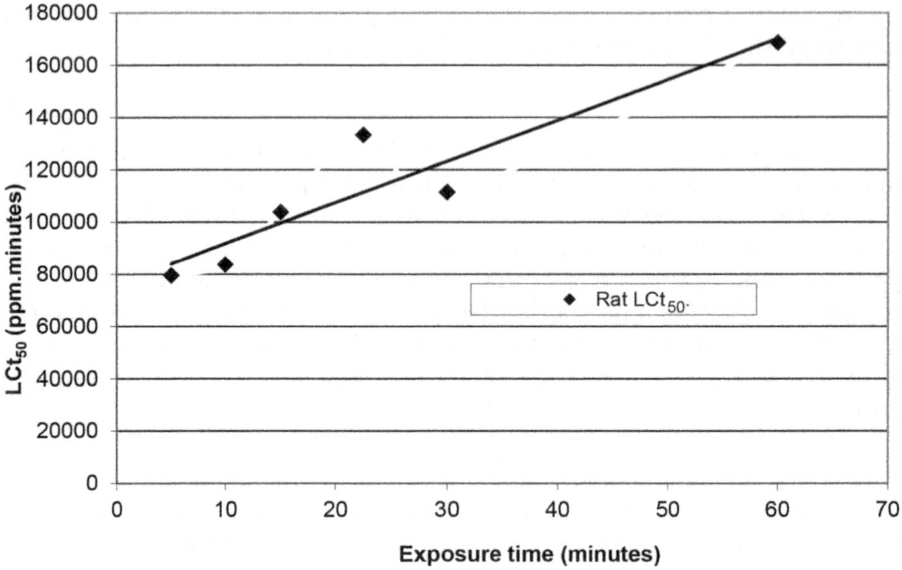

Figure 8.1 *Ct* exposure doses for HCl lethality in rats for different exposure times. A perfect fit would provide a horizontal line.[6]

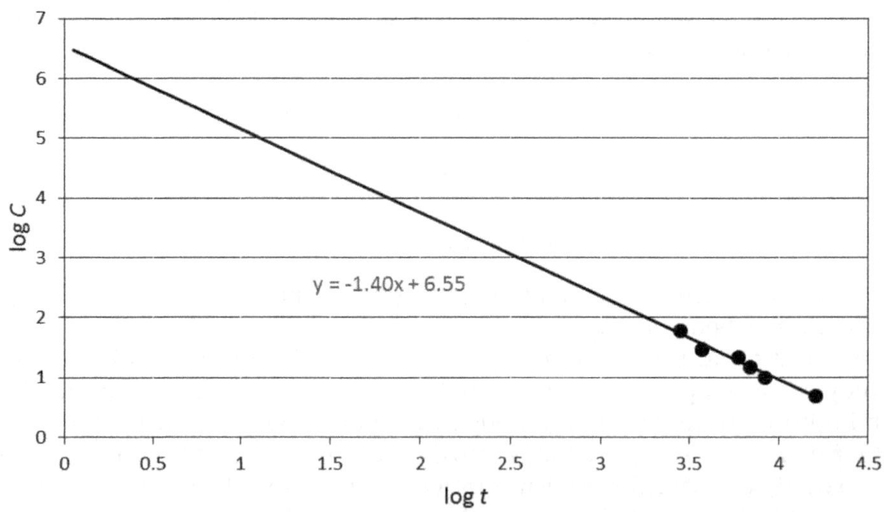

Figure 8.2 Plot of $\log C$ *versus* $\log t$ for the HCl rat lethality data. The linear fit gives a value of n of 1.4 and a $\log k$ of 6.55.

However, in this case the improvement is modest so that overall it seems that for the acute lethal effects of irritants on the lung for periods of up to an hour or so, the $Ct = k$ model is reasonable and an average LCt_{50} value for HCl in the rat over the times measured is approximately 114 000 ppm min.

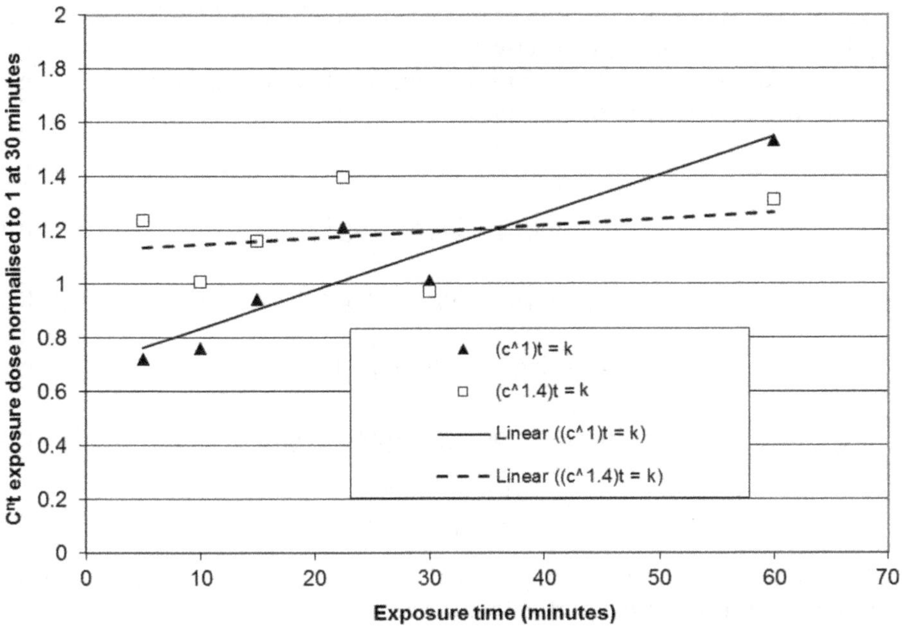

Figure 8.3 Comparison of normalised Ct and $C^{1.4}t$ exposure HCl dose models for rats. A perfect fit would provide a horizontal line at an ordinate value of 1.

Table 8.3 SLOT DTL and SLOD DTL values for various substances.

Substance name	'n' value	SLOT DTL (ppm^n min)	SLOD DTL (ppm^n min)
Acrolein	1	420	1680
Ammonia	2	3.78×10^{8}	1.03×10^{9}
Arsine	2	3706	5.9×10^{4}
Bromine	2	2.5×10^{5}	8.67×10^{5}
Carbon dioxide	8	1.5×10^{40}	1.5×10^{41}
Hydrogen cyanide	2	1.92×10^{5}	4.32×10^{5}
Hydrogen sulfide	4	2×10^{12}	1.5×10^{13}
Methanol	1	8.02×10^{5}	2.67×10^{7}
Methyl isocyanate	1	750	1680
Ozone	1	1980	3600
Sulfur dioxide	2	4.655×10^{6}	7.448×10^{7}
Vinyl chloride	1	3.39×10^{6}	1.36×10^{7}

Much effort has been put into defining n for different compounds. In 2007 the UK Health and Safety Executive (HSE) published a document, "*Assessment of the Dangerous Toxic Load (DTL) for Specified Level of Toxicity (SLOT) and Significant Likelihood of Death (SLOD)*".[13] This lists a large range of chemicals and an extract from the table is reproduced in Table 8.3.

Assumptions regarding the value of n also underlie the US AEGLs.[7] Two examples (HCN and tabun) are reproduced in Table 8.2. AEGL-1, 2 and 3 imply minimal risk, risk of significant health effects or impaired capacity to escape from a scene of release of the chemical concerned, and a significant risk of death, respectively.

n can be derived from data defining effects at two concentrations of C and t (assuming it is constant!) as follows:

$$C^n t = k$$

$$n \log C_1 + \log t_1 = \log k \tag{8.10}$$

$$n \log C_2 + \log t_2 = \log k \tag{8.11}$$

$$n \log C_1 + \log t_1 - n \log C_2 - \log t_2 = 0 \tag{8.12}$$

$$n\ (\log C_1 - \log C_2) = \log t_2 - \log t_1 \tag{8.13}$$

$$n = \frac{(\log t_2 - \log t_1)}{(\log C_1 - \log C_2)}. \tag{8.14}$$

Applying this to Table 8.2 for HCN and tabun shows that the AEGL committee used $n = 2$ for both substances.

A critical assumption is that n **is** constant between different endpoints, between species and, in particular, that it does not vary with t.

Recent work by Bide and Risk (2004)[14] has suggested that for sarin this may not be true. Additionally, and less surprisingly, n seems to vary with the endpoint considered. Professor T. C. Marrs has pointed out that work by Bide *et al.* (2005)[15] suggests that for human lethality (based on extrapolation from animal data), $n = 1.40$ for exposures in the range of 0.17 to 30 minutes. For the mouse, $n = 1.8$ for lethality over the range 1 to 6 hours and for human miosis, n approaches 4. Miller *et al.* (2000)[16] have taken the analysis further and have pointed out that a three dimensional surface is needed to illustrate the equation:

$$(C - C_0)^\alpha\ t^\beta = k, \tag{8.15}$$

or in a non-threshold case:

$$C^\alpha\ t^\beta = k. \tag{8.16}$$

This leads to a mathematical problem! In the words of the authors:[17]

"The introduction of an exponent on either C *or* t *into Haber's rule generates what is termed a power function, which represents a family of curves parameterized by the values of* γ*, the exponent, and* k*. Given two variables* X *and* Y*, one must select either* $X^\gamma Y = k$ *or* $XY^\gamma = k$ *since the number of parameters to be estimated cannot exceed the number of variables. When dealing with concentration and time as the two variables of interest, the type of response outcome led Bliss (1940)*[11] *to emphasize* C *while Druckery (1967)*[17] *emphasized* t*."*

A way out of this difficulty is provided by a three dimensional plot (see Figure 8.4).

Using data from studies of the effects of NO_2 on bacterial infectivity in mice and a probit model to fit the data (probit value $= y$), the authors fitted eqn (8.17).

$$y = m + \alpha \ln C + \beta \ln t \tag{8.17}$$

From this they made parameter estimates and derived confidence limits for a probit model to fit murine inactivity data from Gardner *et al.* (1979).[18] The derived results were; for m 2.6 (2.0–3.1), α 0.95 (0.74–1.19) and β 0.29 (0.22–0.39).

The results of the model are compared with the real data used to derive the model in the pair of graphs shown in Figure 8.4 (A shows the real data, B shows the model results). The fit is not perfect! It will be seen that the model

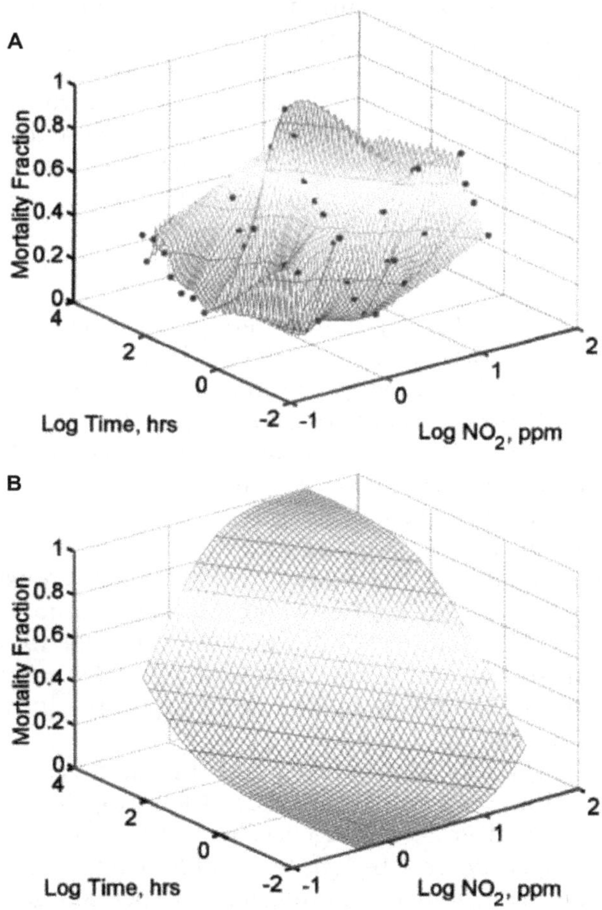

Figure 8.4 Relationships between mortality fraction, log NO_2 concentration and log exposure time. A: Results from experimental data. B: Predicted results using calculation model.

surface resembles the sigmoid curve familiar to toxicologists; it begins fairly flat, steepens and then flattens off again at high concentrations and long time periods. But the key lies in the slope of the surface: very different levels of effect (mortality fraction) can be produced by similar products of concentration and time.

8.5 Conclusions

Haber's Law has long played an important role in toxicological thinking and, in its original ($Ct = k$) or modified ($C^n t = k$) forms, underlies much work in the standard setting field. Its enduring popularity testifies to its utility and generic validity in providing a simple basis for evaluating approximate toxic hazard ranges for inhaled substances.

Recent work has shown that even in its modified form it may not be completely reliable. This is especially the case for predicting the acute effects of irritant and asphyxiant combustion products on escape capability and time to incapacitation, where more sophisticated models may be required, taking into account the dynamics of uptake, target organ concentration and activity, metabolism and excretion. Further work to explore the applicability of Haber's Law to a range of compounds is definitely still needed: a remarkable conclusion after about 93 years of use!

References

1. H.-J. Klimisch, R. Bretz, J. E. Doe and D. A. Purser, Classification of Dangerous Substances and Pesticides in the European Economic Community Directives: A Proposed Revision of Criteria for Inhalation Toxicity, *Regul. Toxicol. Pharmacol.*, 1987, **7**, 21–34.
2. F. Flury and F. Zernik, 1931, Shaedliche Gase, Daempfe, Nebel, Rauch- und Staubarten, Springer-Verlag, Berlin, 1931/Nachdruck, 1969.
3. F. Haber, *Funf Vortrange aus den jaren 1920–1923*, Verlag von Julius Springer, Berlin, 1924.
4. R. F. Coburn, R. E. Forster and P. B. Kane, Consideration of Physiological Variables That Determine the Blood Carboxyhaemoglobin Concentrations in Man, *J. Clin. Invest.*, 1965, **44**, 1899–1910.
5. E. Peterson and R. D. Stewart, Predicting the Carboxyhaemoglobin Levels Resulting from Carbon Monoxide Exposures, *J. Appl. Physiol.*, 1975, **39**, 633–638.
6. D. A. Purser, Hazards from smoke and irritants, in *Fire Toxicity*, ed. A. Stec and R. Hull, Woodhead, Cambridge UK, 2010, ch. 3, pp. 51–117.
7. US Environmental Protection Agency, *Acute Emergency Guideline Levels for Hazardous Substances (AEGLs)*, National Research Council.
8. F. Flury, Ueber Kampfgasvergiftungen: I. Ueber Reizgase, *Z. gesante experimentelle Medizin*, 1921, **13**, 1–15.
9. F. Flury and W. Wirth, Zur Toxikologie der Loesungsmittel, *Arch. Gewerbepath. Gewerbehyg.*, 1934, **5**, 1–90.

10. C. I. Bliss, The relation between exposure time, concentration and toxicity in experiments on insecticides, *Ann. Entomol. Soc. Am.*, 1940, **33**, 721–766.
11. W. Ostwald and A. Dernoschek, Uber die Bezichungen zwischen Adsorption und Giftigkeit, *Kolloid-Zeitschr.*, 1910, **6**, 297–307.
12. G. E. Hartzell, S. C. Packham, A. F. Grand and W. G. Switzer, Modeling of Toxicological Effects of Fire Gases: III. Quantification of Post-Exposure Lethality of Rats From Exposure To HCl Atmospheres, *J. Fire Sci.*, 1985, **3**(3), 195–207, DOI: 10.1177/073490418500300304.
13. Health and Safety Executive, 2007, *Assessment of the Dangerous Toxic Load (DTL) for Specified Level of Toxicity (SLOT) and Significant Likelihood of Death (SLOD)*, Available at: http://www.hse.gov.uk/hid/haztox.htm.
14. R. W. Bide and D. J. Risk, Inhalation toxicity in mice exposed to sarin (GB) for 20-720 min, *J. Appl. Toxicol.*, 2004, **24**, 459–467.
15. R. W. Bide, S. J. Armour and E. Yee, GB toxicity reassessed using newer techniques for estimation of human toxicity from animal inhalation toxicity data; new method for estimating acute human toxicity (GB), *J. Appl. Toxicol.*, 2005, **25**, 393–409.
16. F. J. Miller, P. M. Schlosser and D. B. Janszen, Haber's rule: a special case in a family of curves relating concentration and duration of exposure to a fixed level of response for a given endpoint, *Toxicology*, 2000, **149**, 21–34.
17. H. Druckery, Quantitative aspects in chemical carcinogenicity, in *Potential Carcinogenic Hazard from Drugs. Evaluation of Risk*, ed. Truhaut, R., UICC Monograph Series, Springer, Berlin, 1967, vol. 7, pp. 60–78.
18. D. E. Gardner, F. J. Miller, E. J. Blommer and D. L. Coffin, Influence of exposure mode on the toxicity of NO_2, *Environ. Health Perspect.*, 1979, **30**, 23–29.

CHAPTER 9

Carbon Monoxide†

ROBERT L. MAYNARD,* ISABELLA MYERS AND JOHN A. S. ROSS

University of Birmingham, UK
*Email: robertmaynard3@gmail.com

9.1 Introduction

This chapter is intended to provide guidance for scientists, medical staff and others dealing with or advising on how to deal with casualties from fires. It covers the effects of the rather high concentrations of carbon monoxide in the products of combustion from fires. For more comprehensive information, especially about the effects of prolonged and repeated low level exposures to carbon monoxide, the reader is directed to the three large compendia - ref. 1–3. Carbon monoxide is a toxic, colourless gas with a faint odour of garlic; the odour is seldom noticed and the gas is generally regarded as odourless and tasteless. Carbon monoxide is flammable: it burns with a blue flame of low luminosity. In the days when gas produced from coke (coal gas, water gas and producer gas) was used for lighting, higher hydrocarbons (largely ethylene) present in the gas provided the carbon that is the basis of luminosity. Alternatively, a gas mantle was used: this contained a proportion of thorium and cerium oxides and glowed incandescently when the gas flame played upon it. Luminous flames contain nano-particles of carbon: these glow and provide the visible light in the

†*The views expressed here are those of the authors and should not be taken as those of the UK Department of Health or those of the Health Protection Agency (England and Wales).*

Issues in Toxicology No. 23
Toxicology, Survival and Health Hazards of Combustion Products
Edited by David A. Purser, Robert L. Maynard and James C. Wakefield

Published by the Royal Society of Chemistry, www.rsc.org

yellow band of the spectrum. Luminous flames and soot deposition are signs of incomplete combustion and provide a warning of production of carbon monoxide. Blackening of the China clay radiants of gas fires, discolouration of plaster work around gas fires and luminous flames are commonly seen when poorly maintained devices are used. Carbon monoxide is produced when carbon-containing fuel burns with an inadequate supply of air. It is also produced in coal fires by the reduction of carbon dioxide as it passes over glowing coke within the fire.

$$C + O_2 \rightarrow CO_2$$

$$CO_2 + C \rightarrow 2CO$$

Incomplete combustion occurs in petrol engines as the balance of fuel and air moves away from the stoichiometric ratio. This occurs during cold starting (when a fuel-rich mixture is set by the choke control), during idling and slow running. Three way catalytic converters convert carbon monoxide to carbon dioxide but need to reach their operating temperature before this process becomes efficient. A petrol engine car equipped with a catalytic converter, started from cold and allowed to idle in a closed garage, remains a menace to health. In the days of cars without catalytic converters, running a car in a closed garage for 3 minutes generated a dangerous level of CO; running one for 5 minutes generated a potentially lethal concentration.[4] In the indoor environment smoking is a major source of carbon monoxide. This applies to cigars, pipes and cigarettes and also to water pipes (shisha pipes).

9.2 Physical Properties of Carbon Monoxide

Molecular weight	28.01
Critical point	− 145 °C at 43.5 atmospheres
Boiling point	− 191 °C
Specific gravity (SG) relative to air	0.967

Solubility in water

0 °C	3.54 ml/100 ml
20 °C	2.32 ml/100 ml
37 °C	2.14 ml/100 ml

Conversion factors

At 0 °C, 101.3 kPa	1 mg m^{-3} = 0.8 ppm
	1 ppm = 1.25 mg m^{-3}
At 25 °C, 101.3 kPa	1 mg m^{-3} = 0.873 ppm
	1 ppm = 1.145 mg m^{-3}

Of all the common gaseous air pollutants carbon monoxide occurs in the highest ambient concentrations. Its concentration in ambient air is measured in parts per million (ppm); that of ozone, for example, is measured in parts per billion.

Carbon monoxide is produced in fires, indeed it is the major cause of death of those caught in a fire environment. Carbon monoxide is produced slowly during the smouldering phase of a fire, and at very low yields during early, well-ventilated, fires. However, rates of production increase rapidly when fires become large in relation to the available ventilation, and even more once flash-over occurs (see Chapters 2 and 3). Production then depends, just as in a car engine, on the departure of the fuel : air ratio from the stoichiometric ratio:

$$[(\text{kg fuel})/(\text{kg air})]/[(\text{kg fuel})/(\text{kg air})]\text{stoich.} \qquad (9.1)$$

Once the stoichiometric ratio exceeds one, the further the departure in the fuel ≫ air direction, the more carbon monoxide will be produced. This has been set out in detail in Chapter 2 and by Babrauskas *et al.* (1991).[5] The combination of a rising concentration of carbon monoxide and a falling concentration of oxygen is particularly dangerous. Other toxic materials, including hydrogen cyanide, increase the risk to health. For a discussion of the interactions between combustion and the generation of toxicants, see Chapters 2 and 3.

9.3 Measuring and Reporting Concentrations of Carbon Monoxide

It will be known that inhaled carbon monoxide passes from the lungs to the blood and combines with haemoglobin. Haemoglobin is, of course, the protein of the red blood cells that transports oxygen in the blood. As carbon monoxide combines with deoxyhaemoglobin it prevents the uptake of oxygen. Rather remarkably, carbon monoxide combines with haemoglobin in exactly the same way as oxygen does. Each gram of haemoglobin can bind 1.34 ml (quoted values range from 1.306 to 1.38 ml) of either oxygen or carbon monoxide or a mixture of both. The normal haemoglobin concentration is often taken to be about 14 g/100 ml blood, so at 100% saturation, blood can bind 18.76 ml oxygen (or carbon monoxide) per 100 ml. We can define the concentration of carbon monoxide in blood in two ways:

1. The number of ml carbon monoxide per 100 ml blood (*e.g.*, 8 ml/100 ml).
2. The extent to which haemoglobin is saturated with carbon monoxide (*e.g.*, 42.6%).

These two methods cause confusion.

The arithmetic runs as follows:

$$(8 \times 100)/(1.34 \times 14) = 42.6$$

We will consider, later, the Haldane equation:

$$[COHb]/[O_2Hb] = M(PCO/PO_2), \tag{9.2}$$

where M is an affinity constant.

It does not matter whether concentrations or saturations are used on the left hand side of the equation:

$$42.6/57.4 = 0.74$$

$$8/(18.76 - 8) = 0.74.$$

Thus

$$\frac{\%\,[COHb]}{\%\,[O_2Hb]} = \frac{[COHb]}{[O_2Hb]} = M\left(\frac{PCO}{PO_2}\right). \tag{9.3}$$

The concentration of carbon monoxide in blood is usually written as [COHb]. We will take this to mean the concentration expressed in ml/100 ml blood. %[COHb] will be taken to mean the percentage saturation of haemoglobin with carbon monoxide. All this will be discussed in more detail later.

9.3.1 Measurement of Carbon Monoxide in Blood and Expired Air

The partial pressure of carbon monoxide in blood can be measured by headspace analysis using standard techniques. More usually the blood carbon monoxide level is assessed by measurement of the carboxy-haemoglobin (COHb) level since only this estimates the total blood content of carbon monoxide. Venous blood should be analysed[6] and analysis is by spectrophotometry using a CO-oximeter.

The CO-oximeter uses a number of wavelengths to measure the proportion of oxyhaemoglobin (O_2Hb), COHb, reduced (deoxygenated) haemoglobin (Hb) and methaemoglobin (MetHb) in a blood sample.[7] The use of the abbreviation Hb for reduced haemoglobin is confusing: it might well be taken to mean the whole Hb concentration (oxygenated and deoxygenated) and we will not use it again. Note also that the terms oxygenated and deoxygenated, rather than oxidised and reduced, are appropriate when describing the transport of oxygen by haemoglobin because whilst taking up, transporting and releasing oxygen, the iron atoms of the haem groups remain in the reduced, ferrous, Fe^{2+} state throughout. Oxidation of the iron atoms to the ferric, Fe^{3+} state produces methaemoglobin, which does not bind to oxygen. The CO-oximeter method provides a high level of accuracy, especially at %[COHb] > 5% (note that the inclusion of the % sign after the 5 is not strictly necessary because the left hand side of the inequality reads %[COHb]. We have included it for clarity). The method is susceptible to interference: methylene blue (which might have been given in a case of suspected cyanide poisoning, though better treatments are available, or in the treatment of methaemoglobinaemia induced by inhalation of the nitrogen oxides that can be produced in fires) and sulfhaemoglobin (produced as a result of exposure to hydrogen sulfide, which can also be produced in fires) interfere with the measurement of COHb.

It is probably unnecessary to warn against using a measurement of the partial pressure of oxygen in arterial blood (PaO_2) to derive a value for the percentage saturation of arterial blood with oxygen (SaO_2) and to use this as a basis for estimating %[COHb]. PaO_2 is maintained in CO poisoning (there need be no shortage of oxygen; the shortage is in the capacity to transport it!) and so PaO_2 will remain high despite SaO_2 falling to a low level. The lesson is: measure [COHb]. It should be noted that in cases of smoke inhalation co-pathologies, such as the inhalation of vomit, irritant effects of other toxins causing atelectasis, bronchospasm, respiratory depression *etc.*, can all lead to a lowering of the partial pressure of oxygen in the blood.

Carbon monoxide can be measured in expired air using gas chromatography or spectrophotometry, but for clinical use electrochemical methods are generally used.[8] Devices using this principle contain a platinum diffusion electrode that catalyses the following reaction:

$$CO + H_2O \rightarrow CO_2 + 2H^+ + 2e^-$$

The electrons flow in the sensing circuitry and the concentration of CO is measured. A twenty second breath hold is imposed before expiration *via* the device is allowed. Such devices are much used in clinics encouraging people to give up smoking. A falling CO concentration provides evidence of abstention and encouragement based on the accurate perception that the concentration of a toxic chemical in the subject's blood stream is declining.

Draeger tubes can also be used to monitor the concentration of CO in ambient air[1] using a colour change in a chemical reactant. Pumped tubes are used for grab-samples. These tubes contain iodine pentoxide:

$$5CO + I_2O_5 \rightarrow I_2 + 5CO_2$$

The length of the iodine stained layer in the tube reflects the concentration of carbon monoxide in the air.

From such methods the concentration of COHb may be inferred by means of an algorithm of which the Coburn–Forster–Kane equation is a good example (see below). The reader should be aware of the limitations of chemical analysis in exhaled breath, however, and cross sensitivity of the analytical method for other gases, notably ethanol in the context of fire victims may be a particular problem if not specifically countered in some way.

Enthusiasm for measuring [COHb] should not be allowed to defeat a very important clinical principle:

Treat the patient not his/her [COHb]

This is important. [COHb] is neither an accurate indicator of clinical status nor of the patient's prognosis when measured after rescue once exposure has stopped and treatment has started. It is however an indicator of exposure to carbon monoxide. This is discussed in greater detail in Section 9.8.

9.4 Endogenous Production of Carbon Monoxide and its Role in Normal Physiology

"...*this gas is likely to provide fuel to run plenty of labs.*"[9]

Nitric oxide (NO) has been recognised as a physiological transmitter of importance for more than twenty years. Carbon monoxide and hydrogen sulfide (H_2S) have also been recognised as transmitters (for reviews see: Brann *et al.*, 1997;[10] Marilena, 1997[11]). That gases could act as transmitters came as a surprise to many physiologists. Gases pass easily through cell membranes (H_2S to a lesser extent than CO and NO because it is a more polar and therefore a less lipid-soluble molecule) and function as short range transmitters. Their actions are described as autocrine, juxtacrine and paracrine rather than as endocrine. They play roles in linking the nervous, endocrine and immune systems: the new subject of neuro-immuno-endocrinology is attracting much attention.

Haem proteins are involved in the functioning of CO, NO and H_2S. CO is generated along with biliverdin as a product of the breakdown of haem. Rather surprisingly haem oxygenase, the enzyme that catalyses the breakdown, also contains haem and acts as a substrate and a catalyst. Nitric oxide synthase (NOS) is also a haem-containing protein, as is cystathione beta-synthase, a key enzyme in the production of H_2S. Haem proteins also act as sensors, for example, of CO. CO itself may play a role in the sensing of oxygen in the carotid body[12] and this subject has been the subject of a more recent review.[13]

The number of pathways in which CO can play a part is very large. These include anti-inflammatory pathways,[14] pathways controlling the pituitary gland and those involved in modulating the release of NO in the walls of blood vessels. NO links platelet function to leukocyte function and here, too, carbon monoxide can play a role.[15] Carbon monoxide is thought to have an anti-apoptosis role and, very surprisingly, in protecting against focal brain ischaemia. The reader with a sharp mind and an interest in molecular physiology is referred to two reviews: Kim *et al.* (2011)[16] and Kajimura *et al.* (2010).[17] The wealth of detail provided in these reviews is staggering. At present, the field is full of detail and new observations. Unifying hypotheses are beginning to appear but are not yet clear. Given the extraordinary range of functions that CO may play in the body it is hardly surprising that poisoning with CO produces a very wide range of effects. There are two physiological actions of carbon monoxide that are worth brief mention at this stage, however, since they inform the interpretation of findings to which some of the gas's toxic properties are attributed.

Endogenously produced carbon monoxide inhibits cellular respiration. In human embryonic kidney cells CO inhibited cellular respiration moderately and the effect was accompanied by inhibition of cytochrome oxidase activity and markedly increased in hypoxic conditions (1% ambient oxygen). Exogenously applied CO inhibited respiration in a moderate fashion, which was again accentuated in hypoxic conditions of 1% ambient oxygen.[18] Endogenously produced nitric oxide interacts with mitochondrial cytochrome c oxidase in a similar manner. Further work has demonstrated that endogenous carbon monoxide signals *via* the inhibition of cytochrome c oxidase and the generation of mitochondrial reactive oxygen species while maintaining cellular ATP levels and increasing mitochondrial membrane potential.[19] Both reduction in cellular oxygen consumption and generation of reactive oxygen species have been identified in response to the application of low levels of carbon monoxide and this is discussed in Section 9.5, but care must be taken in assuming that these actions reflect toxicity rather than a physiological mechanism.

9.5 Uptake and Kinetics of Carbon Monoxide

In principle, the toxico- or pharmaco-kinetics of carbon monoxide are straightforward.[20] If CO is inhaled, it diffuses across the alveolar air–blood barrier, dissolves in plasma, diffuses across the red cell membrane and binds to haemoglobin. As long as the relative concentrations of CO and O_2 in the inspired air are held constant we might expect equilibrium to be reached and indeed it is. The driving force is the difference between the partial pressures of CO in the alveolar gas and the plasma. Because CO binds avidly to haemoglobin and thus disappears from solution, the partial pressure of CO in plasma remains low. CO competes with oxygen for binding sites (the ferrous atoms of haem) and at first the uptake is rapid. As the number of available binding sites declines the rate of uptake slows as it approaches equilibrium. We should not be surprised to know that COHb rises along a curve of progressively declining gradient rather than along a straight line. The exact position of this curve is controlled by the alveolar ventilation rate, which controls the rate at which CO reaches the alveoli: the faster you breathe, the more rapidly equilibrium is reached. The equilibrium position or state is controlled by the relative affinities of CO and O_2 for Hb. The uptake curve can be modelled in a number of ways. We will discuss some empirical models later but we shall begin with the best model: the Coburn–Forster–Kane (CFK) model that generates the well-known CFK equation.[21–23] Derivation of this important equation is not difficult once the language is mastered: see Section 9.12 for the abbreviations. The CFK equation was derived during work on the endogenous production of CO and adopted for predicting the uptake of inhaled CO.

The rate of change of whole body CO must equal the rate of endogenous CO production minus the rate at which CO is lost from the lungs:

$$\frac{dCO}{dt} = \dot{V}_{CO} - Dl_{CO}(\bar{P}_{c,CO} - P_{A,CO}), \tag{9.4}$$

where $\frac{dCO}{dt}$ is the rate at which the body store of carbon monoxide changes, $\dot{V}_{CO}$ is the rate at which carbon monoxide is lost from the lung, Dl_{co} is the diffusion coefficient of carbon monoxide in the lung, $\bar{P}_{c,CO}$ is the mean partial pressure of carbon monoxide in pulmonary capillary blood and $P_{A,CO}$ is the partial pressure of carbon monoxide in alveolar air.

This is that formidable thing: a differential equation. This means that it deals with the rate at which some quantity changes.

Let us use some purely illustrative figures: if CO were being produced at 10 ml min^{-1} and lost *via* the lung at 5 ml min^{-1}, the whole body CO content would be rising by 5 ml min^{-1}.

The rate at which CO is lost from the lung must equal the difference between the rate at which CO enters and leaves the alveoli.

$$\frac{dCO}{dt} = \dot{V}_{CO} - \frac{\dot{V}_A(P_{A,CO} - P_{I,CO})}{P_B - P_{H_2O}}, \tag{9.5}$$

where $\dot{V}_A$ is the alveolar ventilation rate, $P_{I,CO}$ is the partial pressure of CO in inspired air, P_B is the barometric pressure and P_{H_2O} is the saturated vapour pressure of water at 37 °C. Combining eqn (9.4) and (9.5) gives:

$$Dl_{CO}(\bar{P}_{c,CO} - \bar{P}_{A,CO}) = \frac{\dot{V}_A(P_{A,CO} - P_{I,CO})}{P_B - P_{H_2O}} \tag{9.6}$$

and rearrangement yields:

$$P_{A,CO} = \left(\frac{(\bar{P}_{c,CO} + (P_{I,CO}\dot{V}_A) / (Dl_{CO}(P_B - P_{H_2O})}{\left(\left(\frac{\dot{V}_A}{Dl_{CO}(P_B - P_{H_2O})}\right) + 1\right)} \right). \tag{9.7}$$

We now need a short digression. The Haldane equation has already been mentioned. The amount of CO bound to Hb equals the partial pressure of CO multiplied by the affinity of Hb for CO at that *P*CO. Similarly, the amount of oxygen bound to Hb equals the partial pressure of oxygen multiplied by the affinity of Hb for CO at that PO_2. Of course, the affinity of haemoglobin for oxygen varies with the PO_2: see Section 9.6.1 for a discussion of the oxyhaemoglobin dissociation curve. The fact that the affinity of haemoglobin for carbon monoxide varies with *P*CO in exactly the same way as the affinity of haemoglobin for oxygen varies with PO_2 allowed Haldane to argue:

$$[O_2Hb] = fPO_2 \tag{9.8}$$

fPO_2 means 'a function of PO_2'. The function is complex: recall the sigmoid curve. If you are not familiar with this curve, accept for the moment that it is curved like an S. The reasons for its shape are explained below.

$$[COHb] = 245fPCO \tag{9.9}$$

(Assuming that *P*CO and PO_2 are expressed in the same units.)

This means that the curve for CO has exactly the same shape as the curve for oxygen, but that a PCO of only 1/245th of a specified PO_2 and the specified PO_2 produce the same [COHb] and [O_2Hb].

Combining these equations gives:

$$[\mathrm{COHb}]/[\mathrm{O_2Hb}] = 245\mathrm{f}PCO/\mathrm{f}PO_2 \tag{9.10}$$

and cancelling out f gives:

$$[\mathrm{COHb}]/[\mathrm{O_2Hb}] = 245PCO/PO_2 \tag{9.11}$$

For a general form, M is put instead of 245.

M is described as the affinity constant; in fact, it describes the relative affinities of Hb for CO and O_2.

This is the Haldane equation; the most important equation in CO toxicology. Note that we could write either %[COHb]/%[O_2Hb] or [COHb]/[O_2Hb] [see eqn (9.3)].

The affinity of Hb for CO is much greater than the affinity of Hb for O_2; $M \approx 245$. We can see at once that a mixture of one part CO to 245 parts O_2 would lead to half the Hb being COHb and half O_2Hb. This would be the state of affairs if blood was equilibrated with air containing 21% oxygen (the usual level) and 21/245% CO, *i.e.*, 0.086% CO (1001.35 mg m^{-3} at 25 °C). Or for blood equilibrated with alveolar gas at 13.1% oxygen and 570 ppm CO (see section 9.5.1).

Let us return to the CFK equation.

From the Haldane equation we can write:

$$\bar{P}_{\mathrm{c,CO}} = ([\mathrm{COHb}]\bar{P}_{\mathrm{c,O_2}}) \,/\, ([\mathrm{O_2Hb}]M) \tag{9.12}$$

Looking back to eqn (9.4):

$$\frac{d\mathrm{CO}}{dt} = \dot{V}_{\mathrm{CO}} - Dl_{\mathrm{CO}}(\bar{P}_{\mathrm{c,CO}} - P_{\mathrm{A,CO}}).$$

We can substitute from above to obtain:

$$\frac{d\mathrm{CO}}{dt} = \dot{V}_{\mathrm{CO}} - Dl_{\mathrm{CO}}\left(\left(\frac{[\mathrm{COHb}]\bar{P}_{\mathrm{C,O_2}}}{[\mathrm{O_2Hb}]M}\right) - \frac{(\bar{P}_{\mathrm{C,O_2}}) + (P_{\mathrm{I,CO}}\dot{V}_{\mathrm{A}})/(Dl_{\mathrm{CO}}(P_{\mathrm{B}} - P_{\mathrm{H_2O}}))}{\left(\left(\frac{\dot{V}_{\mathrm{A}}}{Dl_{\mathrm{CO}}(P_{\mathrm{B}} - P_{\mathrm{H_2O}})}\right) + 1\right)}\right) \tag{9.13}$$

Putting 95.1 for $P_{\mathrm{B}} - P_{\mathrm{H_2O}}$ and rearranging:

$$\frac{P_{\mathrm{I,CO}}}{\left(\frac{1}{Dl_{\mathrm{CO}}}\right) + \left(\frac{95.1}{\dot{V}_{\mathrm{A}}}\right)} \tag{9.14}$$

$$\frac{d\mathrm{CO}}{dt} = \dot{V}_{\mathrm{CO}} - \left(\left(\frac{[\mathrm{COHb}]\bar{P}_{\mathrm{C,O_2}}}{[\mathrm{O_2Hb}]M}\right)\frac{1}{\left(\frac{1}{Dl_{\mathrm{CO}}}\right) + \left(\frac{95.1}{\dot{V}_{\mathrm{A}}}\right)}\right) + \frac{P_{\mathrm{I,CO}}}{\left(\frac{1}{Dl_{\mathrm{CO}}}\right) + \left(\frac{95.1}{\dot{V}_{\mathrm{A}}}\right)} \tag{9.15}$$

Assuming that total body CO is proportional to [COHb] and blood volume (V_b) we can substitute

$$\frac{V_b d[\mathrm{COHb}]_t}{\mathrm{d}t} \text{ for } d\mathrm{CO}/dt$$

We can now write:

$$\frac{V_b d[\mathrm{COHb}]_t}{dt} = \dot{V}_{\mathrm{CO}} - \left(\left(\frac{[\mathrm{COHb}]_0 \bar{P}_{\mathrm{c,O_2}}}{[\mathrm{O_2Hb}]M} \right) \frac{1}{\left(\frac{1}{Dl_{\mathrm{CO}}}\right) + \left(\frac{95.1}{\dot{V}_{\mathrm{A}}}\right)} \right) + \frac{P_{\mathrm{I,CO}}}{\left(\frac{1}{Dl_{\mathrm{CO}}}\right) + \left(\frac{95.1}{\dot{V}_{\mathrm{A}}}\right)} \tag{9.16}$$

If we put

$$\beta \text{ for} \left(\frac{1}{Dl_{\mathrm{CO}}}\right) + \left(\frac{95.1}{\dot{V}_{\mathrm{A}}}\right),$$

and divide by V_b,

$$\frac{d[\mathrm{COHb}]_t}{dt} = \left(\frac{\dot{V}_{\mathrm{CO}}}{V_b}\right) - 1/V_b \left(\frac{[\mathrm{COHb}]_0 \bar{P}_{\mathrm{C,O_2}}}{[\mathrm{O_2Hb}]M} \left(\frac{1}{\beta}\right) \right) + \frac{P_{\mathrm{I,CO}}}{V_b \beta} \tag{9.17}$$

Let, for ease of manipulation,

$$x = \frac{[\mathrm{COHb}]_0 \bar{P}_{\mathrm{c,O_2}}}{[\mathrm{O_2Hb}]M},$$

then:

$$\begin{aligned}
\frac{d[\mathrm{COHb}]_t}{dt} &= \frac{\dot{V}_{\mathrm{CO}}}{V_b} + \frac{1}{V_b}\frac{x}{\beta} + P_{\mathrm{I,CO}}/V_b\beta \\
&= \frac{\dot{V}_{\mathrm{CO}}}{V_b} + 1/V_b\beta(P_{\mathrm{I,CO}} - x) \\
&= \frac{\dot{V}_{\mathrm{CO}}}{V_b} + 1/V_b\beta \left(P_{\mathrm{I,CO}} - \frac{[\mathrm{COHb}]_0 \bar{P}_{\mathrm{c,O_2}}}{[\mathrm{O_2Hb}]M} \right).
\end{aligned} \tag{9.18}$$

This equation appears in the US EPA Criteria document on CO (2000)[3] as eqn (5.3) and is described as non-linear. This seems to be incorrect as it is identical with their eqn (5.4), except that β has been substituted for

$$\left(\frac{1}{Dl_{\mathrm{CO}}}\right) + \left(\frac{95.1}{\dot{V}_{\mathrm{A}}}\right).$$

Non-linearity is produced by considering the term [O_2Hb]. In the linear eqn (9.18) this is taken as constant. If we replace [O_2Hb] with (1.38Hb – [COHb]) then the equation does become non-linear. At levels of COHb of less than 6% or so (*i.e.*, %[COHb] = <6%) this correction produces only a $\pm 0.5\%$ difference in the calculated [COHb]. The assumption regarding blood volume might also be questioned. Why all this explanation about linear and non-linear differential equations? The answer is that the linear form can be integrated to produce what is called an explicit solution (see Box 9.1); the non-linear form cannot be so integrated and a computer-based iterative solution is needed. Note, and this is important, that although the linear differential equation is rightly described as linear it describes a curvilinear relationship.

9.5.1 Explicit, General Solution of the CFK Equation

Coburn argued:

$$\frac{d\mathrm{CO}}{dt} = \dot{V}_{\mathrm{CO}} - \left(\left(\frac{[\mathrm{COHb}]\bar{P}_{\mathrm{c,O_2}}}{[\mathrm{O_2Hb}]M} \right) \frac{1}{\left(\frac{1}{Dl_{\mathrm{CO}}}\right) + \left(\frac{95.1}{\dot{V}_{\mathrm{A}}}\right)} \right) + \frac{P_{\mathrm{I,CO}}}{\left(\frac{1}{Dl_{\mathrm{CO}}}\right) + \left(\frac{95.1}{\dot{V}_{\mathrm{A}}}\right)}.$$

Substituting and integrating leads to a large equation. It is large only because the left hand side is so complicated. To simplify, we present the equation in terms of operators '*a*' and '*b*'.

The solution runs:

$$\frac{a}{b} = \mathrm{e}^{-(\bar{P}_{\mathrm{c,O_2}},t)\,/\,MV_{\mathrm{b}}[\mathrm{O_2Hb}]\left(\frac{1}{Dl_{\mathrm{CO}}} + \frac{95.1}{\dot{V}_{\mathrm{A}}}\right)}, \tag{9.19}$$

where

$$a = \frac{[\mathrm{COHb}]\bar{P}_{\mathrm{c,O_2}}}{[\mathrm{O_2Hb}]M} - \dot{V}_{\mathrm{CO}}\left(\frac{1}{Dl_{\mathrm{CO}}} + (P_{\mathrm{B}} - P_{\mathrm{H_2O}})\,/\,\dot{V}_{\mathrm{A}}\right) - P_{\mathrm{I,CO}}, \tag{9.20}$$

and

$$b = \frac{[\mathrm{COHb}]_0\bar{P}_{\mathrm{c,O_2}}}{[\mathrm{O_2Hb}]M} - \dot{V}_{\mathrm{CO}}\left(\frac{1}{Dl_{\mathrm{CO}}} + (P_{\mathrm{B}} - P_{\mathrm{H_2O}})\,/\,\dot{V}_{\mathrm{A}}\right) - P_{\mathrm{I,CO}}, \tag{9.21}$$

where $[\mathrm{COHB}]_0$ is the carboxyhaemoglobin concentration at time zero.

Perhaps the most interesting feature of this is that at equilibrium, when $t = \infty$, the terms on the right hand side of the equation become equal to zero.

Thus, we have:

$$\frac{a}{b} = 0.$$

Thus:

$$a = 0$$

Or

$$\frac{[\mathrm{COHb}]\bar{P}_{\mathrm{c,O_2}}}{[\mathrm{O_2Hb}]M} - \dot{V}_{\mathrm{CO}}\left(\frac{1}{Dl_{\mathrm{CO}}} + (P_{\mathrm{B}} - P_{\mathrm{H_2O}})\,/\,\dot{V}_{\mathrm{A}}\right) - P_{\mathrm{I,CO}} = 0 \tag{9.22}$$

Eqn (9.22) simplifies to eqn (9.23):

$$[\mathrm{COHb}]_{\mathrm{eq}} = P_{\mathrm{I,CO}}M\left(\frac{[\mathrm{O_2Hb}]}{\bar{P}_{\mathrm{c,O_2}}}\right) + \left(M\dot{V}_{\mathrm{CO}}\left(\frac{[\mathrm{O_2Hb}]}{\bar{P}_{\mathrm{c,O_2}}}\right)\left(\frac{1}{Dl_{\mathrm{CO}}} + \frac{(P_{\mathrm{B}} - P_{\mathrm{H_2O}})}{\dot{V}_{\mathrm{A}}}\right)\right) \tag{9.23}$$

(This is Coburn's eqn 7.)[21]

The integrated form of the CFK equation can be arranged in a form suitable for calculation using a pocket calculator. This is set out in Box 9.1 as an operative sequence including a worked example providing adopted constants. This usage of the CFK equation is taken from the work of Hickman 1989.[24] Note that the terminology has been simplified from that used above.

Box 9.1 Model calculation

$$[\mathrm{COHb}]_t = 1/A\left(\frac{AC}{[\mathrm{COHb}]_0}\right) + (1-C)\dot{V}_{\mathrm{CO}}B + (1-C)P_{\mathrm{I,CO}}$$

$$A = P\mathrm{O_2}/M[\mathrm{O_2Hb}]$$

$$B = \frac{1}{D} + P/\dot{V}_{\mathrm{A}}$$

$$C = \mathrm{e}^{(-tA/V_{\mathrm{b}}B)}$$

M	Haldane coefficient = 240
$[\mathrm{O_2Hb}]$	oxyhaemoglobin concentration = 0.2 ml ml^{-1} blood
$[\mathrm{COHb}]_t$	carboxyhaemoglobin concentration at time t in ml CO per ml blood
$[\mathrm{COHb}]_0$	initial carboxyhaemoglobin concentration = 0.001 ml ml^{-1} blood (0.5% COHb, 100% COHb = 20 ml/100 ml blood = 0.2 ml ml^{-1} blood, 0.5% = 0.2×0.5/100 = 0.001 ml ml^{-1} blood
$P\mathrm{O_2}$	partial pressure of oxygen in lung capillaries = 13.3 kPa
$\dot{V}_{\mathrm{CO}}$	endogenous rate of production of CO = 0.007 ml min^{-1}
D	diffusivity of CO = 225 ml min^{-1} kPa (note that in reality this is not a constant but is altered by a number of factors including exercise)
P	Barometric pressure – saturated vapour pressure of water at 37 °C = 95.1
V_{b}	blood volume = 5500 ml
$P_{\mathrm{I,CO}}$	partial pressure of CO in inspired air = 0.0101 kPa = 100 ppm (adopted for the purposes of this example)
$\dot{V}_{\mathrm{A}}$	alveolar ventilation rate = 6000 ml min^{-1}
t	exposure time = 60 minutes (adopted as an example)

$$A = \frac{13.3}{240 \times 0.2} = 0.2771$$

$$B = \frac{1}{225} + \frac{95.1}{6000} = 0.0203$$

$$\begin{aligned} C &= \exp\left(-\left(\frac{60 \times 0.2771}{5500 \times 0.0203}\right)\right) \\ &= \exp(-0.1489) \\ &= 1/\exp(0.1489) \\ &= 1/1.1606 \\ &= 0.8617 \end{aligned}$$

$$AC = 0.2771 \times 0.8617 = 0.2388$$

$$1 - C = 0.1383$$

$$\dot{V}_{CO}B = 0.007 \times 0.0203 = 0.0001421$$

Then:

$$\begin{aligned} [\mathrm{COHb}]_t &= \left(\frac{1}{0.2771}\right)((0.2388 \times 0.001) + (0.1383 \times 0.0001421) \\ &\quad +(0.1383 \times 0.0101)) \\ &= \left(\frac{1}{0.2771}\right)0.001655 \\ &= 0.0060 \end{aligned}$$

$[\mathrm{COHb}]_t = 0.006\ \mathrm{ml\ ml^{-1}}$ blood
$= 0.6$ ml/100 ml blood
20 ml CO/100 ml blood = 100%
0.6 ml/100 ml blood = 0.6 × 100/20 = 3.0%

Breathing 100 ppm CO for 1 hour at rest produces a concentration of carboxyhaemoglobin of 3%.

It will be seen that this figure appears in the tables throughout this chapter.

The reader may feel that the CFK equation, though undeniably impressive, lacks something in terms of ease of use. Other equations are available.

In 1945, Forbes proposed a very simple equation:[25]

$$\%[\mathrm{COHb}]_t = 3 \times [CO]t \qquad (9.24)$$

Where [CO] is the %CO in inspired air (we might prefer to write %[CO]) and t is the duration of the exposure expressed in minutes. The equation applies to adults at rest. For adults undertaking various levels of work the figures 5, 8 or 11 are substituted for 3.

The CO uptake calculation illustrated in Box 9.1, which uses a constant value for HbO_2, and that in Equation 9.24, are both linear uptake models, which give reasonably good results for short CO exposure times. A better application of the CFK equation, incorporating a variable term for HBO_2, and a non-linear uptake curve, was proposed by Peterson and Stewart[26] in 1975 (see Chapter 14). This superseded a simpler non-linear equation by the same authors which took the form:

$$\%[\mathrm{COHb}] = 0.0051[\mathrm{CO}]^{0.858} \times t^{0.63} \tag{9.25}$$

Where [CO] is the concentration of carbon monoxide expressed as ppm and t is in minutes.

This equation works well for calculations involving exposure to constant levels of CO for short periods. Due to the complexity and different parameters taken into account by the various models the reader wishing to carry out CO uptake calculations is recommended to consult the original publications cited.

We might ask, how do we calculate the equilibrium position? The answer, of course, is by use of the Haldane equation. This is easy, though the arithmetic can catch one out:

Recall eqn (9.3):

$$\%[\mathrm{COHb}]/\%[\mathrm{O_2Hb}] = M(P\mathrm{CO}/P\mathrm{O_2}).$$

Remember that what is equilibrating is alveolar gas and haemoglobin. This means we need to think in terms of the partial pressure of oxygen in the alveoli.

$$P_\mathrm{A}\mathrm{O_2} = 13.3\ \mathrm{kPa}$$

$$P_\mathrm{A}\mathrm{CO} = \mathrm{FICO} \times (101.3 - 5.7).$$

Let FICO = 0.0001 (*i.e.*, 0.01%).

Then:

$$P_\mathrm{A}\mathrm{CO} = 0.0001 \times 95.6 = 0.00956\ \mathrm{kPa}.$$

Let $M = 245$.

Then:

$$\begin{aligned}
&245 \times 0.00956/13.3 = 0.1761\\
&\%[\mathrm{COHb}]/\%[\mathrm{O_2Hb}] = 0.1761\\
&\%[\mathrm{O_2Hb}] = 100 - \%[\mathrm{COHb}]\\
&\%[\mathrm{COHb}] = 0.1761(100 - \%[\mathrm{COHb}])\\
&\%[\mathrm{COHb}] = 17.61 - 0.1761\%[\mathrm{COHb}]\\
&1.1761\%[\mathrm{COHb}] = 17.61\\
&\%[\mathrm{COHb}] = 17.61/1.1761 = 14.96.
\end{aligned}$$

A simple example may help with the calculation, see Box 9.2.

Box 9.2

Fred and Bill own a pound. Fred's share is only a third of Bill's; how much does Fred own? Intuitively, we see that there are four shares between Fred and Bill: one share is 25% of the whole. Thus, Fred has 25% of the whole (25p) and Bill has 75% of the whole (75p). Let Fred's share (F) be 0.17 of Bill's share (B). Intuition may fail and simple algebra is needed:

$B+F=1$

$F=0.17B$

$B+0.17B=1$

$B=1/1.17=0.85$ (Bill has 85p)

$F=0.15$ (Fred has 15p).

Check: 85/15 = 0.17.

Values calculated using this method are shown in the first edition of the EHC document on carbon monoxide.[27] The reader will find that the authors have specified the %CO in alveolar air (which is, of course, the same as the percentage in inspired air at equilibrium if a respiratory quotient of 1 is assumed) and have calculated the partial pressure of CO in the alveolar air by multiplying the % figure by 101.3. This is incorrect, the % should be multiplied by 95.1: the total gas pressure in the alveoli excluding the pressure exerted by water vapour.

Table 9.1 shows the results obtained by use of the CFK model.

More interesting than the equilibrium values are the values at time points along the way to equilibrium. These are also shown in Table 9.1.

Figure 9.1 shows uptake curves for 50 and 200 ppm CO under sedentary conditions, light and heavy work, with V_A specified for each level of activity. Note that Dl_{CO}, the diffusion limit coefficient of carbon monoxide in the lung (or "transfer factor"), is not a constant but increases with V_A, enabling the rate of CO uptake to increase at high V_A levels.

The affinity of fetal Hb for oxygen is greater than that of adult Hb and the fetal O_2Hb dissociation curve is displaced to the left of the adult curve. Increased affinity for oxygen enhances oxygen uptake at the placenta: fetus and umbilical vein blood has a partial pressure of oxygen of about 4 kPa. Note that this blood has the highest PO_2 of any blood in the fetus. The fetus [Hb] is about 18 g/100 ml compared with about 14 g/100 ml in the adult. The affinity of fetus Hb for CO is greater than its affinity for O_2 but M is not as great for fetus Hb as for adult Hb: about 170 as compared with about 240. Longo and his colleagues[28] have studied the uptake of CO by the fetus in great detail and have published eqn (9.26):

$$\frac{[\mathrm{COHb}]_\mathrm{F}}{[\mathrm{COHb}]_\mathrm{M}} = \left(\frac{[\mathrm{O_2Hb}]_\mathrm{F}}{P\mathrm{O}_{2,\mathrm{F}}}\right)\left(\frac{P\mathrm{O}_{2,\mathrm{M}}}{[\mathrm{O_2Hb}]_\mathrm{M}}\right)\left(\frac{M_\mathrm{F}}{M_\mathrm{M}}\right), \tag{9.26}$$

where M (subscript) indicates mother and F (subscript) indicates fetus.

In humans the ratio of $[\mathrm{COHb}]_\mathrm{F}$ to $[\mathrm{COHb}]_\mathrm{M}$ is about 1.1.

Table 9.1 [HbCO] values predicted from Coburn *et al.* (1965)[21] model (see ref. 27, pp. 119).[a]

Time	200 ppm S	200 ppm L	200 ppm H	100 ppm S	100 ppm L	100 ppm H	75 ppm S	75 ppm L	75 ppm H	50 ppm S	50 ppm L	50 ppm H
15 min	1.8	3.5	5.2	1.2	2.0	2.8	1.0	1.6	2.2	0.82	1.2	1.6
30 min	3.1	6.2	9.2	1.8	3.3	4.8	1.5	2.6	3.7	1.1	1.9	2.6
45 min	4.3	8.7	12.6	2.4	4.6	6.5	1.9	3.5	4.9	1.4	2.5	3.4
60 min	5.5	11.0	15.5	3.0	5.7	7.9	2.3	4.3	6.0	1.7	3.0	4.1
90 min	7.7	14.9	20.2	4.0	7.6	10.2	3.1	5.8	7.7	2.2	4.0	5.2
2 h	9.7	18.1	23.7	5.0	9.2	11.9	3.9	7.0	9.0	2.7	4.7	6.1
4 h	16.3	26.2	30.4	8.3	13.2	15.3	6.3	10.0	11.5	4.4	6.9	7.7
6 h	21.1	30.0	32.4	10.7	15.1	16.2	8.1	11.3	12.2	5.5	7.6	8.2
8 h	24.5	31.7	32.9	12.4	15.9	16.5	9.4	12.0	12.4	6.4	8.0	8.3
24 h	32.7	33.2	33.2	16.5	16.7	16.6	12.4	12.5	12.5	8.4	8.4	8.3
∞	33.4	33.2	33.2	16.8	16.7	16.6	12.7	12.5	12.5	8.5	8.4	8.3

Time	35 ppm S	35 ppm L	35 ppm H	25 ppm S	25 ppm L	25 ppm H	10 ppm S	10 ppm L	10 ppm H	5 ppm S	5 ppm L	5 ppm H
15 min	0.72	1.0	1.3	0.66	0.84	1.0	0.55	0.61	0.67	0.52	0.54	0.56
30 min	0.93	1.4	1.9	0.80	1.2	1.5	0.61	0.72	0.82	0.54	0.57	0.60
45 min	1.1	1.9	2.5	0.95	1.4	1.9	0.66	0.81	0.95	0.56	0.61	0.64
60 min	1.3	2.2	3.0	1.1	1.7	2.2	0.71	0.90	1.1	0.58	0.63	0.68
90 min	1.7	2.9	3.7	1.3	2.1	2.7	0.80	1.1	1.2	0.62	0.69	0.74
2 h	2.0	3.4	4.3	1.6	2.5	3.1	0.89	1.2	1.4	0.66	0.73	0.78
4 h	3.2	4.7	5.4	2.4	3.4	3.9	1.2	1.5	1.6	0.77	0.84	0.86
6 h	4.0	5.4	5.7	2.9	3.9	4.1	1.4	1.6	1.7	0.85	0.88	0.88
8 h	4.5	5.7	5.8	3.3	4.1	4.2	1.5	1.7	1.7	0.91	0.91	0.89
24 h	5.9	5.9	5.9	4.3	4.2	4.2	1.9	1.8	1.7	1.05	0.93	0.89
∞	6.0	5.9	5.9	4.4	4.2	4.2	1.9	1.8	1.7	1.06	0.93	0.89

[a]S = sedentary subjects, L = light physical work, H = heavy physical work.

Further displacement of the fetus O_2Hb dissociation curve to the left by binding of CO places the fetus at some risk: the PO_2 of foetal blood has to fall to a very low level before significant amounts of oxygen are released to the tissues. This may play a part in explaining the effects of maternal smoking on fetal development. The relatively slow washout of CO from the fetus is the reason why hyperbaric oxygen is often recommended for the treatment of pregnant women who have been exposed to carbon monoxide.

One is sometimes asked to calculate the level of carbon monoxide to which a person has been exposed given that the subject's %[COHb] is measured at some known time after exposure has stopped. If we can assume that equilibrium had been reached during exposure we can estimate the exposure concentration from a graph showing the washout of CO from the body. For example, if the %[COHb] is about 15% at 4 hours after exposure has stopped it is likely that the equilibrium concentration was about 30%. This equilibrium concentration is associated with an exposure to about 300 ppm CO. In practice we do not know that equilibrium had been reached and therefore cannot make this extrapolation using the standard curves.

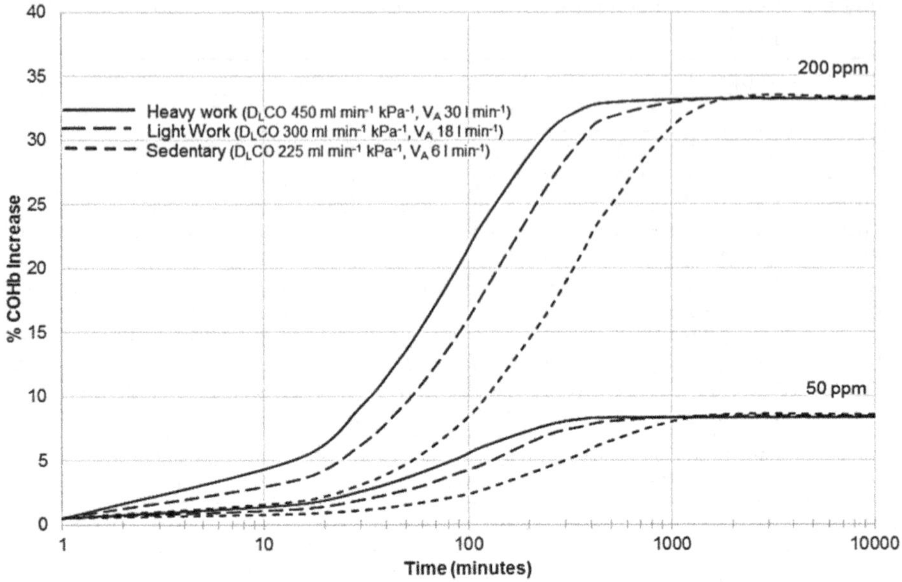

Figure 9.1 Predicted %[COHb] increase for 50 and 200 ppm CO at different activity levels calculated using the CFK model.

This point leads on to a related issue. While these models allow us to think about mechanisms underlying carbon monoxide uptake, they do not allow us accurately to predict what will happen for an individual in any anticipated exposure or to deduce what happened to an individual after an exposure. Indeed, the most comprehensive model, the CFK equation, is based on consideration of the endogenous production of CO in resting subjects and not on exogenous exposure. Also, the values taken for a number of the physiological constants used for the equation in 1965 might now be contested leading to consideration of the effects of variability and chance in retrospectively fitting it to empirically derived data.

9.6 Toxicological Mechanisms of Action of Carbon Monoxide

It will be clear by now that carbon monoxide binds to haemoglobin. This impedes the binding of oxygen to haemoglobin and inhibits the release of such molecules of oxygen that do find binding sites on molecules of haemoglobin not completely saturated with carbon monoxide. Thus, the tissues are deprived of oxygen and a state of hypoxia is induced despite a normal concentration of oxygen in the inspired air. A lack of oxygen leads to a reduction in production of ATP and thus a lack of access to high energy phosphate bonds throughout the energy consuming processes of the cell. Many cellular functions fail, cells die and if the damage is sufficiently severe the patient dies.

This explanation of the toxicological effects of CO has seemed satisfactory for more than a hundred years; indeed it has long been regarded as perhaps the best explanation of the effects of any toxicologically active compound. However, this explanation has been challenged in recent years. In 1976, Goldbaum *et al.* wrote, "Our studies indicate that a high concentration of carboxyhaemoglobin (COHb) does not interfere with the oxygen carrying capacity of the blood".[29] This extraordinary statement, later qualified by the authors, drew attention to their views, based on the results of their experimental work, that binding of CO to Hb was not the key mechanism, on the contrary the presence of CO dissolved in plasma and available for binding to cytochrome enzymes was very much more important. How odd! And yet there were hints that this might be, at least in part, the case much earlier than 1976. In 1927, JBS Haldane reported that rats "living on O_2 dissolved in their blood in the presence of sufficient CO to combine with almost all their haemoglobin are killed by the addition of more CO, which must affect some substance in their tissues, albeit at levels of CO not clinically encountered. It is concluded that cells contain a catalyst of oxidation which is poisoned by CO."[30] It should be noted that JBS Haldane did not deny the importance of binding of CO to Hb; what he added was a further explanation of the effects. The original explanation, binding of CO by Hb, had been put forward by his father, JS Haldane[31] and earlier by Claude Bernard in 1858.[32] Claude Bernard summarised the results of his studies of the mechanisms of action of carbon monoxide in his famous work, '*An introduction to the study of experimental medicine*'.[32] His account is still worth reading; his concluding sentence runs, "so death came through death of the molecules of the blood [binding of CO to Hb], or, in other words, by stopping their exercise of a physiological property [transporting oxygen] essential to life." The brackets [ ] are ours. Recent work on the mechanisms of action of carbon monoxide can be divided into three areas:

1. Exploring the secondary mechanisms triggered by hypoxia, which lead to tissue damage.
2. Exploring aspects of the effects of the binding of CO to molecules other than haemoglobin.
3. Considering the effects of raising concentrations of a molecule (CO) now known to be a physiological transmitter and its effects on other transmitters including nitric oxide.

Work in all these areas has been enlightening. Before turning to these developments we must consider the classical interpretation of the toxicology of carbon monoxide: its binding to haemoglobin.

9.6.1 Binding of CO to Haemoglobin

Haemoglobin comprises four amino acid chains: two alpha chains each of 141 amino acids and four beta chains of 146 amino acids. The chains are folded in a precise way (the quaternary structure of the molecule) and four haem groups lie within the four crevices formed by the folding. The haem groups each contain an iron atom and are attached, *via* that iron atom, to a histidine residue that lies at position 87 in the alpha chain and 92 in the beta chain. The haem group is also held by less firm bonds to other amino acids.[33] This bonding causes the crevices to form; the shape of the crevices controls the access of oxygen to the haem groups and binding of oxygen to those haem groups affects the bonding to amino acids and thus the shape of the crevices. pH, temperature, CO_2 binding and the ionic strength of the fluid in which haemoglobin occurs also affects the bonding and thus alter the affinity of Hb for oxygen. 2,3-Diphosphoglycerate (2,3-DPG) binds to the beta chains of deoxygenated Hb and also alters the affinity of the molecule for oxygen.

As oxygen molecules bind to a haemoglobin molecule, the binding of further oxygen molecules is facilitated: binding of oxygen causes conformational changes affecting the shape of the crevices and this increases the ease of access of further oxygen molecules. Conversely, as oxygen molecules detach from oxyhaemoglobin the further detachment of oxygen molecules is facilitated. These facts underlie the sigmoid shape of the curve, which describes the loading and unloading of oxygen onto and from haemoglobin. As oxygen is taken up the haemoglobin molecules become increasingly 'relaxed', and as oxygen is given off it becomes increasingly 'tense'. Reduction in pH (increase in hydrogen ion concentration), increase in temperature and an increase in PCO_2 all displace the dissociation curve to the right: this is the Bohr shift. The Bohr shift facilitates the release of oxygen in tissues with a high metabolic rate. A right shift of the curve implies a decrease in the affinity of haemoglobin for oxygen. This means that to achieve a given level of saturation a higher PO_2 is needed. In some species, especially fish, acidic conditions lead not only to a Bohr shift but also to a Root shift, which implies a decrease in the capacity of haemoglobin to carry oxygen. 2,3-DPG, produced in increased amounts by tissue with a low PO_2 and an increased rate of glycolysis, also moves the dissociation curve to the right. Stored blood with a low level of 2,3-DPG has a left-shifted curve. Interestingly, binding of haemoglobin to hydrogen sulfide produces sulfhaemoglobin and a shift to the right of the dissociation curve.

The shape of the oxyhaemoglobin dissociation curve has been known for many years: the standard curve is shown in Figure 9.2. The shape of the curve is well described by eqn (9.27)[34] and is shown in Figure 9.2.

$$SO_2 = \left(\frac{100({PO_2}^3 + 2.667PO_2)}{({PO_2}^3 + 2.667PO_2 + 55.4)}\right) \tag{9.27}$$

(SO_2 = saturation of Hb with O_2, expressed as a percentage.)

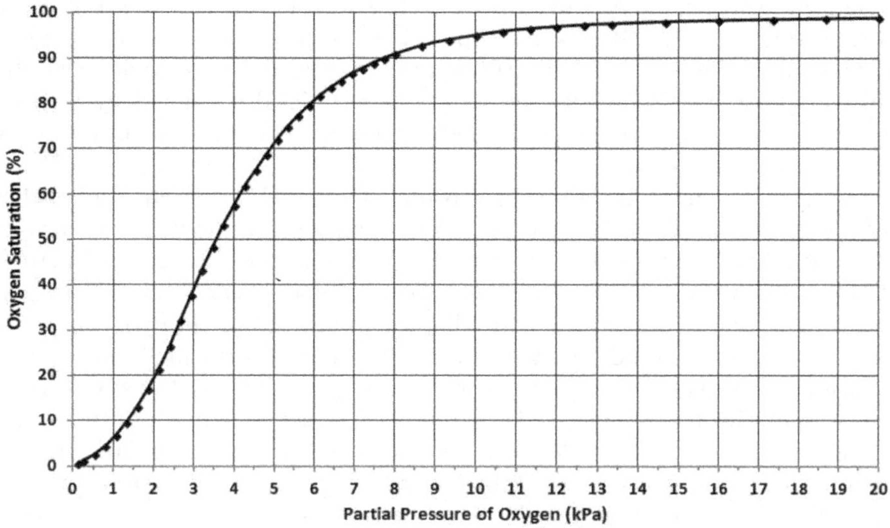

Figure 9.2 Oxygen dissociation curve at 37 °C and pH 7.4. Plotted using data taken from Severinghaus, 1979.[35]

Perhaps the most remarkable thing about carbon monoxide is that it binds to haemoglobin in exactly the same way as oxygen, though with more avidity. The greater affinity of haemoglobin for CO than for O_2 should not be confused with considerations of the comparative rates of uptake and release of these gases by Hb. CO combines with Hb about a tenth as fast as does O_2 and dissociates at about a sixtieth of the rate of O_2.The affinity of haemoglobin for CO is much greater than its affinity for oxygen: if x% saturation with oxygen is produced by exposure to y kPa O_2, x% saturation with CO will be produced by x/M kPa CO. This is Haldane's equation.[36] The identical nature of the binding of CO and O_2 means that we could convert the standard O_2Hb dissociation curve into the COHb dissociation curve simply by writing PCO instead of PO_2 on the x axis and altering the scale. This is sometimes referred to as Haldane's First Law. For practical purposes the law holds well; for a detailed discussion of its inadequacy at high levels of saturation, see Roughton (1954).[37]

We discussed earlier the fact that a mixture of 50% COHb and 50% O_2Hb would be produced by breathing air containing 21/245% CO (0.09% CO). It might be thought that this would produce an effect similar to that which would be produced by reducing the haemoglobin concentration to a half of its usual level. In fact it produces a much more severe effect. A reduction of the haemoglobin concentration from the normal level of 14 g/100 ml to 7 g/100 ml is not uncommon in anaemia. One of us (RLM) recalls seeing a patient with a haemoglobin concentration of about 6 g/100 ml: the patient had 'bled down' to this level as a result of untreated haemorrhoids. Other than a certain and not unexpected pallor the patient was remarkably well, if a little short of breath on exercise. An increase in cardiac output and a lowered blood viscosity were serving him well. The factor that explains the more severe effect of 50% COHb is the imposed left shift of the dissociation curve. This is called the Haldane shift, though the term is also used to describe the shift of the CO_2 dissociation curve (some CO_2 is also carried by haemoglobin) as a result of oxygenation/deoxygenation. The left shift imposed by CO is easily understood.

Consider a subject breathing a mixture of air and CO such that his/her %[COHb] will reach 50%. Now it is very unlikely that some molecules of haemoglobin will take up only CO and others only O_2. More likely, it is a mixture: $(CO \times 4)$Hb, $(CO \times 3)(O_2 \times 1)$Hb, $(CO \times 2)(O_2 \times 2)$Hb, $(CO \times 1)(O_2 \times 3)$Hb and $(O_2 \times 4)$Hb. As the oxygen tension falls, the molecules of Hb with a mixture of CO and O_2 on board do not give up their CO and thus the facilitatory effect of losing gas molecules on the further loss of gas molecules (what we might call cooperative loss) is negated. Losing oxygen becomes more difficult and to achieve a given level of loss the PO_2 must fall further and further. The dissociation curve has moved to the left. As noted above, this shift to the left is particularly dangerous for the fetus, whose O_2Hb dissociation curve is already left-shifted in comparison with the adult curve.

In summary, CO competes very effectively with oxygen for binding sites on haemoglobin. Furthermore, binding of CO impedes the release of oxygen at the tissues (a lower PO_2 is required to release oxygen) and thus the tissues become hypoxic. The very high affinity of Hb for CO, in comparison with its affinity for oxygen, makes even low concentrations of CO very dangerous.

9.6.2 Binding of CO to Myoglobin

Myoglobin (Mb) is a haem protein found in muscle cells in both skeletal and cardiac muscle.[38] Other haem proteins are now known to occur in other cells: neuroglobin (Nb) is, for example, found in neurones[39] and cytoglobin (or histoglobin) is expressed in a wide variety of tissues.[40] Myoglobin is a one chain molecule with one haem group and thus one iron atom. Clearly, there is no possibility here of cooperative action between oxygen molecules and the dissociation curve of O_2Mb is hyperbolic rather than sigmoid in shape. The hyperbolic dissociation curve means that O_2Mb releases oxygen only at very low partial pressures of oxygen. Myoglobin would be of little use as a substitute for haemoglobin as a transporter of oxygen but it plays an important part as a storage site for oxygen in muscle cells and as a facilitator of oxygen diffusion within muscle cells. Myoglobin takes up oxygen that has been released from O_2Hb and which has diffused from the muscle capillaries and across the plasma membrane of the muscle cells (sarcolemma). O_2Mb diffuses within the muscle cell and brings oxygen into close proximity with mitochondria. One might wonder about the diffusion rate of a large molecule like O_2Mb: it is slower than that of oxygen but Mb is present in a high concentration and this compensates for the slow diffusion.[41] The PO_2 at the mitochondrion is low and O_2 is released from O_2Mb. Binding of oxygen to Mb keeps the partial pressure of oxygen at the inner surface of the sarcolemma low: this maintains the PO_2 diffusion gradient from the capillary to the interior of the muscle cells and defends the mitochondria against unduly high levels of oxygen.

Carbon monoxide binds to myoglobin in the same way as oxygen does. However, the ratio of affinities is not 245, as for Hb, but only about 40. Myoglobin is thus likely to bind CO only at low levels of PO_2, though it may store a considerable fraction of total body CO. As PO_2 falls, binding of CO to Mb increases and delivery of oxygen to the mitochondria may be significantly impaired. This is clearly a potential danger in an environment where high concentrations of CO and low concentrations of O_2 are present; such environments occur in fires.

9.6.3 Binding of CO to Other Haem Proteins

Binding of oxygen to both Hb and Mb does not involve oxidation of the iron atoms of the haem group: the iron remains in the ferrous (Fe^{2+}) state. This is not the case for cytochrome c oxidase, the enzyme that catalyses the last step of the electron transport chain.[42] The reactions of the electron transport chain take place, of course, within the mitochondrion. The iron of cytochrome c oxidase shuttles between the ferric (Fe^{3+}), oxidised form, and the ferrous (Fe^{2+}), reduced form. If CO could bind to the iron atom of the haem of cytochrome c oxidase the electron transport chain would be disabled and a leak of electrons from earlier stages of the process (earlier links in the chain) would occur. This sort of disabling of the electron transport chain should not be confused with uncoupling of oxidative phosphorylation. Uncoupling involves breaking the link, proposed in the chemiosmotic theory of Mitchell, between the movement of hydrogen ions and the production of ATP.[43] Uncoupling is produced by compounds such as dinitrophenol but not by cyanide or CO. Disabling the electron transport chain reduces ATP production and, *via* glycolysis (which cuts in, so to speak, when oxidative phosphorylation is disabled), leads to acidosis due to the production of lactic acid.

Does CO bind to cytochrome c oxidase in significant amounts? Table 9.2 shows that the affinity of cytochrome c oxidase for CO, in comparison with its affinity for oxygen, is low when compared with that of haemoglobin.

It will be immediately clear that cytochrome c oxidase binds to O_2 with greater affinity than it does to CO: the reverse of the case as regards Hb or even Mb.

Table 9.2 Warburg coefficient (R) and Haldane coefficient (M) values for haem proteins from different sources (modified from EHC, 1999, pp. 152).[1]

Haem protein	Source	R^a	M^b	Temp./°C
Haemoglobin	Human	0.0045	218	37
Myoglobin	Sperm whale	0.025–0.040	25–40	25
Cytochrome c oxidase	Bovine heart	5–15	0.1–0.2	25
Cytochrome P450	Bovine adrenal	0.1–12	0.1–10	30–37

$^a R = PCO/PO_2$ at 50% inhibition.
$^b M = 1/R$.

Interestingly, CO can only bind to reduced cytochrome c oxidase (to Fe^{2+}) and under normal conditions the iron is in the oxidised (Fe^{3+}) form, though the PO_2 at the mitochondrion can be very low (<130 Pa). Brown and Piantadosi (1990)[44] demonstrated that binding could occur in the rat brain. Iheagwara *et al.* (2007)[45] reported a reduction in cytochrome c oxidase activity in the mouse myocardium on exposure to 0.1% CO. Very interestingly, these authors found that the reduction in activity was due to a reduction in the amount of enzyme present despite normal levels of cytochrome c oxidase mRNA. This suggests some effect of CO on protein synthesis or an up-regulation of haem oxygenase that leads to an increased rate of breakdown of haem. Binding of CO to reduced cytochrome c oxidase can retard binding of oxygen when, on reperfusion, oxygen becomes available in increased amounts. Piantadosi (1999)[46] pointed out that this occurred in isolated mitochondria on exposure to concentrations of CO as low as 100 ppm; see also Miro *et al.*, 1998.[47]

Castoldi *et al.* (2006)[48] showed that 500 ppm CO, 6 hours per day, 5 days per week for 4 weeks did not appear to affect rat brain or lymphocyte cytochrome c oxidase activity. The authors did, however, show that CO exposure increased NO-stimulated cyclic GMP activity in brain tissue. This was a delayed effect and is particularly interesting because increased cGMP activity leads to activation of receptors and triggering of apoptosis in some cells. We shall return to these findings later.

As a final point on effects on cytochrome c oxidase, JBS Haldane's suggestion,[30] made in the 1920s, of an effect of a 'catalyst of oxidation' seems well up to his usual standard of genius!

9.6.4 Recap

CO binds to haem. This reduces oxygen transport and delivery to the tissues and the capacity of cells to use such oxygen as is delivered. The latter point, about use of oxygen, was not included in the classical explanation of the toxicity of CO: that rested on the effect on transport and delivery to the tissues. This rather limited explanation was attacked by Goldbaum, whose work has been widely quoted.

9.6.5 Goldbaum's Experiments

Goldbaum *et al.* undertook a series of studies using dogs in the 1970s.[49] Though undertaken more than 40 years ago, the results are still interesting.

9.6.5.1 First Series

In essence, this series involved looking at the survival of dogs in the following groups:

- Those whose Hb had been reduced by about 70% by bleeding: anaemic dogs.

- Those whose Hb had been reduced by about 70% but then 'topped up' to normal with red cells in which the HB was 100% saturated with CO. These dogs had a %[COHb] of 70%. We shall call this Group X.
- Those whose red cells had been completely replaced with red cells in which the Hb was 70% saturated with CO. These dogs also had a %[COHb] of 70%. We shall call this Group Y.
- Controls.

The authors had thus rather cleverly produced two preparations each with 70% [COHb] but arrived at it, so to speak, in two different ways. None of the controls or anaemic dogs died. None in Group X died, while deaths in Group Y were frequent. The authors argued that if the effects of CO were due to binding to red cell Hb there should have been as many deaths in Group X as in Group Y. They argued that the effects had little to do with Hb and that diffusion of CO dissolved in the plasma into cells where it might affect the use of oxygen was the key to the process. A little later the authors realised that another explanation was possible: in Group Y the release of oxygen might be impaired by a left shift of the O_2Hb dissociation curve. In Group X we must assume that about 30% of the Hb was functioning normally. Drabkin had pointed this out in 1943.[50]

9.6.5.2 Second Series

The authors then undertook another series but with a further, fifth group. In this group the animals were exposed to CO introduced into the peritoneal cavity. This was very innovative of the authors: the CO was taken up and bound to Hb, the blood then passed through the lungs and any CO dissolved in the plasma was lost by diffusion into the alveoli. Thus, they had created an animal with a %[COHb] of 70% and, they argued, with no CO in the plasma. If release of oxygen from Hb was the cause of the effects of CO then deaths in this group should have been the same as deaths in Group Y above. Remarkably, no deaths occurred in what we might call the peritoneal group. This confirmed their view that binding of CO to Hb was not the mechanism of effect but that the activity of plasma-borne CO was. In fact, the level of CO in plasma would have been in the order of 5×10^{-5} ml L^{-1}(245 times less than for oxygen at a SaO_2 of 70%) as dictated by the CO/Hb dissociation curve due to release of CO from the red cells into plasma after passage through the pulmonary capillaries. This is less than would be expected in arterial blood had CO been absorbed by the lungs, but not zero as claimed by Goldbaum. These results have remained unexplained and, as far as we know, unrepeated.

Coburn (1979) commented on Goldbaum's results.[51] He made several points:

- The use of 13% CO to raise the %[COHb] could have caused death from anoxia in some of animals studied.
- 70% [COHb], if reached slowly, is compatible with survival.
- The physiological monitoring undertaken by Goldbaum *et al.* was limited.

Coburn also argued that rapid equilibration between CO in plasma and that attached to Hb was likely. Goldbaum had relied on the plasma CO being rather high as a result of incomplete equilibration with Hb. Coburn considered the suggestion that CO was binding to cytochrome c oxidase was important but argued that binding would not be significant unless mitochondrial oxygen tension dropped to about 13 Pa. This was a low, but not an impossibly low figure.

Some have accepted Goldbaum's findings. Gorman and colleagues[52–54] have stressed that oxygen uptake by the brain is maintained by increased blood flow during at least some levels of CO poisoning.[55] They have argued that it is not the lack of oxygen that causes the damage. This takes the argument a stage further: now it is being suggested that it is not the impairment of oxygen transport or release OR even the use of oxygen by cells that is the problem. If not these, then what? The answer may lie in the production of oxidative free radicals. Thom[56] showed that glutathione was released from red cells and that markers of oxidative stress appeared in rats exposed to CO, but the effects were not related to %[COHb]. It seems unlikely that the effects of CO have nothing whatever to do with a reduction in oxygen availability at the mitochondrion. But that this reduction leads to damage due to increased production of oxidative free radicals is possible. The role of free radicals seems especially likely during the reperfusion phase of CO poisoning.

9.7 Carbon Monoxide, Free Radicals and Oxidative Stress

Free radicals figure in the explanations of a wide range of toxicological effects and pathological processes. Reperfusion injury, occurring in the myocardium after infarction and in the brain after stroke, is thought to be free radical mediated. Some of the effects of CO appear very similar to those seen in reperfusion injury. This area cannot be understood without a digression on the subject of free radicals.

9.7.1 Free Radicals

Many atoms, molecules or, indeed, ions can form free radicals. A free radical is an atom or ion that has an unpaired electron in its outer electron shell. In the case of a molecule, at least one of the atoms of the molecule has to have an unpaired electron in its outer shell. Consider the oxygen molecule. This is usually represented as O_2 but we might have written $O_2^{\bullet}$, the superscript dots representing unpaired electrons in the outer shells of each oxygen atom. How surprising! Is oxygen, then a double free radical? It could be described as one but it lacks the reactivity usually conferred by free electrons and is stable, apparently because the orbits of the two unpaired electrons are parallel to each other. This is clearly a complex area of chemistry! The para-magnetic properties of oxygen are due to these two free electrons. Gain of an electron by the oxygen molecule leaves one of the oxygen atoms with an unpaired electron (the other has a pair), now the molecule is a free radical and not stable. This form may be written as $O_2^{\bullet -}$, this is the superoxide radical or anion. One might wonder why all ions are not free radicals: the presence of a negative charge might suggest the presence of a free electron. This confusion is easily resolved: the charge (positive in cations and negative in anions) is due to an imbalance between the number of electrons present and the number of protons present in the atomic nucleus. Consider the electrons of fluorine: there are two shells and the number of electrons in these shells can be written: 2,7. This means seven electrons in the outer shell. If fluorine accepts an electron it achieves a stable electron structure: 2,8. In fact it has the same electron structure as neon. But it now has one more electron than protons and it has a negative charge.

This can be summed up by an equation explaining the formation of sodium fluoride:

$$Na\ (2,8,1) + F\ (2,7) \rightarrow Na^+\ (2,8) + F^-\ (2,8)$$

The electron configuration is shown in the brackets. Note that Na^+ and F^- have the same electron configuration but opposite charges. The formation of sodium fluoride is an example of the formation of an ionic bond. The hydroxyl anion has no free electron in the outer electron shell of the oxygen atom but it is negatively charged. The hydroxyl free radical, *per contra*, does indeed have a free electron in the outer electron shell of the oxygen atom but is not negatively charged. The hydroxyl anion is not toxicologically important; the hydroxyl free radical is! The hydroxyl free radical is better written as $HO^\bullet$ because the free electron is on the oxygen atom.

The superoxide radical is the starting point for the formation of a range of other free radicals:

$$O_2^{\bullet -} + H^+ \rightarrow HO_2^\bullet \text{ (the hydroperoxyl radical)}$$

$$O_2^{\bullet -} + H_2O_2 \rightarrow HO^\bullet + OH^- + O_2 \text{ (this is the Haber–Weiss reaction)}$$

$$H_2O_2 + Fe^{2+} \rightarrow Fe^{3+} + HO^\bullet + OH^- \text{ (this is the Fenton reaction)}$$

Nitric oxide is a free radical: the unpaired electron is on the nitrogen atom. Nitrogen dioxide is unusual: apparently 50% of the odd electron density is on the N atom (see Pryor, 1986).[57] More complex free radicals, with the unpaired (odd) electron on the carbon atom, are formed from fatty acids; semiquinone free radicals are formed from polycyclic aromatic hydrocarbons found in ambient air and may be, in part, responsible for the toxicological effects of inhaled ambient particles. Free radicals are produced in cells as a result of a leak of electrons from the electron transport chain of the mitochondrion. Removing these radicals is essential for survival of cells and free radical scavengers undertake this task. Reduced glutathione is a good example: it is oxidised by oxidative free radicals. Superoxide dismutase catalyses the following reaction:

$$O_2^{\bullet -} + 2H^+ \rightarrow H_2O_2 + O_2$$

All this is interesting but what has it to do with reperfusion injury? A period of hypoxia is associated with a decline in production of ATP. When ATP levels fall, the following reaction catalysed by adenylate kinase takes place:

$$2ADP \rightarrow ATP + AMP$$

AMP is converted to adenosine and a series of reactions is triggered. The sequence runs:

adenosine → inosine → hypoxanthine → xanthine → uric acid.

The danger for the cell is in the last reaction.

Xanthine oxidase is formed from xanthine dehydrogenase, the conversion being driven by the non-lysosomal enzyme calpain, which is activated by calcium ions. The intracellular concentration of calcium ions rises under hypoxic conditions. This increase is caused by a decrease in activity of ionic pumps in the cell membrane: these pumps are powered by ATP. The inward movement of calcium triggers, in some parts of the brain, a release of excitatory amino acids such as aspartate and glutamate. Activation of further receptors causes further influxes of calcium ions.

Calcium also activates lipoperoxidase and cell membrane lipids are attacked. Free fatty acids are released and chain reactions leading to free radical accumulation occur. The sudden appearance of oxygen during reperfusion, at a time when levels of free radical scavengers will be low due to reduced production, increases free radical formation. Cell damage releases an array of mediators of inflammation. Leukocytes begin to stick to the endothelium of capillaries and then leave these vessels and enter the tissue. Activated leukocytes release a further array of cytokines and other messenger molecules. An inflammatory response is under way. The reader will understand and perhaps appreciate that space does not permit a digression on the role of cytokines in inflammation. Complicated diagrams showing the many pathways and reactions involved in these processes have been produced. A word of caution is necessary. Drawing diagrams is fun and may help to understand the processes at a qualitative level. Providing quantitative detail is much more difficult: the relative importance of the various pathways is difficult to define. It should be recalled that much of this theory has been worked out using animal models; qualitative extrapolation to man may be justified but quantitative extrapolation is less easy and, at present, all but impossible.

For detailed accounts of the link between carbon monoxide and free radical formation, see Piantadosi *et al.* (2006),[58] Raub and Benignus (2002),[59] Thom (1993),[60] Thom *et al.* (1997),[61] Piantadosi *et al.* (1995),[62] Traystman *et al.* (1991)[63] and Kristian and Siesjo (1998).[64] Each of these papers makes a strong case for free radicals being involved in the toxicological effects of CO, especially as regards the very important delayed effects on the brain. The papers also deal with points not covered here, for example, see Piantadosi *et al.* (2006)[58] for a fascinating, though formidably detailed, exploration of the link between CO and mitochondrial pore transition. Here too, calcium seems to be the key factor.

It will be appreciated that all of the above is based on the premise that carbon monoxide leads to hypoxia at a cellular level and, perhaps, if this is granted all else follows. CO, however, in the absence of hypoxia has regulatory effects on cellular respiration and reactive oxygen species (ROS) production as discussed above regarding the functions of the endogenously produced gas. These effects are accentuated in hypoxia but may fall into the category of physiological actions rather than toxic effects. Indeed, reduced cerebral and myocardial oxygen consumption are early features of exposure to carbon monoxide but at levels that may not be associated with the symptoms of toxicity in man.[65,66,‡]

9.8 Clinical Features of Carbon Monoxide Poisoning

Discussion of the clinical effects of CO tends to focus on two patterns of exposure: long term exposure to low concentrations, as might occur as a result of a leak of CO from a malfunctioning heating device; and short term exposure to high concentrations, as are likely to occur in a combustion environment or, again, after failure of a heating device. In addition, the effects of exposure to rather low concentrations in ambient air might be considered.

9.8.1 Exposure–response Relationship

Table 9.3 sets out information of concentrations of carbon monoxide, the %[COHb] reached at equilibrium at the various concentrations and the effects that might be expected. Such tables should be used with caution. Effects depend not only on [COHb] but also on the pattern of exposure, see below. The carboxyhaemoglobin concentration is a poor basis for deciding on therapy on admission to hospital and when measured often does not seem to fit with the patient's symptoms and signs. The patient may not have been exposed for long enough to reach equilibrium; the patient may have reached a much higher %[COHb] than is recorded, he/she will be in the washout phase.

[‡]RLM would like to thank Dr Claude Piantadosi for his willingness to explain the binding of CO to cytochrome c oxidase whilst the above sections were being prepared. Needless to say, any residual errors are the fault of the present author.

Table 9.3 Concentrations of carbon monoxide, the %[COHb] reached at equilibrium at the various concentrations and the effects that might be expected.

[CO] in air/ppm	Inspired/ %	COHb at equilibrium (approx.)	COHb by ranges (approx.)		Toxic effects[a]
1	0.0001	0.49	0	A	None. COHb normal range: 1–1.5%. Note that inhaled CO **adds** to the baseline value.
3		0.81			
5		1.14			
7		1.46			
9		1.78	2		Vasodilation, slight reduction in time to exhaustion, possibly minor central nervous system (CNS) effects.
10	0.001	1.94			
30		5.03	5		
50		8		B	Reduced performance of complex tasks.
70		11	10		
90		13			Headache, nausea, impaired judgement, lethargy, dizziness, shortness of breath (not marked early on).
100	0.01	14	20		
300		33		C	Severe headache, vomiting, confusion. Collapse. Death may occur.
500		45			
700		54	50	D	Fits.
900		60			Coma, weak pulse, slow breathing.
1000	0.1	62			Death.
3000		83	80		Rapidly fatal.
10 000	1.0	94			
100 000	10.0	99			

[a]Provided as a **guide** only. Effects vary with pattern of exposure and from subject to subject.

Table 9.3 is too complicated to be easily memorised. The following rule of thumb guide may be useful:

0.001% CO is safe;
0.01% CO is worrying;
0.1% CO is really dangerous;
1.0% CO is deadly.

These percentages are indicated, broadly, by the arbitrary scale of clinical effects shown in the Table 9.3: A, B, C and D. In the case of CO exposure in homes, patients often complain of effects that fall into group B. Headache is common, nausea, perhaps some dizziness, weakness and confusion are often reported. These symptoms are by no means restricted to the effects of exposure to CO; on the contrary they are often seen in viral infections, such as influenza, and in food poisoning, which may be viral or bacterial in origin. And herein lies a danger: unless the possibility of exposure to CO is considered the symptoms and signs may well be attributed to some other cause. Of course headache is a very common symptom, we all suffer from headaches now and again, and most headaches are not caused by CO. But some are. The great danger is that the diagnosis of exposure to CO is missed and the patient, thought to be incubating some infection, is sent home with advice to keep warm and go to bed. Such advice can be lethal: the source of CO may well be a heating device. As exposure continues the level of confusion increases and the patient may slip into sleep from which he/she will not wake. How can this be avoided?

Some standard clinical questions will often reveal the true diagnosis:

- How long have you had these symptoms?
- Are they worse indoors than out?
- Do they disappear when you are away from home, for example, at work or at school?
- Has anybody else in your family got the same symptoms?

Patients suffering from coronary artery disease may report unexpected anginal pain. Patients with pulmonary disease may report shortness of breath. Physical signs tend to be rare in mild to moderate poisoning. Another sign sometimes mentioned is pink colouration of the skin. This is rarely seen (Box 9.3). Table 9.4 sets out a long list of symptoms and signs.

Box 9.3 Colour

It is probably unnecessary to mention the cherry pink colouration of the skin and mucous membranes often spoken of in association with CO poisoning as this is rarely seen. Cyanosis is more likely in cases of severe CO poisoning. The pink colour is due to carboxyhaemoglobin, which has a cherry red (or pink) colour rather than the scarlet of oxyhaemoglobin. Nitric oxide also combines with haemoglobin to produce a ferrohaem–NO complex. This is pink in colour and was discovered by JS Haldane to be the reason for the pink colour, after cooking of bacon (JBS Haldane, *Science and Everyday Life*, 1939, London, Lawrence and Wishart Ltd). The effect is produced by the potassium nitrate used in curing bacon. The complex is apparently stable on heating. Nitrites in general oxidise the iron of haemoglobin to the ferric (Fe^{3+}) form: methaemoglobin. Sodium nitrite is used in the treatment of cyanide poisoning: cyanide ions bind only to ferric (Fe^{3+}) iron and thus do not bind to normal haemoglobin and do not interfere with oxygen transport. Cyanide ions do, however, bind to ferric iron of cytochrome enzymes and therefore interfere with oxygen usage. Methaemoglobin has a brown colour. Hydrogen sulfide also combines with haemoglobin, to produce sulfhaemoglobin. This is brown in bulk but looks green in thin layers: it accounts for the discolouration of dead bodies, the H_2S being produced by bacteria.

Table 9.4 Symptoms and signs of carbon monoxide poisoning.

Symptoms	Signs	Neurological effects
Headache (90%)	Vomiting	Confabulation
Dizziness (82%)	Tachycardia	Seizures, fits, convulsions
Weakness (53%)	Tachypnoea	Coma
Nausea	Hypotension	Cog-wheel rigidity
Shortness of breath	Flame shaped retinal haemorrhages	Peripheral neuropathy
Defective concentration	Retinal venous congestion	Agnosia[a]
Poor memory	Papilloedema	Apraxia[b]
Poor judgement	Bullous skin lesions	Amnesia
Confusion	Vesicular skin lesions	Parkinsonian tremor
Disorientation	Sweat gland necrosis	Cortical blindness
Anginal pain	Rhabdomyolysis	Incontinence
Hallucinations	Renal failure	Depression
	Localised alopecia	Personality changes
		Intellectual decline
		Gait disturbances
		Cerebellar signs
		Speech impairment

[a]Loss of ability to recognise sensory stimuli. For example, not knowing the name of a common object.
[b]Inability to carry out a purposeful movement.

This is a formidable list! The signs relating to effects on the nervous system may appear in the recovery period and may be delayed by some days or a week or so. The latent period between CO poisoning and apparent recovery and the development of signs of damage to the nervous system is very characteristic. The percentage figures in the first column of the table are taken from the paper by Lowe-Ponsford and Henry (1989),[67] which is an excellent review.

9.8.2 Effect of Short Duration Exposure to High Concentrations of Carbon Monoxide

JS Haldane, his son JBS Haldane and colleagues frequently exposed themselves to sufficient CO to produce %[COHb] of over 30%. In the case of the Haldanes, exposure was often continued to close to the point of collapse. The exposure was then terminated and the subject recovered on breathing air. Table 9.5, taken from the work on JS Haldane, sets out the effects.[31,36,68] This work comprised JS Haldane's first experimental studies. They were undertaken to investigate the effects of carbon monoxide in mining accidents and Table 9.5 presents the findings.

Having noted that his own %[COHb] had reached 56%, Haldane noted that the level found in some miners who had died had been 60%.

He then went on to make a critically important point. He discussed the difference between the effects of carbon monoxide exposure and those of a

Table 9.5 The effects of carbon monoxide inhalation.

%[COHb]	Effects
$<20\%$	No particular effects
20%	Extra exertion, such as running up stairs, produced a slight feeling of dizziness and some extra palpitations and hyperpnoea
30%	Very slight symptoms (slight increase in pulse rate, deeper breathing, slight palpitation) at rest. Running up stairs followed, in half a minute, by dizziness, dimness of vision and abnormally increased breathing and pulse
40%	Symptoms more marked. Any exertion had to be made with caution for fear of fainting
50%	No real discomfort at rest. Vision and hearing impaired, intelligence probably greatly impaired. Unable to rise from chair without assistance. Writing very bad, spelling uncertain. Difficult to judge distances. Attempts to go any distance led to collapse
56%	Reached this level in one experiment. Hardly able to stand, impossible to walk
Recovery, breathing air	
25–35%	Nausea and headache, 'extreme depression', similar to symptoms of mountain sickness

lack of oxygen: the key difference was that in the case of CO exposure, fainting, especially on exertion, was much more common. Haldane suggested that the heart was not able to respond to the increased blood flow through the muscles, the blood pressure fell and fainting occurred. He made a second key point: the respiratory centres of the brain were much less stimulated by hypoxia caused by exposure to CO than by exposure to a low oxygen tension. The reason suggested for this is that the PO_2 of the blood is substantially maintained during exposure to CO but depressed by a lack of oxygen. In general terms, but see below, the sensors of the carotid body drive the respiratory centres of the brain to respond to a reduced PO_2, rather than to the oxygen content of the blood. More recently, however, it has been established that CO is a direct inhibitor of carotid body activity.[12,13] The lack of warning before fainting is a very dangerous characteristic of CO poisoning and clearly a danger in a combustion environment. Haldane concluded his discussion with an explanation of the left shift of the O_2Hb dissociation curve. The reader will understand that JS Haldane is rightly regarded as one of the greatest physiologists of the 20th century.

Haldane's studies were, in general, of short duration exposures. Would different effects have been produced if the pattern of exposure had been different? Haber's Law (or Rule) is much used in inhalation toxicology. It states that the effect of an inhaled compound is proportional to the total number of molecules taken into the target tissues. This is expressed by the following equation:

$$Ct = K, \tag{9.28}$$

where C is the concentration of the inhaled compound and t the duration of exposure. K is a constant. Haber introduced this concept during his work on chemical warfare agents, especially chlorine and phosgene, during World War I.[69] He defined K as the lethal product (of C and t). His point was that as long as K was constant (C and t could vary reciprocally: $Ct = K$, $0.5C2t = K$, $0.1C10t = K$, $10C0.1t = K$) then the effect produced would remain the same. It is known that this simple relationship does not apply to all compounds; indeed it does not apply precisely even to phosgene. A modified equation is now used:

$$C^n t = K \tag{9.29}$$

The exponent n defined the relative importance of C as compared with t. For example, if the equation were:

$$C^4 t = K \tag{9.30}$$

then C would be the dominant variable. Values of n of >1 are well known: irritants have n values of >1; for example, sulfur dioxide has an n value of 2. The effects of exposure to sulfur dioxide are very dependent on concentration and not very much affected by the duration of exposure. Carbon monoxide is not an irritant and we should not expect n to exceed 1. This was accepted by the UK Health and Safety Executive in its published list of values indicating significant likelihood of death (SLOD figures).[70] Many refinements of Haber's Law have been proposed.[71] The product of C and t can be described as the toxic load: the amount of material taken on board, so to speak, during an exposure. The perception that the toxic load controls the response is likely to break down at:

a. Very long duration exposure to low concentrations, because the toxic material might well be metabolised and excreted without causing harm. Adaptation might also occur.
b. Very short duration exposures to very high concentrations. It is possible that at very high concentrations different effects might be seen: overloading of a metabolic pathway or diversion of the toxicant along a pathway that produces a more toxic product might occur. Of course, the exposure must be of sufficient duration to allow a breath to be drawn!

Interestingly, the conditions defined in a. are seen in smokers. Cigarette smokers often have %[COHb] of 4–10%. Some adaptation occurs and smokers tend to have a higher Hb concentration than non-smokers. The question of whether smokers are affected by their raised levels of COHb is a large one. That CO plays a part in accelerating atherosclerotic changes seems likely. But smokers do not, in general, seem to suffer from symptoms of CO poisoning. This has been seen by some to indicate that long term exposure to low environmental concentrations of CO should not produce ill effects. Time series studies have shown an association between peak hourly levels of CO and the occurrence of heart attacks.[72] Whether this is a causal association might be doubted, whether ambient particles are a more likely candidate for the active agent might be argued, but the fact that CO is related to development of coronary artery disease suggests at least some plausibility for the assertion that CO is playing a part.

The effects of short duration exposure to high concentration of CO have been studied in rats by Gu *et al.* (2005).[73] This important work was undertaken specifically with effects on fire-fighters and military personnel in mind and is very relevant to this chapter. Forty two rats, six per group, were used for studies of effects on minute volume (referred to by the authors as respiratory minute volume, RMV), %[COHb] and blood pH. Thirty six rats were used for the study of lethality only.

The authors reported hyperventilation in three rats during the last minute of exposure to 10 000 ppm CO. In two rats exposed to 12 000 ppm CO there was a marked decline in minute volume during the last minute of exposure. The authors suggested that exposure for 5 minutes to 10 000 ppm CO produced a different effect from 5 minutes exposure to 12 000 ppm CO: hyperventilation in the former, respiratory depression in the latter. The mechanism underlying the hyperventilation is not clear. Hypoxia (though PO_2 would have been largely maintained) and acidosis were suggested. Note that in JS Haldane's work there is mention of increased breathing even at comparatively low concentrations of CO. The authors suggested that both a respiratory acidosis and a metabolic acidosis could occur. Respiratory acidosis implies a failure of ventilation and retention of CO_2: this is not associated with hyperventilation. Hyperventilation is, however, a physiological response to metabolic acidosis. No measures of indices designed to differentiate between the different forms of acidosis, for example, measurement of standard bicarbonate concentration, were reported. The study showed that %[COHb] rose in line with the adverse clinical effects.

This is the pattern seen in the uptake phase of carbon monoxide poisoning during exposure to carbon monoxide or soon after recovery when pulmonary and blood levels of the gas are higher than those in tissues. It is not what is seen during the excretion phase of poisoning after rescue when tissue levels exceed those in the blood and lungs and when the clinical picture results from a mixture of ongoing intoxication and any established tissue damage, possibly irreversible, sustained during the period of intoxication. This difference in presentation has practical applications. While the level of poisoning cannot be adduced from the condition or blood carboxyhaemoglobin level of a survivor on admission to hospital, a reasonable assessment of exposure can be made from the symptoms of victims during an acute exposure.

In considering the effects of CO on respiration it is instructive to consider the analysis provided by Comroe (1975).[74] Comroe considered conditions in arterial blood and in end-capillary blood in the carotid body. Table 9.6 is a modified from Comroe's work. Note that Comroe assumed that 0.5 ml O_2 was extracted from each 100 ml blood passing through the carotid body.

Table 9.6 repays careful study. Note that in the example of hypoxaemia the amount of oxygen being transported is very similar to that in the example of

Table 9.6 Oxygen extraction in the carotid body and the effects on breathing in hypoxia, anaemia and carbon monoxide poisoning.

			Arterial blood				Carotid end-capillary blood			
Conditions	Hb (g/100 ml)	COHb (g/100 ml)	S_AO_2 (%)	P_AO_2 (kPa)	O_2 content (ml/100 ml; O_2 bound to Hb + dissolved O_2)	O_2 removed (ml O_2/100 ml)	O_2 content	$S_{ec}O_2$ (%)	$P_{ec}O_2$ (kPa)	Effect on breathing
Normal	15	0	98.5	13.3	19.7 + 0.3 = 20.0	0.5	19.5	96.0	11.7	None
Hypoxaemia	15	0	50.0	3.6	10.0 + 0.08 = 10.08	0.5	9.58	47.5	3.3	Marked
Anaemia	7.5	0	98.5	13.3	9.85 + 0.3 = 10.15	0.5	9.65	94.5	10.7	None
Anaemia	3	0	98.5	13.3	3.94 + 0.3 = 4.24	0.5	3.74	87.5	7.3	Slight/marginal
CO poisoning	7.5 (active)	7.5	98.5	13.3	9.85 + 0.3 = 10.15	0.5	9.65	94.5	10.7	None

CO exposure, but that the former has a marked effect on respiration, whilst the latter has not. This is because the end-capillary PO_2 is so low in the hypoxaemia example (3.3 kPa) but essentially normal in the CO case. A word of caution is necessary in comparing the examples of anaemia with CO poisoning. The end-capillary PO_2 shown for CO exposure is a little high. This is because it was calculated without allowing for the left shift of the O_2Hb dissociation curve that would be produced by CO exposure. Comroe recognised this and put the figures 10.7–6.7 kPa in place of 10.7 kPa for end--capillary PO_2 in the CO example. Even with an end-capillary PO_2 of 6.7 kPa stimulation of respiration would be only marginal. Note that in the example of severe anaemia it is suggested that there is only a slight effect on respiration at rest, though people suffering from severe anaemia do become breathless on exercise.

A second paper of obvious relevance is that of Purser and Berrill (1983)[75] on the effects of carbon monoxide on behaviour in monkeys in relation to human fire hazard. Cynomolgus monkeys were trained to respond to a buzzing noise by pressing a lever on one side of a cage and then to cross quickly to the other side to receive a reward. The time taken between pressing the lever and collecting the reward was taken as a performance indicator. (Note that performance might imply either performance of central processing functions, recalling that pressing the lever preceded the availability of a reward, or/and motor performance needed to cross the cage.) Animals were exposed to 900 or 1000 ppm CO for a 30 minute period. The onset and offset of exposure were rapid: a square-wave pattern was used. Five studies at 1000 ppm (the preliminary studies) and 10 at 900 ppm were undertaken. Monkeys were exposed more than once, a long recovery period being allowed between exposures.

Clinical observations: the findings from the preliminary studies (three animals, five exposures) were dramatic. No effects were noted until 18–20 minutes into the 30 minute exposure period, when the animals became less active. Then, without much warning (consisting of minor signs such as eye closure, head shaking, yawning), the animals went through a period of around 20 seconds with signs of severe intoxication, involving a swaying or staggering gait, followed by collapse with loss of consciousness. Severe incapacitation lasted for up to 30 minutes and vomiting occurred on three occasions. Delay in responding to the buzzer was marked.

In the second series of exposures (nine exposures, three animals: three exposures per animal, plus one trial run) animals were exposed to 900 ppm and significantly less response was seen. In six of nine exposures, collapse occurred but recovery was more rapid than in the preliminary studies.

Measurements: in all the studies involving these actively moving animals, the time taken to cross the chamber rose. This increase occurred from around the middle of the exposure period and the usual time taken to cross the chamber was trebled just before the stage of intoxication. Response decreased progressively during the recovery period. In addition, CO_2 production declined during the exposure period and rose again, rather slowly, during the recovery period. Depression of respiration had been produced.

The authors measured the %[COHb] immediately after the end of the 30 minute exposures to 900 ppm CO as 33%[COHb], representing approximately 30%[COHb] at the time of collapse in these actively moving animals. In earlier studies, sedentary animals in chairs accumulated CO more slowly, but higher levels in excess of 40%[COHb] were achieved before collapse. The early depression of respiration in the active animals is striking in that it would have occurred when the %[COHb] was only 10–15%. The behavioural reaction time was reported to be a more sensitive indicator (detected effects at lower %[COHb]) than physiological monitoring, including electroencephalography, used in earlier studies.

Perhaps the most important finding of this key study was that collapse occurred essentially without warning. The authors suggested that compensatory factor, such as an increase in brain blood flow, could 'cope' up to a point but once that point had been reached, collapse followed rapidly. This is really important for people exposed to high concentrations of CO in a combustion environment. That activity speeds the onset of collapse is also important: rescuers not wearing closed circuit breathing apparatus are at great risk.

We will return to the effects of CO on performance after considering, in more detail, effects on the brain.

9.9 The Neurological Effects of CO Poisoning

The wide range of neurological symptoms and signs that may be observed in or reported by patients exposed to CO has already been noted.[76] These are likely to be particularly marked if a period of unconsciousness has occurred. Effects on the brain (the spinal cord has been rather ignored by researchers) can be studied in a variety of ways:

- Neuro-physiological techniques: electroencephalography (EEG) and visual evoked potentials (VEPs). The interpretation of changes involves a highly specialised technique and will not be discussed here. Lengthening of VEP latency seems to be a consistent finding. The interested reader is referred to Feldman's account.[77]
- Neuro-psychological techniques: behavioural test batteries.
- Neuro-imaging techniques: magnetic resonance imaging (MRI), computer aided tomography (CAT scan), single photon emission topography (SPECT, PET scan), magnetic resonance spectroscopy (MRS).
- Neuro-pathological techniques: sectioning and staining for light and electron microscopy.

An excellent account of all these approaches has been provided by RG Feldman in his *Occupational and Environmental Neuro-toxicology*.[77] The following summary draws heavily on Feldman's work.

9.9.1 Neuro-pathology

The essential cause of damage to the brain in CO poisoning is hypoxia: hypoxia due to impaired transport, delivery and perhaps usage of oxygen. Hypoxia and, later, reperfusion lead to the formation and release of oxidative free radicals and inflammation: the leukoencephalopathy spoken of by some workers. Stasis of blood in capillaries, leak of red cells and fluid from capillaries, leading to peri-vascular oedema are all features of this response. Demyelination follows and microglial cells accumulate lipid. Severely damaged areas may become necrotic.

These changes are seen in CO poisoning and have long been recognised, see Hill and Semerak (1918), and Grinker (1926).[78,79] White matter, with its rather exiguous blood supply, is more susceptible to damage than grey matter. Some areas of grey matter are also prone to damage.

Damage tends to be most marked in the hippocampus, the globus pallidus (a part of the basal ganglia), the internal capsule and that great mass of white matter the centrum ovale. These areas are all susceptible to damage by other causes of hypoxia. Damage also occurs in the cerebral cortex: blindness due to damage to the temporal cortex (as occurs in the Klüver–Bucy syndrome) may occur; personality changes due to damage to the frontal cortex have been reported. Damage to Purkinje cells of the cerebellum may lead to incoordination. Clearly, almost any part of the brain may be damaged if hypoxia is sufficiently severe and prolonged. One reason given for the increased susceptibility of some areas is that they have a poorer than average, so to speak, blood supply. Meyer (1936)[80] provided a detailed analysis of this theory. The globus pallidus, unlike the remainder of the basal ganglia is, in part, supplied with blood *via* the anterior choroidal artery (from the internal carotid artery); the hippocampus also receives its blood supply from the anterior choroidal artery; these tend to be of very fine calibre and follow a tortuous course. This seems to be especially the case for those vessels supplying that part of the hippocampus most affected by CO poisoning: the Sommer sector or region CA1. The centrum ovale is supplied by a long penetrating branch of the pial arterial supply: it is particularly long in man due to the thick cerebral cortex (Box 9.4).

The cortex is sensitive to general and severe hypoxia, indeed more sensitive than the white matter. Perhaps, when the supply of oxygen is completely removed, the cortex 'fails' before lesions develop in the white matter. Hypoxia may develop insidiously in CO poisoning and damage to the white matter has time to develop. A very detailed account of the neuropathological effects of CO has been provided by Lapresle and Fardeau (1967).[81]

The potential for widespread damage to the brain explains the range of neurological effects reported in CO poisoning. An aspect that is not well explained is the latent period that often extends from the recovery of consciousness to the appearance of neurological signs. It is generally accepted that this is caused by the more slowly developing effects of reperfusion injury.

Box 9.4 Basal ganglia

The basal ganglia are a group of masses of grey matter deep in the forebrain. Developing mainly from the forebrain but connected to the substantia nigra of the midbrain, they play a key role in controlling muscle tone and are important in the control and coordination of voluntary movement. The basal ganglia comprise the following nuclei (the word nucleus is used in neuro-anatomy to indicate a mass of neurone cell bodies): caudate, putamen, globus pallidus, and the amygdaloid nucleus. The basal ganglia are widely connected, for example, to the thalamus and substantia nigra. Damage to the basal ganglia may cause involuntary movements: the characteristic tremor of Parkinson's disease is due to malfunction in the substantia nigra. The basal ganglia used to be said to comprise a key part of the extra-pyramidal system, though this term has been discarded in recent years.

Species difference complicate the problem: Meyer pointed out that the globus pallidus of the omnivorous rat did not seem susceptible to necrosis as a result of exposure of the animal to CO, but that of carnivores was.

9.9.2 Neuro-imaging

The various techniques now available predict the location of lesions seen at post-mortem examination. Lesions of the globus pallidus are particularly well seen on an MRI scan. Jaeckle and Nasrallah (1985)[82] noted that detection of lesions of the globus pallidus correlated, perhaps unsurprisingly, with the development of parkinsonian signs, which developed after the latent period had elapsed. MRS techniques have demonstrated an increase in excitatory amino acids, *e.g.*, aspartate, in areas of damage: this fits well with theories of mechanisms of reperfusion injury, see Section 9.7.

9.9.3 Neuro-psychological Methods

This is a difficult area to summarise, especially for one not trained in psychological techniques. The reader is advised to consult Raub and Benignus (2002)[59] for a critical review of published work. The majority of studies have been of patients or volunteers exposed to rather low concentrations of CO. The question has been: does low level exposure produce subtle effects that can only be detected by use of test batteries including, for example, tests of visual perception, manual dexterity, learning capacity, attention level and driving performance? Raub and Benignus's analysis suggests that a %[COHb] of 15–20% is needed before a 10% decrement in performance in these tests can be recognised with confidence. JS Haldane's aforementioned report of effects confirms this (Table 9.5). Speaking of effects at lower %[COHb] they concluded, "It is [also] reasonable [to conclude on the basis of the evidence] that none of the reports of CO behavioural effects in humans are, without further work, entirely credible." (Brackets [ ] are ours.) This view was put in the EHC report on CO (1999),[1] to which Raub contributed. The reader is referred to this report in which the evidence of behavioural effects is reported and discussed in detail (see also Gordon and Mercandetti, 1989).[83] In this chapter we have not been concerned, particularly, with the effects of exposure to low concentrations of CO and will not pursue their possible behavioural consequences any further.

9.10 Management of Carbon Monoxide Poisoning in Association with Smoke Inhalation

Most deaths due to fire are caused by inhalation of smoke or fumes with carbon monoxide being identified as one of the most important, if not the most significant toxin. This status is due to the generation by incomplete combustion in fires of high concentrations of carbon monoxide and its well-established toxicity. The perceived importance of carbon monoxide in the toxicology of fire smoke is also undoubtedly due, at least in part, to its relative ease and speed of analysis both as a gas and in the blood stream bound to haemoglobin with bed-side monitoring being possible. The reaction in blood and the formation of carboxyhaemoglobin reflects the reduction of the blood's oxygen carrying capacity, to which is attributed the major degree of the toxic effects of the gas. As such an obvious and easily monitored therapeutic target, it is not surprising that carbon monoxide poisoning in smoke inhalation victims attracts the attention of the medical and nursing team.

The treatment of patients poisoned by carbon monoxide is, in principle, extremely simple: remove the patient from the source of exposure and give oxygen, lots of oxygen.[84,85] What could be more straightforward? The treatment of carbon monoxide poisoning in the context of smoke inhalation, unfortunately, is complicated by a number of concurrent threats for the victim. While carbon monoxide poisoning where the source of the gas is an indoor charcoal barbecue may be relatively straightforward, that sustained by victims of a major conflagration may be complicated by thermal burns, the inhalation of corrosive gases and the possibility of prior intoxication, usually with alcohol. Whatever the situation, however, it is important to give high concentration oxygen *via* a close-fitting face mask and high flow oxygen source or by endotracheal tube. This will reduce the half-life of COHb from 4 hours or so (breathing air) to about 50 minutes. Oxygen should be given until the patient is alert.

The interaction of oxygen and carbon monoxide for binding sites on haemoglobin is competitive but carbon monoxide is the dominant ligand with an affinity for haemoglobin some 245 times greater than that of oxygen. Removal of the victim from the smoke allows oxygen to replace carbon monoxide, which is excreted in the exhaled breath. This is the most important event in rescue and treatment but, because of the binding characteristics of carbon monoxide, its half-life as carboxyhaemoglobin is 4.5–5.0 hours when breathing air, and the toxic effects of carbon monoxide persist for some time after rescue. This interlude of persistent toxicity can therefore be shortened by the administration of oxygen to accelerate the excretion of carbon monoxide. Theoretically, administration of 100% oxygen should reduce the half-life of carboxyhaemoglobin by a factor of five to 60 minutes. In practice, half-lives vary depending on the efficiency of oxygen delivery, from around a minimum of 50 minutes, but an average half-life of 74 minutes is observed.[86] Again, however, oxygen delivery to the tissues remains impaired for some time after rescue and the initiation of treatment. The competition of oxygen for binding sites on haemoglobin can be further amplified by increasing the partial pressure of inhaled oxygen to higher than ambient pressure. This requires compressing the patient who then breathes 100% oxygen; so-called hyperbaric oxygen therapy. At three atmospheres absolute (300 kPa, 2250 mm Hg or two atmospheres of pressure above ambient) the patient can breathe 100% oxygen without undue toxicity and the half-life of carboxyhaemoglobin is reduced to about 20 minutes. This effectively reverses the oxygen delivery impairment of carbon monoxide poisoning in a little over 60 minutes. Hyperbaric oxygen, however, does more than just accelerate the excretion of carbon monoxide.

While the bulk of oxygen is transported in the blood bound to haemoglobin, some is carried dissolved in plasma. The solubility of oxygen in plasma at 37 °C is about 0.225 ml kPa^{-1}. The relative quantities of oxygen carried in solution and on haemoglobin in blood are indicated in Figure 9.2. At rest, with a blood haemoglobin level of 150 g L^{-1}, the tissue extraction of oxygen from blood is in the region of 40 ml L^{-1} as the oxyhaemoglobin saturation falls from 95% to 75%. It can be seen that this amount of oxygen can be held in solution alone while breathing 100% oxygen at an

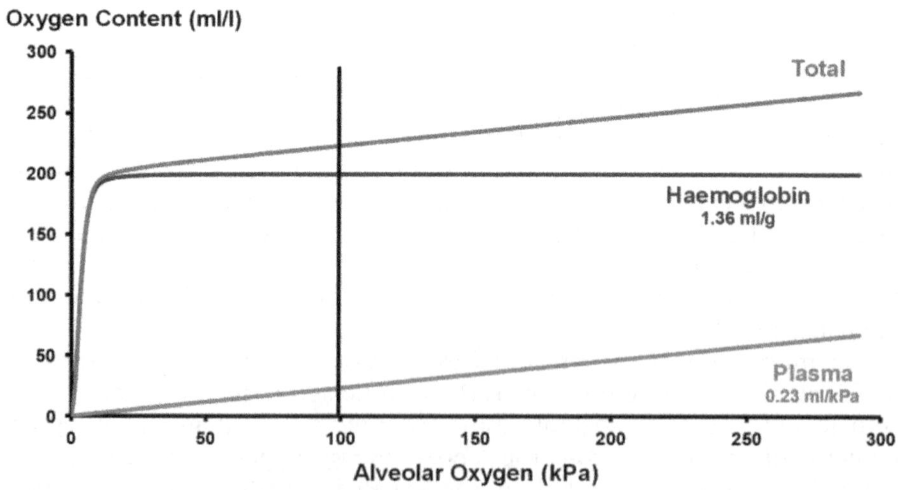

Figure 9.3 Oxygen content and alveolar oxygen level.

ambient pressure of above two atmospheres absolute (Figure 9.3). At this pressure and above, and with oxygen breathing, haemoglobin is not required for oxygen transport to the tissues. So, at three atmospheres absolute any tissue hypoxia due to reduced oxygen transfer by haemoglobin can be completely reversed in as short a time as it takes to compress the patient. The attractive option of both maximally accelerating carbon monoxide excretion and immediate reversal of hypoxia was first put into practice in Glasgow, where first two[87] and then 22 comatose or semi-comatose cases[88] of acute carbon monoxide poisoning were successfully treated with hyperbaric oxygen at a pressure of two atmospheres absolute. On average, cases were treated within 30 minutes of rescue with rapid and seemingly complete short term recovery. One case, however, treated several hours after rescue did not respond—the patient going on to die some days after rescue. No controlled trial was conducted, but it was considered at the time that the treatment was a potential life-saver with, therefore, better than expected short term recovery.

Blood dissolved O_2 content $= PO_2$ (kPa)$\times 0.225$ ml L^{-1};
$13.4\times 0.225 = 3.02$ ml L^{-1} in air;
$88.6\times 0.225 = 19.93$ ml L^{-1} 100% oxygen;
$188.6\times 0.225 = 42.435$ ml L^{-1} 2 atm oxygen;
$288.6\times 0.225 = 64.94$ ml L^{-1} 3 atm oxygen.

The effect of hyperbaric oxygen treatment on the long term neurological outcome for cases of carbon monoxide poisoning has now been the subject of several more recent randomised controlled clinical trials that have been subjected to metanalytical review within the Cochrane collaboration. This found "conflicting, potentially biased, and generally weak evidence regarding the usefulness of hyperbaric oxygen for the prevention of neurological injury".[89] It is fair to say, however, that the original application of the technique, to abbreviate short term treatment and shorten recovery time after poisoning, has not been subjected to randomised controlled trials. Because of this, a health technology assessment exercise concluded rather "that the evidence available would indicate that it is appropriate to include the use of hyperbaric oxygen therapy in algorithms for the management of carbon monoxide poisoning".[90] How, therefore, might hyperbaric oxygen therapy be integrated into an algorithm for the treatment of smoke inhalation?

Available information supports the concept that rapid hyperbaric oxygen therapy may speed recovery from carbon monoxide poisoning but does not produce any long term benefit. It follows from this that if hyperbaric therapy is immediately available it might be indicated as a useful option in patients with an indication for its use. In 2002 the UK Department of Health noted that there was debate about the added value provided by hyperbaric oxygen in the treatment of carbon monoxide poisoning, but that a COHb concentration of >20% might be reason to consider hyperbaric oxygen for victims who had lost consciousness at any stage, who had signs of neurological deficit, who had cardiac ischaemia or arrhythmia on electrocardiography (ECG) or who were pregnant. If it is to be used, hyperbaric therapy should dovetail into the overall management of the fire victim who may be subject to a number of other problems that may be life threatening in the short to medium term and that need to be controlled at the time when hyperbaric therapy might be most useful. Facial burns are an indication for early endotracheal intubation to secure the upper airway since the oedema and swelling that they cause may close the upper airway. Glottic or supra-glottic oedema can also compromise the upper airway and lead to death by suffocation. If this is caused by the inhalation of hot gases there may be perioral burns since the dry skin of the face is more likely to burn that the moist intra-oral mucosa. Corrosive gases in fire smoke, such as hydrogen chloride (released by low temperature pyrolysis of various materials but importantly electrical insulation), can cause chemical burns in the upper airway with high water solubility allowing them to target moist mucosal surfaces. There may be no sign of thermal injury but progressive upper airway obstruction occurs. Upper airway obstruction is a feature of the first few hours after rescue. Later on, the manifestations of deeper lung injury manifest with symptoms of cough with breathing difficulty and signs of bronchospasm, and blood stained sputum with progressive hypoxia as a chemical pneumonitis develops caused by the irritant material in smoke. At this stage, hypoxia is associated with increasing degrees of pulmonary shunting. This greatly reduces the effects of supplementing inhaled oxygen and any benefit that might be gained from hyperbaric oxygen is degraded. In addition, airway obstruction in these cases may make pulmonary barotrauma a risk of hyperbaric therapy, which then would be best avoided.

In the decision to treat carbon monoxide poisoning with hyperbaric oxygen other causes of neurological impairment associated with fire need to be considered. Intoxication with alcohol is common in fire victims and other agents may also be implicated, although less frequently. If carbon monoxide exposure is identified then hydrogen cyanide exposure is also highly likely, although laboratory assessment of any resulting intoxication is not yet available for the treating physician. As for carbon monoxide, there are relatively specific antidotes that can be used for cyanide intoxication with intravenous hydroxycobalamin being the latest advocated having a better therapeutic index than other agents.[91] Also in common with carbon monoxide, there is no evidence base indicating any clinical benefit from such treatment in fire victims.

9.11 Summary

In summary, carbon monoxide poisoning is an important cause of death and morbidity in the victims of fire. Acute toxicity is caused by conversion of haemoglobin to carboxyhaemoglobin with reduction of oxygen delivery to tissues, with acute loss of function most importantly affecting the brain and heart. Acute adaptation to poisoning occurs with vasodilatation and an increase in tissue blood flow, which offsets reduced blood oxygen content. Decompensation occurs when the cardiac output required can no longer be sustained and sustained poisoning leads to hypoxic tissue injury. More recent research has shown that carbon monoxide is an important cell messenger and is involved in cellular free radical metabolism. It is clear that exogenously applied carbon monoxide can disrupt such systems at moderate and low doses but the toxicological significance of this is as yet unclear. The treatment of carbon monoxide poisoning for surviving fire victims lies in the provision of general supportive care and must be integrated with management of the other well recognised problems faced by these patients. Early airway management is critical, as may be the management of thermal burns and poisoning with other agents such as alcohol and cyanide. Since intensive care management must be immediately available for fire victims, long transport times to specialist facilities should be avoided if possible. If such journeys are unavoidable, a critical care team should manage the transfer. Hyperbaric oxygen offers a theoretically attractive mode of treatment for patients with significant carbon monoxide poisoning, but only if delivered soon after rescue and before the onset of any smoke-induced pneumonitis. While such treatment may speed recovery, there is no evidence that there is any long term benefit. The treatment, therefore, does not justify transfer to a distant unit or any reduced access to critical care for the patient.

9.12 Abbreviations

O_2	Oxygen
CO	Carbon monoxide
CO_2	Carbon dioxide
M	Haldane coefficient
R	Warburg coefficient
COHb	Carboxyhaemoglobin
O_2HB	Oxyhaemoglobin
H^+	Hydrogen ion
e^-	The electron, negatively charged
H_2O	Water
PaO_2	The partial pressure of oxygen in arterial blood
PO_2	The partial pressure of oxygen
P_AO_2	The partial pressure of oxygen in alveolar air
$\bar{P}_{c,O_2}$	The mean partial pressure of oxygen in pulmonary capillary blood
PCO	The partial pressure of carbon monoxide
SaO_2	The percentage saturation of arterial blood with oxygen
$P_{A,CO}$	The partial pressure of carbon monoxide in alveolar air
$\bar{P}_{c,CO}$	The mean partial pressure of carbon monoxide in pulmonary capillary blood
MetHb	Metahemoglobin
NHb	Neuroglobin
Fe^{2+}	Ferrous ion

Fe^{3+}	Ferric ion
H_2S	Hydrogen sulfide
I_2O_5	Iodine pentoxide
I_2	Iodine
NO	Nitric oxide
NOS	Nitric oxide synthase
$\dot{V}_{CO}$	The rate at which carbon monoxide is lost from the lung
Dl_{co}	The diffusion coefficient of carbon monoxide in the lung
P_{I,O_2}	The partial pressure of oxygen in inspired air
P_B	The barometric pressure
$\bar{P}_{H_2O}$	The saturated vapour pressure of water at 37 °C
dCO/dt	The rate at which the body store of carbon monoxide changes
$\dot{V}_A$	The alveolar ventilation rate
V_b	Blood volume
t	Time
Mb	Myoglobin
ATP	Adenosine triphosphate
ADP	Adenosine diphosphate
AMP	Adenosine monophosphate
cGMP	Guanosine 3,5-monophosphate
Ca^{2+}	Calcium ion
Na	Sodium
F	Fluorine
$O_2^{\bullet-}$	Superoxide free radical
H_2O_2	Hydrogen peroxide

References

1. International Programme on Chemical Safety Environmental Health Criteria 213: Carbon Monoxide, 2nd edn, World Health Organization, Geneva, 1999, Available at http://www.inchem.org/documents/ehc/ehc/ehc213.htm, Accessed July 2012.
2. D. G. Penney, *Carbon Monoxide Poisoning*, ed. D. G. Penney, CRC Press, 2007.
3. US Environmental Protection Agency Air Quality Criteria for Carbon Monoxide (Final Report) http://cfpub.epa.gov/ncea/isa/recordisplay.cfm?deid=18163, Accessed February 2013.
4. T. Sollman, *A Manual of Pharmacology*, W B Saunders Company, 8th edn, 1957, pp. 973–979.
5. V. Babrauskis, B. C. Levin, T. C. Gann, M. Paabo, R. H. Harris, R. D. Peacock and S. Yusa, Toxic potency measurement for fire hazard analysis, U S National Institute of Standards and Technology, Gaithersburg, MD. NIST Special Publication, No. 827, 1991.
6. M. Touger, E. J. Gallagher and J. Tyrell, Relationship between venous and arterial carboxyhaemoglobin levels in patients with suspected carbon monoxide poisoning, *Ann. Emerg. Med.*, 1995, **25**, 481–483.

7. R. W. Mayes, Measurement of carbon monoxide and cyanide in blood, *J. Clin. Pathol.*, 1993, **46**, 982–988.
8. H. J. Vreman, J. J. Mahoney and M. D. Stevenson, Carbon monoxide and carboxyhaemoglobin, *Adv. Pediatr.*, 1995, **42**, 303–334.
9. M. Barinaga, Carbon monoxide: killer to brain messenger in one step, *Science*, 1993, **259**, 309.
10. D. W. Brann, G. K. Bhat, C. A. Lamar and V. B. Mahesh, Gaseous transmitters and neuroendocrine regulation, *Neuroendocrinology*, 1997, **65**, 385–395.
11. G. Marilena, New physiological importance of two classic residual products: carbon monoxide and bilirubin, *Biochem. Mol. Pharmacol.*, 1997, **61**, 136–142.
12. N. Prabhakar, NO and CO as second messengers in oxygen sensing in the carotid body, *Respir. Physiol.*, 1999, **115**, 161–168.
13. N. R. Prabhakar and G. L. Semenza, Gaseous messengers in oxygen sensing, *J. Mol. Med.*, 2012, **90**, 265–272.
14. T. Dolinay, M. Szilasi and M. Liu, Choi AMK. Inhaled carbon monoxide confers anti-inflammatory effects against ventilator-induced lung injury, *Am. J. Respir. Crit. Care Med.*, 2004, **170**, 613–620.
15. S. R. Thom, S. T. Ohnishi and H. Ischiropoulos, Nitric oxide released by platelets inhibits neutrophil B2 integrin function following acute carbon monoxide poisoning, *Toxicol. Appl. Pharmacol.*, 1994, **128**, 105–110.
16. Y.-M. Kim, H.-O. Pae, J. E. Park, Y. C. Lee, J. M. Woo, N.-H. Kim, Y. K. Choi, B.-S. Lee, S. R. Kim and H.-T. Chung, Hemeoxygenase in the regulation of vascular biology: from molecular mechanisms to therapeutic opportunities, *Antioxid. Redox Signalling*, 2011, **14**, 137–167.
17. M. Kajimura, R. Fukuda, R. M. Bateman, T. Yamamoto and M. Suemastu, Interactions of multiple gas-transducing systems: hallmarks and uncertainties of CO, NO and H_2S gas biology, *Antioxid. Redox Signaling*, 2010, **13**, 157–192.
18. G. D'Amico, F. Lam, T. Hagen and S. Moncada, Inhibition of cellular respiration by endogenously produced carbon monoxide, *J. Cell Sci.*, 2006, **119**, 2291–2298.
19. B. S. Zuckerman, B. K. Chin, M. Bilban, J. d'Avila, J. Rao, T. R. Billiar and L. F. Otterbein, Carbon monoxide signals via inhibition of cytochrome c oxidase and generation of mitochondrial reactive oxygen species, *FASEB J.*, 2007, **21**, 1099–1106.
20. J. B. S. Haldane, the dissociation of oxyhaemoglobin in human blood during partial CO poisoning, *J. Physiol.* 1912, **45**, pp. xxii–xxiv.
21. R. F. Coburn, R. E. Forster and P. B. Kane, Considerations of the physiological variables that determine the blood carboxyhaemoglobin concentration in man, *J. Clin. Invest.*, 1965, **44**, 1899–1910.
22. Biological effects of carbon monoxide, ed. R. F. Coburn, *Annals of New York Academy of Sciences,* 1970, vol. 174 , pp. 1–430.
23. R. F. Coburn, The carbon monoxide body stores, *Ann. N. Y. Acad. Sci.*, 1970, **174**, 11–22.

24. A. J. Hickman, Personal exposures to carbon monoxide and oxides of nitrogen, Transport and Road Research Laboratory, Research Report 206, 1989.
25. W. H. Forbes, F. Sargent and F. J. W. Roughton, The rate of carbon monoxide uptake by normal men, *J. Appl. Physiol.*, 1945, **143**, 594–608.
26. J. E. Peterson and R. D. Stewart, Predicting the carboxyhaemoglobin levels resulting from carbon monoxide exposures, *J. Appl. Physiol.*, 1975, **39**, 633–638.
27. International Programme on Chemical Safety Environmental Health Criteria 13: Carbon Monoxide, 1st edn, Available at http://www.inchem.org/documents/ehc/ehc/ehc013.htm, Accessed February 2013.
28. L. D. Longo, Carbon monoxide in the pregnant mother and fetus and its exchange across the placenta, *Ann. N. Y. Acad. Sci*, 1970, **174**, 313–341.
29. L. R. Goldbaum, T. Orellano and E. Dergal, Mechanism of the toxic action of carbon monoxide, *Ann. Clin. Lab. Sci.*, 1976, **6**, 372–376.
30. J. B. S. Haldane, Carbon monoxide as a tissue poison, *Biochem. J.*, 1927, **21**, 1068–1075.
31. J. S. Haldane, The action of carbonic oxide on man, *J. Physiol.*, 1895, **18**, 430–462.
32. Bernard Claude, An Introduction to the Study of Experimental Medicine, Dover edition, Dover Publications Inc., 1957, This is a reprint of the first English translation of 1927, Published originally in French in 1865.
33. J. F. Nunn, Hyperoxia and oxygen toxicity, Nunn's Applied Respiratory Physiology, Butterworth Heinemann, 4th edn, 1993, ch. 32, pp. 537–555.
34. J. W. Severinghaus, M. Stafford and A. M. Thunstrom, Estimation of skin metabolism and blood flow with $tcPO_2$ and tc PCO_2 electrodes by cuff occlusion of the circulation, *Actaanaesthscand.*, 1978, **22**(suppl. 68), 9–15.
35. J. W. Severinghaus, Simple accurate equations for human blood O_2 dissociation computations, *J Appl Physiol: Respir., Environ. Exercise Physiol.*, 1979, **46**, 599–602.
36. J. S. Haldane, The relation of the action of carbonic oxide to oxygen tension, *J. Physiol.*, 1895, **18**, 218–229.
37. F. J. W. Roughton, The equilibrium between carbon monoxide and sheep haemoglobin at very high percentage saturations, *J. Physiol.*, 1954, **126**, 359–383.
38. K. D. Jurgens, S. Papadopoulos, T. Peters and G. Gros, Myoglobin: just an oxygen store or also an oxygen transporter?, *News Physiol. Sci.*, 2000, **15**, 269–274.
39. T. Burmester, B. Weich, S. Reinhardt and T. Hankeln, A vertebrate globin expressed in the brain, *Nature*, 2000, **407**, 520–523.
40. J. T. Trent and M. S. Hargrove, A ubiquitously expressed human hexacoordinate haemoglobin, *J. BiolChem.*, 2002, **277**, 19538–19548.
41. J. B. Whittenberg and B. A. Wittenberg, Myoglobin function reassessed, *J. Exp. Biol.*, 2003, **206**, 2011–2020.

42. E. Grainger, B. Lassnig, A. Kuznetsov, G. Rieger and R. Margreiter, Mitochondrial oxygen affinity, respiratory flux control and excess capacity of cytochrome c oxidase, *J. Exp. Biol.*, 1998, **201**, 1129–1139.
43. P. Mitchell, Keilin's respiratory chain concept and its chemiosmotic consequences, *Science*, 1979, **206**, 1148–1159.
44. S. D. Brown and C. A. Piantadosi, In vivo binding of carbon monoxide to cytochrome c oxidase in rat brain, *J. Appl. Physiol.*, 1990, **68**, 604–610.
45. K. N. Iheagwara, S. R. Thom, C. S. Deutschman and R. J. Levy, Myocardial cytochrome oxidase activity is decreased following carbon monoxide exposure, *Biochim. Biophys. Acta*, 2007, **1772**, 1112–1116.
46. C. A. Piantadosi, Diagnosis and treatment of carbon monoxide poisoning, *Respir. Care Clin. North Am.*, 1999, **5**, 183–202.
47. O. Miro, J. Casademont, A. Barrientos, A. Urbano-Marquez and F. Cardellach, Mitochondrial cytochrome c oxidase inhibition during acute carbon monoxide poisoning, *Pharmacol. Toxicol.*, 1998, **82**, 199–202.
48. A. F. Castoldi, T. Coccini, G. Randine, M. Hernandez-Viadel, V. Felipo and L. Manzo, Lymphocyte cytochrome c oxidase, cyclic GMP and cholinergic receptors as peripheral indicators of carbon monoxide neurotoxicity after acute and repeated exposure in the rat, *Life Sci.*, 2006, **78**, 1915–1924.
49. L. R. Goldbaum, T. Orellano and E. Dergal, Studies on the relation between carboxyhemoglobin concentration and toxicity, *Aviat., Space Environ. Med.*, 1977, **48**, 969–970.
50. D. L. Drabkin, F. H. Lewey, S. Bellet and W. H. Ehrich, The effect of replacement of normal blood by erythrocytes saturated with carbon monoxide, *Am. J. Med. Sci*, 1943, **205**, 755–756.
51. R. F. Coburn, Mechanisms of carbon monoxide toxicity, *Prev. Med.*, 1979, **8**, 310–322.
52. D. F. Gorman and W. B. Runciman, Carbon monoxide poisoning, *Anaesth., Intensive Care*, 1991, **19**, 506–511.
53. D. Gorman, A. Drewry, Y. L. Huang and C. Sames, The clinical toxicology of carbon monoxide, *Toxicology*, 2003, **187**, 25–38.
54. W. W. Runciman and D. F. Gorman, Carbon monoxide poisoning: from old dogma to new uncertainties, *Med. J. Aust.*, 1993, **158**, 439–440.
55. A. K. Sinha, J. Klein, P. Schultze, J. Weiss and H. R. Weiss, Cerebral regional capillary perfusion and blood flow after carbon monoxide exposure, *J. Appl. Physiol.*, 1991, **71**, 1196–1200.
56. S. R. Thom, M. Kang, D. Fisher and H. Ischiropoulos, Release of glutathione from erythrocytes and other markers of oxidative stress in carbon monoxide poisoning, *J. Appl. Physiol*, 1997, **82**, 1424–1432.
57. A. A. Pryor, Oxy-radicals and related species: their formation, lifetimes and reactions, *Ann. Rev. Physiol*, 1986, **48**, 657–667.
58. C. A. Piantadosi, M. S. Carraway and H. B. Suliman, Carbon monoxide, oxidative stress, and the mitochondrial permeability pore transition, *Free Radical Biol. Med.*, 2006, **40**, 1332–1339.

59. J. A. Raub and V. A. Benignus, Carbon monoxide and the nervous system, *Neurosci. Biobehav. Rev.*, 2002, **26**, 925–940.
60. S. R. Thom, Leukocytes in carbon monoxide-mediated brain oxidase injury, *Toxicol. Appl. Pharmacol.*, 1993, **123**, 234–247.
61. S. R. Thom, Y. A. Xu and H. Ischiropoulos, Vascular endothelial cells generate peroxynitrite in response to carbon monoxide exposure, *Chem. Res. Toxicol*, 1997, **10**, 1023–1031.
62. C. A. Piantadosi, L. Tatro and J. Zhang, Hydroxyl radical production in the brain after CO hypoxia in rats, *Free Radical Biol. Med.*, 1995, **18**, 603–609.
63. R. J. Traystman, J. R. Kirsch and R. C. Koehler, Oxygen free mechanisms of brain injury following ischaea and reperfusion, *J. Appl. Physiol.*, 1991, **71**, 1185–1195.
64. T. Kristian and B. K. Siesjo, Calcium and ischemic cell death, *Stroke*, 1998, **29**, 705–718.
65. P. Langston, D. Gorman, W. Runciman and R. Upton, The effect of carbon monoxide on oxygen metabolism in the brains of awake sheep, *Toxicol*, 1996, **114**, 223–232.
66. S. M. Ayres, S. Giannelli and H. Mueller, Myocardial and systemic responses to carboxyhemoglobin, *Ann. N. Y. Acad. Sci.*, 1970, **174**, 268–293.
67. F. L. Lowe-Ponsford and J. A. Henry, Clinical aspects of carbon monoxide poisoning, *Adverse Drug React. Acute Poisoning Rev.*, 1989, **8**, 217–240.
68. J. S. Haldane and J. G. Priestly, *Respiration*, Oxford, Clarendon Press, 1935.
69. F. J. Miller, P. M. Schlosser and D. B. Janszen, Haber's rule: a special case in a family of curves relating concentration and duration of exposure to a fixed level of response for a given endpoint, *Toxicology*, 2000, **149**, 21–34.
70. Health and Safety Executive, Toxicity levels of chemicals: Assessment of the Dangerous Toxic Load (DTL) for Specified Level of Toxicity (SLOT) and Significant Likelihood of Death (SLOD), Available at http://www.hse.gov.uk/chemicals/haztox.htm (last accessed 12 February 2013).
71. D. R. Sommerville, K. H. Park, M. O. Kierzewski, M. D. Dunkel, M. I. Hutton, N. A. Pinto, Toxic Load Modelling, in *Inhalation Toxicology*, ed. H. Salem and S. A. Katz, Taylor and Francis, 2nd edn, 2006, pp. 137–158.
72. J. J. McGrath, Biological plausibility for carbon monoxide as a copollutant in PM epidemiological studies, *Inhalation Toxicol.*, 2000, **12**, 91–107.
73. Z. Gu, A. J. Januszkiewicz, M. A. Mayorga, G. D. Coleman and C. R. Morrissette, Consequences of brief exposure to high concentrations of carbon monoxide in conscious rats, *Inhalation Toxicol.*, 2005, **17**, 755–764.
74. J. H. Comroe Jr., Physiology of Respiration. *Year Book Medical Publishers Inc.* 2nd edn, 1975.
75. D. A. Purser and K. R Berrill, Effects of carbon monoxide on behaviour in monkeys in relation to human fire hazard, *Arxh. Environ. Health*, 1983, **38**, 308–315.

76. M. D. Ginsberg, Carbon monoxide intoxication: clinical features, neuropathology and mechanism of injury, *Clinical. Toxicol.*, 1985, **23**, 282–288.
77. R. G. Feldman, Occupational and Environmental Neurotoxicology, Lippincott-Raven, 1999, ch. 22, Carbon monoxide, pp. 378–399.
78. E. Hill and C. B. Semerak, Changes in the brain in gas (carbon monoxide) poisoning, *JAMA*, 1918, **71**, 644–648.
79. R. R. Grinker, Parkinsonism following carbon monoxide poisoning, *J. Nerve Ment. Dis.*, 1926, **64**, 18–28.
80. A. Meyer, The selective regional vulnerability of the brain and its relation to psychiatric problems, *Proc Roy Soc Med*, 1936, **29**, 1175–1181.
81. J. Lapresle and M. Fardeau, The Central Nervous System and Carbon Monoxide Poisoning, II, Anatomical study of brain lesions following intoxication with carbon monoxide (22 cases), *Progr. Brain Res.*, 1967, **24**, 31–75.
82. R. S. Jaeckle and H. A. Nasrallah, Single case study: major depression and carbon monoxide induced parkinsonism: diagnosis, computerised axial tomography, and response to L-dopa, *J. Nerve Ment. Dis.*, 1985, **173**, 503–508.
83. M. F. Gordon and M. Mercandetti, Carbon monoxide poisoning producing purely cognitive and behavioural sequelae. Neuropsychiatry, *Neuropsychol. Behav. Neurol.*, 1989, **2**, 145–152.
84. M. V. Balzan, G. Agius and A. G. Debono, Carbon monoxide poisoning: easy to treat but difficult to recognise, *Postgrad Med Journal*, 1996, **72**, 470–473.
85. K. R. Hardy and S. R. Thom, Pathophysiology and treatment of carbon monoxide poisoning, *Clin. Toxicol*, 1994, **32**, 613–629.
86. L. K. Weaver, S. Howe, R. Hopkins and K. J. Chan, Carboxyhemoglobin half-life in carbon monoxide-poisoned patients treated with 100% oxygen at atmospheric pressure, *Chest*, 2000, **117**, 801–808.
87. G. Smith and G. R. Sharp, Treatment of carbon-monoxide poisoning with oxygen under pressure, *Lancet*, 1960, 905–906.
88. G. Smith, I. M. Ledingham, G. R. Sharp, J. N. Norman and M. E. Bates, Treatment of coal-gas poisoning with oxygen at 2 atmospheres absolute, *Lancet*, 1962, 816–819.
89. N. A. Buckley, D. N. Juurlink, G. Isbister, M. H. Bennett, E. J. Lavonas, Hyperbaric oxygen for carbon monoxide poisoning (Review), *The Cochrane Collaboration*, John Wiley & Sons, Ltd. 2011.
90. K. Ritchie, S. Baxter, J. Craig, K. Macpherson, L. Mandava, H. McIntosh and S. Wilson, The clinical and cost effectiveness of hyperbaric oxygen therapy HTA programme: Systematic Review 2 – July 2008, NHS Quality Improvement Scotland, 2008, pp. 14–16.
91. P. Lawson-Smith, E. C. Jansen and O. Hyldegaard, Cyanide intoxication as part of smoke inhalation – a review on diagnosis and treatment from the emergency perspective, Scandinavian J. Trauma Resusc. Emerg. Med. 2011; 19: pp. 14–18, Available at http://www.sjtrem.com/content/19/1/14 (accessed May 2015).

CHAPTER 10

Hydrogen Cyanide—Physiological Effects of Acute Exposure during Fires

DAVID A. PURSER

Hartford Environmental Research, Hatfield, UK
Email: david-purser@ntlworld.com

10.1 Introduction

10.1.1 Contribution of Hydrogen Cyanide to Incapacitation and Death in Fires

The two most important asphyxiant gases produced in fires are hydrogen cyanide (HCN) and carbon monoxide (CO). Because CO is always present in fires it is useful to consider the contribution of HCN in relation to that of CO.

Almost all fuels involved in fires have a high carbon content, mostly in the 50–90% range by mass,[1] so that large amounts of carbon monoxide are produced, especially in the under-ventilated (vitiated) combustion conditions occurring during fires in buildings (see Chapters 2 and 3).[2–4] Carbon monoxide is a major cause of asphyxia and incapacitation during fires, and most fire fatalities have lethal levels of carboxyhaemoglobin (%COHb) in their blood.[5] The distribution of %COHb levels in fire victims is similar to that measured post-mortem in non-fire victims of carbon monoxide gas exposure[6] (see Chapter 14), indicating that carbon monoxide is the main ultimate cause of acute deaths in smoke victims at a fire scene or immediately after rescue.

Issues in Toxicology No. 23
Toxicology, Survival and Health Hazards of Combustion Products
Edited by David A. Purser, Robert L. Maynard and James C. Wakefield

Published by the Royal Society of Chemistry, www.rsc.org

In contrast to this, nitrogen is absent from many fuels involved in fires, and when present is typically at low levels compared to carbon (from around 0.1% by mass in wood to 26% in polyacrylonitrile).[7] Cyanide is not routinely measured in the blood of fire victims, but when found is usually at low concentrations, and when at potentially toxic levels is always found in combination with high blood %COHb levels.[8–10]

As a result of the above, the importance of HCN in fires is often unrecognised, but in reality it is likely to make a major contribution to incapacitation and deaths in many fires for the following reasons:

- Certain materials commonly used in the contents of buildings and transport vehicles, especially for upholstered furniture and bedding (fabrics used for covers and polyurethane foam cushioning materials) often have a significant fuel nitrogen content (~10–26%).[2,7]
- When these materials are burned (especially in under-ventilated fires), the extent of conversion of fuel nitrogen to hydrogen cyanide is similar to the extent of conversion of fuel carbon to CO.[2,5,7,11]
- HCN has been measured at high concentrations of 1000 ppm† or more during the first few minutes of typical domestic fires and larger conflagrations such as aircraft fires.[2,12,13]
- Hydrogen cyanide causes loss of consciousness from asphyxia within a minute or so at concentrations above ~200 ppm.[14] It shows considerable deviations from Haber's rule so that *Ct* exposure doses for incapacitation are between 10–100 times lower than that from CO (depending on the HCN concentration).[5]
- Whereas most CO remains in the blood as carboxyhaemoglobin and is then very stable in cadavers and stored blood, the dynamics of uptake into the blood and dispersal into the tissues result in blood cyanide concentrations after exposure often being much lower than at the time of incapacitation.[15,16] Blood and tissue cyanide levels then decrease rapidly (~50% within 24 hours) in cadavers, and over periods of weeks in stored blood (depending on the conditions of storage), so are likely to be significantly lower by the time of measurement than at the time of death, and much lower than at the time of incapacitation.[5,16,17]
- Although the distribution of %COHb levels in fire smoke fatalities and CO poisoning victims are similar, the mode of the fire data set is lower than that for the CO data set, indicating some contribution to smoke toxicity in addition to the CO, which may include the effects of cyanide and other toxic fire gases and particulates (see Chapter 7).

In full-scale reconstructions of house fires, fires in large buildings and aircraft fires, the high HCN concentrations achieved within a few minutes of ignition are capable of causing loss of consciousness in exposed subjects

†Throughout this chapter, "ppm" represents the volume fraction (mL HCN per mL air)$\times 10^{-6}$.

within seconds, several minutes before blood carboxyhaemoglobin levels become high enough to cause incapacitation.[2] HCN can therefore be considered a major cause of rapid incapacitation in fires, preventing escape, so that subjects remain at the fire scene to die a few minutes later as a result of burns or asphyxia (due mainly due to CO poisoning with a contribution from HCN).

Although blood cyanide levels measured post-mortem in fire victims are generally rather low compared with levels associated with lethality in animal studies, in one particular study, high blood cyanide levels were measured when fresh blood samples were taken from fire survivors of domestic fires as they were rescued at the fire scene.[18] High cyanide concentration levels were also measured in the blood of fatalities from the Manchester Airtours aircraft fire.[19] Other studies have shown loss of cyanide in stored blood,[17,20,21] so that the importance of HCN in fires may be generally underestimated.

There are therefore problems with the interpretation of blood cyanide data because the dynamics of uptake, distribution and removal from the blood are not well understood, so that it is difficult to relate clinical signs to blood levels, especially some time after an exposure has occurred.[14,15] A further complication with assessing the effects of different blood cyanide levels is that, whereas with CO the main toxic effects are related to the formation of carboxyhaemoglobin and its effects on the carriage and delivery of oxygen in the blood, with cyanide the target is mitochondrial cytochrome in the tissues of the affected organs (particularly the brain and heart).

10.2 Mechanism of HCN Toxicity and Interactions with Carbon Monoxide and Oxygen

The main effect of hydrogen cyanide is inhibition of mitochondrial electron transport by binding with cytochrome ass3 and c + c1.[22,23] This causes a strong inhibition of tissue oxygen consumption and oxidative phosphorylation. The resultant hypoxia therefore relates to an inability to use oxygen rather than its absence. As long as respiration is maintained the blood remains highly oxygenated, giving venous blood a bright red appearance and oxygen content close to that of arterial blood.[14] The active agent is free cyanide both as cyanide ions and covalent hydrogen cyanide (CN^- and HCN) in plasma and tissue fluids (both extracellular and intracellular fluids).

The severity of toxic effects may not be directly related to whole blood cyanide concentrations.[5,14,22] HCN penetrates erythrocytes and has some affinity with them. Cyanide binds with methaemoglobin (also known as "ferrihaemoglobin") to form cyanomethaemoglobin, which removes some free cyanide from circulation, and some cyanide is considered to be protein-bound in the plasma.

If the main toxic effects of CO are on oxygen transport and delivery, while cyanide impairs oxygen consumption, it is not immediately obvious why the toxicities of CO and HCN should be additive when the two gases are inhaled together as a mixture. The observed extent of asphyxia might depend on which of the two gases was exerting the greater toxicological effect and thus acting as the limiting factor restraining oxidative metabolism. However, some degree of interaction might be predicted. If the main component was CO then tissue hypoxia should result from impaired oxygen delivery and low tissue PO_2. Introduction of cyanide may then further limit the metabolism of this limited oxygen supply, giving an overall additive effect on oxidative metabolism although increasing tissue PO_2. If cyanide were the main component then there would be an excess of oxygen present due to impaired consumption by cyanide, with high venous oxyhaemoglobin concentrations and high tissue PO_2. The introduction of CO may initially have little effect, but with increasing impairment of oxyhaemoglobin dissociation, it is possible that a decrease in tissue PO_2 may also result in an additive effect on the impairment of tissue respiration.

In practice, results of experimental exposures in animals to CO and HCN mixtures do show a considerable degree of additivity.[24–29] Experiments in rats involved a series of exposures to different concentrations of CO and HCN individually to determine the relationships between exposure concentrations and times to incapacitation and death. From these, *Ct* fractional effective dose (FED) expressions were derived to predict incapacitating and lethal exposure doses. The models were then used to calculate incapacitating and lethal doses for combinations of both gases for exposure periods of 1 to 30 minutes, which were then compared with experimentally determined incapacitating and lethal doses for the same mixtures of the two gases. The results show direct additivity. For example, exposure to $0.5\times$a lethal dose of CO plus $0.5\times$a lethal dose of HCN resulted in an approximate lethal concentration for 50% of the subjects (LC_{50}) for the mixture. This was also true for incapacitating doses and for different combination mixtures, providing the sum of fractional doses was 1.[26–28]

Another reason why CO and HCN may show additivity in terms of time and dose with regards to incapacitation and death may be that the hyperventilation induced by HCN inhalation increases the rate of uptake of any CO present, resulting in increased carboxyhaemoglobin concentrations. Another may be some degree of interaction at the mitochondrial level. Due to its high affinity for haemoglobin and myoglobin, most inhaled CO is sequestered in the blood, but to the extent that some penetration of free CO into the mitochondria occurs; this does interfere with cytochrome electron transport as does HCN.[23] There could therefore be some additivity between effects of CO and HCN on oxygen metabolism.

Other factors may be related to effects on carbon dioxide, blood pH and cerebral circulation. As discussed in Section 10.6.3, the initial episode of hyperventilation occurring during inhalation of HCN can result in a transient hypocapnia and respiratory alkalosis, which subsequently gives way to metabolic acidosis as the hyperventilation subsides and cyanide-induced inhibition of oxidative metabolism results in increased lactate formation.

Another interaction of interest is that between oxygen and cyanide. Since the PO_2 of venous blood is high during cyanide poisoning from inhalation of

HCN in air, the administration of normobaric 100% oxygen would not be expected to be of clinical benefit following cyanide inhalation. In practice, administration of oxygen has been found to improve the clinical condition of cynomolgus monkeys during recovery following incapacitating exposures to HCN,[11,30] and other studies have shown improved survival, and beneficial effects on cerebral and heart function.[31] Normobaric oxygen has not been shown to displace cyanide from cytochrome oxidase and did not improve mitochondrial oxidative processes in the renal cortex of rabbits during cyanide poisoning.[32] However, Klimmek *et al.*[31] suggest that normobaric oxygen does improve circulation and aerobic metabolism in the brain or other parts of the body after poisoning by cyanide. Administration of hyperbaric oxygen has been found to displace cyanide from the tissues back into the blood after cyanide poisoning, as well as improving mitochondrial oxidative processes.[32] Another finding is that administration of cyanide even at lethal doses resulted in only a limited extent of inhibition of cytochrome c oxidase, to maximal levels in the brain of 40% and heart of 60%.[33] Also, the extent of inhibition of cytochrome oxidase does not equate with the extent of inhibition of cellular respiration. In *in vitro* systems, large effects on cellular respiration required more than 50% inhibition of cytochrome oxidase activity. It has been suggested that inhibition of cytochrome c oxidase may not be the ultimate biochemical lesion in cyanide-induced lethality.[33]

These findings indicate that the relationship between tissue PO_2 and both the biochemical and physiological effects of cyanide are not simple, or indeed fully understood. From the perspective of combustion toxicology this is important because the atmospheres to which fire victims are exposed contain a mixture of CO, CO_2, HCN and low oxygen. The extent to which CO and low inspired oxygen combine with the effects of PCO_2, pH and circulatory changes to produce tissue hypoxia, and the interactions of the resultant PO_2 with tissue cyanide levels on tissue respiration, may have important implications for the onset of incapacitation during fires. The effects of normobaric and hyperbaric oxygen may be important with respect to post-exposure emergency treatment.

As described in the following sections, the uptake of HCN tends to be self-limiting, and the dynamics of distribution within the body may result in severe but relatively short-lived incapacitating effects over time-scales of a few minutes during exposures typical of fires, being followed by less severe effects over longer time-scales.[5,11,14,15,25,30] For these reasons it is likely that while HCN is a major cause of incapacitation during exposure it may be less important as a direct cause of death.

10.3 Physiological Effects of Acute HCN Exposure

The physiology of incapacitation by HCN inhalation was studied in a series of experiments on cynomolgus monkeys (*Macaca fascicularis*) seated in chairs and exposed for 30 minute periods *via* a face-mask to sub-lethal concentrations of HCN (60–200 ppm). These were either generated from a cylinder or as thermal decomposition atmospheres containing HCN (from polyacrylonitrile and polyurethane foams—see Chapter 7).[14,25,30] A set of physiological parameters was measured continuously during three successive 30 minute periods (pre-exposure, exposure and recovery). Respiratory rate, flow and volume were measured and used to calculate respiratory minute volume ($\dot{V}_E$). The electrocorticogram (ECoG) was monitored as an indication of cerebral function and the electrocardiogram for measurement of heart rate and rhythm, with waveform changes used as an indication of myocardial hypoxia. Clinical signs and reflexes were also monitored at intervals. During some experiments blood samples were taken at intervals from a femoral artery for blood cyanide analysis.

Figure 10.1 shows an example of a typical response pattern. Initially, on inhalation of these HCN mixtures, there was no response from the animal (no signs of irritancy or other effects), then at some stage during the exposure, an episode of hyperventilation occurred consisting of increases in both respiratory rate and tidal volume. The hyperventilation increased the rate of HCN uptake, further stimulating respiration, so that $\dot{V}_E$ increased by a factor of two to three. At the peak of this episode the animal lost consciousness, showing a large increase in slow-wave (delta) brain activity, a decrease in fast (beta) activity, a 50% decrease in heart rate, cardiac arrhythmias and T-wave changes consistent with myocardial hypoxia. As the subject became unconscious the hyperventilation decreased, thereby decreasing the rate of HCN uptake. Over the next few minutes the condition of the animal improved and it partly recovered consciousness, with improved reflexes, despite the continued HCN exposure. After this stage it went into a slow decline and deeper unconsciousness, with gradually decreasing respiration towards the end of the exposure period (which was considered to be leading to eventual apnoea if the exposures had been continued). During the recovery period, the animal fully recovered consciousness after a few minutes breathing air, and physiological parameters [ECoG, heart rate and electrocardiography (ECG)], returned to normal. There was a rebound hyperventilation throughout most of the recovery period.

The lowest exposure concentration producing loss of consciousness within the 30 minute exposure period was 90 ppm and the highest exposure was to 200 ppm. At the higher concentrations of around 150–200 ppm the hyperventilatory episode was marked and started within a minute or so of the beginning of the exposure, as in Figure 10.1. At lower concentrations the increase in ventilation was more gradual, with a brief hyperventilatory episode occurring towards the end of the exposure.

This pattern of developing incapacitation from inhaled HCN is rather different from that induced by oral or intravenous cyanide, in that during the early stages the rate of uptake is enhanced by the positive feedback episode of hyperventilation, but during the later stages the exposure becomes, to some extent, self-limiting, in that the rate of uptake decreases progressively as $\dot{V}_E$ decreases. It is for this reason that HCN may present a cause of rapid incapacitation during fires, but is not obviously a major cause

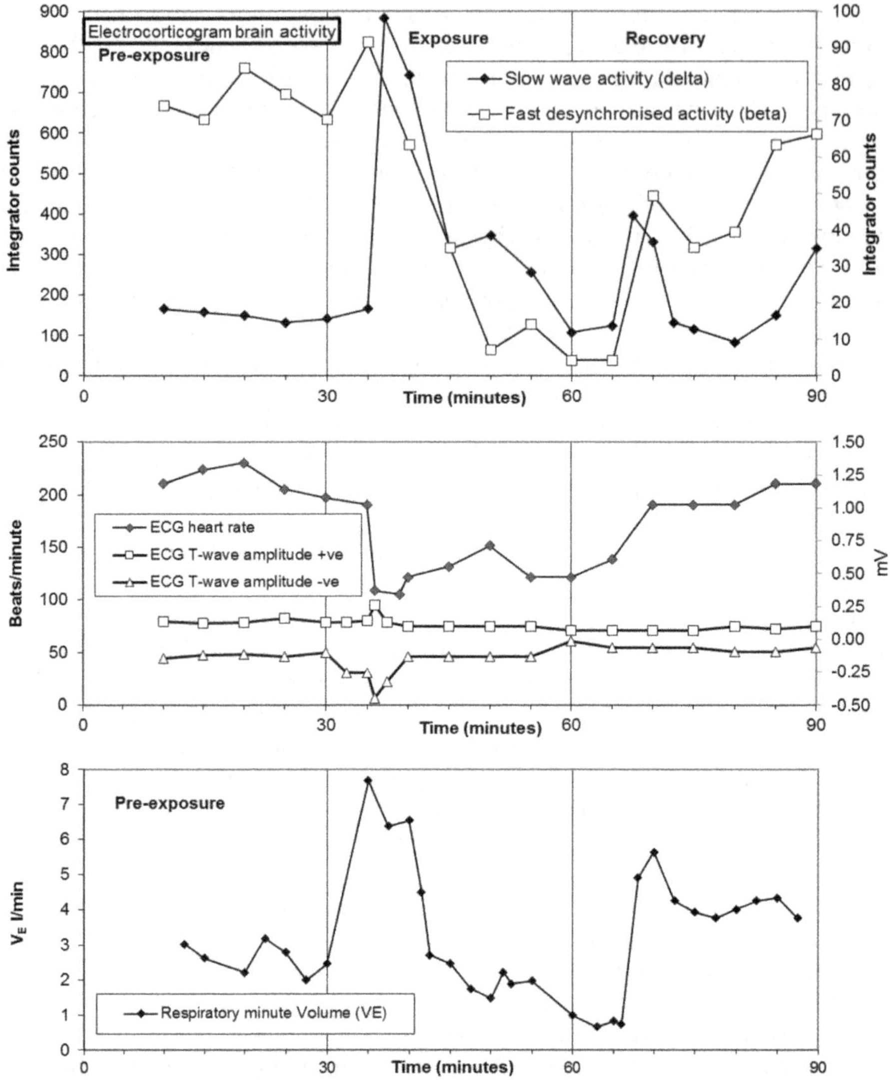

Figure 10.1 Physiological effects of exposure to HCN gas (147 ppm) on a macaque monkey.[14]

of death during fires. A likely scenario is that fire victims are overcome, principally by HCN during the early minutes of a fire, and collapse unconscious. They then continue to inhale both HCN and CO until the %COHb level becomes sufficient (with an additive HCN component) to cause apnoea, circulatory collapse and death. In the relative concentration ranges for CO and HCN typical of fires (CO ≈ 1000–20 000 ppm and HCN ≈ 0–1000 ppm), although HCN may cause incapacitation within a few minutes, approximately 30 minutes or more may be required for the inhaled cyanide dose to reach levels causing circulatory and respiratory failure. When these 4–5 kg cynomolgus macaque monkeys were exposed to different concentrations of HCN between 87 and 200 ppm, the times to incapacitation varied from 30 to 0.9 minutes (*Ct* exposure doses from 2610–270 ppm min), but all were alive, although unconscious, after 30 minute exposure doses of up to ~6000 ppm min. (See Chapter 7, Figure 7.25.) For CO, incapacitation occurred for any concentration above 1000 ppm after inhalation of a constant exposure dose of 27 000 ppm min in active animals (39 000 ppm min in resting animals). At the high CO concentrations typical of fires it is therefore likely that although HCN may cause incapacitation before CO, lethal doses of CO are likely to be inhaled before lethal doses of cyanide.

10.4 Dose–Response Relationships for HCN Incapacitation

If incapacitation by HCN followed Haber's rule, then the development of a calculation model for uptake and time to incapacitation would be a simple matter of determining the appropriate constant *Ct* exposure dose for each animal species. In practice, as noted by Haber in his 1920s lectures,[34] the inhalation toxicity of HCN shows a considerable departure from this ideal behaviour. Figure 10.2 shows data for time to incapacitation as a function of inhaled HCN concentration for resting 4 kg primates (*Macaca fascicularis*). The data set plotted includes exposure to HCN gas in air, HCN generated by thermal decomposition of polyacrylonitrile,[14] and both flexible and rigid polyurethane foams.[25] The similarity of the results for HCN gas and HCN generated from the polymers indicates that HCN is the principle toxic agent producing incapacitation in these thermal decomposition atmospheres. Also plotted are data for time to incapacitation from HCN exposure in rats resting[26] or exercising in a wheel.[35,36] For resting rats, time to incapacitation at a given HCN concentrations is similar to, but somewhat longer than in primates, while for exercising rats it is slightly shorter than for the resting primates. Since rats have a much smaller body size than primates (and therefore a greater rate of oxygen consumption and $\dot{V}_E$ per gram bodyweight), it might be predicted that rats should be overcome more rapidly than primates at a given HCN concentration. In practice rats are more resistant to hypoxia than primates, so that, as for intoxication by CO and simple low oxygen hypoxia, they are able to tolerate higher inhaled doses (see discussion in Chapter 7). When rats are forced to exercise, the resulting increase in $\dot{V}_E$ and metabolism results in them being overcome considerably earlier, and it is predicted that similar differences would occur between resting and exercising humans and other primates.

Figure 10.2 is plotted with time to incapacitation as the dependent variable (y axis), enabling a curve to be fitted for calculation of time to incapacitation in monkeys as follows:

$$t_{IN} = \frac{1.2 \times 10^6}{[CN]^{2.36}}, \tag{10.1}$$

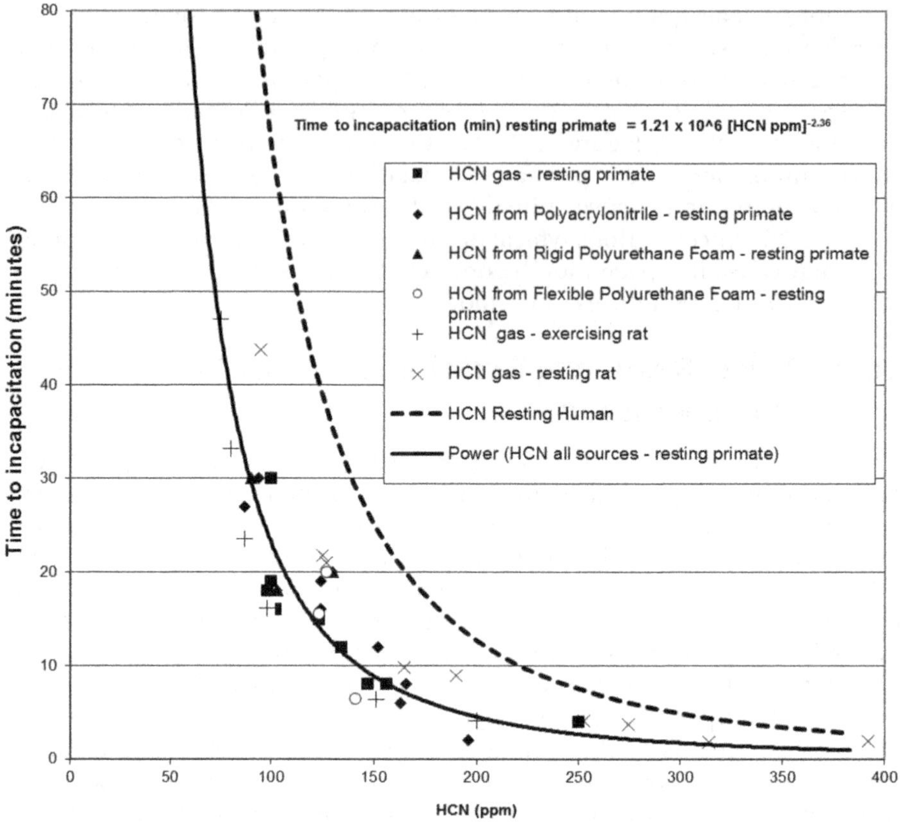

Figure 10.2 Measured times to incapacitation as a function of HCN concentration in macaque monkeys, resting and exercising rats, and predicted for humans.[14,25,26,30,35]

where t_{IN} is the time to incapacitation (minutes) and [CN] is the inhaled HCN concentration. This exponential expression gives a reasonably good fit with the data (regression coefficient = 0.84).

In order to derive an expression for time to incapacitation for humans it is considered that the monkey data provide the most useful model. Because the monkeys were much smaller animals than adult humans (approximately 4 kg *versus* 70 kg for humans), at a given level of physical activity and respiration it is predicted that the rate of uptake and therefore time to incapacitation would be somewhat shorter than for an adult human. Comparing the resting metabolic rates for monkeys and humans as a function of body size it is estimated that the time required for a resting adult human to inhale the same dose as a monkey would be increased by approximately a factor of 2.83 (plotted as the dashed line in Figure 10.2), although for a human infant the relationship would be approximately the same as for the monkey.

Table 10.1 Incapacitation and death by inhaled hydrogen cyanide.

HCN/ppm	Species and effect	Author
100	Loss of consciousness after 20–30 minutes in primates and humans	Purser[14], Kimmerle[37]
200	Loss of consciousness after approximately 2 minutes	Purser[14], Kimmerle[37]
300 +	Death occurs "rapidly"	Kimmerle[37]
444	A man survived an accidental exposure	Bonsall[38]
530	A man, Barcroft, survived a 1.5 minute exposure – his dog exposed at the same time died	Barcroft[39]
539	Suggested 10 minute LC_{50} in humans	McNamara[40]
1000	One breath may cause loss of consciousness	Purser[5]

Data on human exposures to HCN are limited but Kimmerle[37] quotes some approximate data showing a similar effect in humans to that in monkeys, with incapacitation occurring after 20 to 30 minutes at 100 ppm HCN and after 2 minutes at 200 ppm, death occurring rapidly at concentrations exceeding approximately 300 ppm (Table 10.1)

Other data suggest that human victims might be able to survive higher concentrations of HCN for short periods. McNamara[40] suggests 539 ppm as the 10 minute LC_{50} concentration for humans, and there is a report of a survival from an accidental exposure to 444 ppm.[38] An experimental human exposure (by Barcroft) to 530 ppm HCN was survived without immediate symptoms for 1.5 minutes, although a dog exposed at the same time suffered respiratory arrest.[39] Dogs are considered to be particularly susceptible to cyanide poisoning, but it does seem likely that, to some extent, with HCN (as with CO) this represents the influence of body size on uptake rate and time to incapacitation, so that, as proposed, a human would be able to tolerate exposure to a given concentration longer than a cynomolgus monkey. However, due to the rapid absorption and transport of inhaled cyanide from the lung to the heart and brain, and the known departures from Haber's rule, it seems possible that inhalation of high concentrations of around 1000 ppm and above may result in almost immediate collapse and apnoea in adult humans.

Another issue with respect to calculation of time to incapacitation in human subjects is the effect of the level of physical activity on $\dot{V}_E$. For an adult 70 kg human at rest, $\dot{V}_E$ is approximately 7.5 L min^{-1}, and during fast walking (for example, to evacuate a building during a fire incident) $\dot{V}_E$ would be around 20–25 L min^{-1}, an increase of approximately a factor of three. On this basis, the expression for time to incapacitation in a resting 4 kg monkey should provide a reasonable prediction for time to incapacitation in a moderately exercising adult human. For a more general expression for humans $\dot{V}_E$ can be introduced as a variable term as follows:

$$t_{IN} = \frac{2.43 \times 10^7}{[CN]^{2.36} \times \dot{V}_E}, \tag{10.2}$$

where t_{IN} is the time to incapacitation (minutes), [CN] is the inhaled HCN concentration (ppm), $\dot{V}_E$ is the respiratory ventilation (L min^{-1}) and 2.43×10^7 is the trend line constant $1.21 \times 10^6 \times 2.83$ (ratio of basal metabolic rate kJ h^{-1} kg^{-1} bodyweight between monkey and human)$\times$7.1 L min^{-1} resting human $\dot{V}_E$.

10.4.1 Deviation from Haber's Rule

The deviation from Haber's rule is illustrated by the *Ct* products (HCN ppm min) shown as a function of exposure concentration in Figure 10.3. If HCN toxicity followed Haber's rule there would be a constant *Ct* product dose for incapacitation and a higher constant dose for lethality. In practice, for incapacitation in monkeys the exposure dose increases from around 800 ppm min at 200 ppm to 5100 ppm min at 60 ppm, with a similar but slightly smaller differential in resting rats. This also applies to lethal exposure doses in rats over a similar concentration range.

For rats, the *Ct* product ratio between incapacitation and lethality is around a factor of four. At 127 ppm, time to incapacitation is 21 minutes and time to death 83 minutes, while at 600 ppm time to incapacitation is

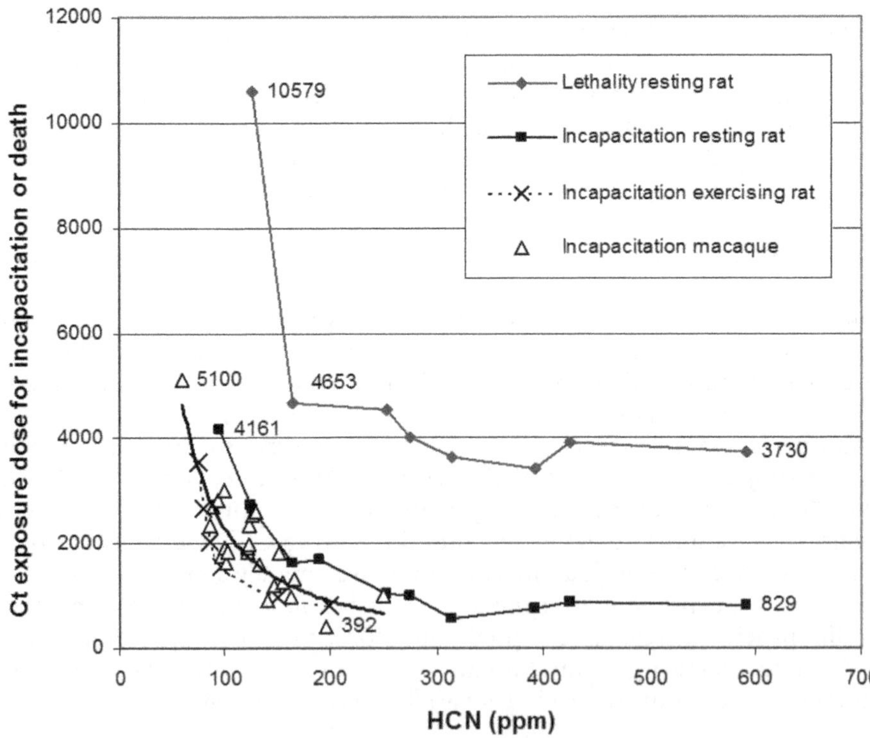

Figure 10.3 Deviation from Haber's rule for HCN: *Ct* product incapacitating and lethal exposure doses for different HCN concentrations.[26,35]

1.4 minutes and time to death 6.3 minutes. For primates, the lowest concentration at which incapacitation was obtained was 87 ppm (after 27 minutes), while signs were minimal following exposure to 60 ppm for an hour. For rats, the lowest concentration leading to deaths was 127 ppm after 83 minutes (standard deviation 32.1).

10.5 Lethal Doses of HCN, NaCN and KCN by Different Routes of Administration

In Section 10.4, the incapacitating and lethal effects of HCN intoxication following inhalation have been described in terms of exposure concentrations and *Ct* product doses. In practice, the toxic effects are some function of the actual dose of cyanide inhaled (mg CN kg^{-1}). It is possible to calculate the inhaled dose of cyanide from the exposure concentration if the volume of air breathed each minute ($\dot{V}_E$) and retained fraction are measured. Unfortunately, few attempts have been made to measure cyanide uptake from inhalation, although some estimates for cynomolgus monkey exposures are described in Section 10.6. An advantage of calculating inhaled doses would be to compare the results of HCN inhalation experiments directly with the results of studies of cyanide poisoning following administration by other routes.

Figure 10.4 shows a comparison of the range of LD_{50} results obtained for administration of HCN, NaCN and KCN by different routes. Most data are for rabbits, mice, rats or guinea pigs, tabulated by Ballantyne,[33] except where

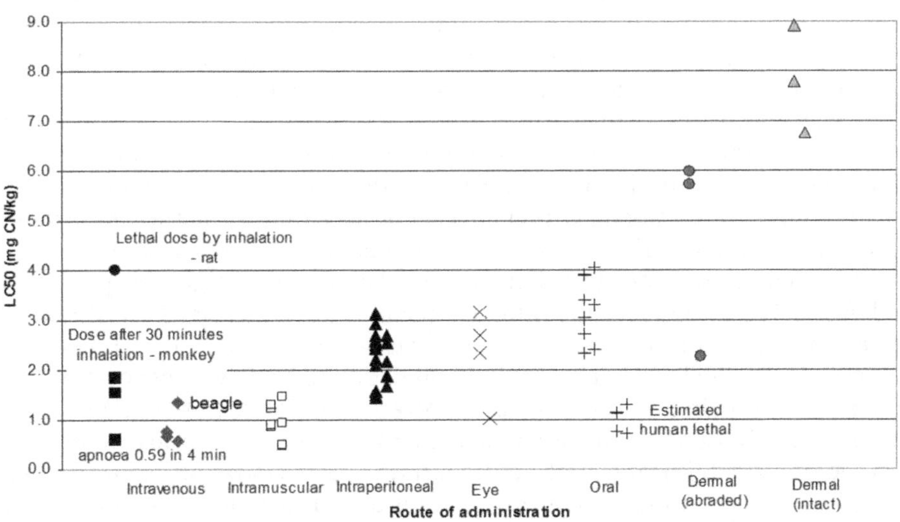

Figure 10.4 Lethal cyanide doses (mg CN kg^{-1}) of HCN, NaCN and KCN in rodents (rats, mice, guinea pigs) and rabbits, and in beagles, monkeys and humans, where indicated, by different routes of administration.[15,33,41,42]

indicated for monkeys,[15] beagles[41] and humans.[42] For each route of administration the data for HCN are plotted slightly to the right of the data for NaCN and KCN for comparison. According to Ballantyne, HCN can be somewhat more toxic by some routes than NaCN and KCN, because HCN is initially absorbed in covalent form and is able to penetrate rapidly into both aqueous and lipid tissue components, while cyanide salts are absorbed in the ionised state. This may be indicated for the LD_{50} following intravenous bolus administration, which was measured in rabbits as 0.57 mg kg^{-1} for CN in the form of HCN, 0.65 for NaCN and 0.75 for KCN. For eye and dermal administration there is also some indication of a lower LD_{50} for HCN, possibly due to more efficient absorption. For intramuscular, intraperitoneal and oral administration there are no obvious differences between the different compounds so that toxicity is dependent on the dose of CN^-.

The main and quite marked differences in toxicity are found between the different routes of cyanide administration. One factor involved may be differences in the efficiency of absorption by the different routes, especially for percutaneous (dermal) administration. Also, for oral administration, there may be somewhat more exposure of the liver than by other routes. But in practice cyanide is quite efficiently absorbed by all routes, excluding dermal and including oral. The differences in toxicity are considered to be due not so much to the efficiency of absorption by different routes but mainly due to the rates of absorption.

The most potent method is by rapid administration of an intravenous "bolus" of cyanide. This results in a transient high blood cyanide concentration (especially blood plasma cyanide), delivering a high dose to key target tissues, such as the myocardium and brain, over a short time-scale of around a minute. At lethal doses this results in severe signs of intoxication within a minute or so, and death within a few minutes.[33,43] Intramuscular administration also results in rapid systemic absorption, with LD_{50} concentrations only slightly higher than for intravenous infusion, at an average value of around 1.04 mg kg^{-1} for rabbits compared with 0.57 mg kg^{-1} intravenous. Intraperitoneal administration and instillation into the conjunctival sac of the eye requires a somewhat higher dose of 2.19 mg kg^{-1} for several rodent species and rabbits. The oral LD_{50} for rabbits and rats was 3.25 mg kg^{-1}.

This then raises the question of what lethal dose might be predicted following HCN inhalation. From the results obtained for these different routes of administration and observed times to severe toxic effects and death it is to be predicted that following inhalation of a low concentration of cyanide (approximately 60–150 ppm) over a prolonged period of up to approximately an hour, the lethal dose might be somewhere close to or even exceeding that following oral administration, especially for exposure concentrations low enough for incapacitation and death after 30–60 minutes, which may allow time for significant cyanide metabolism and detoxification during the period of exposure. On the other hand, exposure to high inhaled concentrations above approximately 150–200 ppm (those sufficient to cause collapse and dyspnoea within a few minutes) might be expected to show lethal inhaled doses similar to those causing rapid death following intravenous bolus administration.

During experimental exposures of HCN to rats over periods of up to 30 minutes, in which $\dot{V}_E$ was measured, the approximate lethal dose was estimated to have been 4 mg kg^{-1},[44] which is comparable to that measured during oral administration to rats. During two of the sub-lethal primate exposures described in the previous section, when animals were exposed to concentrations of 150 ppm HCN, incapacitation occurred within a few minutes and animals were unconscious, with severely reduced respiration approaching apnoea, by the end of the a 30 minute exposure.[15] The calculated doses inhaled by this time were 1.52 and 1.84 mg CN kg^{-1} (taking into account the measured $\dot{V}_E$ and measured fraction of inhaled HCN absorbed from the lungs). It is estimated that these exposure conditions, with gradual continuous uptake over 30 minutes, could produce a pattern of absorption somewhat similar to that occurring following oral administration, as shown in Figure 10.4, and that a lethal exposure dose under these conditions might have been in the region of 2–3 mg CN kg^{-1}.

In one such experiment a brief exposure occurred for 3 minutes to approximately 289 ppm HCN, resulting in apnoea after 4 minutes when a dose of approximately 0.59 mg CN kg^{-1} had been inhaled. It is considered that this exposure would have been fatal within a minute or so if the exposure had not been halted. This exposure was therefore similar to that occurring when an intravenous bolus was injected in rabbits giving a lethal dose of 0.57 mg kg^{-1} CN.

Taking these results in the context of the rodent LD_{50} data it is therefore likely that, for cynomolgus monkeys of around 4 kg bodyweight, the approximate lethal dose of HCN inhaled at concentrations between approximately 100–150 ppm is likely to be similar to that following oral dosing of rodents at around 2–3 mg kg^{-1} bodyweight, while at higher concentrations, especially above 200 ppm, it is likely to be closer to that obtained following intravenous administration at 0.5–1.0 mg kg^{-1}.

These differences in cyanide toxicity relating to the dose and rate of absorption are indicative that cyanide toxicity is not a simple function of dose absorbed, but is more a function of the intracellular cyanide concentration over any period in key target tissues including heart and brain. The following sections examine the relationships between incapacitating effects and the dynamics of uptake from the lungs and partitioning between blood, interstitial and intracellular tissue fluid compartments during and following cyanide intoxication.

The approximate intravenous bolus LD_{50} in dogs is 1.75 mg kg^{-1} and when three beagles were dosed intravenously with a bolus of 2.58 to 2.77 mg kg^{-1}, all died within 1–5 minutes.[41] The intravenous bolus LD_{50} in dogs is therefore slightly higher than that measured for rabbits (0.57 mg kg^{-1}). For humans, the estimated oral lethal dose is approximately 1.0 mg kg^{-1},[42] which is somewhat lower than the measured oral LD_{50} for rodents of 3.25 mg kg^{-1}.[33]

10.6 Relationship Between HCN Toxicity and Blood Cyanide Concentration During and After Exposure

10.6.1 Blood Cyanide Uptake and Incapacitation in Cynomolgus Monkeys

To examine the relationship between HCN exposure concentration, exposure time, blood cyanide concentration and physiological effects, three macaque monkeys were exposed *via* a face-mask to HCN on two occasions while arterial blood samples were taken at intervals from a femoral artery catheter.[15] Following a 20 minute pre-exposure period, two animals were exposed at each of three HCN exposure concentrations (~53, 99 and 149 ppm HCN) for 30 minutes, followed by a 60 minute post-exposure recovery period. Respiratory minute volume ($\dot{V}_E$), the electrocardiogram and clinical signs were monitored throughout. In separate experiments, an HCN mixture was inhaled *via* a three-way valve to enable exhaled air to be collected so that the difference between inhaled and exhaled HCN concentration could be measured. On several occasions measurements were made of the HCN content of exhaled air during the first few minutes of the recovery period.

Figure 10.5 shows the whole blood cyanide concentrations for each pair of exposures at the three HCN concentrations. At 52 and 54 ppm there were no

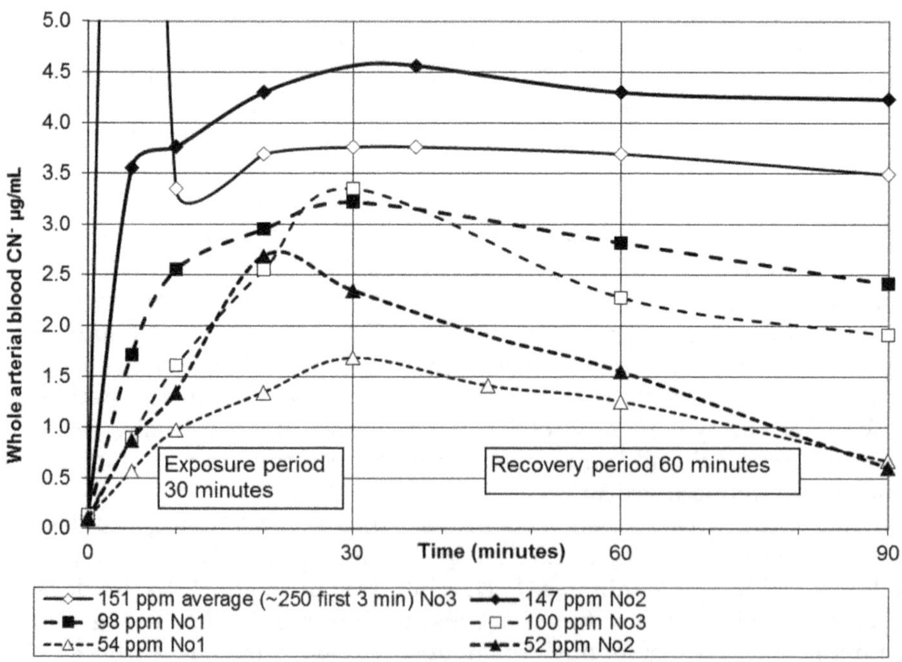

Figure 10.5 Uptake and loss of arterial blood cyanide during and after 30 minute exposures to HCN at three different exposure concentrations with two macaques exposed at each concentration.[15]

obvious clinical effects on the animals, apart from a mild hyperventilation during exposure and minor rebound hyperventilation after exposure. For one animal the arterial blood cyanide concentration increased steadily throughout the exposure period, but somewhat more steeply during the first 10 minutes. For the other animal there was a more marked hyperventilation from 20 minutes when the blood concentration peaked before levelling off. At 98 and 100 ppm there was a gradually developing hyperventilation and more rapid rate of increase in blood cyanide during the first 10 minutes or so. One animal became semiconscious after 14 minutes and unconscious after 27 minutes. The other animal became unconscious at 30 minutes. At 147 ppm for animal No. 2 a marked hyperventilation occurred immediately on exposure and during the first 6 minutes, accompanied by a rapid increase in arterial blood cyanide, after which the hyperventilation subsided and the blood cyanide concentration stabilised with a subsequent slow gradual increase. The animal was semiconscious from around 12 minutes and unconscious from 24 minutes. For animal No. 3 at the high exposure, an error resulted in a high initial exposure concentration of around 250–289 ppm, with a very rapid initial increase in arterial blood cyanide concentration, but after adjustment the average exposure concentration over the exposure was 151 ppm, and after the initial peak the arterial blood concentration levelled off during the second half of the 30 minute exposure period (the exposure pattern for this animal is shown in more detail in Figure 10.7).

It is evident that uptake of cyanide into the blood is not a simple linear function of exposure concentration and exposure time. Figure 10.6 shows

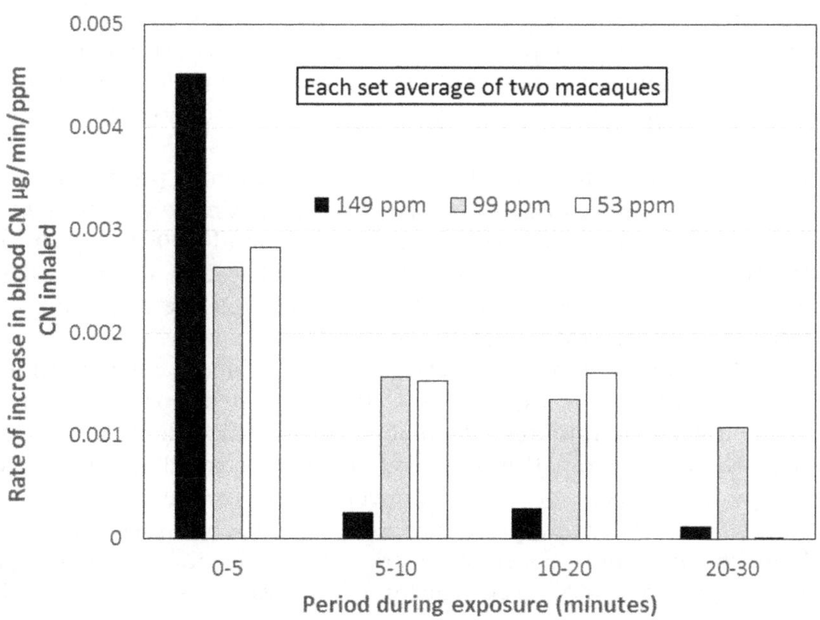

Figure 10.6 Rate of increase of blood cyanide during exposure (µg min^{-1} ppm^{-1} HCN).

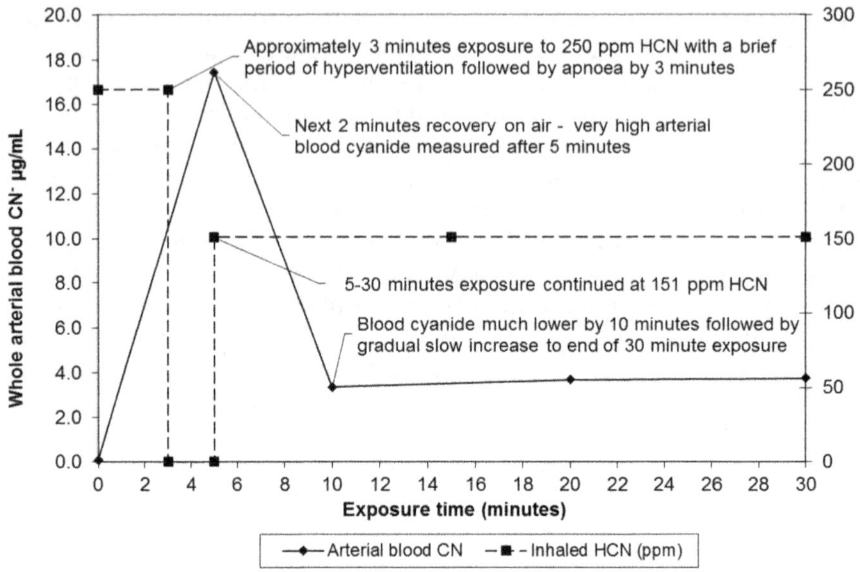

Figure 10.7 Arterial whole blood cyanide during and after exposure to ~250 ppm for 3 minutes, then 151 ppm HCN from 5 to 30 minutes.

the normalised blood uptake expressed as micrograms increase in blood cyanide per minute per ppm HCN inhaled. At 149 ppm HCN the initial rate of cyanide uptake into the blood is very rapid due to the high inhaled concentration and the immediate marked hyperventilation, but by 10 minutes into the exposure the rate of further increase in blood cyanide concentration is very low, partly due to the reduced ventilation, but possibly because a higher proportion is passing out of the blood into the tissues. At 53 and 99 ppm the rate of uptake is somewhat higher initially, but is generally maintained at a relatively high rate throughout the exposure period.

A more extreme example of this dynamic situation is presented by the exposure in which an animal was exposed for approximately 3 minutes to a higher HCN concentration than intended at around 250–289 ppm HCN (Figure 10.7). This resulted in a brief period of strong hyperventilation followed by apnoea (cessation of breathing). The exposure was paused and followed by exposure to air for 2 minutes, after which a blood sample taken at 5 minutes showed an extremely high blood concentration of 17.45 μg CN^- mL^{-1}. The exposure was then resumed from 5 minutes at 151 ppm, but at 10 minutes the blood concentration had decreased to 3.35 μg CN^- mL^{-1}. From this it is concluded that rapid uptake of HCN resulted in a transient very high concentration in the systemic blood compartment, which then rapidly decreased, despite continued exposure, as the high concentration systemic blood mixed with that in less rapidly perfused regions and a large proportion of the cyanide was dispersed into the tissues.

By calculating the mass of HCN inhaled during exposure as $0.6 \times \dot{V}_E$ (see 10.6.4 for measurement of inhaled fraction absorbed) and relating this to the total blood volume, it is possible to calculate the proportion of the inhaled cyanide dose in the whole blood compartment of the body, the remainder therefore being in the extracellular and intracellular tissue fluids.

Figure 10.8 shows that during the early stages of all the exposures the proportion of the total inhaled dose in the whole blood compartment was 25–38%, but that the proportion in the blood then decreased throughout the remainder of the exposure period until only about 10% remained in the blood at the end of the exposure. Almost all the remaining 90% was therefore in the extracellular and intracellular tissue fluid compartments, with a small amount possibly having been metabolised.

There is therefore some relationship between whole blood cyanide and the onset of incapacitation, which occurred after the cyanide concentration exceeded approximately 3 $\mu g\, mL^{-1}$ (0.115 $mmol\, L^{-1}$), but the relationship is not a simple one since the animals recovered rapidly after HCN intake ceased without an appreciable decrease in the blood cyanide concentration. Since the main toxic site of action of cyanide is in the body tissues where it combines with mitochondrial cytochrome a3 and impairs oxidation, the intracellular cyanide concentration is more relevant than the whole blood concentration. In the blood it is known that a significant proportion of cyanide is taken up by the red cells and that it combines reversibly with methaemoglobin in the red cell to form cyanomethaemoglobin.[22] This therefore effectively detoxifies some of the cyanide and also means that whole blood cyanide may not be representative of the cyanide concentration

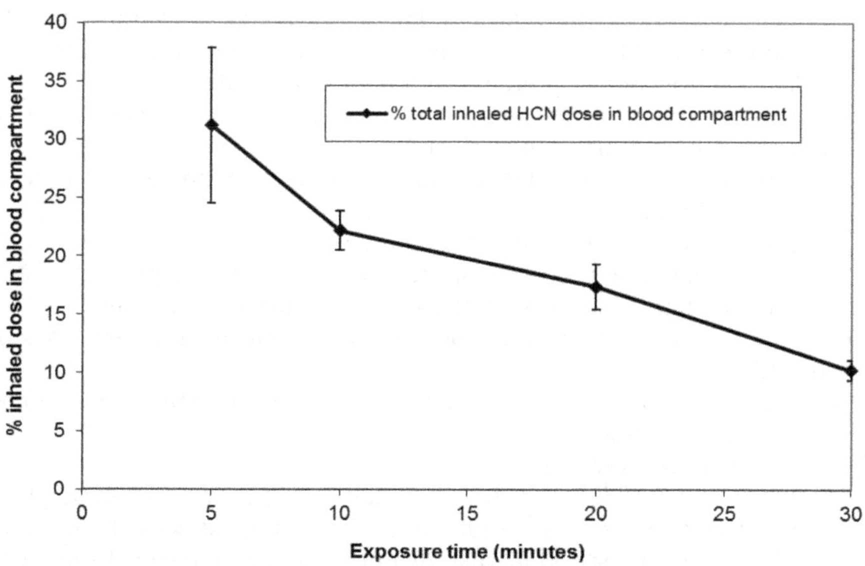

Figure 10.8 Percentage of total inhaled HCN dose in the blood compartment throughout exposure for six primate exposures.

in other body fluid compartments. In the plasma some cyanide is known to be protein bound and some free,[45] and the dynamics of uptake into the blood, and dispersal and partitioning between plasma, interstitial and intracellular fluid spaces, are all relevant to predicted toxic activity. A further complication is that the episode of hyperventilation induced by HCN uptake results in a transient hypocapnia and respiratory alkalosis, which subsidies as ventilation reduces and is succeeded by metabolic acidosis as oxidative metabolism is inhibited and lactate increases as a result of glycolysis.[31] These changes in PCO_2 and blood pH affect oxygen dissociation and capillary PO_2, as well as causing sequential changes in cerebral blood flow.

In order to understand and be able to predict the relationships between inhaled HCN, whole blood cyanide and toxicity, it is therefore necessary to consider a set of parameters as follows:

- The relationship between clinical condition and whole blood cyanide concentration during exposure and recovery.
- The dynamics of cyanide uptake in terms of:
  - The inhaled HCN dose [HCN concentration$\times \dot{V}_E$ or $\dot{V}_A$ (alveolar ventilation)].
  - The fraction of inhaled HCN absorbed during inhalation.
  - Excretion of HCN in expired air.
  - The effects of hypocapnia, alkalosis and acidosis.
- The partitioning of cyanide in the blood between red cells and plasma:
  - The rate of uptake from plasma into red cells.
  - The methaemoglobin content of blood.
  - The proportion of red cell cyanide combined with methaemoglobin.
  - The rate of reaction of cyanide with methaemoglobin and the equilibrium ratio between cyanomethaemoglobin and cyanide.
  - The significance of protein binding of cyanide in plasma.
- The dynamics of dispersal and partitioning of cyanide between plasma and other body fluid compartments:
  - Into interstitial and intracellular fluid compartments (volume of distribution).
  - The extent of binding to cytochrome a3.
  - The extent that plasma cyanide concentration is representative of intracellular tissue concentration and predictive of toxicity.
- The rate of loss of cyanide from the blood during the post-exposure period:
  - Rhodanese activity and conversion to thiocyanate (and other metabolic pathways).
  - Excretion *via* the lungs.
  - Effects of cyanide antidotes (methaemoglobin formers) on displacement of cyanide from tissues into blood and red cells. Preferential binding of cyanide to methaemoglobin rather than cytochrome a3.
  - Competition between oxygen and cyanide in the mitochondrion and extent to which oxygen displaces intracellular cyanide.

10.6.2 Relationship Between Whole Blood Cyanide and Clinical Condition

During HCN exposures in monkeys, loss of consciousness occurs following an episode of hyperventilation. At higher HCN concentrations of around 150 ppm and above, this episode tends to occur within a few minutes of the start of an exposure and consists of a 2–3 times increase in minute volume for a few minutes, leading to loss of consciousness. This was generally followed by a partial recovery as the hyperventilation subsided, despite the continued exposure. Figure 10.5 shows that the initial loss of consciousness occurred in the high dose animals when whole blood levels exceeded approximately 3.0 $\mu g\ ml^{-1}$, but the blood level then continued to increase to around 4 $\mu g\ mL^{-1}$. Although the animals were unconscious their clinical condition did not deteriorate greatly during this period, and after exposure they recovered full consciousness within 10 minutes of the end of the exposure, despite their blood level remaining close to 4.0 $\mu g\ mL^{-1}$ for up to an hour after the exposure. Figure 10.9 shows a more detailed analysis for animal No. 2. During the first 15 minutes of exposure the measured HCN concentration was 134 ppm. The upper part of the figure shows that the animal started to hyperventilate during the third minute of exposure, and the lower part of the figure shows that there was a very rapid increase in arterial whole blood cyanide to 3.6 $\mu g\ mL^{-1}$ by 5 minutes, but with subsequent increases to 3.8 $\mu g\ mL^{-1}$ by 10 minutes and 4.6 $\mu g\ mL^{-1}$ by 30 minutes. There was then a very slow decline post-exposure to 4.2 $\mu g\ mL^{-1}$ an hour later.

For both high dose animals the whole blood cyanide concentration remained close to 4 $\mu g\ mL^{-1}$ for up to an hour after the exposure (4.23 for the animal shown in Figure 10.9), although the animals made a good recovery within a few minutes. For the intermediate dose pair the blood level remained just above 3.0 $\mu g\ mL^{-1}$ for around 10 minutes after exposure but the animals also recovered within a few minutes. For the low dose group there were no significant adverse signs and the maximum blood level was 2.2 $\mu g\ mL^{-1}$.

It is therefore necessary to explain why, following a hyperventilatory episode and loss of consciousness at around 3.0 $\mu g\ mL^{-1}$, animals make some degree of recovery during the next few minutes of exposure and a rapid recovery to normal levels of consciousness soon after exposure when the whole blood cyanide concentration is increasing further during the exposure period and maintained at levels as high as 4.0 $\mu g\ mL^{-1}$ for up to an hour after exposure. There are several possible explanations for this finding relating to:

- The dynamics of uptake and dispersal of cyanide in the body.
- The partitioning of blood cyanide between erythrocytes (red blood cells), plasma, and extracellular and intracellular fluids.
- The effect of hyperventilation on PCO_2 and blood pH (transient respiratory alkalosis followed by metabolic acidosis).
- Biochemical interactions in the mitochondria.

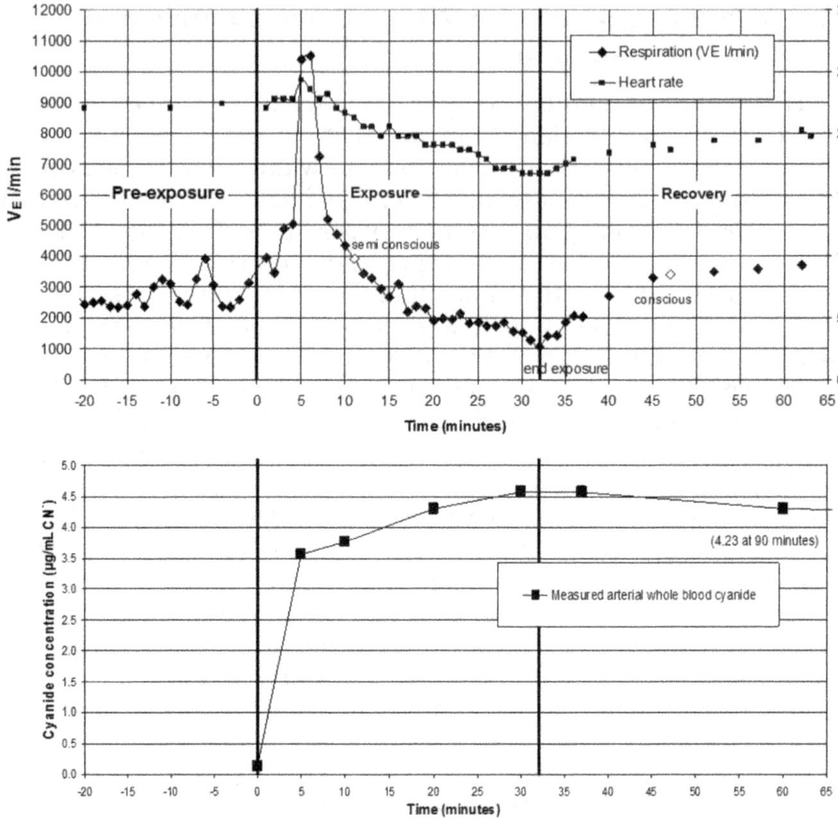

Figure 10.9 Top: minute volume ($\dot{V}_A$, L min^{-1}) and heart rate during pre-exposure, exposure and recovery period for 32 minutes' exposure to HCN at ~147 ppm (actual concentrations: 134 ppm 0–15 minutes, 160 ppm 15–30 minutes). Bottom: measured arterial whole blood cyanide (CN^-, µg mL^{-1}).

When cyanide is taken up by the inhalation route it is absorbed from the lungs directly into systemic circulation, passing within seconds through the heart and brain. In the blood, cyanide exists partly as free cyanide in solution in the plasma, partly bound to plasma proteins, but up to a certain level the largest proportion penetrates into the red cells where most or all of it reacts with methaemoglobin to form cyanomethaemoglobin.[22,31,45,46] Of these three components the red cell fraction is effectively detoxified, while a proportion of that in the plasma is free to disperse into the extracellular and intracellular fluid compartments, initially in the heart and brain, to exert toxic effects. One possibility in relation to the primate experiments is that during the hyperventilatory episode the uptake dynamics are such that a significant proportion of blood cyanide is in the form of free cyanide and is carried rapidly to the heart and brain, while at later stages of exposure, when ventilation has reduced, a greater proportion of blood cyanide is in a bound form in the red cells. This may result in relatively high whole blood cyanide, of which only a relatively small proportion is toxicologically active. During the long recovery period there may be a decrease in the free cyanide concentration before significant loss of red cell cyanide from cyanomethaemoglobin. The decrease in whole blood cyanide may therefore be small but the decrease in the free component may be sufficient for the clinical condition of the animals to be improved. The detoxifying effect of methaemoglobin is well known, so that induction of additional methaemoglobin is used as an effective treatment for cyanide poisoning.[22] The extent of this "detoxification" mechanism during exposure therefore depends upon the maximum amount of cyanide that can be sequestered by the red cells and the rate of sequestration during rapid uptake from the lungs.

It may also be that during the initial uptake period most of the cyanide is preferentially dispersed in a small blood compartment consisting of the most rapidly perfused tissues, including the cerebral circulation, with a proportion as free cyanide, while over a longer period the cyanide becomes more mixed with less rapidly perfused blood and body fluid compartments, with a higher proportion as bound cyanide, as indicated in Figures 10.7 and 10.8. Another factor is that during cyanide poisoning more red cells are released from the spleen into the circulation, so that more are available to take up cyanide from the plasma.

10.6.3 Effects of Hypocapnia

During this hyperventilatory episode ventilation increases by approximately a factor of two to three, which results in a temporary respiratory alkalosis with a reduction in blood PCO_2 (hypocapnia). The pH change results in a left shift in the oxygen dissociation curve and the PCO_2 reduction in a reduced cerebral blood flow, effects known to result in syncope (loss of consciousness) in humans.[47] This is likely to be enhanced by the direct effects of cyanide on heart rate, rhythm and cardiac output, as well as on cerebral metabolism. The result is the reported transient loss of consciousness.

The subsequent decrease in respiration and circulation then enables PCO_2 and pH to recover towards normal levels, the direct effect of cyanide on metabolism then resulting in a metabolic acidosis, with a right shift in the oxygen dissociation curve and increased cerebral blood flow. This may explain why the animals are seen to make a partial recovery during the middle part of the exposure, followed by a slow decline as blood and tissue cyanide concentrations increase towards the end of an exposure. Arterial PCO_2 and pH were not measured during these primate exposures, but effects on PCO_2, pH and cerebral blood flow were measured in anaesthetised dogs by Klimmek *et al.*[31] Intravenous infusion of potassium cyanide induced hyperventilation accompanied by hypocapnia, alkalosis and reduced cerebral blood flow. Combinations of initial hypercapnia, alkalosis and then metabolic acidosis have also been reported by Levine.[48] Djerad *et al.*[49] examined the effects of controlled respiratory alkalosis, a normal pH state

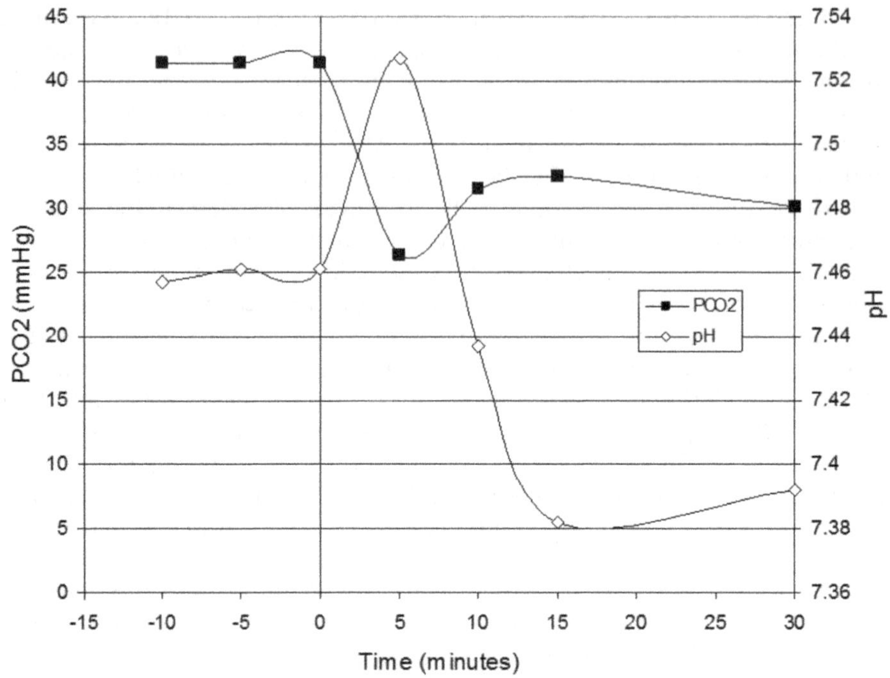

Figure 10.10 Arterial blood PCO_2 and pH following an intravenous bolus of 8.4 mg kg^{-1} NaCN (CN^- 4.5 mg kg^{-1}) in cynomolgus monkeys.[43]

and acidosis on cerebral blood flow and distribution of cyanide to cerebral tissue. Brain uptake of cyanide was reduced under alkalosis but increased under acidosis, presumably due to the effects of PCO_2 on cerebral circulation. For these three studies the animals were anaesthetised, unlike in the primate studies, so the relationship between state of consciousness, hyperventilation, cyanide uptake, and alkalosis followed by acidosis could not be examined.

When Stemler *et al.*[43] infused 8.5 mg kg^{-1} NaCN intravenously into anaesthetised cynomolgus monkeys they also observed hyperventilation, a transient arterial alkalosis and a decrease in PCO_2, as shown in Figure 10.10. Since these animals were also treated with sodium nitrite, the effects of the cyanide were ameliorated after approximately 10 minutes, but there were signs of a recovery in PCO_2 and pH, followed by lowering of pH. This lowered pH is consistent with an increase in venous lactate. Klimmek *et al.*[31] obtained a 500% increase in lactate during intravenous infusion of cyanide in dogs.

10.6.4 Cyanide Uptake from Inhaled Hydrogen Cyanide and Pulmonary HCN Excretion

When 125 ppm HCN mixture in air was inhaled by cynomolgus monkeys *via* a face-mask and three-way valve, the exhaled mixture was found to contain approximately 55–58% less cyanide than the inhaled mixture.[15] This is considered consistent with other findings of 60% (range 58–77%) inhaled HCN retention in humans,[50] taking into account the additional dead space in the apparatus used. Ballantyne used a somewhat higher figure of 70% retention in dogs[33] (which may reflect the much higher collateral ventilation in the dog lung).

Measurements of HCN in exhaled air during the first few minutes of the recovery period showed low HCN concentrations (<1 ppm) but the concentration was found to be approximately proportional to the blood cyanide concentrations, which were in the 2–4 $\mu g\,mL^{-1}$ range.[15] This is also consistent with published data.[33]

From these results it is considered that HCN uptake represents approximately $0.6\times\dot{V}_E$ and is approximately equivalent to full clearance of alveolar ventilation ($\dot{V}_A$) into the blood. Also, the body fluids at the pH ranges occurring during HCN inhalation have a high solubility for cyanide (which can be absorbed both in associated form as HCN into lipids and ionised to CN^- in body fluids, as well as becoming bound to proteins, methaemoglobin and cytochromes). The result is that over short time-scales of up to 30 minutes, little cyanide is excreted *via* the lungs at blood cyanide concentrations resulting in incapacitation, and cyanide uptake can be considered to be a linear function of the inhaled dose as follows:

$$[\mathrm{CN}]_D = 6.48\times10^{-4}\ [\mathrm{HCN}]\ \dot{V}_E\ t, \qquad (10.3)$$

where $[\mathrm{CN}]_D$ is the inhaled cyanide dose retained (mg), [HCN] is the inhaled HCN volume fraction in inspired air expressed as ppm at 20 °C, $\dot{V}_E$ is the respiratory minute volume in litres per minute at 20 °C, t is the exposure time in minutes and 6.48×10^{-4} is 0.00108 volume fraction (ppm) to mass concentration ($mg\ L^{-1}$) conversion times 0.6 uptake fraction.

10.6.5 Partitioning Between Red Cell Methaemoglobin and Blood Plasma

Cyanide has differing degrees of binding to different blood components. Cyanide added to a blood sample *in vitro*, and blood samples taken from animals dosed with cyanide intravenously or by intramuscular injection had a proportion in free solution in plasma or serum, a proportion protein-bound in plasma and the greatest proportion recovered from the red cells.[45] A proportion of red cell haemoglobin is in the form of methaemoglobin (normal range 0.01–0.5 g per 100 mL; average 0.16–0.32 g per 100 mL in humans).[51] In beagle experiments[41] the methaemoglobin concentration in untreated animals was 1.5%, which would be 0.225 g per 100 mL at a haemoglobin concentration of 15 g per 100 mL. For experiments in cynomolgus monkeys the mean methaemoglobin concentration was measured as 25.6 $\mu mol\ L^{-1}$ ± 6.0 SD, which corresponds to 0.163 g per 100 mL.[43] Methaemoglobin reacts with cyanide to form cyanomethaemoglobin. The ferric haem group competes with cytochrome a3 for cyanide by complexing in the ionic form.[22] The affinity of methaemoglobin for cyanide exceeds that of cytochrome a3. Cyanomethaemoglobin is a dissociable complex so some free cyanide is present.[22] Erythrocytes have a high affinity for cyanide and there is rapid uptake of plasma cyanide by erythrocytes.[33,52–55] McMillan and Svoboda[56] found rapid assimilation of cyanide by red cells, passing through the red cell membrane as HCN, where it is sequestered, but that red cells play no role in cyanide degradation. It is therefore likely that most if not all red cell cyanide is in the form of cyanomethaemoglobin, as demonstrated *in vitro* by Lundquist *et al.*[57] over a range of whole blood concentrations (although some may exist as free cyanide within the red cell interior fluid).

The methaemoglobin content of blood includes a component occurring as a result of normal oxidative metabolism and the nitrite or nitrate content of the blood (where nitrate becomes active once reduced to nitrite). Nitrites and nitrates may be derived from dietary intake, but during exposure to cyanide in fires it is likely that some may be inhaled in the form of NO and NO_2 fire gases. When nitrogen-containing fuels are burned in fires, a small proportion of fuel nitrogen is oxidised to NO and NO_2, especially during the early (well-ventilated) phase of a flaming fire. As the fire develops and becomes under-ventilated, there is a large increase in the proportion of fuel nitrogen release as HCN. It is therefore possible that fire victims may have slightly higher blood nitrates and methaemoglobin than unexposed subjects (see Chapter 2).

The maximum binding capacity of human blood for oxygen (or methaemoglobin for cyanide) depends on the haemoglobin content of the blood (13.5–18.0 g dL^{-1} in adult males and 11.5 g dL^{-1} in adult females[51]) and the molecular mass of haemoglobin, which can vary between 64 000–67 000 Da.[58] Taking a molecular mass of 64 458 Da gives a value of 0.0155 mmol g^{-1} Hb. Since each molecule of Hb combines with four molecules of oxygen, this gives a theoretical maximum calculated value of 0.0621 mmol oxygen per gram of haemoglobin or 1.39 mL oxygen at standard temperature and pressure (STP). Literature estimates vary between 1.32 to 1.39 mL O_2 g^{-1} Hb but a commonly used value is 1.36 mL O_2 g^{-1} Hb (representing 0.061 mmol O_2 or 1.94 mg).

The same calculations apply to the binding capacity of methaemoglobin for cyanide so that:

1 g methaemoglobin (0.0152 mmol) can bind 0.061 mmol $CN^- = 1.58$ mg or 1.36 mL CN at STP.

Methaemoglobin has a very high affinity for cyanide,[45] $K_{obs} = 10^{5.7}$, so may be induced as a treatment for cyanide intoxication by infusion of antidotes such as sodium nitrite or 4-dimethylaminophenol (DMAP). Experiments in which beagle bitches were dosed with a bolus of HCN by intra-jugular injection following prior oral administration of DMAP provide an example of the binding capacity of methaemoglobin for cyanide.[41]

Table 10.2 shows data from two experiments in which administration of DMAP provided 9.1 and 9.5% conversion of Hb to metHb, representing 0.0137 and 0.0143 g metHb mL^{-1}, assuming a haemoglobin concentration of 15 g per 100 mL blood. If the binding capacity of metHb for beagles is similar

Table 10.2 Comparison of calculated methaemoglobin CN with measured whole blood and plasma CN for two beagles dosed with DMAP and intravenous cyanide.

HCN dose/mg kg^{-1}	Beagle 1 2.56	Beagle 2 2.49
Methaemoglobin (metHb)/%	9.1	9.5
MetHb mL^{-1} blood/g mL^{-1a}	0.0137	0.0143
Calculated CN binding capacity/μg mL^{-1b}	21.55	22.50
Measured CN (μg mL^{-1}) per mL blood	25.04	24.80
Measured plasma CN (μg mL^{-1} plasma)	0.19	0.42
Plasma component of whole blood/μg mL blood^{c}	0.10	0.23

^{a}Assuming Hb 15 g per 100 mL blood.
^{b}At 1.58 mg CN^{-} g^{-1} metHb.
^{c}Assuming 0.55 mL plasma mL^{-1} whole blood.

to that calculated for humans, the calculated maximum cyanide binding capacity as cyanomethaemoglobin gives 21.55 and 22.50 μg CN mL^{-1} blood, which are close to the measured values during these experiments of 25.04 and 24.80 μg mL^{-1}. Of these levels, the plasma components were very low at 0.01 and 0.23 μg mL^{-1} blood (assuming 0.55 mL plasma mL^{-1} whole blood).

These experiments therefore confirm that the binding capacity of methaemoglobin for cyanide is approximately 1.58 mg CN^{-} g^{-1} metHb. They also confirm the high affinity of methaemoglobin for cyanide, since in these experiments most of the oral dose was sequestered into the blood. At an average 85 mL blood kg^{-1}, the calculated maximum blood cyanide concentration is 28–30 μg mL^{-1}, assuming 100% absorption. The plasma CN concentration is also 60–130 times lower than the whole blood cyanide concentration, indicating that almost all CN is in the red cell fraction as cyanomethaemoglobin.

Because the affinity of methaemoglobin for cyanide is higher than that of cytochrome a3, following administration of a methaemoglobin-forming antidote, almost all tissue cyanide is sequestered back into the blood as cyanomethaemaoglobin.

The affinity of metHb for cyanide was measured by Christel *et al.*[45] in beagles in terms of an equilibrium constant (K_L) for the reaction between ferrihaemoglobin and free plasma cyanide as:

$$K_L = [HbFe^{3+}CN^-]/(HbFe^{3+}H_2O] \times [CN^-] = 10^{7.9} \quad (10.4)$$

In these experiments the ratio of whole blood cyanide to plasma cyanide was around 60–94 : 1 in the presence of methaemoglobin excess. They also found that the reaction between cyanide and methaemoglobin *in vivo* approached equilibrium within a few minutes.

The efficacy of methaemoglobin for rapid sequestration of cyanide is also demonstrated by an experiment in which a lethal bolus dose of sodium cyanide (8.4 mg kg^{-1}) was administered intravenously to anaesthetised cynomolgus monkeys followed 1 minute later by intravenous 4-DMAP (5 mg kg^{-1}).[43] In these experiments the calculated blood cyanide concentration assuming all cyanide was sequestered into the blood was equal to the measured blood cyanide concentration.

The importance of these findings with respect to intoxication during inhalation of HCN is that since blood has a natural methaemoglobin content, it can be predicted that during the early stages of exposure to incapacitating concentrations of HCN, cyanide taken up initially into the pulmonary circulation will first enter the blood plasma, but then rapidly enter the red cells and combine with methaemoglobin, leaving a low residue in the plasma. Once the methaemoglobin is saturated, any additional cyanide should accumulate in the plasma and other body fluid compartments. It also means that the whole blood cyanide concentration is likely to be significantly greater than the plasma and tissue cyanide concentration, depending on the blood methaemoglobin content.

From the data on normal blood methaemoglobin concentrations it is likely that during cyanide intoxication approximately 2–3 $\mu g\,mL^{-1}$ CN^- whole blood cyanide will be in the form of cyanomethaemoglobin and essentially non-toxic. The toxicologically important fraction is therefore the plasma and tissue cyanide component.

10.6.6 The Dynamics of Dispersal and Partitioning of Cyanide between Plasma and Other Body Fluid Compartments

During inhalation, hydrogen cyanide entering the blood circulates rapidly to key target organs including the heart, the aortic and carotid bodies, and the brain. The acute physiological effects therefore depend upon the changing cyanide concentrations in the tissue of these key organs. Since creation of additional methaemoglobin is an effective treatment for cyanide poisoning, and cyanide is tightly bound to methaemoglobin, it is likely that normal methaemoglobin acts to some extent as a protective buffer during the early stages of cyanide poisoning and that cyanide sequestered into the red cell is rendered relatively non-toxic. The severity of acute cyanide poisoning is therefore likely to be more closely related to plasma cyanide concentrations than whole blood cyanide[33,53] and the penetration of cyanide into tissue extracellular and intracellular fluid compartments is likely to relate to plasma cyanide or at least to the concentration of free cyanide in plasma, since protein-bound cyanide may also be at least temporarily unavailable for penetration to the mitochondria in affected tissues. There is evidence that brain exposure to cyanide is related to plasma cyanide levels, in that Ballantyne found similar average cyanide concentrations in blood plasma or serum and cerebrospinal fluid (CSF) in sheep dying following intramuscular cyanide injection.[59]

If plasma cyanide is the most important parameter, then it is important to determine the dynamics of uptake and partitioning of cyanide between red cell and plasma with time during an acute exposure, and possibly also the dynamics of partitioning between protein-bound and free cyanide in plasma. If, effectively, all cyanide in blood is rapidly sequestered by methaemoglobin up to a high saturation level, so that only above this does it appear in the plasma (as proposed by Lundquist *et al.*[57]), then cyanide should be relatively non-toxic up to a critical dose level. If there is a more dynamic partitioning between red cell and plasma cyanide, then red cell sequestration may offer some degree of protection, but the plasma concentration may increase in parallel with the red cell concentration until the plasma and tissue concentrations become high enough to exert toxic effects. Once intake ceases, it is possible that the red blood cell cyanide may remain relatively stable for a period while the plasma and cyanide dispersed throughout the body tissues is detoxified, so that the plasma concentration decreases more rapidly than the whole blood or red cell cyanide. The cyanide may then be gradually released from the red cells, maintaining a low plasma level until detoxification is complete (based upon the high affinity for methaemoglobin for cyanide and depending on the dynamics of the rate of release from the red cells compared with the rate of detoxification by rhodanese and other pathways). The key problem is to determine the time–concentration relationships between exposure conditions, plasma and tissue cyanide.

Rapid mixing is predicted between free blood cyanide (essentially the plasma component) and extra-blood fluids. For humans the circulating blood volume is 70 mL kg^{-1} and the resting cardiac output is 5–7 L min^{-1} (adult male average = 5.6 L min^{-1}).[60] For a 70 kg man this represents approximately 80 mL min^{-1} kg^{-1}, or 1.14 × the circulating blood volume per minute. For cynomolgus monkeys such as those used for the data shown in Figure 10.5, the circulating blood volume has been measured as between 55 and 95 mL kg^{-1} and the cardiac output as 211–316 mL kg^{-1} min^{-1}, or approximately 3.5 × the circulating blood volume each minute.[61,62] When a bolus of 2.38 mg kg^{-1} intravenous sodium cyanide was introduced into cynomolgus monkeys, the half-life for the initial dispersal was approximately 1 minute[43] and has been measured as 20 seconds in rats.[49]

For continuous slow uptake by inhalation or intravenous perfusion, or several minutes after a bolus dose, it is therefore likely that tissue fluid concentrations will be similar to plasma levels. During or soon after administration of a bolus dose or during the rapid inhalation uptake of a hyperventilation episode it is likely that a transient peak of a minute or so will occur in plasma levels and possibly in well-perfused tissues such as the brain and heart.

A number of studies provide some data relating to these issues. A simple method for estimating the distribution of cyanide to body fluids is to calculate the volume of distribution as:

$$\text{Volume of distribution } (V_d) = \text{Cyanide dose (mg kg}^{-1}\text{ bodyweight)/ blood cyanide concentration (mg mL}^{-1}) \qquad (10.5)$$

Figure 10.11 shows the volume of distribution in cynomolgus monkeys following an intravenous bolus dose of 3.37 mg NaCN kg^{-1} bodyweight

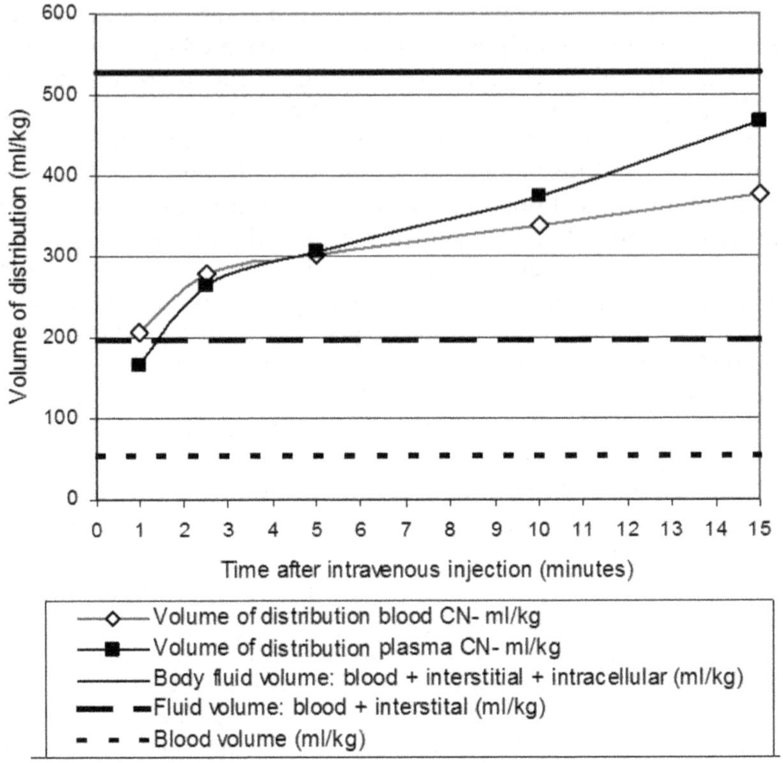

Figure 10.11 Volumes of distribution for whole blood and plasma cyanide following intravenous injection of 3.37 mg kg^{-1} NaCN (1.79 mg CN^{-} kg^{-1}).

(1.79 mg CN^{-} kg^{-1}). There is very rapid dispersal of injected cyanide beyond the blood compartment, since the volume of distribution at 1 minute is 200 mL kg^{-1} (a volume equal to the blood plus interstitial fluid volume) compared with the blood volume of 55 mL kg^{-1}. The volume of distribution then increases steadily to a maximum of 376 mL after 15 minutes (a time-scale over which cyanide metabolism is considered to have been negligible). This is approaching the total fluid volume (blood + interstitial + intracellular) of 528 mL, indicating distribution into the intracellular fluid spaces of most tissues. One factor limiting the volume of distribution is the fraction of cyanide in the red cells bound as methaemoglobin. Based upon the measured blood methaemoglobin concentration of 25.6 μmol L^{-1}, the calculated maximum blood cyanide concentration in the form of cyanomethaemoglobin is 2.6 μg CN^{-} mL^{-1}. Subtracting this, it is possible to calculate the volume of distribution in terms of plasma cyanide, which at 467 mL comes closer to the total fluid volume of 528 mL. For the case of inhaled HCN shown in Figure 10.9, the calculated volume of distribution for whole blood cyanide was 419 mL at the end of the 30 minute exposure period.

Table 10.3 Partitioning of cyanide following intravenous (IV) cyanide bolus and continuous infusion in cynomolgus monkeys and beagle dogs.

	Cynomolgus	Cynomolgus	Dog	Dog
	IV bolus	IV bolus	IV continuous infusion	IV bolus
Time/minutes after dose	15	15	40	1.5
Dose CN^-/mg kg^{-1}	1.79	1.00	1.12	2.67
Blood concentration CN^-/µg mL^{-1}	4.76	3.38	3.25	8.90
Volume of distribution/ L kg^{-1}	0.38	0.30	0.34	0.29
Calculated body fluids CN^-/µg mL^{-1}	2.88	1.50	1.57[a]	3.77
Calculated red cell CN^-/µg mL^{-1} blood	3.12	2.52	2.78[b]	2.70
Measured methaemoglobin/g per 100 mL	0.16	0.16		0.24
Maximum CN^- as cyanomethaemoglobin/ µg mL^{-1}	2.58	2.58		3.73

[a]Measured plasma concentration.
[b]Measured red cell concentration.

Table 10.3 compares the partitioning of cyanide in two of the cynomolgus monkey experiments 15 minutes after an intravenous bolus of 1.79 mg and 1.0 mg CN kg^{-1},[43] 40 minutes into an experiment involving continuous intravenous infusion in a dog at 0.028 mg CN kg^{-1} min^{-1} (total dose 1.12 mg CN kg^{-1})[45] and during an experiment involving an intravenous bolus of 2.67 mg CN kg^{-1}, which was lethal after 1.5 minutes.[41] From the administered doses and the whole blood concentrations it is possible to calculate the volumes of distribution and predicted concentrations in the body fluids, including plasma, for the two cynomolgus experiments on the assumption of even distribution between plasma, interstitial and intracellular fluid volumes, and to estimate the red cell CN concentrations. The calculated concentrations for body fluids, including plasma, are 2.88 and 1.50 µg mL^{-1}. For the intravenous infusion experiment, involving a total dose similar to the second cynomolgus experiment (1.12 compared with 1.0 µg mL^{-1}), the measured plasma concentration of 1.57 µg ml^{-1} is similar to the cynomolgus value of 1.50 µg mL^{-1}. The calculated and measured red cell CN concentrations are also similar at 2.78 and 2.52 µg mL^{-1}. For the cynomolgus experiments, in which the measured methaemoglobin concentration was 25.6 µmol L^{-1} (0.16 g per 100 mL) it is also possible to calculate the maximum predicted cyanomethaemoglobin concentration of 2.58 µg mL^{-1}, which is comparable with the calculated values for total red cell cyanide and similar to the 2.78 µg mL^{-1} red cell cyanide measured in the intravenous infused dog. For the intravenous bolus dog, the calculated red cell cyanide was 2.70 μg mL^{-1} and the maximum calculated CN as cyanomethaemoglobin for the measured methaemoglobin of 0.24 g per 100 mL was 3.73 μg mL^{-1}. The measured red cell cyanide is therefore somewhat lower than predicted despite a high plasma concentration, most likely because the animals died within 1.5 minutes, before there was time for full conversion of the methaemoglobin.

An important study of the distribution of cyanide during continuous uptake at sub-lethal concentrations was carried out by Christel *et al.*[45] This experiment used continuous intravenous infusion, which provides a useful analogue for the continuous uptake during HCN inhalation. Beagle dogs were maintained under anaesthesia with nitrous oxide and halothane. Radio-labelled $K^{14}CN$ was introduced at a continuous rate of 0.07 mg kg^{-1} min^{-1} into a leg vein, while blood samples were taken for analysis from a catheter placed in the superior vena cava. After 45 minutes, when the whole blood concentration reached 3.9 μg mL^{-1} and the plasma concentration was 1.04 μg mL^{-1} (40 μmol L^{-1}), there was a marked decrease in respiration and heart rate, at which point sodium thiosulfate was administered as an antidote.

The relationship between whole blood, red cell and plasma cyanide is shown in Figure 10.12. Over 45 minutes, there was a steady increase in whole blood cyanide up to 3.9 μg mL^{-1}. For the first 15 minutes 92% of whole blood cyanide is taken up into the red cells, but as the whole blood level exceeds approximately 1.5–2.0 μg mL^{-1}, the plasma component gradually increases up to 0.58 μg mL^{-1}, representing 15% of whole blood cyanide.

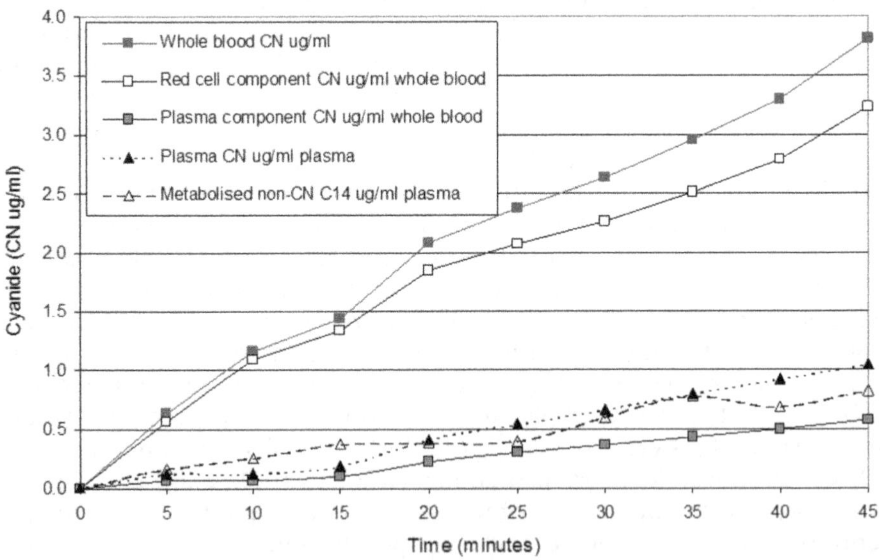

Figure 10.12 Cyanide concentrations in red cells, plasma and whole blood during continuous intravenous infusion of 0.07 mg kg^{-1} min^{-1} KCN in a dog. Concentrations are expressed as fractions of whole blood volumes or plasma.[45]

The actual plasma cyanide concentration is almost twice this value at 1.05 μg mL^{-1} (since plasma represents 0.55 mL mL^{-1} whole blood). Since 60% of plasma cyanide was found to be protein-bound, the free cyanide concentration at this point was approximately 0.24 μg mL^{-1}. Also shown is the metabolised cyanide present (calculated as the total ^{14}C–^{14}C in the form of CN). The concentration of cyanide metabolites is similar to the plasma cyanide concentration at 0.83 μg mL^{-1} plasma after 45 minutes.

This pattern of uptake is consistent with the concept that, initially, most cyanide absorbed into the blood rapidly penetrates into the red cells, where it combines with methaemoglobin, a small proportion remaining in the plasma. Since in this case uptake is continuous there is likely to be a small plasma component still being absorbed. Unfortunately, the methaemoglobin content of the blood was not measured for this experiment, but based upon a level of 0.24 g per 100 mL reported in dogs by Bright[41] there could be sufficient present to combine with approximately 3.7 μg mL^{-1} (as shown in Table 10.3). If this was the case it is predicted that, if the infusion had continued, red cell CN uptake (or at least red cell binding to methaemoglobin) would have ceased and there would have been a greater increase in plasma CN. In practice, the infusion was stopped at this point because apnoea occurred. The accumulation of ^{14}C cyanide metabolites in the plasma shows a slow, continuous cyanide metabolism at a rate resulting in plasma levels increasing at a rate of 0.0192 μg mL^{-1} plasma min^{-1}.

This experiment therefore confirms that, during steady continuous uptake of cyanide, similar to that occurring during HCN exposure by inhalation, some plasma cyanide is always present in addition to red cell cyanide. Also, no saturation point was reached for red cell cyanide in this experiment, so either the methaemoglobin was not exhausted or red cells contain cyanide in addition to that reacting with methaemoglobin (for example, as free cyanide inside the red cell), although there was an increase in the proportion of cyanide in plasma. A question arising from this experiment is what would have happened if the KCN injection had been stopped. Would the amounts and proportions of cyanide in the cell and plasma have remained approximately constant before gradually decreasing, or would the plasma cyanide have preferentially decreased?

For another dog experiment, performed by Vesey and Wilson,[54] KCN (16.77 μmol kg^{-1}, representing 0.174 mg CN kg^{-1}) was injected instantaneously. Figure 10.13 shows that even for this low cyanide dose the initial effect of the injected bolus is a rapid increase in both red cell and plasma cyanide, with the greater proportion in the red cells. However, the rate of increase to toxic levels is very rapid and at 2 minutes the proportion in the plasma is greater than in the previous experiment, at approximately 30% of whole blood cyanide. After the peak at 2 minutes, red cell cyanide decreases gradually to 47% of the peak level over 15 minutes, due to dispersal and mixing throughout the different body blood compartments, losses into the tissues fluid compartments and metabolism to thiocyanate.

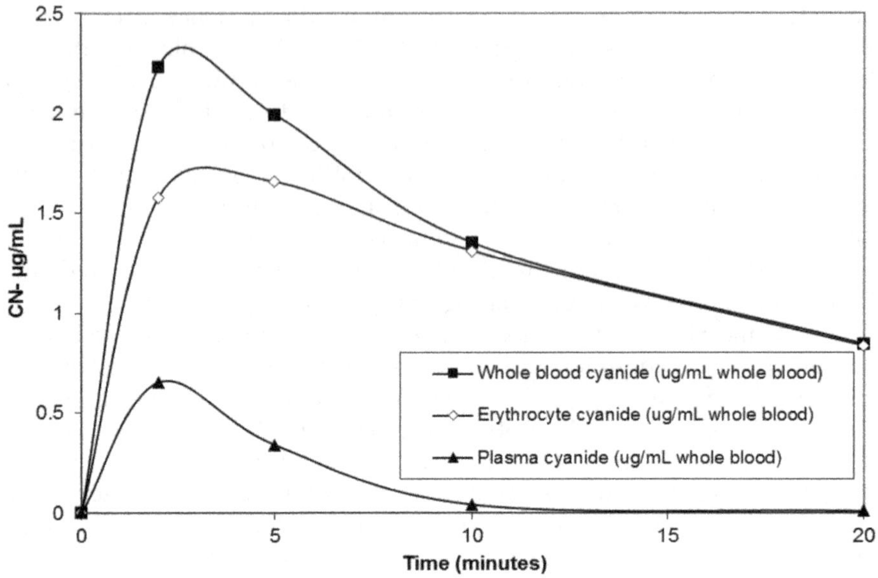

Figure 10.13 Cyanide concentrations in erythrocytes, plasma and whole blood during single intravenous injection of KCN (16.77 μmol kg^{-1}) in a dog. Concentrations are expressed as fractions of whole blood volumes.[54]

The plasma level decreases rapidly to near zero after 10 minutes and this is presumably also due to a combination of dispersal into the extracellular and intracellular body water and detoxification. This experiment also shows red cell and plasma cyanide increasing in tandem during the first 2 minutes, without any obvious lag before red cell uptake. From the results of this experiment it would appear that uptake into both plasma and red cells is rapid, but that plasma cyanide disperses rapidly out of the blood compartment when uptake ceases.

Ballantyne *et al.* carried out a series of experiments in sheep and rabbits.[63,64] Anaesthetised animals were given lethal intramuscular injections of KCN and the plasma, serum and whole blood cyanide levels were measured in samples taken from the heart and other tissues after death. The cyanide distribution recorded is therefore that at death, which occurred 11–26 minutes after administration. The results show some scatter, but are useful in indicating the partitioning between red cells and plasma at high (lethal) blood concentrations, providing further data on the upper limits (if any) of red cell cyanide capacity. Figure 10.14 shows data from Ballantyne's sheep and rabbit experiments, and the dog experiments plotted against whole blood cyanide concentrations. Also plotted are the results of Lundquist *et al.*'s *in vitro* experiments,[57] in which KCN was added to 3 mL blood samples. Cyanide in the different fractions is expressed per mL whole blood, so that the whole blood concentration represents the sum of the red cell and plasma concentrations. Whole blood cyanide is plotted on the

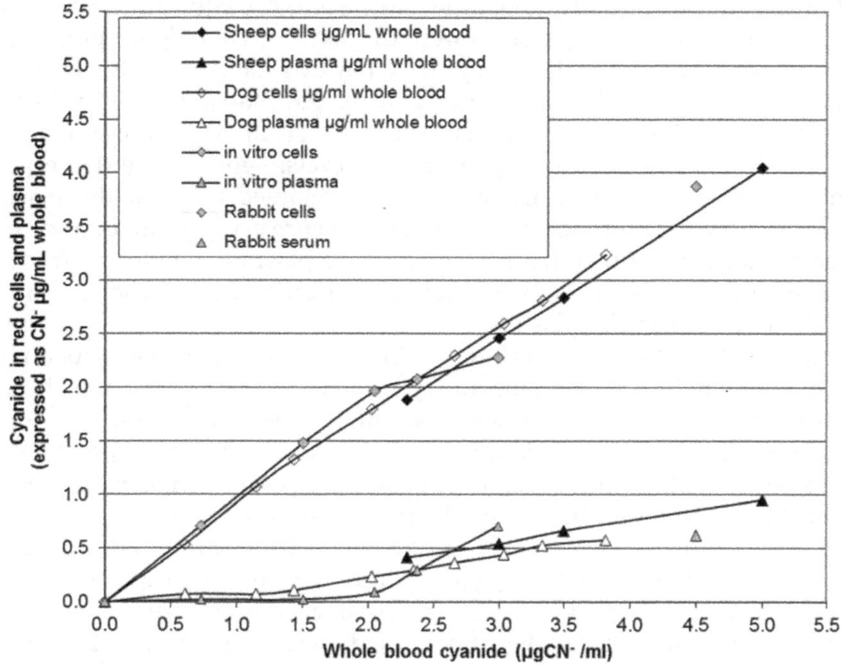

Figure 10.14 Cyanide concentrations in different blood fractions, including whole blood, erythrocytes (cells), plasma or serum, from different *in vivo* and *in vitro* experiments plotted against whole blood concentrations. The concentration for each component is expressed as µg CN per mL whole blood.[57,63,64]

x-axis, while the *y*-axis shows the concentration in the specific blood compartments, but expressed as concentrations per mL whole blood. For example, the right hand side of the figure illustrates that when the amount in the sheep plasma represented a concentration of 0.95 $\mu g\, mL^{-1}$ whole blood that in the red cells represented 4.05 $\mu g\, mL^{-1}$ whole blood. Summing these fractions gives the whole blood concentration of 5 $\mu g\, mL^{-1}$.

For the dog, sheep and rabbit experiments there appears to be no upper limit for red cell cyanide for whole blood concentrations of up to 5.0 $\mu g\, mL^{-1}$ containing a red cell concentration of 4.05 $\mu g\, mL^{-1}$ and a plasma concentration of 0.95 $\mu g\, mL^{-1}$. In none of these experiments was the background methaemoglobin concentration stated, but based upon human data with average values of 0.16–0.32 g per 100 mL, this would be sufficient to sequester approximately as much as 2.7–5.3 µg CN^- mL^{-1} blood. This contrasts with the *in vitro* study, in which the measured methaemoglobin of 0.11 g per 100 mL became saturated with cyanide at 2.31 $\mu g\, mL^{-1}$. Plasma levels were minimal below whole blood concentrations of 2 $\mu g\, mL^{-1}$ in the *in vitro* experiment, although this measured only free cyanide. The levels in the dog experiment, which included total plasma cyanide, were somewhat higher. At 3 μg mL^{-1} whole blood, the plasma and cell concentrations in the *in vitro* and sheep lethality studies were the same, with a higher proportion in the plasma than in the dog study. This may be indicative of the dynamics of plasma levels. In the sheep experiment the whole blood levels had increased to lethal levels, so that the rate of uptake of cyanide was sufficiently rapid to produce lethal plasma levels, while in the dog experiment the slower rate of infusion may have enabled more of the plasma cyanide to flow away into the tissues. For the *in vitro* experiment, cyanide was added to blood contained in a Petri dish so the plasma cyanide was trapped in the dish and once the cells had reached saturation point the rest of the cyanide remained in the plasma.

Experiments by Marrs and Bright[46] using dogs provide further data on the dynamics of whole blood and plasma cyanide. For these experiments, beagle bitches were injected with intravenous single doses of 0.67 mg kg^{-1} HCN (similar to the low dose primate inhalation exposures of approximately 0.96 mg kg^{-1} HCN). Whole blood and plasma cyanide were measured 5, 15 and 45 minutes after injection. The results, summarised in Figure 10.15, show that whole blood cyanide peaked at 2.14 μg mL^{-1} cyanide after 5 minutes (compared to 2.02 μg mL^{-1} after 30 minutes exposure in the primates). At this time in the dogs the plasma cyanide concentration was approximately one tenth of the whole blood concentration at 0.22 μg mL^{-1} (representing a whole blood concentration of 0.12 μg mL^{-1} for this component). The right hand axis shows plasma plotted as a fraction of whole blood concentration rather than the directly measured plasma

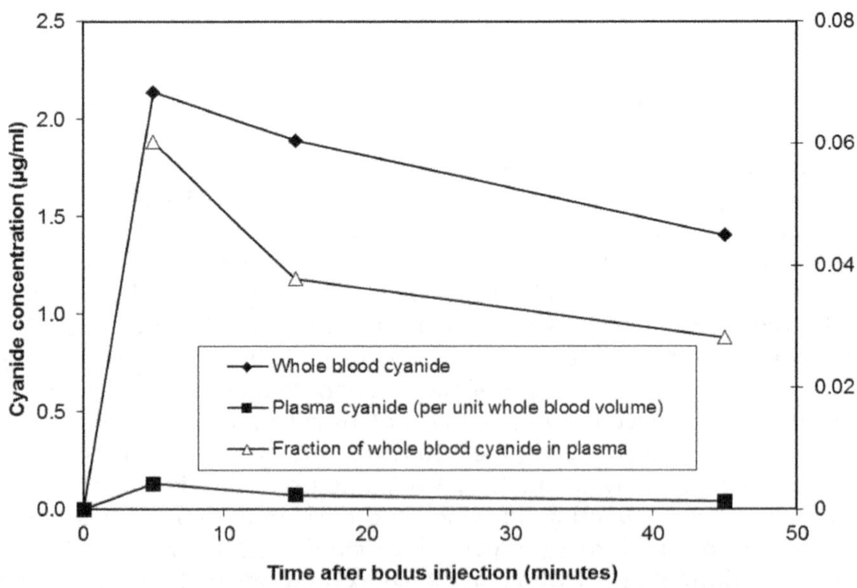

Figure 10.15 Cyanide concentrations in whole blood and plasma and during intravenous bolus injection of HCN (0.67 mg kg^{-1}) in a dog.[46]

concentrations on the left hand axis. From these peak values the whole blood cyanide decreased gradually over the next 40 minutes to 1.4 μg mL^{-1}, with a half-life of 71 minutes (compared to 74 minutes in primates). However, the plasma concentration decreased much more rapidly, to 0.07 μg mL^{-1} after 40 minutes, with a half-life of only 22 minutes. At 5 minutes the plasma cyanide represented 6% of the whole blood cyanide, while 10 minutes later it was 4% and after 40 minutes it was 3%.

10.6.7 Summary of Blood Partition and Distribution Findings

Although the results of these experiments reveal the complexity of the parameters influencing the uptake, partitioning and dispersal of cyanide, and that some aspects are not fully resolved, the basic findings may be summarised as follows:

- Inhaled or intravenously injected cyanide is initially dispersed into plasma but penetrates rapidly into red cells over time-scales of seconds to a minute or so.
- Red cells contain methaemoglobin, for which the normal range for humans is given as 0.01–0.5 g per 100 mL blood. An average value would be around 0.225 g per 100 mL. Whole blood levels of 0.16 g per 100 mL were reported for cynomolgus monkeys and 0.225 g per 100 mL in beagles. These levels are capable of binding 2.6–3.5 μg CN mL^{-1} whole blood as cyanomethaemoglobin. Up to the level of saturation, the bound component represents approximately 90–95% of whole blood cyanide, but the proportions can vary over short periods depending upon the dynamics of uptake and dispersal. Despite the limits on red cell methaemoglobin, no upper limits on red cell cyanide concentration have been apparent up to whole blood cyanide concentrations of approximately 5 μg mL^{-1}. Since red cells have a significant fluid content it is likely that high concentrations of red cell cyanide consist of a mixture of cyanide bound to methaemoglobin and essentially detoxified, and free cyanide in solution or possibly loosely bound to normal haemoglobin.[65]
- Cyanide in plasma rapidly mixes and reaches equilibrium with interstitial fluid and intracellular fluids. Equilibrium times may vary somewhat between different tissues depending on the perfusion of different organs, but have a half-life of around 30–60 seconds for the major organs including the brain and heart.
- Actual plasma levels vary depending upon the dynamics of uptake and dispersal into and from the blood. When HCN is introduced gradually by inhalation or intravenously at relatively low concentrations, by far the major proportion is initially taken up into red cells, so that plasma and tissue fluid cyanide remain low until methaemoglobin approaches saturation. When HCN is introduced rapidly by inhalation during an

episode of hyperventilation, especially at higher concentrations above approximately 150 ppm, or by intravenous injection of a bolus dose of cyanide around 1 mg CN kg^{-1} bodyweight or more, then a transient plasma peak is likely to occur and then subside over a period of a few minutes as the cyanide disperses into the tissues.

- Uptake of inhaled HCN is approximately $0.6\dot{V}_E F_{iHCN}$ (where F_{iHCN} is the inhaled HCN concentration in mg L^{-1}). Excretion in exhaled breath is minor over time-scales of minutes. At blood levels up to around 3 μg mL^{-1} in dogs, the rate of metabolism of ^{14}C cyanide was equivalent to a reduction in plasma cyanide of 0.0192 μg mL^{-1} min^{-1}. However, in cynomolgus monkeys, post-exposure the rate of decrease in blood cyanide was slow at blood levels above 3 μg mL^{-1}.
- In anaesthetised dogs exposed continuously to intravenous cyanide, apnoea occurred at a plasma concentration of 1.04 μg mL^{-1} (whole blood = 3.3 μg mL^{-1}) and ventricular extrasystoles at 1.8 μg mL^{-1}. Since these dogs were anaesthetised with halothane and nitrous oxide it is possible that somewhat higher concentrations might have been tolerated in waking animals. In cynomolgus monkeys exposed to HCN by inhalation, incapacitation with syncope and reduced respiration occurred at whole blood cyanide concentrations above approximately 3.0 μg mL^{-1}.

10.6.8 A Calculation Model for Uptake and Dispersal of Inhaled HCN in Cynomolgus Monkeys and Humans

The results of the experiments in which waking cynomolgus monkeys were exposed for 30 minutes periods to three concentrations of HCN (Figures 10.5 and 10.9) gave rise to a number of issues relating to the blood levels and clinical effects. In particular, for the higher concentration exposures, why the whole blood cyanide increased rapidly during the first few minutes and then more or less levelled out, despite the continued cyanide exposure, and why the animals showed an initial period of incapacitation as the whole blood level exceeded approximately 3.0 μg mL^{-1}, but then little change or a partial recovery during the remainder of the exposure, and a full recovery soon after the end of the exposure period, despite the whole blood cyanide concentrations being maintained at around 4.0 μg mL^{-1} throughout this period.

Since the inhaled cyanide concentration and $\dot{V}_E$ were measured in addition to the arterial blood cyanide concentration it is possible to calculate the uptake dose of CN with time and the volume of distribution. From the data on methaemoglobin concentration in cynomolgus monkey blood,[43] and the volumes of blood, interstitial and intracellular body fluid compartments, it is also possible to estimate the uptake and dispersal dynamics, assuming an appropriate half-life for mixing between blood and the other fluid compartments.

Figure 10.16 shows the measured arterial whole blood cyanide concentration and calculated cyanide concentrations (CN^- μg mL^{-1}) in different

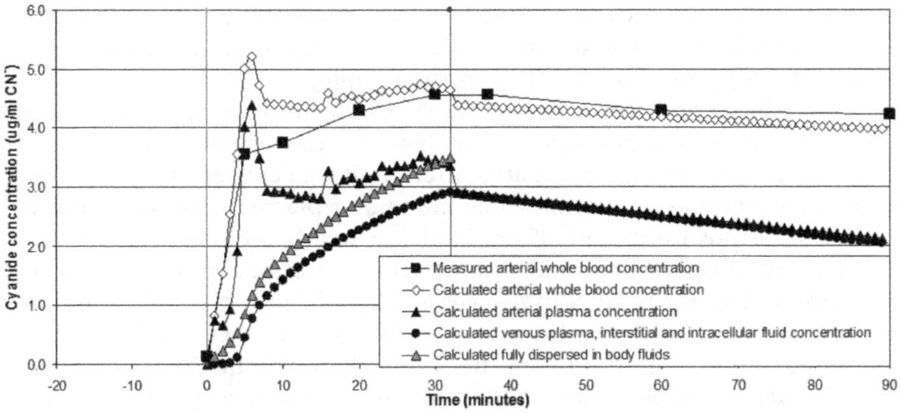

Figure 10.16 Comparison of calculated with measured arterial whole blood CN, and calculated CN concentrations in arterial plasma and venous plasma (also representing interstitial and intracellular CN) for the cynomolgus monkey HCN exposure shown in Figure 10.9.

compartments for the waking cynomolgus monkey exposure detailed in Figure 10.9. By the end of the 32 minute exposure period to 147 ppm HCN, the calculated inhaled dose was 1.84 mg $CN^- kg^{-1}$. If the inhaled CN is assumed to be evenly dispersed throughout the total blood, interstitial and intracellular body fluid volume of 2.6 L, the predicted concentration at the end of the exposure period is 3.49 $\mu g\,mL^{-1}$. The measured arterial blood concentration was 4.48 $\mu g\,mL^{-1}$ and the volume of distribution 420 $mL\,kg^{-1}$.

During the first few minutes of the exposure there is a very rapid increase in both calculated and measured whole arterial blood cyanide, with a peak after 5–6 minutes during the episode of hyperventilation. For the first 3 minutes, most of the inhaled cyanide combines with methaemoglobin up to a saturation level of 2.8 μg CN mL^{-1} whole blood, after which all subsequent cyanide remains in the plasma and begins to distribute to the interstitial and intracellular body fluids with an assumed half-life of 1 minute. Due to the rapid HCN uptake during the first few minutes, an arterial plasma peak of around 4 $\mu g\,mL^{-1}$ is predicted, the difference between this and the whole blood concentration representing the constant red cell cyanomethaemoglobin component. Also plotted is the calculated cyanide concentration in the venous return plasma, which is considered to be the same as the concentration in the interstitial and intracellular fluids. This is predicted to increase gradually after around 5 minutes as significant amounts of cyanide are mixed into the tissue fluids.

At 11 minutes into the exposure, just after this calculated plasma CN peak, the animal became semi-conscious and the respiration and heart rate decreased. At this point the calculated inhaled dose was 1.02 $mg\,kg^{-1}$ and the calculated venous plasma CN was 1.5 $\mu g\,mL^{-1}$, which compares with the intravenous dose in the continuously infused beagle of 1.1 $mg\,kg^{-1}$, giving a plasma concentration 0.92 μg CN mL^{-1} or 1.57 $\mu g\,mL^{-1}$ in terms of

CN-derived ^{14}C. The measured venous whole blood concentration in the dog was 3.3 $\mu g\, mL^{-1}$ compared with a calculated venous whole blood concentration of 3.8 $\mu g\, mL^{-1}$ at this time in the monkey.

During the period from approximately 10–20 minutes into the exposure when the rate of cyanide uptake was low due to the reduced ventilation, there is a decrease in the calculated arterial plasma concentration, followed by a small increase in arterial whole blood and plasma cyanide toward the end of the exposure period. There is also a steady increase in the venous plasma and intracellular fluid cyanide concentration. At the end of the exposure period as the HCN intake ceases there is a rapid small drop in the arterial whole blood and plasma concentrations as they equilibrate with the venous and tissue fluid concentrations. Throughout the exposure and recovery periods it is assumed that plasma and tissue cyanide is slowly metabolised at a rate similar to that measured in the beagle. The result is a gradual decrease in plasma and tissue cyanide, as shown in Figure 10.16, to a concentration of 2 $\mu g\, mL^{-1}$ after 60 minutes recovery. This coincides with a smaller decrease in the whole blood cyanide concentration, due to the presence of maintained red cell cyanide (and because the plasma component is only 55% of whole blood).

Although this calculation model contains some approximations it provides a reasonably good prediction of the actual HCN uptake and measured arterial blood cyanide concentrations during the exposure and recovery periods in the cynomolgus monkey exposed to 147 ppm HCN. The measured and calculated dose, blood and plasma concentrations after 11 minutes exposure are also consistent with those measured following continuous infusion of a similar dose in the dog by Christel *et al.*[45]

The model demonstrates why the whole blood concentration profile was as measured, with the rapid initial increase, plateau and slow decline during the recovery period. It also provides a possible explanation of the rapid onset of incapacitation during the early stages of the exposure, followed by partial recovery and rapid recovery during the post-exposure period. During the period of HCN inhalation it may be that some well-perfused tissues, such as the heart and brain, are exposed to cyanide concentrations closer to the arterial plasma concentrations than the venous concentrations. From the calculated profile, incapacitation may occur in response to the early arterial plasma cyanide peak (perhaps exacerbated by the transient hypocapnia at this time). The subsequent decrease in respiration and arterial plasma cyanide (and recovery from hypocapnia) may then lead to the observed partial recovery and improvement in clinical signs. This is then followed by a slow decline in clinical condition as the cyanide concentrations in blood and tissues gradually increase towards the end of exposure.

When exposure ceases there is a rapid small decrease in arterial plasma cyanide followed by gradual slow decrease in tissue cyanide due to cyanide metabolism, leading to a recovery of consciousness as levels decrease below a critical threshold.

These calculations may provide the basis for a better predictive calculation model for time and dose to incapacitation in humans exposed to HCN inhalation and help to explain the deviations from Haber's rule for time and dose to incapacitation during HCN inhalation. It is proposed that a key predictor of incapacitation is the arterial plasma cyanide concentration, or the whole blood cyanide concentration less a constant for CN combined with the background methaemoglobin. Loss of consciousness, bradycardia and hypoventilation are likely to occur when the arterial plasma cyanide exceeds a critical value of approximately 1–2 $\mu g\,mL^{-1}$ (whole blood level above approximately 3.0 $\mu g\,mL^{-1}$) and recovery from a brief period of intoxication when it falls below this level.

When the inhaled HCN concentration exceeds approximately 140 ppm and induces an episode of hyperventilation then blood levels are predicted to exceed this critical level within a few minutes. For lower HCN concentrations, providing less of a ventilatory stimulus, the rate of uptake is slower, so that an arterial plasma spike is avoided and incapacitation occurs after a longer period as the accumulated concentration in plasma and tissue fluids exceeds 1–2 $\mu g\,mL^{-1}$.

For the two animals exposed to the intermediate concentration of around 99 ppm, with a more gradual hyperventilation, incapacitation occurred as the whole blood level reached 3.0 $\mu g\,mL^{-1}$ near the end of the exposure. As with the high exposure group, they recovered consciousness within a few minutes despite their whole blood level remaining above 3.0 $\mu g\,mL^{-1}$ for more than 10 minutes. For the low dose pair there were no significant adverse signs and the maximum whole blood level was 2.2 $\mu g\,mL^{-1}$.

10.7 Cyanide Metabolism and Rate of Cyanide Clearance from the Blood Post-exposure

The main mechanism for loss of cyanide is following metabolism through a number of processes of which the major pathway (responsible for about 80% of cyanide metabolism) is by conversion to thiocyanate by rhodanese in the presence of a sulfur donor (mainly disulfate) and then excretion *via* the urine.[66] Thiosulfate can also play this role as a sulfur donor when it is used as an antidote in cases of cyanide poisoning. The rate of metabolism is relatively slow in relation to incapacitating exposures of a few minutes to HCN (for example, during fires) so that, during continuous intravenous infusion in dogs, the plasma clearance of ^{14}CN was 0.0192 $\mu g\,mL^{-1}$ plasma min^{-1}.[45] A complicating factor is that cyanide metabolism is an oxidative process, which is itself inhibited by the effects of cyanide on oxidation and acidosis.[32]

The effects of blood cyanide concentration on cyanide metabolism are evident in the cynomolgus HCN inhalation experiments shown in Figure 10.5, in which the decrease in whole blood cyanide over a period of 1 hour post-exposure was much more rapid for the low dose animals than

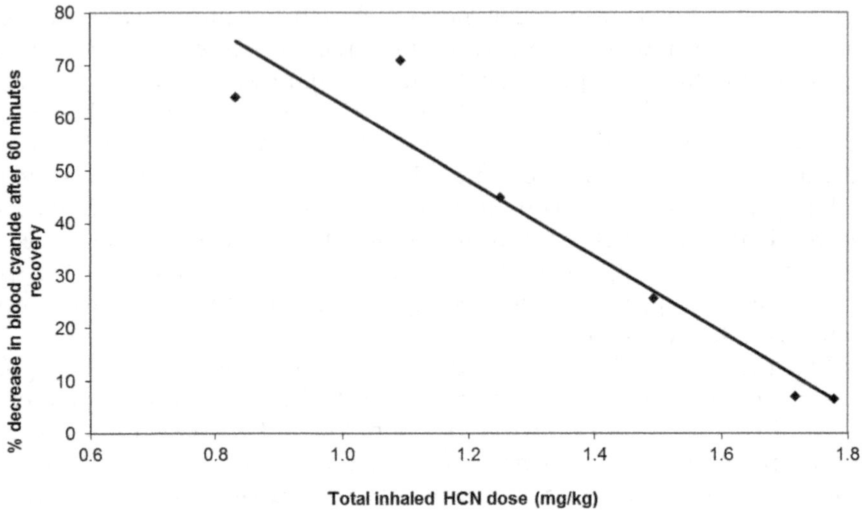

Figure 10.17 Relationship between percentage decrease in blood CN^- concentration and total inhaled HCN body dose (mg CN kg^{-1}) after 60 minutes recovery.

the intermediate and high dose animals. By the end of the exposure period, approximately 10% of the inhaled HCN dose remained in the blood while 90% was in the tissues. This total body burden of cyanide then gradually decreased as it was metabolised (presumably to thiocyanate and other metabolites) and excreted. Figure 10.17 shows the percentage decrease in blood concentration after 60 minutes as a function of total inhaled dose. For the two high concentration exposures at 149 ppm HCN, the total inhaled dose was 8 mg cyanide ($\sim$1.8 mg kg^{-1}) and the blood concentration decreased by only 5% over 60 minutes. A greater decrease occurred for the intermediate doses, and around 70% for the 53 ppm exposures with doses of around 4–5 mg ($\sim$1 mg kg^{-1}).

Table 10.4 summarises the blood decrease data. As shown in Figure 10.5, the rate of decrease in blood cyanide during the recovery period was approximately linear at all blood levels. This suggests that at these blood levels the enzyme system was saturated. On this basis, the half-life for the high dose animals was approximately 417 minutes, decreasing to 74 minutes for the low dose animals.

Assuming the decrease in whole blood cyanide is proportional to the overall decrease in body cyanide, the mass of cyanide metabolised in 60 minutes was approximately five times greater for the low dose animals than for the high dose animals. It is considered likely that the greater degree of hypoxic stress in the high dose animals inhibited the metabolism of HCN in the liver or other sites. These figures compare with reported two-phase half-lives in dogs, for which the first phase was 18.4–24 minutes, and 330 minutes for the second phase.[46] Djerad *et al.*[49] reported a three-phase half-life for low intravenous instantaneous doses of radio-labelled cyanide in

Table 10.4 Decreases in blood cyanide during 60 minute recovery period and half-lives.

HCN	Blood CN^- after 30 minute exposure/ $\mu g\,mL^{-1}$	Blood CN^- after 60 minute recovery/ $\mu g\,mL^{-1}$	Decrease after 60 minutes/ $\mu g\,mL^{-1}$	Rate of decrease/ $\mu g\,mL^{-1}\,h^{-1}$	Half-life/ min	Cyanide loss in 1 hour[a]/ $\mu g\,kg^{-1}$	Loss rate/ mg $CN^-\,kg^{-1}\,min^{-1}$
149	4.16	3.89	0.30	0.30	417	158	0.003
99	3.29	2.16	1.13	1.13	115	597	0.010
53	2.02	0.64	1.38	1.38	74	729	0.012

[a]Assuming a distribution throughout plasma, interstitial and intracellular volume of 528 mL kg^{-1}.

rats, involving an initial rapid distribution phase with a half-life of 21.6 seconds, a second phase with a half-life of 2.4 minutes and a long phase with a half-life of 87.4 minutes. These reports appear to be consistent with the findings from the primate work in that there is obviously a very rapid distribution phase occurring during continuous inhalation dosing over time-scales of a few seconds to a minute or so, and a longer detoxification phase following the end of dosing with a half-life of around 74 minutes at relatively low dose levels, which is similar to that reported for rats of 87.4 minutes.

The rate of detoxification (removal) at the low dose level ($\sim$0.012 mg kg^{-1} bodyweight min^{-1}) is also comparable with that of 0.017 mg kg^{-1} min^{-1} estimated for humans by McNamara.[40] However, the long detoxification time-scales associated with high dose levels in the monkeys do not appear to have been reported previously.

10.8 Forensic Considerations with Respect to HCN Exposure and Post-mortem Blood Cyanide

Following fire incidents, measurements of blood carboxyhaemoglobin and blood cyanide may be important in order to establish the nature and extent of exposure to fire effluents during incidents and possible causes of incapacitation and death. A high blood carboxyhaemoglobin concentration in the blood of a fire survivor is indicative of a significant exposure to smoke and toxic gases during a fire, and since carboxyhaemoglobin is stable in the blood of cadavers this also applies to fire fatalities. Since all burning fuels contain carbon but only some contain nitrogen, it is possible to find fire survivors and fatalities with high blood carboxyhaemoglobin concentrations but low blood cyanide. However, most domestic fires involve upholstered furnishing with relatively high nitrogen content and HCN concentrations in such fires are typically around 500 ppm. For this reason it might be predicted that for some fire victims there would be a relationship between %COHb and blood cyanide concentrations, and this does seem to be the case, in that high blood cyanide concentrations are associated with high blood %COHb concentrations.[8–10,67] High concentrations of both HCN and

CO have been measured during aircraft cabin fire tests[12,13] and high blood cyanide concentrations were measured in the blood of the Manchester Airtours fire victims.[19] Cyanide is unstable in cadavers and stored blood, so that the high blood levels measured in fresh blood samples from domestic fire survivors and domestic fire fatalities demonstrate that HCN is an important toxic agent. For one study,[18] fresh blood samples were taken from survivors and victims at the fire scene. The range in survivors was 0–1.2 μg CN mL^{-1} and in fatalities was 0.9–5.4 μg CN mL^{-1}.

As stated, the uptake and effects of HCN inhalation tend to be somewhat self-limiting during fire incidents, so that cyanide is considered more important as a cause of rapid incapacitation during a fire rather than as a primary cause of death. Based upon experience with primate exposures it is considered that exposed subjects surviving long enough to reach hospital are likely to recover from the effects of HCN inhalation at the fire scene with supportive therapy. A number of treatments are available for cyanide poisoning, but most have certain limitations. Treatment may be indicated if cyanide can be smelled on the breath and, as stated, it has been found that the HCN concentration in exhaled breath correlates well with blood cyanide concentration. Of course, such subjects are also likely to have a high blood %COHb concentration, requiring normobaric or hyperbaric oxygen therapy. From the findings reported here it is considered that such treatments will also be beneficial in ameliorating the effects of any cyanide inhaled.

From the perspective of establishing the extent of exposure to HCN at the fire scene and the likely contribution of HCN exposure to incapacitation and death during an incident, it is necessary to establish the relationship between blood cyanide and tissue cyanide measurements made from samples taken at autopsy and the likely blood cyanide concentrations at the time of exposure. Unlike the relationship between post-mortem blood carboxyhaemoglobin concentration and carbon monoxide exposure at a fire scene, which can be assessed quantitatively, interpretation of blood cyanide data is much more challenging. A particular set of issues relates to changes in cyanide levels in cadavers and in stored blood samples. These have been reviewed by McAllister *et al.*[17] Essentially, the main issues with respect to interpretation of blood cyanide data are as follows:

- There is a low natural background concentration of cyanide in blood, which is slightly elevated in smokers and following ingestion of certain foodstuffs (non-smokers 0.075 μg mL^{-1}, smokers 0.184 μg mL^{-1}).[8]
- As described in Section 10.7, short exposure to high cyanide concentrations can cause incapacitation at a fire scene within a few minutes, in association with transient high blood cyanide concentrations, which can then decrease to much lower levels once incapacitation has occurred, so that blood concentrations at a fire scene at the time of death may be considerably lower than during the critical exposure period when incapacitation occurred.

- Cyanide is unstable in the blood of cadavers and there can be a considerable decrease between death at a fire scene and the time when a blood sample is taken at autopsy. Curry[68] found in a case of death by inhalation of cyanide vapour that the blood level taken at the moment of death was 3.5 $\mu g\,mL^{-1}$, while samples taken at autopsy the next day were 1.0 $\mu g\,mL^{-1}$ (femoral) and 0.5 $\mu g\,mL^{-1}$ (carotid), a decrease of approximately 79%. Ballantyne *et al.*[69] took samples at intervals from 15 rabbits killed by intravenous KCN (8 mg $CN^-\,kg^{-1}$ bodyweight) stored at a temperature of 10–15 °C. The blood concentrations decreased from 5.41 $\mu g\,mL^{-1}$ immediately after death, by 30% after 1 day and 62% after 3 days, and then to near zero after 14 days.
- Another problem is that cyanide levels continue to decrease in blood stored after removal. Ballantyne[70] studied the effect of storage under three conditions: room temperature (20 °C), refrigerated (4 °C) and frozen (−22 °C). The results were that at room temperature and 4 °C there were considerable decreases of blood level, which fluctuated with time. After 3 months at 4 °C there was an approximate 20% decrease in concentration.
- Change in cyanide concentration in frozen blood samples. In the above study, small changes occurred in blood samples following freezing and thawing. Frozen samples containing significant cyanide concentrations were generally the same when thawed and measured after 3 months as when they were first taken. The exception was the control blood, which showed a small increase of approximately 0.1 $\mu g\,mL^{-1}$ within a few days of freezing, the levels then remaining constant for the remainder of the 3 months. Although this was a small increase it represented a large percentage change in the control value of 0.03 $\mu g\,mL^{-1}$. The increase is considered to be due to re-conversion of small amounts of thiocyanate to cyanide. This effect is considered likely to be significant only when examining low concentrations of cyanide, such as is found in the blood of smokers, but minor in the context of blood cyanide concentrations relevant to cyanide poisoning during fires.
- In some situations large increases in blood cyanide have been found to occur in stored blood samples. Although not fully understood, this phenomenon is considered to be due to contamination of samples by cyanogenic bacteria. However, in several large studies of post-mortem blood samples taken from fire victims[8,18,20] such increases have not been reported.
- Sodium fluoride has been found to be an effective stabiliser for cyanide in stored blood, both from the perspective of decay and increases.[71]

An example of how changes in blood cyanide can affect the interpretation of conditions during a major fire incident is presented in Figure 10.18, which summarises opinion on likely blood cyanide concentration during exposure of decedents in the San Juan Dupont Plaza Hotel fire.[72] In this incident the bodies of decedents contained different concentrations of

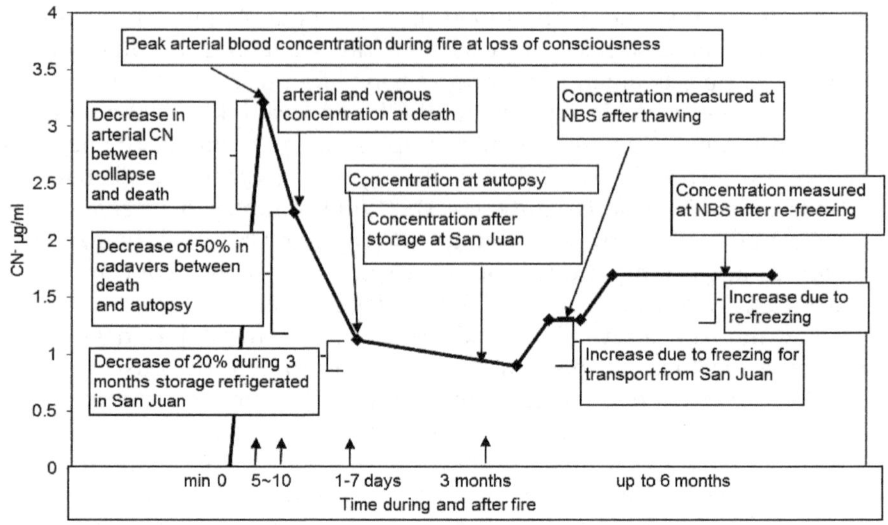

Figure 10.18 Predicted changes in blood cyanide concentrations in a sample between exposure and analysis. From the San Juan Dupont Plaza hotel fire. NBS, National Bureau of Standards.

carboxyhaemoglobin depending upon their location during the fire. Some victims exposed to intense flaming conditions had relatively low concentrations, while those exposed to fire effluent in more protected locations had much higher carboxyhaemoglobin concentrations consistent with incapacitation and death from asphyxia. The blood samples also contained cyanide, but there was some question as to whether the measured concentrations were sufficient for HCN to have made a significant contribution to incapacitation and death at the scene. When frozen samples shipped to the National Bureau of Standards were thawed and tested, the average cyanide concentration was 1.3 $\mu g\,mL^{-1}$, but when samples were refrozen and subsequently re-thawed an increase of 0.4 $\mu g\,mL^{-1}$ occurred, most likely due to re-conversion of thiocyanate to cyanide.[73] If this is subtracted from the original measured value it gives a "true" average value of 0.9 $\mu g\,mL^{-1}$ in the frozen samples sent from San Juan. However, at San Juan, the blood sample taken from the decedents had been stored under refrigeration for 3 months, during which time it is estimated that cyanide concentrations would have decreased by approximately 20%, giving a value of 1.125 $\mu g\,mL^{-1}$ at the time the samples were taken at autopsy. Also, due to the large number or bodies involved, the autopsies and blood samples were taken over a period of 6 days, so that over this period it is estimated that blood cyanide concentration would have decreased by at least 50% on average in cadavers, giving a value of 2.25 $\mu g\,mL^{-1}$ at time of death during the fire. Based upon the primate work it is likely that the concentration at death would have been significantly lower than that at the time of

Table 10.5 Blood cyanide concentrations and signs.

Blood cyanide/ $\mu g\,mL^{-1}$	Sign
0.2–0.5	No signs (Hall[74])
0.5–1.0	Flushing, tachycardia (Hall)
1.0–2.5	Intoxicated, narcotised (Hall)
1.0–2.0	Lowest level suggested as compatible with death from acute cyanide poisoning (Ballantyne and Marrs[75])
1.7	Human fatality (Ballantyne and Marrs)
2.0	Concentration immediately after death in rabbits (Ballantyne[64])
2.15	Rapid loss of consciousness in dogs (Ballantyne and Marrs)
2.4	Human fatality (Ballantyne and Marrs)
2.5	Rapid loss of consciousness in primates (Purser *et al.*[14])
2.5–3.0	Coma (Hall)
3.4	Rapid loss of consciousness in primates (Purser *et al.*[14])
3.5	Human case at death, 0.75 $\mu g\,\mu L^{-1}$ at autopsy next day (Curry[20])
3.6	Rapid death in dogs (Ballantyne and Marrs)
5.0	Human fatality (Ballantyne and Marrs)
In summary:	
~2.0–2.5	Concentration in fresh post-mortem blood consistent with incapacitation due to HCN
~3.0	Concentration in fresh post-mortem blood consistent with death due to HCN
~1.0–1.25	Concentration consistent with incapacitation due to HCN in post-mortem blood not taken and analysed immediately after exposure
~1.5	Concentration consistent with death due to HCN in post-mortem blood not taken and analysed immediately after exposure

incapacitation. Assuming even a small decrease of 30% means that the blood concentration at the time of incapacitation could have been at least as high as 3.2 $\mu g\,mL^{-1}$, which is a typical concentration associated with loss of consciousness. Based upon these data and the fuels involved in the fire, it was concluded that although the actual blood concentrations measured at the National Bureau of Standards were relatively low, they were sufficient to indicate the presence of HCN inhaled by the victims (since they were well in excess of control levels) and that, due to the effects described, they were consistent with concentrations at the fire scene sufficient to have caused incapacitation.

Table 10.5 summarises reports and opinion on blood cyanide concentrations associated with signs of different severity, although these should be taken as only a very approximate guide for the reasons stated.

References

1. A. Tewarson, Generation of Heat and Chemical Compounds in Fires, in *SFPE Handbook of Fire Protection Engineering*, ed. P. J. DiNenno *et al.*, National Fire Protection Association, Quincy, MA, 3rd edn, 2002, pp. 82–161.

2. D. A. Purser, Toxic Product Yield and Hazard Assessment for Fully Enclosed Design Fires Involving Fire Retarded Materials, *Polym. Int.*, 2000, **47**, 1232–1255.
3. D. A. Purser, A. A. Stec and T. R. Hull, Fire scenarios and combustion conditions, in *Fire Toxicity*, ed. A. Stec and R. Hull, Woodhead, Cambridge, UK, 2010, ch. 2, pp. 26–50.
4. D. A. Purser, A. A. Stec and T. R. Hull, Effects of material and fire conditions on toxic product yields, in *Fire Toxicity*, ed. A. Stec and R. Hull, Woodhead, Cambridge, UK, 2010, ch. 14, pp. 515–540.
5. D. A. Purser, Assessment of Hazards to Occupants from Smoke, Toxic Gases and Heat, in *The SFPE Handbook of Fire Protection Engineering*, ed. DiNenno P. J., National Fire Protection Association, Quincy, MA 02269, 4th edn, 2008, pp. 96–193.
6. G. L. Nelson, Carbon monoxide and fire toxicity: A review and analysis of recent work, *Fire Technol.*, 1998, **34**, 38–58.
7. D. A. Purser and J. A. Purser, HCN yields and fate of fuel nitrogen for materials under different combustion conditions in the ISO 19700 tube furnace. Fire Safety Science – Proceedings of the ninth international symposium, 2008, International Association for Fire Safety Science, pp. 1117–11128.
8. R. A. Anderson, I. Thompson and W. A. Harland, The Importance of Cyanide and Organic Nitriles in Fire Fatalities, *Fire Mater.*, 1979, **3**, 91–99.
9. R. A. Anderson and W. A. Harland, Fire deaths in the Glasgow area III – The role of hydrogen cyanide, *Med. Sci. Law*, 1982, **22**, 35–40.
10. R. A. Anderson, A. A. Watson and W. A. Harland, Fire Deaths in the Glasgow Area: 1. General Conclusions and Pathology, *Med. Sci. Law*, 1981, **21**, 175–183.
11. D. A. Purser and W. D. Woolley, Biological Studies of Combustion Atmospheres, *J. Fire Sci.*, 1983, **1**, 118–145.
12. D. A. Purser, Behavioural Impairment in Smoke Environments, *Toxicology*, 1996, **115**, 25–40.
13. D. A. Purser, Modeling Time to Incapacitation and Death from Toxic and Physical Hazards in Aircraft Fires, in *Conference Proceedings No. 467, Aircraft Fire Safety*, NATO-AGARD, Sintra, Portugal, May 22–26, 1989, pp. 1–13.
14. D. A. Purser, P. Grimshaw and K. R. Berrill, Intoxication by Cyanide in Fires: A Study in Monkeys Using Polyacrylonitrile, *Arch. Environ. Health*, 1984, **39**, 394–400.
15. D. A. Purser, *Determination of Blood Cyanide and Its Role in Producing Incapacitation in Fire Victims*, Royal Society of Chemistry Meeting, Huntingdon Research Centre, 1984, 6th June 1984.
16. D. A. Purser, Asphyxiant components of fire effluents, in *Fire Toxicity*, ed. A. Stec and R. Hull, Woodhead, Cambridge, UK, ch. 4, 2010, pp. 118–198.
17. J. L. McAllister, R. J. Roby, B. Levine and D. Purser, Stability of cyanide in cadavers and in postmortem stored tissue specimens: A review, *J. Anal. Toxicol.*, 2008, **32**, 1–9.

18. F. J. Baud, P. Barriot, V. Toffis *et al.*, Elevated Blood Cyanide Concentrations in Victims of Smoke Inhalation, *N. Engl. J. Med.*, 1991, **325**, 1761–1766.
19. D. F. King, *Aircraft Accident Report 8/88*, UK Department if Transport, Air Accidents Investigation Branch, HMSO, London, UK, 1988.
20. A. S. Curry, Cyanide poisoning, *Acta Pharmacol. Toxicol.*, 1963, **20**, 291–294.
21. B. Ballantyne, The forensic diagnosis of acute cyanide poisoning, in *Forensic Toxicology*, Bristol, Wright, 1974, pp. 99–113.
22. R. P. Smith, Toxic Responses of the Blood, in *Casarett and Doull's Toxicology*, ed. J. Doull, C. Klassen, M. Amdur, 2nd edn, 1975, ch. 14, pp. 311–331.
23. C. A. Piantadosi, A. L. Sylvia and F. Jőbis-Vandervliet, Differences in brain cytochrome responses to carbon monoxide and cyanide in vivo, *J. Appl. Physiol.*, 1987, **62**, 1277–1284.
24. F. L. Rodkey, A mechanism for the conversion of oxyhaemoglobin to methaemoglobin by nitrite, *Clin. Chem.*, 1976, **22**, 1986–1990.
25. D. A. Purser and P. Grimshaw, The Incapacitative Effects of Exposure to the Thermal Decomposition Products of Polyurethane Foams, *Fire Mater.*, 1984, **8**, 10–16.
26. G. E. Hartzell, D. N. Priest and W. G. Switzer, Modeling of Toxicological Effects of Fire Gases: II. Mathematical Modeling of Intoxication of Rats By Carbon Monoxide and Hydrogen Cyanide, *J. Fire Sci.*, 1985, **3**(2), 115–128, DOI: 10.1177/073490418500300204.
27. G. E. Hartzell, W. G. Switzer and D. N. Priest, Modeling of Toxicological Effects of Fire Gases: V. Mathematical Modeling of Intoxication of Rats By Combined Carbon Monoxide and Hydrogen Cyanide Atmospheres, *J. Fire Sci.*, 1985, **3**(5), 330–342, DOI: 10.1177/073490418500300504.
28. B. C. Levin, J. L. Gurman, M. Paabo, L. Baier and T. Holt Toxicological effects of different time exposures to the fire gases carbon monoxide or hydrogen cyanide or to carbon monoxide with hydrogen cyanide or carbon dioxide. US-Japan Panel of Fire Research and Safety May 1987.
29. J. C. Norris, S. J. Moore and A. S. Hume, Synergistic lethality induced by the combination of carbon monoxide and cyanide, *Toxicology*, 1986, **40**, 121–129.
30. D. A. Purser, A bioassay model for testing the incapacitating effects of exposure to combustion product atmospheres using cynomolgus monkeys, *J. Fire Sci.*, 1984, **2**, 20–36.
31. R. Klimmek, C. Roddewig, H. Fladerer and N. Weger, Cerebral blood flow, circulation, and blood homeostasis of dogs during slow cyanide poisoning and after treatment with 4-dimethylaminophenol, *Arch. Toxicol.*, 1982, **50**, 65–76.
32. P. Plawson-Smith, E. C. Jansen, L. Hilsted, A. H. Johnsen and O. Hyldegaard, Effect of acute and delayed hyperbaric oxygen therapy on cyanide whole blood levels during acute cyanide intoxication, *Undersea Hyperbaric Med.*, 2011, **38**, 17–26.

33. B. Ballantyne, Toxicology of cyanides, in *Clinical and Experimental Toxicology of Cyanides*. ed. B. Ballantyne and T. C. Marrs, Wright, Bristol, 1987, ch. 3, pp. 41–125.
34. F. Haber, *Funf Vortrange aus den jaren 1920–1923*, Verlag von Julius Springer, Berlin, 1924.
35. C. R. Crane, D. C. Sanders and B. R. Endecott, *Inhalation toxicology; IX. Times to incapacitation for rats exposed to carbon monoxide alone, to hydrogen cyanide alone, and to mixtures of carbon monoxide and hydrogen cyanide*, Department of Transportation/Federal Aviation Administration, Washington, DC, 1989, Publication No. DOT/FAA/AM.89/4. Available from National Technical Information Service, Springfield, VA 22161, Order No. ADA208195.
36. L. C. Speitel, Fractional effective dose model for post-crash aircraft survivability, *Toxicology*, 1996, **115**, 167–177.
37. G. Kimmerle, Aspects and Methodology for the Evaluation of Toxicological Parameters during Fire Exposure, *J. Combust. Toxicol.*, 1974, **1**, 4.
38. J. L. Bonsall, Survival Without Sequelae Following Exposure to 500 mg/m^3 of Hydrogen Cyanide, *Hum. Toxicol.*, 1984, **3**, 57–60.
39. J. Barcroft, The Toxicity of Atmospheres Containing Hydrocyanic Acid Gas, *J. Hyg.*, 1931, **31**, 1–34.
40. B. P. McNamara, Estimates of the Toxicity of Hydrocyanic Acid Vapors in Man, *EB-TR-76023*, Edgewood Arsenal, Aberdeen Proving Ground, MD, 1976.
41. J. E. Bright, A prophylaxis for cyanide poisoning, in *Clinical and Experimental Toxicology of Cyanides*, ed. B. Ballantyne and T. C. Marrs, Wright, Bristol, 1987, ch. 15, pp. 359–382.
42. J. D. Pritchard, Hydrogen Cyanide – Toxicological Overview, Health Protection Agency, 2007, Version 2.
43. F. W. Stemler, W. A. Groff, A. Kaminskis, R. P. Johnson, H. L. Froehlish and S. F. Hawkins, *A Comparison of the Treatment of Cyanide Poisoning in the Cynomolgus Monkey with Sodium Nitrite or 4-Dimethylaminophenol (4-DMAP), with and without Sodium Thiosulfate*, U.S. Army medical research institute of chemical defense UAARMRICD-TR-94_01, Aberdeen Proving Ground MD, USA. February 1994.
44. J. L. McAllister, Fire Victim Blood Cyanide Stability and the Development of a Cyanide Uptake Model, PhD Thesis, 2010, University of Maryland, Baltimore.
45. D. Christel, P. Eyer, M. Hegemann, M. Kliese, W. Lörcher and N. Weger, Pharmacokinetics of cyanide in poisoning of dogs, and the effect of 4-Di-methylaminophenol or thiosulfate, *Arch. Toxicol.*, 1977, **38**, 177–189.
46. T. C. Marrs and J. E. Bright, Effects on blood and plasma cyanide levels and on methaemoglobin levels of cyanide administered with and without previous protection using PAPP, *Hum. Toxicol.*, 1987, **6**, 139–145.

47. J. F. Nunn, *Applied Respiratory Physiology*, Butterworth and Co, London, 1969.
48. S. Levine, Nonperipheral chemoreceptor stimulation of ventilation by cyanide, *J. Appl. Physiol.*, 1975, **39**, 199–204.
49. A. Djerad, C. Monier, P. Houzé, S. W. Borron, J.-M. Lefauconnier and R. J. Baud, Effects of respiratory acidosis and alkalosis on the distribution of cyanide into the rat brain, *Toxicol. Sci.*, 2001, **61**, 273–282.
50. H. D. Landahl and R. G. Herrmann, Retention of vapors and gases in the human nose and lung, *Arch. Ind. Hyg. Occup. Med.*, 1950, **1**(1), 36–45.
51. R. D. Eastman, *Clinical Haematology*, John Wright, Bristol, 5th edn, 1977.
52. S. Barr, The micro determination of cyanide: Its application to the analysis of whole blood, *Analyst*, 1966, **91**, 268–272.
53. C. J. Vesey, V. P. Cole and P. J. Simpson, Cyanide and thiocyanate concentrations following sodium nitroprusside infusion in man, *Br. J. Anaesth.*, 1976, **48**, 651–660.
54. C. J. Vesey and J. Wilson, Red cell cyanide, *J. Pharm. Pharmacol.*, 1978, **30**, 20–26.
55. Schultz, Clinical pharmacokinetics of nitropusside, cyanide, thiosulphate and thiocyanate, *Clin. Pharmacokinet.*, 1984, **9**, 239–251.
56. D. E. McMillan and A. C. Svoboda, The role of the erythrocyte in cyanide detoxification, *J. Pharmacol. Exp.Ther.*, 1982, **221**, 37–42.
57. P. Lundquist, H. Rosling and B. Sörbo, Determination of cyanide in whole blood, erythrocytes and plasma, *Clin. Chem.*, 1985, **31**, 591–595.
58. C. Ronco, R. Bellomo and J. A. Kellum, Arterial and Venous blood gases, in *Critical Care Nephrology*, Saunders, Philadelphia, 2000, ch. 116, pp. 607-610.
59. B. Ballantyne, Blood, brain and cerebrospinal fluid cyanide concentrations in experimental acute cyanide poisoning, *J. Forensic Sci. Soc.*, 1975, **15**, 51–57.
60. E. Asmussen, On muscular exercise, in *A companion to Medical Studies*, ed. R. Passmore and J. S. Robson, Blackwell, 1973, ch. 42, vol. 1 , pp. 1–11.
61. R. P. Forsyth, B. I. Hoffbrand and K. Melmon, Redistribution of cardiac output during hemorrhage in unanaesthetized monkey, *Circ. Res.*, 1970, **27**, 311–320.
62. J. H. Stevens, P. O'Hanley, J. M. Shapiro, F. G. Mihm, P. S. Satoh, J. A. Collings and T. A. Raffin, Effects of Anti-C5a antibodies on adult respiratory distress syndrome in septic primates, *J. Clin. Invest.*, 1986, **77**, 1812–1816.
63. B. Ballantyne, Blood, brain and cerebrospinal fluid cyanide concentrations in experimental acute cyanide poisoning, *J. Forensic Sci. Soc.*, 1975, **15**, 51–57.
64. B. Ballantyne, Artifacts in the definition of toxicity by cyanides and cyanogens, *Fundam. Appl. Toxicol.*, 1983, **3**, 400–408.
65. M. Milani, Y. Ouellet, H. Ouellet *et al.*, Cyanide binding to truncated hemoglobins: A crystallographic and kinetic study, *Biochemistry*, 2004, **43**, 5213–5221.

66. B. A. Logue and D. M. Hinkens, The analysis of cyanide and its metabolites in biological samples, ed. Bruckner, A. W. and G. Roberts, *Proceedings of the International Cyanide Detection Testing Workshop*, NOAA Technical Memorandum NMFS-OPR-40, Silver Spring, MD, 2008, pp. 80–94.
67. C. J. Clark, D. Campbell and W. H. Reid, Blood Carboxyhaemoglobin and Cyanide Levels in Fire Survivors, *Lancet*, June 20, 1981, 1332–1335.
68. A. S. Curry, Cyanide poisoning, *Acta Pharmacol. Toxicol.*, 1963, **20**, 291–294.
69. B. Ballantyne, The forensic diagnosis of acute cyanide poisoning, in *Forensic Toxicology*, Wright, Bristol, 1974, pp. 99–113.
70. B. Ballantyne, Changes in blood cyanide as a function of storage time and temperature, *J. Forensic Sci. Soc.*, 1976, **16**, 305–310.
71. J. L. McAllister, R. J. Roby, B. Levine and D. A. Purser, The Effect of Sodium Fluoride on the Stability of Cyanide in Postmortem Blood Samples from Fire Victims, *Forensic Sci. Int.*, 2011, **209**, 29–33.
72. D. A. Purser, Trial documentation and testimony in relation to San Juan Dupont Plaza hotel fire.
73. B. C. Levin *et al.*, Report of test on analysis of carboxyhaemoglobin and cyanide in blood from victims of the Dupont Plaza hotel fire in Puerto Rico, Report No. FR 3975, 1988.
74. A. Hall, B. H. Rumack, M. I. Schaffer and C. H. Linden, Clinical toxicology of cyanide: North American clinical experiences, in *Clinical and Experimental Toxicology of Cyanides*, 1987, ch. 12. pp. 312–333.
75. B. Ballantyne and T. C. Marrs, Post-mortem features and criteria for the diagnosis of acute lethal cyanide poisoning, in *Clinical and Experimental Toxicology of Cyanides*, 1987, ch. 9, pp. 217–247.

CHAPTER 11

Mechanism of Action of Combustion-derived Nanoparticles

KEN DONALDSON,*[a] AMANDA HUNTER,[b] CRAIG POLAND[c] AND STEVE SMITH[d]

[a] Queens Medical Research Institute, University of Edinburgh, UK; [b] University/BHF Centre for Cardiovascular Sciences, University of Edinburgh, UK; [c] Institute of Occupational Medicine, Edinburgh, UK; [d] Royal Holloway, University of London, UK
*Email: ken.donaldson@ed.ac.uk

11.1 Introduction

Epidemiological studies reveal the potential health risks from inhalation of combustion-derived nanoparticles (CDNP) from various types of fuels. This review deals with the mechanisms of adverse effects caused by CDNP using the examples of diesel exhaust particles and wood smoke for traffic-derived CDNP and for domestic smoke exposure from wood-burning stoves respectively. The target populations that are affected by these sources of exposure differ. Almost all individuals, especially those in conurbations, are exposed to low levels of diesel exhaust particles (DEP) as the predominant nanoparticle type in the ambient urban aerosol. In addition to this exposure, some individuals will be exposed to wood smoke in their home if they have a wood-burning stove. Exposure to wood smoke may be the more important form of exposure to CDNP in some parts of the developing world.

Issues in Toxicology No. 23
Toxicology, Survival and Health Hazards of Combustion Products
Edited by David A. Purser, Robert L. Maynard and James C. Wakefield

Published by the Royal Society of Chemistry, www.rsc.org

11.2 Health Impacts of CDNP

Epidemiological studies have shown clear associations between both short and long term average ambient concentrations of particles monitored as PM2.5 and the occurrence of respiratory and cardiovascular disease. PM2.5 (the mass concentration of particles of generally less than 2.5 μm aerodynamic diameter) is dominated in numerical terms by particles of less than 100 nm diameter: nanoparticles. That the nanoparticles component of PM2.5 explains a significant proportion of the effects associated with PM2.5 is plausible especially as regards effects on the cardiovascular system. Short term exposure to PM2.5 is associated with deaths from both respiratory and cardiovascular diseases; long term exposure is associated with an increased risk of death from cardiovascular disease and from lung cancer but, perhaps surprisingly, not from other respiratory diseases (Table 11.1). It seems likely that the effect on lung cancer might be associated with exposure to CDNP derived from burning fossil fuels: petroleum derivatives, including diesel oil, and coal.[1,2]

Without doubt, most of the work on the mechanism of combustion-derived nanoparticles has been carried out using DEP, reflecting the wide exposure of people to DEP as a consequence of traffic pollution reaching its highest concentrations in conurbations where the traffic is heaviest and the population is greatest. The impact of DEP and other traffic-derived nanoparticles has therefore been a point of focus for the research community. Less is known regarding the mechanism of action of wood smoke but wood burning is widespread and so this is also dealt with in Sections 11.8 and 11.9.

11.3 Toxicological Considerations in the Physicochemical Composition of Combustion-derived Nanoparticles

Exposures to combustion-derived particles can occur in a number of scenarios, as described in the previous sections, and different combustion conditions and fuels produce different emissions. Therefore, particles from diesel engines have characteristics that may be quite different from those derived from the burning of biomass in an open fire. However, they will have some similarities, for example, a graphenic carbon core and various metal and organic components. Each of these components is capable of giving rise to oxidative stress in tissue (Figure 11.1).

Table 11.1 Adverse health responses of CDNP in susceptible populations.

Target	Susceptible population	Response
Lungs	Individuals with asthma or chronic obstructive pulmonary disease (COPD)	Exacerbations; mortality
Cardiovascular system	Individuals with coronary heart diseases	Heart attack/ stroke; mortality

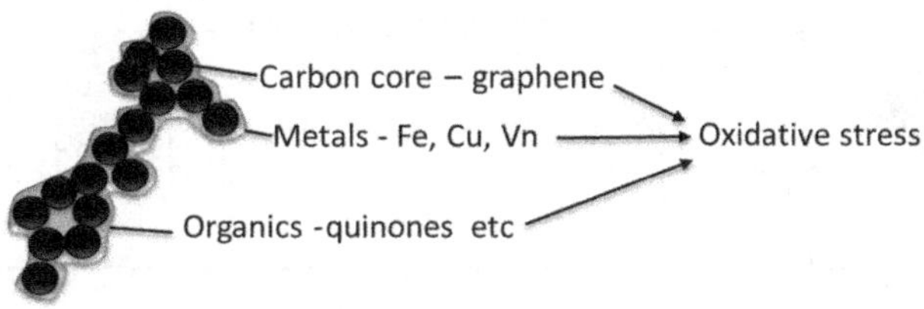

Figure 11.1 The 3 components of CDNP that cause oxidative stress.

Emissions from diesel-powered vehicles are a mixture of particles and gases and may be taken as an exemplar for CDNP. Diesel particles consist of chains of nanoparticles of carbon, as shown in Figure 11.1, which in turn are made up of a central core that is characterised by non-aligned graphene sheets. The individual nanoparticles are typically 20–80 nm in diameter. The chains of nanoparticles are of the order of 100–200 μm long and their aerodynamic diameters are largely sub-micron, predominantly 0.1–0.3 μm. Soot nanoparticles therefore make a significant contribution to the accumulation mode of the ambient aerosol and deposit with high efficiency in the lungs. CDNP are formed in the combustion process and emitted into cooler regions of exhaust systems and the ambient air, where metals, inorganic anions and myriad organic compounds condense and adsorb to the soot nodules.

Figure 11.1 shows that combustion-derived particles are chains and aggregates of carbon (graphene)-cored nanoparticles surrounded by metals and organics of variable solubility in water and lipid. These three components can induce oxidative stress and inflammation in the lung environment. In addition, the organic components and the oxidative stress produced by these and other components can cause organic/oxidative adducts of DNA and protein, which can also be involved in pathogenesis.[3] This is well understood for cancer, where the adduction of guanine is a common feature after exposure to metabolites of polycyclic aromatic hydrocarbons. In addition, the adduction of proteins and their oxidation is profoundly involved in changing the redox balance of cells and in the induction of gene expression for components of inflammatory pathways. These changes affect cell function. The result of increased inflammation in susceptible individuals with ongoing inflammatory lung disease includes exacerbations of asthma and chronic obstructive pulmonary disease (COPD). Chronic inflammation may lead to remodelling of airway walls and fibrosis, and may play a part in the induction of lung cancer. As regards to the cardiovascular impacts, organic compounds, metals and particles, to varying degrees not yet established, may cross into the circulation and enter the plasma. These components may pass *via* the blood to points in the circulation where atherosclerotic plaques are present and cause direct effects on them leading to their destabilisation and rupture. These effects, in turn, may lead to atherothrombotic events cumulating in heart attack or stroke. In addition, the oxidative stress and inflammatory events in the lung can contribute effects *via* host-derived mediators, which

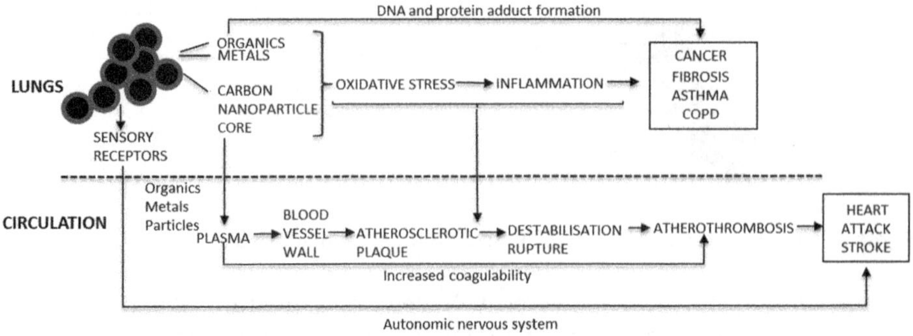

Figure 11.2 Summary of the pathobiological processes that lead to the adverse health effects of CDNP.

pass into the blood to atherosclerotic plaques and contribute to their destabilisation and rupture. Additionally, changes in the plasma related to increased coagulability may occur. This increased coagulability of the blood means that if an atherothrombotic event occurs following destabilisation or rupture of a plaque, the thrombus formed is likely to be propagated and increased in size compared with that likely to be produced when increased coagulability was not present[4] (see Figure 11.2 for a summary of these events).

Organic carbon compounds found in diesel emissions are characterised by a wide range of vapour pressures and are spread across the volatile, semi-volatile and non-volatile fractions. The volatile fraction includes representatives from aliphatic, aromatic and molecules with polar groups. Carbonyl compounds comprise a major component of the gas-phase fraction and include formaldehyde, acetaldehyde and more reactive unsaturated species including acrolein and crotonaldehyde. Non-volatile and semi-volatile compounds include aliphatic and polycyclic aromatic hydrocarbons (PAHs), as well as oxygenated species. These include straight chain organic acids, ketones and esters, as well as polar forms of 4- and 5-ringed PAHs, typically ketone, quinone, carboxyaldehyde and hydroxy derivatives. An unresolved complex mixture (UCM) or "hump" is a prominent feature of gas chromatography-mass spectrometry (GCMS) separations and consists of many aliphatic branched and cyclic hydrocarbons that do not resolve as individual compounds.

The semi-volatile fraction is dynamic and distributed between the gaseous and particulate phases. Vapour pressure is a key driver but other factors influence the distribution between the gaseous and particulate phases and they include surface area, available liquid phase and the adsorbent characteristics of the solid matter.[5] The dynamic nature of this fraction is highlighted by the fact that dilution of vehicular emissions with relatively uncontaminated air, or under-controlled dilution regimes used by equipment sampling emissions, alters the balance of partitioning of semi-volatiles between gas and particulate phases: higher dilution or greater mixing ratios result in an increase of semi-volatiles in the gaseous phase and a significant loss of organic carbon mass from the particle phase.[6,7]

At least in terms of the organic fraction of DEP, no two samples are completely alike and this provides for difficulty in making generalisations as to the contributions of different components following toxicological assessment. The contrasting composition of DEP samples is exemplified by comparing the US National Institute of Standards and Technology standard sample of diesel soot, NIST 2975, with A-DEP (produced at the National Institute for Health Sciences, Japan). Both these samples have been used for toxicological investigations and in some cases compared. When compared, NIST 2975 contains an organic extractable fraction of 1.5% and is described as low in alkane and PAH content and very little in the way of an unresolved complex mixture, whilst A-DEP is rich in alkanes and PAHs, a more evident unresolved complex mixture and an overall extractable fraction of 68.6%.[5]

11.4 The Nature of Cellular Oxidative Stress

As described in Section 11.3, oxidative stress is seen as a key mechanism in the induction of inflammation and propagation of a range of adverse effects relating to the respiratory and cardiovascular systems. The role of oxidative stress in many disease states has been described, ranging from neurodegenerative conditions, such as Alzheimer's disease,[8] to drug-induced liver injury from acetaminophen (paracetamol).[9] Whilst not dealt with in this chapter, it has been observed that children and young adults exposed to high levels of air pollution in Mexico City showed significant up-regulation of inflammatory markers in several regions of the brain, as well as endothelial activation and oxidative stress.[10] The recent outdoor air pollution and brain health workshop convened by the National Institute of Environmental Health Sciences/National Institute of Health reported in a publication by Block and colleagues provides an excellent review of the neurological effects of air pollution.[11]

In terms of the disease endpoints dealt with in this chapter, the basis and role of oxidative stress is considered here in more detail.

Biological systems are, in terms of the balance between reduction and oxidation, more reducing than the surrounding environment. There is thus a constant battle to maintain the normal reducing conditions found in tissues and cells. Such conditions are optimal for cellular and biochemical processes, yet there is a constant onslaught of intracellular and externally generated oxidants and electrophiles.

In a reduction–oxidation (redox) chemical reaction, an oxidant removes electrons from another reactant thereby oxidising it, but at the same time the oxidant is reduced as it takes on the electrons. Oxidations can occur by one or two electron transfers; one electron oxidations are characteristic of free radical reactions whilst two electron oxidants are typified by hydrogen peroxide (H_2O_2), which is a major component of the oxidant burden in cells.

Electrophiles ("electron loving") are chemical species that are electron poor and accept pairs of electrons to form new covalent bonds, whereas nucleophiles are electron rich and provide a pair of electrons to form new covalent bonds. Electrophiles are synonymous with Lewis acids and nucleophiles with Lewis bases. Examples of electrophiles therefore include metals, Hg^{2+}, Cd^{2+} and Zn^{2+}, and organic compounds polarised by electron-withdrawing functional groups, of note are α,β-unsaturated carbonyls and nitroalkenes. In biological systems, nucleophilic nucleic acids and amino acids (cysteine, lysine and histidine) react with electrophiles produced endogenously (products of lipid peroxidation) and from external sources (smoking and air pollution). Electrophiles form reversible covalent adducts with thiolate (RS^-) groups of sulfur-proteins and small molecules (*e.g.*, glutathione, GSH). These represent post-translational changes and are associated with changes in enzyme activity and cellular signalling pathways, including those involved in inflammatory processes.

In terms of oxidation–reduction potentials, cellular compartments are more negative than their surrounding environment and this requires a tightly controlled redox environment in order to maintain normal physiological function and homeostasis. The consequences of an imbalance in the redox balance in favour of oxidation, resulting in oxidative stress, arises from a range of cellular injuries, including those that derive from increased concentrations of xenobiotics in tissue. The consequences of oxidative stress depend on its extent and the cell type that is under stress. Figure 11.3 shows the general outcome of different degrees of oxidative stress in epithelial cells and macrophages, which are the main types of cell likely to be exposed to diesel nanoparticles deposited in the lung.[12]

The intracellular redox environment is governed mainly by the balance between glutathione (GSH) and its oxidised disulfide form (GSSG), reactive oxygen species (ROS), thioredoxins, nicotinamide adenine dinucleotide (NAD/NADH) and nicotinamide adenine dinucleotide phosphate (NADPH/NADP). The reductive cellular environment is maintained by a flow of electrons channelled through NADPH from the respiratory cycle; the normal redox potential of cells is maintained within a narrow range by the GSH redox buffer system, which is present in a high concentration. Regulation of the redox environment in this way maintains the functional oxidation state of macromolecules of proteins, lipids and DNA, and hence cellular function. Oxidative stress in terms of reduction potentials is characterised by a shift to more positive values and a reduction in the ratio of GSH to GSSG (more GSSG). This change can be accompanied by post-translational and functional changes of proteins, and oxidation of lipids and DNA resulting in adverse effects on cellular health and function. There are a host of other

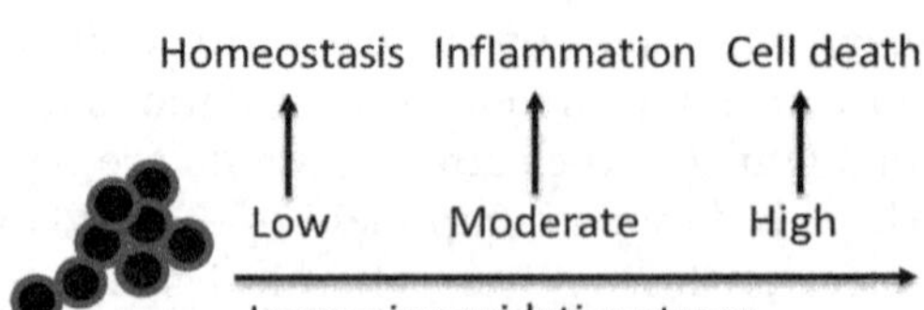

Figure 11.3 Impact of various degrees of oxidative stress from CDNP on pathological processes.

substances that are also involved in the antioxidant defence system of cells, including bilirubin, uric acid, ascorbate, vitamins A and E, ergothioneine and possibly melatonin.[14]

The reactive oxygen species (ROS) and reactive nitrogen species (RNS) arising from particles and from the damage they cause to cells include:

- superoxide anion ($O_2^{\bullet-}$);
- hydroxyl radical ($^{\bullet}OH$);
- singlet oxygen ($^1O_2$);
- hydrogen peroxide (H_2O_2).

$O_2^{\bullet-}$ and $^{\bullet}OH$ are free radicals whilst $^1O_2$ is an excited state of O_2. H_2O_2 is an oxidant and not a radical but on reaction with ferrous iron (Fe^{2+}), $^{\bullet}OH$ can be produced by way of the well-known Fenton reaction pathway. ROS are products of O_2 reduction (oxygen acts as an oxidising agent and is itself reduced) and several enzymes are involved in their generation, with NAD(P)H oxidases (NO_x) probably being the main contributors to cellular ROS, along with others including xanthine oxidase and peroxidases. $O_2^{\bullet-}$ radicals are the initial products of O_2 reduction and these are effectively dismutated to H_2O_2 by superoxide dismutases (SODs). In turn, catalase, glutathione peroxidase and peroxiredoxins swiftly and efficiently reduce H_2O_2. Increased generation of ROS is associated with bursts of activity of macrophages and neutrophils, especially in the lung, as a response to phagocytosis of particles, bacterial infection or other tissue injury. NO rapidly diffuses through tissues to red blood cells where reaction with oxyhaemoglobin rapidly removes NO through the generation of nitrate. Whilst SODs quickly remove $O_2^{\bullet-}$, in situations where NO and $O_2^{\bullet-}$ come into close contact, they rapidly react together to form the much more powerful oxidant peroxynitrite ($ONOO^-$), which is thought to be a key element in the pathological processes associated with NO.

Oxidants and electrophiles are therefore major drivers in the generation of oxidative stress. CNDP contain a range of electrophiles and oxidants, as well as potential electrophiles that may be formed by metabolic transformations, as in the case of P450 mono-oxygenases.

11.5 Oxidative Stress Caused by the Specific Components of Combustion-derived Particles

Several studies have shown that DEP produce oxidant and pro-inflammatory effects in the lung, as well as mutagenic effects, effects on the balance between cell proliferation and apoptosis, and responses linked to cardiovascular disease. The nature of the effects differ depending on the types of DEP and thus on the conditions under which the DEP were generated. For example, a comparison of DEP NIST 2975 and the organic-rich A-DEP showed that A-DEP was the more mutagenic, whilst the nature of the pulmonary toxicities differed between the two types,[15,16] most likely reflecting their different compositions. DEP exposure has an effect on NO generation; this appears to be through dysfunction of endothelial nitric oxide synthase. Oxidation of the cofactor BH_4 (5,6,7,8-tetrahydrobioperin) results in an uncoupling of the enzyme such that $O_2^{\bullet -}$ is generated instead of NO.[17] DEP promote signalling and transcription pathways involved in cellular antioxidant homeostasis (*e.g.*, Keap1/Nrf2) and the balance between cell proliferation and apoptosis.[18]

DEP from different production processes seem to activate pro-inflammatory signalling in lung epithelia by more than one mechanism. The effect does not seem to be a function of the amount of organic carbon in the DEP samples and whist the organic fraction is strongly implicated in certain types of DEP, in others it has been related to the metals. These findings illustrate the complexity of DEP-mediated toxicity. To address this complexity and better explain the basis for DEP-mediated toxicity, the contribution of each of these components of DEP towards oxidative stress is considered in the following sub-sections separately.

11.5.1 Oxidative Effects of the Carbon Core

The naked carbon core of DEP and other combustion particles is formed of disjointed graphene sheets. Graphene is formed as a sheet of carbon atoms bound together with double electron bonds in a thin film only one atom thick. It has important and interesting properties, including the capacity to cause oxidative stress. Pure graphene can cause cell injury by mechanisms that include oxidative stress[19] and can also directly oxidise glutathione.[20] However, the state of the graphene in diesel soot is unknown and it could certainly be oxidised or derivatised in various ways due to the combustion process: carboxyl-functionalised graphene is less oxidative than underivatised graphene.[21] Carbon black (CB) is very similar to the carbon core of DEP in that it is formed of disjointed sheets of graphene but, in contrast, has little metal or organic contamination. However, CB has been reported to cause oxidative stress in cells *in vitro*[22,23] and directly oxidises antioxidants like glutathione,[24] as well as causing oxidative stress after deposition in the lung.[25,26] On balance, CB is less potent the DEP in causing oxidative stress. This is borne out by electron paramagnetic resonance (EPR): studies using tempone as a spin trap show that DEP (2975) produces more $O_2^{\bullet -}$ than CB on a weight for weight basis.

11.5.2 Oxidative Effects of Transition Metals

Diesel exhaust particles contain a number of metals, including the redox-active transition metals, iron (Fe) and copper (Cu), which can be important in the oxidative and pro-inflammatory effects of DEP; electrophilic (Lewis acids), Cd, Hg, Pb and Ni, as well as Zn. Diesel exhaust particles can be shown to generate hydroxyl radicals *via* a mechanism that involves Cu and

Fe[27] and in support of this, the DEP-mediated generation of ROS can be blocked *in vitro* by the use of iron chelators.[28] The inflammatory effects of DEP *in vivo* can be ameliorated by iron chelator treatment.[29] In addition, the generation of hydroxyl radicals in DEP-treated rat lungs (detected using a spin trap) was also abolished with treatment with an iron chelator.[30] Zinc is probably in a category of its own because of its importance as a co-factor in over 1000 proteins. Zinc binds to thiol moieties in these proteins and these bonds are sensitive to redox changes. For example, the metallothioneins (MT) are an essential part of Zn homeostasis, minimising "free" Zn ions in the cytoplasm and delivering Zn to proteins. The release of Zn ions occurs by oxidation of thiol groups to sulfide forms. Zinc toxicity is associated with oxidative stress and can promote H_2O_2 accumulation and a shift to more positive reduction potentials in cells.[31] It has been suggested that the toxicity of electrophilic metals is a consequence of displacement of Zn from thiol sites, resulting in the release of Zn^{2+} ions and hence Zn toxicity.

11.5.3 Oxidative Effects of Organic Electrophiles

Studies of the mechanism of increased cytokine release induced by DEP have implicated the organic, rather than the metal, components.[32,33] It has been suggested that two mechanisms are involved: the first being that super oxide radical production occurs from auto-oxidation or metal catalysed oxidation of hydroquinones to quinones; the second being the associated reduction of O_2 to produce $O_2^{\bullet -}$, which on dismutation forms H_2O_2. The second mechanism implies electrophilic reactivity towards nucleophiles in the cell. DE and DEP carry a number of electrophiles that react with active cysteine (S^-) groups of proteins; these include 1,2-naphthoquinone (1,2-NQ), acrolein, *o*-anisaldehyde, 1-nitronaphthalene and formaldehyde. In addition, metabolic activation of neutral hydrocarbons can create electrophiles. For example, an epoxide derivative of benz[*a*]pyrene readily reacts with DNA[34] and naphthoquinones (1,4-NQ and 1,2-NQ) are products of naphthalene metabolism. Photochemical oxidation in ambient air also produces electrophiles from gaseous and condensed hydrocarbons. This section describes how such compounds react and produce products that are involved in cellular oxidative stress.

4-Hydroxynonenal (HNE), an electrophile Michael addition compound endogenously produced through lipid peroxidation, forms adducts with the active cysteine residue (Cyst 280) of SIRT3, which is an NAD^+-dependent protein in mitochondrial matrix, and inhibits its activity.[35] Martyniuk *et al.*[52] (2011) examined the effects of three α,β-unsaturated carbonyls [acrylamide (ACR), acrolein and methylvinyl ketone (MVK)] on the activity of glyceraldehyde-3-phosphate dehydrogenase (GAPDH). The order of enzyme inhibition (ACR < MVK < acrolein) was related to the electrophilic reactivity of each compound and the specific adducts formed at the active site of the enzyme (Cyst 152). It was concluded that these results demonstrated a general cytotoxic mechanism of electrophilic α,β-unsaturated carbonyls.[36]

In cell culture experiments, measured redox changes and H_2O_2 production in real time using fluorescence reporters in response to exposure to 1,2-naphthoquinone (a component of DEP) showed a rapid loss of reducing potential (more positive reducing potential) and an increased production of H_2O_2 of mitochondrial origin. H_2O_2 production was not the cause of the redox changes and 1,2-NQ induced HO-1 expression was not mediated by H_2O_2. In explanation of the results, it is suggested that the electrophilic 1,2-NQ reacted covalently with cysteine thiolate groups. A recent series of studies have shown that DEP and airborne particulate matter (APM) and gas-phase constituents in polluted air contain components with electrophilic activity.[37–39]

Oxidant and electrophile interactions with nucleophilic centres can result in the activation of proteins and signalling pathways. Transient receptor potential (TRP) ion channels are a family of receptors that are activated by a variety of natural plant products to produce pain and inflammation. Hazari *et al.*[40] examined the effects of DE on cardiac arrhythmias and on activation of TRP channels. It was concluded that a single exposure to DE increased the sensitivity of the heart to arrhythmias; it was suggested this was mediated by activation of TRPA1, mainly by gaseous components in DE. Deering-Rice[41] reported that TRPA1 channels were strongly activated by DEP from scrapings of an in-service and poorly functioning diesel engine and the NIST diesel soot standard, DEP-2975. Most, if not all, the activation was found in ethanol (EtOH) extracts of the PM samples and related to 1,2-naphthoquinone and acrolein, followed by *o*-anisaldehyde, 1-nitronaphthalene and formaldehyde; although acrolein and formaldehyde, in particular, have high vapour pressures and are associated with the gas phase component.

An example of interaction between transition metals and organic electrophiles is shown in Figure 11.4, which highlights the possible interaction between redox-active transition metals and quinones, leading to the generation of damaging hydroxyl radicals ($HO^•$). Cellular reducing equivalents (NADH and ascorbate) can reduce a quinone to its semi-quinone radical and

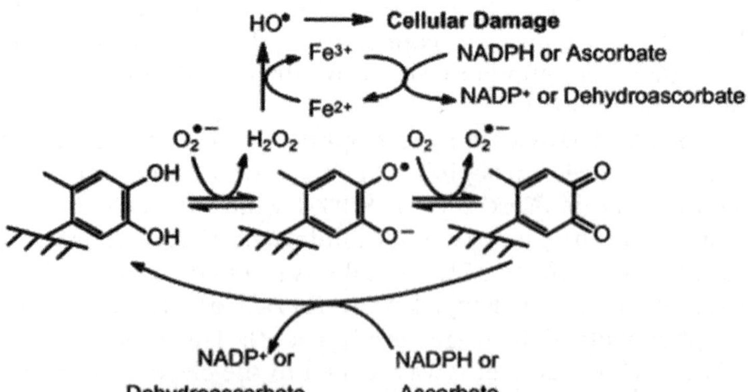

Figure 11.4 Potential interactions between transition metals and quinones in CDNP in producing harmful hydroxyl radicals. Drawn after Squadrito *et al.*[42]

then to the hydroquinone form. Re-oxidation to the quinone can occur with O_2 as the oxidant, which is itself reduced to produce superoxide anions ($O_2^{\bullet -}$). It is a redox cycle, catalysed by redox-active transition metals, and Fenton chemistry produces $HO^{\bullet}$ radicals. It is not clear whether these reactions *in vivo* are significant. They have been promoted as a chain reaction and therefore as a mechanism of sustained free radical production; however, in the cell it is thought that the ubiquitous presence of proteins would terminate such reactions. The more redox-active quinones are also reactive electrophiles and are likely to form covalent adducts with cysteine residues of proteins.

11.6 Oxidative Stress Signalling Causing Inflammation

The role of oxidative stress-sensitive signalling pathways in the transcription of pro-inflammatory genes is generally well understood,[43] and for the specific instance of DEP[44,45] is shown diagrammatically in Figure 11.5. These pathways are stimulated principally in epithelial cells and macrophages, the cells that receive the greatest doses of deposited particles, leading to the expression of pro-inflammatory genes. A range of pathways are activated during expression of such genes, including lipid oxidations (*e.g.*, 4-hydroxynonenal), Ca^{++} flux, kinases of the mitogen-activated protein kinases

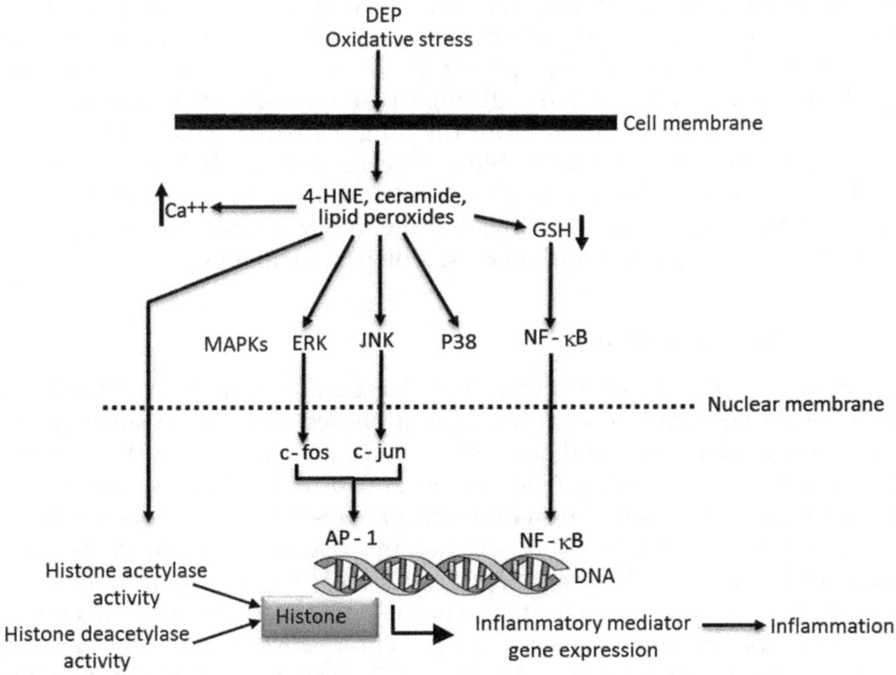

Figure 11.5 Signaling pathway involved n oxidative stress-responsive pro-inflammatory gene expression by CDNP.

(MAPK) family and redox-regulated transcription factors, such as activator protein 1 (AP-1) and nuclear factor kappa-light-chain-enhancer of activated B cells (NF-κB); genes expressed include those controlling expression of cytokines, chemokines and adhesion molecules. Typical cytokine genes expressed in the lungs *via* this mechanism after DEP exposure include those for granulocyte-macrophage colony-stimulating factor (GM-CSF),[46] interleukin 1β (IL-1β) and IL-8.[47]

11.7 Mechanism of Cardiovascular Effects Following Inhalation of CDNP

There are, in general, three hypothetical mechanistic pathways whereby nanoparticles depositing in the lungs might cause impacts on the cardiovascular system. These are *via* inflammation, direct translocation of nanoparticles into the cardiovascular system (especially into atherosclerotic plaques) and *via* the autonomic nervous system. These pathways are considered in the following sub-sections.

11.7.1 Inflammation

This implies that CDNP cause an inflammatory response in the lung and that inflammatory mediators pass from the lung and enter the circulation where they enhance inflammation-induced damage to the cardiovascular system. Oxidative stress accompanies inflammation and so oxidants and cytokines generated in the lungs during CDNP-induced inflammation could have effects in tissues remote from the lung. Cytokines could enhance inflammatory events in walls of blood vessels and in their endothelium, leading to decreased plaque stability whilst oxidants like superoxide could scavenge NO leading to endothelial dysfunction, which may promote plaque formation/destabilisation and increased blood coagulability.

11.7.2 Translocation

An alternative mechanism featuring translocation of particles from the lungs into the cardiovascular system has been proposed. Due to the small size of nanoparticles, it is possible that nanoparticles penetrate through the alveolar–capillary barrier becoming blood-borne and directly affecting the cardiovascular cells. It has been shown in several *in vivo* studies that nanoparticles of differing sizes and compositions can become systemically available after deposition in the lung. For example, inhalation of 20 nm iridium particles by rats led to a small percentage (1–2%) of the deposited dose accumulating in the liver, spleen, kidneys, heart, brain, reproductive organs and skeleton (bone marrow).[48] In addition, it has been fairly well established that particles can penetrate into several layers of the alveolar epithelium[49] and cross similar structural barriers, including those of the olfactory system.[50]

11.7.3 Autonomic Nervous System

The third hypothesis suggests involvement of the autonomic nervous system. Particles are hypothesised to stimulate sensory receptors within the lung, which signal to the central nervous system leading to changes in cardiac function, including, for example, changes in heart rate variability (HRV) *via* changes in the activity of the autonomic nervous system.[51]

It is of course possible that no single pathway is responsible for such complex and varied and diverse adverse outcomes that have been reported in epidemiological studies and that all three pathways are involved. Figure 11.2 summarises the overall hypothesised mechanisms for the action of CDNP on the lungs and cardiovascular system.

11.8 Wood Smoke

Wood smoke has historically been viewed as a relatively benign substance owing to its natural origins. However, it is now well established that wood-burning fires emit a range of damaging pollutants, including free radicals, carbon monoxide (CO), nitrogen oxides and particulate matter (PM). Wood smoke particles are generally smaller than 1 μm, with a size distribution of 0.15 μm to 0.4 μm.[1] Leonard *et al.*[53] demonstrated that coarse particles in wood smoke contained more carbon radicals per unit mass than the CDNP; however, the CDNP generated more hydroxyl radicals in an acellular system *via* a Fenton-like reaction. The CDNP also caused significant increases in H_2O_2 production by monocytes and lipid peroxidation. Combustion of wood is typically inefficient and partially oxidised organic chemicals are produced: wood smoke particles have a higher content of organic carbon and PAHs than do DEP.

11.9 Health Effects of Wood Smoke

In contrast to urban PM, which is dominated by CDNP such as DEP, the respiratory effects of wood smoke appear to be greater than the cardiovascular effects.[26] Studies have reported increases in admissions with upper and lower respiratory tract illnesses, including asthma and respiratory symptoms, following acute exposure. Following the Californian wild-land fires in 2003, average increases in PM2.5 conditions were associated with 34% increases in asthma admissions.[31] There were also increases in hospital admissions for treatment of acute bronchitis, COPD and pneumonias. There was, however, limited evidence of a small impact of wildfire-related PM2.5 on cardiovascular admissions. There are conflicting results on the effects on lung function. A study has shown that there was no difference in lung function as monitored by spirometry between fire-fighters who had been exposed to smoke from wild-land fires and those who had not.[32] Repeated exposures have been shown to result in chronic respiratory disease, with exposure to wood for heating or cooking during female adolescence having recently been shown to be associated with COPD later in life.[33] Installation of a chimney, thereby reducing smoke exposure from indoor biomass combustion and PM2.5 exposure by half, resulted in lower diastolic blood pressure[29] and less non-specific ST depression on electrocardiograms in Guatemalan women. However, there was no change in the apparent effect on heart rate variability.[30] Further research needs to be undertaken in this field.

In contrast to urban air pollution, the mechanisms underlying the adverse health effects of wood smoke inhalation have not been intensively studied. However, the predominant component of both these types of air pollution is CDNP. The large surface area to mass ratio of these particles enables the transport of large amounts of adsorbed toxic agents to internal targets. Additionally, quinones are particularly reactive organic components of PM with potential to produce ROS and to induce oxidative stress *via* their redox capacity[8] (see Section 11.5.3). Some of the gases adsorbed on the surfaces of PM, such as volatile organic compounds, are mutagenic with some having as much as 100 times the activity of some well-known carcinogens.[9]

Several *in vitro* studies of cultured cells have shown wood smoke PM increases expression and production of pro-inflammatory cytokines, and causes oxidative stress and oxidative damage to DNA.[10–12] Furthermore, particles from incomplete combustion of wood caused more severe cytotoxic effects and DNA damage *in vitro* than particles produced under conditions allowing more complete combustion.[8] A number of animal studies have shown that short term inhalation can compromise pulmonary immune defence mechanisms, mainly through impairment of alveolar macrophage activity. Inhalation of wood smoke can also alter pulmonary structure causing bronchiolitis, congestion of parenchymal blood vessels and mild emphysema.[6,13–15]

A number of exposure studies in healthy human subjects have shown similar increases in markers of systemic and lung inflammation. In one such study, controlled exposure to wood smoke was followed by minor increases in markers of systemic inflammation, including acute-phase protein serum amyloid A, to some extent serum C-reactive protein, and in levels of factor VIII, which is involved in blood coagulation. Increases in urinary excretion of the isoprostane 8-iso-PGF2, which is a marker of lipid peroxidation, were also seen.[16] The oxidative DNA damage and related repair capacity in peripheral blood mononuclear cells was also investigated and although wood smoke exposure was followed by significant up-regulation of the repair gene hOGG1, no direct genotoxic effects were observed.[10] In addition, there was an increase in exhaled nitric oxide and levels of Clara cell protein in serum: both are markers of inflammatory effects on distal airways.[17]

Fire-fighters assessed for evidence of pulmonary and systemic inflammation following wild-land fire suppression showed significant increases in sputum neutrophils, circulating white blood cells and band cells (immature neutrophils).[18] Serum IL-6, IL-8 and monocyte chemotactic protein-1 levels also increased significantly following fire-fighting: this was not reported following strenuous exertion without exposure to wood smoke. Urinary biomarkers of oxidative stress, 8-hydroxy-2′-deoxyguanosine (8-OHdG) and malondialdehyde (MDA) were measured in wild-land fire-fighters, revealing no significant change from pre- to post-shift levels. However, a change in 8-OHdG, corrected for renal function, from pre- to post-shift was associated with the length of fire-fighting career and the subject's age, suggesting that the oxidative stress response may be dependent on age or the length of fire-fighting career, and thus perhaps on cumulative exposure to toxicologically active materials.[19] By contrast, Sehlstedt *et al.* reported virtually no effects on inflammation or oxidative stress in the airways after exposure to relatively high concentrations of PM2.5.[20] Although there was an increase in documented respiratory symptoms, neither exhaled nitric oxide, systemic and airway inflammatory parameters in bronchoalveolar lavage nor bronchial mucosal biopsies were significantly affected. Moreover, in atopic individuals there was no effect on markers of oxidative stress, DNA damage, cell adhesion or cytokines in response to clean air, and low and high dose wood smoke.[21] Microvascular function assessed at 6 hours post-exposure in the same study remained unchanged.

Traffic-derived air pollution is known to cause relatively acute changes in vascular function,[22] yet this does not appear to occur with wood smoke. An exposure study in healthy fire-fighters comprehensively assessed cardiovascular endpoints in response to wood smoke at a level analogous to their exposure during wild-land fire suppression. There was no significant change in arterial stiffness, endothelial function, *ex vivo* thrombus formation or platelet activation when compared with filtered air. Additionally, there was no change in lung function or exhaled nitric oxide concentration.[23] Despite not having acute effects, exposure to wood smoke may have a cumulative effect in the genesis of atherosclerosis *via* low-level systemic inflammation and/or oxidative stress. Long term intermittent exposure to ultrafine particles with diameter <0.18 μm is implicated in the development of atherosclerosis in mice at an exposure[24] comparable in terms of particle number concentration with levels found in fires. More studies assessing the long term effects of wood smoke exposure are warranted. Physical and chemical properties are likely to influence the biological potency of the particles and differences in this may account for the differing effects of particles of fire smoke and those from other sources such as traffic pollution. Such factors may also account for the differing effects of wood smoke between studies: this is likely to be due to different combustion conditions and appliances used.

11.10 Summary of Mechanisms of Action of Combustion-derived Particles

We have considered particles produced by two very different sources: diesel particles produced by diesel engines and the particles found in wood smoke. Particles of both types include CDNP, though they differ in terms of their chemical composition. In addition, patterns of human exposure to these two types of particle may also differ significantly. Samples of CDNP vary in structure and composition depending on the material burnt and on the exact conditions under which it is burnt. Despite this variability, which can hamper comparisons and the drawing of conclusions, the impact of various physicochemical components of CDNP on biological systems has been described in the previous sections. Overall, it has been shown that CDNP, as exemplified by DEP, have components capable of causing oxidative stress in the lungs following deposition and these components include a carbon core, metals and a range of organic species. Alterations in the relative concentrations of these components of CDNP due to differences in combustion efficiency, fuel *etc.* alter the potency of differing samples and this has been noted even between standard research samples. Of great interest are the mechanisms by which CDNP can cause adverse health effects in exposed individuals and, as mentioned, oxidative stress is at the centre of the hypothesised patho-biological pathway.

Bullet point list of mechanisms of action of combustion-derived particles

- CDNP exemplified by DEP have components capable of causing oxidative stress in the lungs following deposition: carbon core, metals and a range of organics.
- Oxidative stress is measurable in cells, animals and humans exposed to DEP.
- Oxidative stress initiates inflammation and pro-inflammatory effects are detectable in cells, animals and humans exposed to CDNP.
- Inflammation underlies the major pulmonary effects of DEP and PM in human populations—exacerbations of COPD and asthma.
- Oxidative stress and inflammation are plausibly involved in atherosclerosis development and in plaque destabilisation and rupture, leading to heart attacks and strokes.
- Wood smoke appears to have less ability to cause pulmonary oxidative stress and inflammation than DEP and impacts little on the cardiovascular system.
- Difference in composition between DEP and wood smoke probably explains the differences in potency between DEP and wood smoke. This implies, generally, that CDNP from various sources will have different potency depending on their composition.

References

1. Committee on the Medical Efects of Air Pollutants, *The Mortality Effects of Long-Term Exposure to Particulate Air Pollution in the United Kingdom*, Health Protection Agency, Chilcot, 2010, Crown Copyright, p. 108.

2. Committee on the Medical Effects of Air Pollutants. *Long-Term Exposure to Air Pollution: Effect on Mortality*. Health Protection Agency, Crown Copyright, Chilton, 2009, p. 186.
3. K. Donaldson, L. Tran, L. A. Jimenez, R. Duffin, D. E. Newby, N. Mills, W. MacNee and V. Stone V, Combustion-derived nanoparticles: a review of their toxicology following inhalation exposure, *Part Fibre Toxicol.*, 2005, **2**, 10.
4. N. L. Mills, K. Donaldson, P. W. Hadoke, N. A. Boon, W. MacNee, F. R. Cassee *et al.*, Adverse cardiovascular effects of air pollution, *Nat. Clin. Pract. Cardiovasc. Med.*, 2009, **6**, 36–44.
5. B. J. Turpin, P. Saxena and E. Andrews, Measuring and simulating particulate organics in the atmosphere: Problems and prospects, *Atmos. Environ.*, 2000, **34**, 2983–3013.
6. R. M. Harrison, M. Dall'Osto, D. C. S. Beddows, A. J. Thorpe, W. J. Bloss, J. D. Allan, *et al.*, Atmospheric chemistry and physics in the atmosphere of a developed megacity (London): an overview of the REPARTEE experiment and its conclusions, *Atmos. Chem. Phys.*, 2012, **12**, 3065–3114.
7. A. L. Robinson, N. M. Donahue, M. K. Shrivastava, E. A. Weitkamp, A. M. Sage, A. P. Grieshop *et al.*, Rethinking organic aerosols: semivolatile emissions and photochemical aging, *Science*, 2007, **315**, 1259–1262.
8. A. Butterfield, A. M. Swomley and R. Sultana, Amyloid beta-Peptide 1-42-induced Oxidative Stress in Alzheimer Disease: Importance in Disease Pathogenesis and Progression, *Antioxid. Redox Signaling*, 2013, **19**, 823–835.
9. T. A. Baillie and A. E. Rettie, Role of biotransformation in drug-induced toxicity: influence of intra- and inter-species differences in drug metabolism, *Drug Metab. Pharmacokinet.*, 2011, **26**, 15–29.
10. L. Calderon-Garciduenas, A. C. Solt, C. Henriquez-Roldan, R. Torres-Jardon, B. Nuse, L. Herritt *et al.*, Long-term air pollution exposure is associated with neuroinflammation, an altered innate immune response, disruption of the blood-brain barrier, ultrafine particulate deposition, and accumulation of amyloid beta-42 and alpha-synuclein in children and young adults, *Toxicol. Pathol.*, 2008, **36**, 289–310.
11. M. L. Block, A. Elder, R. L. Auten, S. D. Bilbo, H. Chen, J. C. Chen *et al.*, The outdoor air pollution and brain health workshop, *Neurotoxicology*, 2012, **33**, 972–984.
12. N. Li, T. Xia and A. E. Nel, The role of oxidative stress in ambient particulate matter-induced lung diseases and its implications in the toxicity of engineered nanoparticles, *Free Radical Biol. Med.*, 2008, **44**, 1689–1699.
13. A. Nel, T. Xia, L. Madler and N. Li, Toxic potential of materials at the nanolevel, *Science*, 2006, **311**, 622–627.
14. T. W. Sedlak, M. Saleh, D. S. Higginson, B. D. Paul, K. R. Juluri and S. H. Snyder, Bilirubin and glutathione have complementary antioxidant and cytoprotective roles, *Proc. Natl. Acad. Sci. U. S. A.*, 2009, **106**, 5171–5176.

15. D. M. DeMarini, L. R. Brooks, S. H. Warren, T. Kobayashi, M. I. Gilmour and P. Singh, Bioassay-directed fractionation and salmonella mutagenicity of automobile and forklift diesel exhaust particles, *Environ. Health Perspect.*, 2004, **112**, 814–819.
16. P. Singh, D. M. DeMarini, C. A. Dick, D. G. Tabor, J. V. Ryan, W. P. Linak *et al.*, Sample characterization of automobile and forklift diesel exhaust particles and comparative pulmonary toxicity in mice, *Environ. Health Perspect.*, 2004, **112**, 820–825.
17. T. W. Cherng, M. L. Paffett, O. Jackson-Weaver, M. J. Campen, B. R. Walker and N. L. Kanagy, Mechanisms of diesel-induced endothelial nitric oxide synthase dysfunction in coronary arterioles, *Environ. Health Perspect.*, 2011, **119**, 98–103.
18. D. Cao, P. A. Bromberg and J. M. Samet, Diesel particle-induced transcriptional expression of p21 involves activation of EGFR, Src, and Stat3, *Am. J. Respir. Cell Mol. Biol.*, 2010, **42**, 88–95.
19. O. Akhavan, E. Ghaderi and A. Akhavan, Size-dependent genotoxicity of graphene nanoplatelets in human stem cells, *Biomaterials*, 2012, **33**, 8017–8025.
20. M. Sehlstedt, R. Dove, C. Boman, J. Pagels, E. Swietlicki, J. Löndahl, R. Westerholm, J. Bosson, S. Barath, A. F. Behndig, J. Pourazar, T. Sandström, I. S. Mudway and A. Blomberg, Antioxidant airway responses following experimental exposure to wood smoke in man, *Part Fibre Toxicol.*, 2010, **7**, 21.
21. L. Forchhammer, P. Møller, I. S. Riddervold, J. Bønløkke, A. Massling, T. Sigsgaard and S. Loft, Controlled human wood smoke exposure: oxidative stress, inflammation and microvascular function, *Part Fibre Toxicol.*, 2012, **9**, 7.
22. V. Stone, J. Shaw, D. M. Brown, W. MacNee, S. P. Faux and K. Donaldson, The role of oxidative stress in the prolonged inhibitory effect of ultrafine carbon black on epithelial cell function, *Toxicol. in Vitro.*, 1998, **12**, 649–659.
23. L. Foucaud, M. R. Wilson, D. M. Brown and V. Stone, Measurement of reactive species production by nanoparticles prepared in biologically relevant media, *Toxicol. Lett.*, 2007, **174**, 1–9.
24. H. Zielinski, I. S. Mudway, K. A. Berube, S. Murphy, R. Richards and F. J. Kelly, Modeling the interactions of particulates with epithelial lining fluid antioxidants, *Am. J. Physiol.*, 1999, **277**, L719–L726.
25. X. Y. Li, D. Brown, S. Smith, W. MacNee and K. Donaldson, Short-Term Inflammatory Responses Following Intratracheal Instillation of Fine and Ultrafine Carbon Black in Rats, *Inhalation Toxicol.*, 1999, **11**, 709–731.
26. C. A. Dick, D. M. Brown, K. Donaldson and V. Stone, The role of free radicals in the toxic and inflammatory effects of four different ultrafine particle types, *Inhalation Toxicol.*, 2003, **15**, 39–52.
27. E. DiStefano, A. Eiguren-Fernandez, R. J. Delfino, C. Sioutas, J. R. Froines and A. K. Cho, Determination of metal-based hydroxyl radical generating

capacity of ambient and diesel exhaust particles, *Inhalation Toxicol.*, 2009, **21**, 731–738.

28. S. Park, H. Nam, N. Chung, J. D. Park and Y. Lim, The role of iron in reactive oxygen species generation from diesel exhaust particles, *Toxicol in Vitro*, 2006, **20**, 851–857.
29. T. Arimoto, M. B. Kadiiska, K. Sato, J. Corbett and R. P. Mason, Synergistic production of lung free radicals by diesel exhaust particles and endotoxin, *Am. J. Respir. Crit. Care Med.*, 2005, **171**, 379–387.
30. J. Y. Han, K. Takeshita and H. Utsumi, Noninvasive detection of hydroxyl radical generation in lung by diesel exhaust particles, *Free Radical Biol. Med.*, 2001, **30**, 516–525.
31. W. Y. Cheng, H. Tong, E. W. Miller, C. J. Chang, J. Remington, R. M. Zucker *et al.*, An integrated imaging approach to the study of oxidative stress generation by mitochondrial dysfunction in living cells, *Environ. Health Perspect.*, 2010, **118**, 902–908.
32. S. Boland, V. Bonvallot, T. Fournier, A. Baeza-Squiban, M. Aubier and F. Marano, Mechanisms of GM-CSF increase by diesel exhaust particles in human airway epithelial cells, *Am. J. Physiol.: Lung Cell. Mol. Physiol.*, 2000, **278**, L25–L32.
33. M. Dergham, C. Lepers, A. Verdin, S. Billet, F. Cazier, D. Courcot *et al.*, Prooxidant and proinflammatory potency of air pollution particulate matter (PM(2).(5)(-)(0).(3)) produced in rural, urban, or industrial surroundings in human bronchial epithelial cells (BEAS-2B), *Chem. Res. Toxicol.*, 2012, **25**, 904–919.
34. T. M. Penning, M. E. Burczynski, C. F. Hung, K. D. McCoull, N. T. Palackal and L. S. Tsuruda, Dihydrodiol dehydrogenases and polycyclic aromatic hydrocarbon activation: generation of reactive and redox active o-quinones, *Chem. Res. Toxicol.*, 1999, **12**, 1–18.
35. K. S. Fritz and D. R. Petersen, Exploring the biology of lipid peroxidation-derived protein carbonylation, *Chem. Res. Toxicol.*, 2011, **24**, 1411–1419.
36. W. Y. Cheng, J. Currier, P. A. Bromberg, R. Silbajoris, S. O. Simmons and J. M. Samet, Linking oxidative events to inflammatory and adaptive gene expression induced by exposure to an organic particulate matter component, *Environ. Health Perspect.*, 2012, **120**, 267–274.
37. M. Shinyashiki, A. Eiguren-Fernandez, D. A. Schmitz, S. E. Di, N. Li, W. P. Linak *et al.*, Electrophilic and redox properties of diesel exhaust particles, *Environ. Res.*, 2009, **109**, 239–244.
38. A. Eiguren-Fernandez, M. Shinyashiki, D. A. Schmitz, E. DiStefano, W. Hinds, Y. Kumagai *et al.*, Redox and electrophilic properties of vapor- and particle-phase components of ambient aerosols, *Environ. Res.*, 2010, **110**, 207–212.
39. N. Iwamoto, A. Nishiyama, A. Eiguren-Fernandez, W. Hinds, Y. Kumagai, J. R. Froines *et al.*, Biochemical and cellular effects of electrophiles present in ambient air samples, *Atmos. Environ.*, 2010, **44**, 1483–1489.

40. M. S. Hazari, N. Haykal-Coates, D. W. Winsett, Q. T. Krantz, C. king, D. L. Costa *et al.*, TRPA1 and sympathetic activation contribute to increased risk of triggered cardiac arrhythmias in hypertensive rats exposed to diesel exhaust, *Environ. Health Perspect.*, 2011, **119**, 951–957.
41. C. E. Deering-Rice, E. G. Romero, D. Shapiro, R. W. Hughen, A. R. Light, G. S. Yost *et al.*, Electrophilic components of diesel exhaust particles (DEP) activate transient receptor potential ankyrin-1 (TRPA1): a probable mechanism of acute pulmonary toxicity for DEP, *Chem. Res. Toxicol.*, 2011, **24**, 950–959.
42. G. L. Squadrito, R. Cueto, B. Dellinger and W. A. Pryor, Quinoid redox cycling as a mechanism for sustained free radical generation by inhaled airborne particulate matter, *Free Radical Biol. Med.*, 2001, **31**, 1132–1138.
43. Y. W. Janssen, S. Matalon and B. T. Mossman, Differential induction of c-fos, c-jun, and apoptosis in lung epithelial cells exposed to ROS or RNS, *Am. J. Physiol.: Lung Cell. Mol. Physiol.*, 1997, **17**, L789–L796.
44. G. G. Xiao, M. Wang, N. Li, J. A. Loo and A. E. Nel, Use of proteomics to demonstrate a hierarchical oxidative stress response to diesel exhaust particle chemicals in a macrophage cell line, *J. Biol. Chem.*, 2003, **278**, 50781–50790.
45. S. K. Donaldson, N. Mills, W. MacNee, S. Robinson and D. E. Newby, Role of inflammation in cardiopulmonary health effects of PM, *Toxicol. Appl. Pharmacol.*, 2006, **207**(Suppl), 483–488.
46. S. Boland, V. Bonvallot, T. Fournier, A. Baeza-Squiban, M. Aubier and F. Marano, Mechanisms of GM-CSF increase by diesel exhaust particles in human airway epithelial cells, *Am. J. Physiol.: Lung Cell. Mol. Physiol.*, 2000, **278**, L25–L32.
47. S. Boland, A. Baeza-Squiban, T. Fournier, O. Houcine, M. C. Gendron, M. Chevrier *et al.*, Diesel exhaust particles are taken up by human airway epithelial cells in vitro and alter cytokine production, *Am. J. Physiol.*, 1999, **276**, L604–L613.
48. W. G. Kreyling, M. Semmler-Behnke, J. Seitz, W. Scymczak, A. Wenk, P. Mayer *et al.*, Size dependence of the translocation of inhaled iridium and carbon nanoparticle aggregates from the lung of rats to the blood and secondary target organs, *Inhalation Toxicol.*, 2009, **21**(Suppl. 1), 55–60.
49. J. Ferin, G. Oberdorster and D. P. Penney, Pulmonary retention of ultrafine and fine particles in rats, *Am. J. Respir. Cell Mol. Biol.*, 1992, **6**, 535–542.
50. G. Oberdorster, Z. Sharp, A. P. Elder, R. Gelein, W. Kreyling and C. Cox, Translocation of inhaled ultrafine particles to the brain, *Inhal. Toxicol.*, 2004, **16**, 437–445.
51. R. D. Brook, B. Urch, J. T. Dvonch, R. L. Bard, M. Speck, G. Keeler *et al.*, Insights into the mechanisms and mediators of the effects of air

pollution exposure on blood pressure and vascular function in healthy humans, *Hypertension*, 2009, **54**, 659–667.

52. C. J. Martyniuk, B. Fang, J. M. Koomen, T. Gavin, L. Zhang, D. S. Barber and R. M. Lopachin, Molecular mechanism of glyceraldehyde-3-phosphate dehydrogenase inactivation by α,β-unsaturated carbonyl derivatives, *Chem. Res. Toxicol.*, 2011, **24**, 2302–2311.
53. S. S. Leonard, S. Wang, X. Shi, B. S. Jordan, V. Castranova and M. A. Dubick, Wood smoke particles generate free radicals and cause lipid peroxidation, DNA damage, NFκB activation and TNF-α release in macrophages, *Toxicology*, 2000, **150**, 147–157.

CHAPTER 12

Dioxins and Other Carcinogens

DAVID A. PURSER

Hartford Environmental Research, Hatfield, UK
Email: david-purser@ntlworld.com

12.1 Introduction

An important area of concern with respect to combustion products is the extent to which they contain carcinogens, including halogenated dioxins. Dioxins are also associated with other toxic effects in humans, including chloracne, effects on the liver, reproductive system and thyroid stimulating hormone, and a wider range of toxic effects in animal studies.[1,2] Given the huge range of different substances occurring in combustion product mixtures, it is not surprising that quite a number are known or suspected carcinogens. Among these, where halogens are present in fuels, are halogenated organic substances, including dibenzodioxins, dibenzofurans and polychlorinated biphenyls (PCBs), all of which show similar toxic effects, including carcinogenicity. The main concerns are therefore the yields and mass releases of these substances in combustion products, their bioavailability and potential effects on exposed subjects.

A number of references have been made to these substances and some of their effects in different chapters. The identities and approximate yields of toxic combustion products, including some carcinogens and dioxins, have been discussed in Chapters 2 and 3, while Chapter 17 in particular examines the extent to which groups, such as firefighters, have an increased occupational cancer risk from exposure to combustion products. Chapter 19 has considered health effects post-World Trade, including cancer incidence.

Issues in Toxicology No. 23
Toxicology, Survival and Health Hazards of Combustion Products
Edited by David A. Purser, Robert L. Maynard and James C. Wakefield

Published by the Royal Society of Chemistry, www.rsc.org

Comprehensive coverage of the identification and measurement of these substances in combustion products, their environmental dispersal and bioavailability, toxicology and epidemiology with respect to disease risk is too large a subject area for this book: this chapter presents an overview highlighting the main aspects.

Exposure to combustion-derived dioxins and other carcinogens occurs in several ways in occupational contexts and in indoor or outdoor environments including:

- Acute inhalation exposure to a fire smoke plume.
- Acute dermal and inhalation exposures at fire sites (during post-fire investigations or clean-up procedures) or through contact with contaminated clothing.
- Acute oral ingestion (eating with soot-stained hands or clothing in a contaminated area).
- Long term exposure due to repeated acute occupational exposures to one or more of the above sources (amongst, for example, firefighters, fire scene investigators or workers exposed to diesel fumes in confined workspace environments).
- Long term workplace exposure to industrial combustion processes.
- Long term exposure to aerosolised soot and volatile organic substances in buildings that have been inadequately decontaminated during post-fire restoration (resulting in a "sick building").
- Long term indoor exposures to fumes from poorly vented solid fuel cooking or heating fires.
- Long term outdoor exposures to air pollutants derived from combustion processes.
- Long term oral exposures to foodstuffs or drinking water contaminated by combustion-derived dioxins and other carcinogens as a result of environmental releases.

One of these sources of exposure about which concerns are sometimes expressed relates to the hazards from acute exposures to diluted smoke plumes during fire incidents. As discussed in Chapters 2 and 3, it is considered that, due to the low yields of carcinogenic products and the high dilutions in smoke plumes, the exposure concentrations and doses from any specific incident are unlikely to result in a significant increase in body burden of dioxins or other carcinogens in comparison with the chronic exposures from oral ingestion and inhalation of air pollutants experienced by the entire population. The only exceptions have been exposures to very high concentrations of dioxins from the Seveso incident and possibly exposures to high concentrations of PCBs from fires in electrical transformers in which these compounds were used as coolants and insulating fluids. Of much greater concern are the increased societal risk and incidence of cancer and other diseases resulting from repeated and chronic exposures to low concentrations of dioxins and other carcinogens.

Evidence for the carcinogenicity from combustion products is obtained from three main sources:

- Measurement of known or probable carcinogens in combustion product mixtures.
- Epidemiological evidence for increased cancer risk in specific occupational or environmental exposure populations.
- Animal inhalation and other toxicity studies.

Individual carcinogens have been measured in combustion products from a range of different fuels. As discussed in Chapters 2 and 3, the combustion product mixtures from different fuels usually contain a set of substances common to most fuels plus other sets of substances more related to the elemental and molecular composition of the specific fuels. The yields of different substances, including different carcinogens, vary with the fuel type and the combustion conditions. Some liquid fuels with high vapour pressure, such as diesel and petrol, already contain carcinogens that can be vaporised during combustion, with a fraction escaping the flame zone, so that the resulting effluent plume may contain carcinogens both from the original fuel and from its combustion products.

Epidemiological studies have shown evidence for increased cancer risk (especially lung cancer) from several combustion product sources. The classic case with the strongest effect is for tobacco smokers, but evidence for increased cancer risk has also been demonstrated for environmental tobacco smoke.[3] Diesel engine exhaust fumes have been identified as increasing cancer risk based on studies of occupationally exposed populations such as miners.[4,5] Household use of solid fuels and high-temperature frying also present an increased carcinogenic risk.[6] For even larger populations, an increased lung cancer risk has been identified in relation to the extent of exposure to particulate and other air pollutants.[7–10]

The findings of these epidemiological studies have been reinforced by animal exposure and mutagenicity studies for direct and environmental tobacco smoke,[3] extracts from coal fired domestic furnaces and automobile exhaust particulates.[6,11]

12.2 Carcinogens Identified in Combustion Products

12.2.1 Classification and Identification of Carcinogenic Substances in Combustion Products

Carcinogenic chemicals are classified by the International Agency for Research on Cancer (IARC) according to the scheme shown in Table 12.1 for the substances for which studies have been reported in IARC monographs.[12]

For most of these substances, carcinogenic activity has been identified by human epidemiology studies and/or by experimental studies in animals, usually rodents. For human epidemiology studies, possible carcinogens may

Table 12.1 IARC classification scheme for substances covered by IARC monographs.

Group	Activity	Number identified
Group 1	Carcinogenic to humans	108
Group 2A	Probably carcinogenic to humans	64
Group 2B	Possibly carcinogenic to humans	272
Group 3	Not classifiable as to its carcinogenicity to humans	508
Group 4	Probably not carcinogenic to humans	1

be identified as a result of a statistically significant increase in the incidence of specific cancers in exposed populations compared with unexposed populations or populations exposed to lower levels. For individual chemicals, an increased human cancer risk has generally been identified in groups of workers exposed in particular workplace environments, often to relatively high concentrations over many years. Where animal studies are used, the identification of a substance as a carcinogen generally relies on a statistically significant increase in the incidence of particular cancer tumours in rodents over lifetime studies, compared with the background incidence in unexposed control groups. The case for a particular chemical being a carcinogen is strengthened when positive evidence is obtained from both human epidemiology and animal experimental studies, and where plausibly causal mechanisms, such as mutagenicity, and particular biochemical activity can be identified. For human epidemiological and workplace studies, findings are often complicated by incomplete knowledge of the exposure history and of confounding due to other factors such as possible exposure to carcinogens in addition to the substance under investigation.

For any particular substance the cancer risk (probability of a cancer occurring) depends on a number of factors, including the carcinogenic potency of the substance and the extent of exposure, which for an inhaled substance depends upon the exposure concentration and the period of time over which the exposure occurs.

The IARC website[12] includes a list of agents classified by the IARC Monographs 1–109. Combustion products include tobacco smoke (direct and second hand exposure) and diesel exhaust in Group 1. Indoor exposure to emissions from household combustion of biomass fuel (primarily wood) is classified in Group 2A and engine exhaust gasoline as 2B. Almost the entire population is exposed to these substances, many to a significant extent, and in recent years awareness of the risk has prompted measures to reduce exposures in some cases.

Carcinogens associated with combustion products consist of low molecular mass organic vapours such as formaldehyde, 1,3-butadiene and benzene, and high molecular mass, mostly aromatic organic compounds such as polycyclic aromatic hydrocarbons (PAHs) and metals (including arsenic, nickel, chromium, antimony, cobalt, molybdenum, lead, beryllium and cadmium).[5,13–17] If nitrogen is present, carcinogens including acrylonitrile and nitrosamines can also be formed, and if halogens such as chlorine are present, then dichloromethane, PCBs, chlorinated dibenzodioxins and dibenzofurans are also formed. For inhaled combustion products the main target is the lung. This is especially the case for higher molecular mass, aromatic organics and metals, which are mainly associated with smoke particles deposited in the lung. Some substances are absorbed systemically and target other organs such as the blood, lymphatic system, liver (where many substances are metabolised, sometimes into carcinogenic substances), the kidneys and the bladder (where substances tend to become concentrated before excretion). Some dermal absorption may occur when soot collects on clothing during handling or from frequent contact with dirty clothing impregnated with soot (testicular or scrotal cancer risk).

Table 12.2 lists some of the carcinogens measured in combustion products from different fuel sources, classified according to their elemental composition. In this context, "fuel" includes any material or substance, or mixture of materials, burned in a fire. The table is in no sense intended to be exhaustive but illustrates the general categories of some carcinogens identified in combustion products. Hecht[17] lists 55 carcinogens from cigarette smoke that have been evaluated by the IARC, of which 20 listed in Table 12.2 have been found to induce lung tumours in at least one animal species, but it is likely that others are also present in cigarette smoke (the fuel most studied as a source for combustion product carcinogens) and in smoke from other fuels. The table shows the IARC classification for each substance and indicates some examples of some fuel sources.

12.2.2 Fuels Containing Carbon, Hydrogen and Oxygen

The first section of Table 12.2 lists substances produced by combustion of fuels containing carbon and hydrogen. Some of these also contain oxygen, but all of them combine with oxygen when they burn. The products therefore consist of series of aliphatic and aromatic hydrocarbons, some of which are partly oxidised to form alcohols, carbonyls and organic acids.[18,19] Some of these are low molecular mass substances with high vapour pressures, which are present partly in the vapour phase of the smoke and also partly adsorbed onto carbonaceous soot particles. These include substances such as saturated and unsaturated low molecular mass aldehydes (formaldehyde, acrolein, acetaldehyde and crotonaldehyde). These are highly reactive and irritant substances, of which formaldehyde has been established as a carcinogen and acetaldehyde as a possible carcinogen in humans, while the carcinogenic potential of acrolein and crotonaldehyde is currently unknown.[12] However, all form DNA adducts and are therefore potential carcinogens.[17,20] The vapour phase of cigarette smoke contains free radicals, causing lipid peroxidation, which is another route to DNA damage. Ethylene oxide occurs in the vapour phase of tobacco smoke, and 1,3-butadiene.[20] A major flame reaction in smoke formation is ring cyclisation. This leads to

Table 12.2 List of carcinogens found in combustion products.[a]

Carcinogen	IARC classification	Source examples
Substances containing carbon and hydrogen with or without oxygen		
Formaldehyde	1	Any organic fuel or aldehydes, especially from cellulosic materials
Acrolein	3	
Acetaldehyde	2B	
Ethylene oxide	1	
1,3-Butadiene	1	
Benzene	1	
Styrene	2B	
Methylstyrene	2B	
Polyaromatic hydrocarbons:	1	
Benzo[*a*]pyrene	2B	
Benzo[*b*]fluoranthane	2B	
Benzo[*j*]fluoranthane	2B	
Benzo[*k*]fluoranthane	2B	
Dibenzo[*a,i*]pyrene	2B	
Indeno[1,2,3-*cd*]pyrene	2A	
Dibenz[*a,h*]anthracene	2B	
5-Methylchrysene	2B	
Substances containing nitrogen in addition to C, H and O		
Acrylonitrile	2B	Any nitrogen-containing fuel
Ethyl carbamate	2A	Tobacco
Hydrazine	2B	Tobacco
N-Nitrosamines:		
N-Nitrosodiethylamine	2A	
4-(Methylnitrosamino)-1-(3-pyridyl)-1-butanone (NNK)	1	Diesel exhaust particulates
Nitroarenes, for example:		
3-Nitrobenzanthrone	2B	
6-Nitrochrysene	2A	
1-Nitropyrene	2A	
Aza-arenes		
Dibenz[*a,h*]acridine	2B	Polyurethanes
7*H*-Dibenzo[*c,g*]carbazole	2B	
Toluene diisocyanate	2B	
Substances containing halogens in addition to C, H and O		
1,2-Dichloromethane	2B	Any fuels containing chorine or bromine, *e.g.*, pentachlorophenol from treated wood
Pentachlorophenol	2B	
PCBs	1	
Chlorinated di-benzo dioxins	1	
Chlorinated di-benzo furans	3	
Brominated di-benzo dioxins	3	
Brominated di-benzo furans	3	
Substances containing phosphorus in addition to C, H and O		
Tris-(2,3-dibromopropyl) phosphate	2A	Flame-retardant treatment for children's nightwear that is released during use, but may not be released in combustion products

Table 12.2 (*Continued*)

Carcinogen	IARC classification	Source examples
Metals		
Chromium (hexavalent)	1	Fuels containing these metals, tobacco, vehicle exhaust emissions, treated wood, coal
Nickel	1	
Molybdenum		
Antimony (trioxide)	2B	
Arsenic	1	
Lead	2B	
Radioactive metals		
Polonium-210		Tobacco smoke
Mineral fibres		
Asbestos fibres	1	Fires involving asbestos or ceramic fibres (but not glass fibre or mineral wool)
Ceramic fibres	2B	

[a]Compiled from references cited in accompanying text.

the formation of aromatic substances in the combustion products from all organic fuels, the yields depending partly on the molecular structure of the fuel and partly upon the combustion conditions (see Chapters 2 and 3). The lower molecular mass aromatic substances (including benzene, styrene and methylstyrene) are prominent in the series of volatile organic compounds (VOCs) produced from most burning fuels.[21] Given this set of products, it is not surprising that the vapour phase fraction or tobacco smoke and smoke from burning coal and other fuels have been found to be carcinogenic.[3,5,6,17] However, the main source of carcinogens in tobacco smoke and smoke from sources such as diesel and other fuels is lower vapour pressure substances principally associated with the particulate phase of smoke. As discussed in Chapter 2, ring cyclisation in flames progresses by increased condensation to form larger and larger ring compounds with progressive elimination of hydrogen. From benzene and related compounds this progresses to form polyaromatic hydrocarbons, including benzo[*a*]pyrene and others listed in Table 12.2, which are considered to be the main carcinogens in smoke from tobacco and other sources.

12.2.3 Fuels Containing Nitrogen

In addition to C, H and O, many fuels also have significant nitrogen contents. This can range from around 0.1–3% by mass in natural products such as coal, wood or tobacco to approximately 5–26% in some common polymers such as polyamide, polyacrylonitrile and polyurethane foams.[21] Nitrogen is also often present as a fire retardant additive in polymers. When these burn, the combustion products include nitrogen-containing substances, including nitrogen gas, ammonia, nitric oxide (which oxidises gradually to nitrogen dioxide and may be implicated in cancer), hydrogen cyanide and a range of aliphatic and aromatic nitrogen-containing substances, including carcinogenic nitriles, hydrazine, nitrosamines, nitroarenes and aza-arenes.[17,21,22]

Flexible polyurethane foams decompose on heating into isocyanates released into the airborne phase and a polyol, which remains as a flaming pool fire.[21,23] The isocyanates combine to form a highly irritant particulate "yellow smoke", but a proportion remains as free toluene diisocyanate, which is known to form DNA adducts and is classified as possibly carcinogenic to humans.

12.2.4 Fuels Containing Halogens

Toxic halogenated combustion products are of particular concern with respect to environmental contamination. Halogens, especially chlorine, are present at low concentrations of around 1 mg g^{-1} in natural fuels such as wood or straw. A measurement of tobacco gave 12 and 6 μg g^{-1} for bromine and chlorine, respectively.[24] Treated materials, such as food waste, wood or newspaper, have higher halogen contents and may release specific toxic substances such as pentachlorophenol (PCP) and its combustion products. In particular, chlorinated and brominated compounds are used as flame-retardant treatments for many polymeric products. Polyvinylchloride (PVC) contains approximately 50% chorine, with about half this content in plasticised PVCs (for example, as used typically as cable insulation and jacket materials).[21] Environmental contamination by release of brominated polyether flame retardants by mechanisms other than fires, and their bioavailability and toxicity have been particular issues, but beyond this is the potential release of carcinogenic and otherwise toxic substances when products treated with these compounds are involved in fires or disposed of by incineration. The most important toxic and carcinogenic substances, which are released at low yields by any fuels containing halogens, are the polychlorinated bi-phenyls (PCBs), chlorinated di-benzo dioxins, chlorinated di-benzo furans, brominated di-benzo dioxins, brominated di-benzo furans and those with both chlorine and bromine on the same molecules. As with other products, the yields depend upon the elemental and molecular composition of the fuels and very much upon the combustion conditions.

12.2.5 Fuels Containing Phosphorus

Elemental phosphorus, inorganic phosphorus compounds and a wide variety of aliphatic and aromatic phosphorus compounds are used as flame-retardant additives in many products. These additives often also contain chlorine or bromine.[25] As with the polybrominated flame retardants there are concerns regarding the extent to which some of these additive compounds may be released from treated products during use or disposal, the extent to which they may therefore be released into the indoor and outdoor environments, and their bioavailability and toxicity. The most cited case of this kind involved the use of tris-(2,3-dibromopropyl) phosphate as a flame retardant for children's nightwear, which was introduced as a result of the high incidence of serious and sometimes fatal burns resulting from children standing too close to fires while wearing loose fitting and flammable nightwear. The additive was found to be released, particularly when the nightwear became damp from contact with fluids, such as urine, resulting in dermal absorption, or orally by "mouthing". The additive is classified as a human carcinogen and was withdrawn from this application in 1977.[26,27] This case and the issues related to the polybrominated ethers have led to campaigns to ban the use of all flame retardants, irrespective of the fact that their release, bioavailability and toxicity differs considerably for different additive compounds and systems. A developing problem is that not only is there pressure to remove the additives from products, but also to reduce the flammability performance requirements for products (the extent to which they are required to resist catching fire), which is likely to result in an increase in fire injuries and deaths. The main applications of these flame-retardant additives are to reduce the flammability of polymer systems and products to acceptable levels and to meet regulatory test requirements. Rather than relaxing the test requirements, the response should be to use alternative materials and products that do not require additives, or use flame-retardant additives with low environmental releases and bioavailability.

The aspect addressed here is the potential formation and release of toxic substances containing phosphorus during combustion. Most phosphorus compounds are readily decomposed to release phosphorous pentoxide when burned, so that the release of potentially toxic or carcinogenic phosphorus compounds in fires is likely to be minimal,[28] but cannot be excluded.

12.2.6 Metals

Natural products, such as wood or tobacco, contain traces of a variety of carcinogenic metals depending upon aspects such as the composition of the soils they are grown in. Metals may also be present as additives, for example, arsenical wood preservatives, or in other fuels such as lead added to petrol or various metals in other products. Those listed in Table 12.2 have all been measured in combustion products from tobacco and from heating and cooking fuels.[6,17] In fires, metals may also be released from other materials involved, in addition to the combustible fuels (see Chapter 19). One particular radioactive metal identified in tobacco smoke is polonium-210.[17]

12.2.7 Mineral Fibres

Another hazard, particularly from large fire incidents, is the airborne release of carcinogenic asbestos or ceramic fibres. Asbestos fibres have been found dispersed over a wide area following large fire incidents.[29] Health hazards from mineral dust following the World Trade Center incident are discussed in Chapter 19. Of the different types of fibres used in construction products, glass fibres and mineral wool fibres are not considered to present a cancer risk, but ceramic fibres may present a potential problem and have an IARC classification of 2B.

Table 12.3 illustrates the carcinogens known or predicted to exist in the combustion products from some common fuel materials, with some indication of the cancer target organs associated with these individual carcinogens and/or occupational exposures to combustion products including them. All fuel materials produce a basic similar range of carcinogens when they burn, plus specific additional carcinogens depending upon their composition. The yields of certain carcinogens can be higher from some materials than others. For example, benzene and PAHs are likely to be produced at higher yields from petroleum-based fuels, while halogenated PCBs are likely to be produced at higher yields from PVC and materials treated with halogenated flame retardants. For all materials, the yields vary considerably depending upon the combustion conditions.

12.3 Polychlorinated Biphenyls, Dioxins and Furans

12.3.1 Range of Different Substances and Toxicity

PCBs are mixtures of 209 individual chlorinated congeners, manufactured as coolants, lubricants and insulating fluids for use in electrical equipment, especially large transformers and capacitors. They are also formed during combustion of fuels or fuel mixtures containing halogens, but considerably higher yields occur in combustion products from fires involving the parent compounds. Different congeners exist as solids, liquids or vapours.[30] The manufacture of PCBs stopped in the US and UK in 1977, but use in the UK in certain applications continued up to 2000. Commercial products contain mixtures of different congeners and also dibenzofurans as contaminants. Human intake can occur *via* inhalation, skin absorption or, more typically, *via* water and food. The Environmental Protection Agency (EPA) standard for PCBs in drinking water is 0.5 ppb. The Food and Drug Administration (FDA) limits for food are from 0.2 ppm in infant and junior foods up to 0.3 ppm in eggs, poultry and red meat (fat basis). The toxic effects include increased cancer risk, which has been established for workers exposed during capacitor manufacture and in case-controlled studies of the general population, particularly for cancers of the liver, biliary tract, intestines and skin. PCBs have been shown to cause liver cancer in animals. In addition, toxic effects on the skin, the foetus, the endocrine system, the liver and the immune and nervous systems have been described. PCBs are considered as "dioxin-like" compounds and from the perspective of health hazards are considered in conjunction with dioxins.

Dioxins are polychlorinated dibenzo-*p*-dioxins (PCDDs); furans are polychlorinated dibenzo-*p*-furans (PCDFs). These PCDDs and PCDFs are halogenated aromatic compounds substituted in several positions by one or more chlorine atoms. They are formed as trace by-products in processes involving chlorine and organic compounds. Many possible types can be formed, including some 75 different chlorinated dioxins and 135 chlorinated furans. Brominated dioxins are also formed; some variants contain

Table 12.3 List of common fuel materials involved in fires, carcinogens in the fuels and combustion products, and cancer target organs for specific chemical carcinogens.

Product	Carcinogens in fuel	Carcinogens in combustion products	Cancer target organ/condition
Petroleum products	Benzene	Formaldehyde	Nasal
	Polycyclic aromatics	Benzene	Blood, leukaemia, (lymphoma, myeloma?)
	1,3-Butadiene		
	1,2-Dichloroethane	Polycyclic aromatic hydrocarbons	Lung and skin
	Tetraethyl and tetramethyl lead	1,3-Butadiene	Lymphohaematopoietic cancers (human) Heart, lung, stomach, liver, eye, mammary glands and ovaries (in mice)
	Dibromoethane	Styrene	
		Methylstyrene	
		1,2-Dichloroethane	Skin in animals
		Tetraethyl and tetramethyl lead	Digestive tract
		Dibromoethane	Skin in animals?
		Halogenated PCBs	Liver, biliary tract, intestines and skin (melanoma)
		Dioxins	Liver, thyroid and other in rats and mice
Diesel	Benzene	Formaldehyde	As for the same compounds above
	PAHs	Benzene	
		PAHs	
		1,3-Butadiene	

Used sump oil	Benzene	Formaldehyde	As for the same compounds above
	Polycyclic aromatics compounds	Benzene	
	1,3-Butadiene	Polycyclic aromatics compounds	
	1,2-Dichloroethane	1,3-Butadiene	
	Tetraethyl and tetramethyl lead	1,2-Dichloroethane	
	Dibromoethane	Tetraethyl and tetramethyl lead	
	Halogenated PCBs Dioxins	Dibromoethane	
	Chromium (hexavalent)	Halogenated PCBs Dioxins	
	Nickel	Chromium (hexavalent)	Nasal, lung
	Molybdenum	Nickel	Nasal, lung
	Antimony	Molybdenum	Lung
		Antimony	Lung
Aliphatic hydrocarbons and CHO polymers such as polyethylene Polymethylmethacrylate, polystyrene	None	Formaldehyde	As for the same compounds above
		Benzene	
		1,3-Butadiene	
		Styrene	
		Methylstyrene	
		PAHs	
Solid fuels, wood including treated wood pallets	Arsenic	Arsenic	Stomach, kidney, lung
	Chromium	Chromium	Nasal, lung
	Pentachlorophenol	Pentachlorophenol	Hodgkin's disease, soft tissue sarcoma and acute leukaemia, occupational exposure
	Aldrin, dieldrin	Aldrin, dieldrin	Liver in mice
		Formaldehyde	
		Benzene	As for the same compounds above
		PAHs	
		1,3-Butadiene	
		1,2-Dichloroethane	
		Dibromoethane	
		Halogenated PCBs Dioxins	

Table 12.3 (*Continued*)

Product	Carcinogens in fuel	Carcinogens in combustion products	Cancer target organ/condition
Tobacco		Fifty-five known carcinogens, including those listed in Table 12.2	Oral cavity, pharynx, nasopharynx, oesophagus stomach, colon and rectum, liver, pancreas, lung, kidney, bladder, lymphohaematopoietic cancers
Vehicle tyres	Traces of cadmium and lead	Formaldehyde Benzene	As for the same compounds above, plus lung and prostate
		Acrylonitrile Polycyclic aromatics compounds Nitrosamines? 1,3-Butadiene 1,2-Dichloroethane Halogenated PCBs Dioxins Cadmium Lead	Lung, prostate
Halogenated materials		As for CHO materials plus higher yields of PCBs, dioxins and furans	As for the same compounds from most materials
Nitrogen-containing materials		As for CHO materials plus nitrogen-containing carcinogens, including nitrosamine and nitroarenes	As for the same compounds from most materials

both chlorine and bromine. The toxic potencies of the different substances vary over a wide range, and 17 are of concern as the most toxic. In order to simplify this complex picture, the toxicity of a mixture of PCBs, dioxins and furans is commonly expressed as a "toxic equivalent" (TEQ). This is obtained by multiplying the concentration of individual dioxin and furan species ("congeners") by a suitable toxic equivalent factor (TEF) and summing the results to obtain an overall toxic potency. TEF values are calculated relative to the most toxic congener 2,3,7,8-tetrachlorodibenzo-*p*-dioxin (2,3,7,8-TCDD), which is assigned a TEF of 1; others congeners may have TEFs as low as 0.01 or even 0.0001 or less for some PCBs.[31] Some modifications were made to the calculation of TEFs by the World Health Organization (WHO) in 2005.[32] The relative TEFs are important when considering reported levels of contamination by dioxins and furans, since the total mass of compounds is likely to be much greater than the toxic equivalent mass.

12.3.2 Daily Intakes from Different Sources

Dioxins are fat soluble so levels are quoted for body fat or for the fat content of blood. A mean background level of 57 pg TEQ g^{-1} fat (1.4 ng total PCDD/Fs) has been reported in human fat tissue in the Welsh population.[33]

Dioxins have very long half-lives in the body. The elimination half-life for 2,3,7,8-TCDD in humans is 8 years and the terminal half-lives for some other congeners are 12–132 years.[34]

Routes of intake for PCBs and dioxins include food, air and skin contact. By far the largest source for the general population is in food, especially meat, fish and dairy products. There has been a considerable decrease in total daily intake from food in the UK from peak levels occurring during the 1980s as illustrated by the following figures:[35]

- 1982: 240 pg TEQ $person^{-1}$ day^{-1}.
- 1988: 125 pg TEQ $person^{-1}$ day^{-1}.
- 1992: 69 pg TEQ $person^{-1}$ day^{-1}.

According to the Ministry of Agriculture, Fisheries and Food (MAFF), this decrease in intake was caused mainly by changes in dietary habits and a fall in the average fat content of many foodstuffs, rather than a decline in environmental PCDD/F levels.[39] However, as will be discussed in Section 12.3.6, there is strong evidence for a considerable decline in levels of dioxin contamination of land and food since the mid-1980s. Water is a very minor source of dioxin intake. Air is also a relatively minor source of intake[31,36,37] as follows:

- Four UK air sampling sites (averages for Bowland, London, Manchester and Middlesbrough in 1993): 6.8 pg m^{-3} for 17 dioxin and furan congeners.
- Hamburg: 0.02 pg 2,3,7,8-TCDD m^{-3} and total of 0.1 pg TEQ m^{-3}.

Assuming 20 m^3 air breathed each day and 100% absorption this represents a total possible intake of:

- UK: 140 pg $person^{-1}$ day^{-1} of these congeners ≡ 1.4–14 pg TEQ depending upon the congeners present.
- Hamburg: total 2.1 pg TEQ $person^{-1}$ day^{-1}.[31]

Cigarette smoke contains dioxins and furans with a TEQ value of 1.81 ng TEQ m^{-3} sufficient to deliver 4.3 pg TCDD kg^{-1} bodyweight (bw) day^{-1}, equivalent to approximately 0.08 pg TEQ kg^{-1} bw day^{-1} or 5.8 pg TEQ $person^{-1}$ day^{-1} smoking 20 cigarettes.[38]

Due to their poor aqueous solubility only minute quantities would be absorbed following prolonged skin contact.[31] For example, a baby wearing six nappies per day made from chlorine-bleached paper is estimated to absorb less than 0.001 pg kg^{-1} bw day^{-1}. Therefore, firefighters, fire investigators or fire victims should absorb little through dermal contact with soot contaminated with dioxins.

On the basis of the figures quoted in the previous sections, the average daily intake of dioxins and furans as TEQ for a 70 kg adult can be estimated, as shown in Table 12.4.

In their 2001 report, the UK Committee on Toxicology estimated the daily intake for the average consumer as 1.8 pg WHO-TEQ kg^{-1} bw day^{-1} and for the 95th percentile consumer as 3.1 pg WHO-TEQ kg^{-1} bw day^{-1}, but that dietary intake was decreasing.[39] The 2012 addendum to the US ATSDR report[40] gives somewhat lower levels for dietary intake from meat, poultry and fish of 0.64–0.78 pg TEQ kg^{-1} bw day^{-1}, but this does not appear to include dairy products. In their previous report, dairy products represented 28% of oral intake, representing approximately 0.33 pg TEQ kg^{-1} bw day^{-1} of a total intake of 0.7 pg TEQ kg^{-1} bw day^{-1} in 1989[30] (assuming an average bodyweight of 70 kg). The data in Table 12.4 show that, with the possible exception of smokers and occupationally exposed groups, by far the greatest source of human dioxin and furan intake is *via* food. This, therefore, raises the questions of the hazards associated with such intake and in the context of this chapter the extent to which food content arises as a result of combustion processes.

Table 12.4 Average daily intake of dioxins and furans from all sources for a 70 kg adult.[31]

Source	pg TEQ kg^{-1} bw day^{-1}	pg TEQ $person^{-1}$ day^{-1}
Food	1	69
Air	0.03	2.1
Consumer products	0.01	0.7
Water	very little	
Total	1.04	71.8

12.3.3 Toxic Effects of Dioxins and Recommended Maximum Daily Intake Levels

Information on the toxic effects of dioxins and furans are derived from epidemiological studies in man and detailed experimental studies in animals.[1,30,31,37,41] Studies in man include measurements of dioxin levels in individuals that try to link these with the predicted effects of the compounds on health. A variety of populations have been studied, including those working in industries using compounds known to be contaminated with dioxins and furans and those accidentally exposed such as at the Seveso incident.[41] By taking results from studies in a range of animal species together with the human data it is possible to make estimates of likely effects in humans. Safety factors of one or more orders of magnitude are used to set recommended maximum daily intake levels, which can be used to provide guidance on the significance of environmental exposure levels. As more information becomes available these recommended levels are reviewed and updated.

The best indicator of exposure to high levels of PCDD/Fs in humans is the skin disease chloracne, which is associated with transient effects on liver enzymes. This is associated with very high levels of dioxins in the body, usually as a result of skin contamination by lubricants, and it has been several decades since major incidents have occurred.[1,30,31,40,41] In addition, there are concerns about three major toxic effects found in animal studies: immunotoxicity, reproductive toxicity and carcinogenicity. So far, it has not been possible to determine whether PCDD/Fs are immunotoxic to humans. Studies have produced conflicting findings.[42] Effects on the endocrine system include decreased levels of testosterone reported in male workers manufacturing 2,4,5-trichlorophenol.[37] Reproductive toxic effects on fertility and offspring bodyweight have been reported in rhesus monkeys. Animal studies indicate that dioxins are amongst the most potent carcinogens known, but epidemiological evidence linking dioxins with cancer in the general human population is less clear, although there is evidence for elevated cancer risk in workers using phenoxy herbicide preparations (which contain dioxins as contaminants).[1,37]

12.3.4 Exposure Limits and Guidelines

Dioxins and furans produce chloracne at high doses, may harm human metabolism, development and reproduction, and may constitute a human cancer hazard.[1,31,37,42–44] These adverse effects may occur at levels less than ten times greater than current average body burdens,[37] but effects at current average body burdens remain unproven.[45] The UK Committee on Toxicology (COT)[31,37] recommendations were to reduce human exposures, identify major sources and reduce environmental inputs to reduce levels in food and human tissues.[31]

The currently recommended exposure limits and guidelines for dioxins are expressed as acceptable daily intake (ADI) or tolerable daily

Table 12.5 Exposure limits and guidelines for dioxins and furans.

Agency	Exposure limit	Exposure limit per person (assuming 70 kg)
UK: COT (2001)[1]	2 pg TEQ kg^{-1} day^{-1} TDI	140 pg TEQ day^{-1} TDI
WHO (2010)[46]	2 pg TEQ kg^{-1} day^{-1} ADI	140 pg TEQ day^{-1} ADI
US: EPA (2012)[48]	0.7 pg TCDD kg^{-1} day^{-1} TDI	49 pg TEQ day^{-1} TDI

intake (TDI) values. The UK COT (1995) accepted the World Health Organization recommendations,[44] which were based upon the lowest observed adverse effect in animals for carcinogenic, hepatotoxic, immunotoxic and reproductive toxicity endpoints.[33] Based upon a no effect level of 1000 pg kg^{-1} bw day^{-1} for animal studies, a factor of 100 was applied to take into account toxicokinetic differences and uncertainties regarding reproductive toxicity, giving a recommended value (ADI) of 10 pg kg^{-1} bw day^{-1} (WHO, 1991).[44] This was approximately five to ten times the total daily intake in the UK in 2001 of approximately 1–2 pg TEQ kg^{-1} bw day^{-1}. The recommended limits were later reduced to 2 TEQ kg^{-1} bw day^{-1} TDI.[46] The US EPA at one stage considered a lower level based upon assumptions regarding possible carcinogenicity of 6.4 fg TCDD kg^{-1} bw day^{-1} TDI[47] and in 2012 set its safe threshold at 0.7 pg TEQ kg^{-1} bw day^{-1} in relation to non-cancer health effects,[48] with updated guidance on thresholds for cancer to follow. These limits and guidelines are summarised in Table 12.5.

Based upon the data from Tables 12.4 and 12.5, the average UK citizen in 2001 was receiving close to the maximum acceptable daily intake recommended by the WHO and UK COT, but somewhat more than the EPA recommended levels. It was, however, noted that oral intake (the main source) was gradually declining in the UK.

12.3.5 Environmental Contamination by Dioxins and Furans Resulting from Combustion of Halogenated Materials

PCBs, dioxins and furans constitute a fraction of the combustion products from combustion of any fuels containing halogens. Local indoor contamination of buildings results from fire incidents, smoking tobacco, cooking or from open fires. Contamination of the outdoor environment results from non-combustion sources and a range of combustion sources, including large fire incidents, industrial processes, waste incineration, wildfires and stubble burning.

With respect to the use of halogenated materials, especially flame retardants, in the built environment and unwanted fires there are two particular issues:

1. The extent to which the production, use and disposal of organohalogen compounds and other fuel mixtures containing halogens leads to general levels of environmental contamination by toxic flame-retardant

additives, PCBs, dioxins and dibenzofurans that constitute a risk to public health.
2. The extent to which exposure of people to combustion products from halogenated materials during and after fires in buildings constitutes a risk to long term health.

An important consideration with respect to the use of PVC and halogenated flame retardants is the extent to which they contribute to background levels of environmental dioxin and furan contamination. Reviews of sources of dioxins and furans entering the environment include Eduljee (1988),[49] DOE Pollution Paper No. 27 (1989),[31] including the advice of the COT for the Department of the Environment, and Alcock and Jones (1996).[50] These found many sources of dioxin input into the environment, some resulting from man-made organochlorine compounds and some from natural processes involving the combustion of organic materials with inorganic chlorine salts (for example, coal fires, forest fires and stubble burning). Up to approximately the late 1980s, probably the three main sources were dioxins as contaminants in polychlorinated biphenyls (PCBs) used in transformers, penta-chlorinated phenols (used as wood preservatives) and municipal waste incineration (MSW). Other sources were industrial and hospital incinerators, crematoria, steel processing, coal fires, leaded petrol combustion in motor vehicles, certain herbicides and occasional accidents such as Seveso.[41] Two more personal sources to the individual were hexachlorophene (a once widely used antiseptic also incorporated in toothpaste) and cigarette smoke. For the majority of these chlorinated chemicals, small amounts of dioxins often occurred as contaminants from the manufacturing process, or could be formed when the material was burned.

Of these sources, PCBs were banned in the 1970s and PCP (pentachlorophenol) manufacture ceased in the 1980s. PVC and other halogenated flame-retardant materials may contribute significantly to total chlorine input and dioxin release from municipal and other waste incineration. This depends upon the dioxin output from incinerators and its relationship to the amount and nature (organic or inorganic) of the chlorine load in the incinerator fuel. In old style incinerators, even when high temperatures were used, dioxins could not only be released from the fuel, but could be formed secondarily from hydrogen chloride and organic residues in the flue: this was often the main route of formation. There was felt to be a poor relationship between the chlorine content of the fuel, in particular the amount of PVC and related compounds present, and the dioxin content of the flue gases, although this view was challenged in later work.[51] Modern incinerators are designed to remove dioxins from the flue gases, or prevent their secondary formation, and since the modifications were made, the stack emissions from incinerators are greatly reduced by factors of between 100 and 1000.[52] This compares with a factor of ten reduction achieved by removing PVC in a situation where it constituted 50% of chlorinated waste.[52]

12.3.6 Trends in Dioxin Sources and Levels of Environmental Contamination

Time series examination of environmental dioxin levels in a variety of places and materials reveals a reasonably consistent pattern.[50] These are illustrated in Figure 12.1.[53] Studies of lake bed sediments in remote locations in the US and in park grass samples collected since 1860 in north London show constant low levels of contamination up to around 1950. This is followed by a very large increase over the period from approximately 1955 to 1980, followed by a large and rapid decline during the late 1980s and early 1990s. This has been accompanied by decreases in levels of air contamination (measured in Germany), decreases in contamination of sewage sludge, decreases in levels in wild life, such as sea birds, decreases in livestock tissue and especially large decreases in food fats (including cow's milk) and in human milk as shown in Figure 12.1.

This markedly improving picture runs counter to the data for PVC production and use, which started in the later 1950s. Figure 12.1 also shows the greatly increasing world production since 1950. Consumption in Europe has risen from low levels in the early 1960s to 600 000 tonnes in the early 1990s and continues to rise, while PCDD/F concentrations in a wide range of environmental media are falling. This enormously increasing tonnage of PVC and materials containing chlorinated and brominated fire retardants in use presents an increasing disposal and recycling problem. Just one example of this kind of problem is presented by the huge increase in the amount of electrical cable in building plenum spaces. New cable, much or which consists of PVC or other halogenated materials, is continually being introduced into both new and existing buildings, and in the latter case, old cable is seldom removed. It is estimated that in buildings in the US, the total amount

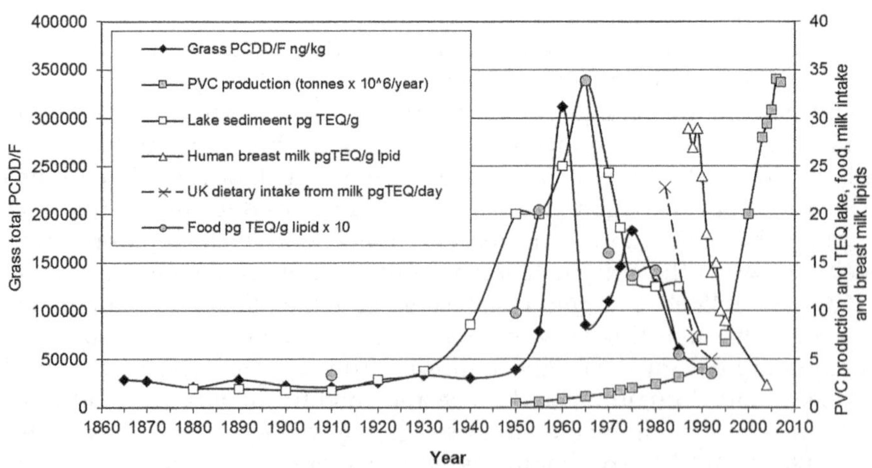

Figure 12.1 Trends in environmental dioxins compared with annual PVC production.

of plenum cable present (expressed in terms of millions of feet) increased from 1615 to 4246 between 1991 and 1996. This represents an average annual growth rate of 46% and the total was predicted to reach 8804 million feet during 2000 compared with 357 million feet in 1983 (a 25 fold increase). All of this material (and that derived from halogenated polymers from other sources) will eventually have to be disposed of or recycled and it is possible that, as the previous major sources of dioxin contamination are eliminated, the amount produced by the combustion of PVC and other halogenated materials in landfill, accidental fires and modern incinerators will come more into prominence. However, data show emissions from such sources continue to be low.[50,52]

The general picture with regard to the environmental hazards from PVC and halogenated fire retardants is therefore not a simple one. Certainly, as far as dioxin release is concerned, there seems to be little evidence that PVC, and chlorinated or brominated flame retardants have been major sources of dioxins up to recent times. However, the release and combustion of increasingly large tonnages of PVC and other halogenated materials does need consideration.

12.3.7 Extent to Which the Exposure of People to Combustion Products from PVC and Materials Containing Halogenated Fire Retardants During and After Fires in Buildings Constitutes a Risk to their Long Term Health

Although general environmental contamination by dioxins and furans arises from a number of sources, a possible concern is that significant local environmental contamination may result from particular fire incidents in buildings and that this might represent a health hazard. Building occupants during fires, the population living in areas close to fires, members of the emergency services attending fires and those involved in post-fire examination and clean-up operations could all be exposed to risk as a result of exposure to dioxins and furans. These risks from exposure to dioxins arising from the combustion of halogenated materials can only be considered realistically in relation to the health hazards presented by other combustion products from those materials in building fires and the health hazards from other materials burning in typical fires. It is also important to consider the nature, circumstances and frequency of exposures, as well as to compare the intake of dioxins and other toxic substances from fires with the likely levels of intake from other sources.

The most important possible routes by which building occupants or the population around building fires are likely to be exposed to dioxins and furans is by inhalation of smoke particulates during a fire, aerosolised soot afterwards and oral ingestion from contamination of food due to contact with soot deposits.

There are a number of potential sources of dioxins in most building fires, but is likely that PVC is the major chlorine source in most buildings and it has been noted that dioxin yield from decomposition of copper-covered PVC wiring is high.[52,54] The yields of dioxins and furans depend very much on the decomposition conditions in the fire. Incinerator and other studies have shown[52,54] that the yields can be much greater (by a factor of five) when a furnace is first started and relatively cold[52] than when it is hot, and that 300–400 °C is the optimum formation temperature. It is therefore to be expected that dioxin yields will vary considerably depending upon the type and size of fire and its ventilation. Vitiated fires of the kind found in buildings, particularly during stages before serious structural breaching, are likely to produce high dioxin yields, while yields may be somewhat lower in very large conflagrations. Some data are available from a few accidental fires and some small scale experiments.[52,54,55]

There is evidence from an EPA study[55] that yields of dioxins and furans from small, inefficiently burning fires are much greater than those obtained from incinerators. In this study, samples of household waste were burned in a 55 gallon oil drum, a common method of waste disposal. In one experiment involving waste with a 4.5% PVC content the total yield of PCDDs and PCDFs was 493 ng g^{-1} waste decomposed. This compares with a figure of 0.0035 ng g^{-1} waste decomposed in a modern municipal waste incinerator, a factor of more than 100 000 times greater yield. On this basis the total dioxin and furan output of a household using this method to dispose of its waste could be approximately equivalent to the entire output of a municipal waste incinerator. Assuming a factor of ×0.1 for the TEQ congeners present, this could represent 49.3 ng TEQ g^{-1} waste decomposed.

Two major considerations in terms of human exposure are the concentrations in the effluent plume and the concentrations in soot deposits. It is very difficult to obtain good quantitative data in such situations due to difficulties in sample collection and difficulties in knowing the mass of PVC, halogenated flame retardants and other fuels involved.

Soot samples from a number of accidental fires studied involving PVC and other materials seem to have dioxin contamination in a range from approximately 5–400 ng TEQ g^{-1} soot.[52,54] A high value of 390 ng g^{-1} was obtained in some soot samples taken from a PVC warehouse fire in Canada, while other parts of the same fire had levels of only 0.3 ng g^{-1}.[52] The levels reported for the Dusseldorf airport fire of 42.6 ng TEQ g^{-1} might be considered fairly typical,[56] the German authorities citing levels in fires of up to 200 ng TEQ g^{-1}. The dioxins produced in the Dusseldorf fire are considered not to have been derived only from PVC, but from halogenated additives in other materials. The composition of soot deposits sampled from the airport departure hall after the fire is shown in Table 12.6. While chlorinated dioxins predominate, a significant proportion was brominated and mixed bromine–chlorine dioxins and furans.

The PCB content of the soot was considerably higher than the dioxin content; however, the WHO-TEFs for PCBs are generally considerably lower

Table 12.6 Dioxin and PAH composition of soot samples from the Dusseldorf airport fire.

Total TEQs	PCDD/F – ITE[a]	42.6 ng TEQ g^{-1}
Total chlorinated dioxins and furans	PCDD/F	1.4 μg g^{-1}
Total brominated dioxins and furans	PBDD/F	0.2 μg g^{-1}
Total chlorine and bromine dioxins and furans	PCBDD/F[b]	0.7 μg g^{-1}
Total PCBs	PCB (6 DIN)[c]	38 μg g^{-1}
Total polyaromatic hydrocarbons	PAH (EPA)	5000 μg g^{-1}
Benzo[*a*]pyrene	BaP[a]	260 μg g^{-1}

[a]ITE, international toxic equivalents.
[b]PCBDD/F, polychlorinated/polybrominated dibenzo-*p*-dioxins and polychlorinated/polybrominated dibenzo-*p*-furans; BaP, benzo[*a*]pyrene.
[c]6 DIN, sum of six congeners according to Deutsches Institut für Normung.

than those for chlorinated dioxins. The PAH content of the soot was also high, including a significant benzo[*a*]pyrene content.

Taking the EPA household waste fire data, which are likely to be similar to the yield in a small domestic fire, it is possible to estimate likely soot dioxin concentrations (from the dioxin and smoke particulate yields in several different experiments) to lie in the range of approximately 700–7000 ng TEQ g^{-1} soot, which are one to two orders of magnitude greater than those measured in soot recovered from fire incidents. It is estimated that this could represent a concentration of around 20 ng TEQ m^{-3} in a dilute smoke plume [optical density (OD) m^{-1} = 0.01] in the vicinity of such an oil barrel fire. This compares to a figure of 5–40 ng TEQ m^{-3} measured in the effluent plume from old fashioned incinerators in Sweden, which if diluted by a factor of 100 would give a maximum of 0.4 ng TEQ m^{-3} in a diluted plume.

Based on these data it is possible to estimate possible hazards to a person working in the vicinity of a burning building during a fire and exposed to a dilute smoke plume or inside a building after a fire and exposed to soot (Table 12.7).[53] Assuming a person was exposed for a 1 hour period to the dilute smoke plume and inhaled 1 m^3 of air, then the inhaled total dose of dioxins and furans could be 400 pg TEQ for a 1/100 diluted incinerator plume or between 1980 and 19 800 pg TEQ for a similar dilution of the household waste fire case (dilution to give a visibility of 100 m through the smoke). This can be compared to the maximum tolerable daily intake of 140 pg TEQ per person (from Table 12.5).

Another potential source of inhalation exposure is the inhalation of soot dust while working in a damaged area after a fire without respiratory protection. Assuming a level of 1 mg m^{-3}, this might result in the inhalation of up to 5 mg soot during a 5 hour period, representing 3304–33 043 pg TEQ for the household waste fire case. Using the German figure of 200 ng TEQ g^{-1} gives a dust intake figure of 1000 pg TEQ. A proportion of this will be removed from the lungs by mucociliary clearance, but most of this is likely to be swallowed, adding to any oral dose.

Table 12.7 Possible total dioxin and furan intake (pg TEQ) for a person near or inside a building during or after a fire.

Dioxin concentration of smoke particles and soot/ng TEQ g^{-1}	Intake	Dose received/pg TEQ	Fraction of daily intake limit (140 pg TEQ day^{-1})
Old Swedish incinerator plume (assuming 1/100 plume dilution factor)	0.05–0.40 ng TEQ m^{-3} diluted smoke inhalation for 1 hour (1 m^3)	50–400	0.35–2.86
EPA household waste study, 700–7000[a]	Inhalation, 100 m visibility, smoke dilution 2–20 ng TEQ m^{-3} for 1 hour (1 m^3)	2000–20 000	14.3–143
EPA household waste study, 700–7000[a]	Dust inhalation 1 mg m^{-3} for 5 hours (5 m^3)	3300–33 000	23.6–235.7
German fire residue, maximum 200	Dust inhalation 1 mg m^{-3} for 5 hours (5 m^3)	1000	7.1
EPA household waste study, 700–7000[a]	Oral intake 0.01 g soot	7000–70 000	50–500
German fire residue, maximum 200	Oral intake 0.01 g soot	2000	14.3

[a]Assuming TEQ = total dioxin and furan content × 0.1. Copyright HER.

The other main route of ingestion is likely to be by oral ingestion through contamination of food by contact with dirty hands or clothing. If 10 mg of soot (a very small pinch) was ingested, this would represent a dose of 6610–66 100 pg TEQ for the household waste case or 2000 pg TEQ for the German data.

The data are summarised in Table 12.7. They indicate that a diluted smoke plume from a fire is likely to contain considerably more dioxins and furans than the diluted smoke plume from even an old style incinerator, and that being enveloped in smoke from a fire dilute enough to provide 100 m visibility could result in the inhalation of up to 140 times the daily maximum acceptable intake. The inhalation of soot dust after a fire could provide a similar or somewhat greater hazard depending upon the dust concentration and duration of exposure. However, oral intake with food is likely to be the most hazardous for someone working in a fire contaminated area. This could provide up to 500 times the maximum acceptable daily intake (more than a year's normal intake) if just 0.01 g of soot was ingested (for example, by eating with soot-stained hands), but the hazard could be minimised by simple hygiene precautions.

These findings demonstrate that a potential hazard may exist at fire scenes and that personnel should take care to avoid dust inhalation or ingestion by the use of simple hygiene precautions (dust particulate masks and gloves, with careful hand washing after exposure).

Potential hazards from dioxins, furans and PAHs in relation to the health and safety of workers at the scenes of the Dusseldorf and Lengerich fires were examined in some details following these incidents.[57–59] The cancer risk from PHDD/Fs was substantially lower than that from PAHs. Both types of pollutants are strongly bound to soot and have a low bioavailability. No chronic toxic effects have been reported from individuals accidentally involved in single fire incidents and a German study has shown professional fire-fighters to have no higher dioxin blood levels than the dioxin background levels in the general population.[60] These findings support the conclusion that exposure at a single fire incident is unlikely to present a serious chronic health hazard from dioxins, furans and PAHs, and that simple hygiene precautions can protect professionals from the potential effects of repeated exposures.

An important consideration with regard to dioxin intake is that it is a long term accumulation problem. Considering the very long half-lives in the body, exposure during any one day or even over a week is less important than the overall cumulative exposure over a long period extending to years. The health significance would therefore depend on how often a person was exposed to such levels. It is also important to remember that the recommended maximum exposure levels include a safety factor of around an order of magnitude or more. On this basis it would seem that there is a potential small but significant risk to health for a person in contact with dioxins and other carcinogens from contaminated soot from fires on a regular basis. The hazard from dioxins and furans due to contact with soot and the combustion process was recognised in the Department of the Environment pollution paper No. 37.[30] Section 8.5, page 41 states, "occupational exposure may also occur for persons involved with combustion processes. These may include the operation of all types of combustion plant and incinerators, including the handling of ash; the burning of chlorinated materials *e.g.*, PCP-treated timber and chlorinated plastics; or open fires; and the smelting of plastics-coated scrap metal. Similarly entry into areas contaminated by soot as a result of fire engulfment of PCP-containing materials may give rise to exposure to PCDDs and PCDFs."

With regard to health hazards and, in particular, cancer risk to the general population, it seems that reduction of hazards from dioxin intake since the 1980s has been quite a success story. Exposure to all smoke toxins from tobacco smoke (both direct and environmental) has also decreased significantly in Western countries (although still responsible for 28% of all cancer deaths[61]).

However, exposure to smoke carcinogens other than dioxins due to air pollution from industrial and vehicle emissions, especially diesel smoke particulates, remains a serious and possibly increasing societal health risk. Studies reviewed by the Committee on the Medical Effects of Air Pollutants (COMEAP) in 2009 and 2010[62,63] led to the conclusion that current levels of fine particles (PM2.5) in the UK are responsible for 29 000 attributable deaths each year from several diseases. Although cardiovascular diseases are a major component, a proportion is attributed to cancer each year, with air pollutants regarded as a cause of lung cancer. According to IARC, 223 000 deaths worldwide from lung cancer are attributable to air pollution.[64]

12.4 Conclusions

Combustion products from all burning carbonaceous fuels contain a wide range of aliphatic and aromatic organic substances, including many known or suspected carcinogens. These include a range of common substances produced by most fuels, plus others specific to the elemental composition and molecular structure (including nitrogen- and halogen-containing substances, and carcinogenic metals). The yields of different substances depend partly on the fuel composition, but greatly upon the combustion conditions. Combustion products from a number of sources have been identified as human carcinogens by IARC.

Where halogens are present in the burning fuel the combustion products include PCBs, halogenated dioxins and furans, which are associated with a range of toxic effects, including carcinogenicity. As with other combustion products the yields in fires are very dependent upon the combustion conditions, being relatively high in poorly ventilated fires and very low in emissions from modern incinerators. Environmental contamination from dioxins has occurred historically from a range of sources, but there has been a considerable decrease since the peak in the 1970s to low present day levels, despite increasing production and use of halogenated materials such as PVC.

Exposure to carcinogenic combustion products can be *via* inhalation, through dermal contact or by oral ingestion, of which inhalation is the route of main concern. For dioxins, the main route of exposure is *via* oral ingestion in food.

It is considered that potential health hazards from acute exposure to airborne carcinogens and dioxins are likely to be minor during most fire incidents and exposure to contaminated soot at post-fire sites can be minimised by simple hygiene precautions.

Of more concern, especially with regard to carcinogens, is the long term risk arising from repeated and continuous exposures in the domestic environment (from open fires, cooking fumes and tobacco smoke), in the workplace from combustion processes and especially from ambient air pollution (especially that arising from vehicle emissions).

References

1. Department of Health, COT statement on the tolerable daily intake for dioxins and dioxin-like polychlorinated biphenyls. Committee on toxicity of chemicals in food, consumer products and the environment. COT/2001.07, October 2001.
2. *EPA's Reanalysis of Key Issues Related to Dioxin Toxicity and Response to NAS Comments*, US Environmental Protection Agency, Washington, DC, February 2012, vol. 1.
3. S. S. Hecht, Tobacco smoke carcinogens and lung cancer, *J. Natl. Cancer Inst.*, 1999, **91**, 1194–1210.

4. D. T. Silverman, C. S. Samaic, J. H. Lubin, A. E. Blair *et al.*, The diesel exhaust in miners study: A nested case-control study of lung cancer and diesel exhaust, *J. Natl. Cancer Inst.*, 2012, **104**, 855–868.
5. IARC, Diesel engine exhaust carcinogenic, Press Release No. 213 12, International Agency for Research on Cancer, June 2012.
6. IARC, Household use of solid fuels and high-temperature frying, *IARC Monographs on the Evaluation of Carcinogenic Risks in Humans*, Lyon, France, 2010, vol. 95.
7. Committee on the Medical Effects of Air Pollutants (COMEAP), *Long-term Exposure to Air Pollution: Effect on Mortality*, Department of Health, 2009.
8. Committee on the Medical Effects of Air Pollutants (COMEAP), *The Mortality Effects of Long-Term Exposure to Particulate Air Pollution in the United Kingdom*, Department of Health, 2010.
9. S. H. L. Yim and Barrett, Public health impacts of combustion emissions in the United Kingdom, *Environ. Sci. Technol.*, 96, 2012, **46**, 4291–42.
10. Y. Grosse, B. Lauby-Secretan, F. El Ghissassi, V. Bouvard, N. Guha, D. Loomis, K. Straif and International Agency for Research on Cancer Monograph Working Group, Carcinogenicity of diesel-engine and gasoline-engine exhausts and some nitroarenes, *Lancet Oncol.*, 2012, **13**(7), 663–664.
11. F. Pott and W. Stober, Carcinogenicity of airborne combustion products observed in subcutaneous tissue and lungs of laboratory rodents, *Environ. Health Perspect.*, 1883, **47**, 293–303.
12. IARC, Agents classified by the *IARC Monographs*, vol. 1–109, Available at http://monographs.iarc.fr/ENG/Classification/, Accessed 28th June 2014.
13. J. Lewtas, Impact of fuel choice on comparative cancer risk of emissions, *Energy & Fuels*, ACS Publications, 1993.
14. F. Pott and W. Stobert, Carcinogenicity of Airborne Combustion Products Observed in Subcutaneous Tissue and Lungs of Laboratory Rodents, *Environ. Health Perspect.*, 1983, **47**, 293–303.
15. T. Silverman, C. Samanic, J. Lubin, A. Blair, P. Stewart, R. Vermeulen, J. Coble, N. Rothman, T. W. Schleiff, R. Ziegler, S. Wacholder and M. Attfield, The Diesel Exhaust in Miners Study: A Nested Case – Control Study of Lung Cancer and Diesel Exhaust, *J. Natl. Cancer Inst.*, 2012, **104**, 855–868.
16. R. B. Hayes, The carcinogenicity of metals in humans, *Cancer, Causes Control*, 1997, **8**, 371–385.
17. S. S. Hecht, Tobacco Smoke Carcinogens and Lung Cancer, *J. Natl. Cancer Inst.*, July 21, 1999, **91**(14).
18. W. D. Woolley and P. J. Fardell, Basic Aspects of Combustion Toxicology, *Fire Saf. J.*, 1982, **5**, 29.
19. W. D. Woolley, S. A. Ames and P. J. Fardell, Chemical Aspects of Combustion Toxicology of Fire, *Fire Mater.*, 1979, **3**, 110–120.
20. J. C. Stavridis, *Oxidation: the Cornerstone of Carcinogenesis*, Springer, 2008.
21. D. A. Purser, Assessment of Hazards to Occupants from Smoke, Toxic Gases and Heat, in *The SFPE Handbook of Fire Protection Engineering*, ed.

P. J. DiNenno, National Fire Protection Association, Quincy, MA 02269, 4th edn, 2009, pp. 96–193.
22. D. A. Purser and J. A. Purser HCN yields and fate of fuel nitrogen for materials under different combustion conditions in the ISO 19700 tube furnace and large-scale fires. *9th International Symposium on Fire Safety Science*, University of Karlsruhe, Germany, 21st–26th September 2008.
23. W. D. Woolley, Nitrogen-containing Products from the Thermal Decomposition of Flexible Polyurethane Foams, *Br. Polym. J.*, 1972, **4**, 27–43.
24. A. I. Muller, C. A. Bizzi, J. S. F. Pereira, M. F. Mesco *et al.*, Bromine and chlorine determination in cigarette tobacco using microwave-induced combustion and inductively coupled plasma optical emission spectrometry, *J. Braz. Chem. Soc.*, 2011, **33**, 1649–1655.
25. D. A. Purser Toxicity of fire retardants in relation to life safety and environmental hazards. in *Fire Retardant Materials*, ed. A. R. Horrocks and D. Price, Woodhead Publishing Ltd, Cambridge, UK, 2001, ch. 3, pp. 69–127.
26. Consumer Product Safety Commission, CPSC Bans TRIS-treated Children's Garments. April 7, 1977, Release No. 77-030.
27. Council Directive 79/663 EEC, *Off. J. Eur. Communities*, 1979, L197/37.
28. D. A. Purser, Combustion toxicology of anticholinesterases, in *Clinical and Experimental Toxicology of Organophosphates and Carbamates*, Butterworth-Heinemann, Oxford, UK, 1992, pp. 386–395.
29. S. Bridgman, Community health risk assessment after a fire with asbestos containing fallout, *J. Epidemiol. Community Health*, 2001, **55**, 921–927.
30. Agency for toxic substances and disease registry (ATSDR), *Toxicological Profile for Chlorinated dibenzo-*p*-dioxins*, U.S. Department of Health and Human Services, December 1998.
31. DOE report of an interdepartmental working group on polychlorinated dibenzo-*para*-dioxins (PCDDs) and polychlorinated dibenzofurans (PCDFs), Pollution Paper No. 27, Department of the Environment, HMSO, London, UK, 1989.
32. M. K. Kotz, A. K. Gerteisen, R. Tritschler and H. Womterhalter, Influence of new WHO-TEFs on TEQ-based results of food samples and analytical opportunities, *Organohalogen Compd.*, 2007, **69**, 99–101.
33. R. Duarte-Davidson *et al.*, The relative contribution of individual PCBs, PCDDs and PCDFs to toxic equivalent values derived for bulked human adipose tissue samples from Wales, United Kingdom, *Arch. Environ. Contam. Toxicol.*, 1993, **24**, 100–127.
34. H. J. Geyer, K.-W. Schramm, E. A. Feicht, A. Behechti *et al.*, Half-lives of tetra-, penta-, hepta-, and octachlorodibenzo-*p*-dioxin in rats, monkeys and humans—a critical review, *Chemosphere*, 2002, **48**, 631–644.
35. Ministry of Agriculture, Fisheries and Food (MAFF), *Dioxins in Food – UK Dietary Intakes*, Food Surveillance Information Sheet No. 71, MAFF, The Stationery Office, London, 1995.

36. G. H. Eduljee, Dioxins in the environment, *Chem. Br.*, December 1988, 1223–1226.
37. C. Humfrey, M. Taylor, K. Amaning, IEH report on health effect of waste combustion products. Report R7, Medical Research Council, Institute for Environmental Health, 1997.
38. H. Muto and Y. Takizawa, Dioxins in cigarette smoke, *Arch. Environ. Health*, 1989, **44**, 171–174.
39. Department of Health, COT statement on the tolerable daily intake for dioxins and dioxin-like polychlorinated biphenyls. Committee on toxicity of chemicals in food, consumer products and the environment. COT/2001.07, October 2001.
40. Agency for toxic substances and disease registry (ATSDR), *Addendum to the Toxicological Profile for Chlorinated Dibenzo*-p-*dioxins (CDDs)*, U.S. Division of Toxicology and Environmental Medicine, November 2012.
41. F. Tschirley, Dioxin, *Sci. Am.*, 1986, **254**, 29–35.
42. Agency for toxic substances and disease registry, US Department of Health and Human Services, Dioxin toxicity, *Am. Fam. Physician*, 1993, **47**, 855–861.
43. COT statement by the committee on toxicity of chemicals in food, consumer products and the environment on the EPA draft health assessment document for 2,3,7,8-tetrachloro-*p*-dioxin and related compounds, available from the Department of Health, Skipton House, 80 London Road, London, 1995.
44. WHO, Consultation on Tolerable Daily Intake from Food of PCCDs and PCDFs, World Health Organisation Regional Office for Europe, Bilthoven, Netherlands, 1991.
45. S. Safe, Polychlorinated dibenzofurans: environmental impact, toxicology, and risk assessment, *Toxic Subst. J.*, 1991, **11**, 177–222.
46. World Health Organization, Dioxins and their effects on human health, Fact sheet No. 225, May 2010.
47. ENVIRON, An expert panel review of the US environmental protection agency's draft risk characterisation of the potential health risks of 2,3,7,8-tetrachloro-p-dioxin (TCDD) and related compounds, available from the Science Advisory Board of the US Environmental Protection Agency, Washington DC, USA, 1995.
48. *EPA, Reanalysis of Key Issues Related to Dioxin Toxicity and Response to NAS Comments*, US Environmental Protection Agency, Washington, DC, February 2012, vol. 1.
49. G. H. Eduljee, Dioxins in the environment, *Chem. Br.*, December 1988, 1223–1226.
50. R. E. Alcock and K. C. Jones, Dioxins in the environment: A review of trend data, *Environ. Sci. Technol.*, 1996, **30**, 3133–3143.
51. V. Thomas and T. J. Spiro, An estimation of dioxin emissions in the United States, *Toxicol. Environ. Chem.*, 1995, **50**, 1–37.

52. S. Maklund. Dioxin emissions and environmental emissions. A study of polychlorinated dibenzodioxins and dibenzofurans in combustion processes, PhD Thesis, University of Umea, Umea, Sweden, 1990.
53. D. A. Purser, Influence of fire retardants on toxic and environmental hazards from fires, in *Fire Retardancy of Polymers. New Strategies and Mechanisms*, ed. T. R. Hull and B. K. Kandola, RSC Publishing, Cambridge, 2009, ch. 24, pp. 381–404.
54. W. Carroll, Is PVC in house fires the great unknown source of dioxin?, *Fire Mater.*, 1996, **20**, 161–166.
55. United States Environmental Protection Agency, Evaluation of emissions from the burning of household waste in barrels, United States Environmental Protection Agency Technical Report EPA-600/R-97-134a, vol. 1, pp. 1–69, 1997.
56. P.-M. Weinspach, J. Gundlach, H.-G. Klingelhöfer, K. Nitscheke, R. Ries and U. Schneider Analyse des Brandes am, 11th April 1996, Empfehlungen und Konsequenzen für den Rhein-Rhur-Flughafen Düsseldorf, Bericht – Teil 1, 14th April 1997, Staatskanzlei Nordrhein-Westfalen, Germany.
57. J. H. Troitzsch Fire gas toxicity and pollutants in fires: the role of flame retardants, FRPM07, 11th European Meeting on Fire Retardant Polymers, Bolton 3rd–6th July 2007.
58. Independent Experts Commission for the Minister President of the State of North Rhine Westphalia for Investigating the Consequences from the Dusseldorf Airport Fire, Report Part I, Analysis of the 11th April 1996 Fire, Recommendations and Consequences for the Dusseldorf Airport, 14 April 1997.
59. Major Fire in a Plastics Warehouse in Lengerich, October 1992—Documentation. Ministry for the Environment and the Ministry of the Interior of the State of North Rhine Westphalia and the town of Lengerich, June 1994 (in German).
60. Environmental Medical Studies on Fire Fighters, Ruhr-University Bochum and Heinrich-Heine University Dusseldorf on Behalf of the Ministry of Labour, Health and Social Affairs of the State of North Rhine Westphalia, October 1992 (in German).
61. Estimated premature deaths for each jurisdiction are as follows: England, 79 100; Scotland, 13 000; Wales, 5600; Northern Ireland, 2300, Sources: Statistics on smoking: England, 2012; The NHS Information Centre for Health and Social Care, 2012; ScotPHO Smoking Ready Reckoner, 2011 edn; Welsh Government, Health Improvement; NIdirect, smoking.
62. Committee on the Medical Effects of Air Pollutants (COMEAP), *Long-term Exposure to Air Pollution: Effect on Mortality*, Department of Health, 2009.
63. Committee on the Medical Effects of Air Pollutants (COMEAP), *The Mortality Effects of Long-term exposure to particulate air pollution in the United Kingdom*, Department of Health, 2010.
64. Air pollution and cancer, *IARC Scientific Publication No. 161*, ed. K. Straif, A. Cohen and J. Samet, 2014.

CHAPTER 13

Irritant Gases

JAMES C. WAKEFIELD

Public Health England, Centre for Radiation, Chemical and Environmental Hazards, Chilton, Didcot, Oxfordshire OX11 0R, UK
Email: James.Wakefield@delphichse.com

13.1 Introduction

The combustion of most commonly used materials, ranging from natural sources, such as wood, to synthetic plastics and polymers, will result in the generation of irritant gases. Therefore, irritant gases are present in most fire atmospheres, irrespective of whether the combustion is smouldering or flaming (see Chapters 2 and 3). The irritant gases evolved and the rate of generation may, however, depend upon the fuel composition and the mode of combustion, relative to non-flaming or flaming, decomposition, temperature, especially for flaming fires, and the ventilation conditions. Irritant gases produced during combustion can be divided into two main classes relating to their chemical composition, either as inorganic acid gases or as organic irritants. For acid gases the conversion efficiencies of fuel elements, such as halogens to acid gases, in the combustion products is generally high and similar for all combustion conditions. The yields of nitrogen oxides from fuel nitrogen are higher under well-ventilated flaming combustion conditions, which also involve some oxidation of atmospheric nitrogen. The yields of organic irritants are very dependent on the combustion conditions. Under non-flaming decomposition conditions in air, yields of organic irritants (mostly partly oxidised substances such as aldehydes and ketones) are high. Under flaming conditions, the yields depend on the combustion efficiency. Under well-ventilated flaming conditions, for fuels low in gas phase

Issues in Toxicology No. 23
Toxicology, Survival and Health Hazards of Combustion Products
Edited by David A. Purser, Robert L. Maynard and James C. Wakefield

Published by the Royal Society of Chemistry, www.rsc.org

flame inhibitors (especially halogens), combustion efficiency is high and yields of irritants are generally low. Where combustion efficiency is low, due to fuel-rich ventilation conditions or the presence of flame inhibitors, yields of organic irritants are high (see Chapters 2, 3 and 7).[1]

The exact nature of the injury following exposure to an irritant gas depends upon the chemical involved, its concentration, the exposure duration and its solubility. However, the initial effect of exposure to these irritant gases is likely to be sensory irritation.[1–4] Irritation of the eyes will cause pain and stinging of the eyes, initiation of a blinking reflex and lacrimation.[1,2,4] The severity of sensory irritation is largely dependent upon the concentration of the irritant present, and is independent of the exposure duration. An individual exposed to irritant gases in a combustion atmosphere with the effect of stinging or burning of the eyes and throat may shut their eyes and hold their breath to alleviate the irritation, and thus hinder their ability to escape from the hazard.[1,2] An additional characteristic sign of exposure to irritant gases is a burning sensation of the mucous membranes of the upper respiratory tract, including the nose, mouth and throat.[1–3] Pulmonary irritation will commonly occur following sensory irritation, due to inhalation of the irritant gas into the lungs. This irritation of the lungs gives rise to bronchoconstriction, coughing and breathing difficulties. Unlike sensory irritation, the severity of pulmonary irritation is dependent upon both the concentration and the duration of exposure to the irritant gas. Exposure to high concentrations of irritant gases can cause inflammation of the lung tissues, pulmonary oedema and could potentially be fatal in a period of between 6 and 48 hours after removal from the exposure.[1,2]

Some irritant products commonly occurring in combustion products and their physiological and toxic effects are reviewed in this chapter. The results of human and animal exposures studies on individual irritant substances and irritant combustion product mixtures are reviewed in Chapter 7. Aspects related to the yields, rates of production and chemistry of combustion product irritant are reviewed in Chapters 2 and 3. The derivation and application of calculation models for predicting the effects of irritant exposures are presented in Chapter 14.

Before discussing individual irritant compounds it will be useful to consider the physiology of irritation and irritant receptors in a little more detail. Irritation is a rather general term but is used by toxicologists and physiologists to describe a range of responses to certain types of chemical and physical insult. Anybody who has been exposed to smoke from a bonfire will know about irritation of the eyes, nose and possibly the airways. Pain, smarting, coughing, sneezing, gasping and a choking sensation are all parts of the irritant response to smoke. Such effects are produced by the interaction of chemicals and particles in the smoke with what are described as irritant receptors. The exact structure of these receptors is not as well-known as it might be and they are thought to involve naked nerve endings that lie between epithelial cells near the surface of mucous membranes. It is thought that no specialised receptors, such as those involved in touch and perhaps in detecting changes of temperature, are involved. Attempts to distinguish between irritant receptors, nociceptors and pain receptors are rather pointless: nociceptors being simply a term invented to describe receptors that respond to unpleasant or potentially damaging stimuli. Pain is certainly produced by many irritant compounds: the example of smoke from a bonfire and its effects on the eyes makes this clear. Of course when the term "receptors" is used in this context we mean something a little different from the pharmacological use of the term. No molecular structure is specified; what is intended is that there is some means by which irritant compounds are recognised by the body. Recognition is the first stage of responding to irritants. Impulses are transmitted from the irritant receptors to the central nervous system and there, *via* inter-neurons, the link with the efferent limb of the "irritant reflex" is made. Impulses travel to the central nervous system *via* a number of nerves: the trigeminal nerves (the fifth cranial nerve) supply the mucosa of the nose; the vagus nerves (tenth cranial nerve) supply the airways of the lung and also the larynx. Afferent nerve fibres found in the vagus nerve are myelinated, as are those of the trigeminal nerve. Pain, on the other hand, is said to be mediated *via* fine unmyelinated fibres, the C fibres. The efferent limb of the irritant reflex also involves a number of nerves belonging to both the autonomic and somatic nervous systems. Narrowing of the airways and lacrimation are autonomic reflexes; gasping or cessation of breathing are mediated *via* nerves supplying the muscles of inspiration, including the phrenic nerves and the intercostal nerves.

Irritant reflexes are, like all reflexes, protective. In teleological terms the reflexes act to minimise further exposure and to remove material from mucosal surfaces. Lacrimation washes out the conjunctival sac, narrowing of the airways prevents irritants reaching the lungs, sneezing expels material from the nose and coughing expels material from the airways. Explaining the responses of the cardiovascular system to exposure to irritants is less easy as both bradycardia and hypertension may occur. The reflexes of the upper airways are complex: cold water placed in the nose can inhibit breathing and produce a marked slowing of the heart. This is believed to be related to the so called diving reflex seen in some aquatic mammals. Sudden death due to cardiac arrest may occur on inhalation of cold water. Irritant receptors may be stimulated by chemicals and particles but also respond to distortion of mucosal surfaces. Contraction of airway smooth muscle provoked by histamine, for example, leads to stimulation of airway irritant receptors. It is interesting that this response can be blocked by isoprenaline, suggesting that it is the contraction of the smooth muscle rather than the histamine that stimulates the receptors. Inter-species differences in airway responses to irritants have been described. The marked inhibition of respiration seen in the mouse forms the basis of a test for potential irritants.[5] The same reflex occurs in most if not all mammals, but is much less obvious in primates, including man. The physiological responses in primates and rodents are described and compared in Chapter 7. In primates the initial trigeminal reflex inhibition of respiration, consisting of breath holding at end inspiration, develops within a few seconds into a hyperventilation resulting from stimulation of vagal airway receptors.[1,4]

Despite the fact that the response to irritants is very well known and has been studied for many years, the exact mechanism by which chemicals stimulate irritant receptors remains curiously unknown. The range of compounds that can stimulate the receptors suggests that they are notably unselective. This in turn suggests that perhaps the receptors do not interact with the stimulating molecules *per se*, but perhaps respond to some change in their local environment produced by the stimulating molecules. For example, it might be expected that if the receptors respond to local hydrogen ion concentration then responses to a wide range of acids and alkalis might be expected. As will be seen, irritant receptors do respond to a range of inorganic acids and also to alkaline compounds such as ammonia. Unfortunately, this simple theory does not explain the responses of irritant receptors to compounds as disparate in structure and behaviour as formaldehyde and the so called tear gases. Aldehyde groupings may play a part: aldehydes, such as formaldehyde, glutaraldehyde and acrolein, are notable irritants.

One characteristic of the response to irritants is the extent to which it departs from Haber's Law. Haber's Law states that the response to inhaled toxicants is directly proportional to the product of the exposure concentration and the duration of exposure ($R = kCt$). Interestingly, Haber was led to formulate his law by his work on compounds, including chlorine and phosgene, during the First World War. Haber's Law may be adapted for irritants as:

$$R = kC^n t, \tag{13.1}$$

where n varies, depending on the compound, from 1 (non-irritants) to about 4 (classical irritants). It will be clear that if n is about 4 then concentration (C) will dominate the relationship and time (t) will be relatively unimportant (k is a constant). This is the case as regards, for example, the response to sulfur dioxide. The response depends largely on concentration, develops rapidly once exposure begins, soon reaches a maximal level and does not further increase as the duration of exposure is extended. The exact mechanism by which irritant receptors respond to, for example, the fourth power of concentration remains unknown. These aspects of responses to irritant concentrations and exposure doses are discussed in Chapters 7 and 8. The development and application of irritancy calculation models to evaluation of human fire hazards is presented in Chapter 14.

13.2 Inorganic Acid Gases

The most common inorganic acid gases evolved during combustion include the halogen acids (HCl, HF, HBr) and oxides of sulfur, nitrogen and phosphorous.[1,6] The degree of generation of these gases during combustion will, however, be dependent upon the chemical composition of the materials involved (see Chapter 3).

13.2.1 Hydrogen Chloride (HCl)

Hydrogen chloride (HCl) is considered to be the most important halogen acid gas that may be evolved during combustion.[1] Due to the chlorine content of many commonly used materials, including plastics and polymers such as polyvinylchloride (PVC), combustion of these materials will be likely to result in the generation of HCl. In an experimental situation, PVC has been shown to release as much as 50% of its weight as HCl upon combustion.[1,7]

HCl is a strong sensory and respiratory irritant, with the main targets being the eyes, skin, nose, mouth, throat and trachea. Exposure to HCl at 35 ppm has been reported by one study to induce sneezing, chest pain, hoarseness, laryngitis and a feeling of suffocation.[8,9] Inhalation of HCl at about 50 ppm has been reported to be strongly irritating to the eyes, nose and throat, including pain, coughing, inflammation and oedema of the upper respiratory tract, and concentrations of around 100 ppm have been described as being extremely irritating and excruciatingly painful to the upper respiratory tract.[1–3] Exposure to HCl in the range of around 50 to 100 ppm is considered to be barely tolerable.[9] Due to the high water solubility of HCl, irritation of the upper respiratory tract is the major effect. However, pulmonary irritation can occur following exposure to higher concentrations of HCl resulting in damage to the alveolae and production of pulmonary oedema.[3] Inhalation of high concentrations of HCl has also been associated with corrosive burns to the eyes, nose, mouth and throat, ulceration of the nasal septum, tachypnoea, bronchoconstriction and laryngeal spasm, which may lead to suffocation.[2,6,8] Exposure to high levels that are sufficient to produce pulmonary toxicity may lead to the development of reactive airways dysfunction syndrome (RADS), which is a form of irritant-induced asthma. Brief exposure of humans to concentrations of HCl in the region of 1000 to 2000 ppm can be regarded as dangerous and potentially fatal[6] (see Chapter 7).

13.2.2 Hydrogen Bromide (HBr)

Hydrogen bromide (HBr) is another gas in the group of halogen acids that may be a product generated during combustion, particularly of synthetic polymeric materials. Flame-retardant compounds containing halogens are commonly added to many plastics and polymers to reduce their flammability, an example being brominated flame retardants such as decabromodiphenyl ether (deca-BDPE) used in high impact polystyrene (*e.g.*, television casings).[7,8] Upon combustion of such flame-retarded polymeric materials, the bromine present is likely to be liberated, yielding the irritant gas HBr[1,10,11] and some bromine gas.

Limited data are available on the irritant effects of HBr, as it has not been studied as extensively as HCl. The effects are expected to be similar to those caused by HCl at comparable concentrations.[1,2] Exposure of six volunteers to HBr at 5 ppm resulted in nasal irritation in all of the subjects and throat irritation in one of the subjects.[12] Exposure to concentrations between 1300 and 2000 ppm has resulted in death.[13] It has been suggested that the sensory irritation produced by exposure to HBr at 200 ppm may be sufficient to slow escape from a hazardous combustion environment[1] (see Chapters 7 and 14).

13.2.3 Hydrogen Fluoride (HF)

The combustion of synthetic fluorine-containing polymers, such as polytetrafluoroethylene (PTFE), perfluoroalkoxy (PFA), fluorinated ethylene propylene (FEP) and polyvinylidene fluoride (PVDF), leading to the liberation of fluorine on decomposition, is likely to result in the production of the halogen acid gas hydrogen fluoride (HF)[14–16] (see Chapter 7).

As in the cases of other halogen acid gases (HCl and HBr), HF is a strong sensory irritant and corrosive gas, which upon inhalation for 1 hour at concentrations of just 0.5 ppm has been shown to cause irritation and corrosion of the mucous membranes of the nose, mouth and throat. Inhalation of higher concentrations is likely to result in injury to the lungs, with the onset of pulmonary oedema, which may be delayed 24–48 hours after exposure.[14,17,18] Studies in volunteers showed some relatively low sensory and lower airways irritancy at 0.2–2.9 ppm, including chest tightness, soreness, coughing, expectoration and wheezing. At 3.0–6.3 ppm more severe effects were seen.[19] HF is the most potent irritant of the halogen acid gases based on animal lethality data (Acute Emergency Guideline Level 3, AEGL-3), but has equivalent potency to HCl and HBr based on sensory irritancy (AEGL-2; Table 13.1), or has been reported as having a slightly lower sensory irritancy.

Table 13.1 Worker exposure limits (WEL) and Acute Emergency Guideline Levels (AEGL) for inorganic acid combustion products.

	WEL[38]/ppm		AEGL-2/AEGL-3[39]/ppm[a]				
	LTEL[a]	STEL[a]	10 min	30 min	60 min	4 h	8 h
HCl	1	5	100/620	43/210	22/100	11/26	11/26
HBr[b]	–	3	150/740	50/250	25/120	13/31	13/31
HF	1.8	3	95/170	34/62	24/44	12/22	12/22
SO_2	–	–	0.75/30	0.75/30	0.75/30	0.75/19	0.75/9.6
NO[c]	NR[d]						
NO_2	–	–	20/34	15/25	12/20	8.2/14	6.7/11
P_2O_5	Data not available						

[a]LTEL = long term exposure limit (8 hour time weighted average); STEL = short term exposure limit (15 minute reference period); AEGL-2 = Acute Exposure Guideline Level 2 (the level of the chemical in air at or above which there may be irreversible or other serious long-lasting effects or impaired ability to escape); AEGL-3 = Acute Exposure Guideline Level 3 (the level of the chemical in air at or above which the general population could experience life-threatening health effects or death).
[b]Interim AEGL.
[c]Values for NO_2 should be used for emergency planning.
[d]NR = not recommended due to insufficient data.

The generation of HF during combustion occurs only when fluorocarbons are present in the burning fuels, so is likely to be present at considerably smaller quantities than the production of HCl in most fires involving common fuels.

13.2.4 Sulfur Dioxide (SO_2)

Sulfur dioxide (SO_2) is an inorganic irritant gas that is commonly produced upon the combustion of fossil fuels, but may also be formed during the thermal decomposition of any sulfur-containing compounds such as vulcanised rubber used in the manufacture of tyres.

SO_2 is a respiratory irritant and may cause narrowing and tightening of the airways. Individuals with asthma are significantly more sensitive to SO_2 than people who do not have the condition. The World Health Organization (WHO) suggests that exposure to 0.4 ppm may lead to significant narrowing of the airways in those suffering from asthma. In most, the effect would not be expected to be large, but some individuals may be clinically affected.[20] Inhalation of SO_2 at 1 ppm for 1–6 hours has been shown to increase airway resistance and decreased forced expiratory volume and forced expiratory flow in healthy patients.[21] SO_2 is readily absorbed by the mucosa of the upper respiratory tract, with irritation of the nose and mouth being the most common effect following inhalation.[21,22] At concentrations of around 10 ppm, SO_2 causes moderate to severe eye irritation with lacrimation.[21] The irritation effect of SO_2 is due to its conversion to sulfuric acid in the presence of water upon the mucous membranes of the upper respiratory tract and eyes. Sulfur dioxide also stimulates pain/irritant receptors that are present as fibres in the epithelium.[23] Exposure to high concentrations of SO_2 can be fatal, due to asphyxiation caused by blockage of the upper respiratory tract as a result of severe irritation.[9]

13.2.5 Oxides of Nitrogen (NO_x)

Nitrogen oxides, such as nitric oxide (NO) and nitrogen dioxide (NO_2), are commonly present as mixtures in combustion atmospheres and can be denoted collectively as NO_x[1,24–26] (see Chapters 2 and 3). Oxides of nitrogen are likely to be generated during the combustion of any nitrogen-containing materials, with the formation of each being dependent upon the availability of oxygen in the fire environment. Nitrogen present as N_2 is a major constituent of air, therefore, during combustion, NO_x may also be formed due to oxidation of this atmospheric nitrogen provided that the temperature of combustion is sufficient.[1,10,24,25,27,28] Burning coal, wood, tobacco, oil and gas all generate nitrogen oxides. Nitrogen oxide is first formed and then oxidised to nitrogen dioxide.

Nitrogen oxides are less soluble than most irritant gases and are therefore more likely to reach the bronchioles and alveoli following inhalation, giving rise to pulmonary damage.[25]

13.2.5.1 Nitric Oxide (NO)

Nitric oxide is the main oxide of nitrogen formed during well-ventilated combustion in fires,[24] but gradually oxidises to nitrogen dioxide over time-scales of minutes as the fire effluent plume entrains air and ages.

The possible health effects arising from exposure to NO have been investigated in relatively few studies, but it is recognised to be significantly less active as an airways irritant than NO_2.[29] NO has a low sensory irritant potency to the upper respiratory tract, but some difficulty in evaluating the effect experimentally arises from the frequent presence of NO_2 as a contaminant. The assessment of sensory irritancy for human exposures is discussed in Chapter 14. Deep inhalation can result in the delayed onset of pulmonary oedema occurring a few hours post-exposure and may be aggravated by physical exertion.[30] Controlled exposure of healthy human volunteers to NO at concentrations above approximately 20 ppm (24.6 mg m^{-3}) has demonstrated a significant increase ($\sim$10%) in total airway resistance.[25]

13.2.5.2 Nitrogen Dioxide (NO_2)

Nitrogen dioxide (NO_2) is likely to increase in concentration in smoke as it drifts away from the fire and when sufficient oxygen is present because NO will gradually be converted to NO_2 in air. Nitrogen dioxide is an irritant and an oxidant, which produces inflammation and oedema of the lungs if inhaled in high concentrations.

Due to its relatively low aqueous solubility, NO_2 has a relatively low upper respiratory tract irritancy, being carried through the upper airways with little absorption, but in the lung it is converted in the presence of water into nitric acid (HNO_3) and nitrous acid (HNO_2) in the mucous membranes of the respiratory tract.[1,10,28] Low concentrations of NO_2 may cause cough, headache, difficulty breathing, nausea, vertigo and fatigue.[31] NO_2 has been shown to cause significant increases in airway resistance in healthy individuals at exposure concentrations as low as 2.5 ppm.[25] Exposure to high concentrations of NO_2 have resulted in sudden death due to severe constriction of the airways and larynx.[32] Severe pulmonary oedema may occur within a few hours following removal from exposure to NO_2.[31,32] During pulmonary oedema, a decrease in the partial pressure of arterial oxygen may also be observed due to impairment of the diffusion capacity.[32] When CO_2 is present in conjunction with NO_2 there is an increase in the toxicity when compared to NO_2 alone, perhaps due to increased respiration and an increase in the intake of nitrogen dioxide. Studies in healthy individuals have shown a threshold level for effects of around 2 ppm. Individuals with asthma are more sensitive, with a threshold being around 0.2 ppm.[33]

13.2.6 Phosphorus Pentoxide (P_2O_5)

The combustion of any compounds containing phosphorus may lead to the formation of phosphorus pentoxide (P_2O_5) and phosphoric acid. Many traditional flame retardants contain halogen groups, such as bromine, however, there has been an increasing drive to develop halogen-free flame retardants.[34,35] Phosphorus is a common constituent of many halogen-free flame retardants such as isopropylated triphenyl phosphates. These are incorporated into polyurethane foams to increase the resistance of the material to burning. Some halogen-containing flame retardants, such as tris(2-chloropropyl) phosphate, may also incorporate phosphorus to enhance the flame-retardant effect.[35] Such phosphorus-containing flame retardants are incorporated into casings and housing of electronic components (*e.g.*, televisions) in place of halogenated compounds.

Phosphorus pentoxide is corrosive to the eyes and upper respiratory tract, as it readily reacts with moisture in the mucous membranes to form orthophosphoric acid. Inhalation of P_2O_5 may also lead to pulmonary oedema, the onset of which may be delayed for a few hours following removal from the exposure.[36,37]

13.3 Organic Irritants

13.3.1 Formation of Organic Irritants

The combustion of organic compounds results in the formation of organic irritant products in the fire effluent. Many organic irritants may be evolved upon combustion; however, the formation of these products will depend largely upon the nature of the material involved and the combustion conditions (see Chapters 2 and 3). The incomplete combustion or pyrolysis of materials including wood, fossil fuels, synthetic and natural polymers, and foodstuffs will give rise to the formation of a range of aldehydes.[1,40] Of this group, acrolein and formaldehyde are most commonly considered, being high potency sensory irritants, although crotonaldehyde is another highly potent unsaturated sensory irritant that has been detected in combustion products, in addition to less potent irritant aldehydes such as acetaldehyde and butyraldehyde.[1,9] The effects of organic irritants individually and in combustion product mixtures are described in Chapter 7, with application to human hazard analysis in Chapter 14. The effects of the two most important known organic irritants occurring in combustion products, acrolein and formaldehyde, are described in the flowing paragraphs.

13.3.2 Acrolein

The combustion of cellulose-based materials, such as wood, cotton and paper, has been demonstrated to evolve significant quantities of acrolein in fire smoke[1,41] (see Chapter 3). Acrolein generated during the combustion of wood may represent up to 13% of the total aldehydes in the fire effluent.[40] Studies to investigate the amount of acrolein generated during combustion have measured acrolein at up to 50 ppm in wood smoke and 60 ppm in smoke generated during the combustion of cotton[9] (see Chapter 2). Acrolein has also been shown to be generated from polyurethane foams in much greater quantities during incomplete combustion. The complete combustion of polyethylene foam was shown to evolve acrolein in the range of 2–23 ppm, whilst incomplete combustion (smouldering fires) generated between 76–180 ppm acrolein.[42]

Acrolein is the most potent of the irritants. It is severely irritating to the respiratory tract and eyes and is a potent lachrymatory agent. Concentrations of acrolein in the region of 0.5 to 5 ppm have been shown to cause the onset of lacrimation and eye irritation in humans within a 10 minute exposure period.[43] Acrolein at 1.2 ppm has also been reported to cause lacrimation in humans within just 5 seconds of commencing exposure.[44] Volunteers exposed to a steadily increasing concentration of acrolein reported nose irritation at 0.26 ppm, throat irritation at 0.43 ppm and a reduction in respiratory rate at 0.6 ppm.[45] Exposure to acrolein may result in the onset of pulmonary oedema, which may not manifest until a few hours post-exposure.[46,47] The onset of pulmonary oedema is due to inflammation in the respiratory mucosa resulting from protein denaturation caused by acrolein.[3] Eye irritation resulting from exposure to acrolein has been observed at concentrations as low as 0.06 ppm and nasal irritation has resulted from exposure to 0.15 ppm. A reduction in respiratory rate has been observed in male volunteers exposed to acrolein at 0.3 ppm for 40 minutes. Respiratory effects, including coughing, chest pain and difficulty breathing, have been reported following exposure to 0.26 ppm acrolein. Most individuals are unable to tolerate acrolein in air at above 2 ppm for more than 2 minutes.[48] Exposure to acrolein at concentrations above 10 ppm has been shown to be fatal within just a few minutes.[9]

13.3.3 Formaldehyde

Formaldehyde is likely to be generated under similar conditions of combustion to acrolein[1] (see Chapter 3). The combustion of wood has been shown to evolve formaldehyde in the fire effluent at up to 80 ppm and the combustion of cotton has been shown to generate up to 70 ppm formaldehyde.[9] Formaldehyde is also commonly present as a smoke product from the combustion of many polymers and plastics, particularly during incomplete combustion.[49]

Formaldehyde is a potent sensory irritant, causing mild to moderate irritation of the upper respiratory tract and eyes at concentrations in the range of 0.2 to 3 ppm (see Table 13.2).[50] In both normal subjects

Table 13.2 Worker exposure limits (WEL) and Acute Emergency Guideline Levels (AEGL) for organic irritant gas combustion products.

	WEL[38]/ppm		AEGL-2/AEGL-3[39]/ppm				
	LTEL	STEL	10 min	30 min	60 min	4 h	8h
Acrolein	0.1	0.3	0.44/6.2	0.18/2.5	0.10/1.4	0.10/0.48	0.10/0.27
Formaldehyde[a]	2	2	14/100	14/70	14/56	14/35	14/35

[a]Interim AEGL.

and asthmatics, no significant clinically detrimental effects were seen in lung function at up to 3 ppm for 3 hours.[51] Formaldehyde inhalation at high concentrations will give rise to respiratory effects, including bronchospasm, dyspnoea, respiratory depression and laryngeal spasm. Inhalation of formaldehyde at high concentrations may also cause the onset of laryngeal and pulmonary oedema, which may be potentially fatal and may be delayed for a few hours post-exposure.[49,52] Upper respiratory tract irritation has been reported following exposure to formaldehyde in the range of 0.1–25.0 ppm and lower respiratory tract and pulmonary irritation resulting from exposure to 5.0–30.0 ppm. Pulmonary oedema, inflammation and pneumonia have been reported following exposure to formaldehyde at 50–100 ppm and concentrations greater than 100 ppm may be fatal.[29]

13.4 Other Inorganic Irritants

The combustion of many polymeric materials will also be likely to generate additional irritants to those previously described. These irritants are inorganic by nature, but do not fall within the previous category of inorganic acid gases. Such inorganic irritants commonly derived upon combustion of natural and synthetic materials include ammonia, chlorine and phosgene.

13.4.1 Ammonia (NH_3)

Ammonia (NH_3) is likely to be present in the fire effluent upon the combustion of any nitrogen-containing materials, including wood, coal, paper and household waste. Ammonia may be generated in significant amounts during the incomplete combustion (pyrolysis) of low rank coals (containing water and impurities due to inadequate pressure, heat or time during formation) and biomass.[27]

Ammonia is severely irritating and corrosive to the respiratory tract and eyes and causes lacrimation and respiratory distress.[53] Ammonia reacts with the water present in the eyes and mucous membranes of the respiratory tract to form ammonium hydroxide, resulting in necrosis of the cells with which it comes into contact due to its alkaline properties.[54] Exposure to ammonia at 50 ppm results in the immediate onset of moderate irritation to the eyes, nose and throat.[54] Exposure to higher concentrations of ammonia in the region of 400 ppm have been demonstrated to cause severe irritation of the nose and throat, 500 ppm ammonia has been shown to increase the rate of breathing and concentrations of 1700 ppm are reported to induce coughing.[53,55] Exposure to ammonia at high concentrations above 1500 ppm may be associated with the onset of pulmonary oedema, which may be delayed for up to 24 hours post-exposure.[54,56] Inhalation of very high concentrations of ammonia ($\sim$5000 ppm and above) may be rapidly fatal due to obstruction of the airways.[54]

13.4.2 Chlorine

The generation of chlorine during combustion may occur in similar situations to those in which HCl is formed, with both products likely to be present simultaneously, but the evidence for chlorine release in combustion products from chorine-containing products is limited. The combustion of any chlorine-containing compounds, including plastics, polymers and synthetic rubbers such as polyvinylchloride, result in the presence of chlorine in the fire effluent almost exclusively in the form of hydrogen chloride. However, it is possible that chorine may be formed under particular combustion conditions and from bleaches, such as sodium hypochlorite, when involved in fires, or during major incidents such as rail crashes involving propane and chlorine tankers, when exploding propane tankers have ruptured the chlorine tankers, thereby releasing chlorine into the fire plume (as in the Mississauga incident in 1979).

Exposure to chlorine can cause severe irritation of the eyes, nose, throat and upper respiratory tract. Concentrations of chlorine in the range of 1 to 3 ppm are associated with mild irritation of the eyes and the mucous membranes of the nose and throat, coughing, difficulty breathing and headache.[8,57,58] Moderate irritation of the upper respiratory tract and eyes is caused by exposure to chlorine concentrations of 5 to 15 ppm. Exposure to higher concentrations of chlorine in the region of 30 ppm will result in immediate chest pain, vomiting and coughing, with the onset of toxic pneumonitis and pulmonary oedema resulting from exposure to 40 to 60 ppm chlorine.[57,58] The onset of pulmonary oedema may be delayed for several hours post-exposure.[59] Exposure to chlorine at concentrations greater than 430 ppm for 30 minutes can be fatal, whilst concentrations above 1000 ppm may be fatal in just a few minutes due to respiratory failure.[57–59]

13.4.3 Phosgene ($COCl_2$)

Phosgene ($COCl_2$) may to be evolved from the combustion of any chlorinated organic compounds, for example, chlorinated solvents, such as chloroform, and plastics and polymers such as polyvinylchloride. The requirement for oxygen for the formation of phosgene during the combustion of materials containing chlorine and carbon suggest that it is likely to be formed in the greatest quantities during fire situations where ventilation is not a limiting factor. However, there is limited evidence that it is actually formed in significant quantities in most fires involving chlorinated fuels.

Phosgene is a potent irritant, particularly of the lower respiratory tract. An acute exposure to a concentration of 3 ppm results in irritation of the eyes and upper respiratory tract.[60] Phosgene will also penetrate into and irritate the deep lung tissues following inhalation, due to its relative insolubility in the mucous membranes of the upper respiratory tract, with onset of lung damage at exposures of greater than 300 ppm min^{-1}.[59] Exposure to

Table 13.3 Worker exposure limits (WEL) and Acute Emergency Guideline Levels (AEGL) for inorganic irritant gas combustion products.

	WEL[38]/ppm		AEGL-2/AEGL-3[39]/ppm				
	LTEL	STEL	10 min	30 min	60 min	4 h	8h
Ammonia	25	35	220/2700	220/1600	160/1100	110/550	110/390
Chlorine	0.5	1	2.8/50	2.8/28	2.0/20	1.0/10	0.71/7.1
Phosgene	0.02	0.06	0.6/3.6	0.6/1.5	0.3/0.75	0.08/0.20	0.04/0.09

phosgene can cause fatal pulmonary oedema at concentrations above 150 ppm min^{-1}.[60] The toxicity of phosgene following an acute exposure in the range of 30 to 300 ppm min^{-1} will typically display three phases. The first stage is the reflex syndrome characterised by irritation to the eyes and respiratory tract, pain, difficulty breathing and coughing, which occur during exposure and usually subside once withdrawn from exposure. The second phase is a clinical latent phase in which the exposed individual experiences no symptoms and generally feels recovered. The duration of the latent phase may be between 30 minutes to 24 hours and depends upon the severity of phosgene exposure, with a shorter latent period as the concentration or duration increase. The final phase of phosgene toxicity is the clinical oedema stage, which involves the manifestation of pulmonary oedema and is associated with shortness of breath, coughing, cyanosis, shock and respiratory arrest.[59,60] In cases of phosgene exposure in the lethal range (above 300 ppm min^{-1}), the latent period may be truncated or non-existent, with immediate onset of pulmonary oedema,[60] but exposure to lethal concentrations can be associated with a latent period of 8 or more hours. This was the classic experience in World War I.[60] Exercise is particularly dangerous in people exposed to phosgene as it exacerbates the adverse effects. Thus, the exertion required to escape a combustion atmosphere where phosgene is present can cause collapse, as can helping other casualties such as carrying people (Table 13.3).

13.5 Conclusions

Irritant gases are formed in the majority of combustion scenarios and are responsible for the sensory effects encountered during exposure to smoke. The irritant gases evolved and the rate of generation depend upon the elemental and organic composition of the fuels involved, but for organic irritants very much upon the combustion conditions, with high yields from non-flaming decomposition and under ventilated flaming combustion.

The most important inorganic irritants released in fires are acid gases including hydrogen halides, nitrogen oxides, sulfur dioxide and phosphoric acid. A wide range of organic irritants occur in combustion products, but those generally considered the most important are the low molecular mass aldehydes, formaldehyde, acrolein and crotonaldehyde, due to their high sensory irritant potency.

The primary initial effect on fire victims is immediate sensory irritation of the eyes, nose and throat, which may hinder escape attempts and cause incapacitation, the severity of irritation being dependent only upon the concentration of the irritant present and independent of the exposure duration.

Pulmonary irritation commonly follows sensory irritation, due to inhalation of the irritant gas into the lungs resulting in bronchoconstriction, coughing and breathing difficulties. Unlike sensory irritation, the severity of pulmonary irritation is dependent upon both the concentration and the duration of exposure. Exposure to high concentrations of irritant gases can potentially be fatal in a period of between 6 and 48 hours after removal from the exposure due to the development of pulmonary oedema. Continued monitoring of exposed individuals is, therefore, important during this period.

References

1. D. A. Purser, Assessment of Hazards to Occupants from Smoke, Toxic Gases and Heat, in *The SFPE Handbook of Fire Protection Engineering,* ed. P. J. DiNenno, National Fire Protection Association, Quincy, MA 02269, 4th ed, 2009, pp. 96–193.
2. G. E. Hartzell, Overview of combustion toxicology, *Toxicology*, 1996, **115**(1–3), 7.
3. T. Prien and D. L. Traber, Toxic smoke compounds and inhalation injury—a review, *Burns*, 1988, **14**(6), 451–460.
4. D. A. Purser, Hazards from smoke and irritants, in *Fire Toxicity*, ed A. Stec and R. Hull, Woodhead, Cambridge UK, 2010, ch. 3, pp. 51–117.
5. Y. Alarie, Irritating properties of airborne materials to the upper respiratory tract, *Arch. Environ. Health*, 1966, **13**, 433–449.
6. Department of Health (DoH), Health advisory group on chemical contamination incidents, Smoke toxins, 1996.
7. A. Marongiu *et al.*, Thermal degradation of poly(vinyl chloride), *J. Anal. Appl. Pyrsolysis*, 2003, **70**(2), 519.
8. International Programme on Chemical Safety (IPCS), Chlorine and Hydrogen Chloride, *Environmental Health Criteria 21*, WHO, Geneva, 1982.
9. I. N. Einhorn, Physiological and toxicological aspects of smoke produced during the combustion of polymeric materials, *Environ. Health Perspect.*, 1975, **11**, 163–189.
10. B. C. Levin and E. D. Kuligowski, Toxicology of fire and smoke, in *Inhalation Toxicology*, ed. H. Salem and S. A. Katz, Taylor & Francis, Boca Raton, 2nd edn, 2006, pp. 205–228.
11. F. B. Clarke, Effects of brominated flame retardants on the elements of fire hazard: A re-examination of earlier results, *Fire Mater.*, 1999, **23**(3), 109.
12. American Conference of Governmental and Industrial Hygienists (ACGIH), *Documentation of the Threshold Limit Values and Biological Exposure Indices*, 6th edn, Cincinnati, OH, 1991.

13. National Library of Medicine (NLM), *Hazardous Substance Data Bank: Hydrogen Bromide*, National Library of Medicine, Bethesda, MD, 1992.
14. D. A. Purser, Recent Developments in Understanding the Toxicity of PTFE Thermal Decomposition Products, *Fire Mater.*, 1992, **16**, 67–75.
15. G. Oberdorster, Toxicology of ultrafine particles: in vivo studies, *Philos. Trans. R. Soci., A*, 2000, **358**(1775), 2719–2740.
16. W. Young *et al.*, A study of the toxicity of pyrolysis gases from synthetic polymers, *J. Combust. Toxicol.*, 1976, **3**, 157–165.
17. International Programme on Chemical Safety (IPCS), Hydrogen Fluoride, *Poisons Information Monograph. PIM 268*, WHO, Geneva, 1990.
18. K. Lund *et al.*, Exposure to hydrogen fluoride: an experimental study in humans of concentrations of fluoride in plasma, symptoms.
19. Expert Panel on Air Quality Standards (EPAQS), Guidelines for Halogens and Hydrogen Halides in Ambient Air for Protecting Human Health against Acute Irritancy Effects, 2006, Department for Environment, Food and Rural Affairs, Scottish Executive, National Assembly of Wales, Department of the Environment in Northern Ireland.
20. Committee on the Medical Effects of Air Pollutants (COMEAP), COMEAP statement on the banding of air quality, 1998, Department of Health.
21. Agency for Toxic Substances and Disease Registry (ATSDR), *Toxicological Profile for Sulfur dioxide*, US department of Health and Human Services, Atlanta, US, 1998.
22. International Programme on Chemical Safety (IPCS), Sulfur Oxides and Suspended Particulate Matter, *Environmental Health Criteria 8.*, WHO, Geneva, 1979.
23. J. G. Widdicombe, Pulmonary and respiratory tract receptors, *J. Exp. Biol.*, 1982, **100**, 41.
24. D. A. Purser and J. A. Purser, HCN yields and fate of fuel nitrogen for materials under different combustion conditions in the ISO 19700 tube furnace. Fire Safety Science – Proceedings of the 9th international symposium, International Association for Fire Safety Science, 2008, pp. 1117–1128.
25. International Programme on Chemical Safety (IPCS), *Nitrogen, Oxides of Environmental Health Criteria 188*, WHO, Geneva, 2nd edn, 1997.
26. K. M. Thomas, The release of nitrogen oxides during char combustion, *Fuel*, 1997, **76**(6), 457.
27. P. Glarborg, A. D. Jensen and J. E. Johnsson, Fuel nitrogen conversion in solid fuel fired systems, *Prog. Energy Combust. Sci.*, 2003, **29**(2), 89.
28. International Programme on Chemical Safety (IPCS), Nitrogen Oxides, *Poisons Information Monograph: PIM G017*, WHO, Geneva, 1992.
29. Committee on the Medical Effects of Air Pollutants (COMEAP), *Guidance on the Health Effects of Indoor Air Pollutants*, Department of Health, 2004.
30. International Programme on Chemical Safety (IPCS), Nitric Oxide, *International Chemical Safety Card: 1311*, WHO, Geneva, 1998.
31. G. D. Leikauf and D. R. Prows, Inorganic compounds of carbon, nitrogen, and oxygen, in *Patty's Toxicology*, ed. E. Bingham, B. Cohrssen and

C. H. Powell, John Wiley & Sons, Inc, New York, 5th edn, 2001, ch. 47, vol. 3.
32. E. P. Horvath *et al.*, Nitrogen dioxide-induced pulmonary disease: five new cases and a review of the literature, *J. Occup. Med.*, 1978, **20**(2), 103–110.
33. Expert Panel on Air Quality Standards (EPAQS), Nitrogen Dioxide, 1996, Department for Environment, Food and Rural Affairs.
34. R.-J. Jeng *et al.*, Flame retardant epoxy polymers based on all phosphorus-containing components, *Eur. Polym. J.*, 2002, **38**(4), 683.
35. S. Zhang and A. R. Horrocks, A review of flame retardant polypropylene fibres, *Progress Polym. Sci.*, 2003, **28**(11), 1517.
36. International Programme on Chemical Safety (IPCS), Phosphorus Pentoxide. *International Chemical Safety Card: 0545*, WHO, Geneva, 1997.
37. E. Bingham, Phosphorous, selenium, tellurium, and sulfur, in *Patty's Toxicology*, ed. E. Bingham, B. Cohrssen and C. H. Powell, John Wiley and Sons, Inc, New York, 5th edn, 2001, ch. 44, vol. 3.
38. Health and Safety Executive, EH40/2005, Workplace Exposure Limits, 2nd edn, 2011, The Stationary Office, London.
39. U.S. Environmental Protection Agency. Acute exposure guideline levels. Available at http://www.epa.gov/oppt/aegl/pubs/chemlist.htm, 2013, accessed November 2013.
40. International Programme on Chemical Safety (IPCS), Acrolein, *Environmental Health Criteria 127*, WHO, Geneva, 1991.
41. B. A. Zikria, J. M. Ferrer and H. F. Floch, The chemical factors contributing to pulmonary damage in "smoke poisoning", *Surgery*, 1972, **71**(5), 704–709.
42. W. J. Potts, T. S. Lederer and J. F. Quast, A study of the inhalation toxicity of smoke produced upon pyrolysis and combustion of polyethylene foams, Part I, Laboratory studies, *J. Combust. Toxicol.*, 1978, **5**, 408–433.
43. L. E. Kane and Y. Alarie, Sensory irritation to formaldehyde and acrolein during single and repeated exposures in mice, *Am. Ind. Hyg. Assoc. J.*, 1977, **38**(10), 509–522.
44. V. M. Sim and R. E. Pattle, Effect of possible smog irritants on human subjects, *J. Am. Med. Assoc.*, 1957, **165**(15), 1908–1913.
45. Agency for Toxic Substances and Disease Registry (ATSDR), *Toxicological Profile for Acrolein*, US Department of Health and Human Services, Atlanta, CS, 2007.
46. C. A. Hales *et al.*, Smoke aldehyde component influences pulmonary edema, *J. Appl. Physiol.*, 1992, **72**(2), 555.
47. International Programme on Chemical Safety (IPCS), Acrolein, *International Chemical Safety Card: 0090*, WHO, Geneva, 2001.
48. International Programme on Chemical Safety (IPCS), Acrolein, *Concise International Chemical Assessment Document 43*, WHO, Geneva, 2002.
49. International Programme on Chemical Safety (IPCS), Formaldehyde, *Environmental Health Criteria 89*, WHO, Geneva, 1989.
50. Agency for Toxic Substances and Disease Registry (ATSDR), *Toxicological Profile for Formaldehyde*, US Department of Health and Human Services, Atlanta, US, 1999.

51. International Programme on Chemical Safety (IPCS), Formaldehyde, *Concise International Chemical Assessment Document 40*, WHO, Geneva, 2002.
52. International Programme on Chemical Safety (IPCS), Formaldehyde, *International Chemical Safety Card: 0275*, WHO, Geneva, 2004.
53. International Programme on Chemical Safety (IPCS), Ammonia, *Environmental Health Criteria 54*, WHO, Geneva, 1986.
54. Agency for Toxic Substances and Disease Registry (ATSDR), *Toxicological Profile for Ammonia*, US Department of Health and Human Services, Atlanta, US, 2004.
55. L. Silverman, J. L. Whittenberger and J. Muller, Physiological response of man to ammonia in low concentrations, *J. Ind. Hyg. Toxicol.*, 1949, **31**(2), 74–78.
56. K. E. Flury *et al.*, Airway obstruction due to inhalation of ammonia, *Mayo Clin. Proc.*, 1983, **58**(6), 389.
57. International Programme on Chemical Safety (IPCS), Chlorine, *Poisons Information Monograph: PIM 947*, WHO, Geneva, 1996.
58. Health Protection Agency (HPA), Chlorine, Toxicological Overview, in *HPA Compendium of Chemical Hazards*, 2007.
59. D. T. Teitelbaum, The halogens, in *Patty's Toxicology*, ed. E. Bingham, B. Cohrssen and C. H. Powell, John Wiley & Sons, Inc, New York, 5th edn, 2001, ch. 48, vol. 3.
60. International Programme on Chemical Safety (IPCS), Phosgene, *Environmental Health Criteria 193*, WHO, Geneva, 1997.

CHAPTER 14

Acute Effects of Combinations of Toxicologically Active Substances and Heat on Fire Victims in Buildings and during Exposures to Outdoor Smoke Plumes

DAVID A. PURSER

Hartford Environmental Research, Hatfield, UK
Email: david-purser@ntlworld.com

14.1 Introduction

Hazard to life from fires in buildings (or other occupied enclosures) depends upon the performance of a dynamic system involving interactions between the building, the fire and the occupants.[1] Since fire hazards are essentially time-based phenomena, the design objective is to ensure, as far as possible, that should a fire occur, the occupants are warned and have time to escape before conditions become untenable. Once a fire occurs in a building or other enclosure, the outcome in terms of fire safety and hazard depends on two parallel time-based processes: the growing fire hazard and the escape performance of the occupants.

Issues in Toxicology No. 23
Toxicology, Survival and Health Hazards of Combustion Products
Edited by David A. Purser, Robert L. Maynard and James C. Wakefield

Published by the Royal Society of Chemistry, www.rsc.org

In fire safety engineering terms it is necessary to ensure that available safe escape time (ASET) is greater than required safe escape time (RSET) by an acceptable margin of safety.[2,3]

- ASET is the time from ignition to that when conditions become untenable to occupants such that they are no longer able to escape without assistance.
- RSET is the time from ignition to that when affected occupants are able to reach a place of safety (this involves a series of stages, including times for detection, warnings and occupant evacuation behaviours).[2]

In situations designed such that occupants are not expected to escape (defend in place strategy), then conditions in occupied enclosures should not develop so as to threaten safety.

The hazards from toxicity and heat are therefore the main determinants of ASET, but they also affect escape time due to the effects of exposure to smoke on occupant behaviour in terms of deciding to enter and continue through smoke or turn back and seek refuge, and on walking speed and "wayfinding" ability in smoke.

Toxicity, or more realistically toxic hazard, in fires is therefore a system property of the full-scale fire scenario. It consists of a sequence of time-varying effects depending upon the changing combustion conditions and the varying rates of development of a sequence of different physiological effects, caused by different toxic fire effluent components. For this reason it is not possible to make meaningful measurements of "toxicity" for any material or manufactured product in any small or even large-scale fire "toxicity" test. Rather, it is necessary to determine toxic hazard in terms of calculated time to incapacitation, injury or death for specified full-scale fire scenarios.

Exposure to toxic smoke is the main cause of death and injuries from fires and the main cause of incapacitation during fire incidents, preventing escape and leading to fire deaths from all causes. Severe injury and death from burns occurs mostly when victims are intimate with the fire (for example, setting fire to their clothing or being in the same room as, and close to, a fire). The incidence of smoke deaths and injuries has changed considerably during the post-war period, as shown in Chapter 1, Figures 1.1 and 1.2.[4] A four-fold increase in smoke deaths between the late 1950s and early 1970s from 1.94 to 8.12 per million (mostly in domestic dwellings) has generally been attributed mainly to the increased use of modern materials in upholstered furniture and bedding. At this time typical furniture was easily ignited, resulting in a rapidly growing fire releasing large quantities of dense, irritant and particulate rich smoke containing high concentrations of carbon monoxide and hydrogen cyanide. A regulatory requirement was introduced in 1988 for domestic upholstered furniture covers, fillings and composites to pass a cigarette and small flame ignition test.[5] The flame-retarded polyurethane foam fillings and covering materials used since this time have resulted in fewer and slower-growing fires, and this coupled with the introduction and wide-spread use of domestic smoke detectors, the decrease in the incidence of smoking in the home and other factors has since resulted in a considerable overall decrease in the incidence of fire deaths and injuries (see Chapters 1 and 2).

The time required for occupants to escape from a fire depends upon a set of parameters related to fire detection and warning times, the behaviour of occupants in response to alarms in terms of starting (pre-movement) times and exit choice, and the time required to travel through escape routes and out of a building. In situations where evacuating occupants see or are exposed to smoke, the exit choice and movement speed (and hence their travel time) can be affected. These effects need to be considered in evacuation calculations, and calculation methods for these parameters are presented in this chapter in Section 14.3.

The main tenability limits for ASET are conditions that cause incapacitation of occupants such that they are unable to escape.

For the toxic effects of exposure to combustion products in fire effluent, particularly in relation to the survival, injury or death of fire victims, it is mainly the more immediate physiological effects that dominate at the fire scene. This is because toxic physiological effects occur rapidly, often within a few seconds, while pathological changes tend to occur over time scales of hours, days or weeks. Physiological effects also occur in response to heat and smoke exposure in fires, interacting to some extent with the effects of toxic products. Burns represent pathological effects rapidly following the physiological effects of painful heat exposures.

Box 14.1 lists the acute fire hazards affecting escape capability, post-exposure condition, survival and longer-term health. These tend to be encountered more or less in the order shown, with exposure first to smoke, which is likely to be irritant, followed by asphyxia or burns, depending upon the type of fire scenario and the proximity of the person to the fire. Once a victim has become trapped or incapacitated in a fire, then conditions usually become lethal within a further few seconds or minutes. This is because flaming fires grow exponentially, so that concentrations of smoke and toxic gases, and the heat intensity, increase rapidly, resulting in death either from asphyxiation or heat exposure. The key determinant of survival is, therefore, incapacitation so that the lethal potency of fire effluent is of limited relevance, since essentially all fires are potentially lethal if the victim remains at the fire scene.

Figure 14.1 illustrates the sequence and time scales over which physiological and pathological toxic effects occur during and after a fire. A common scenario is for a fire involving room contents, such as an upholstered armchair, occurring on the lower floor of an enclosed two-story house with the bedroom door closed but the lounge door left open. In experiments re-creating this scenario[6] the fire self-extinguished after approximately 10 minutes due to oxygen depletion and the effluents then mixed evenly throughout the available space, giving the maintained toxic gas

Box 14.1 Acute survival hazards during fires and longer-term health hazards

- Behavioural effects of seeing fire or smoke (reluctance to enter smoke-logged escape routes or move past flames and other disruptive effects on evacuation behaviour).
- Physiological, behavioural and pathological effects of direct exposure to optically obscure, irritant smoke containing high concentrations of toxic combustion products:
  - Difficulty of finding escape routes and slow movement speed due to effects of smoke obscuration on visibility.
  - Further impairment of vision and eye pain due to immediate, concentration-related effects of sensory irritant smoke products on eyes—blepharospasm (reflex closure of eyes due to pain from exposure of cornea to irritants).
  - Impairment and eventual prevention of evacuation due to immediate, concentration-related, painful sensory irritant effects of smoke on upper respiratory tract (mouth, nose, throat) and airways (bronchoconstriction, chest pain).
  - Uptake over a period of several minutes of asphyxiant gases leading to incapacitation and collapse once a critical exposure dose has been inhaled.
  - Impairment and eventual prevention of evacuation due to pain from heat exposure, with incapacitation due to heat exposure or burns once critical exposure dose of heat has been received.
  - Death during exposure at the fire scene (or after rescue, usually within a few hours) resulting from exposure to asphyxiant gases, exacerbated by upper respiratory tract irritancy and effects of heat or burns on the upper respiratory tract.
  - Death during exposure at the fire scene (or within a few days of rescue) due primarily to heat exposure and/or body surface burns.
  - Death within a few hours of rescue from lung oedema and inflammation due to inhalation of irritant fire effluent gases and particulates into the lung (also sometimes complicated by thermal airway damage).
  - Death within a few days of rescue due to bronchopneumonia.
  - Increased risk of heart attack or stroke (usually between a few hours to several days after rescue) possibly resulting from stress and/or effects of fine particles and asphyxiant gases on blood circulation (cardiac infarct and/or cerebrovascular incidents).
  - Long-term health problems, including neuropathology, sensitisation and reactive airways disease syndrome.

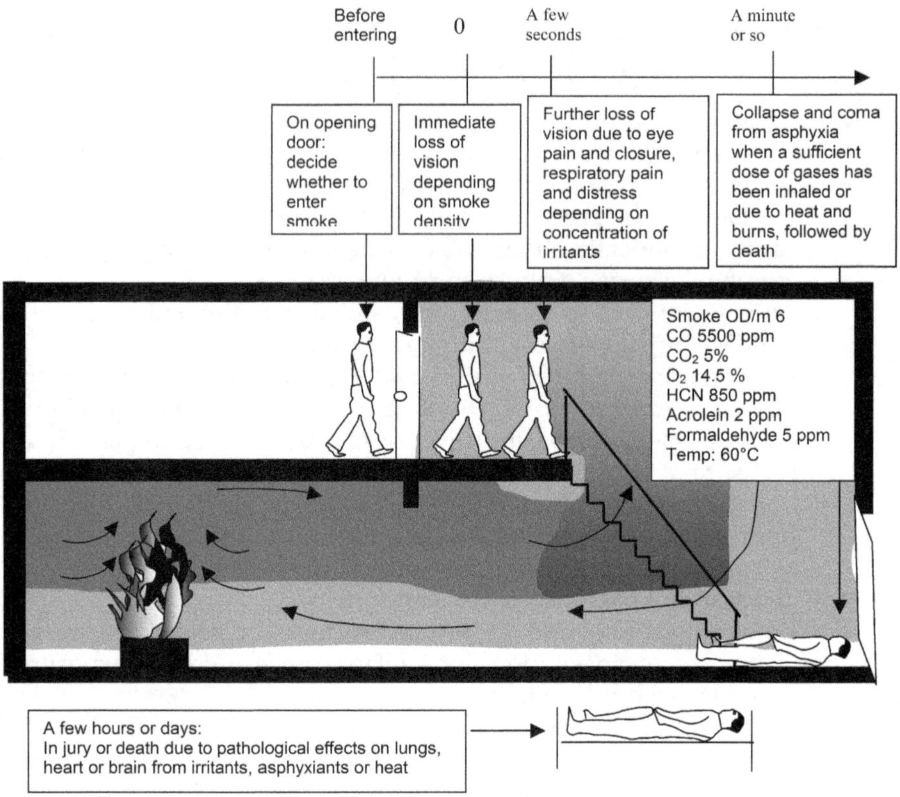

Figure 14.1 Sequence of hazards affecting a subject escaping from a bedroom of an enclosed two-storey dwelling with a fire on the floor below.

concentrations and temperature listed in Figure 14.1. Consider the sequence of hazards faced by a bedroom occupant if they awoke and opened the bedroom door at this time. On opening the bedroom door an occupant attempting to escape must first decide whether to enter the dense smoke, or close the door and remain in the bedroom. If they step out onto the landing the first physiological effect is immediate loss of vision due to obscuration by smoke, which is followed within a few seconds by further visual impairment and eye pain from the effects of smoke irritants and pain to the nose and chest with breathing difficulties. As the subject inhales asphyxiant gases, especially CO and HCN, a dose builds up in the body. There are no immediate effects, but when the inhaled dose reaches a critical level (after a few seconds or minutes) this leads to collapse and coma, followed by death within a further few minutes. In this example scenario the smoke was hot (60 °C), but not sufficient to cause pain or distress within a few minutes. If the victim is rescued alive after collapse, then they may recover, or may suffer a further set of health problems, including brain damage, heart attack or

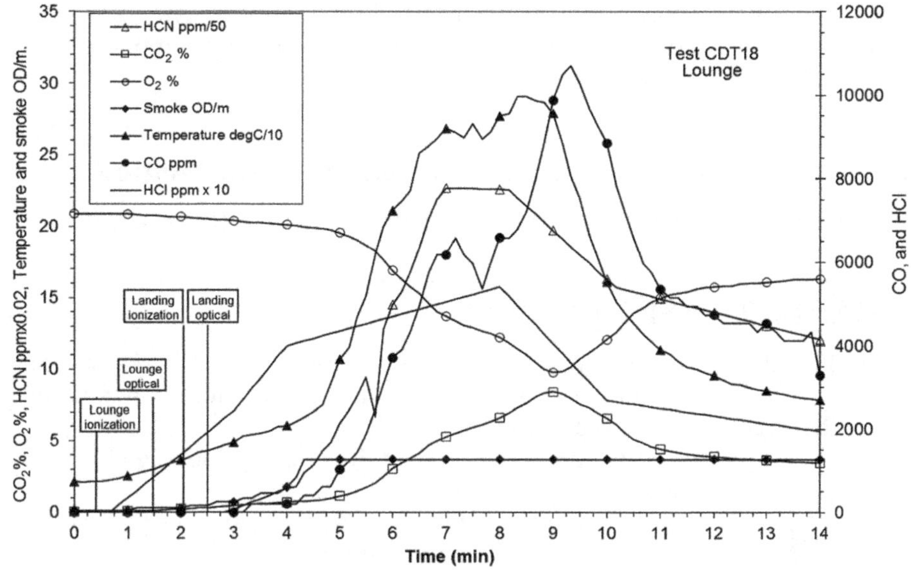

Figure 14.2 Example of time-concentration curves for smoke, toxic gases and temperature at head height in the domestic lounge of a house during an armchair fire. The doorway to the hall is open but the house is otherwise enclosed. The time for activation of smoke detectors is shown.[6]

lung inflammation over a period of a few hours to a few days, all of which can be fatal.

Not only do the different physiological effects during the fire occur over a time sequence as in this example, but in most cases the fire conditions also change rapidly as the hazards develop. Figure 14.2 shows the time–concentration curves of the main fire gases, smoke and heat in the lower floor fire room during the early stages of the same fire. In order to determine the developing hazard and time to incapacitation it is therefore also necessary to evaluate these time-dependent processes.

14.2 Fractional Effective Dose Methods and Application to Fire Hazard Analysis

14.2.1 Fire Effluent Mixtures and the Fractional Effective Dose Hazard Analysis

Because fire effluents contain a mixture of toxic species with varying potencies, and the concentrations and ratios of the various components change dynamically throughout a fire, it has been necessary to develop an assessment method capable of dealing with these issues. This method developed for the analysis of fire hazards and time to incapacitation in fires is the fractional effective dose (FED) method, the basis and application of which is described in the following sections.

14.2.2 The Basis of ASET Analysis and the FED Method

ASET for any system is determined by the parameters in (Box 14.2):[2]

Box 14.2 ASET parameters

- The time–concentration (or time–intensity) curves for the major toxic products, smoke and heat in the fire at the breathing zone of the occupants are depend upon:
  - Fire growth curve [mass loss rate of the burning fuel (kg s^{-1}) and its dispersal volume (kg m^{-3}) with time].
  - The yields of the major toxic products (kg kg^{-1}) and heat (kJ kg^{-1}), for example, kg CO per kg of material burned.
- These terms can be measured directly in full-scale tests or calculated using appropriate fire dynamics computations with appropriate input data, including reaction-to-fire properties and data on product yields under a range of fire conditions. Guidance on calculation methods for these terms is given in BS7974[7] and the *SFPE Handbook of Fire Safety Engineering*.[1] Information is provided in Chapters 2, 3 and 7 on toxic product yields under a range of fire conditions and also in Gottuk and Lattimer[8] and Tewarson.[9] Data on toxic product yields can be obtained using the ISO TS19700 tube furnace[10] or the ASTM E2058 flammability apparatus.[9] An example of a fire profile in shown in Figure 14.2.
- The concentration/time/physiological effect relationships of these products in terms of the physiological/toxic potency of the products and heat [the exposure concentration (kg m^{-3})], or exposure dose (kg m^{-3} min or ppm min) causing toxic effects (and the equivalent effects for heat and smoke obscuration). The important endpoints are:
  - Concentrations and doses (or heat intensity) likely to impair escape efficiency due to behavioural and/or physiological effects.
  - Exposure concentrations or doses likely to cause incapacitation or prevent escape due to behavioural and/or physiological effects.
  - Exposure doses likely to result in post-exposure injury such as lung inflammation or neuropathology.
  - Lethal exposure concentrations or doses.

These terms can be calculated by the application of appropriate physiological methods [fractional effective dose methods (FED equations)] presented in this chapter and in ISO 13571.[11]

Some toxic or physiological effects of exposure to combustion products occur almost immediately on exposure and the severity of the effect is proportional to the concentration of the substance and its potency. This applies to obscuration of vision by smoke and to the painful effects of exposure to irritants. For example, irritant smoke in the eyes or nose immediately causes pain (sensory irritation), reflex closure of the eyes (blepharospasm) and breathing difficulties.[1,12,13]

For other substances, such as asphyxiant gases, the effect depends upon the dose inhaled. The effects therefore take some time to develop and depend upon the concentration inhaled and the time over which it is inhaled. The effects tend to be more persistent that those of sensory irritants, since it takes some time for the toxic material inhaled to be metabolised and/or excreted (for example, carbon monoxide or hydrogen cyanide).

In practice, for asphyxiant substances (and to some extent also for heat) a distinct threshold concentration or exposure dose can be identified at which serious effects are predicted and there is a sudden transition from minor to severe effects. Examples include the dose of CO at which a transition from a headache to collapsing unconscious occurs, or the "dose" of heat to the skin [incident heat intensity (kW m^{-2}) × exposure time] at which a sensation of heat becomes one of severe skin pain. For application to toxic hazard calculations the concept of fractional effective concentration (FEC) or dose is used, whereby the exposure concentration or dose at any point during a fire is expressed as a fraction of the exposure concentration or dose predicted to produce a given effect. For example, the concentration of smoke present at any time during a fire can be expressed as a fraction of the concentration required to seriously impair escape capability:

FEC_{smoke} = Concentration (of smoke) present in a fire at any specific time divided by the concentration considered to significantly affect escape efficiency.

If at any time during exposure the FEC_{smoke} exceeds 1 then escape impairment is predicted. If the smoke density increases further, the impairment increases, but if the smoke clears then the impairment is reduced or eliminated.

A form of fractional effective dose for CO can be expressed as a fraction of the exposure dose predicted to cause incapacitation.

Thus F_{Ico} (fraction of an incapacitating dose of carbon monoxide) can be expressed as:

$$F_{Ico} = \frac{\text{conc. gas present} \times \text{time}}{\text{conc.} \times \text{time for incapacitation}} \tag{14.1}$$

e.g.,

$$F_{Ico} = \frac{1000\ \text{ppm CO} \times 20\ \text{min}}{35\,000\ \text{ppm min}} = 0.57 \tag{14.2}$$

For this simplified example the exposure dose of CO required to cause incapacitation over any short exposure period up to approximately 1 hour is 35 000 ppm min[11] (see Chapter 7). For exposure to a constant concentration of 1000 ppm for 20 minutes the F_{Ico} is therefore 0.57. This would also apply to a scenario with a varying CO concentration in which the average concentration was 1000 ppm, providing the inhaled CO concentration was always well in excess of equilibrium with the blood concentration.

Although the FED calculations for different asphyxiant gases are based upon this simple concept, they are in practice more complex for a variety of physiological reasons, but the user is required to know only the concentration and exposure duration as input data to perform the FED exposure dose calculation.

One reason for expressing the concentration or dose as a fraction of an effective concentration or dose for each toxic product is that the fire atmosphere contains a mixture of toxic products of differing potencies. In order to sum the effects of the different effluent components it is necessary to normalise them in terms of the effective dose. Furthermore, the concentrations of toxic products change with time during a fire, so that for constituents whose effects are dose-related, it is necessary to calculate the effective doses received, based upon the concentrations averaged over short periods of time and then sum these over successive periods. The aim is to determine the time at which the summed effective fractional doses reach unity, at which point the endpoint (such as incapacitation) is predicted to occur for a subject of average sensitivity. For most growing fires, continuing the analysis provides FED values well in excess of unity within a minute or so, indicative of increasing severity of effects and incapacitation of less sensitive subjects.

Effective concentrations and exposure doses for defined endpoints for smoke, heat and toxic effluent mixtures are presented in the following sections. Examples are shown of how the method is applied to fire data to estimate times to different hazard endpoints. A full worked example is given to illustrate the calculations method in detail.

14.2.3 Application of FEC and FED to Full-scale Compartment Fire Data

In order to carry out and FEC and FED analysis of compartment fires, such as those depicted in Figures 14.1 and 14.2, the time–concentration curves for smoke density, toxic gases, temperature and radiation are used to calculate the FEC and FED fractions for the following terms as a function of time during the fire:

- $\mathrm{FEC}_{\mathrm{smoke}}$: Fractional effective smoke concentration, where $\mathrm{FEC}_{\mathrm{smoke}} = 1$ represents a smoke concentration likely to significantly impair escape efficiency.
- $\mathrm{FIC}_{\mathrm{irr}}$: Fractional irritant concentration, where $\mathrm{FIC}_{\mathrm{irr}} = 1$ represents a concentration of organic irritants and/or acid gases likely to

significantly impair escape efficiency. $FIC_{irr} = 5$ represents an approximate incapacitating concentration predicted to cause collapse.

- F_{IN} (or "FED_{IN}"): Fraction of a dose of asphyxiant gases, where $F_{IN} = 1$ represents a dose predicted to cause collapse and loss of consciousness due to asphyxia ($F_{IN} = 2$ represents an approximate lethal dose).
- FED_{heat}: Fraction of an exposure (dose) of radiant and convected heat, where $FED_{heat} = 1$ represents a "dose" sufficient to cause severe pain to unprotected skin (head and hands), $FED_{heat} \sim 8$ represents second degree burns and $FED_{heat} = 12$ represents severe full thickness (third degree) burns with a high risk of lethality.
- FLD_{irr}: Fraction of a lethal dose of irritants, where $FLD_{irr} = 1$ represents an inhaled exposure dose of lung irritants sufficient to cause severe injury or death some hours after exposure.

Expressions for the calculation of each of these terms are presented in Sections 14.3 to 14.7.

Figures 14.3 and 14.4 illustrate the application of this method for the two fire cases in Figures 14.1 and 14.2. Figure 14.3 shows the effects of exposure to fire effluent when the upper floor bedroom occupant steps out on to the landing 10 minutes after ignition of the armchair fire in the open lounge on the floor below. Upon opening the door the occupant is confronted by smoke with an FEC of 12, which means that the smoke density is 12 times the concentration predicted to seriously impair escape attempts. Within a few seconds, exposure to this smoke causes severe irritant pain to the eyes (FIC 15), resulting in reflex eye closure, further impairing and inhibiting escape movement onto the landing. The smoke is also quite hot (60 °C), which may further deter escape, but not hot enough to cause distress. The result at this stage is that the occupant is likely to decide that escape is too hazardous, shut the bedroom door and take refuge in the bedroom. If the subject attempts to descend the stairs, the speed of descent will be slow and the next occurrence depends on the subject's breathing pattern. If the subject holds their breath before being exposed to the smoke, then there is no intake of toxic gases until they are forced to take a breath. At this point the inhaled breath will be a deep one, with immediate incapacitation due to the respiratory tract irritancy. Due to the high concentrations of asphyxiant gases present, inhalation of a few breaths, or even a single deep breath, may result in syncope.

The analysis in Figure 14.3 assumes the subject is breathing normally throughout at a rate associated with a moderate level of activity. This results in uptake of asphyxiant gases (HCN, CO, CO_2 and low oxygen) from the time the subject steps onto the landing. Incapacitation (collapse and loss of consciousness) is predicted when the F_{IN} exceeds unity, which occurs after 0.2 minutes (12 seconds). This is mainly due to the effects of inhaling hydrogen cyanide, which was present at the high concentration of 850 ppm in this experiment. Also considered in the analysis is the calculated time to incapacitation assuming there was no nitrogen in the burning fuel (armchair), in which case the most important asphyxiant gas is carbon

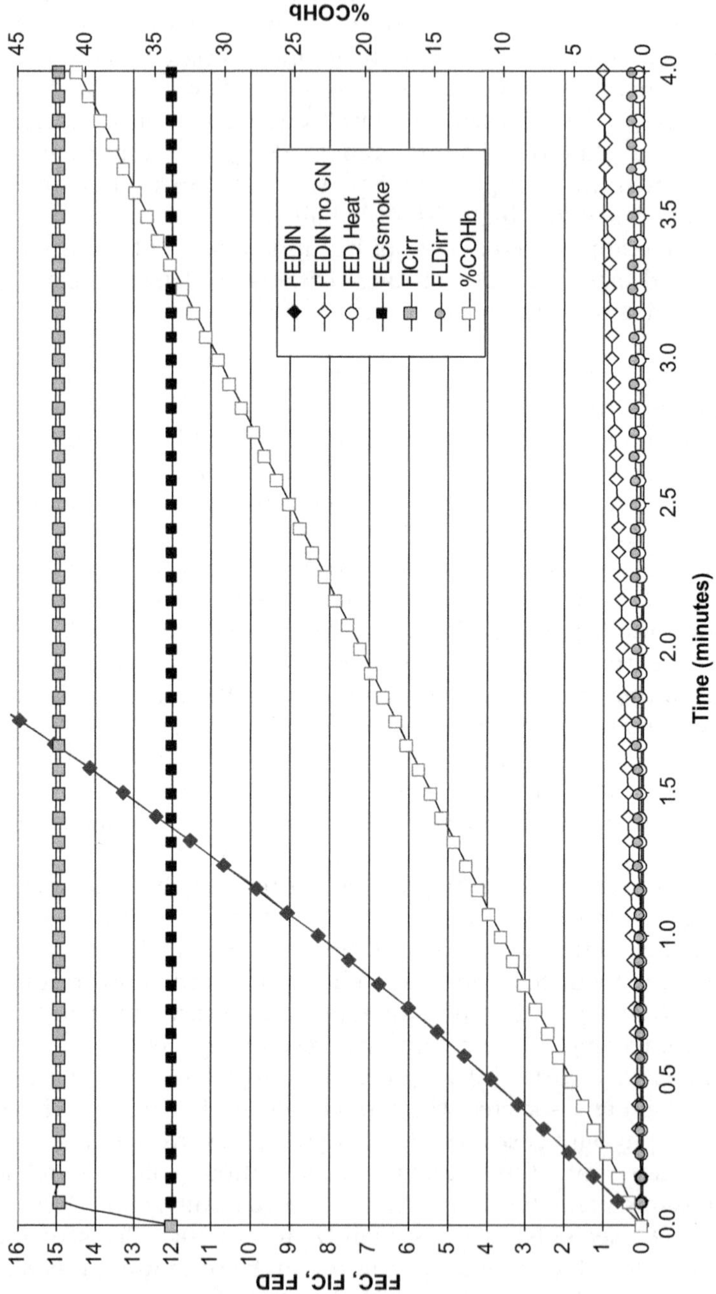

Figure 14.3 Fractional effective dose plots for the static fire hazard condition in Figure 14.1.

monoxide. Uptake of carbon monoxide (increased by the presence of CO_2 due to the stimulatory effect on respiration) results in accumulation of a dose producing an incapacitating level of carboxyhaemoglobin (approximately 30–40% COHb) after 4 minutes of exposure, which represents an FED_{IN} (no CN) of 1. The accumulating %COHb is calculated as part of the analysis and is shown in Figure 14.3. Carbon dioxide also becomes distressing at the exposure concentration of 5% after a minute or so, and would itself be capable of causing incapacitation after approximately 30 minutes.[1]

Also shown in Figure 14.3 are the accumulating doses of heat (FED_{heat}) and of lung irritants, both of which hardly rise above the baseline up to 4 minutes. Since the smoke temperature is only 60 °C, only minor discomfort is predicted over this time scale.

If a subject attempted to escape through these conditions, the predicted results from this analysis are as follows:

- They would find the effects of the smoke exposure to be painful and distressing, and be likely to turn back and take refuge in the bedroom.
- If they were familiar with the building, the distance to the main exit on the lower floor was short and the door easy to open, they might be able to hold their breath long enough to reach the exit and escape before being overcome.
- If they took a breath at any time they are likely to collapse more or less immediately from the effects of the high concentrations of asphyxiant gases, especially hydrogen cyanide.
- If they collapsed on the stairs or near the door, they would die within a further few minutes, mainly from the effects of carbon monoxide, with a blood carboxyhaemoglobin concentration of around 40–60% COHb.
- If they were rescued at 4 minutes and treated with oxygen, they would be likely to make a rapid recovery (depending on their pre-existing health status). They might suffer some eye and throat irritation the next day and develop a minor productive cough, but would be unlikely to suffer significant lung injury. There would be a possibility of suffering a heart attack or stroke due to the inhalation of fine smoke particles and CO and there would be a possibility of suffering long-term brain damage due to cerebral hypoxia.

Figure 14.4 shows an FED analysis for an occupant of the lower floor fire room from the data in Figure 14.2.

This differs from the analysis in Figure 14.3 in that the exposure period starts with ignition of the fire, after which the conditions change rapidly. For a person in the fire room but on the opposite side of the room from the fire, there is no initial exposure to either heat or smoke as the effluent from the early flaming fire collects under the room ceiling. Once the upper layer descends below the soffit of the door to the hall, the smoke starts to flow out into the hall and landing, further reducing the rate of descent of the upper layer in the fire room. For a person standing in the room, the first exposure

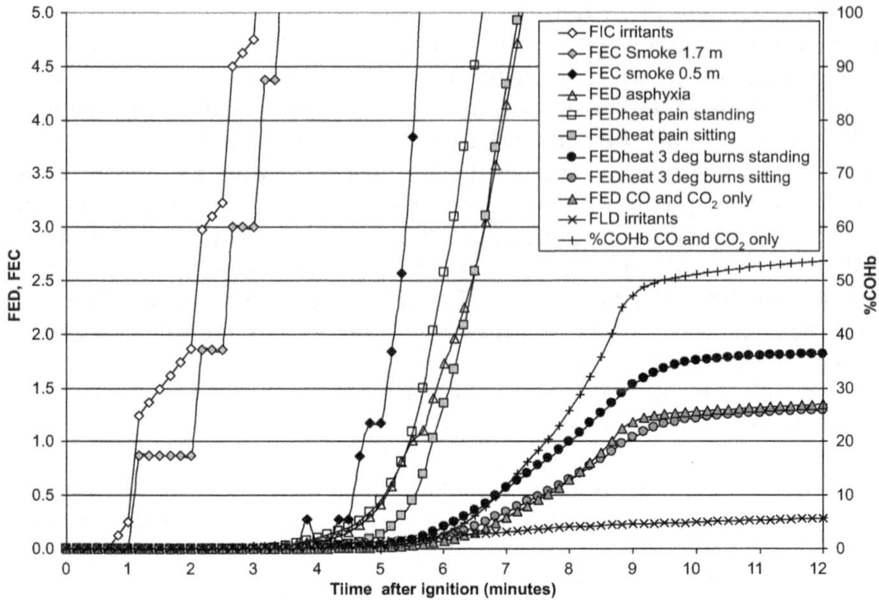

Figure 14.4 Fractional effective dose plots for the changing fire hazard conditions in Figure 14.2.

to smoke is predicted from 1 minute after ignition, when the upper layer descended to head height (1.7 m). FEC = 1 for smoke density and $FIC_{irr} = 1$ for the effects of the irritant gases present in the smoke are predicted at 1 minute, after which a subject is predicted to suffer an impaired escape capability due to the optical obscuration from the smoke and the smoke irritancy, derived from a significant content of both organic irritants and acid gases, including HCl, HBr, HF and SO_2, which were measured in samples from the test using ion chromatography. However, if the subject got down to near floor level (0.5 m), then they would be beneath the smoke layer until 4 minutes. Conditions deteriorated very rapidly from around 4 minutes, so that at 5.5 minutes an FED exceeding 1 is calculated simultaneously for both asphyxia and heat. At this time it is predicted that a standing exposed subject would suffer severe pain to exposed areas of skin, especially the head and hands, and would collapse unconscious due to the effects of inhaled asphyxiant gases and, in particular, hydrogen cyanide. For a person sitting or crouching below the smoke layer up to this time the exposure to hot gases would be somewhat less, so that time to severe pain from heat is predicted at just before 6 minutes. As the exposure to heat continues, an exposed subject would begin to suffer from burns (both to clothed and unclothed areas, depending upon the thickness of the clothing), so that for a standing person fatal extensive full thickness (third degree) burns are predicted after 8 minutes and for a sitting person after 9 minutes. The FED heat curves then level off because by now the fire has self-extinguished and the temperatures in the room are dropping.

The other two curves in Figure 14.4 have been included to consider the situation with regard to asphyxiant gases in the absence of hydrogen cyanide. For these upholstered furniture fires the hydrogen cyanide concentrations increased rapidly to well in excess of those observed to result in loss of consciousness in exposed primates within 2 minutes (see Chapters 7 and 10). Cyanide is therefore the main driver of collapse from asphyxiation in these fires. If the fuels had not contained nitrogen (or $<1\%$ nitrogen by mass) then the main asphyxiant gases present would have been carbon monoxide, carbon dioxide and reduced oxygen levels. Another factor would have been the high concentration of irritant smoke gases and particulates. It is considered that these would have made breathing difficult for exposed occupants, thereby adding to some extent to the asphyxiant effects of the inhaled gases. For the full form of the F_{IN} model for asphyxia, the fraction of a lethal dose of irritants (FLD_{irr}) is included as an additive term, since the breathing difficulties associated are considered likely to add to those of the chemical asphyxia. When this is done for this fire (but without HCN) the main drivers of asphyxia are the inhaled doses of CO and these irritants, increased by the presence of high CO_2 concentrations and with a small contribution for the lowered oxygen concentrations. For this model, the calculated time to FED_{IN} asphyxia due to the effects of CO uptake is 7 minutes. At this time it is calculated that the overall degree of incapacitation is likely to be sufficient to cause collapse and prevent escape. Since the extent of additive effects from the irritants is somewhat uncertain, the final curve shows the situation with this term removed from the analysis. The result shows loss of consciousness after 8.5 minutes as the blood CO level reaches 40% COHb, which happens to coincide with the time by which a person sitting or lying on the floor is likely to receive life threatening burns. After the point of loss of consciousness, a reduction in respiration occurs (see Chapter 7).[1] This, combined with the lower CO concentration occurring around this time, results in a reduction in the subsequent rate of increase in blood %COHb.

After 10 minutes the FLD for lung irritants is 0.25, indicating that severe post-exposure lung inflammation is unlikely if rescued at this time, although elderly subjects in particular have been found to be vulnerable to the development of fatal pneumonia from lung infection. Following rodent and primate exposures to combustion product atmospheres, signs of post-exposure lung inflammation were found to be minor up to a critical exposure dose level, above which severe or fatal oedema and inflammation occurred during the first 12–48 hours.[14]

The predicted results for a subject exposed to these conditions depend very much upon their condition at the time of ignition and their subsequent behaviour:

- A subject awake and ambulant at the time of ignition, and aware of the fire from ignition, should be able to escape without difficulty, suffering no or minimal exposure.

- A waking subject may get into difficulties if they remain at or enter the scene to carry out activities, such as getting dressed, attempting to retrieve belongings or to fight the fire, or if they are unable to escape due to incapacity (for example, due to physical or mental disability, alcohol or drugs). Due to the rapid deterioration in conditions after 4 minutes they may become disorientated and overcome by toxic smoke and heat, collapsing and dying at the fire scene.
- For a sleeping subject the outcome depends upon when they become aware of the fire and their subsequent behaviour. Figure 14.2 shows the activation times for smoke detectors placed on the lounge and upper floor landing ceilings for this experiment. The optical detector placed on the landing ceiling activated 2.5 minutes after ignition, by which time there was already dense smoke at head height in the lounge. Whether or not a subject would escape before being overcome would then depend upon how rapidly they awoke, took in the situation and when they decided to leave the room. Although escape *via* the hallway and main exit would be possible following a rapid response, any delay would be likely to result in the occupant being overcome and collapsing, either in the lounge or hall.
- Unlike the bedroom case it is considered that many occupants in this situation would attempt to escape through the smoke even at high density, since they would be motivated by the need to move away from the source of the fire and effluent.
- A similar analysis has been carried out for the conditions faced by an occupant of an open upper floor bedroom during this fire experiment, both for a situation with the fire room door opened and closed. Once smoke from a fire in an open lounge such as these begins to enter an open bedroom on the floor above, the time available for response and escape is very short (around 1–2 minutes) before incapacitation is predicted. Although smoke leakage occurs around closed doors, the time available for escape from the bedroom is considerably extended if the lounge (fire room) door is closed. When the lounge door was open, conditions in an upstairs bedroom with the bedroom door closed remained tenable for an extended period of around 30 minutes or more.

In the following sections expressions are presented for assessing the physiological effects of each of these hazardous components and their interactions in producing incapacitation and death, including a worked example for a full FED analysis for a set of fire data. In estimating endpoints consideration has been given to relatively conservative estimates that might be used for design purposes, taking into account uncertainties in predicting effects on average human subjects and possible ranges of sensitivity in exposed human populations, and endpoints predicted to produce severe effects in average exposed subjects for forensic application to incident investigations.

14.3 Assessing the Effects of Exposure to Typical Fire Smoke Containing Irritant Particulates and Vapours

Data on the effects of smoke on exposed human subjects are available from reports of fire incidents and a small number of experimental studies (see Chapter 7). For these incidents the smoke composition is unknown but can be considered as "generically irritant" smoke typical of fires in buildings. Experimental studies have been carried out mainly using theatrical smoke (consisting of a non-irritant paraffin aerosol), with one study involving irritant wood smoke (smouldering wood chips in a bee-smoker). For predicting effects of effluent atmospheres containing relatively high concentrations of specific irritants, particularly acid gases, it is necessary to make use of data from animal models, as described in Chapters 7 and 13.

The effects of smoke are partly behavioural, depending upon the situation and nature of the occupant, and partly physiological. The most important considerations are smoke concentrations that are likely to impair, prevent or deter escape, and for occupants attempting to escape through smoke, the effects on walking speed.

The effects of smoke density on escape behaviour during actual fire incidents are reviewed by Bryan.[15] Two studies (in the US and UK)[15–17] report occupant behaviour in terms of distance travelled through smoke of different densities (expressed as visibility distance) and numbers turning back or seeking refuge. Although the irritancy and temperature of the smoke is not mentioned, it can be assumed that the smoke was in most cases somewhat irritant and hot. Both studies showed remarkably similar results (see ref. 15, Table 3-11.27). In the UK study, 91% of those turning back having entered smoke did so when visibility was less than approximately 4 m, representing 62% of those exposed to smoke at this visibility or lower, while only 11% of those exposed to smoke with a visibility >4 m turned back (5% for visibility >9 m). For individual cases, the author has examined or investigated several cases of dwellings fires in which occupants on an upper floor have been confronted with dense, hot smoke from a fire on the floor below when opening a bedroom door. In almost all cases the occupants have decided to close the door and remain in the bedroom or attempt escape *via* a window rather than to attempt to proceed through the smoke-filled escape route, citing the density, irritancy and temperature of the smoke as deterrents. In two incidents, five occupants attempted to escape away from a location where they had opened a landing door and enabled a smoke plume to enter from a fire on the floor below. Three were overcome and died while attempting to escape, while one escaped through a window and one was rescued having taken refuge in a closed room.

During the Dupont Plaza Hotel incident, some witnesses reported occupants being trapped or overcome by smoke, while others successfully travelled a considerable distance and escaped along a long smoke-filled corridor. During the Mont Blanc road tunnel fire, most occupants remained

Table 14.1 Smoke tenability limits.[b]

Smoke density and irritancy/ OD m^{-1} (extinction coefficient shown in brackets)	Approximate visibility (diffuse illumination)	Reported effects
None	Unaffected	Walking speed 1.2 m s^{-1}
0.5 (1.15), non-irritant	2 m	Walking speed 0.3 m s^{-1}
0.2 (0.5), irritant	Reduced	Walking speed 0.3 m s^{-1}
0.33 (0.76), mixed	3 m approx.	30% of people turn back rather than enter
Suggested tenability limits for buildings with:		
- small enclosures and travel distances;		OD m^{-1} = 0.2α_k 0.5 (visibility 5 m)[a]
- large enclosures and travel distances.		OD m^{-1} = 0.08α_k 0.18 (visibility 10 m)[a]

[a]OD m^{-1} is $\log_{10}(I_0/I)$, the logarithm of the ratio of the intensities of light (I_0/I) transmitted over a path length of 1 m from a light source to a receiver in the absence and presence of smoke, respectively. The light extinction coefficient α_k is ln (I_0/I).

[b]For situations where smoke is expressed in terms of particulate mass concentration, these equate to approximately 0.7 and 0.3 g particulates m^{-3}, respectively (where particulates g $m^{-3} \sim 0.356 \times$ OD m^{-1}).

(and died) in their vehicles rather than entering the smoke-filled tunnel, while a small number travelled a considerable distance along the tunnel through smoke before being overcome by toxic gases or heat.[18] The general pattern from these incident investigations is that occupants will generally be unwilling to enter smoke with a visibility of more than approximately 3 m (0.33 OD m^{-1}, extinction coefficient = 0.76), especially if they have an option of turning back or seeking refuge, and women were more likely to turn back than men. If they find themselves enveloped in dense smoke (for example, in a tunnel) they may continue to move slowly through the smoke, often adopting a crouching posture and feeling their way along walls if available.

For exposure to generic irritant smoke it is considered that escape efficiency is likely to be significantly impaired at the concentrations shown in Table 14.1. For application to ASET calculations, different endpoints are recommended for small and large enclosures, since in large enclosures it is necessary to see greater distances to locate escape routes and exits. In this context the limiting concentration for smoke (FEC_{smoke}) at any time during a fire is as follows, where OD_{smoke} is the optical density of the smoke and m is the distance in metres:[1,19]

$$FEC_{smoke} = \frac{OD_{smoke}/m}{0.2 \text{ or } 0.08} \tag{14.3}$$

14.3.1 Movement Speed in Smoke

In experiments where people were asked to walk down a 20 m smoke-logged corridor, Jin[20,21] found that, for non-irritant black smoke (from burning kerosene), walking speed decreased with smoke density and that at an extinction coefficient of 1.15 (OD 0.55 m^{-1}, visibility 2.1 m) walking speed decreased from approximately 1.2 m s^{-1} (no smoke) to approximately 0.3 m s^{-1}. Under these conditions people behaved as if they were in total darkness, feeling their way along the walls or weaving from side to side rather than walking in a straight line.

When people were exposed to irritant smoke, made by heating wood chippings, movement speed was reduced to that in darkness at a much lower smoke density (extinction coefficient 0.5, OD 0.2 m^{-1}) and the experience was found to be more distressing. Jin said that the visibility in irritant smoke decreased sharply at a smoke density exceeding a certain level. "The subjects could only keep their eyes open for a short time and tears ran so heavily that they could not see the words on the signs". Both density and irritancy were found to affect walking speeds, the speed decreasing very rapidly as density and irritancy increased. The subjects could not keep their eyes open and they walked with a zigzag pattern or along the wall.

Figure 14.5 shows plots of Jin's results and that from a study by Frantzich and Nilsson.[22] The figure shows that, although it is possible to suggest a relationship between smoke density and walking speed for both non-irritant and irritant smoke from Jin's data and walking speed was definitely slower in dense smoke, there is a considerable scatter in the data. Also, although the walking speed in the absence of smoke was stated to be approximately 1.2 m s^{-1}, there is no indication of the scatter of individual speeds in clear and low smoke density conditions. The results do appear to show a relationship between smoke density and walking speed for non-irritant smoke over a range of α_k from 0.2–1.13, a linear fit to the data being shown in the figure. For non-irritant smoke with an average density of α_k 0.73, the average walking speed was 0.74 m s^{-1} (standard deviation 0.17). For irritant smoke, although it is possible to obtain a linear fit showing a decrease in speed with smoke density, the data are over a very short density range, $\alpha_k = 0.32$–0.5, and there is a large scatter of walking speeds. For example, at $\alpha_k = 0.42$, speeds range between 0.37 and 1.1 m s^{-1}. For an average density of $\alpha_k = 0.42$, the average speed is 0.75 m s^{-1} (standard deviation 0.21). This is very similar to the average speed in non-irritant smoke, but at a lower density. Taking these data in conjunction with Jin's description of subjects' behaviour, it does seem that a person's ability to walk through irritant smoke is more severely impaired than for non-irritant smoke, but it is less obvious that there is a clear relationship between density and speed. Rather, it seems that some subjects are severely affected at a low smoke density of around $\alpha_k = 0.42$ (OD 0.18 m^{-1}, visibility approximately 5 m), which is close to the suggested design limit for small enclosure of OD 0.2 m^{-1} (see Table 14.1).

The more recent data set for movement in smoke was that obtained by Frantzich and Nilsson.[22] The scenario was somewhat different from that of Jin's work in that subjects were asked to walk through a smoke-logged simulated road tunnel. The "tunnel" contained parked vehicles and had a walkway down the right-hand side from the point at which the subjects

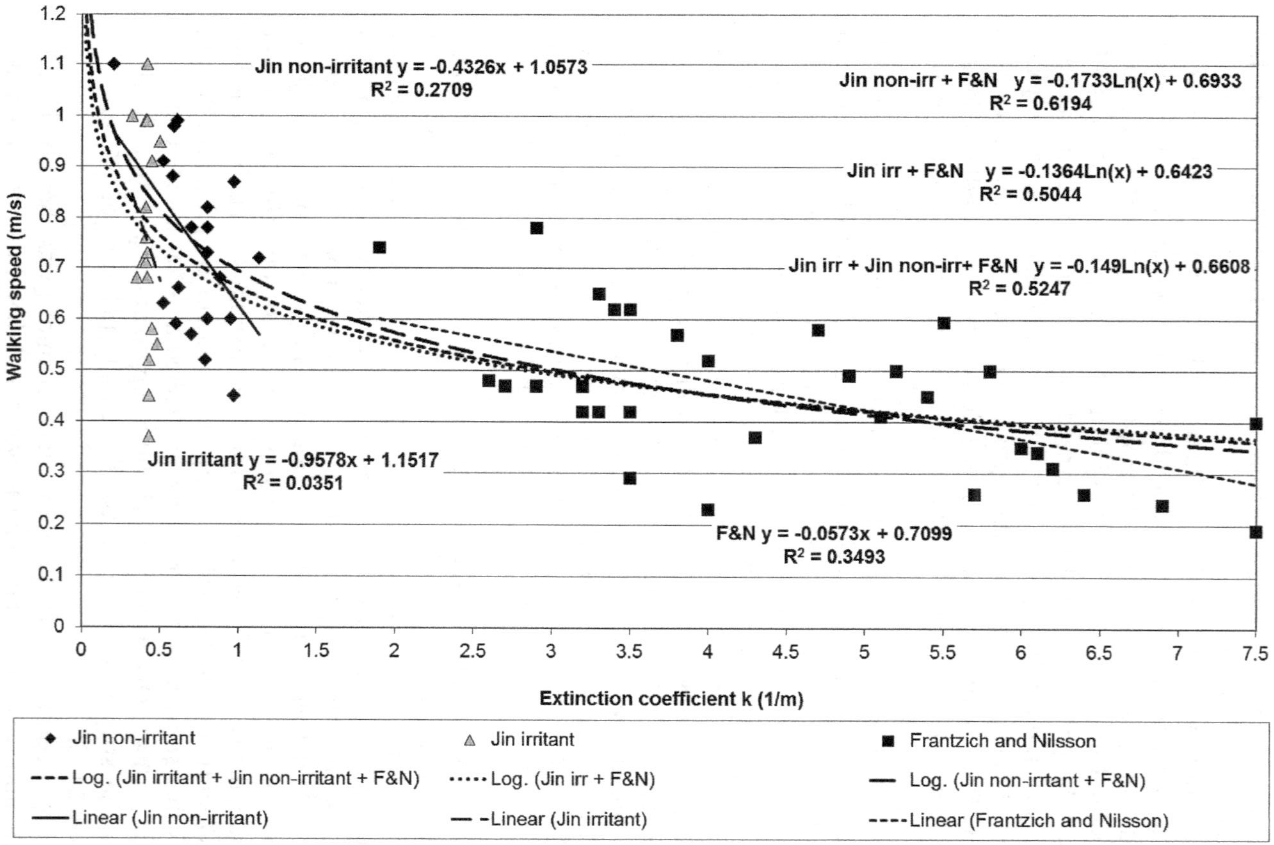

Figure 14.5 Scatter of data and relationship between smoke density [extinction coefficient α_k (1 m^{-1})] and walking speed (m s^{-1}). Data show Jin (non-irritant smoke; Jin non-irr) and Jin (irritant smoke; Jin irr) in a corridor, and Frantzich and Nilsson (F&N; mildly irritant smoke) in a tunnel mock up.

were introduced. There was general lighting (on or off in different tests) and an illuminated sign over an exit near the far end on the left side, the subjects being required at some point to cross the tunnel to reach the exit. Smoke was theatrical smoke made slightly irritant by the addition of some acetic acid vapour.

The results for the case with general lighting are shown in Figure 14.5. Apart from differences in the test scenario, a major difference from Jin's experiments was that much higher smoke densities were used, varying from α_k 1.9 to 7.5. The whole data set therefore sits beyond the density at which Jin found that subjects behaved as if in darkness and walked feeling their way along the wall with their hands. A linear fit to Frantzich and Nilsson's results does show a shallow slope of slower speed with increasing smoke density, but another way of looking at the data set is to consider it all as constituting dense smoke and consider the average walking speed compared to Jin's figure of a speed of 0.3 m s^{-1} (although the slowest measured speed was 0.37 m s^{-1}). For Frantzich and Nilsson's study, the mean walking speed was 0.45 m s^{-1} (standard deviation 0.21), which is quite close to that obtained by Jin in dense smoke. It was also noted in this study that most subjects moved for part of the distance by feeling their way along the walls and that subjects doing this for more than two thirds the total distance had a greater average speed. As with Jin's work, there was a considerable variation in walking speeds between subjects at all smoke densities.

In order to derive calculation expressions for the relationship between smoke density and walking speed for non-irritant or irritant smoke, it is possible to perform a variety of fits to these data sets. Figure 14.5 shows logarithmic curves fitted to three data sets: Jin non-irritant + Frantzich and Nilsson, Jin non-irritant + Jin irritant + Frantzich and Nilsson, and Jin irritant + Frantzich and Nilsson. The three fitted curves are quite similar. For a simple deterministic calculation, an expression fitted to Jin's irritant data and the Frantzich and Nilsson data is suggested as follows:

$$W_{\text{smoke}}\ (\text{m s}^{-1}) = -0.1364 \times \ln \alpha_k + 0.6423, \qquad (14.4)$$

where W_{smoke} is the walking speed in moderately irritant smoke. The standard deviation is 0.157.

Expressions such as eqn (14.4) have important applications to the design of escape routes for buildings, enabling engineers to simulate fire scenarios and evaluate evacuation times and probabilities of escape.

14.4 Tenability Limits and Fractional Irritant Concentrations for Sensory Irritants

As described in Chapter 7, the immediate effects of exposure to sensory irritants occur over a range of increasing severity, from mild eye and respiratory tract irritation to severe pain and breathing difficulties with not obvious threshold levels for different effects. The severity of signs and symptoms is approximately proportional to the logarithm of the exposure concentration. A model relating rate of uptake at the epithelial surface with a concentration-related rate of removal or detoxification has been found to provide a good prediction of the development time and severity of effects on the upper respiratory tract.[23] For humans the eye and upper respiratory tract effects are almost instantaneous and maximal within a few seconds of exposure at painfully irritant exposure concentrations, followed within a short period by lacrimation, mucus secretion and large airway effects, including bronchoconstriction and chest pain.

Although there is no obvious threshold, for practical application to ASET calculations, two threshold concentration endpoints may be calculated for sensory irritants:

- A concentration capable of seriously impairing escape capability and movement speed.
- A concentration capable of causing incapacitation, such that the subject effectively cannot move.

Table 14.2 shows proposed threshold values from several different sources. The threshold concentrations from *The SFPE Handbook of Fire Protection Engineering*[1] and BS7899-2[24] have been suggested by the author as concentrations of common fire irritants likely to severely affect escape capability in most humans (escape impairment), while it is proposed that exposure to the higher concentrations shown in the second column of Table 14.2 is likely to be painful and cause such severe breathing problems as to effectively incapacitate exposed subjects. As described in Chapter 7, the results from human exposure and animal experiments show a wide range of signs so that there is considerable uncertainty in predicting the severity of effects. For the

Table 14.2 SFPE/BS7899-2, ISO 13571 and AEGL guidelines for escape impairment and incapacitation.

	SFPE, escape impaired	Incapacitation: SFPE	Incapacitation: ISO 13571	SFPE, 30 minute, lethal	AEGL-2, 10 minutes, escape impairment	AEGL-3, 30 minutes, life threatening
HCl	200	900	1000	3800	100	210
HBr	200	900	1000	3800	100	250
HF	200	900	500	2900	95	62
SO_2	24	120	150	400–1400	0.75	30
NO_2	70	350	250	63	20	25
NO	—	>1000	—	~1000	—	—
C_3H_4O (acrolein)[a]	4	20	30	150	0.44	2.5
HCHO (formaldehyde)[a]	6	30	250	750	14	70

[a]Where the concentrations of acrolein and formaldehyde (or other important irritants) are unknown, a term derived from smoke density 0.5 OD m^{-1} may be used as an indication of irritancy likely to impair escape efficiency.

ISO 13571 guidance documents on assessment of toxic fire hazards,[11] a considerably less conservative endpoint of incapacitation is proposed. However, for both SFPE and ISO, more conservative design limits are recommended to allow for sensitive members of the exposed population, such as a design FED of 0.3 for the general population or even 0.1 for particularly sensitive groups (such as for those in healthcare premises). For the SFPE guidance this would result in a limiting HCl concentration of 60 ppm to avoid escape impairment in the majority of the exposed population and around 330 ppm for ISO 13571. The Acute Exposure Guidelines Levels (AEGL)[25] for immediate impaired escape capability and long-lasting adverse health effects (AEGL-2) for exposures of up to 10 minutes already include provision for sensitive sub-populations, so that, for example, the proposed limit for HCl of 100 ppm is close to the adjusted SFPE guidance concentration, but somewhat more conservative than the ISO 13571 level, which is closer to the AEGL-3 "life threatening concentrations" for 30 minute exposures.

Since fire effluents contain a mixture of irritants it is necessary to consider how they interact in combination. Based on the data reviewed in Chapter 7 and ref. 23, it is currently recommended that they should be considered additive. The overall FIC for an irritant mixture is then as follows:

$$\begin{aligned} FIC = {} & FIC_{HCl} + FIC_{HBr} + FIC_{HF} + FIC_{SO_2} + FIC_{NO_2} + FIC_{C_3H_4O} \\ & + FIC_{CH_2O} + \Sigma FIC_x, \end{aligned} \tag{14.5}$$

where $\Sigma FICX = FIC$ values summed for any other irritants present.

Each term is a fraction for which the numerator is the exposure concentration of an irritant present at any time during an exposure and the denominator is the concentration predicted to cause impaired escape capability (with use of the SFPE values recommended, with an FED factor of $\times 0.3$ to allow for sensitive individuals). Since these are intended to represent immediate sensory effects, they would be expected to ameliorate if the exposure concentrations decreased (for example, if as a subject moved out of a contaminated area).

14.5 Post-exposure Lung Inflammation and Survival

In addition to incapacitating effects of sensory irritation affecting escape capability, a proportion of inhaled irritants penetrates to the deep lung and when a sufficient exposure dose is accumulated this can lead to inflammatory processes that can be fatal over periods of several hours to several days after exposure (see Chapters 7 and 13, and ref. 23 and 14). For *The SFPE Handbook of Fire Protection Engineering*, the exposure dose considered likely to cause death in an average exposed human subject is estimated as the rat median lethal concentration and time (LCt_{50}) exposure dose for a 30 minute exposure and 14 day post-exposure observation period. A guide to exposure doses of common irritant gases likely to cause incapacitation at the scene or

Table 14.3 Exposure concentrations and exposure doses for incapacitation and lethal lung damage.

Gas	SFPE exposure doses predicted to be lethal to half the population/ppm min	AEGL-3 30 minute life-threatening exposure doses/ppm min
HCl	114 000	6300
HBr	114 000	7500
HF	87 000	1860
SO_2	12 000	900
NO_2	1900	750
NO	~30 000	—
C_3H_4O (acrolein)[a]	4500	75
HCHO (formaldehyde)[a]	22 500	2100

[a]Where the concentrations of acrolein and formaldehyde (or other important irritants) are unknown, a term derived from smoke density 90 OD m^{-1} min^{-1} may be used as an indication of lethal organic irritant exposure dose.

death following exposure is given in Table 14.3. These are not intended as design limits (*i.e.*, maximum exposure limits considered compatible with safe escape from a fire), but as estimates of the extent to which post-exposure deaths from lung oedema and inflammation are likely to occur. The 30 minute AEGL-3 values shown in Table 14.2 are for concentrations above which it is predicted that the general population, including susceptible individuals, could experience life-threatening adverse health effects or death. When these are expressed as exposure doses and compared with the SFPE levels in Table 14.3 it is evident that the AEGL-3 exposure dose limits for this endpoint are much more conservative. The appropriate choice of limits would depend upon the application. Following exposures of primates and rodents to fire effluents, signs of post-exposure inflammatory effects on the lung were minor up to an apparent threshold, above which effects were severe.[1,14]

As with sensory irritancy it is estimated that the effects of lung irritant mixtures are approximately additive. The FED equation for the lethal effects of inhaled irritants (*i.e.*, FLD or fractional lethal dose) is:

$$\begin{aligned} FLD_{irr} = {} & FLD_{HCl} + FLD_{HBr} + FLD_{HF} + FLD_{SO_2} \\ & + FLD_{NO_2} + FLD_{C_3H_4O} + FLD_{HCHO} + \Sigma FLD_x \end{aligned} \tag{14.6}$$

For example, consider a theoretical fire scenario similar to that illustrated in Figure 14.2, in which a fire has self-extinguished leaving a fire effluent atmosphere in an enclosed building with a constant composition containing the concentrations of irritant gases shown in Table 14.4, with an occupant exposed for 20 minutes before rescue. For this mixture of gases, the FDL_{irr} after 20 minutes exposure is 1.00. It is predicted that the occupant is likely to develop fatal lung oedema and inflammation, for which the major driver would be NO_2, with additive components from the other irritant gases.

Table 14.4 Example of an FLD calculation for a theoretical fire effluent atmosphere.

Irritant gas	HCl	HBr	HF	SO_2	NO_2	Acrolein	Formaldehyde
Gas concentration/ppm	1000	500	50	50	55	10	20
Ct exposure dose after 20 minutes/ppm min[a]	20 000	10 000	1000	1000	1000	200	400
Lethal *Ct* exposure dose[a]	114 000	114 000	87 000	12 000	1900	4599	22 500
Individual and total FLD values FLDirr: 1.00	0.18	0.09	0.01	0.08	0.58	0.04	0.02

[a] *Ct* = concentration (*C*) × time (*t*).

14.6 Tenability Limits and Hazard Calculations for Asphyxiant Gases

14.6.1 General FED Expression for Mixed Asphyxiant Gases

The main cause of incapacitation and death during and immediately after fires is exposure to asphyxiant gases. Incapacitation results from loss of consciousness due to the combined hypoxic effects of carbon monoxide, hydrogen cyanide and carbon dioxide, with some additional effects from low oxygen hypoxia, nitrogen oxides and inhaled irritants. Loss of consciousness prevents escape and further uptake of asphyxiants while comatose is likely to result in death within a further minute or so unless the occupant is rescued. The most useful tenability endpoint to work to is therefore considered to be loss of consciousness ($F_{IN} = 1$), with design limits set to prevent this occurring. Since individual susceptibility varies in the population, this is predicted to represent the median of the distribution of exposure doses resulting in collapse. According to one estimate, approximately 11.3% of the population is considered likely to be susceptible below an FED of 0.3 (see ISO 13571).[11] It will be necessary for the designer or regulatory authority to select an FED level suitable to protect vulnerable sub-populations in the chosen application (for example, 0.3 or some other value depending upon the application). Setting such design limits is always a difficult compromise between making acceptable allowance for risks from low probability events, such as fires, and setting limits that can be met within acceptable costs. Using an FED of design limit of 0.3 would represent an upper exposure limit to CO equivalent to 9% COHb.

The effects of combinations of asphyxiant gases causing incapacitation in fires are considered to be approximately additive, but a number of interactions need to be considered (see Chapter 7):

- The FEDs for CO and HCN are considered directly additive (1 : 1) on a fractional dose basis (as found to be the case in rats).[26]
- NO and NO_2 (designated as NO_x in mixtures) also act as asphyxiants, reducing oxygen carriage in the blood due to the conversion of haemoglobin to methaemoglobin. To this extent, their asphyxiant effects can be considered additive with those of HCN and CO. However, methaemoglobin combines with HCN in the blood, thereby reducing its asphyxiant effect. NO_2 is also a potent lung irritant.
- The effects of irritants on lung function also cause some hypoxia and so an additive term is included consisting of the FLD_{irr}.
- The rate of uptake of all these asphyxiant gases (CO, HCN, NO_x and irritants) depends on the respiratory ventilation (V_E) of the subject in relation to body size, which in turn depends upon their level of physical activity. For design purposes it is assumed that the subject is an adult engaged in light work such as walking along an escape route.
- The main effect of carbon dioxide is to increase the breathing rate and thus the rate of uptake of asphyxiant gases (CO, HCN, NO_x and

irritants). A multiplicative term $fMiCO_2$ is used to calculate this effect. Based on experimental data from human exposures it is considered that total ventilation is given by $V_E \times fMiCO_2$.

- Low oxygen hypoxia will be additive with the overall hypoxic effects of CO and HCN, but is not increased by $fMiCO_2$ (in fact, it is improved).
- The beneficial effects of increased CO_2 on the hypoxic effects of CO and low oxygen hypoxia resulting from right shifting of the oxyhaemoglobin dissociation curve are ignored.
- The direct intoxicating effects of CO_2 are considered unlikely to occur before other effects so are normally ignored, but exposure to 7% CO_2 or higher itself causes incapacitation within a few minutes.

On the basis of these considerations the overall FED equation for asphyxiants is:

$$F_{IN} = (F_{ICO} + F_{ICN} + F_{INO_x} + \mathrm{FLD}_{irra}) \times V_E \times fMi_{CO_2} + F_{IO}, \tag{14.7}$$

where:

F_{IN} = Fraction of an incapacitating dose of all asphyxiant gases.
F_{ICO} = Fraction of an incapacitating dose of CO.
F_{ICN} = Fraction of an incapacitating dose of HCN.

Note: If necessary this can be corrected for the presence of other nitriles besides HCN and for the protective effect on cyanide poisoning of NO and NO_2. [CN] can then be calculated as: [CN] = [HCN] + [Total organics nitriles] − 0.67[NO + NO_2].

FI_{NOX} = Fraction of an asphyxiant incapacitating dose of NO + NO_2 [$= (NO_x\ \mathrm{ppm} \times t_{min})/1500$].
FLD_{irra} = Fraction of an irritant dose contributing to hypoxia (this term represents a correction for the effects of irritants on lung function and is derived from the term FLDirr developed in 14.5. This term may be omitted if the effects of asphyxiant gases only are under consideration).

Note: Where significant concentrations of acid gases are present, this term is calculated according to eqn (14.6) divided by 25. Otherwise, it may be expressed in terms of smoke optical density as OD $m^{-1} \times t(min)/(90 \times 25)$.

V_E = Ventilation [L min^{-1}, ambient temperature, ambient pressure, saturated (ATPS)], depending on activity and body size of subject (in the absence of inhaled CO_2).
$fMiCO_2$ = Multiplicative factor for CO_2-induced hyper-ventilation.
F_{IO} = Fraction of an incapacitating dose of low oxygen hypoxia.

For a simple analysis, the direct asphyxiant effects of NO*x* and those of NO*x* on HCN asphyxia may be ignored without significant error.

Both the total volume of air breathed each minute and the efficiency of uptake have upper limits so that a limiting value for $V_E \times fMiCO_2$ of 70 L min^{-1} is recommended.

For each of these gases it is necessary to obtain an expression for the fraction of a dose required to cause incapacitation. The derivation of these expressions has been detailed in Purser.[27,28] Basically, the exposure dose acquired over any period of time during a fire is expressed as a fraction of the dose required to cause incapacitation for each asphyxiant component. These are then summed and corrected for V_E and $fMiCO_2$ to provide an overall F_{IN} for each time period. The fractions are then summed for each successive time period. The time to incapacitation is then the time during the exposure at which the summed F_{IN} exceeds unity. Once incapacitation occurs, the fractional dose continues to increase if exposure continues until a lethal dose has been inhaled. The expressions for the different components of eqn (14.7) are presented in the following sections.

14.6.2 Carbon Monoxide

For the effects of CO, the FED is expressed in terms of %COHb. The denominator is the %COHb predicted to cause loss of consciousness in an active (escaping) person, which is 30% COHb (or 40% COHb for a resting person). The numerator is the increase in %COHb in the subject during each short period of exposure calculated from the inhaled CO concentration in the fire, the exposure time and the volume of air breathed each minute. For a short, high concentration exposure of an adult human the Stewart equation[29] may be used so that:

$$F_{Ico} = (3.317 \times 10^{-5}[CO]^{1.036} V_E t)/D, \quad (14.8)$$

where [CO] is the carbon monoxide concentration (ppm v/v, 20 °C), V_E is the volume of air breathed per minute [L min^{-1} ATPS; for application in eqn (14.7) the V_E term is removed from eqn (14.8)], t is the exposure time in minutes and D is the exposure dose (%COHb) for incapacitation.

The following values may be taken for V_E and D:

Activity level of subject	V_E/L min^{-1}	D/%COHb
Resting or sleeping	8.5	40
Light work: walking to escape	25	30
Heavy work: slow running, walking up stairs	50	20

This expression [the Stewart equation, eqn (14.8)] was obtained from young adult male human volunteers. It is suitable for adults in situations where the CO concentration is high in relation to the blood COHb concentration (as for most flaming fires and short exposure durations), since under these conditions the rate of uptake of CO is almost linear (see Chapters 7 and 9). Where long exposures may lead to near equilibrium conditions, the Coburn Forster Kane (CFK) equation[30,31] should be used since significant deviations from the Stewart equation may then occur. Also, due to limitations on the diffusivity of CO from the lungs into the blood (DL_{CO}), the efficiency of uptake of CO decreases at high V_E levels (*i.e.*, the rate of uptake becomes limited by the maximum rate of absorption into the blood). Figure 14.6 shows a comparison between the calculated rates of increase in %COHb for a 70 kg male subject inhaling CO at 1000 and 4000 ppm at three V_E levels representing sedentary, light activity and fast walking using the Stewart and CFK equations. For 1000 ppm CO, the rates of uptake are similar for both calculation methods up to around 25% COHb, but the CFK values then start to deviate from the linear Stewart model as the blood

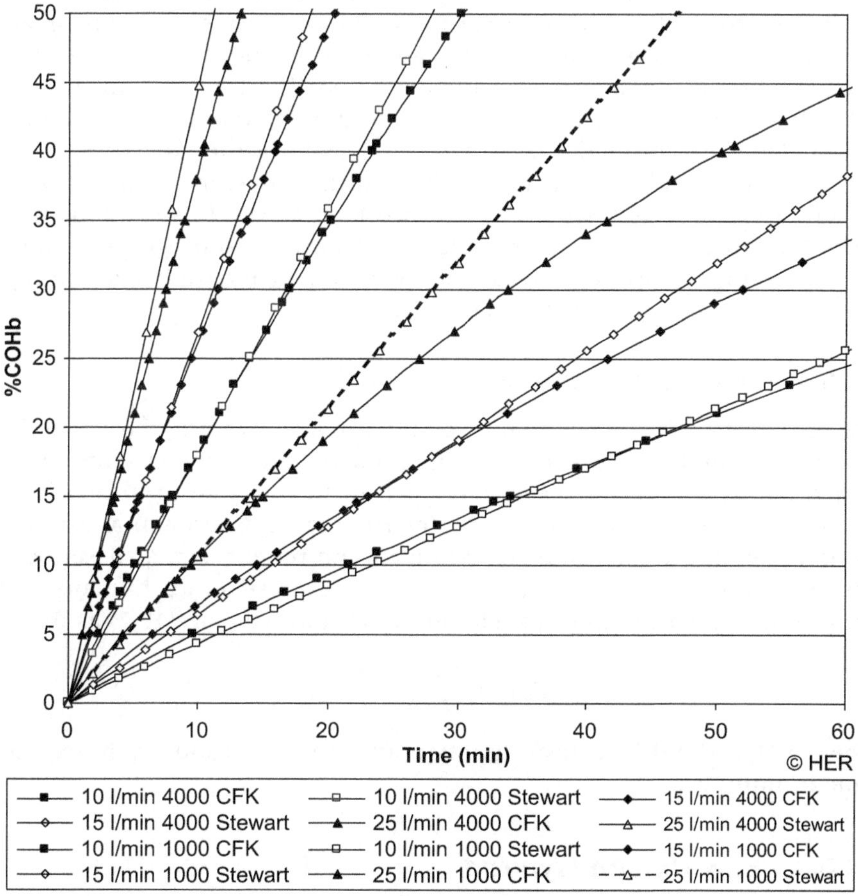

Figure 14.6 Comparison of calculated increases in %COHb for a 70 kg adult make inhaling 1000 or 4000 ppm CO at three V_E levels (10, 15 and 25 L min^{-1}) using the Stewart and CFK equations.

concentration tends to equilibrium, especially at the higher V_E levels. For 4000 ppm CO, a concentration more typical of a serious exposure during a fire, the rates of uptake are very similar and almost linear for all three V_E levels up to around 30–40% COHb, at which dose, loss of consciousness is predicted.

The Stewart equation somewhat underestimates uptake rates for children. Where more precise CO uptake calculations are required, differences in body size (including children) can be accommodated using the CFK equation. For a basic design case, use of the Stewart equation is recommended. Differences in body size and other susceptibilities can be considered as allowed for in the overall safety margin chosen to accommodate more susceptible sub-populations (for example, an FED of 0.3 or other endpoint).

As an alternative to using this expression, the FED_{ICO} may be expressed as a CO exposure dose ratio. This method is used for ISO 13571.[11] For this method FED_{ICO} is expressed as CO ppm × (t/35 000). This is approximately equivalent to a "light work" case for a subject breathing approximately 20 L min^{-1}. The disadvantage of this approach is that it does not allow for consideration of the effects of activity level on V_E and is not linked to the %COHb, which is a commonly used indicator of CO dose and shows a reasonably good correlation with clinical signs for acute exposures.

Since occupants must at least walk in order to escape from a fire, the default case suggested is that for light work. However, this could be varied according to the case. For example, a sleeping person escaping from a basement might start by being at rest, then awaken and walk to a stair (light work) and then climb the stair (heavy work).

14.6.3 Hydrogen Cyanide

For the effects of hydrogen cyanide and low oxygen hypoxia, the expressions are more complex, because the denominators are not constants. An exponential expression has been developed to fit the time to incapacitation *versus* exposure concentration curve for HCN from experimental exposures in non-human primates and some human and rodent data as described in Chapter 7 and Chapter 10. The derived FED expression for humans is the reciprocal of the time to incapacitation in Chapter 10 eqn (10.2) as follows:

$$F_{ICN} = \frac{[CN]^{2.36} \times V_E}{2.43 \times 10^7} t \tag{14.9}$$

where [CN] is the HCN concentration (ppm v/v at 20 °C) and t is the exposure time in minutes.

14.6.4 Low Oxygen Hypoxia

An expression for time to incapacitation by low oxygen hypoxia has been derived from human experiments on time of useful consciousness measured at different simulated altitudes in decompression chambers.[32] For this it is

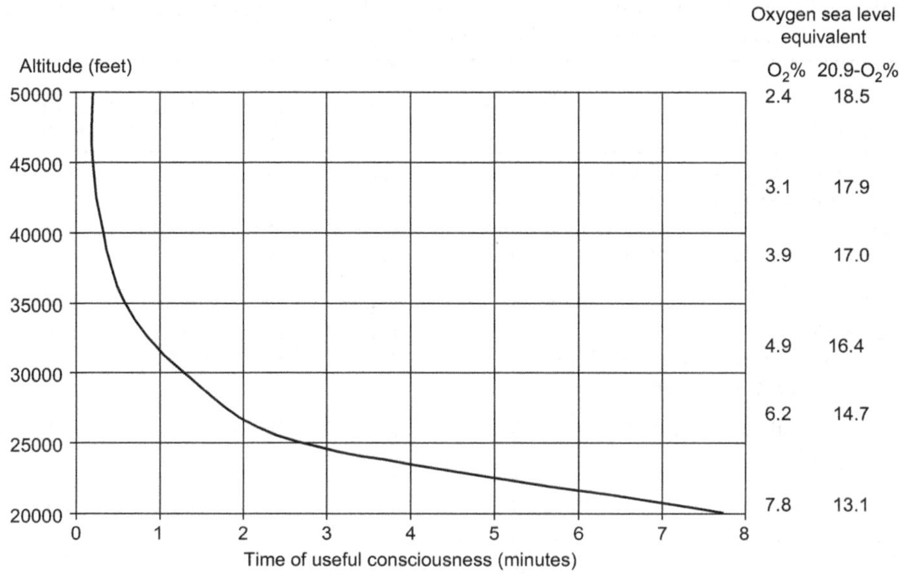

Figure 14.7 Time of useful consciousness for resting subjects exposed to rapid decompression to different simulated altitudes and sea level equivalent oxygen concentration (data from Luft[32]).

estimated that the effects of equivalent sea level exposure at lowered oxygen concentrations are similar (Figure 14.7).

As with the effects of cyanide, this is an exponential relationship, and a fractional incapacitating dose expression for hypoxia (F_{IO}) derived from this curve is as follows:

$$F_{IO} = t/(\exp[8.13 - 0.54(20.9 - [\%\,O_2])]), \tag{14.10}$$

where $[\%\ O_2]$ is the oxygen concentration (% v/v at 20 °C), t is the exposure time in minutes and F_{IO} is the exposure dose for incapacitation.

14.6.5 Effect of Ventilation at Different Activity Levels and Inhaled Carbon Dioxide

Having derived expressions for the main asphyxiant gases it is necessary to consider the effects of ventilation (V_E) at different activity levels and the further stimulatory effects on ventilation of high inhaled CO_2 concentrations on their rate of uptake. For the effects of activity levels, at least between sleeping and light work it is estimated that the rate of uptake of CO, HCN and other atmosphere components is a direct function of V_E as used for the uptake of CO in the empirically derived Stewart equation. Suggested values have been presented for resting, light and heavy work, and Table 14.5 shows a more comprehensive data set relating levels of activity, work, V_{O_2} and V_E. However, it is recognised that, especially at higher V_E levels, the rate of

Table 14.5 Work rate, oxygen consumption and minute ventilation for different activities (for ~70 kg bodyweight).

Activity	Speed/ km h^{-1}	Work/kcal min^{-1}	Work/ Watts	V_{O_2} L min^{-1} STPD[a]	V_E L min^{-1} ATPS[b]
Sleeping and lying, includes turning over, getting up	0	1.1	77	0.22	4.9
Resting	0	1.26	88	0.252	5.6
Sitting, includes reading, eating, desk work	0	1.5	105	0.3	6.7
Sitting	0	1.9	133	0.38	8.5
Standing	0	1.875	131	0.375	8.4
Standing, includes activities like moving between rooms	0	2.5	174	0.5	11.2
Walking, outdoors, and other activities	3	3	209	0.6	13.4
Walking	4	3.8	265	0.76	17.0
Walking	5	4.45	311	0.89	19.9
Light industry	5	5	349	1	22.3
Walking	6	5.15	359	1.03	23.0
Walking	6.4	5.6	391	1.12	25.0
Walking	7	5.8	405	1.16	25.9
Manual labour	7.24	8	558	1.6	35.7
Running	5.6	11.2	782	2.24	49.9
Running, climbing stairs, heavy manual work	8.9	11	768	2.2	49.0

[a]STPD = standard temperature (0 °C) and pressure (760 mmHg), dry.
[b]ATPS = ambient temperature and pressure, saturated.

uptake, especially of CO, is not a simple function of V_E, but increases by a somewhat shallower curve due to absorption inefficiencies (for example, related to the diffusion limit DL_{CO}). This has been taken into account for the deviation of the multiplicative term $fMiCO_2$, which is used to estimate the effect of inhaled CO_2 on the rate of uptake of other gases.[28]

Figure 14.8 shows data from several sources for human subjects and a fitted curve for the relationship between inhaled CO_2 and ventilation and the incapacitating effects of 5–10% CO_2. Data for these curves were pooled from several sources.[33–36] Also shown is a somewhat shallower curve for the effect of increased ventilation on the rate of uptake of CO calculated using the CFK equation.

This data has been used to derive the $fMiCO_2$ term for input into eqn (14.11) as follows:

Ventilatory stimulation by CO_2:

$$fMiCO_2 = \exp([CO_2]/5), \tag{14.11}$$

where $[CO_2]$ is the carbon dioxide concentration (% v/v at 20 °C).

Also shown in Figure 14.8 is a plot of time to incapacitation caused by inhalation of CO_2 itself at concentrations exceeding approximately 5% v/v.

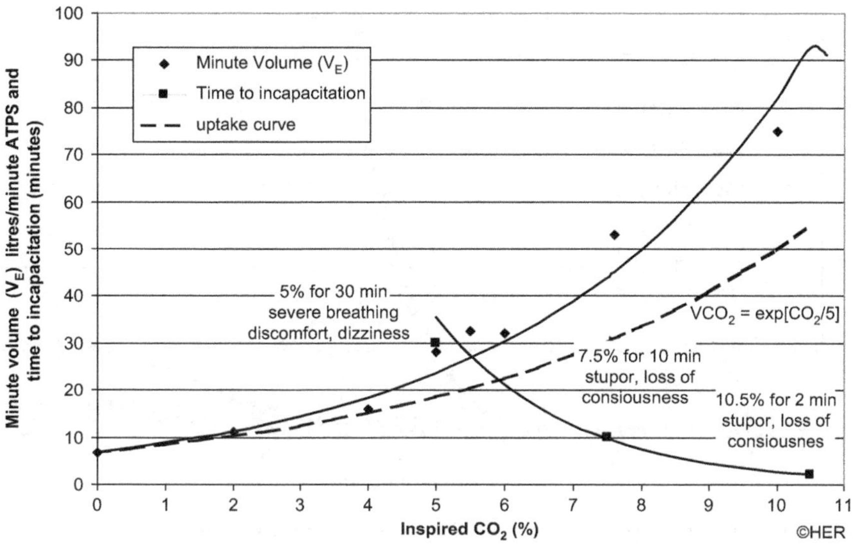

Figure 14.8 Ventilatory response to inhaled carbon dioxide, proposed curve for calculating effects on rate of uptake of CO and HCN, and incapacitating effects of inhaling 5–10% CO_2.

Fitting an expression to this curve provides an FED expression for incapacitation by CO_2:

$$F_{ICO_2} = t/(\exp[6.1623 - 0.5189 \times \% CO_2]) \quad (14.12)$$

14.6.6 Contribution of Lung Irritants to Asphyxia

In order to estimate a contribution to asphyxia from the inhaled dose of irritants the FLD_{irr} term is calculated using eqn (14.6). For the application in eqn (14.7), the term FLD_{irra} is given by eqn (14.13):

$$FLD_{irra} = FLD_{irr}/25 \quad (14.13)$$

14.6.6.1 Contribution from each Asphyxiant Component in a Typical House Fire

Figure 14.9 illustrates an application of the FED calculations for each asphyxiant component to the overall dose and time to predicted incapacitation for the typical furniture house fire time-concentration curves shown in Figure 14.2. The figure shows the plotted solution to eqn (14.7) (F_{IN}), predicting loss of consciousness after 5.5 minutes. For this example it was assumed that the exposed adult subject was at rest in the lounge and became aware of the fire after 2 minutes. The subject then stood up and moved around the room in an agitated state with a V_E of 15 L min^{-1} until he became unconscious from the combined effects of the asphyxiants when $F_{IN} = 1$, after which V_E reduced to 5 L min^{-1}.

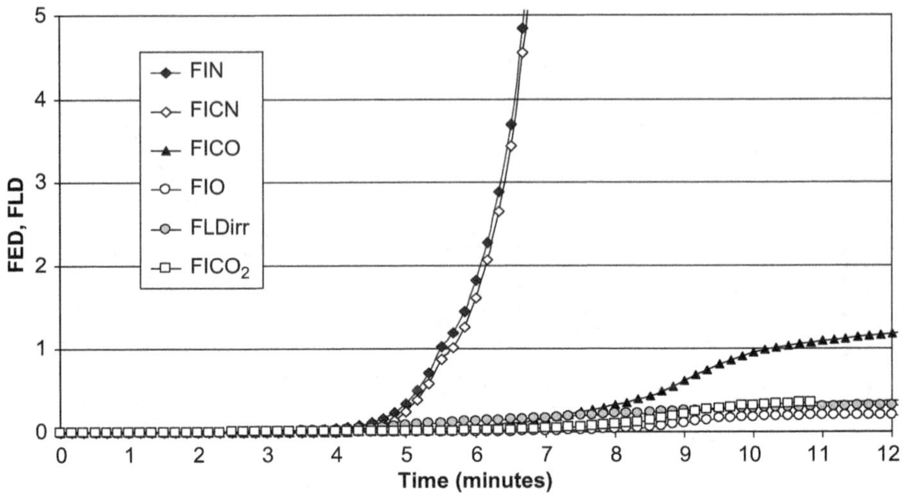

Figure 14.9 Contribution from each asphyxiant component to overall F_{IN} for incapacitation from asphyxia.

The accumulating fractional doses of carbon monoxide, hydrogen cyanide and irritants have been adjusted to reflect the effect on uptake of the subject's total ventilation depending on activity level and inhaled CO_2-induced hyperventilation. The results confirm that the onset of incapacitation is due almost entirely to the uptake of HCN. Once the subject becomes unconscious the rate of uptake of CO is reduced due to the reduced ventilation, but an incapacitating dose of CO at head height is achieved after approximately 11 minutes (representing 30% COHb). In practice, the CO concentration near floor level is somewhat lower, but this has not been adjusted for in this analysis. The contributions to overall hypoxia from the lowered oxygen concentration in the fire (which reached a brief minimum of 9.8% O_2) is small at an FED of 0.2, as is the estimated contribution from the acute effects of inhaled irritants on respiration. The inhaled concentration of CO_2 peaked briefly at 8.4%, but this was not sustained for long enough to produce an incapacitating inhaled exposure dose (F_{ICO2} maximum 0.37). In other experimental fires involving a smoke plume flowing down and open corridor, upper layer oxygen concentrations as low as 1.5% and CO_2 concentrations above 17% have been measured. It is considered that incapacitation would be almost instantaneous if a subject were to put their head into such a layer and inhale a breath.

14.7 Tenability Limits and Hazard Calculations for Radiant and Convective Heat

Similar approaches to those used to calculate the effects of asphyxiant gases have been used to derive FED expressions for effects of radiant and convective heat (see ref. 1).

There are three basic ways in which exposure to heat may lead to incapacitation: through heat stroke (hyperthermia), skin pain and burns, or respiratory tract burns. Thermal burns to the respiratory tract from air containing less than 10 % by volume water vapour do not occur in the absence of burns to facial skin. Therefore, tenability limits with regard to skin pain are lower than for thermal burns to the respiratory tract. Thermal burns to the respiratory tract may occur on inhalation of air above approximately 60 °C when saturated with water vapour, as may occur when water is used to control fires, and 60 °C is the highest temperature at which inhalation of saturated air can be tolerated . The derivation of the following expressions for time to pain and burns from radiant and convective heat is described in Purser.[1]

The tenability limit for exposure of skin to radiant heat is approximately 2.5 kW m^{-2}, below which exposure can be tolerated for at least several minutes. Radiant heat at this level and above causes skin pain followed by burns within a few seconds, but lower fluxes can be tolerated for more than 5 minutes. For situations where occupants are required to pass under a hot smoke layer in order to escape, this radiant flux corresponds approximately to a hot layer temperature of 200 °C. Above this threshold, time (in minutes) to different endpoints for effects of exposure to radiant heat t_{Irad} at a given radiant flux of q in kW m^{-2} is given by eqn (14.14).[37,38]

$$t_{\mathrm{Irad}} = \frac{r}{q^{1.33}}, \tag{14.14}$$

where r is the radiant heat exposure dose [(kW $m^{-2})^{4/3}$ min] required for any given endpoint (Table 14.6).

For convective heat (exposure in a hot air or smoke environment), tolerance time t_{tol} for pain is given by eqn (14.15):

$$t_{\mathrm{tol}} = 2\times10^{31}\times T^{-16.963} + 4\times10^{8}\times T^{-3.7561}, \tag{14.15}$$

where T is the ambient temperature (°C).

Time (in minutes) to serious injury or severe incapacitation is given by eqn (14.16):

$$t_{\mathrm{injury}} = 5\times10^{22}\times T^{-11.783} + 3\times10^{7}\times T^{-2.9636}. \tag{14.16}$$

Time (in minutes) to fatal exposure conditions with extensive third degree burns is then given by eqn (14.17):

$$t_{\mathrm{fatal}} = 2\times10^{18}\times T^{-9.0403} + 10^{8}\times T^{-3.10898} \tag{14.17}$$

Table 14.6 Radiant heat exposure doses for various radiant heat endpoints for exposed skin.

Radiant heat endpoint for exposed skin	r/(kW $m^{-2})^{4/3}$ min
Severe skin pain	1.33–1.67
Second degree burns	4.0–12.2
Third degree (full-thickness) burns	16.7

These empirical expressions are related to exposure to heated air with less than 10% water content by volume.

As with toxic gases, the body of a fire victim may be regarded as acquiring a "dose" of heat over a period of time during exposure, with short exposure to a high radiant flux or temperature being more incapacitating than a longer exposure to a lower temperature or flux. The same fractional incapacitating dose model as with the toxic gases may be applied and, providing that the temperature in the fire is stable or increasing, the fractional dose of heat acquired during exposure can be calculated by summing the radiant and convective fractions using eqn (14.18).

$$\mathrm{FED} = \int_{t_1}^{t_2} \left(\frac{1}{t_{\mathrm{Irad}}} + \frac{1}{t_{\mathrm{Iconv}}} \right) \Delta t. \tag{14.18}$$

Note: t_{Irad} will tend to 0 as q tends to <2.5 kW m^{-2}.

Thermal tolerance data for unprotected skin of humans suggest a limit of about 120 °C for convective heat, above which considerable pain is quickly incurred along with the production of burns within a few minutes. Depending on the length of exposure, convective heat below this temperature may still result in incapacitation due to hyperthermia, especially at high humidity.

14.8 Worked Example of a Fire FED Analysis

FEDs are calculated for successive short periods during the fire and then integrated with time in order to calculate the time when incapacitation is predicted (FED = 1). Flaming fires tend to grow rapidly at rates proportional to the square of the time from ignition expressed in seconds (see Chapter 2). This results in rapid rates of increase in asphyxiant gas concentrations so that variations in individual susceptibility and uncertainties in prediction of incapacitating doses tend to have relatively minor effects on predicted times to incapacitation.

Table 14.7 shows input data and calculation results for a life threat hazard analysis calculation to illustrate use of the methods described. The results of the analysis are presented in Figure 14.10. This figure shows plots of the FEC for smoke, FIC for sensory irritation and FED values for each hazard parameter as they increase with time. The endpoints of escape impairment or loss of tolerability (for smoke obscuration and irritants) and incapacitation (for heat and asphyxiant gases) are reached when the line for each parameter crosses 1 on the y-axis. Higher FECs and FEDs indicate more severe effects. For irritancy, incapacitation is predicted at $\mathrm{FEC_{irr}}$ values of approximately 5–10 and, for asphyxiation, death is predicted at $\mathrm{FED_{IN}}$ values of approximately 2–3. The data used for the worked example are from an actual large-scale room furniture fire, with some additional estimated data for hydrogen cyanide and organic irritants added to illustrate the calculation method.

Table 14.7 Illustration of life threat analysis for the first 6 minutes of a furniture fire based on a single armchair room burn. (The armchair is polystyrene with polyurethane cushions and covers. The room is 39 m^3 with an open doorway. CO, CO_2, O_2, smoke, temperature and radiant flux are as measured in the doorway at 2.1 m height. Other gases are estimates to illustrate the calculation method.)

Gas concentration each minute	1	2	3	4	5	6
Smoke (OD m^{-1})	0.1	0.2	0.5	1.5	3.0	3.5
HCl (ppm)	10	50	150	200	250	200
Acrolein (ppm)	0.4	0.8	2.0	6.0	12.0	14.0
Formaldehyde (ppm)	0.6	1.2	3.0	9.0	18.0	21.0
CO (ppm)	0	0	500	2000	3500	6000
HCN (ppm)	0	0	50	150	250	300
CO_2 (%)	0	0	1.5	3.5	6.0	8.0
O_2 (%)	20.9	20.9	19.0	17.5	15.0	12.0
Temp. (°C)	20	65	125	220	405	405
Heat flux (kW cm^{-2})	0	0.1	0.4	1.0	2.5	2.5
Fractional smoke						
Concentration, FEC_{smoke}	0.50	**1.00**	2.50	7.50	15.00	17.50
Fractional irritant concentration						
FIC_{HCl}	0.05	0.25	0.75	1.00	1.25	1.00
$FIC_{acrolein}$	0.10	0.20	0.50	1.50	3.00	3.50
$FIC_{formaldehyde}$	0.10	0.20	0.50	1.50	3.00	3.50
ΣFIC	0.25	0.65	**1.75**	4.00	7.25	8.00
Fractional lethal dose (irritants)						
FLD_{HCl}	0.00	0.00	0.00	0.00	0.00	0.00
$FLD_{acrolein}$	0.00	0.00	0.00	0.00	0.00	0.00
$FLD_{formaldehyde}$	0.00	0.00	0.00	0.00	0.01	0.01
ΣFLD_{irr}	0.00	0.00	0.00	0.00	0.01	0.02
Fractional asphyxiant dose						
$F_{I_{CO}}$	0.00	0.00	0.02	0.07	0.13	0.23
$F_{I_{CN}}$	0.00	0.00	0.01	0.11	0.38	0.58
FLD_{irr}	0.00	0.00	0.00	0.00	0.01	0.02
$fMiCO_2$	1.00	1.00	1.35	2.01	3.32	4.95
$F_{I_{O_2}}$	0.00	0.00	0.00	0.00	0.01	0.04
F_{IN} (asphyxiants)	0.00	0.00	0.04	0.38	1.72	4.08
ΣFED_{IN}	0.00	0.00	0.04	0.42	**2.14**	6.23
Fractional heat doses						
$FED_{rad\ pain}$	0.00	0.00	0.00	0.00	2.54	2.54
$FED_{conv\ pain}$	0.00	0.02	0.19	1.57	15.55	15.55
$FED_{heat\ pain}$	0.00	0.02	0.19	1.57	18.10	18.10
$\Sigma FED_{heat\ pain}$	0.00	0.02	0.21	**1.78**	19.88	37.98
$FED_{rad\ burns\ full\ thickness}$	0.00	0.00	0.00	0.00	0.20	0.20
$FED_{conv\ burns\ full\ thickness}$	0.00	0.00	0.03	0.19	1.28	1.28
$FED_{heat\ burns\ full\ thickness}$	0.00	0.00	0.03	0.19	1.48	1.48
$\Sigma FED_{heat\ burns\ full\ thickness}$	0.00	0.00	0.03	0.22	**1.70**	3.18

Note: The endpoint, escape impairment (for smoke obscuration and irritancy), or incapacitation (for heat and asphyxiant gases) is reached when the FIC or FED value reaches 1.
Note: Limiting values are in bold. Lethal values are approximately 2–3 times incapacitating levels for dose-related parameters, and incapacitation 5–10 times the FIC.

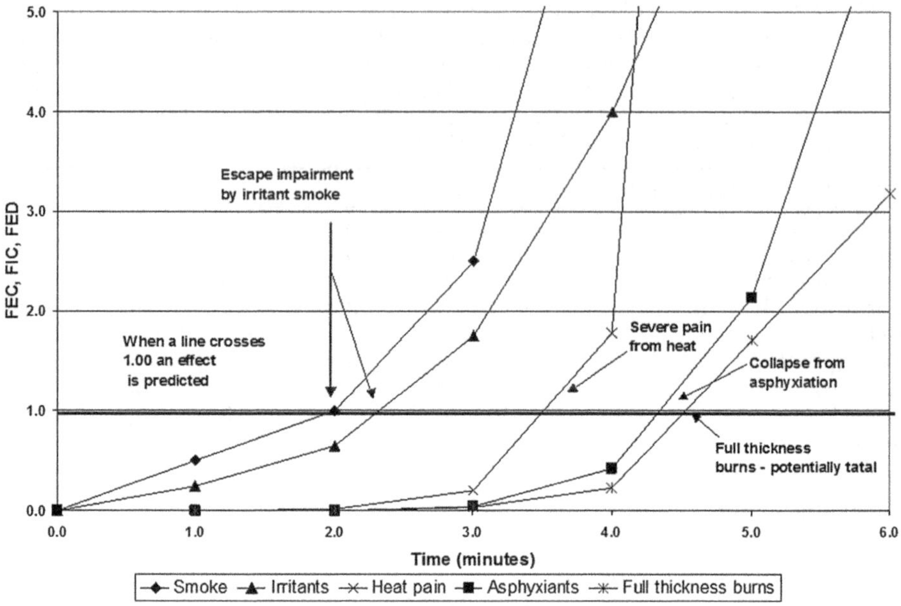

Figure 14.10 Plots of the FED analysis calculated in Table 14.7.

The analysis is designed to predict the severity of each hazard and the time during the fire at which it becomes significant. The toxic gas concentrations, smoke optical density, temperature and radiant heat flux have been averaged over each of the first 6 minutes in the smoke layer at head height in the furniture fire. The analysis shows that the smoke obscuration is the first hazard confronting a room occupant. The level of obscuration exceeds the tenability limit for irritant smoke in a small enclosure after 2 minutes, with an FEC of 1. The second hazard to confront the occupant is irritancy. This becomes significant during the third minute, reaching an FIC of 1 at 3 minutes. The tenability limit designed to protect vulnerable individuals (FIC = 0.3) is exceeded approximately 1 minute earlier. It is therefore predicted that after the second minute the level of obscuration and the irritancy of the smoke would be sufficient to impair and possibly prevent escape from the room due to difficulty in seeing and increasing pain in the eyes and respiratory tract. The effects of radiant and convective heat then become significant, crossing the tenability limit during the fourth minute and reaching an FEDheat value of 1.78, so that it is predicted that a room occupant would suffer severe skin pain due to the effects of convective heat.

During the fifth minute, the radiant flux reaches the tenability limit of 2.5 kW m^{-2}, so that skin pain would be predicted within seconds due to radiation alone were it not that the temperature had already exceeded the limiting exposure dose. Also during the fifth minute the FED_{IN} reaches 2.14, predicting that anyone breathing the smoke would lose consciousness due to asphyxia and might die after 6 minutes. Extensive, potentially fatal full-thickness burns are predicted during the fifth minute.

The level for exposure to asphyxiants considered to provide protection for the most vulnerable sub-populations ($FED_{IN} = 0.1$) is crossed at approximately 3.5 minutes, and that to allow for more normal sensitivity variations ($FED_{IN} = 0.3$) is crossed a few seconds later, only half a minute before the FED_{IN} reaches 1. The exposure dose of irritants is very small during the first 6 minutes of the fire, so that there should be little danger of post-exposure lung damage. An important point about all these parameters is that the FIC and FED curves are rising very steeply after the tenability threshold (FIC or FED of 1) is crossed. This means that if the true exposure concentrations or exposure doses required to cause incapacitation were somewhat higher or lower than the tenability limits chosen there would be little effect on predicted time to incapacitation.

The overall prediction is that for this fire, escape would become difficult during the third minute and incapacitation could occur due to the effects of irritant smoke. A person remaining in the room after this time would suffer severe pain after 4 minutes and full thickness burns, which would probably be lethal a minute or so later. They would lose consciousness during the fifth minute. In this analysis it is assumed that the head of a room occupant would be in the smoke at all times. In practice, if the room doorway were open, the hot, effluent-rich layer would descend from the ceiling to a level probably between 1 and 1.2 m above the floor as the chair reached its peak burning rate. A more sophisticated analysis could allow for the possibility that a room occupant might be at, or move to, a lower level in the room. If the height of the smoke layer with time is measured, then it is possible to allow for this factor in the calculation.

14.9 Application and Validation of FED Methods in Fire Incident Investigation

The FED calculation expressions and method described in the previous sections have been developed and partially validated using a combination of data from human and animal exposures. Full experimental validation for time to and severity of incapacitation effects of exposure to fire effluents cannot be carried out for human subjects, but some validation can be carried out for incidents where detailed information is available on the effects on exposed subjects, combined with full-scale reconstruction of the fire conditions.

The fractional effective dose methods described for calculating exposure doses and times to incapacitation and death from fire data have proved to be a useful tool for investigating fire incidents in which occupants have been injured or died. It has been applied usefully to investigations of several major incidents, including the Dupont Plaza fire, the Mont Blanc Tunnel fire[18] and the Rosepark Care Home fire.[39,40] Where data have been obtained on the fire conditions (time–concentration curves for toxic products and heat in these fires), either from full-scale fire reconstruction experiments of from fire modelling computations, it is possible to calculate the predicted effects on occupants and their timing. Beyond this, when forensic data are available from the victims (primarily blood carboxyhaemoglobin and cyanide concentrations, extent of smoke deposition in the lungs and extent of burns), it has been possible to use FED calculations to validate, at least in part, the fire test or fire modelling calculations.

A difficulty with incident re-creation tests and fire models is that, due to the variability of fires, it can be difficult to establish exactly how close the results are to the conditions in the actual incident. However, from the time–concentration curves of CO in a fire, it is possible to calculate the uptake by exposed subjects and hence the %COHb in the blood at the time of death or rescue (if these times are known). For subjects dying in the fire, the calculated %COHb level can be compared directly with the actual levels measured at post-mortem. For fire survivors, if the post-fire treatment (in terms of air exposure or oxygen therapy) is known, and the time at which a blood sample was taken in hospital is known, then it is possible to back-calculate from the CO wash-out curve to find the %COHb at the time of rescue, which can be compared with the forward-calculated value.[28] If there is reasonably good agreement between the forward- and back-calculated values then this provides a good indication that the combustion conditions and history of the test or modelled fire were close to those occurring during the actual incident. In practice, the CO washout half-life depends on the partial pressure of oxygen achieved and the subject's respiration and body size. Oxygen partial pressures achieved using a face-mask may be somewhat lower than those achieved following intubation (see Chapter 9). For resting adults using a face-mask a typical CO half-life is approximately 60 minutes $\pm$ 10 minutes.[28] Similar assessments can be made for blood cyanide concentrations, burns and smoke deposition. In practice, the FED and forensic approaches tend to be complementary with the fire testing and modelling approaches, in that while some uncertainties often apply to both methods, when used together the level of confidence in the findings can be considerably enhanced.

One example of the use of these methods was for the Mont Blanc Tunnel fire inquiry.[18] During this fire, it was possible to establish the time at which some vehicle occupants had stopped in the tunnel, had attempted to walk back towards the French portal and the locations at which they had collapsed and died. From computational fluid dynamics (CFD) modelling calculations of the vehicle fire, validated against smoke movement data recorded in the tunnel, it was possible to calculate the smoke, heat and toxic gas concentrations in the tunnel. Applying the FED modelling methods described, it was then possible to calculate how long occupants would have been able to walk along the tunnel, how fast they would move in the smoke and when (and especially at what location) they would have been overcome by heat and toxic gases. The results of these calculations showed a good agreement between the actual time and location of collapse obtained from incident data and the times and locations of collapse calculated according to CFD and FED modelling. CFD and FED modelling were therefore validated and the combined approach provided a good understanding of the conditions during the fire and effects on the tunnel occupants.

Another example is application to the Rosepark Nursing Home fire.[39,40] In this incident a brief intense fire involving mixed fuels and some exploding aerosol cans occurred at night in a cupboard opening onto a corridor. Ten elderly sleeping occupants of open rooms off this corridor died at the scene, while two who had their room doors closed were rescued after approximately an hour, but were overcome by toxic smoke exposure and subsequently died in hospital. During the fire, a fire door between the main fire corridor and the next compartment was blown open by the pressure pulses as the aerosols exploded so that a considerable quantity of smoke passed through, resulting in five occupants of open or partly open rooms off this corridor being overcome so that two died after rescue, while three survived. A full-scale reconstruction of the affected part of this building and the fire incident was carried out by the Building Research Establishment (BRE) on behalf of the Scottish Office and the Procurator Fiscal. FED calculations were applied to the fire test data to calculate timing and effects on the occupants, assuming the conditions in the reconstruction were similar to those during the actual incident. Forensic data, principally the %COHb concentrations and extent of burns in the decedents and those rescued from the fire, were also examined. Figure 14.11 shows the time–concentration curves of toxic gases and heat in the fire corridor, which were very similar to those in the open bedrooms off this corridor. Figure 14.12 shows the FED hazard analysis for a standing corridor occupant.

The FED calculations show dense smoke from 4 minutes after ignition, with collapse from the effects of asphyxiant gases predicted after 5.5 minutes due mainly to the effects of HCN and CO, with a lethal (50% COHb) level of

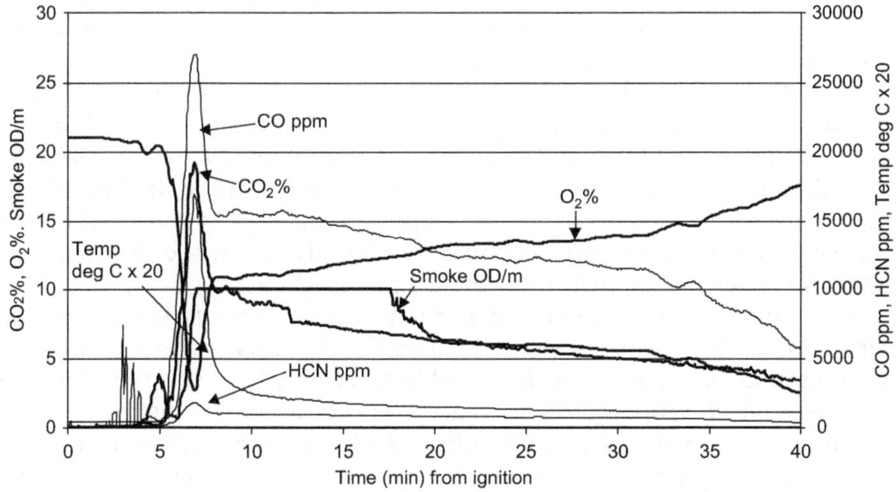

Figure 14.11 Smoke, toxic gases and temperature profile in the fire corridor during the BRE full-scale recreation of the Rosepark Nursing Home fire. Conditions in the open bedrooms off this corridor were similar, except that the temperature at bed height was considerably lower.

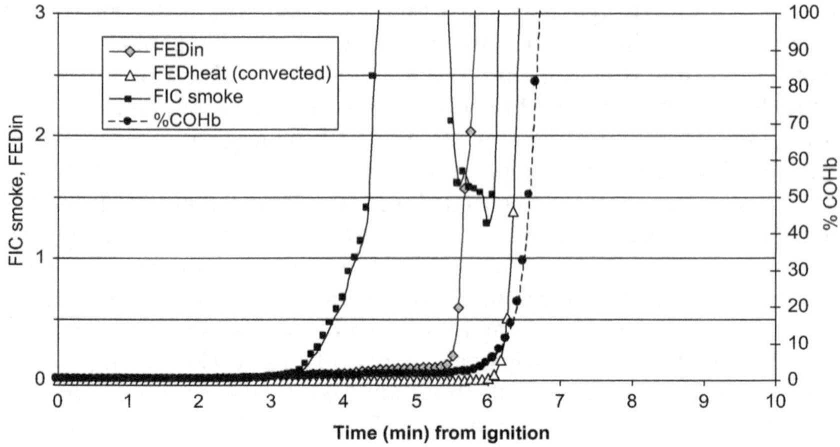

Figure 14.12 Fractional effective dose hazard analysis for the corridor fire profile shown in Figure 14.11.

CO predicted by 6.5 minutes. Severe pain from convective heat exposure is predicted after 6 minutes in the corridor. The smoke and toxic gas conditions at bed height in the open rooms were similar to those in the corridor, but temperatures at bed height were much lower. It was therefore predicted that all open room occupants would have been overcome and died within a few minutes due to the effects of asphyxiant gases, with lethal %COHb blood levels but no burns, and this was the finding from the forensic data, with all room occupants dead at the fire scene when the fire service entered the area. These findings and the pattern of fire damage supported that the reconstructed fire conditions were similar to those in the incident, but since the exact times of death for the room occupants were unknown, these findings could not fully validate the fire conditions.

Figure 14.13 shows the conditions in one of the closed rooms off the fire corridor from which the uptake of CO (%COHb) could be calculated for the room occupant. The time of rescue for this occupant is known, as well as her subsequent treatment and the %COHb concentration in her blood, measured soon after her arrival at hospital. Figure 14.14 shows the FED analysis in terms of the forward-calculated %COHb from the fire data and the back-calculated %COHb from the actual blood data.

As the figures show, there are inevitably some uncertainties in these calculations. The forward-calculated CO uptake depends on the bedroom CO concentration curve, but also on the subject's activity level and therefore the V_E of the subject. From the reported activities of this subject, her average V_E is estimated at between 6 and 8 L min^{-1}. Also, although the exact time she arrived at the hospital and the fact that she received oxygen by face-mask after rescue are known, the exact time at which her blood sample was taken was not recorded, so three estimates were made. Despite these uncertainties, there is a good agreement between the range of forward- and back-calculated values, which therefore validates the test fire conditions in terms of CO

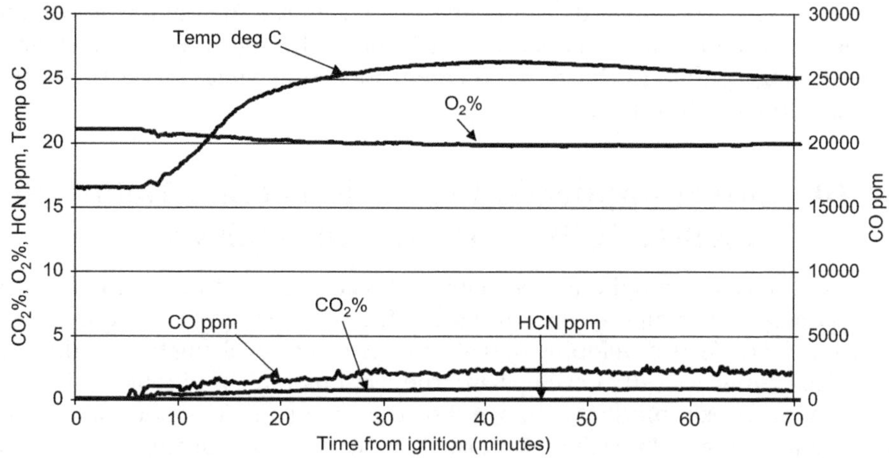

Figure 14.13 Smoke, toxic gases and temperature profile for a closed room off the fire corridor during the BRE full-scale recreation of the Rosepark Nursing Home fire.

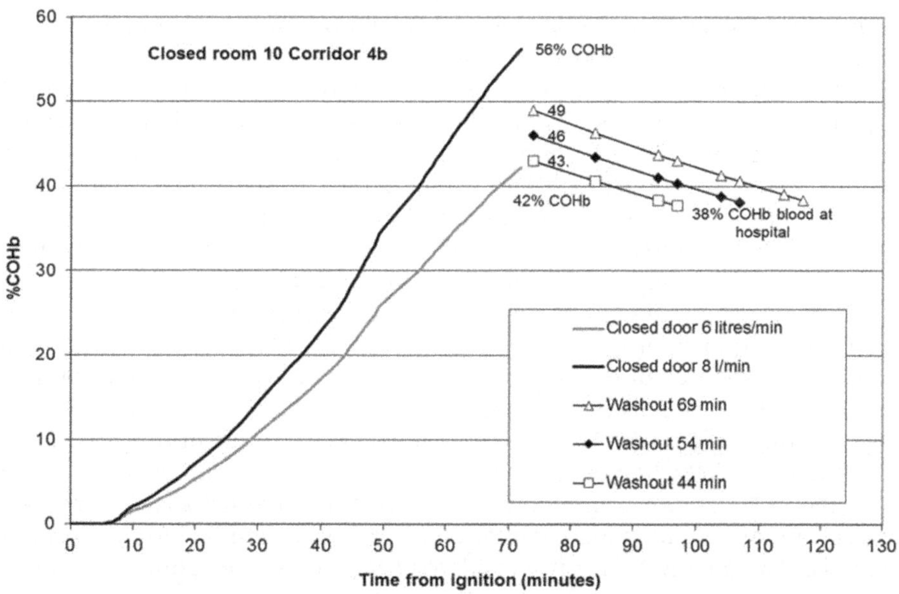

Figure 14.14 Forward-calculated %COHb level for the occupant of a closed room from the CO and CO_2 gas profiles in Figure 14.13, assuming sleeping and standing V_E values of 6 and 8 L min^{-1}, respectively, compared with %COHb levels back-calculated to the time of rescue using three estimates of the time when the blood sample was taken after arrival at hospital.

concentrations against those in the actual incident. For the occupants of rooms off the corridor beyond the fire door, a similar set of calculations showed consistently higher actual %COHb blood levels in the rescued occupants than the forward-calculated values from the fire data, from which it was concluded that the penetration of smoke through the fire door (and other leakage paths) in the actual incident was somewhat greater than that in the reconstruction test.

14.10 Direct Application of Toxic Potency Data from Small-scale Tests to Hazard Analysis

As demonstrated throughout this chapter, it is considered that toxic hazard in fires is a system property of a full-scale fire, consisting of a sequence of physiological effects developing over a time scale and involving fire conditions that also change as a fire develops and the combustion conditions, fuel mass loss rate and toxic products change. For these reasons it is not considered realistic to measure "toxicity" or "toxic hazard" for any burning material in any test, other than one which replicates the full-scale fire scenario. Despite this, small-scale tests have been used to measure the "toxicity" of materials for product specification purposes, although the test conditions have not been related to any form of full-scale fire scenario (see Chapters 2, 3, 5 and 7).[1]

One way in which small-scale test data could be used for a simple form of toxic hazard analysis would be to calculate the overall toxic potency of the mixed combustion products from a test material burned under specified combustion conditions (as described in Chapters 2, 3 and 7). These data could be used to calculate the 30 minute median lethal concentration (LC_{50}) or the LCt_{50} lethal exposure dose (LED) for a material (expressed as g m^{-3} min). This represents the mass loss concentration of the test material calculated to be lethal to 50% of exposed rats. For application to a fire hazard analysis it is then necessary to estimate the mass burning rate of the fuel (kg s^{-1}) using appropriate fire engineering heat release rate curves and the volume into which the products are dispersed with time. This then provides mass loss concentration–time curves for the calculated fire, to which the toxicity data can be applied to calculate the time at which a lethal exposure occurred (representing an FED for lethality of 1). A worked example of this method is presented in ref. 1. Chapter 7, Table 7.7 shows some examples of approximate LC_{50} and LED data for some common classes of polymeric fuels decomposed under different combustion conditions, which can be applied to this form of analysis. For this method it is considered that an incapacitating exposure is represented by approximately 0.3–0.5 times a lethal exposure.

14.11 Estimation of Outdoor Environmental Fire Hazards and Health Effects

14.11.1 Outdoor Hazards

The methods described in the previous sections are intended for assessment of life-threatening exposures to fire effluents for subjects exposed inside buildings or transport systems during fires. Another form of toxic hazard occurs when subjects are exposed to toxic products at concentrations several orders of magnitude lower than for indoor fires, but sometimes for longer periods and on a repeated basis. Such exposures can occur as a result of low level contamination of the indoor environment from poorly ventilated combustion devices, or in outside locations from exposure to dispersing smoke plumes. As with indoor hazards, the toxic hazards to subjects in outdoor locations depend on the concentrations of toxic substances and the exposure doses to which they are exposed. As described in Chapters 2, 3 and 6, the exposure depends upon a range of parameters, including the nature of the burning fuel, the combustion conditions, the proximity of the subject to the fire and the meteorological conditions.

Due to the high dilution factors of $\sim$10–1000 between exposures to fire plumes indoors and outdoors, the hazards are very different outdoors, in that the concentrations of the main life-threatening asphyxiant gases are almost always too low to be toxicologically significant over exposure time scales of hours. In contrast, many of the irritant acid gases and organic species have low odour and irritancy thresholds, so that the main acute hazards to exposed subjects are smoke odour and nuisance, mild eye and respiratory tract irritancy (sometimes resulting in a sore throat the day following exposure), and increased risk of the effects of general air pollution on vulnerable subjects (increased risk of acute cardio-respiratory effects or asthma symptoms). Chronic exposure over time scales of months to years may result in the range of health effects related to air pollution, including increased risk of systemic toxicity such as lung diseases, sensitisation and carcinogenicity. Also of concern (but beyond the scope of this chapter) is secondary oral ingestion from contaminated food and water following environmental deposition of substances such as dioxins and furans.

The potential for these effects can be assessed by calculating the concentrations and doses of toxic species to which subjects may be exposed compared with thresholds for odour and specified health effects from individual toxic vapours and particulates. The exposure concentrations can be estimated from the yields of different toxic species and the mass burning rates of the fuels involved under the estimated fire combustion conditions (see Chapter 3), coupled with appropriate plume dispersal modelling (see Chapter 6). A simpler approach is to determine the smoke density at the point of exposure (from the reported visibility through the smoke) and the relative yields of smoke particulates and species from the source fire. The reported smoke visibility conditions at different locations can then be used to estimate the particulate concentrations and the particulate/toxic product ratios can be used to estimate the likely exposure concentrations for each specific toxic substance. Health effects for exposures can then be assessed by reference to industrial hygiene data and exposure limit values for specific toxicants, or by reference to acute hazard guidelines such as AEGL concentrations.[41]

Large dilution factors occur from air entrainment in fire plumes as they flow away from the source and in heated, low density smoke plumes rising upwards above ground level. For this reason, the acute toxic hazards from most fires are generally minor, with the exception of some industrial fires involving bulk release of specific toxic chemicals. Significant acute hazards may occur for subjects without respiratory protection close to a large fire source if they are exposed to concentrated, down-washed smoke plumes (see Chapters 3, 4 and 6). Examples would be emergency personnel, such as fire officers immediately outside a large burning building during strong winds, or those fighting large wildfires (see Chapters 17, 18 and 19). Under such conditions the concentrations of CO and other toxic products can be sufficient to cause acute effects. For subjects more than approximately a few hundred metres away from most fire sources the main acute hazards are likely to be odour and minor respiratory effects from irritant vapours and particulates, and a small addition to the lifetime exposure to inhaled environmental toxins and carcinogens in air pollutants (most of which are derived from combustion sources; see Chapter 12). Table 14.8 lists some of the toxic species released in low yields in combustion products from mixed waste involving a range of common materials and their reported health effects. Further information is presented in Chapter 4.

14.11.2 Outdoor Hazards from Fires

In terms of the fire scenarios and hazards it is possible to classify outdoor hazards into several different categories:

- Smouldering/non-flaming fires involving large fuel masses inside buildings or in the open.

  The main characteristics of smouldering/non-flaming fires are that the rate of decomposition is slow but the main form of decomposition is oxidative pyrolysis, resulting in high yields of a wide range of irritant organic and inorganic products. The fire effluents tend to consist of somewhat grey smoke, which is highly irritant. The grey colour is indicative of the relatively low carbon soot content and the high content of condensed droplets of organic substances and other particulates. Because the rates of heat release and mass loss are low, the mass production rate of toxic products is low and smoke plumes from such fires tend to be "lazy", flowing slowly away from the source near ground level (depending on the wind conditions). This increases the risk of exposure of subjects standing within approximately a few tens of metres of the source if they remain in the area. Because the mass loss rate is slow the plumes tend to be rapidly diluted to low concentrations away from the immediate location by mixing with outside air.

  A simple example of this type of scenario is self-heating smouldering of a compost heap, or when a large amount of green material is placed on a garden bonfire and extinguishes the flames. The "lazy" grey smoke arising from such sources is extremely irritant to the eyes of a subject standing nearby if they become exposed to the smoke plume close to the source. The immediate, acute health hazards from sensory and

Table 14.8 Common materials likely to be found in mixed waste, their toxic combustion products and toxic effects.

Material and elemental composition	Toxic combustion products	Toxic effect
Cellulosic materials: mainly C, H and O	Carbon monoxide	Asphyxiant
Untreated wood, green waste, cotton and viscose fabrics	Carbon soot particles with adsorbed organics	Lung irritants
	Volatile organic compounds, including:	
	Acrolein	Potent irritant
	Formaldehyde	Potent irritant and carcinogen
	Acetaldehyde	Irritant carcinogen?
	Benzene	Carcinogen
	Styrene monomer	Irritant
	Phenol	Irritant
	Polyaromatic hydrocarbons (PAHs)	Carcinogen, environmental contaminant
	Low yields of dibenzodioxins and dibenzofurans	Environmental toxins, enzyme disruption, immune effects, carcinogen
	Respirable ash	Lung irritant
Treated wood: C, H and O plus metals and other additives, halogens copper, chromium, arsenic (CCA)[a]	Toxic products as for wood plus toxic metal salts	Systemic toxicity, carcinogenicity
Tributyl tin Pentachlorophenol Mainly used for timber used outside	High yields of dibenzodioxins and dibenzofurans, polychlorinated biphenyls (PCBs)	Endocrine disruption, immune system effects, carcinogenicity
Chipboard and veneered, melamine covered chipboard, medium density fibreboard: C, H, O and N	Toxic products as for wood plus:	
	Hydrogen cyanide	Asphyxiant
	Nitrogen oxides	Irritant
Widely used in furniture, fitted kitchens and buildings	Acrylonitrile	Irritant, carcinogen
	Formaldehyde	Irritant carcinogen
Plastics: C, H and O Aliphatic and aromatic solvents Polyethylene Polypropylene Polystyrene Polymethyl methacrylate Epoxy resins Household plastics and plastic bags, plastic kitchen items, building linings, plastic baths	Toxic products generally similar to wood Polystyrene: high yields of particulates and aromatics such as benzene and styrene	Irritant, carcinogen

Table 14.8 (*Continued*)

Material and elemental composition	Toxic combustion products	Toxic effect
Plastics: C, H, O and N plus chlorine, bromine, phosphorus Polyurethane foams in upholstered furniture	Toxic products generally similar to wood plus: Isocyanates Acrylonitrile and other nitriles Hydrogen chloride and bromide Hydrogen cyanide Phosphoric acid Dibenzodioxins, PCBs	 Irritant, sensitisation Irritant, carcinogen Irritant Asphyxiant Irritant As above
C, H, O and N Polyamide (nylon) used in clothing, carpets, kitchen plastic items	As for untreated wood plus hydrogen cyanide and acrylonitrile	Asphyxiant Irritant carcinogen
C, H and Cl Polyvinylchloride (PVC)	As for untreated wood plus High yields of hydrogen chloride	 Irritant
Rubber (tyres) C, H, O, S and other additives	As for untreated wood plus high yields of carbon soot particles and aromatic organic compounds, sulfur dioxide	Irritant

[a]CCA, chromated copper arsenate.

respiratory tract irritancy from such fires can therefore be high, especially if specific toxic chemicals are produced or released (depending upon the composition of the burning fuel), but the effects are generally local and self-limited.

An example of more serious exposure of large populations to such fires is the large area involved in subterranean peat fires in Indonesia, which have resulting in considerable air pollution affecting several countries simultaneously.[42]

- Smoke plume released from a restricted ventilation fire in a building.

  When a fire occurs inside a closed building it grows rapidly until the available ventilation limits the fire size. The fire may then keep burning as a ventilation-controlled fire. If the structure remains intact, then air enters through leakage paths at low levels and the positive pressure at higher levels in the building results in smoke effluent flowing out from under the roof and any small vents. The smoke leaking from such an intact building, formed by inefficient flaming combustion, is rich in smoke particulates and toxic combustion products, including high yields of CO, a wide range of organic aliphatic and aromatic substances [volatile organic compounds (VOCs), PCBs, dioxins and furans], polyaromatic hydrocarbons, isocyanates and acid gases (depending upon the fuel composition).

The emerging smoke plume is likely to be rapidly diluted and dispersed as it is released into the outside air, depending upon the meteorological conditions. The health hazard to people outside the building is therefore relatively low, except if exposed close to the building to relatively high concentration down-washed plumes.

- Smoke plume released from a large, fully developed ventilation-restricted fire.

  If the ventilation to the original fire is increased, for example, by doors or windows being opened, glazing failure or partial structural failure of the building, then the fire grows to a large size, with large, black, energetic (high temperature) smoke plumes flowing from open vents. The mass loss rate of the burning fuel, and therefore the mass release rate, of fire effluents is greatly increased, while the yields of toxic products from the ventilation-controlled combustion conditions remain high. Such fires therefore have the potential to release large amounts of toxic combustion products to the outside environment, but because the plume is hot and has a low density, the smoke plume tends to rise high above the source with considerable outside air entrainment, and is therefore rapidly diluted and carried away from the source area.

  Fires reaching this level of development are therefore capable of dispersing large amounts of toxic products over a wide area and, following air entrainment and dilution away from the source, may present minor health hazards, especially odour and irritancy, to subjects up to several kilometres or more from the source where the plume descends to ground level (depending on the extent of dilution and the exposure duration).

- Smoke plume released from a large, fully developed building fire.

  If a fire continues to burn for some time and to develop beyond the level described in the previous case, then a greater extent of structural failure and therefore increased ventilation can occur, leading to a fully developed post-flashover fire. Depending upon the fuel disposition and ventilation, such fires tend to remain fuel rich inside the building, but as the fire plume leaves the building and entrains more air, the high temperatures result in secondary combustion, so that large flames can be observed in the plumes flowing from the vents. This secondary flaming results in more efficient combustion, destroying much of the organic fuel products released from the fire and reducing the concentrations of toxic species. However, such plumes release very large quantities of black, carbon-rich particulate smoke and, due to flame extinction and to some smoke avoiding secondary combustion, the resulting plume can contain significant yields of organic products in addition to acid gases and other inorganic products (depending on the fuel composition). Such large fires also tend to release and entrain mineral dusts and fibres released from the structure of the building (as during the World Trade Center incident; see Chapter 19).

  Fires such as these have the potential to burn for periods of hours to days, releasing large masses of toxic fire effluents and other materials over a wide area. Due to the highly energetic nature of the fire and smoke plumes, these tend to rise vertically with considerable air entrainment and dispersal, being carried for large distances before cooling, mixing and delivering diluted products to ground level. As with all these different fire types or stages, the meteorological conditions have a large influence on the concentrations at ground level and the areas impacted (see Chapter 18). In general, the toxic hazard to exposed subjects is likely to be low (except where specific bulk chemicals are involved), but there is the potential for exposure to large populations to these low concentration effluents, which then merge with, and add to, the general level of background air pollution (which is itself mainly derived from combustion sources).

- Smoke plumes from outside and wildfires.

  Wildfires have the potential to present a significant health hazard to exposed subjects because they are likely to be very large and occur over very large areas. The hazard depends on the size of the fire and the proximity of subjects (for example, those close to and attempting to fight such fires or the occupants of neighbouring urban areas). Wildfire plumes tend to contain products from areas of smouldering/non-flaming decomposition mixed with those from both poorly ventilated and well ventilated flaming. The resultant mix contains significant yields of CO, VOCs, PAHs and dioxins. Due to the very large masses of fuel consumed and depending upon the conditions, exposure to such effluents may present a significant health hazard to exposed subjects (see Chapter 17).
- Large fuel conflagrations (such as Buncefield, Iraq, Seveso and Teeside).

  Very large conflagrations, although of obvious concern, may in practice prove less hazardous than might be expected. The Buncefield hydrocarbon fire is an example in which very large masses of petroleum products burned for many days (see Chapter 18). Because the fire was very large and energetic, the fire plume entrained very large masses of air and was carried to a high altitude. Also, due to the winter temperature inversion conditions, the plume was carried for many hundreds of miles before grounding. The combustion was relatively efficient and the hydrocarbon fuels were relatively low in precursors for highly toxic potency products. The overall situation was therefore of an additional source term to the enormous mass of combustion products evolved over UK and neighbouring European countries from all sources during the period of the fire. However, if the meteorological conditions had been such that the plume had grounded over London before becoming very widely diluted, it is arguably possible that acute health effects might have shown up in the form of reports of minor health effects (odour and irritancy) affecting large numbers of people and a spike in more serious acute cardio-respiratory hospital admissions, as occurs during episodes of high ambient air pollution. At the time of this fire the author lived and worked within approximately 5 miles of this very large conflagration and observed the large plume rising from the site and passing overhead for several days. It was notable that no irritancy or odour was detectable from this fire until a short period during which the fire was finally extinguished, when the residual, highly diluted, low energy plume dispersed across the surrounding area near ground level and a slight odour was detected.

Where the fuels involved in large conflagration include large quantities of chemical precursors to specific toxic substances with a high potency, as in incidents such as Severso and Bhopal, then of course the health consequences can be more serious.

14.11.3 Odours and Irritants in Outdoor Smoke Plumes

Many substance released in fire plumes have low detection thresholds for odour and irritancy. Odours and sensory irritation by airborne particles and vapours depend upon the concentration of individual substances in air. Odours stimulate the olfactory nerve in the nose, while sensory irritants stimulate mainly pain endings of the trigeminal nerve on the surface of the cornea and the linings of the nose, mouth and throat.[23]

14.11.3.1 Odours

Odours are not generally hazardous, but the effects of intense odours can include vomiting, behaviour changes, such as leaving an affected area, and changes in breathing patterns, all of which are concentration-dependent. The perceived intensity of an odour decreases with duration of exposure. Some substances stimulate both olfactory and irritant receptors in the nose.

The odours of many substances can be detected at very low concentrations in air (based on odour panel studies), but the sensitivity of individuals varies over a wide range of approximately 3–5 orders of magnitude and there can be considerable differences between estimates of different odour panels. Estimates of odour threshold concentrations are therefore somewhat variable. The odour thresholds for different substances also cover a very wide concentration range of approximately seven orders of magnitude. Table 14.9 shows some examples for a range of compounds, some of which are likely to be present in effluent atmospheres.[43] These are shown to one significant figure to reflect the inherent uncertainty of the data. The uncertainty in these estimates is reflected in some of the differences in the figures shown in Table 14.9 and some of those in Table 14.10, which were obtained from different data sources.

The data set shows that a number of aliphatic fatty acids, esters, aldehydes and phenols have very low odour thresholds in the 0.005–0.05 ppm range. Other aldehydes, ketones and aliphatic hydrocarbons have thresholds in the 0.05–0.5 ppm range. It is possible that some of these compounds could be present in combustion products at yields in the 0.01–1% mass range. Based upon the general composition of thermal decomposition product effluents from common materials it is considered that vapours and thermal

Table 14.9 Reported odour threshold concentrations from National Institute of Occupational Health Denmark database.[a]

Substance	Concentration/ppm	Concentration/mg m^{-3}
Methyl mercaptan	0.0001	0.005
Diphenyl ether	0.0003	0.002
2-Methyl phenol	0.0005	0.003
cis-2-Octenal	0.0007	0.004
7-Octene	0.001	0.005
Octanal	0.001	0.007
Nonanoic acid	0.002	0.01
2-Methyl butyl acetate	0.002	0.01
Butanoic acid	0.004	0.01
Octanoic acid	0.004	0.02
Propanal	0.005	0.01
Butanal	0.09	0.03
Butyl acetate	0.07	0.04
Hexanoic acid	0.01	0.06
Acetic acid	0.02	0.04
1-Butanol	0.03	0.1
Propanoic acid	0.04	0.1
Styrene	0.03	0.2
Benzaldehyde	0.04	0.2
1-Octene	0.07	0.3
Pyridene	0.09	0.3
Methyl methacrylate	0.09	0.4
Phenol	0.1	0.4
Crotonaldehyde	0.1	0.4
Ethanol	0.1	0.3
Toluene	0.2	0.6
Acrolein	0.2	0.4
Acetaldehyde	0.2	0.4
2-Propanol	0.4	1
Decane	0.7	4
Formaldehyde	0.9	1
Nonane	1	7
2-Pentanone	2	6
Diethyl Ether	2	7
Octane	6	30
Acetone	5	10
2-Butene	5	10
Heptane	10	40
Benzene	9	30
Hexane	20	80
Butane	200	500
Ethane	10 000	100 000

[a]Jensen *et al.* (1996) VOC database, National Institute of Occupational Health Denmark.

decomposition product mixtures derived from mixed fuels may have odour thresholds in the 1–10 mg m^{-3} concentration range for many individuals. Depending upon the exact product mix and the sensitivity of individual subjects, it may be possible for some subjects to detect the presence of such contaminants at concentrations as low as 0.1 mg m^{-3}.

Table 14.10 Concentrations and exposure doses of common smoke irritants occurring in combustion products causing adverse health effects following acute exposures.

Irritant	Approximate odour threshold/ppm	Occupational exposure limit (OEL)[a]/ppm	Severe sensory irritation in humans/ppm	Approximate acute lethal exposure dose in rodents/ppm min
Toluene diisocyanate	—	=0.1	1	3000
Acrolein	0.1	0.1	4	4000
Formaldehyde	<1	2	8	22 500
Crotonaldehyde	—	2	25	24 000
Acrylonitrile	20	2	>20	129 000
Phenol	0.05	5	>50	165 000
Sulfur dioxide	0.5	2	30	420 000
Ammonia	5	25	1000	141 000
Hydrogen fluoride	<5	3	120	87 000
Hydrogen chloride	<5	5	100	114 000
Hydrogen bromide	<5	3	100	114 000
Nitrogen dioxide	0.5	3	25	5100
Styrene monomer	0.1	100	>700	1 350 000
Acetaldehyde	0.2	100	>1500	2 220 000
Ethanol	10	1000	>5000	12 000 000
Acetone	100	1000	>12 000	56 700 000

[a]These occupational limits were often originally set on the basis mainly of irritancy; some have subsequently been revised for a number of reasons related to chronic toxicity or carcinogenicity risk.

That odour from environmental releases can be detected and affect subjects at great distances from a source is illustrated by the release of methyl mercaptan from the Lubrizol Corporation plant in Rouen on 21st January 2013. This substance is used as a stanching agent and, as listed in Table 14.9, has a very low odour threshold [9.9×10^{-4} ppm (standard deviation 7.2×10^{-4} ppm)].[44] During the night of Monday 21st and throughout Tuesday 22nd January 2013 the diluted plume from this release spread over northern France and southern England, affecting large numbers of people and was reported as far from the source as Northamptonshire and Oxfordshire. The author, who lives in Hatfield, Hertfordshire, experienced a trace of a burning odour and eye irritation indoors at the time and was sufficiently concerned to check the house during the night for sources of smouldering before reading about the release the next day.

14.11.3.2 Irritants at Low Concentrations

Depending upon the exposure concentration, irritants cause mild eye and respiratory tract irritation at low concentrations. At higher concentrations, exposure causes eye and respiratory tract pain with tears, blepharospasm (involuntary closure of the eyes), a burning sensation in the nose, mouth and throat, and mucus secretion in the nose and mouth. Inhalation of irritants stimulates the vagal nerve endings in the lungs resulting in chest pain, bronchospasm and mucus secretion.[1,23,45] (see Chapter 7). These effects are essentially concentration-related, occurring almost immediately upon exposure at concentrations causing strongly irritant effects.[1,13,23] At near threshold concentrations the effects may take longer to develop. For example, when human subjects were exposed to formaldehyde at a concentration of 0.56 ppm, eye irritation was reported after an average of 217 seconds, while at 1.0 ppm irritation was reported after 78 seconds.[46] The intensity of irritant sensation decreases with time, as does that of odour. Thus, when volunteers were exposed to the highly irritant CS riot control agent at 0.08 ppm they found the immediate effects very unpleasant, but after 4–5 minutes were able to play cards. Similarly, one group of subjects was able to endure a relatively high concentration (0.8 ppm) if it was achieved gradually over a 10 minute period, while they were totally unable to bear an immediate exposure to the same concentration.[47]

Data on the effects of inhaled irritants are based upon human experience (usually in the workplace) and some experimental human exposures at low concentrations. These have generally been used to set recommended threshold levels and occupational exposure limits for common irritants. A method often used to evaluate the sensory irritancy of chemicals (particularly those for which there is little or no human data) is the mouse respiratory rate depression (RD_{50}) test (see Chapter 7).[1,23,48] When any mammal (including humans) is exposed to a sensory irritant, a reflex decrease in breathing rate occurs. In mice this reduced breathing rate persists for many minutes and the degree of respiratory rate depression is proportional to the log of the exposure concentration for the chemical.[23,45] Using this test, the concentration causing a 50% decrease in breathing rate is calculated. Such data are available for many substances, including common irritant chemicals for which the effects on humans are relatively well known.[23,45] A relatively good correlation has been established between the mouse RD_{50} concentration and the effects in humans, as shown in Chapter 7, Figure 7.7. In general, at the mouse RD_{50} most people would find the exposure very painful and intolerable, while a workplace tolerance limit (for an 8 hour day) is approximately 0.03 times the mouse RD_{50}.[45] The workplace tolerance limit for irritancy provides a useful indication of the extent to which the general public might be affected by acute or repeated outdoor exposures to individual substances and mixtures in diluted smoke plumes from fires.

This method has been applied to many of the major irritant chemicals occurring in thermal decomposition atmospheres and also to the mixed effluents obtained by non-flaming oxidative thermal decomposition of a range of common materials. The human data and the mouse data for individual substances and effluent mixtures therefore provide a reasonably good indication of the concentration of combustion products likely to prove irritant to subjects exposed to low concentrations in highly diluted smoke plumes. One finding from the thermal decomposition atmospheres was that the observed irritancy was always more severe than that predicted from measurements of the concentrations of individual chemicals present. This was based upon the assumption that different irritants present in a mixture will be approximately additive in their effects (using the FIC approach;[23] see Chapter 7).

When exposure to irritants is prolonged, particularly to those capable of penetrating into the deep lung, then some degree of airway and deep lung inflammation and oedema may develop, usually some hours after exposure. This depends upon the exposure dose of irritant to which the subject is exposed, which can be expressed in terms of *Ct* product concentration (ppm×minutes).[1] Estimation of the potential for irritant atmospheres to cause lung inflammation is based upon acute rodent studies. In general, irritant effects on the lungs resulting from acute exposures are relatively minor and soon resolve up to a threshold exposure dose, above which severe, usually fatal inflammation occurs.[14] The detailed reviews of toxicity data for individual irritants reported in the AEGL publications include data on minor airway and deep lung pathology from acute rodent inhalation studies (mainly 4 hour exposures). These have been used to inform the setting of AEGL-1 values for critical exposure concentrations for specified exposure periods. For fire situations the exposure time evaluated is usually 30 minutes, but effects can generally be predicted for other exposure times on a *pro-rata* basis (*i.e.*, the *Ct* product exposure dose for different endpoints is approximately a constant; see Chapters 7 and 9).

The dynamic concentration range for the effects of odours and irritants is very wide. This means that the threshold concentration for odour detection (or irritation) can be very low and that increasingly stronger effects can be experienced up to high concentration (on a logarithmic scale). Table 14.10 shows some examples of effects of some irritant substances commonly occurring in thermal decomposition product atmospheres. The table shows the approximate odour threshold, the occupational exposure limit, the concentration causing severe irritancy in humans and the approximate lethal exposure dose in rodents.

The odour and irritant potencies of different substances also cover a very wide range (for example, the irritant potency of irritant substances covers approximately six orders of magnitude, as shown in Table 7.1 in Chapter 7). The table also shows a range of irritants, most of which commonly occur in thermal decomposition atmospheres. In this table the reported effects in humans are compared with the mouse RD_{50} concentration. Where human data exist, the mouse data provide a reasonably good indication of the effects in humans (see also Chapter 7, Figure 7.7). Also shown is the LC_{50} concentration, normalised to a 30 minute exposure period (followed by post-exposure observation period of 14 days). The results show that for most irritant substances (especially most organic substances) sensory irritation would be intolerable at concentrations considerably below those required to cause lung inflammation over a 30 minute period. Of the substances shown, the most potent (toluene diisocyanate) has been measured in fire effluents from burning flexible polyurethanes, and all the other substances are common in fire atmospheres, other than the riot control agents CN and CS, which have been included for comparison purposes.

Figure 7.10 in Chapter 7 is particularly relevant because it shows mouse RD_{50} data for the mixed thermal decomposition products from a range of common materials. It is considered that thermal decomposition products from non-flaming decomposition and under-ventilated flaming would have a high irritant potency. Examples are provided by low density polyethylene (RD_{50} 0.11 g m^{-3}) and the phenolic oil used as a binder with fibreglass insulation blankets (RD_{50} 0.05 g m^{-3}). In other experiments polypropylene was found to have an RD_{50} concentration of 0.1 g m^{-3}. Humans are likely to experience some irritancy at these concentrations, but not at intensely painful levels. When primates were exposed to polypropylene fumes, the effects were mild up to concentrations of approximately 2 g m^{-3} mass loss concentration, which was found to cause disruption of a behavioural task.[23]

It is important to note that all these experiments were conducted by prolonged heating at temperatures between approximately 450 °C and 600 °C. Based upon this and other work, it is considered that the results would be similar down to around 350 °C, depending upon the material decomposed. Since organic substances, including irritants, are largely combusted under well-ventilated flaming conditions, it is likely that irritancy is considerably reduced in flaming fires, as shown in Chapter 7, Figure 7.10 for flaming polyurethane, polyamide, acrylonitrile, butadiene, styrene (ABS) and even for PVC. The latter finding is important because the irritancy of PVC combustion products is partly due to the release of hydrogen chloride, which is not destroyed by flaming.

Based upon these findings it is considered that mass loss concentrations of thermal decomposition products would be unpleasantly irritant at concentrations of 0.1 g m^{-3} (0.1 g fuel decomposed for each m^3 diluent air) and could be severely irritant at concentrations of around 2 g m^{-3}. The threshold for odour (or mild irritation) for most substances is considerably below that causing severe sensory irritation.

Another way to estimate the sensory irritant potential of fire effluents is to identify key irritant chemicals likely to be present and estimate the concentrations and effects of these individual irritants. Where the effects of these chemicals on humans are relatively well known, this provides an opportunity to estimate the nature and severity of the effects likely to be experienced at different exposure concentrations. As shown in Table 14.10 and Chapter 7, Table 7.1, two of the most potent organic irritants likely to be produced by the thermal decomposition of oils are formaldehyde and acrolein. Chapter 7, Table 7.1 shows that the RD_{50} concentrations for these are extremely low, at 3.1 and 1.7 ppm, respectively, while the concentrations reported to be highly irritant to humans are 5–10 ppm for formaldehyde and 1–5.5 ppm for

Table 14.11 Acute health effects observed at different formaldehyde concentrations in mg m^{-3} and ppm.

Reported effects	Concentration/ mg m^{-3}	Concentration/ ppm
None	0.0–0.6	0.0–0.5
Minor neurophysiological effects	0.06–1.8	0.05–1.5
Odour threshold 10% of subjects	0.03	0.025
50% of subjects	0.204	0.17
All subjects	1.8	1.5
	0.1–2.4	0.1–2.0
Eye irritation	0.12–30.0	0.10–25.0
AEGL-1: notable discomfort, irritation, or certain asymptomatic non-sensory effects	1.1	0.9
Upper airway irritation	3.72	3.1
Mouse RD_{50}	6–12	5–10
AEGL-2: escape impairment and health effects in a susceptible person	16.8	14
Severe irritation in humans:		
Lower airway and pulmonary effects	6.0–36.0	5.0–30.0
Irritation difficult to tolerate	>6.0	>5
Respiratory difficulties	12–24	10–20
Pulmonary oedema, inflammation, pneumonia[a]	60–120	50–100
AEGL-3 10 minutes, life threatening health effects or death	120	100
Death[a]	>120	>100
30 minute LC_{50} mammal	840–960	700–800

[a]No exposure time quoted.

acrolein. Note that the higher molecular mass unsaturated aldehyde, acetaldehyde, although considered an irritant, is considerably less potent than either formaldehyde or acrolein, with an RD_{50} of 4946 ppm and reported irritancy in humans at concentrations above approximately 1500 ppm. At concentrations above approximately 1000 ppm the central nervous system depressive (narcotic) effects start to become of more concern than odour or irritancy.

The acute effects of exposure to formaldehyde have been particularly extensively studied due to its importance in the workplace and as an indoor air contaminant in the home. The reported effects of formaldehyde[46] are shown in Table 14.11. To put these data into context, concentrations in the range 0.08–0.8 mg m^{-3} have been measured in buildings containing chipboard and a Connecticut study of emissions from urea formaldehyde insulation in homes suggested levels ranged from 0.12–8.4 mg m^{-3}.

The results from human and animal studies show some variations between subjects but no or very minor effects below 0.1 mg m^{-3}, with an odour detection threshold between 0.03 and 1.8 mg m^{-3} depending upon the subject's sensitivity. Eye irritation can occur possibly as low as 0.1 mg m^{-3} and respiratory tract irritation begins at around 0.12 mg m^{-3}, increasing in intensity with concentration. Severe irritancy occurs above

Table 14.12 Irritants effects of exposure to acrolein.

Concentration/ppm	Effect
0.03	AEGL-1: notable discomfort, irritation, or certain asymptomatic non-sensory effects
0.1	Odour threshold and occupational exposure limit (OEL)
0.44	AEGL-2: escape impairment and health effects in a susceptible person[41]
1	Strongly irritant to the eyes and nose in humans[49,50]
1.7	Mouse RD_{50} concentration[45]
2.5	AEGL-3: 30 minutes
4	SFPE escape impairment (average person)
5.5	Severely painful eye and respiratory tract irritation
>10	Suggested as being potentially lethal to humans after a short exposure
20	SFPE incapacitation (average person)
30	ISO/13571 incapacitation (average person)
135	30 minute rat LC_{50}
505	Baboon: no pulmonary effects following 5 minute exposure[51]
1025	Baboon: death occurring following 5 minute exposure[51]

approximately 6 mg m^{-3}, which is considered intolerable, with respiratory difficulties above 12 mg m^{-3}. Pulmonary oedema and inflammation would depend upon both concentration and exposure time (not stated in Table 14.11), but an exposure dose of around 22 500 ppm min is expected to result in lethal pulmonary oedema and inflammation. For more sensitive human subjects some signs of lung inflammation could occur at exposure doses of between a factor of 0.01 or 0.1 of this.

Another important irritant in fire atmospheres is acrolein, the effects of which are summarised in Table 14.12, which is taken from Purser[23] using data compiled from a number of sources.

14.12 Conclusions

In order to evaluate the acute, life-threatening hazards to fire victims it is necessary to evaluate the sequential changes occurring during a developing building fire scenario in terms of the time–concentrations curves for different irritant and asphyxiant substances, smoke particulates and heat, and the sequence of behavioural and physiological effects impairing escape and causing incapacitation and death during (or after) exposure.

The fractional effective dose (FED) calculation methods presented in this chapter are sufficiently well validated to enable prediction of time to and severity of a set of endpoints, including visual obscuration, sensory and lung irritancy, and asphyxiation for the key toxic substance mixtures occurring in combustion products from different types of fires involving different fuels and combustion conditions.

For low level outdoor exposures to fire plumes, the general conclusions are that for most situations in which persons are exposed to grounded smoke plumes from large fires involving mixed fuels, the main toxic hazards are acute minor odour sensation and mild sensory irritancy. It is likely that large numbers of persons exposed to dilute grounded plumes at locations hundreds to thousands of metres from large fire sources may experience such signs and symptoms, and a small proportion of sensitive sub-populations may experience more severe cardio-respiratory health effects. Persons closer to higher concentration, down-washed plumes may experience more severe irritancy, respiratory signs after exposure, other health effects and even some degree of carbon monoxide intoxication (depending upon the fuel composition, the combustion conditions, the plume dispersal and the duration of exposure). More serious health effects may occur where fires or explosions release large masses of specific toxic chemicals in addition to more general combustion products. It is possible to estimate the health hazards from outdoor fire plumes by use of appropriate modelling calculations as described.

References

1. D. A. Purser, Assessment of Hazards to Occupants from Smoke, Toxic Gases and Heat, in *The SFPE Handbook of Fire Protection Engineering*, ed. P. J. DiNenno *et al.*, National Fire Protection Association, Quincy, MA 02269, 4th edn, 2009, pp. 96–193.
2. D. A. Purser, ASET and RSET: Addressing Some Issues in Relation to Occupant Behaviour and Tenability, in *Proceedings of the 7th International Symposium on Fire Safety Science*, International Association for Fire Safety Science, Boston, MA, 2003, pp. 91–102.
3. The application of fire safety engineering principles to fire safety design of buildings. PD7974-6: Human Factors: Life safety strategies—occupant evacuation, behaviour and condition. British Standards Institution, 2004.
4. United Kingdom Fire Statistics, Communities and Local Government, London, published annually.
5. Furniture and Furnishings (Fire Safety) Regulations Statutory Instrument 1988/No. 1324, The Stationery Office Ltd.
6. D. A. Purser, Toxic Product Yield and Hazard Assessment for Fully Enclosed Design Fires Involving Fire Retarded Materials, *Polym. Int.*, 2000, **47**, 1232–1255.
7. The application of fire safety engineering principles to fire safety design of buildings, Code of practice, British Standards Institution, 2001.
8. D. T. Gottuk and B. Y. Lattimer, Effect of Combustion Conditions on Species Production, in *The SFPE Handbook of Fire Protection Engineering*, ed. P. J. DiNenno *et al.*, National Fire Protection Association, Quincy, MA, 4th edn, 2008, pp. 67–95.
9. A. Tewarson, Generation of Heat and Gaseous, Liquid and Solid Products in Fires, in *The SFPE Handbook of Fire Protection Engineering*, ed.

P. J. DiNenno *et al.*, National Fire Protection Association, Quincy, MA, 4th edn, 2008, pp. 109–194.
10. Controlled Equivalence Ratio Method for the Determination of Hazardous Components of Fire Effluents, *ISO/TS 19700*, International Organization for Standardization, Geneva, Switzerland, 2007.
11. ISO 13571, *Life-threatening Components of fire – Guidelines for the Estimation of Time to Compromised Tenability in Fires*, International Organization for Standardization, Geneva, Switzerland, 2nd edn, 2012.
12. D. A. Purser and W. D. Woolley, Biological Studies of Combustion Atmospheres, *J. Fire Sci.*, 1983, **1**, 118–145.
13. D. A. Purser, The application of exposure concentration and dose to evaluation of the effects of irritants as components of fire hazard. *Interflam 2007, 3rd–5th September 2007, Royal Holloway College, Egham, UK, Proceeding*, Interscience Communications, Greenwich, UK, pp. 1033–1041.
14. D. A. Purser and P. Buckley, Lung Irritance and Inflammation During and After Exposure to Thermal Decomposition Products from Polymeric Materials, *Med. Sci. Law*, 1983, **23**, 142–150.
15. J. L. Bryan, Behavioral Response to Fire and Smoke, in *The SFPE Handbook of Fire Protection Engineering*, ed. P. J. DiNenno *et al.*, National Fire Protection Association, Quincy, MA, 4th edn, 2008, pp. 320–354.
16. P. G. Wood, The Behavior of People in Fires, *Fire Research Note 953*, Fire Research Station, Watford, UK, 1972.
17. J. L. Bryan, Smoke As a Determinant of Human Behavior in Fire Situations (Project People), *NBS-GCR-77-94*, U.S. Department of Commerce, National Bureau of Standards, Washington, DC, 1977.
18. D. A. Purser, Application of human behaviour and toxic hazard analysis to the validation of CFD modelling for the Mont Blanc Tunnel fire incident, *Advanced Research Workshop: Fire Protection and Life Safety in Buildings and Transport Systems*, University of Cantabria, Spain, 17 October 2009, Proceedings, pp. 23–57.
19. BS 7974-6 The application of fire safety engineering principles to fire safety design of buildings – Part 6: Human factors: Life safety strategies – Occupant evacuation, behaviour and condition (Sub-system 6). British Standards Institution, 2004.
20. T. Jin, Visibility and Human Behavior in Fire Smoke, in *The SFPE Handbook of Fire Protection Engineering*, ed. P. J. DiNenno *et al.*, National Fire Protection Association, Quincy, MA, 4th edn, 2008, pp. 54–66.
21. T. Jin, Visibility through Fire Smoke—Part 5, Allowable Smoke Density for Escape from Fire, *Report No. 42*, Fire Research Institute of Japan, 1976, p. 12.
22. H. Frantzich and D. Nilsson, Evacuation Experiments in a Smoke Filled Tunnel, Human Behaviour in Fire, Proceedings of the 3rd International Symposium, Belfast, UK, pp. 229–238.
23. D. A. Purser, Hazards from smoke and irritants, in *Fire Toxicity*, ed. A. Stec and R. Hull, Woodhead, Cambridge, UK, 2010, ch. 3, pp. 51–117.

24. British Standard Code of Practice for Assessment of Hazard to Life and Health from Fire, Part 2: Guidance on Methods for the Quantification of Hazards to Life and Health and Estimation of Time to Incapacitation and Death in Fires, *BS 7899-2,* British Standards, London, 1999.
25. NAC/AEGL, Acute Exposure Guidelines for Selected Airborne Chemicals, *National Advisory Committee for Acute Exposure Guideline Levels for Hazardous Substances*, Subcommittee on Acute Exposure Guideline Levels, Committee on Toxicology, Board on Environmental Studies and Toxicology, National Research Council of the National Academies, National Academies Press, Washington, DC, 2004, Note: Reviews of new substances are being added on a continuing basis.
26. G. E. Hartzell, D. N. Priest and W. G. Switzer, Modeling of toxicological effects of fire gases: II. Mathematical modelling of intoxication of rats by carbon monoxide and hydrogen cyanide, *J. Fire Sci.*, 1985, **3**, 195–207.
27. D. A. Purser, Behavioural Impairment in Smoke Environments, *Toxicology*, 1996, **115**, 25–40.
28. D. A. Purser, Asphyxiant components of fire effluents, in *Fire Toxicity* ed. A. Stec and R. Hull, Woodhead, Cambridge, UK, 2010, ch. 4, pp. 118–198.
29. R. D. Stewart, The effects of carbon monoxide on man, *J. Combust. Toxicol.*, 1974, **1**, 167–176.
30. R. F. Coburn, R. E. Forster and P. B. Kane, Consideration of the physiological variables that determine the blood carboxyhaemoglobin concentration in man, *J. Clin. Invest.*, 1965, **44**, 1899.
31. J. E. Peterson and R. D. Stewart, Predicting the carboxyhaemoglobin levels resulting from carbon monoxide exposures, *J. Appl. Physiol.*, 1975, **4**, 633.
32. U. C. Luft, Aviation Physiology—The Effects of Altitude, in *Handbook of Physiology*, American Physiology Society, Washington, DC, pp. 1099–1145, 1965.
33. C. J. Lambertson, Carbon Dioxide and Respiration in Acid-Base Homeostasis, *Anaesthesiology*, 1960, **21**, 642–651.
34. J. H. Comroe, R. E. Forster, A. B. Dubois, W. A. Briscoe, and E. Carlsen, Pulmonary Ventilation, in *The Lung*, Year Book Medical Publishers, Chicago, 1962, pp. 52–53.
35. *Environmental Biology*, ed. P. Altman and D. S. Ditter, Federation of American Societies for Experimental Biology, Bethesda, MD, 1966.
36. B. G. King, High Concentration–Short Time Exposures and Toxicity, *J. Ind. Hyg. Toxicol.*, 1949, **31**, 365–375.
37. S. M. Hockey and P. J. Rew, Human Response to Thermal Radiation, Contract Research Report No. 97/1996, HSE Books, Sudbury Suffolk, UK, 1996.
38. D. A. Purser, Review of Human Response to Thermal Radiation, *Fire Saf. J.*, 1997, **28**, 290–291.
39. Experimental research for Scottish Building Standards Agency following the fire at the Rosepark Care Home, Glasgow, 31st January 2004. BRE Project Report No. 219132, October 2004.

40. D. A. Purser, Fire safety and evacuation implications from behaviours and hazard development in two fatal care home incidents: Rosepark and Frampton House, *Proceedings of the 5th International Symposium on Human Behaviour in Fire*, Downing College, Cambridge, UK, Interscience, London, 19–21 September 2012, pp. 251–262.
41. National Advisory Committee for Acute Exposure Guideline Levels for Hazardous Substances (NAC/AEGL), Acute Exposure Guideline Levels for Selected Airborne Chemicals Subcommittee on Acute Exposure Guideline Levels, Committee on Toxicology Board on Environmental Studies and Toxicology, National Academic Press, Washington.
42. M. E. Harrison, S. E. Page and S. H. Limin, The global impact of Indonesian forest fires, *Biologist*, 2009, **56**, 156–163.
43. B. Jensen, P. Wolkoff, *VOCBASE: Database with Properties of 808 VOCs*, National Institute of Occupational Health, Denmark, 1996.
44. Interim acute exposure guideline levels (AEGLs): Methyl mercaptan, Interim 2: 10/2008, National Research Council of the National Academies, National Academies Press, Washington, DC.
45. Y. Alarie, Bioassay for evaluating the potency of airborne sensory irritants and predicting acceptable levels of exposure in man, *Food Cosmet. Toxicol.*, 1981, **19**, 623–626.
46. C. Humfrey, S. Shuker and P. Harrison, *Indoor Air Quality in the Home*, Institute of Environmental Health, Leicester, Assessment A2, 1996.
47. F. W. Beswick, P. Holland and K. H. Kemp, Acute effects of exposure to orthochlorobenzylidene malonitrile (CS) and the development of tolerance, *Br. J. Ind. Med.*, 1972, **29**, 298–306.
48. M. M. Hirschler and D. A. Purser, Irritancy of the smoke (non-flaming mode) from materials used for coating wire and cable products, both in the presence and absence of halogens in their chemical composition, *Fire Mater.*, 1993, **17**, 7.
49. Documentation of the Threshold Limit Values for Substances in Workroom Air, *American Conference of Governmental Industrial Hygienists*, Cincinnati, 1980.
50. *Registry of Toxic Effects of Chemical Substances*, National Institute for Occupational Safety and Health, Washington, DC, 1982.
51. H. L. Kaplan, A. F. Grand, W. G. Switzer, D. S. Mitchell, W. R. Rogers and G. E. Hartzell, Effects of Combustion Gases On Escape Performance of the Baboon and the Rat, *J. Fire Sci.*, 1985, **3**(4), 228–244, DOI: 10.1177/073490418500300401.

Clinical Toxicology and Management of Combustion Product Exposures

CHAPTER 15

Treatment of Cyanide Poisoning Associated with Fires

TIMOTHY C. MARRS

Edentox Associates, Edenbridge, UK
Email: timothymarrs05@aol.com

15.1 Introduction

The production of hydrogen cyanide (HCN) and other nitrogen-containing substances in fires is discussed in Chapters 2 and 3, and in other texts, *e.g.*, Ballantyne.[1] Erdman stated data are currently unclear about whether, or to what degree, cyanide generates or contributes to morbidity and mortality in human victims of smoke inhalation.[2] HCN is believed to be very rapidly absorbed when inhaled[3] and, provided that the dose of hydrogen cyanide has not been sufficient to produce lethality, blood levels start to fall when exposure ceases.[4–6] Thus, the measure of the greatest importance is to stop further exposure of the poisoned individual. The physiological effects of HCN exposure and dynamics of uptake and distribution and excretion are described in Chapter 10.

It should be recognised that much of the experimental and clinical data related to cyanide poisoning is based not on exposure to HCN, but upon oral poisoning with cyanide salts, where absorption is likely to be slower, being dependant largely on stomach acid releasing hydrogen cyanide, although there are no pharmacokinetic data to confirm this (see Chapter 10).

Issues in Toxicology No. 23
Toxicology, Survival and Health Hazards of Combustion Products
Edited by David A. Purser, Robert L. Maynard and James C. Wakefield

Published by the Royal Society of Chemistry, www.rsc.org

15.2 General Measures and First Aid

The most important measure is to stop further exposure to hydrogen cyanide; thus the casualty needs to be removed from the source of HCN. Because of the volatility of HCN, decontamination of casualties is unlikely to be necessary. It has often been said that oxygen is of no value in cyanide poisoning because the blood is fully oxygenated; however, a number of studies have shown slight, but significant evidence that oxygen is of value in cyanide poisoning and the use of oxygen is now frequently recommended, *e.g.*, by Mégarbane and colleagues (see also Section 15.5 below).[7]

15.3 Antidotes

A number of antidotes have been used for cyanide poisoning in humans, while a great many more have been studied experimentally in animals and/or *in vitro* (see Boxes 15.1 and 15.2) and there have been differences in therapeutic attitudes in different countries. A recent development has been widespread advocacy of the use of hydroxocobalamin.

It has been said that the reason for the number of antidotes that are available and/or have been studied experimentally is that no antidote is totally satisfactory. Many of the antidotes listed in Box 15.2 are satisfactory in studies in experimental animals but the onset of poisoning with cyanides is rapid and this is particularly likely to be the case with HCN. In many cases, the experimental studies have used protocols that are totally unrealistic particularly with respect to the time between experimental poisoning and antidotal therapy and some have even used prophylactic protocols. The antidotes used clinically follow one of three approaches: enhanced enzymic detoxication (sodium thiosulfate), direct binding of cyanide (cobalt compounds, dicobalt edetate, hydroxocobalamin) or indirect binding (the methaemoglobin generators such as sodium nitrite and 4-dimethylaminophenol).

Box 15.1 Antidotes to hydrogen cyanide that act by hastening the enzymic detoxication of cyanide

- a) Sulfur donors
  - i) Sodium thiosulfate
  - ii) Sodium ethanethiosulfonate
  - iii) Sodium propanethiosulfonate
  - iv) Sodium mercaptopyruvate
  - v) Sodium tetrathionate
- b) Exogenous sulfurtransferases
  - i) Rhodanese
    - α) Bovine heart
    - β) Bacterial
  - ii) β-Mercaptopyruvate cyanide sulfurtransferase

Box 15.2 Possible antidotes to HCN

Antidotes to hydrogen cyanide that act by complex formation with heavy metals.

- a) Cobalt
  - i) Dicobalt edetate
  - ii) Hydroxocobalamin
  - iii) Cobaltous chloride
  - iv) Cobaltous acetate
  - v) Cobalt histidine
  - vi) Sodium cobaltinitrite (also a methaemoglobin former)
- b) Iron
  - i) Endogenous methaemoglobin generated by
    - α) Sodium nitrite
    - β) Amyl nitrite
    - γ) 4-Dimethylaminophenol
    - δ) Compounds primarily of interest in prophylaxis
      - A) 4-Aminopropiophenone (*p*-aminopropiophenone, PAPP)
      - B) Hydroxylamine
  - ii) Exogenous methaemoglobin

Carbohydrates and related compounds.

- a) Glucose
- b) Sodium pyruvate
- c) α-Oxoglutaric acid
- d) Glyceraldehyde
- e) Pyridoxal-5-phosphate

Drugs that do not directly interact with cyanide.

- a) Chlorpromazine
- b) Etomidate
- c) Naloxone
- d) Meclofenoxate (centrophenoxone)
- e) Phenoxybenzamine

15.3.1 Sulfur Donors and Rhodanese

15.3.1.1 Sodium Thiosulfate

The main pathway of reaction of detoxication of cyanide is catalysed by the enzyme rhodanese (thiosulfate/cyanide sulfurtransferase, EC 2.8.1.1), which converts cyanide to thiocyanate using endogenous sulfane sulfur. Exogenous sulfur donors, such as sodium thiosulfate, hasten this enzymic reaction by supplying additional sulfane sulfur, the rate of activity of rhodanese normally being limited by the availability of sulfane sulfur.[8] Notwithstanding the poor access of the drug to the mitochondria and the location of rhodanese in that organelle, sodium thiosulfate, which is not the natural co-substrate for the enzyme, can increase the rate of transulfuration in dogs up to 30-fold.[9,10] The reaction of cyanide with sodium thiosulfate can be represented as follows:

$$NaS_2O_3 + CN^- \rightarrow SCN^- + NaSO_4$$

Thiocyanates are much less toxic than cyanides [oral median lethal dose (LD_{50}) for mice is 350–800 mg g^{-1} bodyweight (bw) and for rats is 200–1200 mg kg^{-1} bw].[11] The LD_{50} of sodium thiosulfate intravenously was reported as 3 g kg^{-1} bw in dogs,[12] meaning that sodium thiosulfate would be expected to be of low toxicity in humans.

Used by itself, thiosulfate causes blood cyanide levels to fall only slowly and this antidote is therefore used as a second-line cyanide antidote, usually with a methaemoglobin generator. The usual adult dose of sodium thiosulfate is 50 mL of the 25% solution administered intravenously over 10–15 min.[13]

Sodium thiosulfate forms, with sodium nitrite (see Section 15.3.2), the "classical therapy" for cyanide poisoning.

15.3.1.2 Other Sulfur Donors and Rhodanese

Sulfur donors other than sodium thiosulfate have been studied, partly in an attempt to bring about the collocation of the sulfur donor and the enzyme in the same fluid compartment. These have included sodium salts of ethanethiosulfonic acid and propanethiosulfonic acid. Superiority to sodium thiosulfate in antidotal capability observed was small in a mouse model but combined with rhodanese a greater advantage was seen.[14] However, these alternative sulfur donors have not entered into clinical practice. Another substance that has been studied experimentally is sodium tetrathionate [IUPAC name: (sulfonatodisulfanyl)sulfonate][15] and a large number of sulfur donors were studied by Baskin and colleagues.[16] A related approach along these lines was the use of rhodanese and sodium thiosulfate micro-encapsulated in mouse erythrocytes.[17] Another enzymatic pathway known to play a part in the detoxication of cyanide has been exploited experimentally, namely the β-mercaptopyruvate:cyanide sulfurtransferase route of detoxification; β-mercaptopyruvate:cyanide sulfurtransferase (3-mercaptopyruvate sulfurtransferase, EC 2.8.1.2) is at least partially cytoplasmic in distribution.[18] Sodium 3-mercaptopyruvate has been investigated as a cyanide antidote.[18,19] Prodrugs for 3-mercaptopyruvate have also been studied,[20] including a water-soluble one, 3-mercaptopyruvate dithiane (sulfanegen sodium).[21,22] Another approach has been the parenteral injection of rhodanese, often of beef liver origin, so that it is present in the extracellular fluid.[23]

15.3.2 Complexation with Heavy Metals

15.3.2.1 Introduction

Cyanide forms stable complexes of low toxicity with many heavy metal ions, such as iron, cobalt, silver, chromium and rhodium.[24] Only complexation with iron and cobalt have been used clinically. The "classical treatment" for cyanide poisoning (sodium thiosulfate and sodium nitrite) uses iron in methaemoglobin generated by the latter salt (see Section 15.3.2.2).

15.3.2.2 Treatment of HCN Poisoning by Methaemoglobin Production

The usual source of iron in the treatment of cyanide poisoning is methaemoglobin generated by chemical means from the patient's own haemoglobin using compounds such as the nitrites or 4-dimethylaminophenol. Methaemoglobinaemia is a condition in which the Fe^{2+} in erythrocytic haemoglobin is oxidised to Fe^{3+}. Methaemoglobin is brown. High levels of methaemoglobin are generally undesirable, because haemoglobin in the oxidised form is unable reversibly to carry oxygen; consequently, very high levels are life-threatening as they produce tissue hypoxia (see review by Skold and colleagues).[25] Happily, however, methaemoglobin has the property of combining with some harmful anions, including cyanide, causing them to be less toxic, and methaemoglobinaemias of 30–40% seem to be fairly safe. It has been suggested, however, that with co-exposure to carbon monoxide and cyanide, as may occur in fires, the loss of additional reversible oxygen-carrying capacity in the blood might be undesirable[13] and there is some experimental evidence for that.[26] Cyanide forms an intraerythrocytic complex with methaemoglobin called cyanmethaemoglobin, which has different spectral characteristics from methaemoglobin and is relatively non-toxic. If methaemoglobin production is monitored spectroscopically, misleading results may be obtained if cyanide is present in the subject. The effect of methaemoglobin binding to cyanide is to sequestrate the cyanide inside the red blood cell (see also below). Small amounts of methaemoglobin are present in normal erythrocytes, but levels are kept low by methaemoglobin reductase, an NADH-dependent enzyme present mainly in the erythrocyte; certain drugs and genetic conditions can give rise to abnormally high levels of methaemoglobin (see review by Percy and colleagues)[27] and individuals with pre-existing abnormally high levels of methaemoglobin would be expected to react to methaemoglobin generators by producing high levels of methaemoglobin (see below). Because of the reaction catalysed by methaemoglobin reductase, chemically generated methaemoglobinaemia is reversible so that the binding of cyanide is of limited duration. This would result in cyanide diffusing out of the erythrocyte and a second antidote, usually sodium thiosulfate, is necessary as a result.

Repeated administration of methaemoglobin formers may result in oxidant changes in the red blood cell in addition to methaemoglobin formation; these may include Heinz body formation, as well as reduced red cell survival time.[28]

It is likely that oxidation of Fe^{2+} to Fe^{3+} in other iron-containing body constituents, such as myoglobin, adds to the beneficial effects of methaemoglobin in cyanide poisoning.

In order to use chemically-induced methaemoglobinaemia safely to treat cyanide poisoning, it is important to understand that many commonly used methods for measuring that pigment measure cyanmethaemoglobin as ordinary haemoglobin, not methaemoglobin, which will give a falsely optimistic level for the remaining fraction of blood pigment, which can reversibly carry oxygen. Another consideration is that there are individuals with elevated levels of methaemoglobin because of deficient activity of methaemoglobin reductase[29] or because of abnormal haemoglobins.[27,30] Individuals with these rare conditions may run the risk of developing abnormally high levels of methaemoglobin in response to methaemoglobin generators used therapeutically.

The first nitrite whose efficacy in cyanide poisoning was noted was amyl nitrite and this was as early as 1888.[31] Inhalation of amyl nitrite can give rise to significant levels of methaemoglobin in dogs[32] and successfully antagonised several multiples of a lethal dose of cyanide in dogs.[33] However, rather low levels of methaemoglobin seem generally to be produced in humans,[34] although some case reports suggest otherwise.[35] And it has been suggested that production of methaemoglobin may not be the only or main mechanism of action of amyl nitrite: vasodilation may be at least as important.[13] Most people find the inhalation of amyl nitrite unpleasant and the probable reason for this is that it is not possible to persuade human volunteers to inhale doses comparable to those that can be administered to experimental animals (or even moribund humans). Despite misgivings over its worth, amyl nitrite ampoules (0.2–0.4 mL) are frequently used in acute cyanide poisoning. The use of amyl nitrite in cyanide poisoning has been reviewed.[13,36]

The combination of sodium nitrite and thiosulfate was shown to be effective in cyanide poisoning in the rabbit by Hug[37] in the 1930s. The combination was also demonstrated to be effective in dogs[38,39] and came to be known as the "classical therapy" for cyanide poisoning (see also above). The combined effects of these two drugs, whereby the antidotal effect is much greater for the two together than when they are administered separately, have often been described as synergy. The combination of sodium nitrite and sodium thiosulfate remained standard treatment for many years throughout most of the world, the usual adult dose of sodium nitrite being 300 mg (10 mL 3% w/v) given intravenously by infusion; whether this should be over 5 or 20 min is a source of disagreement.[13] It has been stated that sodium nitrite does not exert its action entirely through the generation of methaemoglobin.[13,40] The evidence for this is based on work that suggested that similar concentrations of methaemoglobin were more effective in mice in protecting against cyanide poisoning than those produced by

4-dimethylaminophenol (DMAP) or hydroxylamine.[41] Suggestions for the mechanism of action of sodium nitrite have included vasoactivity[13,42] and antagonism of cyanide inhibition of cytochrome c oxidase by nitrous oxide.[43]

It has been known for many years that aminophenols are methaemoglobin generators. The mechanism of action seems to be oxidation of the aminophenol to its corresponding quinoneimine. The quinoneimine then oxidises the Fe^{2+} in haemoglobin to Fe^{3+}, the quinoneimine being reduced to the aminophenol, thus establishing a catalytic cycle. Klimmek and colleagues[44] found that, in dogs and humans, DMAP rapidly produced therapeutic levels of methaemoglobin. In fact DMAP acted faster in producing methaemoglobin than other aminophenol derivatives, as well as sodium nitrite, in several species.[45–47] DMAP was not widely used clinically outside Germany and there were some case reports of unfavourable clinical experiences.[48–50] The main toxic effect of DMAP is the one it shares with other methaemoglobin producers, namely the production of hypoxia. DMAP, like other aminophenols, is nephrotoxic,[51] but this would be unlikely to affect its use in humans as a single dose using the advised regimen and, unlike sodium nitrite, DMAP does not cause vasodilatation. The main puzzle with DMAP is why it has proved unsatisfactory in clinical use in some instances, whereas sodium nitrite has been used for many years without major problems. A possible reason is that peak levels of methaemoglobin generated by the recommended doses of the two compounds are very different, being much higher with DMAP.[13] Also, there is the possibility (see above) that nitrites have beneficial activity by mechanisms in addition to methaemoglobin formation. The use of DMAP in cyanide poisoning has been reviewed.[52] The recommended dose is 3.25 mg kg^{-1} intravenously. As with sodium nitrite, use of DMAP should be followed by sodium thiosulfate.[13]

The aminophenones, particularly 4-aminopropiophenone (*p*-aminopropiophenone, PAPP), have been studied in cyanide poisoning. However, because aminophenones need metabolic activation to act as methaemoglobin producers it has been considered that aminophenones would be more suitable for prophylaxis than for *post facto* treatment. The mechanism of action of the proximate methaemoglobin former (probably the *N*-hydroxyaminophenone) is rather different from that of the aminophenols, but like the action of the aminophenols, it involves the establishment of a catalytic cycle: it is thought that the oxidation of Fe^{2+} to Fe^{3+} is accompanied by conversion of the hydroxyaminophenone to the nitroso derivative, which is reduced in the red cell back to the hydroxyaminophenone by NADPH.[53,54]

Another experimental approach to the use of methaemoglobin to combine with cyanide has been the use of exogenous stroma free methaemoglobin and this has been studied in experimental animals.[55–57]

15.3.2.3 Cobalt Compounds

It has long been known that cyanides could form complexes of low toxicity with cobalt. The exact nature of the complexes formed is not always clear.

Evans studied the complexation of cyanide by cobalt compounds and found that cobalt acetate was effective at a molar cyanide cobalt ratio of 5.[58] Evans proposed that cobalt reacted with cyanides to form the cobaltocyanide ion, $[Co(CN)_6]^{4-}$ and then the cobalticyanide ion, $[Co(CN)_6]^{3-}$. The LD_{50} of cobalticyanides is thought to be in the region of 1 g kg^{-1} bw.[13] With dicobalt edetate the effective molar ratio was considered to be about two by Paulet[59] (see also IPCS/CEC).[13]

Although cobalt is an essential component of the diet in animals, including humans,[60] the salts are somewhat toxic.[58,59,61,62] This fact militated against their use as cyanide antidotes, but in 1952 Mushett and colleagues[63] showed that hydroxocobalamin (vitamin B12a), a complex cobalt-containing compound, could antagonise cyanide poisoning in mice. A number of investigators, including notably Paulet,[59] have studied the antidotal efficacy and toxicity of various cobalt compounds, both inorganic and organic salts and organic complexes in animal models.

On the basis of animal studies of both toxicity and efficacy against experimental cyanide poisoning, Paulet concluded that dicobalt edetate was the most satisfactory cobalt compound for the treatment of cyanide poisoning.[59] The usual preparation of dicobalt edetate is a 1.5% solution, known as Kelocyanor [Societe d'études et de recherches biologiques (SERB), 53 Rue Villiers de l'Isle Adam, 75020 Paris].[64] The 20 mL ampoules contain a purple solution of 1.96–2.40 g L^{-1} free cobalt and 13.5–16.5 g L^{-1} dicobalt edetate, as well as 4 g glucose per ampoule.[13] The structure of dicobalt edetate was assumed to be a monocobalt salt of a cobalt edetate anion,[13,58] but this is still not completely clear[13] and crystallised dicobalt edetate is reported to be polymeric, half of the Co^{2+} ions being bound to the edetate^{2-} ion while the remaining Co^{2+} ions are bound to four water ligands as well as to the carboxylate ligands on the $[Co(edetate)]^{2-}$ moiety.[65]

Dicobalt edetate only appears to antagonise cyanide at molar ratios of up to about two (see above). The use of dicobalt edetate has particularly been advocated for use in cyanide poisoning in industrial contexts, because of its rapidity of action.[66] The main adverse effects of dicobalt edetate include urticaria, oedema of the face, hypotension, chest pains and dyspnoea. These effects seem most likely to occur when dicobalt edetate has been administered to subjects where there was a lack of objective evidence that cyanide poisoning was present, although no instance of death resulting from dicobalt edetate has been reported.[67] These effects are often described as allergic type responses but it seems unlikely that there is a true sensitisation basis for the effects. Hypertonic glucose has been reported to increase the LD_{50} of dicobalt edetate in mice.[13,59] One advantage of dicobalt edetate over the methaemoglobin-producing cyanide antidotes is that dicobalt edetate is reported to cross the blood brain barrier.[13]

As was discussed above, Mushett and colleagues[63] showed that hydroxocobalamin, a complex cobalt-containing compound, could antagonise cyanide poisoning in mice in 1952. Hydroxocobalamin is a complex natural substance, also known as vitamin B12a, the IUPAC systematic name being

Coα-[α-(5,6-dimethylbenzimidazolyl)]-Coβ-hydroxocobamide. It is a high molecular mass compound (1346.37 g mol^{-1}), containing cobalt, a corrinoid ring, phosphate, 5,6-dimethyl-1-α-D-ribofuranose and a benzimidazole moiety. In animal experiments hydroxocobalamin is an efficacious antidote. The pharmacodynamics and pharmacokinetics of hydroxcobalamin have been reviewed.[68–71] Hydroxocobalamin has been used to treat pernicious anaemia, although the usual treatment is cyanocobalamin; hydroxocobalamin is sometimes used for pernicious anaemia, particularly with ocular complications, and other conditions with vitamin B12 deficiency, as well as certain metabolic disorders (see review by Schiff and colleagues).[72] In all these cases the preparations used are much too dilute for use as cyanide antidotes. However, there is a preparation specifically designed for the treatment of cyanide poisoning: this is the Cyanokit®.[73] The Cyanokit® comprises two vials, both containing hydroxocobalamin lyophilizate 2.5 g, which require reconstitution with 100 mL 0.9% saline. It is recommended that an initial dose of 5 g be administered over 15 min, while a further 5 g over 15 min to 2 hours may be given if the status of the patient so warrants (Cyanokit® package insert).

Hydroxocobalamin has been shown to be efficacious in a number of species, including mice,[63,74] rats,[75] rabbits,[58] guinea pigs,[76,77] dogs[78,79] and sows.[80–82] Some of the older of these studies made use of somewhat artificial experimental conditions, particularly with respect to the relationship between the time of poisoning and therapy, and in some cases the antidote was administered prophylactically. However, in overview there seems little doubt that hydroxocobalamin can effectively antagonise cyanide under experimental conditions. Guidance issued by the UK College of Emergency Medicine and the UK National Poisons Information Service (NPIS) is that hydroxocobalamin should be immediately available in all emergency departments.[83]

The attractiveness of hydroxocobalamin in smoke exposure over methaemoglobin-producing cyanide antidotes has been discussed, as have doubts as to the importance of cyanide in fire casualties.[2] The evidence base for the efficacy of hydroxocobalamin in fires is based upon case reports and case series. Case reports included one by Fortin and colleagues of a nine month old infant with smoke inhalation to whom 2.5 g hydroxocobalamin was administered intraosseously using a bone-injection gun in a pre-hospital environment. The patient recovered without sequelae.[84] In another case reported by Fortin and colleagues an adult prisoner was discovered about 30 min after exposure to smoke from a burning mattress in cardiorespiratory arrest. The fire was apparently a suicide attempt. There were raised blood cyanide and lactate concentrations. Hydroxocobalamin (5 g intravenously) was administered pre-hospital and again on arrival at the hospital, hyperbaric oxygen being given after the second dose of hydroxocobalamin. The patient recovered without residual effects.[85] Roderique *et al.*[86] reported a case of a 32-year-old woman who was 37 weeks pregnant and exposed to smoke in a house fire. She received hydroxocobalamin, 5 g intravenously.

Despite having burns and injury from smoke inhalation, she recovered well and neither she nor her offspring had significant complications due to smoke inhalation after 6 months follow-up.

There are several case series of the use of hydroxocobalamin in fire victims, none very large.

Houeto *et al.*[87] reported 12 victims of fires (six men and six women), of whom five died. The authors compared blood cyanide concentrations with plasma cyanocobalamin concentrations after intravenous administration of hydroxocobalamin, 5 g. There was correlation between cyanocobalamin in plasma and blood cyanide ($r = 0.87$; $p < 0.0001$). Where the initial blood cyanide concentration was below 40 μmol L^{-1} (1.04 mg L^{-1}), the relationship between cyanide and cyanocobalamin was nearly linear. In one patient with an initial blood cyanide of 96 μmol L^{-1} (2.5 mg L^{-1}), plasma cyanocobalamin approximately doubled after a second dose of 5 g hydroxocobalamin, intimating that where blood concentrations of cyanide are more than 40 μmol L^{-1}, doses of hydroxocobalamin greater than 5 g may be required.

A retrospective study of the pre-hospital use of hydroxocobalamin by the Paris Fire Brigade has been reported by Fortin and colleagues.[88] Hydroxocobalamin was administered to 101 patients over an 8 year period, 30 surviving and 42 dying (17 at the fire scene and 25 at the intensive care unit). The survival status of the other 29 patients was not established. Of the patients found in cardiac arrest ($N = 38$), 21 had a return of spontaneous circulation before admission to hospital. The survival status was known for 72 patients and, of these, the survival rate was 41.7%. The overall survival rate could have been as high as 59% or as low as 30%.

Borron and colleagues[89] undertook a prospective observational study between 1987 and 1994 to assess outcomes in patients treated with hydroxocobalamin at the scene or in the intensive care unit ($N = 69$). Adult victims of smoke inhalation, suspected of being poisoned by cyanide, as determined by soot on the face, or in the mouth, nose or expectorations, together with neurological impairment, received an intravenous infusion of 5 g hydroxocobalamin (maximum 15 g), blood samples for cyanide estimation having been collected before administration of hydroxocobalamin. Cyanide poisoning (predefined) was considered to be present when blood cyanide was greater than or equal to 39 μmol L^{-1} (1.01 mg L^{-1}). Fifty of the 69 patients (72%) admitted to the intensive care unit survived after administration of hydroxocobalamin. In the 42 individuals where cyanide poisoning was confirmed, 28 (67%) survived after hydroxocobalamin administration. Among the 15 adult patients who presented with cardiac arrest, two survived without permanent neurological deficit. Hydroxocobalamin was well tolerated whether cyanide was present or not.

A retrospective review of cases from the Emergency Medical Assistance Unit [Service d'Aide Médical d'Urgence (SAMU)] in France with 161 fire survivors with suspected or confirmed cyanide poisoning was carried out by Fortin and colleagues.[90] This review placed special emphasis on cardiac complications. From the text (but not the abstract or tabular data) it seems that some poisonings by cyanide salts were included. Cardiac arrest was seen in 61 cases (58 asystole, three ventricular fibrillation), while cardiac rhythm disorders were seen in 57 cases. All patients were given hydroxocobalamin (initial dose 5 g) and, if the response was poor, a second dose of 5 g hydroxocobalamin was administered. Of the patients initially in cardiac arrest, 30 died at the scene, 24 died in hospital and five survived arrest.

A serious confounder in some of the above studies and reports may be hyperbaric oxygen, as was pointed out by Desola.[91] Thus, 34 out of 42 patients reported by Borron *et al.*[89] who had blood cyanide concentrations above 39 μmol L^{-1} (1.01 mg L^{-1}) received hyperbaric oxygen therapy. Also, there are animal data reported by Lawson-Smith *et al.*[75] that compared hydroxocobalamin and hyperbaric oxygen therapy in the experimental treatment of cyanide poisoning of rats; the two therapies both had definite and comparable beneficial effects. One study[90] placed particular emphasis on the resuscitation of individuals with cardiac arrest, but Wing *et al.*[92] observed that, in smoke inhalation victims, there are reasons for cardiac arrest other than cyanide, including airway obstruction and hypoxia.

Some studies in dogs have shown an increase in blood pressure and Riou *et al.*[93,94] in one study showed that one of nine dogs developed generalised urticaria.[93] In humans most of the studies discussed above found few adverse reactions. In human volunteers, elevations in blood pressure were seen.[95,96] Uhl *et al.*[96] additionally reported itching, facial erythema and papulae, swelling, reddening of an eye and shivering in one volunteer, and dyspnoea, facial oedema, spontaneous exanthema and urticaria with wheals in another. A pustular rash was also noted in 11 of 66 volunteers treated with 5 g hydroxocobalamin. Some of the adverse effects seen with hydroxocobalamin are not dissimilar to those seen with dicobalt edetate. Høvding[97] reported a patient with no history of allergic disease, who had been treated for pernicious anaemia for many years with 0.2 mg cyanocobalamin at intervals of two to eight weeks. The treatment was changed to hydroxocobalamin and after the third injection the patient developed anaphylactic shock, with peripheral circulatory collapse, together with generalised itching. He was close to losing consciousness when adrenaline was administered, after which he slowly recovered. There have been a number of similar reports.[98–107] In the cases reported by Vidal and Lorenzo,[102] and Djuric and colleagues,[107] the patient was clearly tolerant of cyanocobalamin but not of hydroxocobalamin. In one case, there was a positive reaction to cyanocobalamin intradermal testing, suggesting type I hypersensitivity.[104] Doses of hydroxocobalamin used in pernicious anaemia and similar conditions are typically three orders of magnitude less than those used in cyanide poisoning, so these individuals would clearly be at risk if they were exposed to larger doses of hydroxocobalamin. Clearly hydroxocobalmin can cause adverse effects and the paucity of reports in the case series of use of this antidote is surprising and difficult to explain.

Erdman[2,108] has suggested that there is, as yet, insufficient evidence to demonstrate either the safety or clinical efficacy of hydroxocobalamin when administered in the post-inhalational exposure setting, bearing in mind that cyanide is rapidly absorbed and distributed, and essentially acts immediately. Nevertheless, the use of hydroxocobalamin in smoke victims has often been recommended, *e.g.*, by Borron *et al.*,[89] Dueñas-Laita *et al.*,[109] Anseeuw *et al.*[110–112] and others.[113,114]

Paulet[59] studied a number of cobalt compounds in addition to dicobalt edetate, including the chloride, glutamate and gluconate salt, as well as cobalt histidine. Sodium cobaltinitrite, which is a methaemoglobin former as well as a cobalt compound, has also been studied in experimental animals.[115,116] One answer to some of the disadvantages of hydroxocobalamin is to use other organic complexes of cobalt. One such is cobamide, which lacks the 5,6-dimethylbenzimidazole ribonucleotide and which produced better outcomes than saline or hydroxocobalamin in experimental sodium cyanide poisoning in rabbits. This precursor of hydroxocobalamin is more water-soluble than the latter compound and has, on a molar basis, a greater affinity for cyanide.[117,118] Two cobalt porphyrins have been studied: cobalt(III)-tetrakis(4-sulfonatophenyl)porphyrin[119] and cobalt(III)-5,10,15,20-tetrakis(4-sulfonatophenyl) porphyrin,[120] the former *in vitro* and the latter *in vivo* in mice.

15.3.3 Keto and Allied Compounds

A group of compounds contain keto (oxo) groups have been studied experimentally. They include glucose, sodium pyruvate,[121] α-oxoglutaric acid (2-oxoglutaric acid)[122] and pyridoxal-5-phosphate.[123] Unless one includes glucose given with and after dicobalt edetate, none has been used in humans.

15.3.4 Miscellaneous Substances Studied in the Experimental Treatment of Cyanide Poisoning

A mixed bag of other compounds has been studied in experimental animals with a view to ameliorating the clinical signs of cyanide poisoning. These include chlorpromazine,[124] naloxone,[125] meclofenoxate (centrophenoxine)[126] and phenoxybenzamine.[127,128]

15.3.5 Comparison of Cyanide Antidotes

Many studies have been carried out comparing the main cyanide antidotal regimes. Those carried out in rodents, where methaemoglobin-producing antidotes were used, are difficult to interpret as the much higher activity of erythrocytic methaemoglobin reductase in rodents makes the results of studies difficult to extrapolate to man. In that situation, dogs are the only species used in laboratories that resemble humans, with respect to red blood cell methaemoglobin reductase activity. Furthermore, some studies have used prophylactic protocols. Moreover, the pharmacokinetics of cyanide after orally administered cyanide salts (commonly used in experimental studies) is likely to be quite different from that after inhalation of hydrogen cyanide (see Chapter 10). Also, much of the clinical experience of cyanide antidotes has been of poisoning by cyanide salts, usually the potassium salt. Bearing in mind the difficulty of mimicking the situation in acute human poisoning in animal studies, the most that can be said is that all the main clinically used antidotes seem capable of resuscitating acutely poisoned experimental animals and patients. The reader is referred to reviews such as those by Paulet,[59] Marrs,[5] the International Programme on Chemical Safety,[13] Mégarbane *et al.*,[7] Thompson and Marrs,[71] Borron and Baud,[69] and Marraffa and colleagues.[129]

15.4 Treatment of Hydrogen Cyanide Poisoning Pre-hospital

It will have been noted that, with the exception of the perhaps not very effective inhalation of amyl nitrite, all cyanide antidotes have to be administered intravenously and generally in large (>10 mL) volumes. Depending on the number of casualties of a fire, this will present some difficulty. One possibility with hydroxocobalamin might be the use of the intraosseous route.[130–132]

15.5 Oxygen in Cyanide Poisoning

It used to be thought that oxygen had no role in the treatment of poisoning by cyanides as the blood was fully oxygenated in cyanide poisoning and it was utilisation that was stopped. However, a number of studies, for example, those of Way and Sheehy, have shown that oxygen enhances the antidotal effects of the classical cyanide antidotes.[133,134] The mechanism is unclear but oxygen enhancing detoxication or elimination of cyanide are the likely explanations: inhibition of cytochrome oxidase by cyanide is apparently unaffected by oxygen.[135,136] The US Agency for Toxic Substances and Disease Registry and the UK Health and Safety Executive (HSE) both advise that oxygen should be used in cyanide poisoning.[137,138] There is evidence in experimental animals for the beneficial effects of hyperbaric oxygen[75,139,140] and there have been favourable reports in humans where therapy has included hyperbaric oxygen, *e.g.*, by Goodhart.[141] Purser found improved clinical status following administration of normobaric oxygen in primates immediately after intoxication by HCN (see Chapter 10). Of course, in fires, oxygen will be given for reasons other than cyanide exposure such as anoxia and/or carbon monoxide poisoning.

15.6 Conclusion

The most important measure in HCN poisoning is the removal of the casualty from the source of exposure to HCN. There is evidence to suggest that oxygen is beneficial in cyanide poisoning. Very many antidotes for cyanide have been studied experimentally in animals, but a few of these have been used clinically. Those used clinically include sodium thiosulfate, which hastens the enzymatic detoxication of cyanide. Alternatively, hydrogen cyanide can be detoxified by complexation with heavy metals, either iron (Fe^{++}) in the form of drug-induced methaemoglobin or cobalt. The only methaemoglobin formers that have been used clinically are amyl nitrite, sodium nitrite and 4-dimethylaminophenol. Sodium nitrite, together with sodium thiosulfate, form the "classical therapy" for cyanide poisoning. Two cobalt compounds have been used clinically: dicobalt edetate (Kelocyanor) and hydroxocobalamin. Both appear effective, but the adverse effects of the former, particularly when used without substantial cyanide poisoning, have been emphasised. Hydroxocobalamin is not without side effects and the evidence base supporting its use in fire victims is limited. Further data are needed on the efficaciousness of hydroxocobalamin in HCN poisoning associated with fires.

References

1. B. Ballantyne, Combustion toxicology and implications for adverse human health effects, in *General and Applied Toxicology*, ed. B. Ballantyne, T. C. Marrs and T. Syversen, John Wiley and Sons Ltd, Chichester, 2009, pp. 2603–2630.
2. A. R. Erdman, Reply to letter, *Ann. Emerg. Med.*, 2008, **51**, 111.
3. Toxicological review of hydrogen cyanide and cyanide salts (CAS No. various). In support of summary information on the integrated risk information system (IRIS), United Sates Environmental Protection Agency, Washington DC, USA, September 2010, Available at http://www.epa.gov/iris/toxreviews/0060tr.pdf (accessed 14th August 2013).
4. B. Ballantyne, Toxiocology of cyanides, in *Clinical and Experimental Toxicology of Cyanides*, ed. B. Ballanytne and T. C. Marrs, Wright, Bristol, UK, 1987, pp. 41–126.
5. T. C. Marrs, Antidotal treatment of acute cyanide poisoning, *Adverse Drug React. Acute Poisoning Rev.*, 1988, **4**, 179–206.
6. J. E. Bright and T. C. Marrs, The pharmacokinetics of intravenous potassium cyanide, *Hum. Toxicol.*, 1988, **7**, 183–186.
7. B. Mégarbane, A. Delahaye, D. Goldgran-Tolédano and F. J. Baud, Antidotal treatment of cyanide poisoning, *J. Chin. Med. Assoc.*, 2003, **66**, 193–203.
8. G. E. Isom and J. D. Johnson, Sulphur donors in cyanide intoxication, in *Clinical and Experimental Toxicology of Cyanides*, ed. B. Ballantyne and T. C. Marrs, John Wright, Bristol, 1987, pp. 413–426.
9. D. M. Sylvester, W. L. Hayton, W. Schneiderhan and M. Kiese, Effects of thiosulfate on cyanide pharmacokinetics in dogs, *Toxicol. Appl. Pharmacol.*, 1983, **69**, 265–271.

10. D. Christel, P. Eyer, M. Hegemann, M. Kiese, W. Lörcher and M. Weger, Pharmacokinetics of cyanide poisoning in dogs, and the effect of 4-dimethylaminophenol or thiosulfate, *Arch. Toxicol.*, 1977, **38**, 177–189.
11. Material Safety Data Sheet, Sodium thiocyanate, Fisher Scientific, Fair Lawn, NJ, USA, Available at http://fscimage.fishersci.com/msds/21700.htm (accessed 14th August 2013).
12. D. L. Dennis and W. S. Fletcher, Toxicity of sodium thiosulfate (NSC-45624), a nitrogen mustard antagonist, in the dog, *Cancer Chemother. Rep.*, 1966, **50**, 255–257.
13. International Programme on Chemical Safety (IPCS) with the Commission of the European Community (CEC), Evaluation of Antidotes Series, in *Antidotes for Poisoning by Cyanide*, ed. T. J. Meredith, D. Jacobsen, J. A. Haines, J.-C. Berger and A. N. P. Van Heijst, Cambridge University Press, Cambridge, UK, 1993, vol. 2.
14. L. Frankenberg, Enzyme therapy in cyanide poisoning: effect of rhodanese and sulfur compounds, *Arch. Toxicol.*, 1980, **45**, 315–323.
15. S. I. Baskin and S. D. Kirkby, The effect of sodium tetrathionate on cyanide conversion of thiocyanate by enzymatic and non-enzymatic mechanisms, *J. Appl. Toxicol.*, 1990, **10**, 379–382.
16. S. I. Baskin, D. W. Porter, G. A. Rockwood, J. A. Romano, H. C. Patel, R. C. Kiser, C. M. Cook and A. L. Ternay, *In vitro* and *in vivo* comparison of sulfur donors as antidotes to acute cyanide intoxication, *J. Appl. Toxicol.*, 1999, **19**, 173–183.
17. P. Leung, R. W. Davis, C. C. Yao, E. P. Cannon and J. L. Way, Rhodanese and sodium thiosulfate encapsulated in mouse carrier erythrocytes. II *in vivo* survivability and alterations in physiologic and morphologic characteristics, *Fundam. Appl. Toxicol.*, 1984, **16**, 559–566.
18. N. Nagahara, T. Ito and M. Minami, Mercaptopyruvate sulfurtransferase as a defense against cyanide toxication: molecular properties and mode of detoxification, *Histol. Histopathol.*, 1999, **14**, 1277–1286.
19. J. L. Way, R. Holmes and J. L. Way, Cyanide antagonism with mercaptopyruvate, *Fed. Proc.*, 1985, **44**, 718.
20. H. T. Nagasawa, D. J. Goon, D. L. Crankshaw, R. Vince and S. E. Patterson, Novel, orally effective cyanide antidotes, *J. Med. Chem.*, 2007, **50**, 6462–6464.
21. K. G. Belani, H. Singh, D. S. Beebe, P. George, S. E. Patterson, H. T. Nagasawa and R. Vince, Cyanide toxicity in juvenile pigs and its reversal by a new prodrug sulfanegen sodium, *Anesth. Analg.*, 2012, **114**, 956–961.
22. M. Brenner, J. G. Kim, J. Lee, S. B. Mahon, D. Lemor, R. Ahdout, G. R. Boss, W. Blackledge, L. Jann, H. T. Nagasawa and S. E. Patterson, Sulfanegen sodium treatment in a rabbit model of sub-lethal cyanide toxicity, *Toxicol. Appl. Pharmacol.*, 2010, **248**, 269–276.
23. C. J. Clemedson, H. I. Hultman and B. Sörbo, The antidote effect of some sulfur compounds and rhodanese in experimental cyanide poisoning, *Acta Physiol. Scand.*, 1954, **32**, 245–251.

24. P. Hambright. Anti-cyanide drugs. Annual Summary Report, US Department of Defense: US Army Medical Research and Development Command, Fort Detrick, MD, USA, Unclassified Report, 1986.
25. A. Skold, D. L. Cosco and R. Klein, Methemoglobinemia: pathogenesis, diagnosis, and management, *South. Med. J.*, 2011, **104**, 757–761.
26. S. J. Moore, J. C. Norris, D. A. Walsh and A. S. Hume, Antidotal use of methemoglobin forming cyanide antagonists in concurrent carbon monoxide/cyanide intoxication, *J. Pharmacol. Exp. Ther.*, 1987, **242**, 70–73.
27. M. J. Percy, N. V. McFerran and T. R. Lappin, Disorders of oxidised haemoglobin, *Blood Rev.*, 2005, **19**(2), 61–68.
28. T. C. Marrs, S. Warren, Haematology and toxicology, in *General and Applied Toxicology*, ed. B. Ballantyne, T. C. Marrs, and T. Syversen, John Wiley, Chichester, UK, 3rd edn, 2009, pp. 743–762.
29. M. J. Percy, C. Barnes, G. Crighton, R. J. Leventer, R. Wynn and T. R. Lappin, Methemoglobin reductase deficiency: novel mutation is associated with a disease phenotype of intermediate severity, *J. Pediatr. Hematol./Oncol.*, 2012, **34**, 457–460.
30. A. Mansouri and A. A. Lurie, Concise review: methemoglobinemia, *Am. J. Hematol.*, 1993, **42**, 7–12.
31. L. G. Pedigo, Antagonism between amyl nitrite and prussic acid, *Trans. Med. Soc. Va.*, 1888, **19**, 124–131.
32. G. Bastian and H. Mercker, Zur der Frage der Zweckmäßigkeit der Inhalation von Amylnitrit in der Behandlung der Cyanidvergiftung, *Naunyn-Schmiedebergs Arch. Exp. Pathol. Toxikol.*, 1959, **237**, 285–295.
33. K. K. Chen, C. L. Rose and G. H. A. Clowes, Amyl nitrite and cyanide poisoning, *JAMA, J. Am. Med. Assoc.*, 1933, **100**, 1920–1922.
34. G. Paulet, Sur la valeur du nitrite d'amyle dans le traitement de l'intoxication cyanhydrique, *C. R. Soc. Biol.*, 1954, **148**, 1009–1014.
35. J. M. T. Pierce and M. S. Nielsen, Acute acquired methaemoglobinaemia after amyl nitrate poisoning, *Br. Med. J.*, 1989, **298**, 1566.
36. O. Lavon and Y. Bentur, Does amyl nitrite have a role in the management of pre-hospital mass casualty cyanide poisoning?, *Clin. Toxicol.*, 2010, **48**, 477–484.
37. E. Hug, Acción del nitrito de sodio y del hiposulfito de sodio en el tratamiento de la intoxicación provocada por el cianuro de potasio en el conejo, *Rev. Soc. Argent. Biol.*, 1933, **9**, 91–97.
38. K. K. Chen, C. L. Rose and G. H. A. Clowes, Methylene bue, nitrites and sodium thiosulfate against cyanide poisoning, *Proc. Soc. Exp. Biol. Med.*, 1933, **31**, 250–251.
39. K. K. Chen, C. L. Rose and G. H. A. Clowes, Comparative values of several antidotes in cyanide poisoning, *Am. J. Med. Sci.*, 1934, **188**, 767–781.
40. R. K. Holmes and J. L. Way, Mechanism of cyanide antagonism by sodium nitrite, *Pharmacologist*, 1982, **24**, 182.
41. R. Kruszyna, H. Kruszyna and R. P. Smith, Comparison of hydroxylamine, 4-dimethylaminophenol and nitrite protection against cyanide poisoning in mice, *Arch. Toxicol.*, 1982, **49**, 191–202.

42. J. L. Way, D. Sylvester, R. L. Morgan, G. E. Isom, G. E. Burrows, C. B. Tamulinas and J. L. Way, Recent perspectives on the toxicodynamic basis of cyanide antagonism, *Fundam. Appl. Toxicol.*, 1984, **4**(suppl.), S231–S239.
43. L. K. Cambal, M. R. Swanson, Q. Yuan, A. C. Weitz, H. H. Li, B. R. Pitt, L. L. Pearce and J. Peterson, Acute, sublethal cyanide poisoning in mice is ameliorated by nitrite alone: complications arising from concomitant administration of nitrite and thiosulfate as an antidotal combination, *Chem. Res. Toxicol.*, 2011, **24**, 1104–1112.
44. R. Klimmek, C. Krettek, L. Szinicz, P. Eyer and N. Weger, Effects and biotransformation of 4-dimethylaminophenol in man and dog, *Arch. Toxicol.*, 1983, **53**, 275–288.
45. M. Kiese and N. Weger, Formation of ferrihaemoglobin with aminophenols in the human for the treatment of cyanide poisoning, *Eur. J. Pharmacol.*, 1969, **7**, 97–105.
46. S. F. Hawkins, W. A. Groff, R. P. Johnson, H. L. Froehlich, A. Kaminskis and F. W. Stemler, Comparison of the *in vivo* formation of methemoglobin by 4-dimethyl aminophenol and sodium nitrite in the cynomolgus monkey, *Fed. Proc.*, 1981, **40**, 718.
47. M. Kiese, *Methemoglobinemia: A Comprehensive Treatise*, CRC Press, Boca Raton, FL, USA, 1974.
48. A. van Dijk, J. M. C. Douze, A. N. P. van Heijst, J. Glerum, Clinical evaluation of the cyanide antidote 4-DMAP, Proceedings of the 3rd World Congress of the World Federation of Associations of Clinical Toxicology and Poisons Control Centers and the 12th International Congress of the European Association of Poisons Control Centres, Brussels, Belgium, 27–30th Aug 1986, p. 49.
49. A. van Dijk, J. H. Glerum, A. N. P. van Heijst and J. M. C. Douze, Clinical evaluation of the cyanide antagonist 4-DMAP in a lethal cyanide poisoning case, *Vet. Hum. Toxicol.*, 1987, **29**(Suppl 2), 38–39.
50. A. N. P. van Heijst, J. M. C. Douze, R. G. van Kesteren, J. E. A. M. van Bergen and A. van Dijk, Therapeutic problems in cyanide poisoning, *Clin. Toxicol.*, 1987, **25**, 383–398.
51. L. L. Szinicz, Nephrotoxicität von Aminophenolen: Wirkung von 4-Dimethylaminophenol auf isolierte Nierentubuli von Ratten, *Fortschr. Med.*, 1979, **46**, 1206–1208.
52. N. P. Weger, Treatment of cyanide poisoning with 4-dimethylaminophenol (DMAP)—experimental and clinical overview, *Middle East J. Anesthesiol.*, 1990, **10**, 389–412.
53. W. Graffe, M. Kiese and E. Rauscher, The formation *in vivo* of *p*-hydroxylaminopropiophenone from *p*-aminopropiophenone and its action *in vivo* and *in vitro*, *Naunyn-Schmiedebergs Arch. Exp. Pathol. Pharm.*, 1964, **249**, 168–175.
54. T. C. Marrs, R. H. Inns, J. E. Bright and S. G. Wood, The formation of methaemoglobin by 4-aminopropiophenone (PAPP) and 4-(*N*-hydroxy)aminopropiophenone, *Hum. Exp. Toxicol.*, 1991, **10**, 183–188.

55. R. P. Ten Eyck, A. D. Schaerdel and W. E. Ottinger, Stroma-free methemoglobin solution: an effective antidote for acute cyanide poisoning, *Am. J. Emerg. Med.*, 1985, **3**, 519–523.
56. G. W. Boswell, D. E. Brooks, A. J. Murray, A. A. Doye, D. J. Disselhorst, C. L. Chin and C. B. Clifford, Exogenous methemoglobin as a cyanide antidote in rats, *Pharm. Res.*, 1988, **5**, 749–752.
57. P. H. Breen, S. A. Isserles, E. Tabac, M. F. Roizen and U. Z. Taitelman, Protective effect of stroma-free methemoglobin during cyanide poisoning in dogs, *Anesthesiology*, 1996, **85**, 558–564.
58. C. L. Evans, Cobalt compounds as antidotes for hydrocyanic acid, *Br. J. Pharmacol.*, 1964, **23**, 455–475.
59. G. Paulet, *L 'Intoxication Cyanhydrique et son Traitement*, Masson SA., Paris, France, 1960.
60. H. A. Schroeder, A. P. Nason and I. H. Tipton, Essential trace metals in man: cobalt, *J. Chronic Dis.*, 1967, **20**, 869–890.
61. G. J. A. Speijers, E. I. Krajnc, J. M. Berkvens and M. J. van Lolgten, Acute oral toxicity of inorganic cobalt compounds in rats, *Food Chem. Toxicol.*, 1982, **20**, 311–314.
62. J. D. Donaldson; D. Beyersmann, Cobalt and Cobalt Compounds, in *Ullmann's Encyclopedia of Industrial Chemistry*, Wiley-VCH, Weinheim, Germany, 2005. Available at http://onlinelibrary.wiley.com/doi/10.1002/14356007.a07_281.pub2/abstract (accessed 12th August 2013).
63. C. W. Mushett, K. L. Kelly, G. E. Boxer and J. C. Rickards, Antidotal efficacy of vitamin B_{12a} (hydroxo-cobalamin) in experimental cyanide poisoning, *Proc. Soc. Exp. Biol.*, 1952, **81**, 234–237.
64. Product list, Laboratories SERB, 53, rue Villiers de L'Isle-Adam, 75020 Paris France, Available at http://www.serb-labo.com/media/deliacms/media//1/154-36e815.pdf (accessed 12th August 2013).
65. E. F. K. McCandlish, T. K. Michael, J. A. Neal, E. C. Lingafelter and N. J. Rose, Comparison of the structures and aqueous solutions of [(*O*-phenylenediaminetetraacetato(2-)]cobalt(II) and [ethylenediaminetetraacetato(2-)]cobalt(II), *Inorg. Chem.*, 1978, **17**, 1383–1394.
66. D. D. Bryson, Acute industrial cyanide intoxication and its treatment, in *Clinical and Experimental Toxicology of Cyanides*, ed. B. Ballantyne and T. C. Marrs, John Wright, Bristol, 1987, pp. 348–358.
67. R. J. Flanagan, A. L. Jones, *Antidotes*, Taylor and Francis, London, England, 2001, pp. 232–233.
68. G. Shepherd and L. I. Velez, Role of hydroxocobalamin in acute cyanide poisoning, *Ann. Pharmacother.*, 2008, **42**, 661–669.
69. S. W. Borron and F. J. Baud, Antidotes for acute cyanide poisoning, *Curr. Pharm. Biotechnol.*, 2012, **13**, 1940–1948.
70. M. C. Reade, S. R. Davies, P. T. Morley, J. Dennett and I. C. Jacobs, Australian Resuscitation Council. Review article: management of cyanide poisoning, *Emerg. Med. Australas.*, 2012, **24**, 225–238.
71. J. P. Thompson and T. C. Marrs, Hydroxocobalamin in cyanide poisoning, *Clin. Toxicol.*, 2012, **50**, 875–885.

72. M. Schiff, J. F. Benoist, B. Tilea, N. Royer, S. Giraudier and H. Ogier de Baulny, Isolated remethylation disorders: do our treatments benefit patients?, *J. Inherited Metab. Dis.*, 2011, **34**, 37–145.
73. European Medicines Agency, A. Summary of the European Public Assessment Report (EPAR), European Medicines Agency, Cyanokit London, UK, 2012 Apr 23 (cited 2012 Jul 3), Available at http://www.ema.europa.eu/ema/index.jsp?curl = pages/medicines/human/medicines/000806/human_med_000731.jsp&mid = WC0b01ac058001d124/ (accessed 12th August 2012).
74. R. C. Hatch, D. P. Laflamme and A. V. Jain, Effects of various known and potential cyanide antagonists and a glutathione depletor on acute toxicity of cyanide in mice, *Vet. Hum. Toxicol.*, 1990, **32**, 9–16.
75. P. Lawson-Smith, N. V. Olsen and O. Hyldegaard, Hyperbaric oxygen therapy or hydroxycobalamin attenuates surges in brain interstitial lactate and glucose; and hyperbaric oxygen improves respiratory status in cyanide-intoxicated rats, *Undersea Hyperbaric Med.*, 2011, **38**, 223–237.
76. K. Mengel, W. Kramer, B. Isert and K. D. Friedberg, Thiosulfate and hydroxocobalamin prophylaxis in progressive cyanide poisoning in guinea-pigs, *Toxicology*, 1989, **54**, 335–342.
77. M. A. Posner, R. E. Tobey and H. McElroy, Hydroxocobalamin therapy of cyanide intoxication in guinea pigs, *Anesthesiology*, 1976, **44**, 157–160.
78. C. L. Rose, R. M. Worth and K. K. Chen, Hydroxo-cobalamine and acute cyanide poisoning in dogs, *Life Sci.*, 1965, **4**, 1785–1789.
79. S. W. Borron, M. Stonerook and F. Reid, Efficacy of hydroxocobalamin for the treatment of acute cyanide poisoning in adult beagle dogs, *Clin. Toxicol.*, 2006, **44**, 5–15.
80. V. S. Bebarta, D. A. Tanen, J. Lairet, P. S. Dixon, S. Valtier and A. Bush, Hydroxocobalamin and sodium thiosulfate *versus* sodium nitrite and sodium thiosulfate in the treatment of acute cyanide toxicity in a swine (*Sus scrofa*) model, *Ann. Emerg. Med.*, 2010, **55**, 345–351.
81. V. S. Bebarta, R. L. Pitotti, P. S. Dixon, S. Valtier, L. Esquivel, A. Bush and C. M. Little, Hydroxocobalamin and epinephrine both improve survival in a swine model of cyanide-induced cardiac arrest, *Ann. Emerg. Med.*, 2012, DOI: 10.1016/j.annemergmed.2012.02.002.
82. V. S. Bebarta, R. L. Pitotti, P. Dixon, J. R. Lairet, A. Bush and D. A. Tanen, Hydroxocobalamin *versus* sodium thiosulfate for the treatment of acute cyanide toxicity in a swine (*Sus scrofa*) model, *Ann. Emerg. Med.*, 2012, **59**, 532–539.
83. National Poisons Information Service, 2011, Annual Report 2010/2011, Centre for Radiation, Chemical and Environmental Hazards, Health Protection Agency, Chilton, UK, October 2011, 37-38, Available at http://www.hpa.org.uk/Publications/ChemicalsPoisons/NationalPoisonsInformationServiceAnnualReports/1110NPISAnnualreport20102011/ (accessed 12th August 2013).

84. J. L. Fortin, G. Capellier, C. Manzon, J. Giocanti and O. Gall, Intraosseous administration of hydroxocobalamin in the acute treatment of cyanide poisoning, *Burns*, 2009, **35**, S15–S16.
85. J. L. Fortin, V. Judic-Peureux, T. Desmettre, C. Manzon, D. Grimon, U. Hostalek, C. Fetro and G. Capellier, Hydrogen cyanide poisoning in a prison environment: a case report, *J. Correct. Health Care*, 2011, **17**, 29–33.
86. E. J. Roderique, A. A. Gebre-Giorgis, D. H. Stewart, M. J. Feldman and A. L. Pozez, Smoke inhalation injury in a pregnant patient: a literature review of the evidence and current best practices in the setting of a classic case, *J. Burn Care Res.*, 2012, **33**, 624–633.
87. P. Houeto, J. R. Hoffman, M. Imbert, P. Levillain and F. J. Baud, Relation of blood cyanide to plasma cyanocobalamin concentration after a fixed dose of hydroxocobalamin in cyanide poisoning, *Lancet*, 1995, **346**, 605–608.
88. J. L. Fortin, J. P. Giocanti, M. Ruttimann and J. J. Kowalski, Prehospital administration of hydroxocobalamin for smoke inhalation-associated cyanide poisoning: 8 years of experience in the Paris Fire Brigade, *Clin. Toxicol.*, 2006, **44**, 37–44.
89. S. W. Borron, F. J. Baud, P. Barriot, M. Imbert and C. Bismuth, Prospective study of hydroxocobalamin for acute cyanide poisoning in smoke inhalation, *Ann. Emerg. Med.*, 2007, **49**, 794–801.
90. J.-L. Fortin, T. Desmettre, C. Manzon, V. Judic-Peureux, C. Peugeot-Mortier, J.-P. Giocanti, M. Hachelaf, M. Grangeon, U. Hostalek, J. Crouzet and G. Capellier, Cyanide poisoning and cardiac disorders: 161 cases, *J. Emerg. Med.*, 2010, **38**, 467–476.
91. J. Desola, Hydroxycobalamin, hyperbaric oxygen and cyanide poisoning, *Undersea Hyperbaric Med.*, 2011, **38**, 217–220.
92. K. S. Wing, C. Y. Cheung and L. F. Lung, Hydroxocobalamin for acute cyanide poisoning in smoke inhalation, *Ann. Emerg. Med.*, 2008, **51**, 108.
93. B. Riou, J. L. Gerard, C. D. La Rochelle, R. Bourdon, A. Berdeaux and J. F. Giudicelli, Hemodynamic effects of hydroxocobalamin in conscious dogs, *Anesthesiology*, 1991, **74**, 552–558.
94. B. Riou, A. Berdeaux, E. Pussard and J. F. Giudicelli, Comparison of the hemodynamic effects of hydroxocobalamin and cobalt edetate at equipotent cyanide antidotal doses in conscious dogs, *Intensive Care Med.*, 1993, **19**, 26–32.
95. J. C. Forsyth, P. D. Mueller, C. E. Becker, J. Osterloh, N. L. Benowitz, B. H. Rumack and A. H. Hall, Hydroxocobalamin as a cyanide antidote: safety, efficacy and pharmacokinetics in heavily smoking normal volunteers, *J. Toxicol. Clin. Toxicol.*, 1993, **31**, 277–294.
96. W. Uhl, A. Nolting, G. Golor, K. L. Rost and A. Kovar, Safety of hydroxocobalamin in healthy volunteers in a randomized, placebo-controlled study, *Clin. Toxicol.*, 2006, **44**, 17–28.
97. G. Høvding, Anaphylactic reaction after injection of vitamin B12, *Br. Med. J.*, 1968, **3**, 102.

98. S. Dally and M. Gaultier, Choc anaphylactique dû à l'hydroxocobalamine, *Nouv. Presse Med.*, 1976, **5**, 1917.
99. C. N. Ugwu and F. J. Gibbins, Anaphylactic reaction to vitamin B_{12} appearing after several years of therapy, *Age Ageing*, 1981, **10**, 196–197.
100. J. James and R. P. Warin, Sensitivity to cyanocobalamin and hydroxocobalamin, *Br. Med. J.*, 1971, **2**, 262.
101. I. Pevny, A. Hartmann and J. Metz, Vitamin-B12-(Cyanocobalamin). Allergie, *Hautarzt*, 1977, **28**, 600–603.
102. C. Vidal and A. Lorenzo, Anaphylactoid reaction to hydroxycobalamin with tolerance of cyanocobalamin, *Postgrad. Med. J.*, 1998, **74**, 702.
103. F. Bilwani, S. N. Adil, U. Sheikh, A. Humera and M. Khurshid, Anaphylactic reaction after intramuscular injection of cyanocobalamin (vitamin B12): a case report, *J. Pak. Med. Assoc.*, 2005, **55**, 217–219.
104. F. J. Moloney, R. Hughes, D. O'Shea and B. Kirby, Type I immediate hypersensitivity reaction to cyanocobalamin but not hydroxycobalamin, *Clin. Exp. Dermatol.*, 2008, **33**, 412–414.
105. M. Branco-Ferreira, M. H. Clode, M. A. Pereira-Barbosa and A. G. Palma-Carlos, Anaphylactic reaction to hydroxycobalamin, *Allergy*, 1997, **52**, 118–119.
106. R. Tordjman, T. Genereau, M.-T. Guinnepain, A. Weyer, O. Lortholary, I. Royer, P. Casassus and L. Guillevin, Reintroduction of vitamin B_{12} in 2 patients with prior B_{12}-induced anaphylaxis, *Eur. J. Haematol.*, 1998, **60**, 269–270.
107. V. Djuric, M. Bogic, A. P. Popadic, V. T. Spiric and S. Raskovic, Anaphylactic reaction to hydroxycobalamin with tolerance to cyanocobalamin, *Ann. Allergy, Asthma, Immunol.*, 2012, **108**, 207–208.
108. A. R. Erdman, Is hydroxocobalamin safe and effective for smoke inhalation? Searching for guidance in the haze, *Ann. Emerg. Med.*, 2007, **49**, 814–816.
109. A. Dueñas-Laita, G. Burillo Putze, J. R. Alonso, A. Bajo, B. Climent, E. Corral, F. Felices, A. Ferrer, M. P. Hernández Frutos, S. Nogué and J. Puiguriguer, Bases del manejo clínico de la intoxicación por humo de incendios, Docohumo Madrid 2010, *Med. Intensiva*, 2010, **34**, 609–619.
110. K. Anseeuw, N. Delvau, G. Burillo-Putze, F. De Iaco, G. Geldner, P. Holmström, Y. Lambert and M. Sabbe, Cyanide poisoning by fire smoke inhalation: a European expert consensus, *Eur. J. Emerg. Med.*, 2013, **20**, 2–9.
111. V. S. Bebarta, Comment on antidotes for cyanide poisoning, *Eur. J. Emerg. Med.*, 2013, **20**, 65–66.
112. K. Anseeuw, N. Delvau, G. Burillo-Putze, F. De Iaco, G. Geldner, P. Holmström, Y. Lambert, M. Sabbe and K. Delvau, Antidotes for cyanide poisoning – author's reply, *Eur. J. Emerg. Med.*, 2013, **20**, 66–67.
113. D. J. O'Brien, D. W. Walsh, C. M. Terriff and A. H. Hall, Empiric management of cyanide toxicity associated with smoke inhalation, *Prehosp. Disaster Med.*, 2011, **26**, 374–382.

114. G. Geldner, E. M. Koch, U. Gottwald-Hostalek, F. Baud, G. Burillo, J. P. Fauville, F. Levi, C. Locatelli and T. Zilker, Report on a study of fires with smoke gas development: Determination of blood cyanide levels, clinical signs and laboratory values in victims, *Anaesthesist*, 2013, **62**, 69–616.
115. R. P. Smith, Cobalt salts: effects in cyanide and sulfide poisoning and on methemoglobinomia, *Toxicol. Appl. Pharmacol.*, 1969, **15**, 505–516.
116. M. Jianyao, W. Yu and Q. Laifa, Antidotal effect of intramuscularly administered sodium cobaltinitrite on cyanide poisoning, *J. Med. Coll. PLA*, 1987, **2**, 301–305.
117. M. Brenner, S. B. Mahon, J. Lee, J. Kim, D. Mukai, S. Goodman, K. A. Kreuter, R. Ahdout, O. Mohammad, V. S. Sharma, W. Blackledge and G. R. Boss, Comparison of cobinamide to hydroxocobalamin in reversing cyanide physiologic effects in rabbits using diffuse optical spectroscopy monitoring, *J. Biomed. Opt.*, 2010, **15**, 017001.
118. M. Brenner, J. G. Kim, S. B. Mahon, J. Lee, K. A. Kreuter, W. Blackledge, D. Mukai, S. Patterson, O. Mohammad, V. S. Sharma and G. R. Boss, Intramuscular cobinamide sulfite in a rabbit model of sublethal cyanide toxicity, *Ann. Emerg. Med.*, 2010, **55**, 352–363.
119. P. Hambright and R. J. Langley, Cyanide scavengers: kinetics of the reactions of cyanide with a water soluble cobalt(III) porphyrin, *J. Inorg. Biochem.*, 1988, **32**, 197–205.
120. W. D. McGuinn, L. Baxter, L. Pei, I. Petrikovics, E. P. Cannon and J. L. Way, Antagonism of the lethal effects of cyanide by a synthetic water-soluble cobalt(III) porphyrin compound, *Fundam. Appl. Toxicol.*, 1994, **33**, 76–80.
121. C. Schwartz, R. L. Morgan, L. M. Way and J. L. Way, Antagonism of cyanide intoxication with sodium pyruvate, *Toxicol. Appl. Pharmacol.*, 1979, **50**, 437–441.
122. R. Bhattacharya and R. Vijayaraghavan, Promising role of α-ketoglutarate in protecting against the lethal effects of cyanide, *Hum. Exp. Toxicol.*, 2002, **21**, 297–303.
123. R. L. Keniston, S. Calbellon and K. S. Yarborough, Pyridoxal 5′- phosphate as an antidote for cyanide, spermine, gentomycin and dopamine toxicity: an *in vivo* rat study, *Toxicol. Appl. Pharmacol.*, 1987, **88**, 433–441.
124. J. L. Way and G. Burrows, Cyanide intoxication: protection with chlorpromazine, *Toxicol. Appl. Pharmacol.*, 1976, **36**, 93–97.
125. P. Leung, D. M. Sylvester, F. Chiou, L. Way and J. L. Way, Effect of naloxone HCl on cyanide intoxication, *Fed. Proc.*, 1984, **43**, 545.
126. S. Rump and Z. Edelwejn, Effects of centrophenoxone on electrical activity of the rabbit brain in cyanide intoxication, *Int. J. Neuropharmacol.*, 1968, **7**, 103–113.
127. G. E. Burrows and J. L. Way, Antagonism of cyanide toxicity with phenoxybenzamine, *Fed. Proc.*, 1976, **35**, 533.
128. J. A. Vick and H. L. Froehlich, Studies of cyanide poisoning, *Arch. Int. Pharmacocdyn. Ther.*, 1985, **273**, 314–322.

129. J. M. Marraffa, V. Cohen and M. A. Howland, Antidotes for toxicological emergencies: a practical review, *Am. J. Health-Syst. Pharm.*, 2012, **69**, 199–212.
130. S. W. Borron, J. C. Arias, C. R. Bauer, M. Sanchez, M. Fernández and I. Jung, Hemodynamics after intraosseous administration of hydroxocobalamin or normal saline in a goat model, *Am. J. Emerg. Med.*, 2009, **27**, 1065–1071.
131. S. W. Borron, J. C. Arias, C. R. Bauer, T. Philbeck, P. Hass, W. Lawson, D. Montez, M. Fernández, I. Jung and D. J. Gordon, Intraosseous line placement for antidote injection by first responders and receivers wearing personal protective equipment, *Am. J. Emerg. Med.*, 2011, **29**, 373–381.
132. D. B. Murray, M. Eddleston, S. Thomas, R. D. Jefferson, A. Thompson, M. Dunn, D. S. Vidler, R. E. Clutton and P. G. Blain, Rapid and complete bioavailability of antidotes for organophosphorus nerve agent and cyanide poisoning in minipigs after intraosseous administration, *Ann. Emerg. Med.*, 2012, **60**, 424–430.
133. J. L. Way, Cyanide intoxication and its mechanism of antagonism, *Annu. Rev. Pharmacol. Toxicol.*, 1984, **24**, 451–481.
134. M. Sheehy and J. L. Way, Effect of oxygen on cyanide intoxication III mithridate, *J. Pharmacol. Exp. Ther.*, 1968, **161**, 163–168.
135. J. L. Way, S. L. Gibbon and M. Sheehy, Effect of oxygen on cyanide intoxication 1. Prophylactic protection, *J. Pharmacol. Exp. Ther.*, 1966, **153**, 381–385.
136. G. E. Isom and J. L. Way, Effects of oxygen on the antagonism of cyanide intoxication: cytochrome oxidase, *in vitro*, *Toxicol. Appl. Pharmacol.*, 1984, **74**, 57–62.
137. Agency for Toxic Substances and Disease Registry, Medical management guidelines for hydrogen cyanide, March 3, 2011, Agency for Toxic Substances and Disease Registry, 4770 Buford Hwy NE, Atlanta, GA 30341, USA, 2011, Available at http://www.atsdr.cdc.gov/mmg/mmg.asp?id = 1073&tid = 19 (accessed 14th August 2013).
138. Health and Safety Executive (HSE), Cyanide poisoning – New recommendations on first aid treatment, HSE, Bootle, UK, 1997, Available at http://www.hse.gov.uk/pubns/misc076.htm (accessed 14th August 2013).
139. T. Takano, Y. Miyazaki, I. Nashimoto and K. Kobayashi, Effect of hyperbaric oxygen on cyanide intoxication: *in situ* changes in intracellular oxidation reduction, *Undersea Biomed. Res.*, 1980, **7**, 191–197.
140. P. Lawson-Smith, E. C. Jansen, L. Hilsted, A. H. Johnsen and O. Hyldegaard, Effect of acute and delayed hyperbaric oxygen therapy on cyanide whole blood levels during acute cyanide intoxication, *Undersea Hyperbaric Med.*, 2011, **38**, 17–26.
141. G. L. Goodhart, Patient treated with antidote kit and hyperbaric oxygen survives cyanide poisoning, *South. Med. J.*, 1994, **87**, 814–816.

CHAPTER 16

Acute Lung Injury Following Inhalation of Irritant Products of Combustion

DAVID J. BAKER

Emeritus Consultant Anaesthesiologist SAMU de Paris Hopital Necker - Enfants Malades, Paris, France
Email: djbaker@orange.fr

16.1 Introduction

There are many irritant substances that cause acute injury to the respiratory tract and lungs following inhalation. Examples include released toxic industrial chemicals and chemical warfare agents.[1] However, injury from inhalation of smoke that contains both particulates and various irritant gases remains the most familiar cause of inhalational lung injury, and dealing with this is an important part of the management of burn injury.

The management of burns and smoke inhalation is a very large subject. This chapter does not attempt to cover in detail the pathophysiology and medical management of lung injury following the inhalation of irritant gases from combustion, but rather tries to provide an overview of the subject for the non-specialist. Although this chapter concerns inhalation injury from products of combustion, there are lessons from research that has been

Issues in Toxicology No. 23
Toxicology, Survival and Health Hazards of Combustion Products
Edited by David A. Purser, Robert L. Maynard and James C. Wakefield

Published by the Royal Society of Chemistry, www.rsc.org

carried out in inhalation injury from other causes. Recent developments in this area will therefore be considered.

Acute lung injury (ALI) and acute respiratory distress syndrome (ARDS) have been the subject of a large number of publications over the past three decades and there are a number of detailed reviews of developments in management.[2–8]

Specifically, the chapter will consider:

- The nature of irritant products of combustion and where they act on the lung and respiratory tract.
- The pathophysiology of ALI and the respiratory tract caused by irritant gases of combustion, which may lead to ARDS.
- Clinical presentation following inhalation of irritant products of combustion.
- The management of airway injury and ALI at the pre-hospital and hospital levels.
- Gaps in knowledge at the present time.

16.2 Definitions

The term 'acute lung injury' strictly means damage to the lung tissue containing the air sacs (alveoli) and their surrounding structures. However, injury from toxic combustion products can also affect the large and small airways leading to the lung tissue. This should strictly be termed 'respiratory tract injury'.

Toxic inhalants cause injury through a variety of different mechanisms, including direct irritation of the respiratory tract mucosa, leading to asphyxiation. The nature and extent of the acute injury depends on the nature of the inhaled chemical and its properties. Smoke injury is a term used to define the effects of inhalation of products of combustion containing both particulate matter and irritant gases. In smoke injury, victims are exposed to a large number of inhaled toxic chemicals, which cause chemical actions at all lung levels, and may also suffer thermal injury to the respiratory tract.

16.3 Classification of Inhaled Products of Combustion

Combustion produces a wide range of potential irritants.[9,10] The composition of smoke varies widely and depends on the nature of the fire, its temperature and what is burning. The considerations for the damage these cause after inhalation are:

- particulate or gaseous form;
- chemical properties, acid or alkaline, and solubility;
- temperature of inhaled gases—burn injury to the respiratory tract.

16.3.1 Asphyxiant Gases

These may be divided into:

- Active, with a direct toxic action on key systems in the body. These include:
  - Carbon monoxide (see Chapter 9).
  - Hydrogen cyanide (see Chapter 10).
- Passive, where hypoxia is caused without damage to the lung. These include:
  - Carbon dioxide (see Chapter 14).
  - Oxygen depletion in the inhaled atmosphere (see Chapter 14).

16.3.2 Irritant Gases

Irritant gases are defined as gases that are generated by the combustion of a range of products, both natural (*e.g.*, wood) and synthetic (*e.g.*, plastics). Combustion may be smouldering, flaming or ventilation-controlled. The irritant gases evolved during combustion and the rate of generation depend on the mode of combustion relative to temperature and ventilation (see Chapters 2 and 3).

Irritant gases can be divided into two groups:

- Inorganic.
- Organic.

The detailed properties of both these classes of irritant gas are considered in detail elsewhere in the book (see Chapters 7, 13 and 14). A summary of the properties of common specific irritant products of combustion is shown in Box 16.1.

16.4 Actions of Irritant Products of Combustion on the Lung and Respiratory Tract

The gases and smokes produced by combustion will have different actions depending on where they are acting. Thus, thermal burns are mainly found in the upper respiratory tract, whereas chemical effects, such as chemical bronchiolitis and toxic pulmonary oedema, are found in the lower tract and lung parenchyma. The signs and symptoms (toxidromes) following toxic inhalation injury are a function of the action of the chemical on specific sites. These may be divided between the upper and lower respiratory tract.

The signs and symptoms of inhalation of irritant products of combustion may conveniently be divided into immediate and delayed.

16.4.1 Immediate Actions

Immediate actions of inhaled irritant products may be on both the upper and lower respiratory tract (see Chapter 7).

Box 16.1 Properties of irritant gaseous products of combustion (adapted from Wakefield and Maynard, 2012)[9]

Halogen acids

HCl: This is the most important irritant produced during combustion. Key points in the action of inhaled HCl produced by combustion are: (1) the water solubility of HCl means that its effects will predominate in the upper respiratory tract, but pulmonary irritation can occur with alveolar damage and pulmonary oedema following exposure to higher concentrations. (2) In relation to phosgene, another pulmonary irritant that can be produced as a product of combustion (*vide infra*), HCl was thought to be the end stage product causing damage at the parenchymal level; however, this was never proven. There is an interesting comparison with chlorine (not a product of combustion), which also acts on the upper respiratory tract as a result of its water solubility.

HBr: Produced as a result of brominated flame retardant compounds or from the combustion of high impact polystyrene.

There are few data available about the toxicity of HBr but its effects are expected to be the same as another halogenated acid, hydrogen fluoride. A notable point about HBr is that exposure to low concentrations (5 ppm) would alert the victim to the potential danger from exposure. However, higher concentrations (200 ppm) produce disabling sensory irritation sufficient to prevent escape from the hazard area. There are recorded cases of death following exposure to 1300–2000 ppm.

HF: This halogenated acid can be produced as a product of combustion of fluorine-containing polymers, *e.g.*, polytetrafluoroethane (PTFE). In conjunction with perfluoroisobutene (PFIB), there is a severe risk of the production of toxic pulmonary oedema. PFIB was regarded as a potential chemical warfare (CW) agent at one stage of the Cold War. HF is known to be a powerful sensory irritant and 5 ppm for a 1 hour exposure causes irritation and corrosion of the mucous membranes of the upper respiratory tract. Volunteer studies have shown lower airway irritancy levels of 0.2–2.9 ppm. Exposure at this concentration gives rise to chest tightness, coughing and wheezing. At 3–6 ppm these symptoms become more severe. Animal data show that HF is the most potent of the halogenated acid gases but is equivalent to HCl and HBr based upon the endpoint of sensory irritation.

Sulphur dioxide

This gas can be produced from burning fossil fuels, *e.g.*, coke and from sulfur-containing compounds, *e.g.*, vulcanised rubber tyres. It is extremely irritant, particularly in sensitive individuals where severe bronchoconstriction of asthmatic proportions can occur. In known asthmatics a concentration of only 0.4 ppm is sufficient to cause bronchoconstriction.

In addition, longer term inhalation of SO_2 for 1–6 hours has been shown to increase airway resistance and decrease forced expiratory volume in 1 second (FEV1) results and forced expiratory flow.

Exposure to high concentrations can be fatal due to asphyxiation from blockage of the upper and lower airways.

Nitrogen oxides

Various forms of nitrogen oxides are commonly present in combustion products. Hence, the term NO_x is used rather than specifying the most commonly encountered compounds, NO_2 or N_2O_4. Any nitrogen-containing compound can produce NO_x depending on the combustion conditions and temperature. NO is formed first and then oxidised to higher oxides.

NO_2 is known to be an oxidant and irritant, leading directly to acute lung injury and toxic pulmonary oedema. Hydrolysis to HNO_3 is suspected as a pathophysiological pathway.

Respiratory symptoms are apparent from breathing the gas at a level of 2.5 ppm—studies have shown increased airway resistance at this level in volunteers.

Toxic pulmonary oedema may start from exposure as low as 2 ppm. Persons who are known asthmatics are more susceptible to the action of nitrogen oxides.

Phosphorus pentoxide

This is an important product of combustion due to its presence in fire retardant compounds, *e.g.*, isopropylated triphenyl phosphate compounds in polyurethane foam. These compounds are widely used in many domestic appliances. The pathophysiological action may be due to the formation of orthophosphoric acid.

Ammonia, chlorine and phosgene

These compounds are not inorganic acid gases but can all be produced by combustion.

Organic irritants

Combustion produces a number of organic irritant gases such as acrolein and formaldehyde. In addition, there are over 100 listed polyaromatic hydrocarbons (PAHs), which contain at least two fused benzene rings in the molecular structure. This group includes substances such as benzopyrene, napthalene and anthracene. These compounds are known to have longer term carcinogenic properties.

Other compounds, including dioxins, cause longer term problems. However, these compounds do not cause acute lung injury as defined in the text.

Isocyanates can have early and late effects on the respiratory system and are produced by combustion of nitrogen-containing polymers.

The upper tract includes the nose, pharynx, larynx, trachea and the main bronchi. Actions on these sites include irritation, an increase in secretions, sternutation (sneezing), laryngeal spasm, shortness of breath and coughing. In addition there will be actions on the eyes, causing lachrymation.

The lower tract includes the smaller bronchi, bronchioles (both the lower and terminal divisions) and the alveoli. The actions here are bronchospasm and bronchiolitis (inflammation of the bronchioles). These actions occur within a short period of time (short latency).

16.4.2 Delayed Actions

Following the immediate actions of products of combustion on the upper tract there may be clinical effects that occur with a longer latency. These are mainly due to effects on the terminal bronchioles and alveoli of the lung. Actions at these sites lead to the development of acute lung injury (ALI) and acute respiratory distress syndrome (ARDS) (see Section 16.6).

16.5 Respiratory Failure Following Inhalation of Products of Combustion

Respiratory failure may be defined as a failure of oxygen to be transported to or used at the mitochondrial sites in the body cells.

Respiratory failure is classified as: (1) type 1, where there is a failure of transfer of oxygen from the alveoli due to failure of diffusion of oxygen through the alveolar walls. This is usually a later consequence of acute lung injury, which is discussed in more detail in Section 16.6. (2) Type 2, where there is a failure of ventilation (movement of gas in and out of the lung). Type 2 failure occurs when the airway becomes blocked or breathing is reduced due to toxic effects on the medullary control mechanisms. In a clinical setting, type 2 failure is often accompanied by an increase in the alveolar and arterial partial pressures of carbon dioxide ($PACO_2$ and $PaCO_2$, respectively); in type 1 failure, $PaCO_2$ may be, for some time, maintained within normal limits. A further cause of cellular respiratory failure is due to direct actions on the cytochrome system in the mitochondria, as in the case of cyanide inhalation.

16.6 Acute Lung Injury and Acute Respiratory Distress Syndrome

ALI and ARDS are the most serious consequences of the inhalation of irritant products of combustion producing type 1 respiratory failure. Acute lung injury may be defined as a characteristic form of parenchymal lung disease representing a wide range of severity, from short-lived dyspnoea to a rapidly terminal failure of the respiratory system, when the term acute respiratory distress is used.[11]

The following criteria are required for the diagnosis of ALI:[12]

- Acute onset of impaired oxygenation.
- Severe hypoxaemia defined as a ratio of PaO_2 to fractional inspired oxygen (FiO_2) of ≤300 mm Hg.
- Bilateral diffuse infiltration seen on chest X-ray (CXR).
- A normal pulmonary artery wedge pressure of ≤18 mm Hg, which excludes causes of cardiogenic pulmonary oedema.

ARDS is diagnosed in the same way, except that there is a worse impairment of gas exchange with PaO_2:FiO_2 ≤200 mm Hg. The clinical risks of developing ARDS following ALI have been reviewed by Hudson *et al.*[13]

Inhalation of irritant gases is only one of many causes of ALI. These may be grouped into (1) direct lung injury, as with the inhalation of products of combustion or water in the case of near drowning; and (2) indirect lung injury, following severe traumatic shock and sepsis. In the case of ALI from inhalation of products of combustion the cause will be relatively clear. However, it is sometimes difficult to separate ALI from other diseases, such as diffuse pneumonia or fibrosing alveolitis, which may present with similar features.

It has been suggested[14] that ALI should be divided into (1) pulmonary ALI, where there is a direct insult to the lung tissue, as in the case of inhalation of gases; and (2) extrapulmonary ALI, caused by conditions such as severe sepsis and non-thoracic trauma. There are known to be differences in the clinical outcome of ALI from these two overall causes.[15]

ALI has been studied intensively, and risks and management have been largely based upon information from ALI caused by conditions other than the inhalation of irritant gases.[7] It is known that, overall, patients with a single risk factor have a 25% chance of developing ARDS. However, the presence of two factors increases this chance to 42%.[16] There is agreement that the overall mortality from ARDS is high.

16.6.1 Clinical Features of ALI

Four phases of ALI have been described, which are applicable to its cause by inhalation of irritant gases.[11] These are:

1. Dyspnoea and tachypnoea with a normal CXR. This phase usually lasts about 24 hours.
2. Hypoxaemia is present but the arterial ($PaCO_2$) remains normal or subnormal. Minor abnormalities of the CXR may be present. This phase usually lasts 24–48 hours.
3. Evidence of the diagnostic criteria of ALI as given above. There is significant arterial hypoxaemia with slight elevation of the $PaCO_2$. The CXR shows evidence of diffuse bilateral infiltrates and the lungs become stiff. Ventilatory support usually required at this stage.
4. Established ARDS with massive bilateral consolidation and progressive hypoxaemia, even when ventilated with very high oxygen concentrations.

16.6.2 Pathophysiology of ALI

Several aspects of the damage mechanisms causing ALI are clear from clinical level investigation when managing acute inhalational ALI. Information from the CXR may be supplemented by computerised tomography (CT) scans, which reveal a heterogeneous distribution of the opacities, representing collapsed lung.[14] These can move with a change in posture. A major shunt (blood reaching the arterial system without passing through ventilated regions of the lungs) is evident, which usually cannot be corrected by increasing FiO_2. CT scans also show areas of lung overdistension, which may contribute to increased dead space.[16] The lungs themselves show major changes in mechanics. There is a decrease in lung compliance with ALI from pulmonary causes. There is also an increase in mean total resistance.

Histopathological studies in ALI differentiate between an acute and chronic stage. The acute stage is characterised by damage to the blood–gas barrier with extensive damage to the type 1 alveolar epithelial cells.[17] This leads to a leakage of protein-containing fluid into the alveoli, together with red blood cells (RBCs), leucocytes and strands of fibrin. There is intravascular coagulation. In the chronic or proliferative stage there is attempted repair of the damage listed in the acute stage, leading to ARDS. The destroyed type 1 cells are replaced by type 2 cells, which do not differentiate into type 1 cells. After a few days there is fibrosis, which ultimately overwhelms the lung tissue and usually leads to death.

16.6.3 Cellular Mechanisms of ALI

The cellular mechanisms of ALI have been extensively studied. The reader is referred to a number of reviews for details of this complex area.[2,4,18,19]

The following are key points.

Neutrophils have a key role in human ALI. These activate the ultimate liberation of mediators that cause the actual damage of ALI. A large number of such mediators has been identified in experimental studies and detected in broncho-alveolar fluid washout studies. There are considered to be four main groups of substances released from neutrophils that cause damage:

1. Cytokines, which are pro-inflammatory substances such as tumour necrosis factor (TNF) and interleukins. A range of other factors, such as interleukin 1 (IL-1), IL-8 and also TNF, exert feedback on the neutrophils causing further stimulation.[20]
2. Protease enzymes, which lead to extensive damage in the lung. Elastase is the most damaging and has broad proteolytic activity on collagen, fibrinogen and other proteins, as well as elastin.[21] Reactive oxygen species are also released, which can damage the endothelium by lipid peroxidation.
3. Lipid-derived mediators such as prostaglandins, thromboxanes and leukotrienes. These mediators act by amplifying neutrophil activation.

16.7 Clinical Presentation of Patients Following Inhalation of Irritant Products of Combustion

The signs and symptoms of patients exposed to irritant products of combustion will depend on the severity and duration of exposure. This will lead to a spectrum of conditions presenting with variable latency after exposure. Some effects will be almost instantaneous, as with the irritant effects of smoke.

Acute lung injury and the severe sequelae of ARDS will become apparent after a latent period of hours or days.

The signs and symptoms of persons presenting after exposure to irritant gases from combustion can be classified as sensory and pulmonary.

16.7.1 Sensory Irritation

The initial presentation of a person exposed to irritant products of combustion includes:

- Eyes: stinging and lachrymation.
- Nose and upper respiratory tract: coughing and laryngospasm.
- An additional sign of exposure to irritant gases is a burning sensation of the mucous membranes of the upper respiratory tract.

Effects occur immediately on exposure and the severity is a function of the exposure concentration rather than the exposure duration.

16.7.2 Pulmonary Irritation

This follows inhalation of irritant gases into the more peripheral parts of the lung.

The immediate consequences are:

- coughing;
- bronchoconstriction (particularly in sensitive individuals, *e.g.*, asthmatics);
- breathing difficulties [particularly in those with already compromised lungs, *e.g.*, chronic obstructive airway disease (COAD)].

Pulmonary irritation, unlike sensory irritation, depends on the irritant substance, the concentration and the duration of exposure. This mirrors the Haber equation, originally described for chemical warfare agents[22] (see Chapters 7 and 8). From this equation it is now thought that both concentration and duration of exposure are important determinants of damage.

The individual affected by sensory irritation will tend to close their eyes and hold their breath. Both these actions hinder the ability to escape from the hazard and may exacerbate the clinical situation.

16.8 The Management of Patients who are Affected by Irritant Products of Combustion

The management of patients with acute lung injury following inhalation of irritant products of combustion can be divided into early and continuing care. The latter applies to the management of ALI, where special intensive-care unit (ICU) measures are required, particularly in the management of artificial ventilation. This spectrum extends from the point of injury through to emergency room care in the hospital. Further, continuing hospital care will be necessary for badly affected patients and those who have developed signs of toxic pulmonary oedema. This may require treatment in an intensive care unit.[23] The management of smoke inhalation injury, which includes management of irritant gases, has been extensively reviewed over the past decade.[10,24] This management is integrated with the management of the burn patient in general. This discussion focuses on the immediate and continuing care stages of the management of acute lung injury from irritant products of combustion. It covers immediate life saving measures and other steps necessary to limit the potential damage caused by irritant gas inhalation.

16.8.1 Immediate Management

Immediate management measures include:

1. Rescue of the patient from the scene of the exposure to minimise exposure time. This is usually the role of the fire service, who wear self-contained breathing apparatus to avoid being affected themselves by the toxic products.
2. Reduction of carboxyhaemoglobin levels as much as possible by administration of 100% oxygen.
3. Taking a direct history from the patient if possible; otherwise, an indirect history must be sought from the rescuers (usually the Fire and Rescue Service). Details should include the type of fire, likely products of combustion and likely duration of the exposure. Was there an accompanying explosion?
4. Once the patient has been removed from immediate and further environmental danger a rapid (primary) assessment is necessary. This must include:
   - The overall physical state of the patient. Is there external burn trauma? Are there other physical injuries?
   - Inspection for signs of thermal injury to the upper respiratory tract, *e.g.*, singed nasal hair and carbon particles around the nose and mouth.
   - Establish the conscious level using AVPU (is the patient **A**lert, responsive to **V**ocal commands and/or **P**ainful stimulus or is the patient **U**nconscious). This is the quickest assessment, which can be followed by use of the Glasgow coma scale during later assessment.

- Airway and ventilation assessment. The rate and form of the breathing should be assessed quickly. Rate should be >8 and less than 20 breaths per minute. The respiratory form should be normal, with no abdominal see-sawing, indicating airway obstruction at both the upper and lower levels. The airway should be assessed visually and by auscultation to assess air entry to all parts of the lungs.
- The colour of the mucous membranes should be observed to see whether there is any obvious cyanosis.

Following the primary survey, immediate life saving measures must be started if there is evidence of airway obstruction and respiratory failure or arrest.

16.8.2 Emergency Airway Management

Early management of the airway is important in the case of irritant gas injury accompanied by burns. The risk of early developing oedema should be taken into account even if there is no initial dyspnoea on the primary examination. Intubation is desirable if the necessary skills are present. If the tube is in place it should be left uncut since facial oedema may make management difficult later (in anaesthesia and critical care management, endotracheal tubes are usually cut to size before insertion to allow only a few centimetres to protrude beyond the teeth). Airway management following inhalation of products of combustion is very important and may be life saving if the inhalation injury is accompanied by external burn injury with facial and laryngeal oedema. In the case of the latter, an early tracheostomy may be necessary or an emergency cricothyroidotomy at the site of injury. In their review of the management of smoke inhalation injury, Toon *et al.*[24] emphasised the value of the early use of bronchodilators, such as beta 2 agonists, which will decrease airway resistance and peak airway pressures produced during mechanical ventilation. There may also be benefit from the fact the beta 2 agonists are known to decrease inflammatory mediators such as histamine, leukotrienes and TNF-alpha. Salbutamol, a beta 2 agonist, is widely used by the paramedical emergency services and is therefore likely to be available for early use in combustion inhalation injury.

16.8.3 Emergency Artificial Ventilation

Emergency artificial ventilation must be started if the patient is not breathing adequately. This is best done using a portable gas-powered ventilator capable of delivering 100% oxygen, which will be required during the early stages of the management of acute lung injury. There are now concerns that oxygen toxicity may be a consideration in the development of ALI and so the oxygen concentration should be reduced to 50% as soon as possible. Many portable ventilators (such as the Pneupac® range [Smiths Medical International, Luton, UK]) have this option. Ventilation should be according to ILCOR ALS guidelines[25] with a tidal volume of 600 mL and a rate of

12 per minute. The ventilator should be a time-cycled volume preset, flow generating device, which will be able to overcome changes in lung compliance (V/P) and airway resistance caused by irritant gases. Pressure-generating portable ventilators are unsuitable for emergency ventilation since they deliver only a preset pressure before cycling and, in the face of decreased compliance and increased resistance, may deliver an inadequate tidal volume.

Patients who are still breathing spontaneously should be observed carefully during evacuation for signs of deterioration. The patients will usually require oxygen supplementation by face mask. This can either be free flow or include a demand valve device that delivers oxygen only when the patient takes a breath. This causes considerable saving in oxygen in the pre-hospital setting.

16.9 Assessment of Damage and Identification of Patients at Risk from Developing ALI and ARDS

The parameters defining the onset of ALI following inhalation injury have been considered in the previous sections. Following the primary survey and essential life saving measures the patient will require a secondary survey, which may be done near the site of the incident at an advanced medical post or in the emergency department following transfer to hospital. A key problem of ALI is that pathology may take some time to develop (the concept of latency) and therefore the careful assessment of the airway and respiration is necessary, together with early chest X-ray to provide a baseline against which continuing changes can be compared. The greatest risk following inhalation of irritant gases is the development of ALI with toxic pulmonary oedema, which commences with a latency of usually 12–24 hours.

On arrival at the hospital, immediate chest imaging is required (CXR), together with a chemical pathology screen. If available, CT scanning is valuable in early identification of developing ALI and ARDS.

16.10 Therapeutic Measures for the Management of Inhalation of Irritant Products of Combustion

A treatment protocol for the management of smoke injury has been developed by Toon *et al.*[24] The stages of treatment, both primary and secondary, can be divided into (1) supportive measures and (2) pharmacological measures.

Supportive measures include the management of the airway and artificial ventilation, as well as circulatory support. These measures are in accord with the general perceptions of advanced life support.[25]

Pharmacological measures are aimed at reversing changes to the airway, such as bronchoconstriction, and the early management of specific toxic gases, which may accompany irritant products of combustion such as carboxyhaemoglobin and cyanide. The management of intoxication by these gases is covered in detail elsewhere in the book (see Chapters 9 and 15).

16.10.1 Continuing Artificial Ventilation for ALI

The management of artificial ventilation in cases of developing ALI has been the subject a considerable attention over the past two decades and the type of ventilation used has been shown to be a major factor affecting clinical outcome. Dries[26] has reviewed artificial ventilation strategies for the management of ALI. The following points are important:

- Assessment of respiratory function compromised by smoke inhalation goes far in predicting the likelihood and quality of survival after thermal injury.[27]
- Permissive hypercapnia (also known as hypercarbia; *vide infra*) is a product of a better understanding of lung injury induced by mechanical ventilation. In particular, barotrauma is not only caused by hyperinflation of aerated regions of the lung but also by alveolar wall necrosis with persistent collapse and inflammation of dependent lung zones.
- There is likely to be CT evidence of persistent collapse in the dependent regions.[28] The implication is that, because only smaller regions of the lung will be available for ventilation, airway pressures as low as 30 cm H_2O may be associated with lung injury.
- There is evidence to show that hypercarbia is associated with attenuation of ALI.[29] Hypercarbia is also associated with anti-inflammatory effects and a reduction in cytokine signalling.[30]
- Carefully managed artificial ventilation has been shown to be the most successful intervention in the management of ALI and preventing the transition to ARDS.

16.10.2 Rationale for the Ventilation Strategy in ALI

In lungs affected by ALI there are three types of regions:[31]

- Normal areas, usually in the non-dependent region.
- Areas where there has been severe collapse and alveolar flooding and where ventilation will be impossible.
- Intermediate areas where there are alveoli that are partially ventilated or collapsed that are capable of being 'recruited' to improve gas exchange.

The ventilation measures used in the continuing management of inhalation ALI are designed to ventilate the lung carefully and to open up the intermediate zones. This technique is based upon 'opening the lung up' and keeping it open.[32] This is achieved by the use of high levels of positive end-expiratory pressure (PEEP) initially and then reducing it.

Because much of the lung is not being ventilated, calculation of normal tidal volumes based upon weight is not appropriate. Lower tidal volumes must be used. Pressure support ventilation is backed up with the use of positive end-expiratory pressure. In ventilating persons who are at risk of developing ALI a protective ventilation strategy may be used. This involves using small tidal volumes to prevent alveolar over-distension. The initial tidal volumes used should be between 6 and 8 $mL\,kg^{-1}$ in comparison with the more usual 10 $mL\,kg^{-1}$ for adults.

PEEP has been the mainstay of ALI treatment strategies for a number of years. It is thought to increase recruitment of non-aerated alveoli. This leads to an increase in functional residual capacity (FRC) and improved oxygenation. There may also be a reduced venous admixture and shifting of the tidal volume (Vt) to a more compliant portion of the pressure–volume curve. It should be noted, however, that the actions of increased PEEP are sometimes unpredictable.

Identifying the optimal level of PEEP to use has been the subject of considerable debate. The static pressure–volume curve[11] shows two points of inflexion. The lower point is thought to be the point at which recruitment of unventilated alveoli starts. The upper point is where all available alveoli have been recruited and hyperventilation of alveoli occurs leading to possible damage.[33]

For continuing ventilation, pressure-controlled ventilation should be used, while noting that, with decreased compliance and increased airway resistance, this mode of ventilation does not always deliver an adequate tidal volume (Vt). PEEP should be set by using the pressure–volume curve or until PaO_2 is adequate. If the plateau airway pressure exceeds 35 cm H_2O or the FiO_2 exceeds 0.56, a different ventilation mode should be used. These include inverse ratio ventilation or ventilating in the prone position. Another possibility is the use of high frequency ventilation.[34]

16.11 Pharmacological Support Following Inhalation Injury by Irritant Gases of Combustion

Acute lung injury from inhalation causes has been the study of substantial research using large animal models. This research has focused on ALI caused by inhaled chemical warfare agents and toxic industrial chemicals. However, the pathophysiology of ALI caused by toxic organic chemicals (TOC), such as phosgene, may be applicable to irritant products of combustion, such as oxides of nitrogen, which are lipid soluble and produce their effects after dissolving into the cellular membrane.[10]

Jugg *et al.*[35] have reviewed a series of studies conducted during the past decade using a terminally anaesthetised pig model over a period of 24 hours following exposure to phosgene. The object of the studies was to examine a number of standard 'off the shelf' therapies for ALI, which have been considered for use in ALI but for which the evidence base for the individual effectiveness is moderate at best. The results for individual single pharmacological therapies can be summarised as follows.

16.11.1 Steroids

In phosgene poisoning, no direct therapeutic effect could be demonstrated using inhaled or intravenous steroid therapy alone. The authors noted that further study should take place to investigate their effectiveness in combination therapy.[36]

16.11.2 Beta Agonists

Beta agonists have been considered for some time as a possible therapy for ALI given the discovery of beta receptors in the alveolar walls, which are active in countering the development of toxic pulmonary oedema.[37] The wide availability of salbutamol to paramedical and medical personnel alike makes the option attractive. However, in a study of salbutamol in pigs exposed to phosgene[38] the authors found that early treatment with nebulised salbutamol worsens physiological responses and did not improve survival following phosgene-induced lung injury in pigs.

16.11.3 Diuretics

A further study of frusemide, another 'accessible' therapy, showed no effect on survival and a worsening of the $PaO_2:FiO_2$ ratio (a measure of the effectiveness of oxygenation and ventilation in intensive care).

Studies by the same group of oxygenation therapy and protective ventilation strategies were more encouraging however.[35,39] This finding supports the current protective ventilation strategies in human inhalation injury management detailed above.

16.11.4 New Research Directions in the Management of ALI

There is currently intense interest in finding further pharmacological strategies for the treatment of ALI and ARDS.[11] Compounds being considered include pulmonary vasodilators, surfactants, anti-inflammatory compounds, anti-oxidants, such as *N*-acetyl cysteine, and anticoagulants such as heparin. The use of nebulised heparin and acetyl cysteine in children with massive burn and smoke inhalation injury has demonstrated a significant decrease in the need for re-intubation for progressive pulmonary failure, decreased incidence of atelectasis and reduced mortality.[40] The therapeutic strategy of decontaminating the lungs early after smoke exposure to prevent inhalation injury has received little attention and may be of significant value. This could potentially utilise amphoteric, hypertonic chelating agents developed for topical and ocular chemical exposures.[9]

16.12 Cyanide Poisoning: the Use of Hydroxocobalamin

The management of poisoning by hydrogen cyanide is considered in Chapter 15. Here it should be noted that there have been extensive studies on the early management of cyanide poisoning from smoke inhalation using hydroxocobalamin. Early pharmacological intervention for cyanide poisoning is essential since the problem in this form of (cellular) respiratory failure lies in the fact that oxygen is present in the cells but cannot be used by the mitochondria due to blocking of oxygen access to the cytochrome system (see Chapter 10). Hydroxocobalamin has been in use in the Paris Fire Service since 1995 and experience of its use for smoke-associated cyanide poisoning over an eight year period has been reviewed.[41] Overall, hydroxocobalamin has a risk : benefit ratio rendering it suitable for pre-hospital use in the management of acute cyanide poisoning caused by smoke inhalation. The use of hydroxocobalamin in the management of serious neurological complications following smoke inhalation has been reported by Baud *et al.*[42] Although the combustion of natural and synthetic products can yield cyanide, its toxic role in residential fires is unclear (see Chapter 7). The case reported concerns a woman aged over 50 years who presented comatose, pulseless and apnoeic after a domestic fire. Cardiopulmonary resuscitation and on-site administration of 2.5 g hydroxocobalamin as an antidote to cyanide resulted in a return of spontaneous circulation. These data suggest that smoke inhalation in a residential fire may cause cyanide poisoning. This case provides clinical, biological, analytical and brain imaging data supporting the hypothesis of the toxic role of smoke-induced cyanide poisoning, which may result in neurological sequelae.

16.13 A Summary of Therapeutic Strategies in the Management of ALI from Inhalation of Products of Combustion

Although a number of standard strategies for the management of ALI from all causes have been used in the ICU setting for many years, direct evidence concerning their relevance to inhalational ALI has been limited. Against this there have been a number of studies in animals. The problem remains about how relevant these are to human management. As seen in previous sections, a series of studies in the pig have not confirmed the value of common pharmacological strategies when used singly. However, the value of *combined* therapies has yet to be explored. Only large scale multi-centre trials will give the answers to a number of questions. Meanwhile, human studies continue. Perkins *et al.*[37] have considered further the possibility of the use of beta 2 agonists for ALI. Other non-ventilatory strategies have been considered by Caffee and Mathay.[43]

A dilemma therefore remains for the clinician faced with managing a patient with ALI following the inhalation of combustion products. In this situation, good clinical common sense and judgement will have to prevail given the seriousness of the situation and the limited options open to the intensivist.

At present, the following options are available:

- If ALI has been confirmed and is developing into ARDS with requirement for ventilation a protective ventilation strategy should be introduced. This will involve pressure-controlled ventilation with low Vt and high PEEP titrated after an initial higher PEEP period to open the lung alveoli.
- Consideration should be given to the use of steroids in developing ALI. Despite discouraging results from phosgene studies the strategy may be of use in the management of ALI from other causes. Studies of the use of inhaled and parenteral steroids following inhalation of irritant gaseous products are continuing.
- In emergency airway management, control of airway secretions and management of bronchospasm (high dose steroids, beta adrenergic bronchodilators) is important as is prevention of superimposed infection in the airway.
- Where inhalational injury is combined with external burning there must be an integrated strategy for the management of both, since lung injury has been shown to reduce burn survival.

16.14 Conclusions

Inhalation of gaseous products of combustion gives rise to effects at all levels of the respiratory tree and to respiratory failure of either type 1 or type 2. In severe cases combustion product inhalation will lead to ALI and ARDS. There are many causes of these conditions and hospital experience in management may be based upon experience from other causes. A great deal of work has been done on the management of ALI but given that the cause is pulmonary and not extrapulmonary, treatment strategies should be reviewed carefully. The following points are important:

- Acute lung injury management is based upon experience in the ICU after developing ARDS or in established ARDS. ALI and ARDS (particularly ALI) should be defined in the context of an inhalational insult. Although there are many provoking aetiologies for ARDS they appear to lead to a final common clinical pathway with a common pathophysiology involving fibrosis.
- In the case of ALI/ARDS from the most common causes (sepsis and trauma) there have been many large scale multi-centre trials involving broad clinical endpoints such as ventilation-free days in hospital and the length of ICU stay. These have confirmed (1) the effectiveness of the open lung ventilation strategy and (2) the non-effectiveness of high dose systemic steroids. However, these findings may not be relevant to irritant products of combustion.
- Other pharmacological approaches, such as beta 2 agonists, may be appropriate for future therapy.

- A problem arises for inhalation injury in whether evidence gathered from ICU experience with ARDS resulting from multiple aetiologies as a good model for toxic inhalation injury. Toxic inhalational injury is not represented in the collection of predisposing conditions examined in most of the large scale ARDS trials.
- The nature of ALI from inhalational causes requires careful examination and definition in terms of the chemical actions of the substance concerned and evidence from animal models. The time course of ALI is particularly important. Lung washout studies identifying inflammatory mediators would be useful. Intensive work on inflammatory mediators has been done and may lead to new therapeutic pathways.
- The role of steroids in the early, non-clinical phases following toxic inhalational exposure requires examination. This is not necessarily the same as the role of steroids in developing and established ARDS. Equally, the role of protective ventilatory strategies in the latent period of toxic pulmonary oedema also requires examination.
- Large animal studies have provided detailed findings for therapeutic measures taken following phosgene inhalation. This work may provide valuable clinical advice for clinicians involved in the management of human combustion products inhalation.

References

1. S. D. Tuorinsky and A. M. Sciuto, Toxic inhalation injury and toxic industrial chemicals, *Medical Aspects of Chemical and Biological Warfare*, Borden Institute. Office of the Surgeon General, 2007, ch. 10, pp. 339–370.
2. M. A. Matthay and G. A. Zimmerman, Acute lung injury and the acute respiratory distress syndrome: four decades of inquiry into pathogenesis and rational management, *Am. J. Respir. Cell Mol. Biol.*, 2005, 33319–33327.
3. M. A. Matthay, G. A. Zimmerman, C. Esmon *et al.*, Future research directions in acute lung injury: summary of a National Heart, Lung, and Blood Institute working group, *Am. J. Respir. Crit. Care Med.*, 2003, **167**(7), 1027–1035.
4. The Acute Respiratory Distress Syndrome Network, Ventilation with lower tidal volumes as compared with traditional tidal volumes for acute lung injury and the respiratory distress syndrome, *NEJM*, 2000, **342**(18), 1301–1308.
5. R. G. Brower, P. N. Lanken, N. MacIntyre *et al.*, Higher versus lower positive end-expiratory pressures in patients with the acute respiratory distress syndrome. ARDS Clinical Trials Network, *N. Engl. J. Med.*, 2004, **351**(4), 327–336.
6. L. B. Ware and M. A. Matthay, The acute respiratory distress syndrome, *N. Engl. J. Med.*, 2000, **342**, 1334–1349.

7. A. P. Wheeler and G. R. Barnard, Acute lung injury and the acute respiratory distress syndrome: a clinical review, *Lancet*, 2007, **335**, 1553–1565.
8. K. Leaver and T. W. Evans, Acute respiratory distress syndrome, *BMJ*, 2007, **335**, 389–395.
9. J. Wakefield and R. Maynard, Combustion toxicology, in *Essentials of Toxicology for Health Professionals*, Oxford University Press, UK, 2nd edn, 2012, ch. 18, pp. 227–254.
10. T. Prien and D. L. Traber, Toxic smoke compounds and inhalation injury – a review, *Burns Incl. Therm. Inj.*, 1988, **14**, 451–460.
11. A. Lumb, Acute lung injury, in *Nunn's Applied Respiratory Physiology*, London Elsevier Ltd, 7th edn, 2010, ch. 31, pp. 451–461.
12. G. R. Bernard, A. Artigas, K. L. Brigham *et al.*, The Anglo–American consensus conference on ARDS: definitions, mechanisms, relevant outcomes and clinical trial co-ordination, *Am. J. Respir. Crit. Care Med.*, 1994, **149**, 818–824.
13. L. D. Hudson, J. A. Millberg, A. Anardi and R. J. Maunder, Clinical risks for the development of the acute respiratory distress syndrome, *Am. J. Respir. Crit. Care Med.*, 1995, **151**, 293–301.
14. L. Gattinoni, P. Pelosi, P. M. Suter *et al.*, Acute respiratory distress syndrome caused by pulmonary and extra- pulmonary disease. Different syndromes?, *Am. J. Respir. Crit. Care Med.*, 1998, **158**, 3–11.
15. R. Agarwal, R. Srinivas, A. Nath and S. Jindal, Is the mortality higher in the pulmonary vs extrapulmonary ARDS? A meta-analysis, *Chest*, 2008, **133**, 1463–1473.
16. L. Gattinoni, P. Pelosi, G. Vitale *et al.*, Body position changes redistribute lung – computed tomographic density in patients with acute respiratory failure, *Anesthesiology.*, 1991, **74**, 15–23.
17. M. Bachofen and E. R. Weibel, Structural alterations and lung parenchyma in the adult respiratory distress syndrome, *Clin. Chest Med.*, 1982, **3**, 35–43.
18. P. Enkhbaatar and D. L. Traber, Pathophysiology of acute lung injury in combined burn and smoke inhalation injury, *Clin. Sci.*, 2004, **107**, 137–143.
19. J. F. Pittet, C. Mackersie, T. R. Martin and M. A. Matthay, Biological markers of acute lung injury; prognostic and pathogenic significance, *Am. J. Respir. Crit. Care Med.*, 1997, **155**, 1187–1205.
20. M. Bhatia and S. Moochhala, Role of inflammatory mediators in the pathophysiology of acute respiratory distress syndrome, *J. Pathol.*, 2004, **202**, 145–156.
21. P. A. Ward, Role of complement, chemokines and regulatory cytokines in acute lung injury, *Ann. N. Y. Acad. Sci.*, 1996, **796**, 104–112.
22. R. L. Maynard, The physicochemical properties and general toxicology of chemical warfare agents. in *Chemical Warfare Agents: Toxicology and Treatment*, ed. T. C. Marrs, R. L. Maynard and F. R. Sidell, John Wiley, Chichester, UK, Chapter 2, p. 55.

23. D. J. Baker, Critical care requirements after mass toxic agent release, *Crit Care Med*, 2005, **33**, S66–S74.
24. M. H. Toon, M. O. Maybauer, J. E. Greenwood, D. M. Maybauer and J. F. Fraser, Management of acute smoke inhalation injury, *Crit. Care Resusc.*, 2010, **12**(1), 53–61.
25. C. D. Deakin, J. P. Nolan, J. Soar *et al.*, European Resuscitation Council Guidelines for Resuscitation 2010, Section 4: Adult Advanced Life Support, *Resuscitation*, 2010, **81**(10), 1305–1352.
26. D. J. Dries, Key questions in ventilator management of the burn-injured patient (first of two parts), *J. Burn Care Res.*, 2009 Jan–Feb, **30**(1), 128–138.
27. C. M. Ryan, D. A. Schoenfeld, W. P. Thorpe *et al.*, Objective estimates of the probability of death from burn injuries, *N. Engl. J. Med.*, 1998, **338**, 362–366.
28. L. Gattinoni, A. Presenti, A. Torresin *et al.*, Adult respiratory distress syndrome profiles by computed tomography, *J. Thorac. Imaging.*, 1986, **1**(3), 25–30.
29. D. G. Sinclair, S. Braude, P. L. Haslam and T. W. Evancs, Pulmonary permeability in patients with severe lung injury: clinical correlates and anatural history, *Chest*, 1994, **106**, 535–539.
30. D. O'Croinin, M. Ni Chonghaile, B. Higgins and J. G. Laffey, Bench-to-bedside review: Permissive hypercapnia, *Crit. Care*, 2005, **9**(1), 51–59.
31. L. Gattinoni and A. Pesanti, ARDS: the non-homogenous lung: facts and hypotheses, *Intensive Crit. Care Digest*, 1987, **6**, 1–4.
32. M. D. Reis, D. Gommers and B. Lachmann, Open lung management, in *Mechanical Ventilation: Clinical Applications and Pathophysiology*, ed. P. P. Papadakos and B. Lachman, Sanders: Elsevier, Philadelphia PA, 2008, ch. 25, pp. 276–285.
33. N. Marlakkan, N. J. Snook and A. B. Lumb, New aspects of ventilation in acute lung injury, *Anaesthesia*, 2003, **58**, 647–667.
34. K. P. W. Chan, T. E. Stewart and S. Mehta, High frequency oscillatory ventilation for adult patients with ARDS, *Chest*, 2007, **131**, 1907–1916.
35. B. J. A. Jugg, A. J. Smith, S. J. Rudall and P. Rice, The injured lung: clinical issues and experimental models, *Philos. Trans. R. Soc., B*, 2011, **366**, 306–309.
36. A. J. Smith, D. A. Parkhouse, R. F. R. Brown *et al.*, The effect of steroid treatment with inhaled budenoside or intravenous methyl prednisolone on phosgene – induced acute lung injury in a porcine model, *Mil. Med.*, 2009, **174**, 1–8.
37. G. D. Perkins, F. Gao and D. R. Thickett, In vivo and in vitro effects of salbutamol on alveolar epithelial repair in acute lung injury, *Thorax*, 2008, **63**, 215–220.
38. C. Grainge, R. F. R. Brown, B. J. Jugg *et al.*, Early treatment with nebulised salbutamol worsens physiological responses and does not improve survival following phosgene – induced acute lung injury, *J R Army Med Corps*, 2009, **155**, 105–109.

39. S. A. Parkhouse, R. F. R. Brown, B. J. A. Jugg *et al.*, Protective ventilation strategy in the management of phosgene – induced acute lung injury, *Mil. Med.*, **172**, 295–300.
40. M. H. Desai, R. Micak, J. Richardson *et al.*, Reduction in mortality in pediatric patients with inhalation injury with aerolised heparin/N-acetylcysteine therapy, *J. Burn Care Rehabil.*, 1998, **19**, 210–212, [erratum in *J. Burn Care Rehabil.*, 1999, **20**(1 Pt 1), 49].
41. J. L. Fortin, J. P. Giocanti, M. Ruttimann and J. J. Kowalski, Prehospital administration of hydroxocobalamin for smoke inhalation-associated cyanide poisoning: 8 years of experience in the Paris Fire Brigade, *Clin. Toxicol. (Phila)*, 2006, **44**(Suppl. 1), 37–44.
42. F. Baud, M. Boukobza and S. W. Borron, Cyanide: an unreported cause of neurological complications following smoke inhalation, *BMJ Case Rep.*, 2011 Oct 28, DOI: 10.1136/bcr.09.2011.4881, pii: bcr0920114881.
43. C. C. Caffee and M. A. Mathay, Non-ventilatory treatments for acute lung injury and ARDS, *Chest*, 2007, **131**, 013–920.

CHAPTER 17

Health Effects in Groups Exposed to Wildland and Urban Fires

JAMIE L. McALLISTER

Combustion Science & Engineering Inc., 8940 Old Annapolis Road, Columbia 21045, USA
Email: jmcallister@csefire.com

17.1 Introduction

Occupational and environmental exposure hazards from combustion events pose a risk both to those intimate with a fire and those remote from the fire origin. Potential health hazards in these subpopulations include acute and chronic effects. Acute effects may include asphyxia, irritancy, and odor resulting from exposure to combustion products during and immediately after fire incidents. Continued or developing health effects may include respiratory disease and sensitization. An increased long-term cancer risk may also result from exposure to airborne carcinogens in combustion particulate. Studies of high exposure groups should reveal the extent to which such potential health hazards are significant.

This chapter reviews the evidence of developing health effects in four high exposure groups. The first is firefighters, who experience relatively high concentration exposures to combustion products both on an acute basis at individual incidents and chronically from repeated exposures during a working life. The second group, receiving somewhat lower level acute

Issues in Toxicology No. 23
Toxicology, Survival and Health Hazards of Combustion Products
Edited by David A. Purser, Robert L. Maynard and James C. Wakefield

Published by the Royal Society of Chemistry, www.rsc.org

exposures from individual incidents, is populations exposed to smoke plumes from major incidents such as the Buncefield and World Trade Center fires (which are also reviewed individually in Chapters 18 and 19, respectively). The third group, also exposed acutely to diluted smoke plumes from individual incidents, is populations exposed during wildfires. The fourth group considered is military personnel exposed over a limited period to smoke plumes from oil fires during the Gulf War. For this group, consideration has been to both acute effects during the exposure period and the extent of continued effects on health revealed by follow-up studies.

17.2 Occupational Exposure: Fire Service and Fire Investigation Professionals

Percivall Pott,[1] who in 1775 first linked the occupational exposures of chimney sweeps to the development of scrotal cancer, demonstrated the need to better understand the environmental hazards individuals face in their professional lives. A career in the fire service presents a variety of unique occupational hazards. Firefighting involves continuous exposure to toxicants from various sources both inside and outside of the fire compartment. Firefighters actively engaged in firefighting or rescue operations inside burning buildings use self-contained breathing apparatus (SCBA). There are substantial periods of time during structural and wildland firefighting, however, where front-line firefighters and other team members outside the immediate fire area fail to deploy respiratory protection. Firefighters are exposed to carcinogens during firefighting operations, overhaul operations, training, post-fire investigations at fire scenes and even while at the fire station. The products created during a combustion event are numerous and copious and include compounds such as carbon monoxide, hydrogen cyanide, halogenated agents, nitrogen oxides, sulfur dioxides, and formaldehydes.[2] Additionally, specific compounds linked to carcinogenesis, such as polycyclic aromatic hydrocarbons (PAHs), dioxins, polychlorinated biphenyls (PCBs), nitrosamines, and heavy metals are routinely formed as byproducts of combustion.[3] These products can be inhaled by firefighters or absorbed through their skin.

While the firefighting profession has always been viewed as life-threatening, toxicant exposure has not always been identified as one of the main occupational threats. Traditionally, firefighters have viewed thermal exposures from fires as their primary hazard; smoke exposure, aside from irritant effects, was not considered a primary concern. Over the last four decades, the escalated production and use of synthetic materials has introduced a new assortment of combustion toxicants and different exposure conditions for firefighters. Construction materials and building supplies from developing countries, with higher levels of impurities, *e.g.*, asbestos, sulfur, formaldehyde, *etc.*, can cause unexpected toxicant exposure.

In a recent study, Underwriters Laboratories (UL) conducted individual material testing, full-scale fire testing, and live fire scene sampling.[4]

The purpose of the research was to gain a better understanding of the smoke exposures to firefighters during active firefighting activities and the products of combustion produced by different natural and synthetic materials. Individual material tests showed that synthetic materials produce 12.5 times more smoke particles per consumed mass than natural materials. The products of combustion found in the full-scale fire tests correlated to the results of the individual material tests. Recommended National Institute for Occupational Safety and Health (NIOSH) immediately dangerous to life or health (IDLH) levels, short-term exposure limits (STEL) and Occupational Safety and Health Administration (OSHA) time weighted average (TWA) exposure levels were exceeded during fire growth and overhaul for carbon monoxide, hydrogen cyanide, formaldehyde, benzene, and arsenic. UL found that 99% of smoke particles collected during overhaul were smaller than 1 μm. Of these particles, 97% were too small to be seen by the naked eye, suggesting that what is perceived to be clean air is really not clean. As part of the UL study, Chicago firefighters wore gas sampling equipment during firefighting operations and had their gear examined after fire exposures. Carbon monoxide levels frequently exceeded NIOSH and OSHA limits, and heavy metals, including arsenic, cobalt, chromium, lead, and phosphorous, were collected during atmosphere sampling. Gear inspections found that nearly identical chemical compositions were found on firefighting hoods and gloves (some containing lead, mercury, phthalates, and PAHs), but the concentrations on gloves were 100 times greater than on the hoods.

In addition to changes in the types of toxicants within the occupational environment, there have also been significant changes to firefighter personal protective equipment [*e.g.*, bunker coats, pants, self-contained breathing apparatus (SCBA)] over the last few decades; firefighters were not always fully encapsulated in Kevlar® and Nomex®, nor did they breathe fresh air through a mask during firefighting operations. The personal protective equipment (PPE) of choice in the 1960's and 1970's era of the Fire Department of New York was a jean jacket and 3/4 rubber boots. Having deemed themselves as "smoke eaters", most firefighters that were in the profession 25–50 years ago didn't have SCBA, and if they did have it, no one used it. Smoke eaters entered burning dwellings and traveled as deep as they could before they felt too much heat or encountered too much smoke, a threshold that was set when pain was felt on exposed skin surfaces and airway passages. Overall, firefighters a few decades ago spent little time inside the fire compartment, because their bodies could not tolerate it.

Today's PPE has become increasingly more sophisticated and has allowed firefighters to move from primarily defensive, exterior operations to interior, aggressive attacks on the fire. Although there are obvious benefits to this enhanced protective equipment, its use has allowed firefighters to become more intimate with the fire and the byproducts of combustion; as a result, firefighters have become more complacent in the fire environment. It is this complacency, along with the functional limitations of today's PPE, that is currently believed to be the main cause of respiratory exposures.[5]

Although most firefighters assume some level of occupational health risks present in their profession, firefighters should still be afforded protection from the known, endemic hazards to which they are exposed. Issues arise, however, when the ability of the firefighter to perform his/her duties is impeded or hampered by the protective methods available or when the hazard(s) is not clear and obvious to the firefighter. For example, the use of SCBA appears to be a logical solution to protect firefighters from inhaling products of combustion during extinguishment operations. However, due to the bulk and limited air supply of the SCBA, firefighters may wait to attach their regulators to their face pieces until smoke conditions are such that they can no longer breathe without significant irritation. The rationale for this procedure is to ensure the preservation of as much air as possible, which allows for a longer duration of operation in the structure or a longer period of time to complete their assigned task. Moreover, breathing air through the SCBA adds additional physical strain and discomfort to the firefighters during operations, so some firefighters wait until use of the equipment is absolutely necessary. SCBA and PPE can weigh 40–50 pounds,[6] so there is always a desire to minimize the amount of equipment used. In addition to their protective equipment, firefighters will also be weighed down with forcible entry equipment, racks of hose, saws, and various other tools. Although the protective equipment significantly blocks the transfer of heat to the body, it also limits dexterity, traps body heat, and limits vision. The extra weight of the protective equipment and the functional limitations of the equipment often prevent firefighters from utilizing their gear when exposed to potentially toxic environmental conditions. In this case, although the firefighters are likely recognize that a hazard is present, they may disregard the hazard for ease of job performance.

In some cases, it appears that the hazard is not clear and obvious. A recent study completed by the National Fire Protection Association (NFPA) focused on how firefighters assess the use of, or discontinue use of SCBA and other types of respiratory devices, specifically in conditions that are tenable but still present a respiratory hazard.[5] Examples of these hazardous conditions included wildland fires, vehicle fires, fire investigations, and overhaul and post-fire extinguishment operations. In many cases, an outside fire (wildland or vehicle) or operation that takes place in the fire structure after extinguishment (overhaul or investigation) is not perceived as hazardous, even though off-gassing and high airborne particulate concentrations are still present.[7,8] The overhaul period occurs after the bulk of the fire or all of the fire is extinguished and firefighters begin to pull down ceilings and walls, to ensure that small fires have not developed in the hidden void spaces. It is during this period of time that most firefighters will remove their masks or completely remove the SCBA.[9] Furthermore, overhaul is an extremely physical operation requiring firefighters to pull, push, lift, and rip apart the structure with their tools, thus creating increased tissue oxygen demands, which translate into increased respiratory intake of air components (*e.g.*, carcinogens, dusts, *etc.*) of the post-fire scene. Moreover, in combination with the physical activity, pre-emptive exhaustion and dehydration, and weight of their gear, the SCBA equipment is often the first item removed to reduce some of the strain on the firefighter.

The use of respiratory protection in the form of SCBA is, therefore, generally limited to those sections of teams actively engaged in firefighting or rescue operations in close proximity to a fire, typically inside a building while exposed to dense smoke and heat. Before and after these periods, the same team members are not using SCBA or other forms of respiratory protection, while other team members, also without respiratory protection, are engaged in operations beyond the intense fire area. In this context, it is possible to consider three hazard zones during a fire (see Chapters 2 and 3).

Zone 1 is close to or inside a fire enclosure. Conditions in zone 1 include high temperatures and dense smoke containing high concentrations of toxic gases; firefighters are wearing their SCBA in this zone. During overhaul, when respiratory protection may not be used, this zone is likely to be highly contaminated by smoke deposits, airborne smoke and dust particulates, toxic condensates, and vapors.

Zone 2 is the area immediately outside the fire perimeter, either inside the fire structure near zone 1, immediately outside the fire structure, or in proximity to a wildfire. In this area, the primary smoke plume from the fire is considerably diluted by air entrainment, but can contain concentrations of combustion products capable of causing acute toxic effects including irritancy and asphyxia. More typically, zone 2 will contain a wide range of toxic species exceeding workplace limits and presenting a chronic hazard from repeated exposures. Firefighters and other emergency service personnel without respiratory protection are exposed in this area to intermittent downwashed smoke plumes.

Zone 3 is categorized as the area approximately 100 meters from the source fire to possibly several kilometers or more away from the fire depending upon the fire size and meteorological conditions. Zone 3 may encapsulate a large resident population; firefighters and other emergency personnel in this zone may be exposed to highly diluted smoke plumes presenting a potential acute environmental hazard.

Based on these details, it appears that a large portion of carcinogenic exposures to firefighters occurs outside or during the initial entry and post-fire overhaul stages of the incident. It can be hypothesized that the two primary reasons that firefighters continue to expose themselves to respiratory and other fire scene hazards appear to be related to the field limitations of their personal protective equipment and a misinterpretation of the perceived hazard in comparison with the actual hazard. There is technology available to the fire department to quantify the hazards present in the environment in which they operate during and after fire extinguishment. According to the NFPA study, some departments are utilizing equipment to quantify airborne contaminants, such as carbon monoxide, flammable gases, hydrogen sulfide, hydrogen cyanide, sulfur dioxide, nitrous oxide, and other toxic gases. However, these omit the potentially carcinogenic particulate-associated compounds arguably most hazardous to long term health. Additionally, the simple identification of a toxicant in the air does not directly correlate with exposure levels in firefighters; air concentrations tells us nothing about what is absorbed, distributed, metabolized, or excreted in the body. Air measurements also lack detail regarding the synergistic or additive effects of multiple toxicants or the effects of chronic exposure. Although general occupational exposure limits have been set by OSHA and NIOSH for some of the common carcinogens found in fires, no studies have linked specific or combined contaminant exposure to cancer development in firefighters, nor has any study quantified the resultant blood levels in firefighters post-exposure. Since the particular toxicants responsible for the long term adverse health effects identified in firefighters are unclear, the efforts of departments who are quantifying air quality after fire extinguishment appear incomplete.

Over the last three decades, heightened awareness to the issues of firefighter occupational exposure has led to numerous cohort mortality studies in cities across the United States, including Philadelphia, Boston, Seattle, San Francisco, and Honolulu.[10–15] Additional studies in Florida and New Jersey have focused on occupational mortality at a statewide level.[16,17] Similar studies have also been completed for large cities outside of the United States, including Toronto, Paris, Alberta, and Stockholm.[18–21]

The Philadelphia study examined the cause(s) of deaths and illnesses sustained by approximately 7800 firefighters employed between 1925 and 1986. Factors such as duration of employment, unit assignment (engine or ladder company), year hired, total incident responses, and diesel engine exposure were considered. A statistical analysis comparing the mortality of the firefighters to the rates found in the general population concluded that the overall mortality of Philadelphia firefighters was slightly lower. However, an excess risk of colon cancer and ischemic heart disease existed in the cohort, and an increase in mortality was found for firefighters who suffered from colon cancer, kidney cancer, non-Hodgkin's lymphoma, and multiple myeloma. The Boston study examined the mortality of over 5600 firefighters employed after 1915, noting the year of employment and age at the time of death for each firefighter. The researchers concluded that there were no significant differences between the overall mortality rates of firefighters compared to the expected rates of the general population, with the exception of a higher risk in the "accidental death" category. The Seattle study found that the incidence of cancer among the 2500 firefighters examined was similar to that of the general population. There was, however, an elevated risk of prostate cancer, as well as a slightly elevated risk of colon cancer. The San Francisco study of over 3000 firefighters also found that the overall mortality rate was similar, although slightly less to that of the general population. However, deaths from esophageal cancer and cirrhosis and other liver diseases were significantly higher than expected in the cohort. In Honolulu, the cause of death for 205 male firefighters employed for at least one year between 1969 and 1988 was examined. Their risk of death was compared to that of the Hawaiian male population ages 20 or greater. The study found that a significantly greater risk was present for firefighters to die from brain cancer, prostate cancer, and cirrhosis of the liver than the risk present in the general population.

The statewide studies in Florida and New Jersey included a significantly larger sample size. In Florida, the mortality rate of nearly 37 000 firefighters employed between 1972 and 1999 was compared with the general population in Florida. At the time of publication, this study was the largest of its kind, as well as the first to specifically examine the mortality of female firefighters. Male firefighters were found to have a higher mortality rate due to male breast cancer, thyroid cancer, and bladder cancer, while female firefighters had a significantly higher risk for atherosclerotic heart disease. The New Jersey study focused on the deaths of firefighters and police officers between 1974 and 1980. The firefighters showed increases in skin cancer, cirrhosis, and arteriosclerotic heart disease when compared to the New Jersey male population. Compared to police officers, firefighters also had a significant increase in nonmalignant respiratory disease and leukemia.

In Toronto, the study compared over 5400 firefighters to the male Ontario population. Out of 777 firefighter deaths, the study found significant excesses in deaths from brain tumors, "other" malignant neoplasms, and aortic aneurysms. Slight increases above the expected mortality risk were found for several cancers (pharynx, rectum, prostate, testis, bladder, and lymphatic tissue), as well as ischemic heart disease (including acute myocardial infarctions), arteriosclerosis, vein diseases, chronic bronchitis, emphysema, asthma, cirrhosis, and gallbladder diseases. The 1995 mortality cohort study of 830 firefighters in Paris was the first of its kind in France. As found in the majority of the mortality studies throughout the world, the Paris firefighters had a lower overall mortality rate when compared to the average French male. However, as with the studies in the United States and Canada, increased incidences were found for certain diseases. Within the cohort, higher mortality rates were observed for genito-urinary cancer, digestive cancer, respiratory cancer, and cerebrovascular disease. The Alberta study examined firefighter deaths in Edmonton and Calgary employed between 1927 and 1987. Greater than expected mortality rates were found for malignant neoplasms, and cancers of the lungs, bladder, kidney, ureter, colon, rectum, and pancreas, as well as leukemia, lymphoma, myeloma, and obstructive pulmonary diseases. In Stockholm, firefighters between 1931 and 1983 were found to have a greater mortality rate for both stomach and brain cancers. This study was slightly smaller, comparing the mortality of only 1100 firefighters to the general population.

Throughout all the studies completed, two main trends were noted. Overall, firefighters generally have a lower mortality rate compared to their respective populations (healthy worker effect); however, these studies have identified in particular a relationship between cancer mortality and the firefighting occupation. Each study typically found a handful of cancers for which firefighters had a higher risk of mortality, although the types of cancer may vary slightly. More specifically, in an epidemiological study by Howe *et al.*,[22] researchers indicated that firefighters had increased risk of brain tumors, malignant melanoma, and multiple myeloma. In 2006, LeMasters *et al.* performed a meta-analysis on 32 cohort studies, some of which are discussed above, and identified probable cancer risks for multiple myeloma, non-Hodgkin's lymphoma, and prostate and testicular cancers.[23]

Organizations such as the NFPA and the Bureau of Alcohol, Tobacco, Firearms and Explosives (ATF) have also researched the safety issues associated with firefighting and fire investigation. An NFPA study tallied an average of 4585 injuries per year related to smoke inhalation, gas inhalation, and respiratory distress in firefighters between 1996 and 2006.[5] The Alcohol, Tobacco and Firearms Fire Investigation Division also recognized the respiratory and dermal hazards associated with fire investigators sifting through fire debris, and more recently published the results of a study focusing on fire scene uniform contaminations.[24,25] These studies and more have led the fire service to recognize the cancer trends within their profession and have raised questions regarding preventive measurements to reduce the risk of cancer.

In the ATF study, researchers examined the exposures of fire investigators to products of combustion present after the extinguishment of structure fires. Air sampling was conducted at fire incidents in Washington, D.C. and in Prince Georges County, Maryland, as well as at three staged fires in Alexandria, VA. The samples were checked for dusts, metals, hydrogen cyanide, inorganic acids, aldehydes, PAHs, elemental carbon, and volatile organic compounds (VOCs). Most of the compounds that were tested were found in levels between the minimum detectable concentrations and minimum quantifiable concentrations. Compounds included dusts, aliphatic hydrocarbons, acetone, acetic acid, ethyl acetate, isopropanol, styrene, benzene, toluene, xylene, furfural, phenol, and naphthalene. PAHs with carcinogenic potential included benz[*a*]anthracene, benzo[*b*]fluoranthene, and benzo[*a*]pyrene. While all of the previous compounds were found at levels below NIOSH recommended exposure limits, formaldehyde was found in concentrations nearly twice as high as its limit of 0.1 ppm. The report concluded that investigators are exposed to irritants that cause acute effects, and to carcinogens that could have chronic effects. The study also noted that investigators who did not wear respiratory protection experienced eye and respiratory irritation while working, and suggested that respiratory protection and mechanical ventilation could be used to reduce exposure potential.

The rescue efforts during the aftermath of the World Trade Center (WTC) attacks on September 11, 2001, have also shed light on the extreme respiratory hazards faced by firefighters (see Chapter 19). Three weeks after the attacks, the Centers for Disease Control and Prevention (CDC), NIOSH, and the Fire Department City of New York (FDNY) conducted a study that involved the biomonitoring of those who responded to the WTC fire and collapse.[26] Blood and urine samples were taken from 370 members of FDNY,

321 of who were present at various times between the collapse and rescue, recovery, and cleanup (up to two weeks post-collapse). Variables considered were arrival time, number of work days at WTC, and unit assignments. The samples were then analyzed for 110 chemicals. The study found that exposed firefighters had significant differences in nine chemicals when compared to the control samples and those assigned to special operations command had differences in 14 chemicals compared to the other responders. The New York State Department of Health has since created the World Trade Center Responder Fatality Investigations Program to track those affected while working on site. Since the rescue, recovery, and cleanup efforts have subsided, over 600 people have attributed their various forms of cancer to their work following 9/11.[27] According to the WTC Responder Fatality Investigations Program, most of the deadly tumors were located in the lungs and digestive systems of the victims, while other illnesses included blood cancer, heart disease, and circulatory disease.

17.3 Environmental Exposures

17.3.1 Emergency Workers and the General Public during Large-scale Fire Incidents

Many studies have linked increases in environmental pollution and particulate matter (*e.g.*, $PM_{2.5}$ and PM_{10}) to severe health effects and mortality. Hence, events that increase the normal ambient air pollution concentrations have the potential to adversely affect health in surrounding communities. In large-scale fire events, those intimate with the fire are not the only individuals who may be exposed to the products of combustion. Individuals within zones 2 or 3 may also be acutely exposed to the fire environment and traveling fire plume. A number of studies conducted over the last two decades have focused on air pollution and air quality effects from large-scale fires, as well as the resultant health effects on both emergency workers and the surrounding populations.

Bridgman[28] evaluated the health risks of a factory fire in Tranmere, Merseyside, England, which occurred on September 22, 1994. The roof of the factory contained chrysotile and amosite asbestos; the roof burned during the fire and debris was carried to nearby communities in the smoke plume. Fallout affected 16 000 to 48 000 people who lived near the site. A total of 105 calls were received from the public to request fallout removal. The large majority of the calls were within one mile of the factory in a half mile span. A few calls were received as far out as four miles. In estimating the risk of exposure to asbestos, the researchers utilized asbestos concentration measurements taken from the scene and applied them to a hazard assessment model; real data on injuries or illnesses was not obtained. Risk of mesothelioma and lung cancer was determined to be primarily from chrysotile exposure. It was determined that the health risk from exposure was undetectably small, mainly due to the shortness of exposure and the long latency period of asbestosis (*e.g.*, 10 years). It should also be noted that it rained within 48 hours after the incident, which assisted in significantly minimizing the presence of airborne particulates.

The health effects of another factory fire were investigated by Upshur *et al.*[29] The fire occurred at a polyvinyl chloride (PVC) recycling plant on July 9, 1997 in Hamilton, Ontario and burned for more than 3 days. The majority of the plume dispersed into the atmosphere until a thermal inversion event occurred, which blanketed nearby communities in the plume. Various air monitoring sites were established with concerns of exposure to hydrogen chloride and dioxins. Researchers performed health surveys of local residents, analyzed emergency room admissions, and also surveyed for chloracne, which is indicative of severe dioxin exposure. A total of 163 households (488 individuals) responded to the questionnaires, of which 70% reported that they left the neighborhood during some portion of the evacuation period. A total of 62% reported no symptoms, 17% reported throat irritation, and the remainder (from most to least) reported headaches, difficulty breathing, nausea, abdominal pain, vomiting, diarrhea, eye irritation, dizziness, mouth irritation, and nose irritation. Occurrences were highest during the inversion. Very few residents (only 2%) sought medical assistance; hence, it is not surprising that researchers found no notable increase in emergency room visits during the incident. There were also no reported cases of chloracne. Researchers did not find any evidence of increased morbidity or mortality due to the incident.

Another factory-type fire was investigated by Hoek *et al.*[30] and Morgan *et al.*[31] A large explosion occurred on December 11, 2005 in Hemel Hempstead, Hertfordshire, UK at the Buncefield fuel depot (see Chapter 18). The subsequent fire lasted for many days and involved the combustion of more than 20 motor fuel storage tanks. Hoek *et al.* investigated the health impact of this explosion on surrounding communities by analyzing epidemiological data from individuals who went to the hospital as a result of the incident. Researchers reported that 244 people received medical care as a result of the incident. A large number of the 244 people were emergency workers who were advised to seek medical evaluation from their command officers regardless of symptoms of exposure. Approximately half of the 244 people had symptoms that were attributable to the fire (*e.g.*, exacerbation of asthma, shortness of breath, sore throat, and cough); 17 people were oil depot workers, 40 people were from the general public, and 63 people were emergency workers. Researchers concluded that the public health impact was relatively small compared to the duration of the fire and quantity of combustion byproducts. Researchers reasoned that the time of week (*i.e.*, Sunday), high combustion temperatures, and wind and weather conditions were favorable in reducing the health impact of the fire.

Morgan *et al.*[31] focused specifically on the health effects of emergency workers who responded to the Buncefield fuel depot fire. Utilizing questionnaires, researchers asked emergency workers about symptoms that arose during and after the active fire and compared the cohorts' symptoms to those of local residents outside of the area of the smoke plume. Of the 1949 individuals polled, only 815 responded; a large majority of the respondents operated during the active burning phase of the fire and were within 500 meters of the fire. Within this group, 75% reported inhaling smoke, fumes, or particles, and 85% reported smelling smoke and/or chemicals. Unfortunately, many did not use respiratory protection (1/5th of sample), and most of those that did use respiratory protection only used paper masks. The majority of the reported symptoms occurred during the burning phase, with 40% of the population reporting respiratory irritation, eye irritation, and headaches. The symptoms reported after active burning ceased were insignificant and were consistent with those seen in the general public. No major illnesses were reported from the acute exposures, however, it was determined that the respiratory protection available to emergency workers was insufficient to protect them from the fire scene hazards.

17.3.2 Emergency Workers and the General Public Exposed during Wildland Fires

Many studies have also evaluated the health effects of wildland fires. During the summer and fall of 1997, Southeast Asia was affected by forest fires that burned in Indonesia. Emmanuel[32] investigated the impact of the extensive fires on the lung health of residents in Singapore. Based on air monitoring data, they found that the pollutant standard index (PSI; a metric developed by the US Environmental Protection Agency, EPA) reached an "unhealthy range" for numerous days during the pollution event. Additionally, they reported that the normal PM_{10} values doubled with a recorded range of 60–100 μg m^{-3}; however, there was only a slight increase in carbon monoxide, nitrogen dioxide, and ozone levels. Researchers utilized national mortality data, morbidity data from outpatient and inpatient facilities, and accident and emergency department data to evaluate exposure symptoms. It was found that the haze caused a 30% increase in outpatient attendance for related conditions, a 12% increase in upper respiratory tract illness, a 19% increase in asthmatic events, and a 26% increase in rhinitis. There was not, however, a significant increase in hospital admissions or in mortality.

Between August and September of 2002, wildfires stretching 2500 km long and 1000 km wide burned in Eastern Europe. These fires resulted in significant air pollution and caused public health concerns. Hänninen *et al.*[33] utilized data from these incidents to evaluate mortality risks to the exposed population. Air quality monitors measured peak $PM_{2.5}$ and PM_{10} concentrations approaching 200 μg m^{-3} and 100 μg m^{-3}. Analysis determined that the large majority of the particulate matter was $PM_{2.5}$. In order to estimate risk, researchers utilized the air monitoring data and data from epidemiological studies that investigated exposure–response factors to $PM_{2.5}$. In the 11 Southern Finnish provinces evaluated, the mean exposure level to $PM_{2.5}$ for two weeks was 15.7 μg m^{-3}. Using an increase of 1% mortality for each

10 μg m^{-3}, it was estimated that the incident could result in 17 additional mortalities, although this was not substantiated with actual occurrences.

Ovadnevaite *et al.*[34] investigated the impact of the 2002 summer fires in Lithuania. Using air pollution monitoring and meteorological data, researchers found a significant increase in PM_{10}, NO_2, CO, and SO_2 concentrations during the two month fire event. Specifically, the concentrations of PM_{10} and NO_2 were 5.5 times and 2 times higher, respectively, than the 50 μg m^{-3} limit value set forth by the regulating authority. An increase in these measurements coincided with an increase in complaints of respiratory diseases and exacerbation of bronchial asthma; these complaints were up to 20 times higher in comparison to previous years.

A large-scale forest fire occurred in Hoopa, California between August and November of 1999 near the Hoopa Valley National Indian Reservation. Air concentration measurements taken at the reservation showed that EPA limits were exceeded on three occasions. In an attempt to limit adverse effects to those exposed, officials provided filtered and non-filtered masks, hotel vouchers, high-efficiency particulate arrestance (HEPA) cleaners, and did several public service announcements (PSA). Mott *et al.*[35] evaluated the effectiveness of these interventions in preventing adverse health effects. Researchers evaluated the prevalence of clinical visits during the incident (from September to November of 1999) and compared them to the previous year (from September to November of 1998); they found that there were 217 additional visits to the clinic for respiratory problems during the fire incident. A community survey documented self-reported respiratory symptoms in 60% of the respondents (178 out of 289). A total of 140 people reported leaving the reservation during the incident, 100 people reported wearing masks or face coverings, and 98 people used the HEPA air cleaners provided by officials. The use of masks was not significantly associated with a reduction in respiratory symptoms, while the use of a HEPA cleaner did show a significant correlation. Additionally, those who remembered hearing a PSA were less likely to report respiratory symptoms than those who did not remember hearing a PSA.

In October of 2013, Southern California was devastated by 14 large-scale wildland fires, which infiltrated San Diego, Los Angeles, San Bernardino, and Ventura. The fires burned thousands of homes and commercial and accessory structures, and resulted in 16 deaths. Viswanathan *et al.*[36] investigated the health and air quality effects of the fires on the residents of San Diego, California. Researchers investigated the pollutant concentrations in the air prior to the fire and compared them to those during and after the incident. Various monitoring stations were utilized to compile data for $PM_{2.5}$, PM_{10}, ozone (O_3), CO, and NO_2 concentrations. The $PM_{2.5}$ and PM_{10} concentrations exceeded the EPA limits. Ozone concentrations trended downward during the event and never exceed the EPA limits, CO stayed below EPA limits with the exception of one day during which the 8 hour TWA of 9 ppm was exceeded by 1.6 ppm, and NO_2 never exceeded the 1 hour California State limit of 0.25 ppm (no limit is set by the EPA). Researchers compared the air monitoring data against hospital records for patient admissions. The total number of overall emergency room visits decreased during the fire event, which corresponded with days that residents were asked to remain home. There was a significant increase, however, in visits specifically related to asthma in comparison to pre-fire ratios. There was also a higher predominance of visits for bronchitis and other respiratory symptoms. Researchers concluded that there was a direct correlation between the increased occurrence of asthma, other respiratory symptoms, eye irritation, and smoke inhalation as a result of increased PM concentrations, noting a 1–1.5 day lag between exposure and asthmatic symptoms.

Künzli *et al.*[37] also investigated the health effects of the 2003 Southern California wildfires, but specifically focused on children. The researchers utilized data from the Children's Health Study (CHS), which was a long term health study of California schoolchildren not related to the wildland fire event. Researchers sent out a fire questionnaire to two cohorts of CHS participants; the questionnaire evaluated exposure duration, interventions, and health-related problems. Symptoms were compared against monitored air PM_{10} concentrations during the event. Substantial effects within the general population, including eye irritation and respiratory symptoms, were found to be prevalent in children with asthma and non-asthmatic children. The study suggests that wearing masks, spending less time outdoors, and/or using air conditioning were beneficial intervention techniques.

Two wildland fire events occurred in June of 2001 and June of 2002 in Denver, Colorado. Vedal and Dutton[38] investigated the mortality impact of the two fires. Both fires resulted in a sudden increase in PM concentrations. Researchers evaluated daily deaths in Denver in 2001 and 2002 in comparison to daily PM concentrations. The researchers concluded that there was no significant increase in daily mortality rates on days with increased PM concentrations when compared to a control population. The authors noted that an evaluation of less adverse health effects, *e.g.*, asthma, bronchitis, *etc.*, may show a positive correlation.

17.3.3 Gulf War Veterans Exposed during the Kuwait Oil Fires

Unlike the previously discussed research, which focused on the immediate effects of exposure to a population, Lange *et al.*[39] investigated Gulf War veterans' exposed to the Kuwait oil fires 5 years post-war; Iraqi forces ignited over 600 oil wells as they retreated in February of 1991. The study evaluated 1560 veterans. Phone interviews were performed to evaluate exposure conditions, war preparedness, and pre-exposure and post-exposure physical health. It was found that 8.3%, 4.7%, 24.7%, and 8.6% of the population reported asthma, bronchitis, injury, and major depression, respectively. Smokers and those with a low level of preparedness were more likely to experience asthma and bronchitis than nonsmokers with a high level of preparedness; the correlation between level of preparedness and health symptoms is unclear. Geographic Information Systems (GIS) data was utilized to evaluate troop location against exposure conditions. Risks of symptoms did not correlate with the modeled levels of exposure; those with a high level of exposure in the modeled scenario had the same risk of asthma and bronchitis as those with low levels of exposure. However, in the self-reported measures, those with a high self-reported level of exposure had an increased risk of asthma and bronchitis; the authors attributed this finding to recall bias. Ultimately, they concluded that there was no link between exposure to oil-fire smoke and the enhancement or development of respiratory symptoms among the Gulf War veterans.

Smith *et al.*[40] and Kelsall *et al.*[41] also investigated the effects of the Kuwait oil fires on US and Australian military, respectively. Smith *et al.*[40] utilized Gulf War veteran hospitalization data from Department of Defense records for the time period of August 1991 to July of 1999. They compared this data with air emission concentrations within the Gulf War thereafter of operations. An air dispersion model was used to estimate troop exposure concentrations, and troop location was compiled using GIS data. The researchers grouped the hospitalization data into three categories: those admitted for any cause, those admitted for a major diagnosis, and those admitted for symptoms specific to smoke exposure from oil-well fires. Data was available for a total of 405 142 veterans. Subjects were excluded if they had an existing condition prior to the war. Exposed veterans were found to have a statistically significant decreased risk for ischemic heart disease, and an increased (not statistically significant) risk for emphysema, acute bronchitis, various neoplasms, and other diseases of the respiratory system. While some of the models predicted an increased risk for hospitalization, the researchers found no dose–response correlations, and therefore concluded that the Gulf War veterans did not have a higher risk of morbidity.

Kelsall *et al.*[41] investigated the health status of Australian veterans exposed to the oil-well fires and dust storms during the Gulf War. Noting that there was a high occurrence of self-reported respiratory symptoms and diseases in the veteran population, researchers set out to quantify the respiratory conditions and lung functions of veterans. A total of 1456 veterans were included in the study population, and the study was conducted between August 2000 and April 2002. Questionnaires were distributed to gather information about service history, exposure history, prior and current medical conditions, tobacco use, medicine use, and respirator use during active deployment. Participants also underwent health assessments during which respiratory function was documented; doctors also evaluated the accuracy of any self-diagnosed respiratory symptoms. The researchers found that veteran reporting of respiratory symptoms was statistically significant in comparison to the control group (*e.g.*, military personnel who were not deployed to the Gulf War). Researchers concluded that, although the veterans reported symptoms of increased asthma and bronchitis, there was no significant change in lung function. Based on these findings, researchers concluded that there would be no major long term consequences of exposure to the oil fires or dust storms.

In most of the studies discussed above, the researchers attempted to correlate air quality measurements with hospital admission records or self-reported exposures. While increases in asthma and other respiratory symptoms were commonly reported, many cases were self-reported and not medically diagnosed. A large majority of the fires did result in a significant increase in particulate matter concentrations typically coinciding with an insignificant increase in CO, NO_x, and O_3. Statistically significant, direct correlations between large-scale fire exposures and severe respiratory symptoms or mortality were lacking in the above summarized studies. These findings may be a defect of the study methodology or may suggest that acute exposures to PM concentrations exceeding EPA limits do not have a large impact on morbidity or mortality.

References

1. J. Brown and J. Thornton, Percivall Pott (1714–1788) and Chimney Sweepers' Cancer of the Scrotum, *Br. J. Ind. Med.*, January 1957, **14**(1), 68–70.
2. *Products of Combustion: Draft Summary Information*, Health Protection Agency, Chemical Hazards and Poisons Division, London, July 2006.
3. M. Avakian, D. Dellinger, H. Fiedler, B. Gullet, C. Koshland *et al.*, The Origin, Fate, and Health Effects of Combustion By-Products: A Research Framework, *Environ. Health Perspect.*, November 2002, **110**(11).
4. T. Fabian, J. Borgerson, S. Kerber, P. Gandhi, C. Baxter, C. Ross and J. Lockey, Firefighter Exposure to Smoke Particulates, Underwriters Laboratories, Final Report. Project Number: 08CA31673, April 1, 2010.
5. C. Grant, Respiratory Exposure Study for Fire Fighters And Other Emergency Responders, *The Fire Protection Research Foundation*, National Fire Protection Association, Quincy, MA, December 2007.
6. R. Lackore, Firefighter Anthropometric Data White Paper, Fire Apparatus Manufacturer's Association, FAMA Technical Committee, Chassis Subcommittee, October 20, 2007.
7. G. Kinnes and G. Hine, Health Hazard Evaluation Report 96-0171-2692, Bureau of Alcohol, Tobacco, and Firearms, Washington, D.C., May 1998.
8. T. Reinhardt and R. Ottmar, Smoke Exposure at Western Wildfires, United Stated Department of Agriculture, Forest Service, Pacific Northwest Research Station, Research Paper, PNW-RP-525, July 2000.
9. G. Noonan, J. Stobbe, P. Keane, L. Reed, R. Ronk, S. Hendricks, R. McCarthy, Firesmoke: A Field Evaluation of Self-Contained Breathing Apparatus, National Institute for Occupation Safety and Health, Division of Safety Research, Morgantown, West Virginia, Accession number PB93-159119, 1989.
10. D. Baris, T. J. Garrity, J. L. Telles *et al.*, Cohort mortality study of Philadelphia firefighters, *Am. J. Ind. Med.*, 2001, **39**, 463–476.

11. J. J. Beaumont, G. S. Chu, J. R. Jones *et al.*, An epidemiological study on cancer and other causes of mortality in San Francisco firefighters, *Am. J. Ind. Med.*, 1991, **19**, 357–372.
12. P. A. Demers, H. Checkoway, T. L. Vaughan *et al.*, Cancer incidence among firefighters in Seattle and Tacoma, Washington (United States), *Cancer Causes Control*, 1994, **5**, 129–135.
13. A. W. Musk, R. R. Monson, J. M. Peters *et al.*, Mortality among Boston firefighters, 1915–1975, *Br. J. Ind. Med.*, 1978, **35**, 104–108.
14. N. Heyer, N. S. Weiss, P. Demers *et al.*, Cohort mortality study of Seattle fire fighters: 1945–1983, *Am. J. Med.*, 1990, **17**, 493–504.
15. G. Grimes, D. Hirsch and D. Borgeson, Risk of Death Among Honolulu Fire Fighters, *Hawaii Med. J.*, March 1991, **50**(3), 82–85.
16. F. Ma, L. Fleming, D. Lee *et al.*, Mortality in Florida Professional Firefighters, 1972–1999, *Am. J. Ind. Med.*, 2005, **47**, 509–517.
17. E. Feuer and K. Rosenman, Mortality in Police and Firefighters in New Jersey, *Am. J. Ind. Med.*, 1986, **9**, 517–527.
18. K. Aronson, G. Tomlinson and L. Smith, Mortality Among Fire Fighters in Metropolitan Toronto, *Am. J. Ind. Med.*, 1994, **26**, 89–101.
19. S. Deschamps, I. Momas and B. Festy, Mortality amongst Paris firefighters, *Eur. J. Epidemiol.*, 1995, **11**, 643–646.
20. T. Guidotti, Mortality of Urban Firefighters in Alberta, 1927–1987, *Am. J. Ind. Med.*, 1993, **23**, 921–940.
21. G. Tornling, P. Gustavsson and C. Hogstedt, Mortality and Cancer Incidence in Stockholm Fire Fighters, *Am. J. Ind. Med.*, 1994, **25**, 219–228.
22. G. R. Howe and J. D. Burch, Firefighters and risk of cancer: an assessment and overview of the epidemiologic evidence, *Am. J. Epidemiol.*, 1990, **132**, 1039–1050.
23. G. LeMasters, A. Genaidy, P. Succop *et al.*, Cancer Risk Among Firefighters: A Review and Meta-Analysis of 32 Studies, *J. Occup. Environ. Med.*, November 2006, **48**(11), 1189–1202.
24. G. Kinnes and G. Hine, Health Hazard Evaluation Report 96-0171-2692, Bureau of Alcohol, Tobacco, and Firearms, Washington, D.C., May 1998.
25. E. Snyder, Health Hazard Evaluation Report 2004-0368-3030, Bureau of Alcohol, Tobacco, Firearms and Explosives, Austin, TX, January 2007.
26. P. Edelman, J. Osterloh, J. Pirkle *et al.*, Biomonitoring of Chemical Exposure among New York City Firefighters Responding to the World Trade Center Fire and Collapse, *Environ. Health Perspect.*, December 2003, **111**(16), 1906–1911.
27. J. Lite, 360 post-9/11 workers have died, including 80 of cancer, says state, Daily News (New York), May 8, 2008.
28. S. Bridgman, Community health risk assessment after a fire with asbestos containing fallout, *J. Epidemiol. Commun. Health*, 2001, **55**, 921–927.
29. R. Upshur, M. James, E. Richardson *et al.*, Short-Term Adverse Health Effects in a Community Exposed to a Large Polyvinylchloride Plastics Fire, *Arch. Environ. Health*, May/June 2001, **56**(3), 264–270.

30. M. R. Hoek, S. Bracebridge and I. Oliver, Health impact of the Buncefield oil depot fire, December 2005: Study of accident and emergency case records, *J. Public Health*, June 2007, **29**(3), 298–302.
31. O. Morgan, N. Q. Verlander, F. Kennedy, M. Moore, S. Birch *et al.*, Exposures and reported symptoms associated with occupational deployment to the Buncefield fuel depot fire, England 2005, *Occup. Environ. Med.*, 2008, **65**, 404–411.
32. S. Emmanuel, Impact to lung health of haze from forest fires: The Singapore experience, *Respirology*, 2000, **5**, 175–182.
33. O. Hänninen, R. Salonen, K. Koistinen, T. Lanki *et al.*, Population exposure to fine particles and estimated excess mortality in Finland from an East European wildfire episode, *J. Exposure Sci. Environ. Epidemiol.*, 2009, **19**, 414–422.
34. J. Ovadnevaite, K. Kvietkus and A. Marsalka, 2002 summer fires in Lithuania: Impact on the Vilnius city air quality and the inhabitants' health, *Sci. Total Environ.*, 2006, **356**, 11–21.
35. J. Mott, P. Meyer, D. Mannino and S. Redd, Wildland forest fire smoke: health effects and intervention evaluation, Hoopa California, 1999, *Western J. Med.*, May 2002, **176**(3), 157–162.
36. S. Viswanathan, L. Eria, N. Diunugala, J. Johnson and C. McClean, An Analysis of Effects of San Diego Wildfire on Ambient Air Quality, *J. Air and Waste Manag. Assoc.*, January 2006, **56**, 56–67.
37. N. Künzli, E. Avol, J. Wu, J. Gauderman, E. Rappaport *et al.*, Health Effects of the 2003 Southern California Wildfires on Children, *Am. J. Respir. Crit. Care Med.*, 2006, **174**, 1221–1228.
38. S. Vedal and S. Dutton, Wildfire air pollution and daily mortality in a large urban area, *Environ. Res.*, 2006, **102**, 29–35.
39. J. Lange, D. Schwartz, B. Doebbeling, J. Heller and P. Thorne, Exposures to the Kuwait Oil Fire and Their Association with Asthma and Bronchitis among Gulf War Veterans, *Environ. Health Perspect.*, November 2002, **110**(11), 1141–1146.
40. T. Smith, J. Heller, T. Hooper, G. Gackstetter and G. Gray, Are Gulf Veterans Experiencing Illness due to Exposure to Smoke from Kuwaiti Oil Well Fires? Examination of Department of Defense Hospitalization Data, *Am. J. Epidemiol.*, 2002, **155**(10), 908–917.
41. H. L. Kelsall, M. R. Sim, A. B. Forbes, D. P. McKenzie, D. C. Glass, J. F. Ikin *et al.*, Respiratory health status of Australian veterans of the 1991 Gulf War and the effects of exposure to oil fire smoke and dust storms, *Thorax*, 2004, **59**, 897–903.

Examples of Unusual Conflagrations

CHAPTER 18

Buncefield Fire

THOMAS WAITE, CATHERINE KESHISHIAN AND VIRGINIA MURRAY*

Public Health England, Wellington House, 133-155 Waterloo Road, London SE1 8UG, UK
*Email: Virginia.Murray@phe.gov.uk

18.1 The Buncefield Oil Depot

The Buncefield depot is one of the largest oil storage and transfer sites in the UK, handling refined petroleum products and additives before distribution to other facilities. Due to its scale, parts of the Buncefield terminal complex are classified as a top tier COMAH site, meaning they fall under the Control of Major Accident and Hazards Regulations, 1999. Given the location of Buncefield, approximately 3 miles from Hemel Hempstead in Hertfordshire and 25 miles north of London, it is particularly important in the distribution of fuels to London and the south-east of England, and prior to the events of December 2005 the terminal acted as a main pipeline transit point for 40% of Heathrow airport's aviation fuel. In 2005, Buncefield was made up of three different sites that, between them, were permitted to store approximately 190 000 tonnes of motor fuel, heating oil and other fuels. These three sites were:

- Hertfordshire Oil Storage Ltd (HOSL): a Total UK/Chevron joint venture, permitted to store 34 000 tonnes of motor fuel and 15 000 tonnes of heating oil. This site was divided into two sections: east and west.
- British Pipeline Agency Ltd (BPA): a joint Shell/BP venture with assets owned by UK Oil Pipelines Ltd, permitted to store 70 000 tonnes of

Issues in Toxicology No. 23
Toxicology, Survival and Health Hazards of Combustion Products
Edited by David A. Purser, Robert L. Maynard and James C. Wakefield

Published by the Royal Society of Chemistry, www.rsc.org

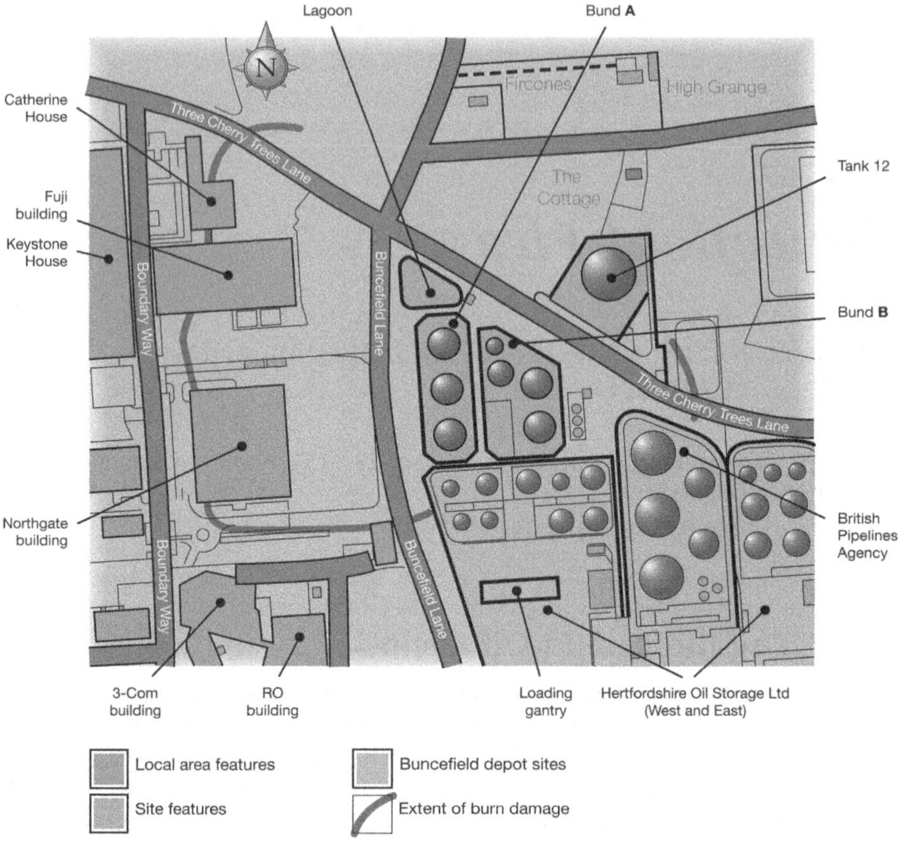

Figure 18.1 Layout of depot and surroundings. The figure also shows the extent of the burn damage.
(Image from BMIIB 2008b.)[3]

motor and other fuels. This site was also divided into two sections: the main site and the north site.
- BP Oil Ltd: with consent to store 75 000 tonnes of motor fuel.

The depot was directly supplied with fuel by three different pipelines: the Merseyside/Buncefield (M/B) pipeline running between Merseyside and the BPA site, the Thames/Kingsbury (T/K) pipeline running between Coryton Refinery and the BPA site and a third line running from Lindsey Oil Refinery in Humberside to the HSOL site (Figure 18.1).[3] On the morning of 11 December 2005, approximately 82 million litres of fuel were stored on site.[1,2]

18.2 Circumstances of the Explosion and Subsequent Fire

At 1900 hours on Saturday 10 December 2005, filling of tank 912 commenced on the HOSL west site at Buncefield with approximately 550 m^3 h^{-1} of unleaded car fuel. Filling continued, although from approximately 0300 hours on Sunday 11 December the level gauge for the tank, which had a capacity of 6 000 000 litres, recorded an unchanged reading. It would probably have been completely full by approximately 0520 hours, but the protection system to automatically close valves failed to operate.[3]

1. From 0520 hours onwards, pumping continued whilst the fuel cascaded down the sides of the tank and into the atmosphere, leading to a rich mixture of fuel and air collecting in the bund around the tank. CCTV footage shows that the vapour from this fuel started to escape in a number of directions, commencing at approximately 0538 hours. Between 0550 hours and 0600 hours the pumping rate into the tank rose gradually from 550 $m^3 h^{-1}$ to around 890 $m^3 h^{-1}$.

Petrol is not a single compound but is a mixture of different hydrocarbons and other organic chemicals. The composition varies throughout the year and in winter is blended with approximately 10% butane in order to boost performance. This was the blend being pumped into tank 912.[5]

In the minutes prior to the explosion, a white mist was seen on CCTV spreading from the area of the overflowing tank 912. It is not entirely certain what the constituents of the white mist were; it has been suggested that a volatile fraction of the hydrocarbon mix, such as butane, may have given rise to it, or ice crystals may have formed from the chilled humid air as a consequence of the rapid evaporation of the escaping fuel. The vapour cloud eventually spread to a diameter of about 360 m, beyond the boundary of the HOSL west site and including the car park of an adjacent business park, as well as the BPA north site, where a tank containing aviation kerosene was sited.

Petrol does not explode easily; however, up to 300 tonnes of petrol had escaped around the tank. It is thought that approximately 10% of this turned to vapour and mixed with air, reaching concentrations capable of supporting combustion. The cloud was extensive enough to be noticed off site by members of the public, as well as by nearby tanker drivers waiting to fill their vehicles. They alerted employees on site who pressed the fire alarm button at 0601 hours, sounding an alarm and starting the firewater pump. The explosion occurred almost immediately, probably caused by a spark when the firewater pump started, as the pump house was within the area covered by the vapour cloud. Ignition of the vapour cloud triggered the first in a series of explosions to occur at the oil depot, measuring 2.4 on the Richter scale. At the time of the explosion, the M/B pipeline was supplying tank 908 with approximately 400 $m^3 h^{-1}$ of diesel oil and a further pipeline was supplying a tank in the same bund as tank 912 with unleaded petrol at a rate of approximately 220 $m^3 h^{-1}$.[3]

18.2.1 The Fire and Aftermath

The resulting fire was of such a scale that a massive smoke plume could clearly be seen across south Hertfordshire, London and the south-east of

England. The explosion was reported to be heard as far away as the Netherlands.

The fire burned for five days, by which time 23 tanks of diesel, kerosene and aviation fuel had been destroyed. More than 33 fire and rescue services were brought in to put out the fire; the largest in Europe since World War II.

The explosions caused significant damage to the site, including catastrophic structural damage or rupturing of 19 tanks and numerous buildings on site, structural damage to all but four of the other tanks, blocked access to two of the three emergency water reservoirs, destruction of the main fire pump houses on site and ignition of adjacent office buildings and several cars. Off site, the explosions caused severe structural damage to properties within 1 km of the depot, broke windows, displaced walls and collapsed ceilings up to 2 km from the depot and caused minor damage as far away as 18.6 km.[4] The cost of this explosion has been put at UK £1 billion.[5]

Petrochemical fires can burn at particularly high temperatures. At Buncefield, the heat was sufficient that for the first two days of the fire a strong thermal image was detected from space, on the Moderate Resolution Imaging Spectroradiometer satellites.[2] Mather *et al.*[2] modelled the energy based on the volume of fuel that burned at Buncefield and conditions at the time, and calculated that a release of thermal energy of this magnitude would give rise to a buoyant plume reaching approximately 2–3 km in height, with an initial plume from the explosion of 3.5 km height. These modelling estimates were found to be consistent with real time satellite imagery.[1,2]

At the time of this explosion, the ground temperature at Buncefield was close to freezing. There was little wind and the weather conditions were predominantly anticyclonic.[6] A strong inversion layer existed in the atmosphere, which the plume, with an initial height of 3.5 km, was able to "punch" through. This, along with the high temperature of the blaze, meant that the plume was then effectively trapped aloft with minimal deposit to the ground;[1,2] this plume can be seen in Figure 18.2.

18.3 Estimation of Emissions and Pollutants

The products of combustion from petrochemical fires are largely predictable and include nitrogen oxides, particulate matter (PM), carbon monoxide, polycyclic aromatic hydrocarbons (PAHs), including benzo[*a*]pyrene, benzene and other volatile organic compounds (VOCs), dioxins and furans. In 2006, The National Environmental Technology Centre (Netcen) estimated that 8000 tonnes of PM_{10} was released by the explosion and fire, equivalent to approximately 6% of the total annual emission of PM_{10} for the UK.[1]

Oil and petrochemical fires that occur outdoors release significant amounts of particulate matter, due in part to the high temperature at which they burn, but also because they have large amounts of oxygen available

Figure 18.2 The smoke plume rising and spreading above the Buncefield site. Courtesy of Hertfordshire Constabulary and Chiltern Air Support Unit.

for combustion[7] (see Chapter 3). They would also be expected to create polycyclic aromatic hydrocarbons and many other organic irritants, although at high enough temperatures these may be completely combusted. The particulate matter and irritants are thought to pose the greatest hazard to the health of those people in the path of the plume.

Targa *et al.*[1] used the known capacity and inventory of the Buncefield site to calculate a suite of scenarios for the quantities of air pollutants emitted by the fire. They suggest that a most realistic scenario is one that assumes 90% of the fuel on the BPA site was lost to combustion along with 60% of the fuel from the HOSL inventory, which encompasses west and east sites. This gives us an estimated 73 000 000 litres of fuel (approximately 58 000 tonnes) that are assumed to have burned in the fire.

Using these assumptions, Targa *et al.* estimated the released quantity of different pollutants, taking into account the emissions factor for each. Emission factors are based on the measured amount of pollutants released from the combustion of different fuels in different contexts, in milligrams per kilogram of fuel burned; in this case, Targa *et al.* mostly used factors relating to "open-burning" oil fires (see ref. 1 for source texts). The pollutants and estimated quantity released are shown in Table 18.1.

As in Table 18.1, an estimated 39.2 kilotonnes of carbon was released in the Buncefield fire. By way of comparison, at the end of the Gulf War, retreating Iraqi forces set fire to 613 Kuwaiti oil wells, burning approximately 61 megatonnes of oil over a nine month period and releasing about 2.1 megatonnes of carbon.[8]

Table 18.1 Estimation of the quantities of pollutants released from the Buncefield fire compared to background 2003 emissions (table adapted from Targa *et al.*, 2006).[1]

Pollutant	Emission factor	Estimated quantity released at Buncefield	Total UK emissions in 2003	Buncefield emissions as % of 2003 total
NO_x	651 mg kg^{-1}	37.2 tonnes	1570 kilotonnes	0.0024
PM_{10}	170 000 mg kg^{-1}	8249.5 tonnes	141 kilotonnes	5.85
$PM_{2.5}$	102 000 mg kg^{-1}	4949.7 tonnes	86.9 kilotonnes	5.70
Dioxins	2.31 ng WHO-TEQ[a] kg^{-1}	1.32 g TEQ g	259 g TeQ g	0.51
Benzo[*a*]pyrene	5 mg kg^{-1}	285.4 kg	4034 kg	7.08
CO	30 000 mg kg^{-1}	1712.7 tonnes	2768 kilotonnes	0.062
Non-methane VOCs	1770 mg kg^{-1}	101 tonnes	1089 kilotonnes	0.0093
Benzene	1022 mg kg^{-1}	58.3 tonnes	13.6 kilotonnes	0.43
CO_2	–	0.144 megatonnes	–	n/a
Carbon	–	39.2 kilotonnes	152,324 kilotonnes	0.026

[a]TEQ, toxic equivalency.

18.4 Public Health Response

As described in Chapter 20, the Health Protection Agency (HPA; Public Health England from April 2013) had a role to play in providing scientific and public health advice at Bronze (such as liaising with staff on the ground for updates, answering emergency service staff queries, providing information and advice), Silver (such as advising on evacuation or sheltering message, describing potential health impacts), and Gold (such as considering longer-term impacts) levels. The HPA had an important role in multi-agency liaison and keeping the public informed about any impact the incident may have on their health, as well as advising and recommending sheltering actions to take to protect themselves, and providing reassurance based on the latest data and evidence. This role extended beyond the immediate acute phase of the fire into the recovery and return to normality.

The HPA's main priority during the incident was to gather information in order to develop a public health risk assessment for the local population and, later, others under the plume within the UK and further afield. Environmental data were required to inform the HPA's dynamic exposure assessment, and investigations were required to identify any health impacts, both in the short and long term, which are discussed in the following sections.

18.5 Environmental Impact Findings

18.5.1 Air Quality Monitoring and Modelling

A core part of the information used by the HPA for exposure and toxicological risk assessment during the blaze was derived from two key sources: air quality monitoring and atmospheric modelling. Monitoring results are input into air quality models, so these methods cannot be entirely disentangled; however, for the purposes of this chapter they will be considered separately. Dispersion modelling of the products of combustion is discussed in Chapter 6.

18.5.2 Atmospheric and Plume Dispersion Modelling

Modelling of the plume consisted of four different sources of information: Chemical Meteorological models (CHEMET), Numerical Atmospheric-dispersion Modelling Environment (NAME), satellite imagery and atmospheric dispersion modelling undertaken by the Environment Agency. Atmospheric modelling is used to predict an area at risk, based on the expected path of the plume, using current and forecasted weather data (allowing for drift).

The Met Office is responsible for providing meteorological advice to the emergency services in the event of accidental release of potentially hazardous substances into the atmosphere.[9] After the explosion, Hertfordshire Fire and Rescue Service requested a CHEMET at approximately 0815 hours on the morning of the fire, after which time the modelling was run every three hours for the duration of the incident.

NAME modelling, also carried out by the Met Office, is used for large scale incidents and allows a prediction of the spread of a plume in three dimensions, including stratification by altitude. In common with the CHEMET, this modelling was undertaken and updated regularly throughout the incident, incorporating additional information as it became available. This was important because, as with any major incident, the initial modelling was undertaking using arbitrary release rates and other information because there was no accurate source-specific data available.[10] An example of a NAME output map can be seen in Figure 18.3.

These two (CHEMET and NAME) models informed the HPA on the areas of potential exposure and allowed rapid delivery of advice at strategic command group, SCG (Gold), level. They were also used to predict plume deposition.[10] These predictions were supported by satellite imagery from the Met Office, which allowed the HPA to confirm which of the fixed air quality monitoring stations on the ground would be most helpful in identifying whether anything from the plume was grounding.

The Environment Agency modelled the short range spread of the plume and expected ground deposition, to inform the HPA grass and soil sampling strategy (see next section).

18.5.3 Air Quality Monitoring

Four main types of monitoring were required in order to provide public health advice:[10] targeted local ground level air quality monitoring, direct

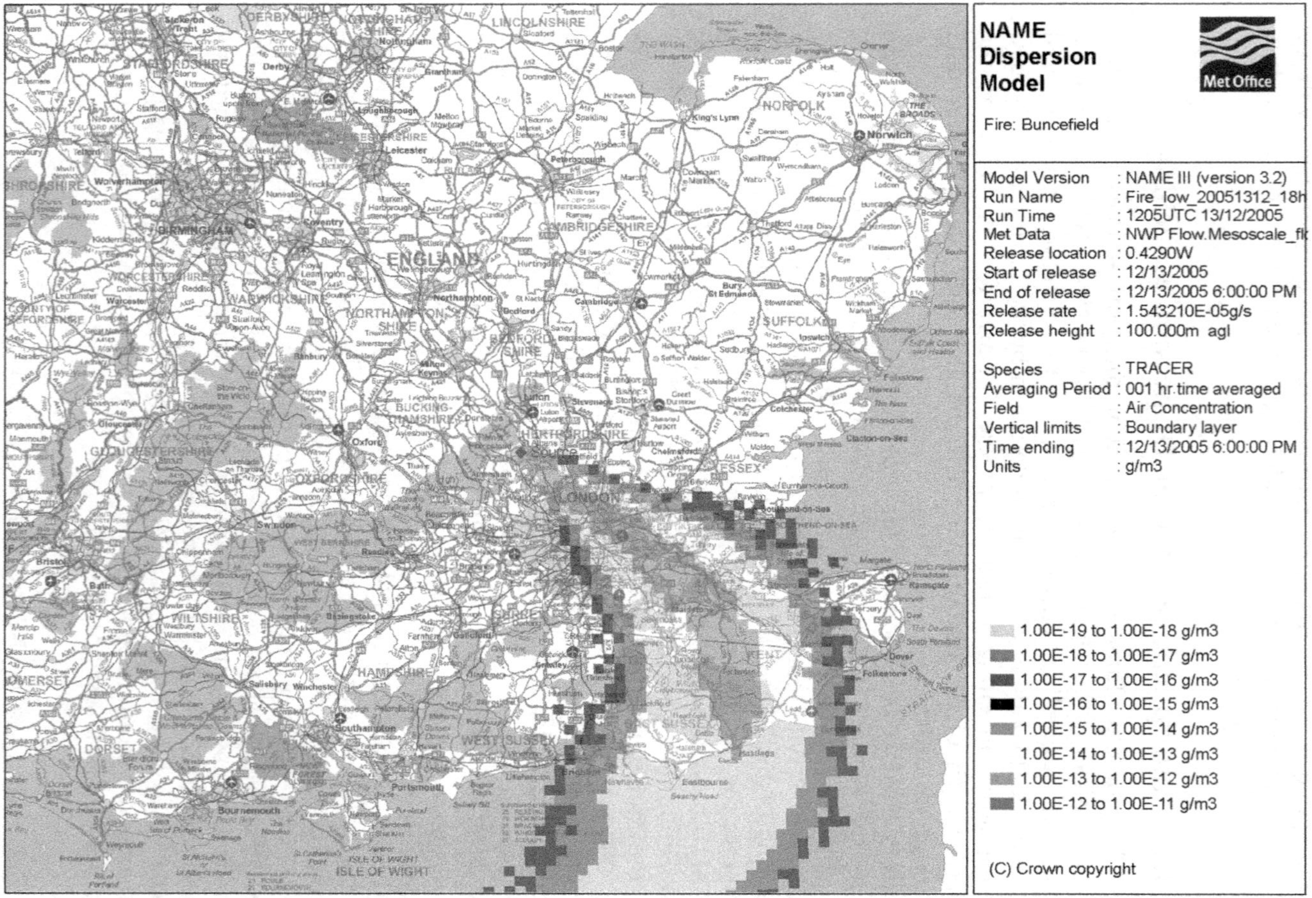

Figure 18.3 Dispersion modelling mapping the density of the plume across the south east of the UK. (Map courtesy of the Met Office – Crown Copyright.)

plume sampling, existing air quality monitoring networks and soil/grass sampling.

1) Locally targeted air quality sampling sites were identified by fire brigade scientific advisers and the Health and Safety Laboratory. A range of chemicals were monitored at these sites, including carbon dioxide, carbon monoxide, sulfur dioxide, hydrocarbons, particulate matter, volatile organic compounds, hydrogen sulfide, ammonia and hydrogen fluoride. The locations chosen were close to the fire, with sampling carried out over the period from 11th to the 13th December 2005. The results did not suggest deterioration in air quality due to the fire.
2) The Facility for Airborne Atmospheric Measurements (FAAM) aircraft was deployed to take *in situ* samples on the 12th and 13th December 2005 in order to determine the location of the plume and ascertain its chemical composition.[10] This identified that the plume existed between 600 m and 1500 m in altitude, thus supporting the view that exposure to the plume by the public at ground level was low. It also provided information that the main constituent of the plume was black carbon (or soot) and that the polycyclic aromatic hydrocarbon content was low.
3) Data from a number of pre-existing air quality networks were analysed. The UK national Automatic Urban and Rural Network (AURN) monitors particulate matter, nitrogen dioxide, sulfur dioxide, carbon monoxide and ozone.[10] These data were available in near-real time to the HPA and SCG.

   In the south-east of England, the Environmental Research Group (ERG) at King's College London manages four regional air quality networks at far greater spatial resolution than the AURN network. ERG reported that the plume appeared to have a small impact on air quality on several occasions on the evening of Sunday 11th December 2005 and early in the morning on 12th December 2005; one reading of 133 $\mu g\ m^{-3}$ of PM_{10} was detected in St Albans in Hertfordshire (King's College London, 2006).[10] The highest readings of PM_{10} detected from the fire were at fixed monitoring stations in East Surrey and Sussex, where "moderate" concentrations, with a maximum of 290 $\mu g\ m^{-3}$, were found.

   ERG, in collaboration with the fire and rescue service scientific advisors and local authority, established monitoring for particulate matter and oxides of nitrogen in Hemel Hempstead, comparing these with nearby St Albans. Throughout the incident, the concentrations of these pollutants remained low and at similar levels to St Albans.
4) Soil and grass samples were also collected after the fire and analysed to confirm that the plume had not grounded locally, as suggested by the air quality data. Seventy-two samples were tested for products of combustion from 33 locations downwind of the fire. These samples were compared with both control locations and known background information (where available). Samples were collected as soon as

possible after the fire and before substantial rainfall in the area, to avoid the possibility of pollution being washed away by the rain.[10]

Analysis was undertaken for a number of pollutants including dioxins, furans, heavy metals and fluorides. The vast majority of results were unremarkable when compared with control site and background information. Historical contamination was thought to be the probable cause of elevated levels at one site, which was supported by a plausible source of contaminated land (a former power station). Aside from this, the rest of the soil and grass samples were unexceptional and did not present a risk to human health.

18.5.4 International Impact

Air dispersion modelling done at the time of the incident subsequently showed that, although pollutants from the fire were long ranging and travelled across France and other countries, atmospheric conditions and combustion parameters meant that the plume was so high that there would be no impact on surface air quality.[1,11] Data from air quality monitoring networks in northern France were assessed and did not show impact from the fire.[1]

18.6 Health Impact Findings

18.6.1 Outcomes of Three Studies of Public Health Impact

In this section we will consider the outcomes of three studies of the public health impact of the incident and the bearing that environmental factors and combustion conditions had on the findings of those studies. The studies are:

1) An emergency medicine department case note review.
2) An occupational health surveillance register.
3) The Buncefield follow up population survey.

18.6.2 Emergency Department Case Note Review

In order to establish the immediate health impact of the fire and explosion, such as injuries or acute effects, the HPA initiated surveillance in emergency departments (EDs). Whilst the fire was still burning, staff registered people presenting with symptoms related to the fire, but this was not sufficient to analyse the overall health impact and results are not presented here.

A retrospective descriptive study of ED attendees was therefore undertaken.[12] A case definition was established: all records that included a clear statement indicating that the attendance was a result of the fire were considered.

Hoek *et al.*[12] reviewed 714 case records of attendances to Hemel Hempstead and Watford EDs for the period Sunday 11th December to

Table 18.2 Emergency department presentations linked to the Buncefield fire, by complaint type. Some patients presented with more than one complaint (adapted from Hoek *et al.*, 2007).[12]

Main presenting complaint	Number of attendees with complaint (%)
Respiratory	73 (62%)
Headache	21 (18%)
Injuries	39 (33%)
Anxiety	8 (7%)
Other	22 (19%)

Wednesday 14th December 2005. In total, 244 people attended as a result of the fire; 117 were symptomatic and the remainder attended for a check-up. Of the 117, 38 were members of the public, presenting with mainly cuts and sprains or respiratory symptoms, including shortness of breath, cough or asthma symptoms; these are shown in Table 18.2. Sixty-three patients were emergency workers who mainly presented with respiratory symptoms; half complained of a sore throat. The remaining 16 patients were workers from the oil depot itself; this group saw the highest proportion of injuries.[12]

Ninety percent of patients were sent home without the need for follow up.[10] Three were admitted to hospital, 15 referred to their general practitioners, three to orthopaedic surgeons and one to a cardiologist.

Despite the scale of the fire and the protracted length for which it burned (four days), the public health impact as measured by ED attendance indicates that significant exposure to hazardous products of combustion was minimal. Similarly, blast injuries and the high casualty numbers expected from a vapour cloud explosion of this sort were not seen.[10] The main burden on EDs was due to asymptomatic emergency personnel, and a number of suggestions were made for the triage and management of the occupational health of emergency workers during major incidents.[12]

18.6.3 Occupational Health Surveillance

It was apparent that, because of ongoing local anxiety about potential toxic hazards from exposure to the plume, it would be important to set up a register of emergency service personnel responding to the fire. The emergency services in Hertfordshire with Hertfordshire County Council and the HPA established an Occupational Health Working Group to set up a health register following the fire.[13,14]

A total of 51 organisations deployed approximately 2100 staff to the incident site; occupational health departments were identified through liaison with public services, government, professional and commercial networks. Key organisations included:

- Thirty-three fire and rescue services.
- Police services.

- Bedfordshire and Hertfordshire Ambulance and Paramedic Service.
- Hertfordshire Country Council.
- Government Office for the East of England.
- Office of the Deputy Prime Minister.
- Health and Safety Executive.
- Ministry of Defence.
- Three Valleys Water.
- Owners of the Buncefield site.
- Health Protection Agency.

The case definition for individuals to be included in the register included occupational deployment to or near the Buncefield Oil Depot fire site between 11 December 2005 and 5 January 2006. This covered the burn phase from 11–14 December, where smoke was being produced, and the clean-up and recovery phase from 15 December to 5 January, where personnel were involved in managing the fire-fighting foam blanket and equipment recovery.

Of the 1949 eligible individuals from participating organisations, 815 (44%) returned the study questionnaire. Of these, 660 (81%) individuals were deployed whilst the fire was burning with almost three-quarters deployed inside the inner safety cordon. Three-quarters of respondents reported inhalation of smoke, fumes or particles, 86% smelled smoke and/or chemicals and about two-fifths (39%) reported using respiratory protective equipment.

Twenty-six respondents reported attending hospital EDs during the burn phase and 12 consulted their general practitioner. An occupational health review was provided for 22 individuals (3%). Reassuringly, only 18 (2%) of respondents took time off work, 15 (2%) noted any subsequent ill effects and 32 (4%) reported feeling anxious about their health.[13,14] The proportion of respondents on the occupational register reporting health symptoms was higher during the burn phase than the control and clean-up phase (see Table 18.3).

18.6.4 Buncefield Follow up Population Survey

In order to assess ongoing health concerns due to the fire, a survey was carried out by the HPA and local public health teams. Over 5000 people in

Table 18.3 Frequency of new symptoms reported by occupational groups exposed during the burn phase. Some respondents reported more than one complaint (Morgan *et al.* 2008).[14]

Main symptom reported	Frequency (percentage of respondents)
Nasal irritation	130 (20%)
Throat irritation	174 (26%)
Eye symptoms	139 (21%)
Cough	138 (21%)
Headaches	105 (16%)

the two local primary care trusts (PCTs), both in affected areas to the south under the plume, and unaffected areas to the north were sent a questionnaire and letter in the post. The survey was intended to identify any concerns about risks to their health from the smoke, contamination of their environment and other fallout from the smoke.

Approximately 40% of those surveyed responded. A telephone survey was carried out to establish why people did not respond; the main reason given was lack of concern about health effects.

The survey found that the level of perceived exposure depended on location, where those who were under the plume were more likely to be worried about exposure. The level of anxiety dropped from approximately 50% during the incident to 13% when the questionnaire was distributed some seven weeks later. The main reason for anxiety during the event was the effect of breathing in toxic smoke, whereas after seven weeks those who were concerned were more anxious about long-term health effects. People were more likely to be distressed or report symptoms due to wider or pre-existing health issues than to perceived exposure to the incident. Throughout the area studies there was a very low rate of psychological distress.

18.7 Discussion

18.7.1 Lack of Major Acute or Chronic Health Impacts

The lack of either major acute or chronic health impacts resulting from the explosion at Buncefield has been well described and stems in no small part from the combination of the environmental conditions described in previous sections and the timing of the incident.

That the incident occurred in the early hours of a Sunday morning was, in the words of Judge Sir David Calvert-Smith, who presided over the operators' trial: "Very, very fortunate. Had the explosion happened during a working day, the loss of life may have been measured in the tens or even hundreds." He went on to observe that, "The failures which led in particular to the explosion were failures which could have combined to produce these consequences at almost any hour of any day. The fact that they did so at 06.01 on a Sunday morning was little short of miraculous."[15]

18.7.2 Public Health Impact

The health impacts of this fire, as with any major incident, can be considered to be acute (immediate or in the short term) or chronic (longer-term) effects. The range of environmental investigations conducted by the many agencies involved concluded that prolonged plume grounding downwind of the fire did not occur. As such, members of the public were not significantly exposed to the products of combustion, and this lack of exposure is reflected in the ED case notes review and public follow up studies. Occupational deployment to the Buncefield fire was associated with minor, rapidly resolving symptoms characteristic of smoke exposure, but was not associated with major acute health symptoms; follow up of these individuals was not recommended because of the mild nature of their acute symptoms.

An evaluation of the risk assessment concluded that the information provided at the time of the incident was accurate and that the public health advice to shelter based on this was correct.

As with all extreme events, public health can be affected by non-toxicological factors. Two thousand people were evacuated, with the local authority, NHS organisations and voluntary sector providing support. More than a year later, some families were still unable to return home, causing considerable disruption and distress. Nearly 100 local businesses were displaced, with 17 forced to relocate permanently and three ceasing to trade, at an estimated cost of £70 million,[16] and more than 1000 people lost their jobs.[17] Some of the community required psychological support, in particular those whose employment, debt or housing was affected; a survey of 721 households showed 55% of respondents had experienced anxiety as a result. The health authorities' view was that the major health impact of Buncefield was the long-term impact on mental health, alcoholism and relationships.[17]

18.7.3 Potential Air Pollution Health Impacts under Alternative Meteorological Scenarios

It was not simply the timing of the incident that affected the mortality and morbidity. The weather governed the distribution of the products of combustion and resultant plume. It was fortunate that the clement weather took the plume, which was particularly buoyant due to the high temperature of the fire, high into the air with little ground deposition.

It was considered that it would be valuable to examine the possible air pollution-related health impacts from the Buncefield Oil Depot fire if alternative meteorological conditions had prevailed.[18] Atmospheric dispersion modelling of the smoke plume was conducted under a range of meteorological conditions that had occurred during 2005. Constant particle emission rates from the fire were assumed. From this, the population exposure to particle concentrations (PM_{10}) was calculated by linking atmospheric dispersion modelling data at 2 km resolution with postcode population data. With this information, the health impacts were estimated using time-series based exposure–response relationships for PM_{10} available from epidemiological literature.

The main outcomes of this assessment were estimations of air pollution-related deaths brought forward, emergency hospital admissions from respiratory problems and emergency hospital admissions from cardiovascular disease. It was identified that the highest four-day population exposure to PM_{10} for meteorological data from 2005 was predicted to have potentially occurred between 5 and 8 August 2005, when northerly winds would have carried the plume towards London and surrounding areas of high population density. If the Buncefield fire had occurred on these days under these conditions, it was estimated that the additional PM_{10} exposure would have resulted in around 12 extra deaths brought forward, around 13 additional emergency hospital admissions and a similar additional number of emergency admissions for cardiovascular disease. These numbers are slightly greater than estimated deaths and emergency admissions attributable to regular anthropogenic PM_{10} concentrations in south-east England over the same four-day period.

From this assessment by Mohan *et al.*,[18] it was found that although the particle pollution-related health impacts of the Buncefield fire could have been higher under different meteorological conditions, it is unlikely that the impacts would be substantially greater than those attributable to regular anthropogenic particle pollution over a similar period.

18.7.4 Public Health Lessons Identified

Public health experts responded to the fire at Buncefield by providing advice and opinion to the public and frontline responders; this advice was often based on information and data provided by other agencies, showing the value of close multi-agency working.

Following the fire at Buncefield, the Major Incident Investigation Board recommended that air quality data in a major incident should be co-ordinated and that public health advice should be available from the outset.[19] As a result, since April 2010 a new, national, multi-agency Air Quality Cell (AQC) has been available, which can deploy mobile teams to monitor air quality in fires or other air pollution incidents. The Environment Agency (EA), with expert advice from key partners in the HPA, Met Office and other responders, use these data, along with air dispersion modelling, to advise tactical and strategic multi-agency groups on the toxicological content of the plume and its potential public health impact.[20]

In 1968, when the site was built, there were approximately nine dwellings on the north periphery of the site. Since that time, there has been gradual development of the adjacent land, particularly from 1965 until the early 1980s. This included construction of residential properties, a number of schools and industrial premises within 3 km to the west of the site.[3] A further recommendation from the investigation board was for local authorities to work with the EA and HPA during the development of emergency off site COMAH plans to ensure public health input and pre-prepared information on chemicals stored and the potential toxic products of combustion.[19]

After the fire and the 2005 London bombings, the HPA and multi-agency partners identified that a health register would be needed to assess the risk properly, as well as short- and long-term health outcomes among affected individuals in a major incident. Formal arrangements for establishing such a register were published by the HPA in 2012.[21]

18.7.5 Vapour Cloud Incidents of Note from Around the World

Large vapour cloud explosions are rare, but noted for their violence. Whilst infrequent, this type of open vapour cloud explosion is not without precedent. The first described explosion of this nature in the UK occurred in Flixborough in 1974 and involved the catastrophic release of superheated cyclohexane, killing 28 and injuring 36.[22] There are, in general, two principal characteristics of open flammable cloud explosions: release of a large quantity of material capable of forming a flammable "cloud" and a delay of undefined duration before ignition.

Other notable incidents when overfilling of a petrochemical or oil tank has resulted in a vapour cloud explosion include:

- Newark, New Jersey. In 1983,overfilling of a storage tank resulted in a spillage of up to 265 tonnes of gasoline within a bunded area. A vapour cloud 450 to 600 m long and 60 to 90 m wide was formed. The explosion caused significant damage on site, damaging storage tanks hundreds of metres from the initial spillage and breaking glass up to 5.6 km away.[23]
- Naples, Italy. In 1985, overfilling of a storage tank resulted in a spillage of approximately 700 tonnes within a bunded area. The explosion resulted in serious damage to structures within 100 m and broke glass up to 1 km away.[24]
- Saint Herblain, France. In 1991, a release of gasoline from a section of pipework inside a bunded area produced a vapour cloud. Ignition of the vapour cloud produced extensive damage, killed an employee and seriously injured up to five others. Structural damage was noted up to 100 m away and broke glass up to 1 km away.[25]

None of these incidents caused damage on the scale of Buncefield or over the same geographic distance.

18.8 Conclusions

One of the first questions asked at every chemical incident and fire, such as the one at Buncefield, is "What is the risk to the public?" Where emergency services on the scene deal with the immediate health and safety of those directly involved, the role of public health and its health protection service is to define the risk to the general public, in terms of both geographical and temporal distance. The HPA has found that an understanding of toxicology and the expected products of combustion is at the heart of providing informed, evidence-based public health advice and guidance. Even before any sampling or modelling has taken place, an initial assessment of the expected health impact of emissions can be undertaken and this can be used to provide early and informed public health messages. At Buncefield, as with many other incidents, refinement of the risk assessment continued as site inventory data and air quality data became available, allowing more accurate evaluation of the contents of the smoke plume. Dynamic meteorological data informed the exposure assessment. Using these multi-agency and multi-disciplinary partnerships, advice can be disseminated and tested.

Some incidents, such as Buncefield, which resulted in high levels of public concern, require the detailed identification, evaluation and follow up of affected individuals. The HPA's role was to co-ordinate the gathering of data and its analysis in order to provide definitive information on the effects of the smoke plume's constituents. This information provided reassurance to the public, emergency responders and oil depot workers affected by this unique incident.

References

1. J. Targa, A. Kent, R. Stewart, G. Coleman, J. Bower, H. Webster, J. Taylor, V. Murray, R. Mohan and C. Aus, Initial review of air quality aspects of the Buncefield oil depot explosion, 2006, AEA/ENV/R/2168 Report, Issue 1.
2. T. Mather, R. Harrison, V. Tsanev, D. Pyle, M. Karumudi, A. Bennett, G. Sawyer and E. Highwood, 2007, Observations of the plume generated by the December 2005 oil depot explosions and prolonged fire at Buncefield, Hertfordshire, UK and associated atmospheric changes, *Proc. R. Soc. A*, May 8; **463**, 1153–1177; doi:10.1098/rspa.2006.1810.
3. Buncefield Major Incident Investigation Board (BMIIB), 2008, The Buncefield Incident 11 December 2005: The final report of the Major Incident Investigation Board, Volume 2, Available at http://www.buncefieldinvestigation.gov.uk/reports/volume2a.pdf, Accessed on 5th November 2011.
4. Hertfordshire Fire and Rescue Service, 2006, Buncefield, Hertfordshire Fire and Rescue Service's Review of the Fire Response, TSO, Norwich UK.
5. Buncefield Major Incident Investigation Board (BMIIB), 2008, The Buncefield Incident 11 December 2005: The final report of the Major Incident Investigation Board, Volume 1, Available at http://www.hse.gov.uk/comah/buncefield/miib-final-volume1.pdf, Accessed on 27th August 2015.
6. A. Jones, M. Hort, A. Manning, D. Ryall, J. Taylor, H. Webster, C. Witham and S. Wortley, The Buncefield oil depot fire: an overview of actual events and the Met Office's dispersion modelling response, *Geophys. Res. Abstr.*, 2006, **8**, 06452.
7. J. Wakefield, 2010, A toxicological review of the products of combustion, Health Protection Agency, HPA Chemical Hazards and Poisons report, June 2010 issue 17, Available at: https://www.gov.uk/government/uploads/system/uploads/attachment_data/file/203572/HPA_Chemical_Hazards_17.pdf.
8. B. Al-Khalaf, Pilot Study: The onset of asthma among the Kuwaiti population during the burning of oil wells after the Gulf War, *Environ. Int.*, 1998, **24**(1/2), 221–225.

9. F. Welch, 2002, What is CHEMET? in Chemical Incident Report, CIRS: London, Available at: http://webarchive.nationalarchives.gov.uk/20140714084352/http://www.hpa.org.uk/webc/HPAwebFile/HPAweb_C/1194947314613, Accessed 27th August 2015.
10. HPA (Health Protection Agency), 2006, The Public Health Impact of the Buncefield Oil Depot Fire, HPA, London, UK, Available at: http://webarchive.nationalarchives.gov.uk/20140714084352/http://www.hpa.org.uk/webc/HPAwebFile/HPAweb_C/1194947321467, Accessed 27th August 2015.
11. R. Vautard, P. Ciais, R. Fisher, D. Lowry, F. M. Breon, F. Vogel, I. Levin, F. Miglietta and E. Nisbet, The dispersion of the Buncefield oil fire plume: An extreme accident without air quality consequences, *Atmos. Environ.*, 2007, **41**(40), 9506–9517, DOI: 10.1016/j.atmosenv.2007.08.055.
12. M. Hoek, S. Bracebridge and I. Oliver, Health Impact of the Buncefield oil depot fire, December 2005, Study of accident and emergency case records, *J. Public Health*, 2007, **29**(3), 298–302.
13. Eastern Region Public Health Observatory (ERPHO) on behalf of the Buncefield Fire Occupational Health Working Group, 2007, Buncefield Fire Occupational Health Working Group, Final Report: ISBN: 1 904389 12 0, Available at http://webarchive.nationalarchives.gov.uk/20140714084352/http://www.hpa.org.uk/webc/HPAwebFile/HPAweb_C/1194947361869, Accessed on 5th November 2011.
14. O. Morgan, N. Q. Verlander, F. Kennedy, M. Moore, S. Birch, J. Kearney, P. Lewthwaite, R. Lewis, S. O'Brian, J. Osman and M. Reacher, Exposures and reported symptoms associated with occupational deployment to the Buncefield fuel depot fire, England 2005, *Occup. Environ. Med.*, 2008, **65**, 404–411, DOI: 10.1136/oem.2007.035303.
15. BBC, 2010, Firms ordered to pay almost £10m over Buncefield blast, Available at http://www.bbc.co.uk/news/uk-england-10660356, Accessed on 6th November 2011.
16. UK HM Government, 2010, Emergency response and recovery non statutory guidance accompanying the Civil Contingencies Act 2004, UK Government, London UK, Available at: http://interim.cabinetoffice.gov.uk/media/353478/err-guidance-050410.pdf, Accessed on 5th November 2011.
17. Dacorum Borough Council, Report of the Buncefield Community Recovery Taskforce, December 2005–December 2007, 2008, Available at: http://www.dacorum.gov.uk/pdf/Buncefield%20report%20lores.pdf, Accessed on 15th August 2012.
18. R. Mohan, H. A. Walton, D. Thomson, H. Webster, P. Wilkinson, C. Grundy, V. Murray and G. Leonardi, The Buncefield Oil Depot Fire of 2005: Potential Air-Pollution Health Impacts Under Alternative Meteorological Scenarios, PLOS Curr. Disasters, 2012 Feb 2 [last modified: 2012 Apr 2] doi: 10.1371/currents.RRN1300.
19. Buncefield Major Incident Investigation Board (BMIIB), 2008, The Buncefield Incident 11 December 2005 Recommendations on the

emergency preparedness for, response to and recovery from incidents, Available at: http://www.buncefieldinvestigation.gov.uk/reports/volume2b.pdf, Accessed on 5th November 2011.

20. *Essentials of Toxicology for Health Protection – a handbook for field professionals*, ed. D. J. Baker, L. Karalliedde, R. Maynard, V. S. G. Murray and N. Parkinson, Oxford University Press, Oxford, UK, 2012.
21. K. Paranthaman, M. Catchpole, J. Simpson, J. Morris, C. R. Muirhead and G. S. Leonardi, Development of a decision framework for establishing a health register following a major incident, *Prehosp. Disaster Med.*, 2012, **27**(6), 524–530.
22. V. Murray, 2001, Flixborough, England, Chemical Disaster 1974, In: Chemical Incident Report, CIRS: London, Available at: http://webarchive.nationalarchives.gov.uk/20140714084352/http://www.hpa.org.uk/webc/HPAwebFile/HPAweb_C/1194947404005, Accessed 27th August 2015.
23. J. K. Bouchard, Gasoline Storage Tank Explosion and Fire: Newark NJ January 7, 1983, National Fire Protection Association (NFPA) Summary Investigation Report (in cooperation with Federal Emergency Management Agency/United States Fire Administration and National Bureau of Standards/Center for Fire Research), referenced in BMIIB initial report.
24. G. Russo, M. Maremonti, E. Salzano, V. Tufano and S. Ditali, Vapour cloud explosion in a fuel storage area; a case study, *Process Saf. Environ. Prot.*, 1999, **77**(6), 310–365.
25. J. Lechaudet and Y. Mouilleau, Assessment of an accidental vapour cloud explosion, *Loss Prev. Saf. Promot. Process Ind., Prepr. Int. Loss Prev. Symp., 1st*, 1995, **314**, 377–378.

CHAPTER 19

The World Trade Centre Disaster

MICHAELA KENDALL,*[a] MITCHELL COHEN[b] AND LUNG-CHI CHEN[b]

[a] Faculty of Medicine, University of Southampton, Southampton, UK;
[b] Department of Environmental Medicine, NYU School of Medicine, Tuxedo NY, USA
*Email: michaela_kendall@yahoo.co.uk

19.1 Introduction

The Twin Towers disaster followed the terrorist attack of 9th September 2001 on New York City's (NYC) World Trade Center (WTC). This man-made environmental disaster, known as 9/11, was caused by two airplanes being flown intentionally into two towers of the WTC and resulted in destruction and debris across a wide urban area of Lower Manhattan, NYC. In the immediate aftermath, during months of continuous fire and subsequent clean-up, there were recorded human exposures to WTC dust and smoke. Hundreds of thousands, perhaps millions, of NYC residents and workers, plus incoming responders to the incident, were exposed to some level of WTC contamination. While the disaster initiated a rapid military response to secure the area, the non-military, post-disaster management continued for years. Defining the occupational and environmental health consequences of this disaster is still a work in progress, as data collection on health effects continues into a second decade.[1]

The 9/11 terrorist attacks generated intense shock, and social and political pressure around the world. As well as the WTC collapse, the attacks included

Issues in Toxicology No. 23
Toxicology, Survival and Health Hazards of Combustion Products
Edited by David A. Purser, Robert L. Maynard and James C. Wakefield

Published by the Royal Society of Chemistry, www.rsc.org

two further airplane attacks on the Pentagon and Pennsylvania, provoking a military response both in the US and abroad. Risks of secondary public health threats in NYC related to sanitation were closely followed by an anthrax release, ratcheting up city-wide tensions related to public and environmental health. In this chapter we limit our focus and review to the WTC dust release from Ground Zero (where the WTC buildings had previously stood) in the NYC environs only. We then summarise the latest findings of ongoing studies linking WTC dust and smoke exposures to specific health consequences. We summarise the events as they unfolded, and describe some of the key lessons that were learned during the disaster.

19.2 WTC Dust Emissions

19.2.1 WTC Dust and Smoke Release

During the WTC disaster, there were three separate air pollution hazards: (1) the initial dust and smoke from the building collapse; (2) smoke from fires burning at Ground Zero until December 2001; and (3) deposited dust at and around the Ground Zero and the Fresh Kills dump site, including in streets and inside buildings, cars, *etc.* The environmental and public health response to the dust and smoke release was rapid. However, the unprecedented scale of the environmental impacts, the size of the affected area, plus the military dimension challenged the normal capacity of the agencies involved. Communication and coordination were key challenges, and practical limitations on access to electricity, water and Manhattan Island itself hampered some efforts. Access into and out of Ground Zero was restricted for months using cordons. New ambient air monitoring networks, additive to existing sites, were activated within hours by local environmental health academics and some agencies. Still, the highest health-relevant exposure measurements were largely missed in the first four days following the attacks, due to onsite security restrictions. The initial dust plume created by the downward implosion of the buildings remains unquantifiable, and the analysis of deposited dust became important in identifying the WTC dust composition.

Regional institutions, including New York University (NYU), responded to an urgent call to support local activities at Ground Zero. The National Institute of Environmental Health Sciences (NIEHS) loosely coordinated one of the key environmental health responses by calling on local funded groups to focus their efforts on dealing with the disaster. Spontaneous groups also formed to share information and resources, but without coordination and following military control of the area, this took time. The Centers for Disease Control and Prevention (CDC) World Trade Centre Working Group and the National Institute for Occupational Safety and Health (NIOSH) later published concerns regarding air quality and public health.[2] Notably, the United States Environmental Protection Agency (US EPA) announced within a few days of the disaster that the air was "safe". The US EPA issued several news releases within days of the attack to reassure the public about air quality in NYC. It became known later that these assurances were being edited by the National Security Council of the White House to protect the security, political and socio-economic priorities of the region and country. Christine Todd Whitman (then head of the US EPA) was later widely condemned for these statements and held to account by an extraordinary audit report.[3] The governing NYC Mayor, Giuliani, and other public figures urged residents to stay at home and remain calm, and for NYC workers to return quickly to work. Wall Street and the surrounding financial district of the city opened less than a week after 9/11.

Participation in the NIEHS WTC Working Group, attending the WTC site and contributing to community outreach activities following the disaster, meant that environmental health scientists were thrust into an unfamiliar and unpredictable situation, with no clear lead agency or individual. The NYC community also proved an exceptionally educated, informed and linked-in population, demanding continuous communication of complex science and risk. These stakeholders provided incisive, challenging questions as a group, and demonstrated powerful activism to get answers, even years on from the event.

19.2.2 Immediate Characterisation and Evaluation of WTC Smoke and Dust

A large body of work details the measurement of airborne dust and gases in the NYC environs following 9/11.[4–7] A dense network of monitors collected air quality data before the disaster and short-term monitoring sites were established within days of 9/11. Studies focused on the ambient "outdoor air", and little monitoring was conducted indoors. Crucially, the earliest ambient concentrations in the first four days remain largely unknown. In hindsight, there was also no clear rationale in selecting the measured parameters, so that regulated air pollutants were adopted as the default. The result is a temporal and spatial patchwork of chemical speciation data for air pollution, with many voids.

The WTC fires persisted until mid-December 2001, releasing combustion emissions into the NYC Metropolitan area. Ambient levels of semi-volatile organic compounds, such as polyaromatic hydrocarbons (PAHs), polychlorinated biphenyls (PCBs), polybrominated diphenyl ethers (PBDEs) and polychlorinated naphthalene (PCN), and other combustion emissions from these fires were not well defined during the monitoring campaign. Ambient PAH measurements were limited despite being extremely high,[8] as were mercury and brominated persistent organic pollutants (POPs: common flame retardant compounds), even though historical samples of particulate matter were available. The analysis of window film samples from Lower Manhattan (acting as passive samplers of combustion emissions) showed deposition occurred within 1 km of the WTC site, and illustrated the chemical complexity of depositing material.[9,10]

Analysis of indoor and outdoor dusts representative of materials that settled immediately following the WTC collapse became crucial in defining WTC dust.[5,11] These samples were collected over the first weeks and months following 9/11. These samples provided some of the only evidence of the dust in the minutes and hours immediately following the WTC collapse. They have become the best indicators of WTC dust exposure in work areas and residential properties around the WTC site.[4,12] The use of settled dust samples to describe the composition of the airborne particles is, however, fraught with difficulty. By definition, settled dust has undergone a size selective process so that smaller particles (and their components) will be under-represented in a settled dust sample. For example, dust components that undergo gas to particle conversions will be in the smallest particle fraction that deposit slower than larger particles. Since certain smaller particle fractions are more likely to deposit in the lung, it is important to define what chemicals are enriched within that fraction in particular. Also, once deposited, the size distribution of the sample will be disturbed and, on a mass basis, the smaller particles contribute little to mass. Therefore, care must be taken in extrapolating deposited dust results to ambient concentrations and inferred human exposures.

The deposited dust comprised large and fine particulate matter and fibres, with a high pH (high alkalinity). The bulk of the material was pulverised building materials and building contents. Asbestos was detected in many samples and glass fibres from the buildings made a significant fraction of deposited dust.[3,5] Substantial concentrations of metals were detected in many samples. The concentration of carcinogenic metals in settled dust and smoke samples was low compared to concentrations of non-carcinogenic metallic elements.[4] Organic constituents included carcinogenic materials, some at elevated levels.

Research scientists from the NYU Particulate Matter (PM) Center immediately began an air pollution monitoring campaign in the area following the WTC collapse.[13] This study ran from 14th September 2001 to January 2002, and involved a number of permanent, trans-located and mobile samplers. Deposited and airborne dust sampling was conducted in the immediate and extended area surrounding WTC Plaza as part of this sampling regime. Large quantities of deposited dust were visible on and in the buildings immediately surrounding the WTC site. Figure 19.1 (taken on 26th November 2001) shows that quantities of dust even remained visible on outdoor building window ledges around the WTC disaster site until at least late November.

Indoor deposited and airborne dust samples were gathered and analysed during visits to buildings within the cordoned off WTC site. Visible quantities of dust remained undisturbed at these locations, thus providing a valuable opportunity to examine the dust depositing in local homes. Figure 19.2 (also taken on 26th November 2001) shows a typical indoor scene from nearby apartments, which faced Tower 1—widespread dust invasion and deposited dust layers up to one inch thick. Visible deposits reduced significantly with short distances from the site. Fibres in these dust deposits

Figure 19.1 Deposited dust and debris remained on the external surfaces of buildings for months after the WTC attack.
This picture was taken from Trinity Street on the 26th November 2001. ©M. Kendall.

were visible during the collection, but processing the dust further emphasised the high fibre content of WTC dust.

Figure 19.3 shows a sample of unseparated WTC dust in the glass vial at the bottom of the picture; the larger plastic sample bag contains the >53 μm fraction, and at the top of this sample, matted balls of fibres can clearly be seen, potentially caused by the separation process. Clearly, then, some of these matted fibres could have been misattributed to the >53 μm size fraction by the separation process itself, suggesting the lower size fractions may have been underestimated in these analyses.

NYU scientists used PM concentrations obtained from their own monitors in Downtown Hospital and those measured at additional monitor sites operated by NY State to estimate the scale of population exposures during this episode.[14] A series of high peaks of $PM_{2.5}$ registered in the Lower

Figure 19.2 Residential properties directly adjacent to the WTC site—such as the property pictured above, on Liberty Street—were particularly badly damaged by the WTC blasts. These buildings suffered moderate structural damage during the WTC collapse: windows were blown out of the properties, allowing dust and debris to enter the homes and damage the interior.
©M. Kendall.

Manhattan area after the 11 September from existing monitoring stations. Outdoor $PM_{2.5}$ concentrations over NYC were assessed by interpolation within a Geographical Information Systems (GIS) environment. For Lower Manhattan the monthly 24 hour average was 20.2 mg m^{-3}, which did not exceed the NAAQS value. A deterministic micro-environmental model (INTAIR) was used to simulate $PM_{2.5}$ concentrations in indoor micro-environments. The mean 12 hour night and day exposures or nightly exposures weighted by residential population were estimated for three NYC population categories - home-makers, office/shop workers and students/children over the period from 14th–30th September.

Average calculated night and day exposures of the Lower Manhattan population were 37.3 mg m^{-3} and 23.6 mg m^{-3} respectively (daily average: 30.5 mg m^{-3}). Consideration was given to group movements and activities, smoking habits of individuals, and special population movements due to access restrictions and risk avoidance behaviours. Findings were that one half of the Lower Manhattan population was exposed to 10–20 mg m^{-3} and 30–60 mg m^{-3}; with one fifth of the population exposed to over 60 mg m^{-3} at

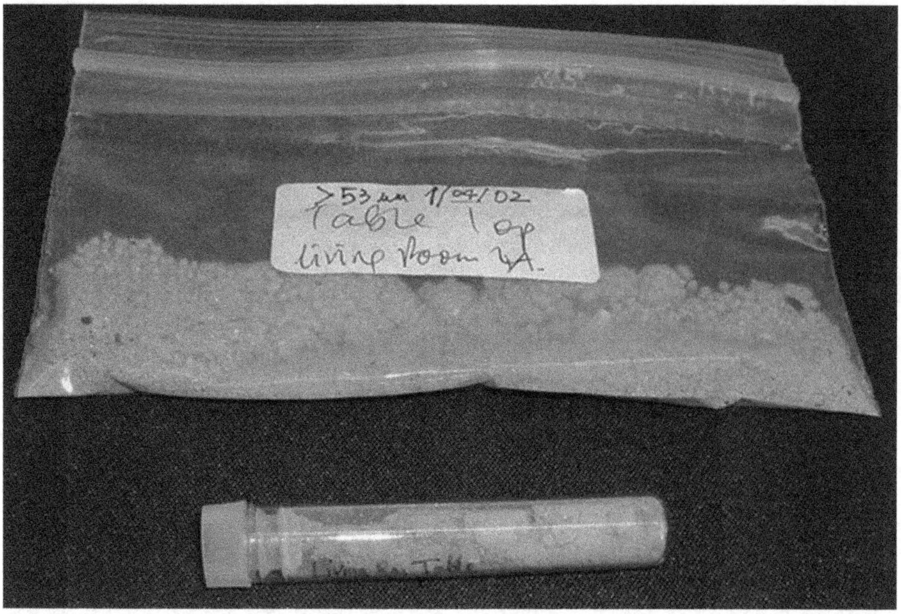

Figure 19.3 The vial at the bottom of the picture shows one of the whole (not size-fractionated) samples of deposited dust collected in the Liberty Street apartments. The sample bag shows the >53 μm dust fraction of the same sample. Two disparate components in the >53 μm dust fraction were visible during handling the sample: a sand-like component at the bottom of the bag and a "wool-like" component (apparent as matted balls, spun together during the separation process) at the top of the sample. ©M. Kendall.

night. Although the outdoor $PM_{2.5}$ concentration was lower than the NAAQS value, personal exposure levels were generally higher so may still be of concern.

19.2.3 Detailed Characterisation of Deposited WTC Dust

The elemental characterisation of WTC dust deposited on surfaces from many studies conducted by different agencies is very consistent.[7] Gravimetric analysis of indoor dust samples collected from inside buildings adjacent to the WTC site in November 2001 was conducted on size separated samples.[13] The size distribution of deposited WTC dust samples collected in Liberty Street was dominated by the 10–53 μm size particles (Figure 19.4). The >53 μm size fraction is sometimes greater due to the large stone, plastic and glass fragments in these samples. Fibres made up to approximately 50% of the mass of deposited particulate matter.

Scanning electron microscopy (SEM) and energy dispersive X-ray analysis (EDAX) was conducted to image and chemically analyse individual fibres and particles. Overall composition results represent the spectra from a large area

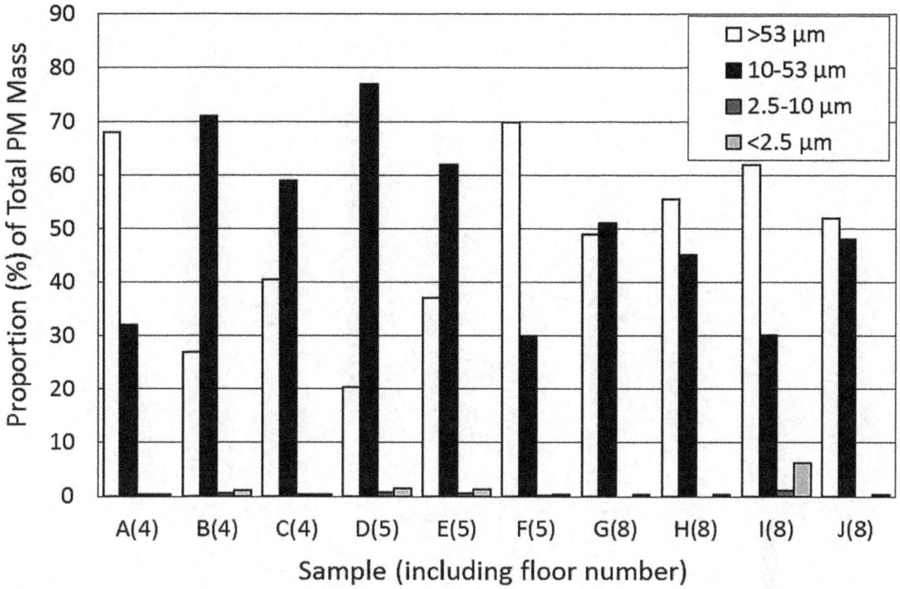

Figure 19.4 Particle size fractions of deposited dust as a percentage of the total dust sample collected.

Table 19.1 The major and minor elements identified by SEM–EDAX analysis of representative individual particles.

Size/μm	Major	Minor
>53	Ca, Si, Fe, O	S, Al, Mg, Mn, Ag, K, Pd, Ti, As, Na, Cl, In
10–53	Fe, Ca, O	Si, Al, Na, Cl, Mg, Pb, C, S, Sn, Mn
2.5–10	S, Si, Pb, Al, O	Fe, Zn, Ca, Mg, Cr, Ni, C
<2.5	Al, Si, Ca, O,	S, K, P, Mn, Fe, Cl, Zn, Mg, Ti, Pd, C

of the Liberty Street sample. Overall EDAX results are presented in Table 19.1. The results show that all samples had a similar overall composition: the major elemental components were Ca, Si and S; Mg, Al, Ti, K, Fe and NaCl were the most common minor elemental components detected. Asbestos was present in some samples, representing 0.8 to 3% by weight of outdoor dusts collected on 16th September 2001 on Cortland Street, Cherry Avenue and Market Street, outside the perimeter of Ground Zero.[4] We also identified one uranium particle using EDAX, which may have been an aerosolised component of the plane wingtips, which reportedly contained uranium. Figure 19.5 shows a selection of these particles and fibres.

Few individual fibres were observed in the >53 μm WTC deposited dust sample, possibly because they had combined to form balls of matted fibres during the size separation process as discussed (Figure 19.3). Figure 19.6 depicts the electron micrographs and individual spectra from EDAX analyses of randomly selected fibres in the 10–53 μm WTC deposited dust sample.

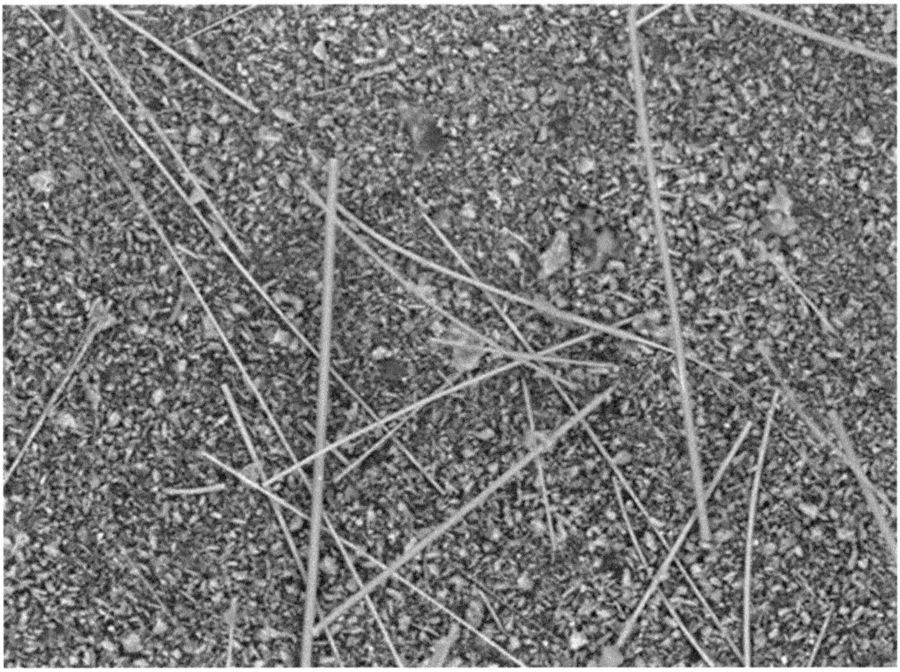

Figure 19.5 SEM image of the 2.5–10 μm WTC deposited dust sample. Individual spectra from EDAX analyses of randomly selected particles and fibres are summarised in Table 19.1.

Maciejczyk *et al.*[15] characterised ambient particulate matter and fallout dust samples collected by the NYU team. This analysis included a general characterisation of the percent distribution of various materials present in each sample and detailed measurement of the inorganic components of the dust mass. They also sought to determine how much of the fine particle mass ($PM_{2.5}$) measured in Lower Manhattan was associated with the WTC dust. The contribution of each aerodynamic size fraction of WTC settled dusts is shown in Figure 19.7. More than 97% of the dust particles' mass was larger than 10 μm aerodynamic diameter. Only $0.43 \pm 0.13\%$ and $0.57 \pm 0.15\%$ of outdoor and indoor dusts were fine particles less than 2.5 μm. Approximately equal percentages of the dust were particles 2.5–10 μm: $1.32 \pm 0.40\%$ and $1.62 \pm 0.46\%$ outdoor and indoor, respectively.

The compositions of the outdoor bulk dust samples, as well as the sized fractionated settled dusts, are shown in Table 19.2. The content and distribution of material were indicative of the complex building debris and combustion plume formed during 9/11. There were considerable differences between the compositions of the bulk samples and the <2.5 μm fractions. In the bulk samples 30–40% was fibreglass, 10% carbon particles and 40% classified as "non-fibrous" dust with the remaining 10–20% consisting of cellulosic material. These samples contained only a trace amount (<1%) of

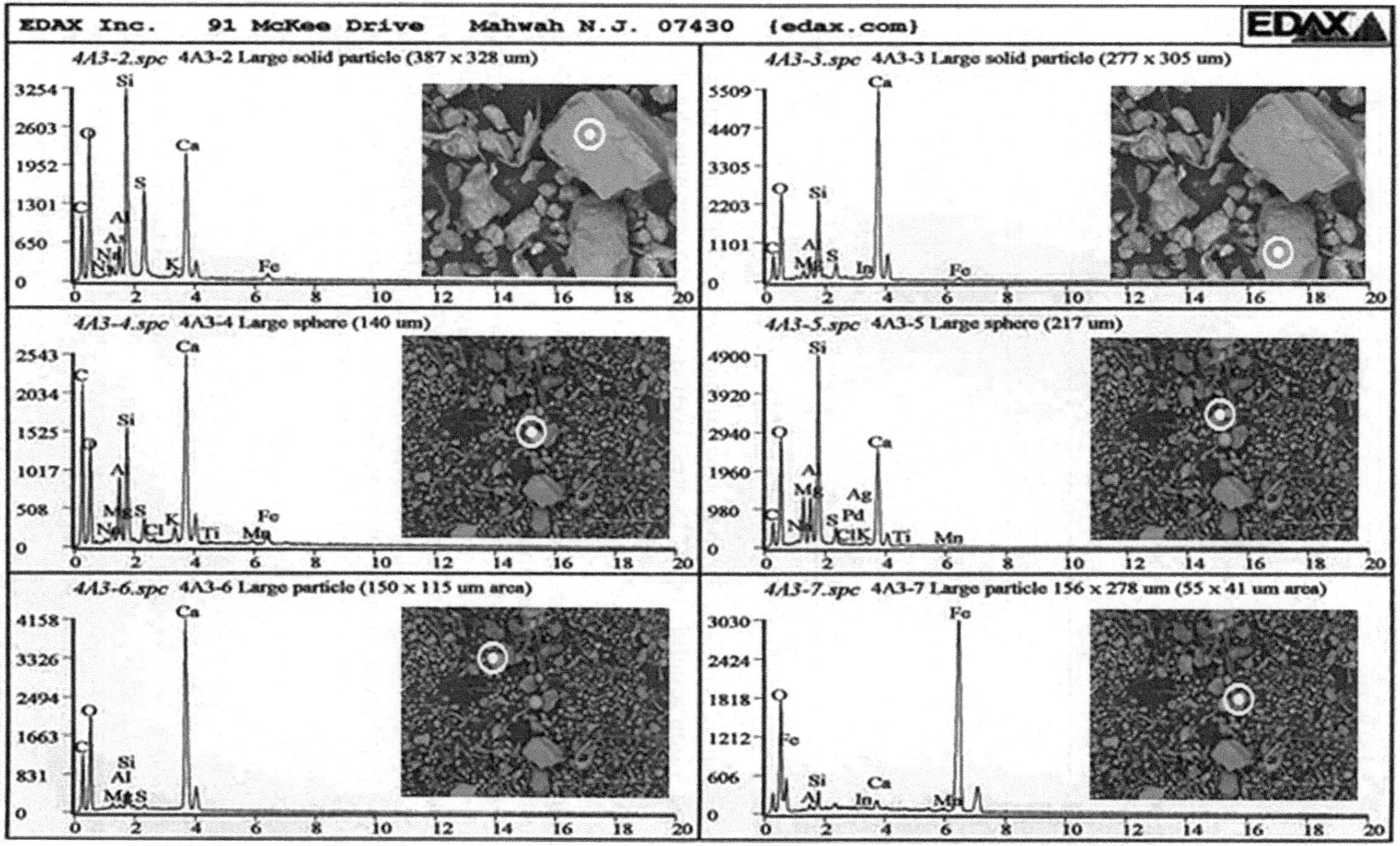

Figure 19.6 Individual SEM–EDAX micrographs and spectra of randomly selected particles in the >53 μm WTC deposited dust sample. The white character ⊙ in each picture marks the approximate spot of analysis on each particle. A brief description of each analysed particle, together with the particle dimensions are provided directly above each spectra.

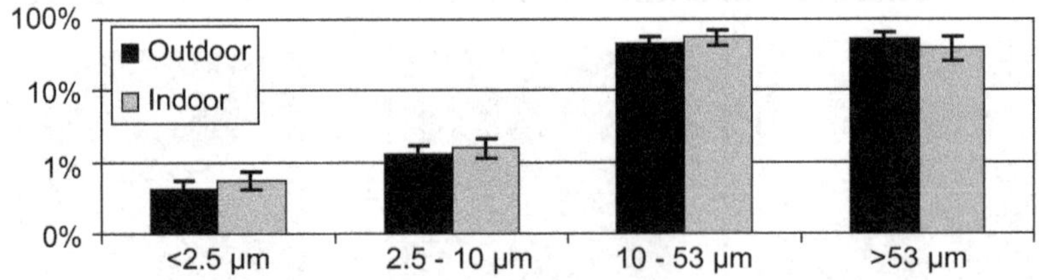

Figure 19.7 Contribution of each size fraction to the total dust.

Table 19.2 Morphological composition of selected dust samples.

Sample[a]	Chrysotile	Amosite	Cellulose	Fiberglass	Non-fibrous	Carbon Particles
#10	0.3%	none	10%	40%	40%	10%
#11	0.3%	none	15%	35%	40%	10%
#13	0.3%	none	20%	30%	40%	10%
#1:>53 μm	none	none	40%	20%	20%	20%
#1:10–53 μm	trace	trace	5%	40%	40%	15%
#1 <2.5 μm	none	none	5%	none	80%	15%
Nassau St.	0 to 0.5%	none	13 ± 10%	30%	57 ± 10%	not done

[a]Samples are bulk dust unless the size fraction is indicated. Nassau St. results are the arithmetic means and standard deviations of six indoor samples.

asbestos. Samples collected by other investigators were similar in composition to those found around the WTC sites, even though the sampling locations were approximately 1 km away.[4] For the PM < 2.5 μm fraction no fibrous materials (including asbestos) were present, the content consisting of 80% non-fibrous dust, 15% carbon particles and 5% cellulose. The dusts collected around the WTC sites were different from those from the apartment building (other than one from the apartment building roof). As with the WTC site samples, the apartment samples contained no asbestos (or only a trace), but less fibreglass, more non-fibrous material and varying amounts of cellulose.

The majority of the suspensions of the bulk WTC settled dusts had a pH greater than 10 (Figure 19.8), but pH decrease with particles size so that for PM 2.5 the pH was close to neutral. Since larger particles are likely to be caught in the upper airway, their alkalinity was partially responsible for the "WTC cough".[16,17]

Figure 19.9 shows ion chromatography results for the concentrations of soluble ions, which varied to some extent in all dust fractions. This indicates either sample heterogeneity or an uneven spatial distribution of the collapsed building materials. The bulk samples contain, on average, 8.2 ± 0.6 and 14.5 ± 1.2 μg g^{-1} of Ca^{2+} and SO_4^{2-}, respectively, which were highly correlated ions (r^2 0.996). This reflects gypsum content in fraction > 53 μm. Soluble calcium and sulfate ion concentrations increased with decreasing particle size and remained correlated. Mean concentrations in dust fraction < 2.5 μm were 150 ± 33 and 385 ± 85 μg g^{-1}, respectively. Figure 19.9 also shows concentrations of other soluble ions (Na^+, K^+, Mg^{2+}, F^-, Cl^- and NO_3^-). Other than

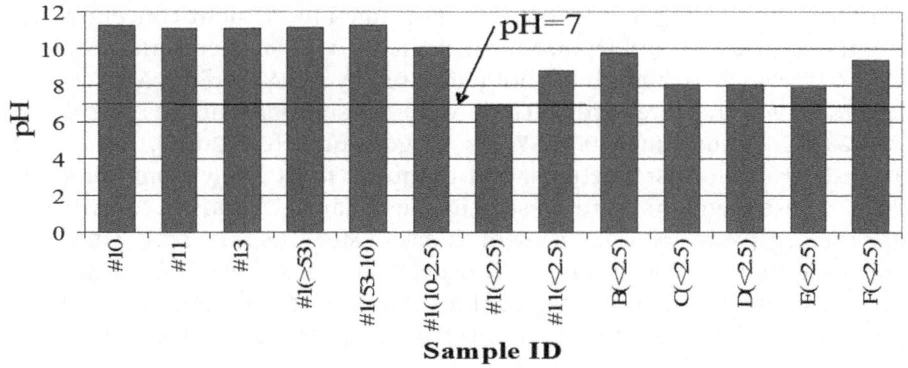

Figure 19.8 pH of the suspensions of selected dust samples. Sample size fraction (in μm) is indicated in parenthesis unless bulk dust.

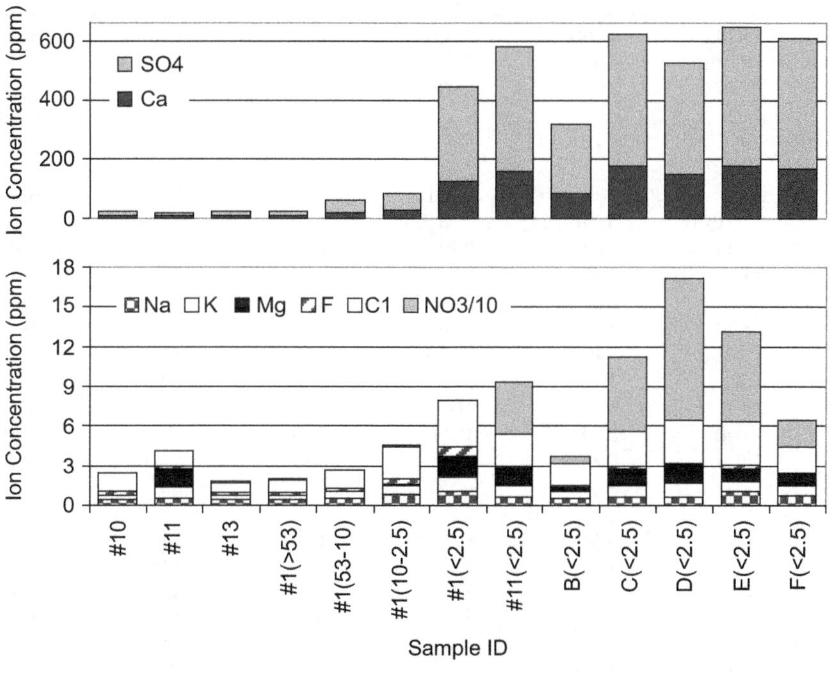

Figure 19.9 Soluble ion concentrations (in $\mu g\ g^{-1}$) in selected dust samples. Sample size fraction (in μm) is indicated in parentheses unless bulk dust.

calcium and sulfate, chlorine is the dominant element in all class sizes. The average molar ratio of Cl^-/Na^+ in <2.5 μm fractions was 2.31 ± 0.55 with poor $r^2 = 0.29$. However, this ratio was 1.49 ± 0.42 in average bulk samples with a higher correlation of 0.952. It is therefore likely that chlorinated compound in addition to coarse sea salt were present in the fine dusts which may have been produced by the fire. Only in fine fractions of dusts contained nitrate ion (highly correlated ($r^2 = 0.98$) with ammonium ion), except for sample #1.

Table 19.3 and Figure 19.10 show mass fraction element concentrations determined *via* XRF and INAA. Not all elements could be determined by both methods, *e.g.*, Si, S, and Pb are not measured by INAA, and our XRF was not calibrated for the rare-earths. There were substantial amounts of calcium (9.7–23%), aluminum (0.7–2.7%), magnesium (0.4–2.8%), and iron (0.2–1.0%) in all dust fractions, and titanium, most likely from paint particles. These elements were present in the collapsed building construction material (gypsum boards, cement blocks, steel beams *etc.*), with progressively higher concentrations in coarser dust fractions. Zn, Cu and Mn were also present in μg g^{-1} quantities. Some elements (Sb and Cl) were produced by the fires rather than mechanical grinding of the collapse, and,

Table 19.3 Element concentrations (in μg g^{-1}) in selected dust samples.[a]

	10–53 μm Mean	St. dev.	2.5–10 μm Mean	St. dev.	<2.5 μm Mean	St. dev.
Na	5300	1500	1300	420	1500	460
Mg	28 000	3000	7100	1400	4400	1800
Al	27 000	2600	8800	1400	7200	2200
Si	–	–	–	–	28 000	6000
S	–	–	–	–	130 000	23 000
Cl	19	3.2	24	7.4	1400	1600
K[c]	300	780	<DL[b]		1300	610
Ca	220 000	5900	230 000	3200	97 000	16 000
Ti	2200	210	930	140	540	400
V	31	2.3	17	2.4	8.3	5.7
Cr	85	22	75	21	30	15
Mn	1200	180	180	49	41	14
Fe	10 000	2400	5700	1200	2300	950
Co	4.4	1.0	3.1	0.60	6.0	4.6
Cu	230	120	160	140	64	35
Zn	1400	370	1100	230	550	290
As	2.7	0.8	2.0	0.4	<DL[b]	
Se	–	–	–	–	20	7.3
Br	–	–	–	–	68	24
Rb	16	3.3	12	1.8	32	10
Sr	–	–	–	–	210	37
Sb	40	14	54	18	3800	630
Ba	–	–	–	–	27	37
Pb	–	–	–	–	65	27
La	38.0	12.0	9.3	3.6	–	–
Sm	9.1	2.5	1.5	0.3	–	–
Ag	1.6	0.7	<DL[b]		–	–
Ce	80	22	17	7.1	–	–
Hf	3.0	0.7	0.6	0.1	–	–
Sc	6.2	1.6	1.5	0.3	–	–
Th	7.2	1.9	1.4	0.3	–	–

[a]NOTE: $n=7$ (10–53 μm), $n=7$ (2.5–10 μm), and $n=28$ (<2.5 μm).
[b]<DL is reported if the value was below the detection limit of the method.
[c]Potassium in size fraction 10–53 μm was detected only in one sample. No values reported if element was not measured by the method.

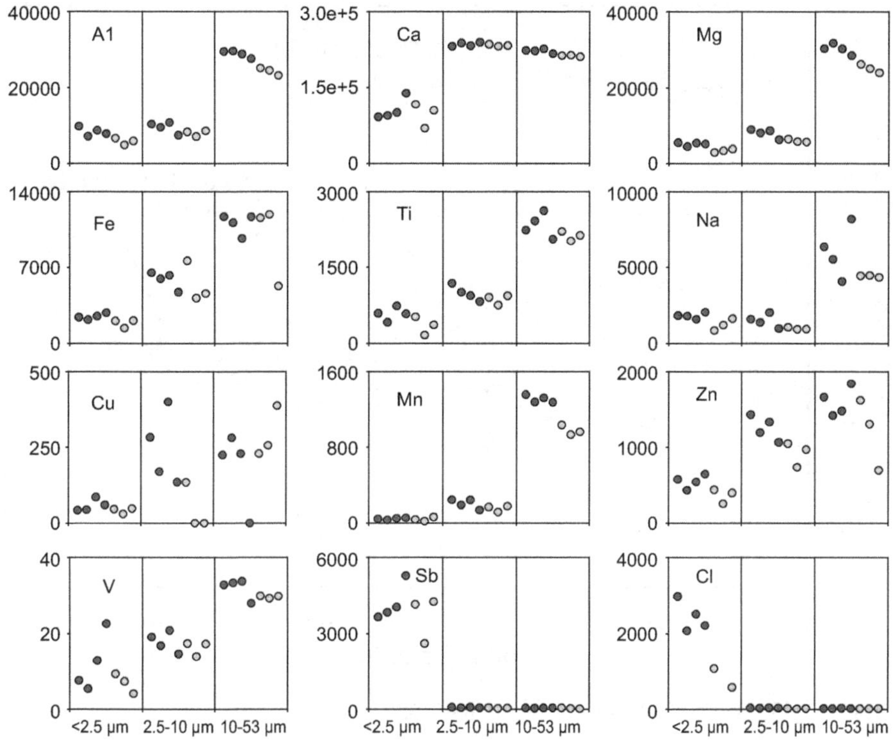

Figure 19.10 Elemental composition (μg g^{-1}) of selected dust samples by size classification. Black (blue online) circles are outdoor; light (yellow online) circles are indoor dust samples.

thus, are much elevated in the fine fraction of dust. Traces of Pb, Cr, V, Ni, elements important for toxicological considerations, were also present.

There is a slight apparent trend for lower element concentration in the indoor than outdoor samples (Figure 19.10). The indoor samples were collected about two months after the collapse so these samples might contain slightly more material (*e.g.* carbon as the result of fires, and pre-existing house dust) that were not measured by either XRF or INAA.

Figure 19.11 shows the results of factor analysis of ambient samples used to predict the factor scores for 28 dust samples of the <2.5 μm fraction. This shows that the factor loadings for elements of combustion origin are clearly separated from crustal elements. Sulfur was associated with both dust and combustion sources, and additionally with regional sulfate aerosol, which is inclusive in category of other source(s). During regional pollution episodes in September and October 2001, elevated sulfur levels were observed at other NY state locations. Following the regression of the $PM_{2.5}$ mass concentrations for three months after the WTC collapse, we were able to predict the relative contribution of WTC dust to ambient $PM_{2.5}$ concentrations.

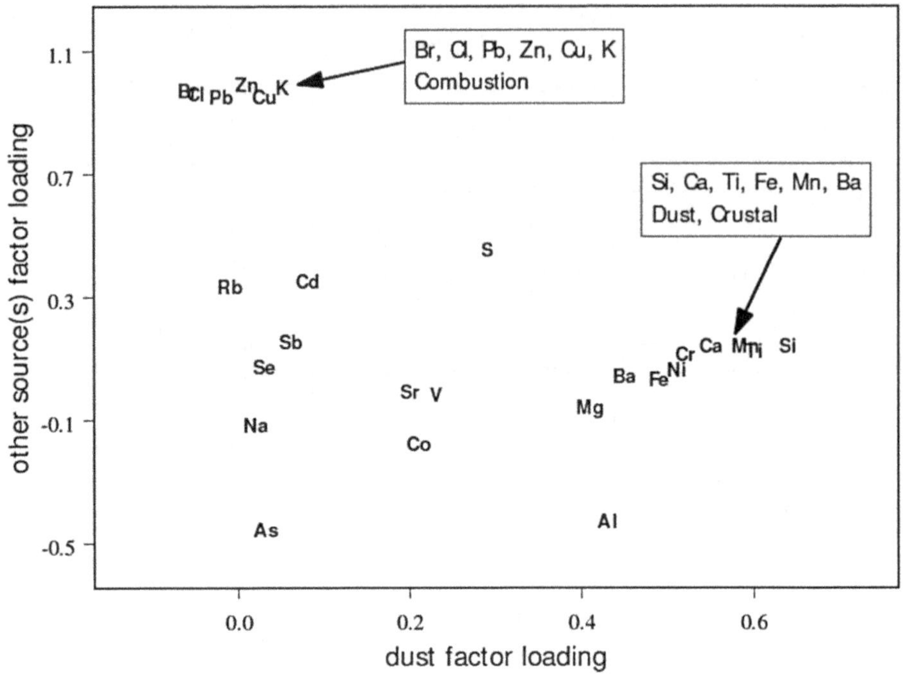

Figure 19.11 New re-calculated factor loadings.

19.3 Toxicology of Deposited WTC Dust

19.3.1 Inflammatory Effects

Samples of the collected, well-characterised WTC deposited dust were subsequently tested for cellular toxicity.[11,18] Mechanistic studies focusing on inflammatory processes in lung cells dominated early evaluations, due to the respiratory illnesses observed in the highly exposed. These acute toxicity assays of WTC dust offer insights into lung cell toxicity that may contribute to the pathophysiology of chronic lung disease seen in rescue workers and residents exposed to WTC dust. Biomarkers of inflammation may help identify pathways producing obstruction after irritant exposure.

Gavett *et al.*[19] found significant neutrophil inflammation and increased airway hyper-responsiveness to methacholine challenge in mouse lungs following a single exposure to WTC dust less than 2.5 μm. Payne *et al.*[20] showed that exposure to high concentrations of WTC dust less than 2.5 μm ($PM_{2.5}$) caused increased formation/release of pro-inflammatory cytokines/chemokines that contribute to inflammation and airway remodeling. Wang *et al.*[21] demonstrated that 2.5 μm WTC dust exposures in lung epithelial cells induced cytokines at lower doses than previous studies and that activation of mitogen-activated protein kinase signaling pathways may be important.

Xu *et al.*[22] examined the acute cytotoxic effects of airborne dust contaminants recovered from the area 24–48 h after the WTC collapse. Two airway cell types, epithelial and smooth muscle cells, most directly exposed to the inhaled dust, were selected. The extent to which the presence of these dusts could modulate the effects of cigarette smoke on these cell types was also determined.

Human cultured airway epithelial cells (BEAS-2B) cells were exposed to WTC dust particles (10–53 μm) (0.01–0.5 μg μl^{-1}), 10% cigarette smoke extract (CSE), or a combination of the two for 2–24 h. Mitochondrial integrity (MTT assays) and apoptosis (PARP immunoblotting) were measured to determine cell viability. Necrotic cell death occurred following exposure to WTC dust or CSE for 2–24 h without evidence of apoptosis. Conditioned cell culture media recovered from the CSE- and/or WTC dust-exposed BEAS-2B cells were then applied to cultured human airway smooth muscle cells that were subsequently assayed for mitochondrial integrity and their ability to synthesize cyclic AMP (a regulator of airway smooth muscle constriction). Cellular toxicity and enhanced cyclic AMP synthesis were observed following exposure to conditioned media from both WTC- and CSE-exposed epithelial cells.

Weiden *et al.*[23] investigated macrophage derived cytokine (MDC) and granulocyte macrophage-colony stimulating factor (GM-CSF) expression in the serum of patients with WTC dust lung injury and *in vitro* using normal alveolar macrophages (AM) exposed to various sizes of WTC-PM. Normal adherent AM from 15 subjects without WTC dust exposure were incubated in media alone, lipopolysaccharide (LPS; 40 ng mL^{-1}), or suspensions of WTC-PM_{10-53} or WTC-$PM_{2.5}$ at concentrations of 10, 50 or 100 mg mL^{-1} for 24 h. Supernatants were assayed for 39 chemokines/cytokines. In addition, sera from WTC dust-exposed subjects who developed lung injury were assayed for the same cytokines. In the *in vitro* studies, cytokines formed two clusters with GM-CSF and MDC as a result of PM_{10-53} and $PM_{2.5}$. GM-CSF clustered with interleukin 6 (IL-6) and IL-12(p70) at baseline after exposure to WTC-PM_{10-53} and in sera of WTC dust-exposed subjects ($n = 70$) with WTC lung injury. Similarly, MDC clustered with growth-regulated oncogene (GRO) and monocyte chemoattractant protein-1 (MCP-1). WTC-PM_{10-53} consistently induced more cytokine release than WTC-$PM_{2.5}$ at 100 mg mL^{-1}. Individual baseline expression correlated with WTC-PM-induced GM-CSF and MDC.

Therefore, WTC-PM_{10-53} induced a stronger inflammatory response by human AM than WTC-$PM_{2.5}$. This large particle exposure may have contributed to the high incidence of lung injury in those exposed to particles at the WTC site. GM-CSF and MDC consistently cluster separately, suggesting a role for differential cytokine release in WTC-PM injury. Subject-specific response to WTC-PM may underlie individual susceptibility to lung injury after irritant dust exposure.

19.3.2 Long-term Effects

Several studies demonstrated the ability of deposited WTC dust from the initial dust and smoke plume to stimulate an inflammatory response. First responders present at Ground Zero during the first 72 h after the WTC collapse suffered significant damage to their respiratory health. This includes chronic disorders like sarcoid-like granulomatous pulmonary disease (SLGPD) and persistent airway hyper-reactivity (AHR; a defining feature of chronic asthma). While a number of animal studies using building-derived WTC dusts have been performed to better understand the etiology of some responder ailments, they primarily used only a dust fraction (fine; $\leq$2.5 μm diameter) likely to be deposited in the lung, dusts collected after 13th September 2001, or regimens that did not reflect exposure scenarios most responders underwent. As significant coarse WTC dust particle deposition was found in the lungs of first responders, and the alkalinity of WTC particles increases with diameter, in an ongoing NIOSH sponsored study, NYU is currently testing the two hypotheses. The first is that an increased presence of large, super coarse (>10 μm) alkaline particles in their lungs contributed to the high incidence/severity of SLGPD and AHR. The second is that a mechanism underlying these two disorders is that (A) the alkalinity of the large particles initially damaged the lung epithelium so that airway remodeling associated with persistent AHR was triggered and particle clearance was reduced, and (B) metals associated with the now retained WTC particles exerted a variety of toxicities including granuloma induction and promotion of ongoing airway remodeling.

A generator was built to generate and deliver super coarse (10–53 μm) WTC particles to a rat model in a manner that mimicked WTC exposure scenarios[24] as an intratracheal inhalation (ITIH) bypassing the nasal passages. Exposed rats had significantly increased lung weight and levels of select tracer metals 1 h post-exposure to $\geq$1200 mg m^{-3}. Using this system, inhalation exposures to determine adverse WTC dust impacts on the respiratory system continue at NYU. As discussed, sample composition may have varied with time and location, and the size separation techniques disturbed the size distribution. Variation in sample storage times before testing may also result in different effects. The studies tend not to account for these variations, but provide an indication of the level of toxicity of specific samples.

19.4 Health Impacts of WTC Dust and Smoke Exposures

19.4.1 Human Exposures to WTC Dust and Smoke

The initial WTC dust plume persisted for several days following the disaster, while the smoke plume following the attack persisted for several months and remains the longest burning commercial fire in US history. During that period, enormous quantities of pulverised building material and smoke were suspended in the atmosphere, deposited on external building surfaces, and in homes and offices.[3] A large proportion of the NYC population was exposed to respirable atmospheric particles resulting from the building collapse, the fires and the subsequent site remediation, over an extended period.[14]

A subset of the exposed NYC population were filmed being exposed to extremely high concentrations of smoke and dust in the WTC collapse plume, including first responders, local workers, local school children and the surrounding residential population. Air concentrations of dust and smoke were estimated to be >100 000 μg m^{-3}.[25,26] Other groups exposed to the longer-term smoke plume and resuspended dust included rescue and military personnel, clean-up workers, and those local workers and residents who returned to work, study or live in the area within days or weeks, following temporary displacement. Many returned to buildings that had been cleaned. However, a number of buildings remained uncleaned or poorly cleaned, allowing continued exposures. Residents, workers and students returning to the area in the aftermath of the disaster were exposed to varying levels of contamination following inadequate remediation.[27] Even if the buildings were cleaned, ventilation systems were often not cleaned, allowing the recirculation of dust into cleaned buildings. The population exposures and the consequences of these exposures have yet to be fully calculated and may never be fully understood through exposure assessment.[28,29]

The highest short-term and cumulative exposures were expected in rescue and recovery workers. Although respirators and breathing masks were available to workers, they were often not used due to practical difficulties in performing tasks and a lack of enforcement onsite. Limited data characterised recovery worker exposures at Ground Zero due to the challenging working conditions.[30] As a consequence, exposure estimates for individuals or specific areas were often self-reported 2–3 years following the disaster and may be subject to self-selection and recall error. Qualitative descriptions of exposure conditions provided anecdotal evidence that elevated exposures were linked to risks for acute and chronic respiratory effects. Individual records of work patterns at Ground Zero, plus filmed exposures to the plume provided further evidence.

In the absence of comprehensive exposure measurements, the high prevalence of acute symptoms and chronic conditions observed, workers and survivors began to attract the attention of rescue worker unions and medical professionals in NYC. Early studies of these groups showed that trace contaminants found in WTC dust were higher in the sera and lungs of first responders than would be normally expected. While ambient PCB levels reduced quickly to "normal" urban levels, certain PCB levels were elevated in the sera of seven first responders compared to the general US population levels.[31] Dioxin-like compounds were elevated in the air immediately after the disaster[29] and 15 dioxin-like compounds in the sera of ~350 firefighters were correspondingly 50% higher than the highest levels in the general population and other firefighters not attending the WTC site.[32] Chrysotile and amosite asbestos fibres, chromium, degraded glass fibres, fly ash and silicates were found in the bronchoalveolar lavage of a firefighter who developed eosinophilic pneumonia after two weeks working at Ground Zero.[33]

Large and irregularly shaped particles, with more metallic inclusions, were found in the sputum of 39 WTC-exposed firefighters compared to firefighters from another location.[34] Aluminium and magnesium silicates, chrysotile asbestos, calcium phosphate and calcium sulfate, small shards of glass and carbon nanotubes were found in seven symptomatic WTC responders.[35] Silica, aluminium silicates, titanium dioxide, talc and metal particles were identified in lung biopsies of 12 WTC-exposed workers and residents who had presented with symptoms and who exhibited interstitial fibrosis, emphysematous change and small airway abnormalities.[36]

Based on ambient outdoor air concentrations, the majority of the NYC population was unlikely exposed to higher than normal particulate levels.[14,29] Except for inhalation exposures during the attacks, the general population was exposed only briefly to elevated levels of toxic materials from resuspended materials and night-time inversions that acted to concentrate the smoke over the city.[6] However, there remained a risk that localised contamination (*e.g.*, asbestos), especially indoors, resulted in higher exposures of some individuals.

19.4.2 Initial Evidence of WTC Dust Health Impacts

The first ailment to be linked to the WTC disaster by several authors was dubbed the "World Trade Center cough".[16,17] The Mount Sinai Medical Center (NYC) estimated that nearly half of those who worked at the site experienced symptoms within days of the disaster, and onsite workers continued to present in line with their cumulative time working at Ground Zero. Skin disorders were also reported in those exposed to WTC dust.[37] Asthma[27] and other respiratory diseases[38,39] soon became the focus of concern amongst site-workers, with several patients presenting with unusual symptoms soon after working at Ground Zero (*e.g.*, see Rom *et al.*, ref. 33). Stress and post-traumatic stress (PTSD) disorders also began to emerge as significant, together with associated conditions such as reflux.[40,41] However, establishing cohort studies took time, partly because of the practical limitations in setting up studies on the difficult, ever-changing site and because the Institutional Review Board (IRB) process that regulates such studies is not designed for rapid turn-around.

19.4.3 Diseases Linked to Exposures of WTC Dust

Over the last decade, several significant adverse health outcomes have been linked to WTC dust and smoke exposures.[42–44] Wisnivesky *et al.*[43] linked exposures to WTC dust in rescue and recovery workers to incidences of persistent multiple illnesses, including asthma, sinusitis, gastro-oesophageal reflux disease, mental health disorders (depression, PTSD and panic disorder) and spirometric abnormalities. Sarcoid-like granulomatous disease—normally arising from inflammatory processes including infection—was observed among firefighters and other first responders.[45,46] Multiple myeloma was also reported.[47]

Respiratory effects were most widespread amongst first responders. The majority of WTC firefighters receiving subspecialty pulmonary evaluation (SPE) within 6.5 years post-9/11 exhibited airway obstruction. In a cohort of 801 never smokers with normal pre-9/11 expiratory volume in one second (FEV1), Nolan *et al.*[48] correlated inflammatory biomarkers and complete blood count (CBC) at monitoring entry within 6 months of 11th September 2001 with a median FEV1 at SPE (34 months; interquartile range, 25–57). Cases of airflow obstruction had FEV1 less than the lower limit of normal (LLN; 100 of 801; 70 of 100 had serum), whereas control subjects had FEV1 greater than or equal to LLN (153 of 801; 124 of 153 had serum). FEV1 declined 12% in these cases, compared to increases of 3% in control subjects. Case subjects had elevated serum MDC, GM-CSF, granulocyte colony-stimulating factor and interferon inducible protein-10 levels. Elevated GM-CSF and MDC increased the risk for subsequent FEV1 less than the LLN by 2.5 fold [95% confidence interval (CI), 1.2–5.3] and 3.0 fold (95% CI, 1.4–6.1) in a logistic model adjusted for exposure, body mass index (BMI), age on 11th September 2001 and polymorphonuclear neutrophils. The model had sensitivity of 38% (95% CI, 27–51) and specificity of 88% (95% CI, 80–93). Elevated serum GM-CSF and MDC levels soon after WTC exposure were associated with increased risk of airflow obstruction in subsequent years.

Loss of lung function in WTC lung injury cases was predicted by biomarkers of inflammation and metabolic syndrome, while pulmonary vascular loss is a known early feature of chronic obstructive pulmonary disease. Weiden *et al.* (2013)[55] investigated if other cardiovascular disease (CVD) biomarkers also predicted WTC lung injury. They carried out a nested case-cohort study used 801 never smoked, WTC exposed firefighters with normal pre-9/11 lung function presenting for subspecialty pulmonary evaluation (SPE) before March 2008, using a sub-cohort of 124/801 with serum drawn within six months of 9/11 to define CVD biomarker distribution. Cases were defined using post incident FEV1 as cases with FEV1 $\leq$ 77% predicted (66/801) and WTC lung injury cases resistant to lung injury with FEV1 $\geq$ 107% (68/801). Susceptible WTC lung injury cases had higher levels of Apo-AII, CRP, and MIP-4 with significant RRs of 3.85, 3.93, and 0.26 respectively, with an area under the curve (AUC) of 0.858, while resistant subjects had significantly higher sVCAM and lower MPO with RRs of 2.24, and 2.89 respectively; AUC 0.830. Six months after the incident biomarkers in serum were found to predict either susceptibility or resistance to WTC lung injury. Pathways causing or protecting subjects from pulmonary vascular disease and associated lung function after an irritant exposure may be defined by these biomarkers.

Zeig-Owens *et al.*[49] reported a modest excess of cancer cases in a WTC dust-exposed cohort of firefighters. Despite cancer latency for solid tumors averaging >20 years, studies of WTC dust- and smoke-exposed firefighters show a statistically significant excess in all-site cancer with less than 10 years follow-up. The aetiological role of WTC dust and smoke exposures in cancer development is unclear, but the earliest and longest serving WTC site-workers were more likely to develop respiratory illness than other groups. Limitations of the study relate to the paucity of data on specific exposures experienced by firefighters, the short time frame since exposure and disease development, the biological plausibility of chronic inflammation as a possible mediator between WTC exposure and cancer, and potential confounders. Additional post-WTC cancer incidence results are expected to come from the non-FDNY WTC Responder Consortium, the WTC registry cohorts and the FDNY Emergency Medical Services (EMS) cohort in the future.

Two deaths were directly attributed to exposures to the dust and smoke. Forty-two-year-old Felicia Dunn died of sarcoidosis in 2002 following WTC dust cloud exposure. New York's chief medical examiner certified that the death was connected to the WTC "with certainty beyond a reasonable doubt", and labeled the death a "homicide". In 2006, 32-year-old James Zadroga, a police officer who had worked 100 h at the site, died of pulmonary fibrosis. According to a New Jersey pathologist, his death was "directly related to the 9/11 incident". The James L. Zadroga 9/11 Health and Compensation Act was signed by President Obama in 2011. This Act provided federal funding to monitor and treat WTC responders/survivors until at least 2015 and reopened the September 11 Victim Compensation Fund, making it accessible to future healthcare claims. It also required a periodic review of the scientific and medical evidence regarding any potential WTC-related health conditions, with specific reference to the findings and recommendations of the Clinical Centers of Excellence (see the next section).

19.4.4 The WTC Health Registry

As the health effects within the emergency responder and NYC community began to emerge, it became clear that a system capable of tracking the exposed population would also be necessary. The lack of centralised medical records in NYC hampered such efforts initially. In addition, many residents moved out of the affected area after 9/11 making follow-up difficult, and public and environmental health investigators were often not able to track WTC dust-exposed residents and responders through work-rosters.

A worker screening and treatment programme for WTC responders was quickly established to provide healthcare for those workers affected by their activities related to the WTC collapse. The WTC Medical Monitoring and Treatment Program (WTC-MMTP) was established by NIOSH through a WTC Centers of Excellence clinic network in NYC. However, it did not cover non-rescue, recovery and clean-up workers, or local residents. The Non-responder WTC Health Registry was established by the Agency for Toxic Substances and Disease Registry and the New York City Health Department, in collaboration with NIOSH. This registry was established to capture all individuals exposed to WTC dust, not just the occupationally exposed, and has more than 70 000 enrollees. It is a rare example of environmental and occupational groups working together on a workplace disaster that affects the wider environment and population. It is implicit that the occupationally exposed may be the first presenting cases. The main drawback with the registry is that approximately 70% of registry enrollees were self-identified and enrollees self-assessed their exposures, potentially creating bias in the dataset. The key advantage is that many of the WTC dust- and smoke-exposed individuals are captured in a centralised database, where co-occurrence, or comorbidity, of respiratory and mental health conditions was experienced by many enrolled.[1] The registry is a valuable tool for establishing the WTC health impacts. Investigations are currently underway to establish the 9/11 impacts related to:[44]

- cardiovascular disease;
- alcohol use;
- PTSD and depression;
- unmet mental health care needs;
- cancer rates among enrollees;
- health of Staten Island landfill and barge recovery workers;
- respiratory and behavioural health of children;
- impact of 9/11 injuries on long-term enrollee health.

19.5 Lessons Learned

The WTC was a unique urban disaster with wide ramifications for environmental health professionals in other disaster scenarios.[50–52] While clean-up and rescue efforts successfully managed many complex risks at the time, we now know that the long-term respiratory health of 9/11 rescue and clean-up workers was compromised. A coordinated air quality response to the WTC disaster was slow to form on the scale required to contain respiratory health risks, and adequate warnings of potential hazards of the dust and smoke emissions were not given. Communication, response coordination and dissemination of safety information to protect workers and residents have long since been strongly criticised.[3] The EPA was quick to declare the air "safe", endangering public trust at a critical time immediately following the disaster. In the haste to clean up the WTC site and revert to business as usual, clean-up practices endangered workers and the public: respiratory protection and other worker protection required vigilance from day one to prevent health consequences, and local residents and workers could have been given more accurate information. Failure to act on known hazards and enforce protective practices can now be seen to have contributed to exposures to smoke and dust that led to long-term health impacts among some groups.

Utilising diverse skills across the NYC community became vital in reducing exposure risks and tracking consequences. The scale and scope of events on 9/11 overextended the abilities of individual health departments and federal environmental managers, requiring occupational and environmental professionals to work together. Forming multi-disciplinary research teams as part of a disaster response may avoid a "groupthink" phenomenon that may unwittingly influence management strategies and health outcomes. In the WTC disaster, academics in NYC also played a vital role, but could not influence the work of federal agencies, only complement it. In the long-term, partnership with community groups and local organisations proved especially crucial and the clear communication of risk and scientific uncertainty introduced specific challenges.[13,52]

For future air quality management in such scenarios, forward planning and ready-to-go systems for emergency monitoring are required. The development of disaster preparedness programs could improve disaster management by providing emergency protocols, to be rapidly deployed. A rapid assessment facility of portable instrumentation (*e.g.*, batteries, solar or fuel cell powered) and "real time" monitors can ideally be deployed to respond to any emergency. Designating environmental parameters to be measured and health standards for short-term exposure assessments would provide important management tools. Having an independent panel of experts ready for assembly, briefings and quick turn-around assessment of i) risk and ii) appropriate actions would also be a valuable resource in such circumstances. Systematic and rapid data gathering would have improved the certainty of long-term health impact studies: increased exposure measurements during the event, together with detailed follow-up of affected individuals, could have provided additional evidence of risk factors and informed screening programmes. Rapidly deployed databases and technology infrastructures capable of population tracking would have been valuable in the immediate aftermath of the disaster.[53,54]

In conclusion, it is now recognised that new strategic approaches are required to deal with an air quality disaster on this scale, and some of the environmental health lessons learned were adopted in subsequent events (*e.g.*, the Deepwater Horizon Gulf oil spill). Some of the lessons from the WTC disaster have therefore informed environmental and occupational health practice during a disaster and influenced how such a disaster might be handled today. Such preparedness would be valuable in responding to a biological or chemical terrorist attack, but also to nuclear or natural disasters.

19.6 Conclusions

Conclusions about the human health effects of long-term exposure or an initial acute exposure to WTC air emissions cannot be directly linked to individual or area exposure data (which is absent). However, it is not unusual that environmental or occupational exposures are not measured directly in exposure cases (*e.g.*, asbestos exposures). The important factor is that substantial risk is proved to such a degree that there is a high probability that subsequent disease with biologically plausible association may have been caused by the *likely* exposure. In those cases, and in the case of the WTC, the health evidence emerges from the population data and this takes decades of monitoring of exposed populations. Risk factors associated with negative health outcomes include exposure to the initial plume, early arrival at the WTC site, work duration at the WTC site and work without a mask or respirator. Emergency response and evacuation plans now accommodate this experience from the WTC disaster. Public and environmental health departments are now seen as being central to preparedness efforts, rather than peripheral. Following 9/11, the impacts of air quality on health, even of short-term exposures, are now more likely to be given due consideration.

References

1. NIOSH, 2011, First Periodic Review of Scientific and Medical Evidence Related to Cancer for the World Trade Center Health Program. NIOSH Publication No. 2011-197. Available at: http://www.cdc.gov/niosh/docs/2011-197/pdfs/2011-197.pdf. Accessed December 18, 2012.
2. Landrigan *et al.*, Health and Environmental Consequences of the World Trade Center Disaster: The NIEHS World Trade Center Working Group, *Environ. Health Perspect.*, 2004, **112**(6), 731–739.
3. U.S. EPA, 2003 *Office of Inspector General. EPA's Response to the World Trade Center Collapse: Challenges, Successes, and Areas for Improvement*, Report No. 2003-P-00012.
4. P. J. Lioy, C. P. Weisel, J. R. Millette, S. Eisenreich, D. Vallero, J. Offenberg, B. Buckley, B. Turpin, M. Zhong, M. D. Cohen, C. Prophete, I. Yang, R. Stiles, G. Chee, W. Johnson, R. Porcja, S. Alimokhtari, R. C. Hale, C. Weschler and L. C. Chen, Characterization of the dust/smoke aerosol that settled east of the World Trade Center (WTC) in lower Manhattan after the collapse of the WTC 11 September 2001, *Environ. Health Perspect.*, 2002, **110**, 703–714.
5. US EPA, 2003 Exposure and Human Health Evaluation of Airborne Pollution from the World Trade Center Disaster (External Review Draft) (EPA/600/R-03/142). US Environmental Protection Agency, Washington, DC.
6. T. A. Cahill, S. S. Cliff, K. D. Perry *et al.*, Analysis of aerosols from the World Trade Center collapse site, New York, October 2–30, 2001, *Aerosol Sci. Technol.*, 2004, **38**, 165–183.
7. L.-M. Yiin, C. P. Weisel, B. Buckley, I. Yang, P. J. Lioy, J. R. Millette, C. Quan, J. Gorczynski, M. Kendall and L. C. Chen, Comparisons of the Dust/Smoke Particulate that Settled Inside the Surrounding Buildings and Outside on the Streets of Southern New York City after the Collapse of the World Trade Center, 11 September 2001, *J. Air Waste Manage. Assoc.*, 2004, **54**(5), 515–528.
8. J. D. Pleil, A. F. Vette, B. A. Johnson and S. M. Rappaport, Air levels of carcinogenic polycyclic aromatic hydrocarbons after the World Trade Center disaster, *Proc. Natl. Acad. Sci. USA*, 2004, **101**(32), 11685–11688.

9. C. M. Butt, M. L. Diamond, J. Truong, M. G. Ikonomou, P. A. Helm and G. A. Stern, Semivolatile organic compounds in window films from lower Manhattan after the September 11th World Trade Center attacks, *Environ. Sci. Technol.*, 2004, **38**(13), 3514–3524.
10. S. Rayne, M. G. Ikonomou, C. M. Butt, M. L. Diamond and J. Truong, Polychlorinated dioxins and furans from the World Trade Center attacks in exterior window films from lower Manhattan in New York City, *Environ. Sci. Technol.*, 2005, **39**(7), 1995–2003.
11. US EPA, 2002, Toxicological Effects of Fine Particulate Matter Derived from the Destruction of the World Trade Center, U.S. Environmental Protection Agency, Washington, DC, EPA/600/R-02/028.
12. M. Kendall, D. Breger and L. C. Chen, SEM characterization of WTC plume particles: Indoor airborne and deposited dust, *Epidemiology*, 2002, **13**(4), 241.
13. L. C. Chen, B. Cohen, M. Cohen, M. Kendall, M. Heikkinen, P. Maciejczyk, G. Thurston "NYU Research and Outreach Activities Related to the WTC Disaster" *Special Paper on WTC Related Activities at NYU*, NYU, February 2002.
14. S. Ng, S. Dimitroulopoulou, A. Grossinho, L. C. Chen and M. Kendall, PM2.5 exposure assessment of the population of New York City Lower Manhattan after the World Trade Center disaster, *Atmos. Environ.*, 2005, **39**, 1979–1992.
15. P. B. Maciejczyk, R. L. Zeisler, J.-S. Hwang, G. D. Thurston, and L. C. Chen, Characterization of size-fractioned World Trade Center dust and estimation of relative dust contributions to ambient particulate concentrations in *Urban Aerosols and Their Impacts: Lessons learned from the World Trade Center Tragedy*, ed. J. S. Gaffney and N. A. Marley, ACS Symposium Series 2006, vol. 919, pp. 114–131.
16. M. D. Scanlon, World Trade Center cough—a lingering legacy and a cautionary tale, *N. Engl. J. Med.*, 2002, **347**, 840–842.
17. L. C. Chen and G. D. Thurston, World Trade Center cough, *Lancet*, 2002, **360**, S37–S38.
18. J. K. McGee, L. C. Chen, M. D. Cohen *et al.*, Chemical analysis of World Trade Center fine particulate matter for use in toxicologic assessment, *Environ. Health Perspect.*, 2003, **111**(7), 972–980.
19. S. H. Gavett, N. Haykal-Coates, J. W. Highfill *et al.*, World Trade Center fine particulate matter causes respiratory tract hyperresponsiveness in mice, *Environ. Health Perspect.*, 2003, **111**(7), 981–991.
20. J. P. Payne, S. J. Kemp, C. A. Dewar, P. Goldstraw, M. Kendall, L. C. Chen and T. D. Tetley, Airborne particulate matter generated by the World Trade Centre (WTC) collapse stimulates cytokine production by human alveolar macrophages (AM) and type II epithelial (TII) cells in vitro, *J. Occup. Environ. Health*, 2004, **46**(5), 420–427.
21. S. Wang, C. Prophete, J. M. Soukup *et al.*, Roles of MAPK pathway activation during cytokine induction in BEAS-2B cells exposed to fine World Trade Center (WTC) dust, *J. Immunotoxicol.*, 2010, **7**(4), 298–307.

22. A. Xu, C. Prophete, L. C. Chen, C. W. Emala and M. D. Cohen, Interactive Effect of Cigarette Smoke Extract and World Trade Center Dust Particles on Airway Cell Cytotoxicity, *J. Toxicol. Environ. Health*, 2011, **74**, 887–902.
23. M. D. Weiden, B. Naveed, S. Kwon, L. N. Segal, S. J. Cho *et al.*, Comparison of WTC Dust Size on Macrophage Inflammatory Cytokine Release In vivo and In vitro, *PLoS One*, 2012, **7**(7), e40016, DOI: 10.1371/journal.pone.0040016.
24. J. M. Vaughan, B. Garrett, C. Prophete, L. Horton, M. Sisco, J. M. Soukup, J. Zelikoff, A. Ghio, R. E. Peltier, L. C. Chen, and M. D. A. Cohen, Novel System to Generate WTC Dust Particles for Inhalation Exposures (Submitted).
25. P. J. Lioy, E. Pellizzari and D. Prezant, The World Trade Center aftermath and its effects on health, *Environ. Sci. Technol.*, 2006, **40**(22), 6876–6885.
26. P. J. Lioy and P. Georgopoulos, The anatomy of the exposures that occurred around the World Trade Center site: 9/11 and beyond, *Ann. N. Y. Acad. Sci.*, 2006, **1076**, 54–79.
27. S. Lin, R. Jones, J. Reibman, J. Bowers, E. F. Fitzgerald and S. A. Hwang, Reported respiratory symptoms and adverse home conditions after 9/11 among residents living near the World Trade Center, *J. Asthma*, May 2007, **44**(4), 325–332.
28. L. Claudio, Environmental aftermath, *Environ. Health Perspect.*, 2001, **109**, 528–536.
29. M. Lorber, H. Gibb, L. Grant, J. Pinto, J. Pleil and D. Cleverly, Assessment of inhalation exposures and potential health risks to the general population that resulted from the collapse of the World Trade Center towers, *Risk Anal.*, 2007, **27**(5), 1203–1221.
30. A. S. Geyh, S. Chillrud, D. L. Williams *et al.*, Assessing truck driver exposure at the World Trade Center disaster site: personal and area monitoring for particulate matter and volatile organic compounds during October 2001 and April 2002, *J. Occup. Environ. Hyg.*, 2005, **2**(3), 179–193.
31. J. Dahlgren, M. Cecchini, H. Takhar and O. Paepke, Persistent organic pollutants in 9/11 world trade center rescue workers: reduction following detoxification, *Chemosphere*, 2007, **69**(8), 1320–1325.
32. P. Edelman, J. Osterloh, J. Pirkle, S. P. Caudill, J. Grainger, R. Jones, B. Blunt, A. Calafat, W. Turner, D. Feldman, S. Baron, B. Bernard, B. D. Lushniak, K. Kelly and D. Prezant, Biomonitoring of chemical exposure among New York City firefighters responding to the World Trade Center fire and collapse, *Environ. Health Perspect.*, 2003, **111**, 1906–1911.
33. W. N. Rom, M. Weiden, R. Garcia *et al.*, Acute eosinophilic pneumonia in a New York City firefighter exposed to World Trade Center dust, *Am. J. Respir. Crit. Care Med.*, 2002, **166**(6), 797–800.
34. E. M. Fireman, Y. Lerman, E. Ganor *et al.*, Induced sputum assessment in New York City firefighters exposed to World Trade Center dust, *Environ. Health Perspect.*, 2004, **112**(15), 1564–1569.

35. M. Wu, R. E. Gordon, R. Herbert *et al.*, Case report: Lung disease in World Trade Center responders exposed to dust and smoke: carbon nanotubes found in the lungs of World Trade Center patients and dust samples, *Environ. Health Perspect.*, 2010, **118**(4), 499–504.
36. C. E. Caplan-Shaw, H. Yee, L. Rogers *et al.*, Lung pathologic findings in a local residential and working community exposed to World Trade Center dust, gas, and fumes, *J. Occup. Environ. Med.*, 2011, **53**(9), 981–991.
37. M. J. Huang, J. Li, J. M. Liff, D. E. Cohen and J. Cone, *J. Occup. Environ. Med.*, 2012, **54**(4), 451–458.
38. D. J. Prezant, M. Weiden, G. I. Banauch *et al.*, Cough and bronchial responsiveness in firefighters at the World Trade Center site, *N. Engl. J. Med.*, 2002, **347**(11), 806–815.
39. J. Weakley, M. P. Webber, J. Gustave *et al.*, Trends in respiratory diagnoses and symptoms of firefighters exposed to the World Trade Center disaster: 2005-2010, *Prev. Med.*, 2011, **53**(6), 364–369.
40. J. M. Stellman, R. P. Smith *et al.*, Enduring mental health morbidity and social function impairment in world trade center rescue, recovery, and cleanup workers: the psychological dimension of an environmental health disaster, *Environ. Health Perspect.*, 2008, **116**(9), 1248–1253.
41. R. M. Brackbill, J. L. Hadler, L. DiGrande *et al.*, Asthma and posttraumatic stress symptoms 5 to 6 years following exposure to the World Trade Center terrorist attack, *JAMA*, 2009, **302**(5), 502–516.
42. S. E. Perlman, S. Friedman, S. Galea *et al.*, Short-term and medium-term health effects of 9/11, *Lancet*, 2011, **378**(9794), 925–934.
43. J. P. Wisnivesky, S. L. Teitelbaum, A. C. Todd *et al.*, Persistence of multiple illnesses in World Trade Center rescue and recovery workers: a cohort study, *Lancet*, 2011, **378**(9794), 888–897.
44. H. T. Jordan, R. M. Brackbill, J. E. Cone, I. Debchoudhury, M. R. Farfel, C. M. Greene, J. L. Hadler, J. Kennedy, J. Li, J. Liff, L. Stayner and S. D. Stellman, Mortality Among Survivors of the Sept 11, 2001, Word Trade Center Disaster: Results from the World Trade Center Health Registry Cohort, *Lancet*, 2011, **378**, 879–887.
45. G. Izbicki, R. Chavko, G. I. Banauch *et al.*, World Trade Center "Sarcoid-Like" Granulomatous Pulmonary Disease in New York City Fire Department Workers, *Chest*, 2007, **3**(131), 1414–1423.
46. L. E. Crowley, R. Herbert, J. M. Moline *et al.*, "Sarcoid like" granulomatous pulmonary disease in World Trade Center disaster responders, *Am. J. Ind. Med.*, 2011, **54**(3), 175–184.
47. J. M. Moline, R. Herbert, L. Crowley *et al.*, Multiple myeloma in World Trade Center responders: a case series, *J. Occup. Environ. Med.*, 2009, **51**(8), 896–902.
48. A. Nolan, B. Naveed, A. L. Comfort, N. Ferrier, C. B. Hall, S. Kwon, K. J. Kasturiarachchi, H. W. Cohen, Z. Zeig-Owens, M. S. Glaser, M. P. Webber, T. K. Aldrich, W. M. Rom, K. Kelly, D. J. Prezant and M. D. Weiden, Inflammatory Biomarkers Predict Airflow Obstruction After Exposure to World Trade Center Dust, *Chest*, 2012, **142**(2), 412–418.

49. R. Zeig-Owens, M. P. Webber, C. B. Hall, T. Schwartz, N. Jaber, J. Weakley, T. E. Rohan, H. W. Cohen, O. Derman, T. K. Aldrich, K. Kelly and D. J. Prezant, Early Assessment of Cancer Outcomes in New York City Firefighters After the 9/11 Attacks: An Observational Cohort Study, *Lancet*, 2001, **378**(9794), 898–905.
50. P. J. Lioy and M. Gochfeld, Lessons learned on environmental, occupational, and residential exposures from the attack on the World Trade Center, *Am. J. Ind. Med.*, 2002, **42**(6), 560–565.
51. F. Lyman Messages in the Dust: What are the Lessons of the Environmental Health Response to the Terrorist Attacks of September 11? The National Environmental Health Association, September 2003. Sponsored by the Chartered Institute of Environmental Health.
52. S. Lin, M. P. Mauer, R. Jones, M. L. Herdt-Losavio, S. A. Hwang and E. F. Fitzgerald, Lessons learned from the September 11th disaster: a state health agency perspective, *Environ. Health Insights*, 2012, **6**, 27–31.
53. D. B. Reissman and J. Howard, Responder safety and health: preparing for future disasters, *Mt. Sinai J. Med.*, 2008, **75**, 135–141.
54. W. N. Rom, J. Reibman, L. Rogers *et al.*, Emerging exposures and respiratory health: World Trade Center dust, *Proc. Am. Thorac. Soc.*, 2010, 7(2), 142–145.
55. M. D. Weiden, B. Naveed, S. Kwon, S. J. Cho, A. L. Comfort, D. J. Prezant, W. N. Rom and A. Nolan, Cardiovascular biomarkers predict susceptibility to lung injury in World Trade Center dust-exposed firefighters, *Eur. Respir. J.*, 2013, **41**, 1023–1030.

Public Health Aspects of Fire Management

CHAPTER 20

Providing Advice to those Exposed to Combustion Products

VIRGINIA MURRAY

Public Health England, Wellington House, 133-155 Waterloo Road, London SE1 8UG, UK
Email: Virginia.Murray@phe.gov.uk

20.1 Introduction

It has been stated that "never has public health been more political or the linkages with government stronger or more demanding."[1] Thus, the importance of coordinating planning and service-delivery activities among local authorities, disaster-response organisations, and governmental agencies is vital to minimise duplication and enable appropriate procedures to implement the best strategies to protect the health of the population.[2]

Emergency planning, preparedness and response are of vital import in the UK and widely supported. The Civil Contingencies Act (CCA) 2004 is an enabling Act of the UK Parliament closely supported by specific regulations[3] and statutory guidance on emergency preparedness. The Act delivers a single framework for civil protection and is regarded as a strong and very useful tool for directing and facilitating emergency planning in England. The Act is separated into two substantive parts: (1) local arrangements for civil protection; and (2) emergency powers. The CCA has increased the legislative requirements on a number of organisations classed as either category-1 or category-2 responders to major incidents, who together

Issues in Toxicology No. 23
Toxicology, Survival and Health Hazards of Combustion Products
Edited by David A. Purser, Robert L. Maynard and James C. Wakefield

Published by the Royal Society of Chemistry, www.rsc.org

provide advice to those exposed to combustion products. Category-1 responders include the emergency services (police, fire, ambulance), primary and acute national health service (NHS) organisations, the Health Protection Agency (HPA) is now Public Health England (PHE) since 2013, the Environment Agency, the Maritime and Coastguard Agency and local authorities; and Category-2 responders include the Health and Safety Executive, strategic health authorities and transport and utility companies.[4]

The functions of the HPA (now PHE) are "to protect the community (or any part of the community) against infectious diseases and other dangers to health."[5] In addition to the agency's role in reducing the dangers to health from infections, chemical and radiation hazards, and extreme events, it also provides support to, and works in partnership with, others who also have health protection responsibilities, and advises, through the Department of Health, all government departments and devolved administrations throughout the UK. In England it provides the local health protection services, which in the rest of the UK are delivered by the three other lead health protection bodies: the National Public Health Service Wales, Health Protection Scotland (HPS), and the Department of Health, Social Services and Public Safety, Northern Ireland. The HPA (now PHE) works closely with all these organisations. At the UK level, the HPA (now PHE) is responsible for providing information and services to support a coordinated and consistent UK public health response to national level emergencies.[6]

The control and coordination of emergency response in England are carried out at three levels of command: strategic (gold command), tactical (silver command) and operational (bronze command).[7] Table 20.1 provides a summary of the level of command level and the coordination of response these levels offer. Although all fires require frontline operational (bronze) response, not all will require other levels of command and control. Where appropriate, public health agencies, including the HPA (now PHE), contribute to support these levels of response.

From other chapters in this book it has been shown that public health advice is of importance to reduce health impacts from exposure to products of combustion. The HPA (now PHE) provides health protection advice on fires and has prepared a non-domestic fire checklist to give guidance to HPA (now PHE) and other responding staff, as non-domestic fires result in significant hazards and risks.[8] The checklist covers many of the issues these fires have presented and recommendations are based on observed best practice and lessons learned from past incidents. The key points recommended for assessment are:

1. Where is the fire?
2. When did it start?
3. What is burning?
4. What are the toxic hazards to air, water and land?
5. In what direction is the plume moving?
6. What is the population at risk?
7. What are the potential health effects?

8. Is advice on sheltering or evacuation required?
9. Alert key groups, including the HPA and chemical experts, NHS organisations [the primary care trust (PCT), the health protection unit (HPU), emergency departments, general practitioners (GPs), NHS Direct, neighbouring HPUs], and local authorities.

Table 20.1 A summary of the control and coordination of emergency response in England, carried out at three levels of command. Source of information: Wright *et al.*, Assessment of health-system crisis preparedness: England, December 2011[7] (adapted from "Emergency response and recovery", Chapter 4. In: *Responding to Emergencies*, version 3. London: HM Government, 2010).

Strategic: GOLD	At the strategic level, the gold command is active in cases where it is clear that the resources, expertise or coordination required are beyond the capacity of the silver command (tactical level). The gold commander of the organisation in charge of the response, typically the police force, sets up a strategic coordinating group, bringing together the gold commanders of other relevant organisations such as the ambulance service, NHS and public health. Each commander is in control of his/her organisation's resources at the scene of the incident. The gold commander is not located at the site of the incident but participates in developing the response strategy in the designated control room.
Tactical: SILVER	At the tactical level, the silver command deals with the overall management and resourcing of frontline response in accordance with the strategy set by the gold command. The most senior officers of each organisation involved in the operations, which include NHS and public health, form a tactical coordinating group to manage the response at that level, defining action to be taken by the bronze command (operational level). The silver commander, who is not normally located at the scene of the incident, reviews the resources at the operational level (bronze command) and works closely with silver commanders of other agencies. The silver command may also take charge of bronze's resources to implement the strategy of the gold command.
Operational: BRONZE	The bronze command (operational level) operates at the scene of the incident and at the receiving NHS hospitals. The bronze commander concentrates efforts and resources on specific tasks within the bronze command's area of responsibility. As management of an incident can become complex, it may be necessary to establish a number of bronze areas so that, for example, at the scene of a train crash, each railway carriage may be under the responsibility of a separate bronze command. If an incident is geographically widespread, the area involved would be divided into different locations, each of which would be under the responsibility of a different bronze command.

This chapter focuses on the evidence relating to the provision of advice to those exposed to combustion products by paying special attention to advice on sheltering and evacuation.

The National Steering Committee on Warning & Informing the Public (NSCWIP) is the UK's primary independent advisory group on warning and informing, and has a web site link on the Cabinet Office national security site.[9] NSCWIP aims "to encourage improvements in the arrangements for warning members of the public of an imminent or actual threat to life, health or property, and to inform them of the appropriate action to take" and considers that the community resilience agenda has moved forward over the last few years and much of the warning and informing activity has been conducted at a local level reflecting the developments from the introduction of the Civil Contingencies Act in November 2004.[3] The four sub-groups of the committee relate to public education, new technology, media issues, and public address. With the support of NSCWIP and many partners, the UK HM Government's leaflet, which was delivered to every address in the UK in 2004, shares the "go in, stay in, tune in" for sheltering and the "get out, stay out and call 999" for evacuation messages very clearly.[10]

Sheltering and evacuation guidance are key to the provision of advice offered to those exposed to combustion products but the evidence to support this advice is limited for major industrial fires and other major events. This chapter considers the content of the current advice within England. By summarising a selection of case studies where sheltering and evacuation were used for health protection from industrial fires and other chemical incidents, this chapter builds a review of what evidence is available. Finally, the issues identified as requiring further work are addressed.

20.2 Current HPA (now PHE) Shelter and Evacuation Guidance

In 2000 the Chemical Incident Response Service (which joined HPA in 2003) reviewed the information on sheltering and evacuation. To support this and to speed decisions in discussion with users, a sheltering or evacuation checklist was developed and is regularly reviewed to make sure it remains current, evidence based and useful (HPA 2009).[11] In this checklist a series of questions to help to facilitate the evacuation/sheltering decisions was developed. In summary these addressed:

- the toxicity of the material to the public, by assessing if it is highly toxic/toxic/irritant/non-irritant, with short-term or long-term effects and if it is explosive or non-explosive;
- if the public are at risk of exposure, by determining if the material can be contained or has the potential for release, if it is capable of dispersal *via* wind, rain, *etc.*, and whether the public are in the path of the projected route and, if so, the distance, plume height, meteorological conditions, and weather conditions that might put the public at risk, and when this might occur;

- if the public might has already been exposed, or might be imminently or in a few hours, and how long such an exposure might last—possibly a few minutes, hours, days, months or years.

The checklist then addresses both the issues of sheltering and evacuation as briefly as possible. Box 20.1 summarises the sheltering advice.

In order to define if evacuation is an appropriate health protection measure, a brief list of the specific instances where evacuation may be appropriate is included.[11] These cover issues that might be relevant:

- Before an incident occurs, this advice for evacuation can be described as precautionary and includes the risk of imminent explosion (*e.g.*, defusing a bomb or making safe an explosive hazard), a small leak likely to escalate rapidly to a release or a threatened release of chemical terrorist or radioactive materials.

Box 20.1 Sheltering: Action to protect the health of the public.[11]

Sheltering is usually the safest option.

GO IN, STAY IN, TUNE IN

- Close windows and doors.
- Minimise drafts by sealing windows and doors with paper/tape or damp towels or newspaper.
- Turn off central heating.
- Turn off mechanical ventilation, including air conditioning.
- Go to an upper floor, if possible to an interior room where ventilation is less.
- Avoid bathrooms and kitchens as they tend to have higher ventilation rates.
- Keep children and pets indoors.
- Have access to a radio and tune in to the local radio station for advice and information.
- Try not to use a telephone or mobile phone unless absolutely vital (to prevent jamming lines).
- Ensure effective communication systems with the public, especially to ensure sheltering is in place as quickly as necessary.
- Ensure effective forms of communication with other emergency services.
- Provide a help-line number, consider using NHS Direct.
- Provide medical assistance post sheltering, especially to those incapacitated.
- Provide post sheltering advice on airing houses.

- During an incident such as the spread of fire or a continuing release of hazard over a prolonged period of time.
- After an incident where gross environmental contamination might present a continuing significant hazard.

If a decision to evacuate is taken then additional issues need to be addressed and these are summarised in Box 20.2.

Box 20.2 Some of the issues requiring consideration on whether to evacuate or not using the findings of a risk assessment.[11]

- Is there sufficient time to evacuate?
- How long will the implementation process take?
  - The emergency services' response time.
  - Public health response time.
- Method being chosen to coordinate and inform the public?
  - Door-to-door.
  - *Via* loudhailers.
  - Radio/TV networks.
  - The need to consider language barriers, the need for translators.
- The time of day (it is more difficult to warn people effectively at 4 a.m. than at 8 p.m.).
- Time to prepare the public:
  - to collect clothes, medication, baby supplies, pets, cheque books, credit cards *etc.* and to secure their homes.
- Time required for the public to move.
- The population profile:
  - Number of elderly, handicapped and immobile.
  - Are any residential homes/nursing homes in the affected area?
  - Are there any people on dialysis machines, or others at special risk?
- The extent of the road network:
  - Transport availability—private and public.
  - Blockage of roads, *e.g.*, flooding or snow.
  - Hazardous travel conditions, *e.g.*, fog, snow, sleet, ice *etc.*
- Consideration of the effects on the evacuees from:
  - outside temperature;
  - psychological trauma/medical risk;
  - cost.
- How large a zone is to be evacuated?
- Possible health risk to the police on the cordon?

One of the issues that has been identified from previous incidents is the need to provide information to those being evacuated, as quickly as possible, about what they need to take with them. One of the major problems that has been identified from previous evacuation is the failure of evacuees to bring their medication with them. As a result, a draft information leaflet for those people being evacuated has been prepared for use in these circumstances, highlighting the need to bring medication (Appendix 20.1).

If individuals have been evacuated then there is a need for local assessment to determine if and when the incident is deemed to be under control and not expected to escalate in order to return people to their homes and places of work or school.[11] The decisions can be guided by environmental sampling and analyses to provide risk assessment information in residential premises after discussion with a medical or scientific toxicologist, where necessary. It may be helpful to prepare a leaflet to explain the situation and actions that should be taken on returning to the premises, such as opening windows and doors to ventilate the premises for an appropriate period of time. Guidance on the content of a "returning home" leaflet should include further advice about whom to contact if any ill health effects develop.[11]

In summary, the HPA (now PHE) shelter and evacuation guidance links to and reflects the current Direct.Gov advice, the UK Government web site that links public services all in one place.[12]

20.3 Incident-related Evidence

Decisions on the appropriateness of sheltering *versus* evacuation have always been difficult and a review of case studies was undertaken to examine how some decisions were made and implemented during incidents.[13] Eight previously published case studies that occurred between 1999 and 2009 (including the fire at Buncefield in 2005, see Chapter 18) demonstrate the decision making process.

Case study 1: Fire at a plastic factory, December 1999

At 10:59 hours the fire service were called to a fire at a factory making plastic jelly shoes and toys in an industrial estate in Paignton in Devon.[14] The initial response of the emergency services was to start to evacuate 472 residents from their homes to a nearby leisure centre. However, this decision was subsequently reviewed by the members of the emergency response team and further evacuation was stopped, with 1278 residents advised to shelter and stay inside their homes. The resultant partial evacuation offered an opportunity to compare the relative health protection offered by these two modes of intervention—probably resulting in the most detailed epidemiological survey comparing sheltering and evacuation following a fire undertaken anywhere in the world at this time.

Kinra *et al.*[15] therefore carried out a cross sectional postal questionnaire survey two weeks after the fire on residents in the affected area and compared the health outcomes among the people evacuated (one third) and sheltered (two thirds). In total, 1096 residents (63%; 299 evacuated, 797 sheltered) provided data for analyses, and the mean symptom score and proportion of cases were higher in evacuated people than in the sheltered population [evacuated: symptom score 1.9, cases 19.7% ($n=59$); sheltered: symptom score 1.0, cases 9.5% ($n=76$); $P<0$]. The difference between the two groups was attenuated markedly at the end of two weeks from the start of the incident. The two main modifiable risk factors for the odds of becoming a case were evacuation [odds ratio 2.5, 95% confidence interval (CI) 1.7 to 3.8] and direct exposure to smoke for more than two hours on the first day of the incident (2.0, CI: 1.7 to 2.3).[13]

The authors state that sheltering may have been a better protective action than evacuation in this chemical incident. Although this study has limitations and many possible sources of bias, it was based on a real event. Evacuations carry their own risks and resource implications; increased awareness may help to reduce unnecessary evacuations in the future.

Case study 2: A chemical release from domestic waste disposal, 1999

Decisions about evacuation and sheltering are often made rapidly without the full toxicological advice to support them. In Kingsbridge in Devon on Thursday, 25 November 1999 an incident occurred involving the inappropriate disposal of rat poison (aluminium phosphide) that had become wet in the rain, leading to production of phosphine gas.[14] Some residents were evacuated immediately at the onset of the event, and later some more were evacuated at the time of clean up of the chemical, with a total of 80 being evacuated. The time line of the event and the major actions involved are summarised below:

- At 13:30 hours, a call from a resident regarding a rubbish fire in the neighbour's backyard to the local fire service, who arrive quickly and apply water but soon stop because the smoke worsens.
- At 13:46 hours, the owner of the house tells the fire service that the smoking black sack contains rat poison (aluminium phosphide) for disposal; the fire service sends a request to search their chemical database for information.
- At 14:02 hours, the police arrive on the scene, evacuate 20 nearby residences exposed to the plume and cordon off the area (100 m radius); others are asked to shelter indoors.
- At 14:45 hours, the fire service receives details about the gas and its effects by fax from head office on discovering that aluminium phosphide, on getting wet, releases phosphine gas and they presume a chemical reaction due to rain and fire water; this results in the agreement that sand should be applied to the chemical while officers use breathing apparatus.

- At 17:00 hours, the public health service is contacted and further expert toxicological advice is sought.
- At 18:36 hours, 12 emergency services personnel attend the local hospital; four of them found to be symptomatic.
- At 19:24 hours, a major incident is declared; silver and gold commands are set up; meetings carry on through the night; no further evacuation decisions are considered possible until air sampling information is available.
- At 04:31 hours, air sampling is commenced; only traces of phosphine in the midst of the plume are found; the decision is taken to evacuate more residences to allow the clean up of the chemical as the process is expected to release more fumes.
- At 06:00 hours, evacuation leaflets are distributed; a decision to evacuate another 100 residences results in only 25 people evacuated as some properties were empty and some residents refused to leave their homes—those who were evacuated were taken to a nearby leisure centre.
- At 14:00 hours, the remaining chemicals were removed from the site by a specialist clean up company.
- At 17:40 hours, residents were returned to their homes and the incident was declared closed.

Case study 3: A warehouse fire, April 2000

At 11:15 hours on Sunday, 30 April 2000 a fire at a warehouse on an industrial estate 4 miles south of St Helens town centre, near the M62 motorway, was reported.[16] The warehouse was thought to contain several hundred different chemicals used in the manufacture of rubber products. The list of chemicals was held in a fire-proof safe within the warehouse itself. The fire was difficult to control, requiring the attendance of 150 fire brigade personnel. Police closed off the roads in the vicinity and advised local residents to remain indoors and close windows. This information was broadcast over a helicopter loudspeaker and on the local radio. At the incident debrief it was noted that it was difficult to encourage the local residents to stay indoors for their own protection because it was a warm sunny day and, despite radio warnings from the police to stay indoors and close windows, these were unheeded by many people and the police noted that use of the helicopter to make announcements to the public was counterproductive as people had to come outside to hear a message telling them to stay indoors and close their doors and windows.

Case study 4: A control of major accident hazards (COMAH) site chemical release, 2000

On 8 March 2000 an incident occurred at the European Vinyls Corporation (EVC) in Runcorn, Cheshire, resulting in a large release of hydrogen chloride

(HCl) gas.[17] The emergency response for this large COMAH site was declared and shelter advice issued to residents and people travelling through the area. After three hours, the leak was contained, the shelter advice lifted, and the incident was stood down. The full health impact can only be estimated: it was reported that two workers from the site were given first aid and taken to the local emergency department, where they were treated and allowed home; an incident team member sustained a back injury and seven members of the public presented with minor symptoms at the emergency department and were allowed home. Following the incident debrief meeting, it was considered that the prompt advice to shelter may have minimised the adverse health effects.

Case study 5: A large plastic recycling plant factory fire, 2002

On Sunday, 21 April 2002, at 22:30 hours, public health services were notified of a fire at a large plastics recycling plant in Creswell, Derbyshire, which had been burning since 17:00 hours.[18] The premises comprised a large yard on which there were substantial quantities of plastic for recycling and a number of brick buildings. The fire had initially produced copious quantities of thick, black smoke, which had dissipated quite rapidly. By 23:30 hours the fire was producing less smoke and was believed to be well under control.

At 01:00 hours, two calls were received from residents of a village 4 miles from the site, with one reporting poor visibility outside the house and some respiratory irritation, and the other reporting worsening of asthmatic symptoms. The person suffering from symptoms of asthma asked whether people were going to be evacuated. At this point the fire brigade had received 12 complaints from the village; however, most were repeated calls from just one or two individuals. The fire brigade were considering evacuation, as they felt it was unlikely that the smoke would disperse before 05:00 hours. However, they were uncertain as to whether the evacuation should involve the village, the site of the fire, where several properties were in the plume, or a nearby village with a community of 4000, where the plume was grounding. At 03:00 hours, the fire incident officer reported that the fire had spread and was once again out of control. The emergency services set up a silver command with a joint incident control point. The fire was brought under control. A site visit at 07:00 hours by the emergency services, the local authority and public health with expert toxicological support showed that the fire was being contained but had been extensive and very hot. The plume was gradually moving away from the village as the wind direction changed, and by 17:00 hours the fire was out.

At the incident debrief it was agreed that it is a difficult process to evacuate people in the middle of the night, and very good reasons for taking this action are needed. In this fire, the burning plastic produced dense clouds of black smoke and generated potentially toxic products of combustion in an open environment, but there was no risk of explosion. Very few complaints of adverse health effects were received at the time and calls to all the health facilities in the area, including GPs, the GP co-op, NHS Direct, and local emergency departments, revealed that there had been no reports of ill health. Additionally, evacuation takes a surprisingly long time and in this fire it was found that it was actually brought under control much more quickly than expected. Finally, it was considered that the "go in, stay in, tune in" message, especially during the night, appears to have been the most appropriate and safest advice.

However, on further investigation it was reported that the "go in, stay in, tune in" message was not formally lifted on 21 April 2002 and a member of the public rang the local authority on Thursday, 24 April to ask if she were allowed to go out and buy food again.

Case study 6: Car tyre depot fire, 2005

Asgari[19] reported a fire where evacuation and sheltering were implemented by the emergency services. In summary, on a Saturday in November 2005, at about 01:00 hours, a fire started in a tyre refitting depot in London, burning approximately 6000 to 8000 tyres. The tyre depot is situated in a light industrial estate, with the refitting yard adjacent to a number of houses on its eastern border. Within the refitting yard a number of gas cylinders were observed and a fireworks warehouse was reported as being present on the same industrial estate. At 02:15 hours, the fire brigade commander at silver command, in conjunction with the police commander on site, decided to evacuate 30 residents, many of whom were elderly, from the street adjacent to the tyre depot because of the risk of fire spreading from the yard, whilst recommending sheltering to other residents in the area. At a silver command meeting held later in the day, it was decided that as the fire was under control and environmental analytical results satisfactory, local residents should be allowed to reoccupy their properties and the sheltering advice for other residents was lifted. Enquiries made to the local emergency medicine department and NHS Direct reported no health concerns as a result of the fire or the related smoke.

Case study 7: Food warehouse fire, 2005

Kirkpatrick *et al.*[20] reported a fire that occurred in November 2005 in a Chinese food warehouse on an industrial estate in London where guidance to shelter was noted as beneficial. They noted that the contents of the warehouse were principally food, oils and plastics, and that the fire started in the early hours of the morning, complicated by two gas mains rupturing, requiring Transco, the then gas transporter (now National Grid) to control the source. Of note, the prompt action of the emergency services had led to the recommendation to local residents to shelter and not evacuate, which was confirmed during a multi-agency site visit that morning. This site visit had been triggered by the scale of the incident, which required eight fire pumps on site and found the presence of asbestos-containing cement dust from roofing, which added to the potential risk and required specialised clean up work.

These case studies illustrate the dilemmas identified by emergency services in providing the best advice to the public, in particular requiring the emergency services to work with their public health partners who can help support these decisions. Of note are the difficulties in making decisions in a multi-agency and timely way, and of implementing these decisions when individuals may refuse to leave their homes. It was also noted that guidance is sometimes not shared or well communicated.

20.4 Discussion

Case studies are widely used in many disciplines, including health care[21,22] to provide records of innovative or good practices and to identify specific problems or issues experienced. Such problems can be documented, as can the actions taken to overcome them.[23,24] Grynszpan *et al.*[24] reported that case studies can validate our understanding and encourage re-evaluation and learning, capturing the complexity of disaster risk and disaster situations. The case studies set out in Section 20.3 describe a series of difficult-to-manage chemical incidents where evacuation and or sheltering were recommended. Building an evidence base from a set of reported industrial fires and other chemical incidents in England is not easy. These events are not part of randomised controlled trials so the evidence needs to be built from critical assessments of the incident reports.

The recently published review by WHO Europe of their *Assessment of Health-system Crisis Preparedness for England* stated in their recommendations that, to enhance the preparedness of the English health-system for crises, the Department of Health may consider taking action to:

1. establish an accessible repository of material relating to all major incidents in England (reports, enquiries, publications *etc.*) to enhance learning and broaden the evidence base;
2. introduce a standardised structure for reporting major incidents, related exercises and lessons learnt, and include the reports in the aforementioned repository of material.[7]

Case studies provide a medium through which to learn practical lessons about successes in disasters risk reduction.[25] The need to document these events and share them nationally and internationally in a standardised manner is increasingly important to build the evidence base. Several projects have identified lessons from case studies.[26–29]

The evaluation of case studies can add understanding to the development of chemical incident response decisions, as demonstrated by the case studies in Section 20.3. In the review of incidents occurring at COMAH sites in England and Wales, January 2009–June 2011, a total of 2227 logged incidents for the period 1 January 2009–6 June 2011 were extracted from the HPA chemical incident documentation system, of which 22 (0.01%) COMAH-associated incidents were identified.[30] In reviewing this data, Stewart-Evans noted that 25% of these incidents resulted in sheltering advice being issued, with evacuation in 14%. He concluded that this review identified a higher percentage of incidents in which shelter advice was issued than in the review between 1 January 2005 and December 2008.[30]

As a result of this and other work, the HPA annual report 2011–2012 recognises the limited evidence and states that the HPA (now PHE) will develop the evidence base for shelter/evacuate messages during radiation and chemical incidents.[43]

20.4.1 Epidemiological Surveillance

Epidemiological surveillance of adverse health effects from chemical incidents can be very valuable, but such studies are hampered by the fact that it is very difficult to quantify exposure and therefore hard to compare shelter and evacuation epidemiologically.

Kinra *et al.*[15] have provided the most widely reported epidemiological study comparing sheltering and evacuation following an industrial fire. Another such study involved a low lying residential area flooded by sewage (a mixture of domestic and trade effluent with a strong chemical odour entered the homes and collected underneath floorboards) on Thursday, 26 November 1998. By comparison of the symptoms of the evacuated residents group, who had been exposed to the odour and, for some, the effluent directly, with those not evacuated (where effluent did not enter their homes), it was shown that both groups suffered similar symptoms, although, the small number of respondents and the absence of a control group made interpretation of the data set difficult.[31] This could imply that the evacuated and the non-evacuated groups might have been equally exposed and at risk, or that the action of evacuation was timely and effectively minimised exposure, and therefore continuing harm to those most at risk.

Because of the paucity of data on sheltering and evacuation, it is relevant to consider data from chemical incidents such as chemical release information. The work of Van Sickle *et al.*[32] has shown that those who had sheltered or who were evacuated after an incident involving the release of chlorine from a derailed train could provide subjective reports that can support rapid public health responses and epidemiologic surveillance. However, the authors recommended that future research should focus on the validity of subjective exposure ratings in comparison with objective exposure markers.

This brief review recommends that further work is undertaken to determine more clearly how to build an epidemiological evidence base on the value of and benefits from sheltering and evacuation guidance from products of combustion to reduce health impacts. The evidence base will be difficult to quantify in terms of exposure and impacts, and it may be complex to compare shelter and evacuation epidemiologically.

20.4.2 Communication

Communication is key in all public health response. The WHO manual for the public health management of chemical incidents (2009) considers communication as a core public health function.[33] The manual states that there are two types of communication where:

- "Risk communication refers to communication about possible incident scenarios, information about possible protective actions and public involvement in siting and licensing of facilities where chemicals are produced, used or stored before an incident occurs."
- "Crisis communication refers to communication about actual risk and appropriate risk-reducing (avoiding) behaviour during an incident."

The manual goes on to state that good risk communication opens communication channels, builds trust, and thereby lays the foundations for effective crisis communication. The cornerstones of effective risk and crisis communication are speed, openness, transparency, and continuity of communication. In particular, the manual notes that mechanisms for interagency communication and public communication are vital for decision making. For sheltering, the manual recommends that guidance should include:

> *"Usually advice to go indoors, shut all doors and windows, disengage mechanical ventilation (shelter-in-place advice) is appropriate for the population downwind of the incident. Building characteristics determine the protectiveness and safe shelter duration, which is the length of time for which those at risk of chemical exposure should remain in a safe shelter. The alert to shelter-in-place should include public alert systems (sirens) supported by continuous and consistent communication through multiple channels including radio, television, websites and telephone transmissions. If the wind direction is expected to shift, preventive evacuation can be considered. After the toxic cloud has passed, there are usually few or no restrictions on outdoor activities."*

However, the evidence base for this information is relatively incomplete. One of the few reports on an inter-agency study on communications addresses findings following two wildfires in Southern California. Taylor *et al.*[34] identified that the community at risk needs "real-time," place-specific information: precise location, severity, size, and direction of spread of the fires. They stated that fire management agencies must contribute real-time, place-specific fire information when it is most needed by the affected public as they try to make sense out of the chaos of a wildland fire. They recommended that disseminating information relating to the fire as widely as possible through multiple pathways will maximise the probability of the public finding the information they need. Palen[35] reported the increasing value of online social media, including the use of photograph sites, after assessing their use in the 2007 wildfires in Southern California.

Although there is extensive information on communication to the public, this brief review recommends that further work is undertaken to determine more clearly how to achieve effective communication from evidence-based knowledge to enhance response for action for sheltering and evacuation guidance from products of combustion.

20.4.3 Compliance with Advice on Sheltering and Evacuation

Risk communication has been described as the process of sharing information and perceptions about risk.[36] Compliance with advice is always complex and relates to perception of risk and peer pressure, and compliance with advice has already been identified as an issue in the case studies 1 and 3. Using risk communication as a two-way interaction, in which experts and non-experts exchange and negotiate perceptions relating to both scientific and community values and preferences, it is considered that compliance with advice may be facilitated.[36]

Again, a limited evidence base exists on compliance and a small US study by Dombroski *et al.*[37] provides some evidence. The authors used focus expert groups and interviews. They reported that these groups (national and local) predicted approximately 80–90% compliance with an order to evacuate workplaces, and 60–70% compliance with an order to shelter in place at home. They went on to report that these experts predicted 10% lower compliance for people asked to shelter at the office or to evacuate their homes. They then went on to report that these groups predicted a further 10% lower compliance if the media were sceptical rather than supportive. These groups considered that preparatory policies could improve public compliance by 20–30%.

This brief review recommends that further work is undertaken to determine more clearly how to achieve compliance to enhance response for action for sheltering and evacuation guidance from products of combustion.

20.4.4 Education

The United Nations and the Organisation for Economic Co-operation and Development (OECD) has introduced guiding principles for chemical accident prevention, preparedness and response, with sections on prevention, emergency preparedness/mitigation, emergency response, follow-up of incidents and some special issues, and education considered to be vital.[38]

Education before an incident is important to ensure a high level of compliance with advice provided during an incident. An example of the work of the NSCWIP is the "go in, stay in, tune in" video, produced for their Public Education Group (Cabinet Office, 2012).[9] This has been specifically designed for 7- to 11-year-old children who are recognised by educational psychologists as being most impressionable for adopting safety lessons for life and who can also take their learning back home. In addition, the "go in, stay in, tune in" video and quiz is available within "Crucial Crew Interactive" for primary school pupils.[39] The idea to develop an interactive safety game for children came out of Crucial Crew from a series of workshops held all over the UK for schoolchildren aged 8–11. At these workshops children were presented with various dangerous or challenging situations, such as a road accident or a gas leak, and they were encouraged to engage in role play exercises in how to deal with these situations.[39] The Crucial Crew's website provides computer game versions of several scenarios that typically confront pupils participating in live LASER programmes.

This brief review recommends that further work is undertaken to determine more clearly how to enhance education to help to inform communication for action for sheltering and evacuation from products of combustion.

20.4.5 Vulnerability of Patients and Health Care Workers during Sheltering

The WHO manual for the public health management of chemical incidents[33] addresses vulnerability and it states that some common examples of populations that must be considered when evaluating population susceptibility are children, pregnant women, elderly persons, hospital patients, and people with low socioeconomic status. The manual goes on to state that the actual list will vary by location and by toxic end-point.[33] The manual considers that, once these groups of more vulnerable individuals have been identified, policy-makers should give special attention to their protection. This might include enforcing building or safety regulations on hospitals, schools or residential facilities for the elderly. This might also include more stringent controls on land-use-planning in areas close to a potentially vulnerable population, as well as specific public education.[33]

Few studies on sheltering and the impact on vulnerability of patients and health care workers have been undertaken, and data from Hurricane Katrina provides a case example.[40] In the week after the hurricane hit, 1749 patients and 7600 others sought shelter at 11 area hospitals. These hospitals required employees to work longer and longer shifts under increasingly desperate conditions. Of note, there was little or no power, as backup generators were failing and temperatures above 37 °C destroyed drugs and equipment. They reported that eyewitness accounts included:

- doctors making rounds by flashlight;
- hospital staff unable to electronically control even basic equipment; and
- no food, clean water, or plumbing.[40]

From the information they collected, these conditions were reported as lasting for days. Without a system of support, individual doctors and nurses were left to decide how to use their limited resources.[40] As a result, ethical and legal inquiries were made about their decisions and actions because many decisions were made individually and not as part of a crisis response plan in the following months.[40]

Such issues for those who are ill, elderly or with disabilities have been identified by others. For example, the WHO/PHE/UNISDR (the United Nations Office for Disaster Risk Reduction) factsheet[41] on disaster risk management for the health of people with disabilities and older people states that, in an emergency response, it is important to ensure physical accessibility of transportation, service delivery points, emergency shelters, latrines, and other services. Additionally, many factors that adversely affect psychological health are related to the way assistance, *e.g.*, food security, shelter, water and sanitation, is provided.[42]

Relatively little evidence exists on the vulnerability of patients and health care workers during sheltering from products of combustion. The American Institute of Medicine in their recent publication on *Crisis Standards of Care: A Systems Framework for Catastrophic Disaster Response*[40] have focused on the care of patients and staff in a disaster, and that they should develop a systems-based framework for catastrophic disaster response, which must be integrated into existing emergency response plans. The authors focused on a different definition, but also important use, of the word "shelter", describing it as shelter-in-place for health care facilities. The authors addressed issues relating to evacuation leading to in-place sheltering that are emergency planning measures used in the US particularly in response to hurricane forecasts. They considered that planning for patients with physical disabilities at alternative care and shelter sites was critical. From the information from the US Census Bureau (2006) they report that 12% of the US population have severe disabilities that should be accommodated, and many more are impacted by lower levels of disability. The authors consider that physical access, assistance with activities of daily living, control of the environment for those with cognitive impairments, and the accommodation of other physical impairments need to be addressed. They concluded that if they are not addressed then these issues can lead to further illness, injury, or unnecessary hospitalisation.

This brief review recommends that further work is undertaken to determine more clearly how to manage the vulnerability of patients and health care workers during sheltering and evacuation from products of combustion.

20.5 Conclusion

In the WHO, PHE and UN International Strategy of Disaster Reduction's disaster risk management for health factsheet on chemical safety, risk management considerations are described.[36] It recommends, because of the complex nature of managing chemical incidents, that governments and communities manage disaster risk related to chemical safety by:

- assessing health risks and impacts in a disaster context;
- considering regulatory action for prevention;
- undertaking emergency preparedness;

- undertaking emergency response;
- being ready to undertake recovery.

This chapter considers the issues of providing advice to those exposed to combustion products. It has focused on the specific advice of sheltering and evacuation. Since there is no systematic literature review of the evidence around the value and efficacy of sheltering and evacuation guidance, this chapter has concentrated on UK Government advice of "go in, stay in, tune in" for sheltering and the "get out, stay out and call 999" message for evacuation for the public. Although this chapter has described the current advice provided to those exposed to combustion products by the HPA (now PHE), the HPA annual report 2011–2012 recognises this limited evidence and it states that the HPA (now PHE) will develop the evidence base for shelter/evacuate messages during radiation and chemical incidents.[43]

Limited evidence was identified. A series of case studies were described to illustrate some of the issues identified in making decisions and to provide the best evidence of how the decisions are made and implemented. This helps in identifying the types of advice relevant for those exposed to combustion products and in specifying how that advice should be provided. Many gaps in research were identified; these included the need for a systematic study of the evidence base on the provision of advice to those exposed to combustion products and the need for standardised documentation of each event to allow improved assessment of the guidance offered and of its efficacy. It would be useful to broaden this systematic literature review to include all incidents, fire, chemical spills, and other events, such as extreme weather events and other natural hazards, where sheltering and evacuation guidance is given to the public.

Appendix 20.1 Draft Information Leaflet for those People being Evacuated (HPA, 2009)

Draft Information Leaflet

IMPORTANT INFORMATION FOR PEOPLE BEING EVACUATED

YOU ARE BEING EVACUATED TO AN EMERGENCY REST CENTRE ESTABLISHED BY THE LOCAL AUTHORITY

The basic facilities you can expect at the emergency rest centre are:

- Safe shelter
- Refreshments
- Accommodation for domestic pets
- Reliable information about the incidents and when it is safe to return home
- First aid

Please consider bringing with you the following:

- **All regular medication, if any**
- Spectacles, hearing aid, walking frame or stick, if any
- Address book and your mobile telephone, etc, if any
- Things for daily living such as tooth brush, towel, change of underwear
- Things you may need for your baby such as warm clothing and milk powder
- Your family and pets, if any
- Your cheque books, credit cards and/or money

Before you leave your home remember to

- Close all your windows and doors
- Turn heating to minimum for winter, off for summer
- Turn off any mechanical ventilation
- Secure your home

YOU WILL BE KEPT INFORMED OF THE SITUATION BY THE LOCAL AUTHORITY.

THANK YOU FOR YOUR CO-OPERATION

References

1. L. M. Jacobs, K. J. Burns and R. I. Gross, Terrorism: a public health threat with a trauma system response, *J. Trauma*, 2003, **55**(6), 1014–1021.

2. C. Bayntun, A health system approach to all-hazards disaster management: A systematic review, PLOS Currents Disasters, 2012 Aug 22, Available at http://currents.plos.org/disasters/article/a-health-system-approach-to-all-hazards-disaster-management-a-systematic-review/ (Accessed on 23 August 2012).
3. UK (United Kingdom), The Civil Contingencies Act 2004, London, HM Government, 2004, Available at http://www.legislation.gov.uk/ukpga/2004/36/pdfs/ukpga_20040036_en.pdf (Accessed on 4 August 2012).
4. *Health emergency planning: a handbook for practitioners*, ed. S. Stuart-Black, J. Stuart-Black and E. Coles, The Stationery Office, London, 2nd edn, 2008.
5. UK (United Kingdom), The Health Protection Agency Act 2004, London, HM Government, 2004, Available at: http://www.legislation.gov.uk/ukpga/2004/17/contents (Accessed on 4 August 2012).
6. HPA (Health Protection Agency), 2012, What the Health Protection Agency does, Available at http://www.hpa.org.uk/AboutTheHPA/WhatTheHealthProtectionAgencyDoes/(Accessed on 6 August 2012).
7. C. Wright, S. Finlay, B. McCloskey, V. Murray, A. Osman, G. Rockenschaub, J. Simpson, L. Soomaroo and C. Stanke, Assessment of health-system crisis preparedness: England - December 2011. World Health Organization. Available at: http://www.euro.who.int/en/what-we-do/health-topics/emergencies/disaster-preparedness-and-response/publications/2012/assessment-of-health-system-crisis-preparedness-england (Accessed on 20 May 2015).
8. HPA (Health Protection Agency), October 2009, Sheltering or Evacuation Checklist, Available at: http://www.hpa.org.uk/webc/HPAwebFile/HPAweb_C/1194947310512 (Accessed on 6 August 2012).
9. Cabinet Office, National Steering Committee on Warning & Informing the Public (NSCWIP), Available at: http://www.cabinetoffice.gov.uk/content/national-steering-committee-warning-informing-public-nscwip/ (Accessed on 4 August 2012).
10. UK HM Government, Preparing for Emergencies, What you need to know, 2004, Available at: http://www.direct.gov.uk/prod_consum_dg/groups/dg_digitalassets/@dg/@en/documents/digitalasset/dg_176618.pdf (Accessed on 4 August 2012).
11. HPA (Health Protection Agency), October 2009. Sheltering or Evacuation Checklist. Available at: http://www.hpa.org.uk/webc/HPAwebFile/HPAweb_C/1194947308210 (Accessed on 5 August 2012).
12. Direct.gov, What you can do to prepare for emergencies, Public services all in one place, Available at: http://www.direct.gov.uk/en/HomeAndCommunity/InYourHome/Dealingwithemergencies/Preparingforemergencies/DG_176651 (Accessed on 15 September 2012).
13. V. Murray, Sheltering versus evacuation Chemical Incident Report July 2000; 17:2–3, Available at: http://www.hpa.org.uk/webc/HPAwebFile/HPAweb_C/1194947330053 (Accessed on 4 August 2012).

14. S. Kinra, Incidents resulting in part evacuation and part sheltering – chemical incidents in Devon, Chemical Incident Report July 2000; 17:12–13. Available at: http://www.hpa.org.uk/webc/HPAwebFile/HPAweb_C/1194947330053 (Accessed on 4 August 2012).
15. S. Kinra, G. Lewendon, R. Nelder, S. Harrison and V. Murray, Evacuation decisions in a chemical air pollution incident: cross-sectional survey, *BMJ*, 2005, **330**, 1471–1474.
16. B. Wiratunga and C. Whiteside, Sheltering to manage acute chemical releases – A fire in a chemicals warehouse in St Helens, Chemical Incident Report July 2000, 17: 5–6. Available at: http://www.hpa.org.uk/webc/HPAwebFile/HPAweb_C/1194947330053 (Accessed on 4 August 2012).
17. S. Woodhouse, Sheltering to manage acute chemical releases – Shelter advice protects against adverse health effects of a hydrogen chloride gas plume, Chemical Incident Report July 2000; 17:4–5. Available at: http://www.hpa.org.uk/webc/HPAwebFile/HPAweb_C/1194947330053 (Accessed on 4 August 2012).
18. R. McNaught and W. Phillips, The Cresswell plastics fire (April 2002), Chemical Incident Report January 2003, 27:14–15, Available at: http://www.hpa.org.uk/webc/HPAwebFile/HPAweb_C/1194947377952 (Accessed on 4 August 2012).
19. N. Asgari, Fire in a tyre depot in Barking, NE London. Chemical Hazards and Poisons Report. 2006, 7: 8, Available at: http://www.hpa.org.uk/webc/HPAwebFile/HPAweb_C/1204542882927#page = 9 (Accessed on 1 August 2012).
20. A. Kirkpatrick, M. Meltzer and D. Thrale, Industrial Fires: What are we missing? Chemical Hazards and Poisons Report. 2006, 7: 9–11, Available at: http://www.hpa.org.uk/webc/HPAwebFile/HPAweb_C/1204542882927#page = 9 (Accessed on 1 August 2012).
21. J. Keen and T. Packwood, Qualitative research: Case study evaluation, *BMJ*, 1995, **311**(1), 444–446.
22. I. R. McWhinney, The value of case studies, *European Journal of General Practice*, 2001, **7**(3), 88–89.
23. V. Murray, G. McBean, M. Bhatt, S. Borsch, T. S. Cheong, W. F. Erian, S. Llosa, F. Nadim, M. Nunez, R. Oyun, and A. G. Suarez, Case studies, in *Managing the Risks of Extreme Events and Disasters to Advance Climate Change Adaptation*, ed. C. B. Field, V. Barros, T. F. Stocker, D. Qin, D. J. Dokken, K. L. Ebi, M. D. Mastrandrea, K. J. Mach, G.-K. Plattner, S. K. Allen, M. Tignor, and P. M. Midgley, A Special Report of Working Groups I and II of the Intergovernmental Panel on Climate Change (IPCC), Cambridge University Press, Cambridge, UK, and New York, NY, USA, 2012, pp. 487–542.
24. D. Grynszpan, V. Murray and S. Llossa, The value of case studies in disaster assessment, *Prehospital Disaster Med.*, 2011, DOI: 10.1017/S1049023X11006406.
25. V. Murray, G. McBean, M. Bhatt, S. Borsch, T. S. Cheong, W. F. Erian, S. Llosa, F. Nadim, M. Nunez, R. Oyun, and A. G. Suarez, Case studies, in

Managing the Risks of Extreme Events and Disasters to Advance Climate Change Adaptation, ed. C. B. Field, V. Barros, T. F. Stocker, D. Qin, D. J. Dokken, K. L. Ebi, M. D. Mastrandrea, K. J. Mach, G.-K. Plattner, S. K. Allen, M. Tignor and P. M. Midgley, A Special Report of Working Groups I and II of the Intergovernmental Panel on Climate Change (IPCC), Cambridge University Press, Cambridge, UK, and New York, NY, USA, 2012, pp. 487–542.

26. P. Kulling, M. Birnbaum, V. Murray and G. Rockenschaub, Guidelines for reports on health crises and critical health events. Prehospital and Disaster, *Medicine*, 2010, **25**(4), 377–383.
27. I. Burton, Forensic disaster investigations in depth: A new case study model, *Environ. Magaz.*, 2011, **52**(5), 36–41.
28. FORIN, 2011, Forensic Disaster Investigations – The FORIN Project, Integrated Research on Disaster Risk, Beijing, China, Available at: www.irdrinternational.org/aboutirdr/scientific-committee/working-group/forensic-investigations/ (Accessed on 5 August 2012).
29. M. Debacker, I. Hubloue, E. Dhondt, G. Rockenschaub, A. Rüter, T. Codreanu, K. L. Koenig, C. Schultz, K. Peleg, P. Halpern, S. Stratton, F. Della Corte, H. Delooz, P. L. Ingrassia, D. Colombo and M. Castrèn, Utstein-Style Template for Uniform Data Reporting of Acute Medical Response in Disasters, *PLOS Curr. Disasters.*, 2012 Mar 23, DOI: 10.1371/4f6cf3e8df15a.
30. J. Stewart-Evans, Review of incidents occurring at COMAH sites in England and Wales, January 2009 – June 2011, Chemical Hazards and Poisons Report 2012; 19–23:12. Available at: http://www.hpa.org.uk/webc/HPAwebFile/HPAweb_C/1317134541578 (Accessed on 5 August 2012).
31. K. MacArthur, J. Gubbins, N. Premaratne, C. Caine, A. Henderson, R. Euripidou and F. Goodfellow, Evacuation from persistent contamination following a sewage spill, Chemical Incident Report July 2000; 17: 8–11, Available at http://www.hpa.org.uk/webc/HPAwebFile/HPAweb_C/1194947330053 (Accessed on 4 August 2012).
32. D. Van Sickle, M. A. Wenck, A. Belflower, D. Drociuk and J. Ferdinands, Panel classification of self-reported exposure histories: a useful exposure index after a mass-casualty event, *Public Health Rep.*, 2007, **122**, 776–783.
33. WHO (World Health Organization) Manual for the Public Health Management of Chemical Incident, 2009, Available at: http://www.who.int/environmental_health_emergencies/publications/FINAL-PHM-Chemical-Incidents_web.pdf (Accessed on 15 September 2012).
34. J. G. Taylor, S. C. Gillette, R. W. Hodgson, J. L. Downing, M. R. Burns, D. J. Chavez and J. T. Hogan, Informing the network: Improving communication with interface communities during wildland fire, *Human Ecol. Rev.*, 2007, **14**(2), 198–211.
35. Leysia Palen, Online Social Media in Crisis Events (short article), *EDUCAUSE Quarterly*, 2008, **31**, 3, (July–September 2008).

36. WHO/HPA/UNISDR Disaster Risk Management for Health factsheets, People with disabilities and older people (2011), Available at: http://www.hpa.org.uk/webc/HPAwebFile/HPAweb_C/1296686249201 (Accessed on 4 August 2012).
37. M. Dombroski, B. Fischhoff and P. Fischbeck, Predicting emergency evacuation and sheltering behavior: a structured analytical approach, *Risk Anal.*, 2006, **26**, 1539–1569.
38. OECD (Organisation for Economic Co-operation and Development) Guiding Principles for Chemical Accident Prevention, Preparedness and Response 2003, Guidance for Industry (including Management and Labour), Public Authorities, Communities, and other Stakeholders Available at: http://www.oecd.org/chemicalsafety/riskmanagementofinstallationsandchemicals/2789820.pdf (Accessed on 15 September 2012).
39. Crucial Interactive Media, 2012, Available at: http://www.crucial-crew.org/interactive-safety-game/ (Accessed on 4 August 2012).
40. IOM (Institute of Medicine), 2012, Crisis Standards of Care: A Systems Framework for Catastrophic Disaster Response, Washington, DC: The National Academies Press.
41. WHO/HPA/UNISDR Disaster Risk Management for Health factsheets, Chemical safety 2011, Available at: http://www.hpa.org.uk/webc/HPAwebFile/HPAweb_C/1296686246437 (Accessed on 15 September 2012).
42. IASC Reference Group for Mental Health and Psychosocial Support in Emergency Settings, 2010, Mental Health and Psychosocial Support in Humanitarian Emergencies: What Should Humanitarian Health Actors Know? Geneva, Available at: http://www.who.int/mental_health/emergencies/what_humanitarian_health_actors_should_know.pdf (Accessed on 4 August 2012).
43. HPA (Health Protection Agency), 2012, Annual Report and Accounts 2011/12, Available at: http://www.hpa.org.uk/webc/HPAwebFile/HPAweb_C/1317134964879 (Accessed on 15 September 2012).

Subject Index